Quantitative Biomedical Optics

This is the textbook and reference resource that instructors, students, and researchers in biomedical optics have been waiting for. Comprehensive and up-to-date, it covers a broad range of areas in biomedical optics, from light interactions at the single-photon and single-biomolecule levels, to the diffusion regime of light propagation in tissue.

Subjects covered include spectroscopic techniques (fluorescence, Raman, infrared, near-infrared, and elastic scattering), imaging techniques (diffuse optical tomography, photoacoustic imaging, several forms of modern microscopy, and optical coherence tomography), and laser-tissue interactions, including optical tweezers.

Topics are developed from the fundamental principles of physical science, with intuitive explanations, while rigorous mathematical formalisms of theoretical treatments are also provided.

For each technique, descriptions of relevant instrumentation and examples of biomedical applications are outlined, and each chapter benefits from references and suggested resources for further reading, and exercise problems with answers to selected problems.

Irving Bigio is Professor of Biomedical Engineering and Electrical Engineering at Boston University. His research activities address the interactions of light with cellular and tissue structures on the microscopic and mesoscopic scales. He pioneered methods of elastic scattering spectroscopy and has developed practical diagnostic and sensing applications that have been demonstrated in large clinical studies. He has co-authored over two hundred scientific publications and is an inventor on nine patents. Trained in optical physics, he gains satisfaction from explaining the fundamentals of complex phenomena in biomedical optics on an intuitive level. He believes that historical developments in physics theory and artistic expression have influenced each other, leading to parallels between the concepts of physical science and the movements in art. He is convinced that Vincent Van Gogh understood the scattering of starlight by interstellar dust (and was aware of spiral galaxies), as evidenced by Starry Night. He is also convinced that the medical field will finally "discover" the benefits of various clinical applications of biomedical optics.

Sergio Fantini is Professor of Biomedical Engineering and Electrical & Computer Engineering at Tufts University. His research interests in biomedical optics are in the area of diffuse spectroscopy and imaging of biological tissue. He contributed to the development of quantitative frequency-domain methods for absolute tissue oximetry, spectral imaging approaches to optical mammography, and the assessment of cerebral hemodynamics in the human brain. He has co-authored about two hundred scientific publications and is an inventor on 11 patents. He thinks that Falstaff and Otello, by Verdi and Boito with due credit to Shakespeare, and Beethoven's Opus 131 are among the greatest expressions of the human mind. He is still waiting to witness Fiorentina win the title in the Italian Serie A. While waiting, he is performing translational research aimed at developing quantitative diffuse optical methods for clinical applications.

"Bigio and Fantini provide a long-needed introduction to the field of biomedical optics and biophotonics, adding spice to the presentation of the basics with historical and etymological gems. The conversational tone of the book is very welcome, and allows room for the clear explanation of subtleties not always clarified in other discussions. The book is a wonderful introduction to the field. It balances rigor with readability. Bravo!"

Steven L. Jacques, Oregon Health & Science University

"This book about biomedical optics provides a remarkably comprehensive introduction to the field. The text is carefully and affectionately developed with quantitative rigor, and it is written in a clear, easy-to-understand style that helps students develop intuition. The subject matter covers basics of linear and nonlinear optical spectroscopy, static and dynamic light scattering and more advanced topics such as light transport through highly scattering tissues, acousto-optics and opto-acoustics, and imaging from microscopy to tomography. The book should prove useful as a textbook for courses targeting both advanced undergraduates and graduate students in science, engineering, and medicine. It will also be a valuable reference for researchers working at the frontiers of knowledge."

Arjun G. Yodh, University of Pennsylvania

"Bigio and Fantini's comprehensive text on Biomedical Optics provides a wonderful blend of accessible theory and practical guidance relevant to the design and application of biomedical optical systems. It should be required reading for all graduate students working in this area."

Rebecca Richards-Kortum, Rice University

CAMBRIDGE TEXTS IN BIOMEDICAL ENGINEERING

Cambridge Texts in Biomedical Engineering provide a forum for high-quality textbooks targeted at undergraduate and graduate courses in biomedical engineering. They cover a broad range of biomedical engineering topics from introductory texts to advanced topics, including biomechanics, physiology, biomedical instrumentation, imaging, signals and systems, cell engineering, and bioinformatics, as well as other relevant subjects, with a blending of theory and practice. While aiming primarily at biomedical engineering students, this series is also suitable for courses in broader disciplines in engineering, the life sciences, and medicine.

Quantitative Biomedical Optics

Theory, Methods, and Applications

Irving J. Bigio
Boston University
Sergio Fantini
Tufts University

Shaftesbury Road, Cambridge CB2 8EA, United Kingdom

One Liberty Plaza, 20th Floor, New York, NY 10006, USA

477 Williamstown Road, Port Melbourne, VIC 3207, Australia

314–321, 3rd Floor, Plot 3, Splendor Forum, Jasola District Centre, New Delhi – 110025, India

103 Penang Road, #05–06/07, Visioncrest Commercial, Singapore 238467

Cambridge University Press is part of Cambridge University Press & Assessment, a department of the University of Cambridge.

We share the University's mission to contribute to society through the pursuit of education, learning and research at the highest international levels of excellence.

www.cambridge.org
Information on this title: www.cambridge.org/9780521876568

First published 2016

A catalogue record for this publication is available from the British Library

ISBN 978-0-521-87656-8 Hardback

Additional resources for this publication at www.cambridge.org/9780521876568

To our families, mentors, teachers, and students

Contents

Preface

A textbook for a new field based on old concepts

Biomedical optics is a field that is both new and ancient. From the vantage point of the natural sciences and engineering, this is a newly developing interdisciplinary field, dealing with the application of optical science and technology to biological and biomedical problems, including clinical applications. On the other hand, the field has been around for thousands of years in a less quantitative way. Physicians' eyes have served as optical spectrographs and sensors, with the brain serving as a database repository and providing the computational power (of a massively parallel computer) for pattern recognition. For example, physicians have known for a long time that a Caucasian patient with yellowing of the skin (or of the sclera of the eye) is likely to be suffering from liver disease. If the patient is flushed red, he/she might be running a fever, and if a local tissue area appears flushed and red, an inflammation is indicated; and the bluish appearance of a patient's lips and nail beds might be indicative of hypoxia. Now that the modern approach has become more quantitative and is developing new technologies, however, the field is growing and beginning to have a major impact on bioscience and healthcare. The emerging field combines the observational with the mathematical and computational, benefits from recent advances in optical technologies, and is coupled with a more rigorous physical-science approach that seeks to understand the basic underlying principles.

This textbook provides a broad survey of the field and covers the basics of a quantitative approach to the subtopics, taking advantage of the powerful tools offered by mathematics, physics and engineering. This quantitative approach and the didactic style, coupled with the description of representative applications and problem sets that accompany each chapter, are designed to serve the needs of students and professionals in engineering and the physical sciences. Students of the biological sciences will also find the text useful, especially if they have a good mathematical background. The basic material about general concepts and methods that are directly relevant to biomedical optics, including some topics of medical statistics, are described at an introductory level, whereas selected topics are covered in greater depth and treated at a more advanced level. Consequently, by proper selection of the material, this textbook can be used for upper-level undergraduate courses, as well as more advanced graduate-level courses on biomedical optics.

The coverage of this book is broader than that of other currently available texts in the field, teaching a broader range of topics under the umbrella of biomedical optics and explaining the interrelationships among them. It also emphasizes aspects in all three areas of theory, instrumentation, and biomedical applications to provide a comprehensive view of each biomedical optics technique that is presented. As a result of its broad coverage, the book can also serve as a reference resource for researchers in biomedical optics. Although the survey is broad, it is by no means exhaustive. Some topics have been bypassed, mainly because they pertain to areas that have not yet developed to the point that they are based on a mathematical formalism or a clear understanding of the physical principles. This book also leaves to others the general field of photonic biosensors, which is broad enough to merit a textbook of its own.

Is it biomedical optics or biophotonics?

One might ask the simplest of questions: what is the proper name for the field, *biomedical optics* or *biophotonics*? The question may be tackled from an etymological point of view by recognizing that the Greek word "οπτική" (optiki) is associated with *vision* while the Greek word "φωτόνιο" (fotonio) is associated with *light*. This seems to suggest that biophotonics may be a more appropriate term for a field in which light is used to interrogate biological systems and to interact with them. One should recognize, however, that the word *optics* has been traditionally used to describe the science of light, which only during the twentieth century was recognized to consist of quanta for which the word *photons* was coined. Therefore, biomedical optics conjures up the more classical elements of lenses, fiber optics, lasers (not so classical) and, perhaps, ophthalmic applications, whereas biophotonics might appear to speak to the quantum-mechanical or statistical nature of light and its interaction with biological tissue. If electronics refers to the generation, manipulation and detection of electrons, then, by analogy, photonics refers to the generation, control and detection of photons, the quantum units of light.

It is arguable that optics and photonics speak to the wave-like and particle-like nature of light, respectively. Although a debate of sorts had been ongoing since the days of Newton as to the true nature of light, modern physicists explain that the two are intertwined (if not entangled!) and the distinction is more philosophical. Historically, the mathematical and physics tools of both approaches have been shown to produce the same results, in most cases, when describing the same phenomena. An instructive view of the matter can be gained from examining how the language of physics itself varies with the frequency (or wavelength) when describing electromagnetic radiation. For low frequencies, such as radio waves, the wave nature completely dominates and enables explanation of all phenomena of interest, given that the photon energy is extremely low ($<10^{-4}$ eV), and a detected radio-wave field is composed of a large number of photons. At very high frequencies, as with "hard" X-rays or γ-radiation, the photon energy is much larger ($>10^4$ eV), and formalisms for interactions focus on the photonic nature of the field (with the exception of the methods for X-ray crystallography). The intersection of the two regimes happens in the range centered on visible light (characterized by photon energies of 1.6–3.1 eV) where both the photoelectric effect and the wave-nature of light are important.

In this textbook, we utilize whichever approach is simpler and more intuitive for understanding a specific concept. Thus, it is easier to think of the wave-nature of light when explaining, for example, the interference effects relevant to optical coherence tomography, whereas it is conceptually simpler to think of photons as particles when describing the random diffusion of photons (also referred to as *photon migration*) in densely scattering media, such as most biological tissues. In the end, we have chosen the word *Optics* in the title of this book, perhaps because one speaks more naturally about, say, "optical diagnostics" or "optical microscopy" or "optical fibers," rather than "photonic diagnostics" or "photonic microscopy" or "photonic fibers."

Why is the field of biomedical optics important?

The dramatic growth of the field in recent years is a consequence of the realization that optical methods offer the potential to have a significant impact on the broad field of health care, and also to provide novel tools for an increasingly quantitative approach to biology. When used for measurement and diagnostic purposes in living systems, light is, under most circumstances, essentially noninvasive. Thus, for biomedical applications there is a growing list of distinct advantages:

- Light (at visible and near-infrared wavelengths) is non-ionizing radiation, and at sub-thermal levels has no cumulative effect on tissue.
- Light can be used to reveal much about tissue that cannot be determined by other imaging or sensing modalities.
- Light can travel farther into tissue than one might think. Although scattering is strong, near-infrared light is only weakly absorbed in tissue, and can diffuse

across several centimeters of tissue, enabling, for example, imaging and sensing of structures and function in solid organs, such as skeletal muscles, breast or brain.
- Optical fibers can be used to deliver and collect light, permitting access to remote sites within the body, mediated by endoscopes, catheters or needles.
- Properties of tissue that are not commonly or readily monitored in real time can be measured with light, enabling new types of diagnostic measurements, e.g., blood oxygenation without drawing blood, cellular nucleus size, etc.
- Optical imaging typically features high temporal resolution, down to the millisecond range.
- The spatial resolution of optical imaging techniques scales with the penetration depth, from sub-micron in microscopy applications at depths up to $\sim 100\ \mu\text{m}$, to several millimeters for diffuse optical imaging at depths of centimeters in tissue. Importantly, throughout the range, optical methods offer functional information not available with other imaging modalities.
- New methods of therapy can be accomplished with light, enabling new ways to treat diseases or repair problems in tissue. The use of light enables interactions to be highly specific as a consequence of wavelength selectivity, spatial selectivity (with tight focusing), temporal selectivity (with ultrashort laser pulses), or cellular and molecular selectivity (with molecular targeting agents).

In short, biomedical optics is ideally suited to serve the trend of modern clinical medicine: the development of noninvasive or minimally invasive diagnostics and therapeutics. It also opens new avenues for biomedical research at the cellular and molecular levels.

Physical modeling in quantitative biomedical optics

In the field of physics, the expression "simplicity is elegance" has been passed down as gospel since Albert Einstein's time. The elegance of simple physical models, however, does not always go hand in hand with the complexity of biological systems. Figure 0.1(a) shows an elegantly simple physicist's view of a chicken, in the spirit of the old joke: "a physicist postulates a chicken as a sphere of uniform density." The chicken is represented in a less simple and more realistic form in Figure 0.1(b). Of course, the simple chicken model of Figure 0.1(a) is far from representative of the complexity of a real chicken, which, some might argue, is more closely represented by the more sophisticated and more complex model of Figure 0.1(b). The question that must often be tackled in quantitative biomedical optics is whether the added complexity of more sophisticated models and

Figure 0.1

(a) An overly simplified view of a chicken; (b) a more sophisticated and complex representation of a chicken. Is it always better to use a more complex model for a biological system? This is a key question in quantitative biomedical optics. (Figure 0.1(b) courtesy of Cliparts.co, http://cliparts.co/clipart/186502.)

treatments is really required by the specific application in hand, and whether it truly teaches more about reality. For example, if one needs to discriminate a chicken from a giraffe, the simple representation of Figure 0.1(a) might be adequate (assuming a physicist would postulate a giraffe as a long cylinder of uniform density!), but it would not be adequate to model shape variations that would distinguish a hen from a rooster, for which the more sophisticated model of Figure 0.1(b) may be necessary. Given the complexity of biological systems, finding a good compromise between the simplicity of physical models and their appropriate representation of the biological parameters of interest is a critical objective to be achieved in quantitative biomedical optics. In writing this book, we have endeavored to strike the right balance, so that the basic physical principles can be understood, while useful quantitative results can be obtained from their application.

Organization of the book

We have strived to present the broad range of topics covered in this book in a consolidated way. We have cross-referenced material from different chapters every time ideas presented in one area had relevance or implications in other areas. We have also used consistent symbols and notations, and we have paid particular attention to the physical dimensions and associated units. For example, even though the scattering phase function (p) is dimensionless, by expressing it in units of sr^{-1} one immediately appreciates its meaning of a probability per unit solid angle. This also results in its discrimination from the scattering probability per unit scattering angle (indicated in this text as p_θ), when the latter is expressed in units of rad^{-1}. As another example, we have explicitly discussed the units of diffuse reflectance and their implications in understanding its physical meaning.

In our descriptions and derivations, our goal is to provide a clear presentation of the basic principles and their implications, and we have included references in the bibliographic sections of each chapter to direct the interested reader to more advanced material or in-depth treatments. For example, we have not presented the solutions to the diffusion equation in a variety of tissue geometries (slab, sphere, cylinder, etc.); rather, we have focused on the two ideal cases of infinite and semi-infinite media, to introduce and discuss the basic parameters in play, their interdependence, and the roles they play in optical measurements in the diffusion regime. Our aim is to strike a good balance between presenting the material in a comprehensive, rigorous, and quantitative fashion, while keeping a descriptive tone with intuitive explanations and illustrative examples.

The chapters of this book are arranged in the following four groupings:

I. **Chapters 1–4.** These provide the broad underpinnings for the field. Chapter 1 defines the ***nomenclature*** that we adopt for this book, including the symbols and units, along with the rationale for many of the choices. Chapter 2 provides a broad overview of the ***optical properties of biological tissues***, and introduces the constituents that contribute to those properties. The quantitative formalisms for representing those optical properties are also presented. Chapter 3 teaches the ***basics of biomedical statistics***, in the language of probability theory, especially as applied to diagnostic tests that may be the translational goal of many of the technologies described in this book. Chapter 4 is an overview of the ***basics of optical spectroscopy***, including a comparison of the merits of different types of spectroscopy as applied to tissue diagnostics; also included are overviews of the optical-science basics of instrumentation for tissue spectroscopy and of optical fibers.

II. **Chapter 5–8.** These chapters present in-depth developments of four classes of tissue spectroscopy, predominantly related to superficial measurements or measurements in the sub-diffuse regime. Chapter 5 covers ***fluorescence spectroscopy and imaging***, starting with the molecular physics of fluorescence, followed by listings of the important endogenous fluorophores and exogenous reporter fluorophores. Then, instrumentation and methods are described, including methods for fluorescence lifetime and polarization measurements. Chapter 6 covers spectroscopic methods for measurement of ***vibrational modes of biomolecules***, starting with the molecular physics concepts of vibrational transitions, followed by discussions of ***IR-absorption spectroscopy*** and ***Raman scattering spectroscopy***. Chapter 7 teaches the physics of ***scattering by single dielectric particles***, including Mie theory, and lays the groundwork for measurements of elastic scattering; also covered are dynamic light scattering from single particles and Doppler flowmetry. Chapter 8 extends the principles established in Chapter 7 to measurements of ***multiply scattered light in the sub-diffuse regime***, and the extraction of tissue optical properties from measurements at short source-detector distances.

III. **Chapters 9–15.** These chapters cover the theoretical formalisms and a range of measurement methods for light transport in the diffuse regime. Chapter 9 introduces the ***Boltzmann transport equation*** and derives its best-known approximations, including the ***diffusion equation*** in various forms and the constraints of boundary conditions. Chapter 10 teaches the basics and formalisms of ***continuous-wave tissue spectroscopy***, with special attention to applications in the diffusion regime. Chapter 11 presents formalism and methods of ***time-domain spectroscopy*** in the diffusion regime, with a special emphasis on the features of the photon time-of-flight distribution. Chapter 12 formulates the ***frequency-domain spectroscopy*** method in the diffusion regime, including a detailed development of the concept of photon-density waves. Chapter 13 presents a broad overview of the types of ***instrumentation and experimental methods for diffuse tissue spectroscopy***. Chapter 14 focuses on methods of ***optical imaging in the diffusion regime***, and covers a broad array of parameter sensitivities and methods for solution of the inverse imaging problem. Chapter 15 concludes this grouping by presenting several exemplary areas of ***applications of diffuse optical methods***, for both clinical and pre-clinical use.

IV. **Chapters 16–19.** These chapters cover methods of higher-resolution imaging in tissue and several advantageous areas of laser-tissue interactions. Chapter 16 introduces the basic concepts of ***acousto-optic and opto-acoustic methods*** for imaging at various length-scales based on the interaction or combination of ultrasound and light in tissue. Chapter 17 starts with the basics of a classical compound microscope, then teaches the basics of several classes of ***modern optical microscopy*** that are important in biomedical science, and concludes with recent developments of super-resolution microscopy. Chapter 18 introduces the fundamentals of ***optical coherence tomography***, implemented with both time-domain and frequency-domain methods, and includes some practical considerations and applications. Finally, Chapter 19 overviews the fundamentals of ***optical tweezers***, with related instrumentation and applications, and presents a survey of the more practical types of ***laser-tissue interactions*** from sub-thermal tissue treatments to microsurgical techniques.

In all chapters, we have included a set of problems. The answers to selected problems (indicated by ∗) are reported at the end of the book, where we also provide tables of symbols and acronyms.

Acknowledgements

We wish to recognize the helpful advice that we received from colleagues on specific topics or sections of the book. Specifically, Angelo Sassaroli (Chapters 9–15), Arvind Saibaba (Chapter 14), Eladio Rodriguez-Diaz (Chapter 3), Eric Miller (Chapter 14), Irene Georgakoudi (Chapter 5), Jean Luc Castagner (Chapter 7), Jerome Mertz (Chapter 17), Mark Cronin-Golomb (optical tweezers), and Misha Kilmer (Chapter 14). In particular, IJB thanks Judith Mourant for her role in the co-development of several concepts presented in the book related to elastic scattering and diffuse reflectance at short distances, and SF is very grateful to Angelo Sassaroli for the many illuminating discussions on diffuse optics and for his help with figures in Chapter 14. We also acknowledge Ousama A'amar for data analysis and instrumentation development relevant to results presented, Satish K. Singh for providing clinical data and images, Katherine Calabro for help with figures and text in Chapter 8 and, together with Roberto Reif, for development of concepts therein, Wei Han Bobby Liu for figures in Chapter 7, and Aysegul Ergin for generating chromophore data. We thank Alper Corlu, Anaïs Leproux, Arjun Yodh, Brian Pogue, Bruce Tromberg, Chris Cooper, David Boas, Jerome Mertz, Joe Culver, Junjie Yao, Kevin McCully, Lihong Wang, Michal Balberg, Regine Choe, Song Hu, Terence Ryan, Ton van Leeuwen, and Turgut Durduran for providing high-resolution images from their publications. We would also like to thank the many current and former students and postdocs in our groups for their input and discussions that helped shape the structure and content of this book; and we recognize our colleagues in the field of biomedical optics, with whom we have had numerous scientific discussions over the last three decades, contributing greatly to shaping this book. Among those colleagues, we wish to explicitly recognize Britton Chance, whose stature, brilliance, charisma, and personality served as inspiration and guidance for the field, who encouraged us personally when presented with our ideas, and whose positive influence on the field in general is still felt years after his death.

IJB: I owe much to my parents Joe and Renée (of blessed memory), especially my mother Renée, who stimulated my intellect and curiosity in so many ways during my youth. Among the many educators who gave selflessly over the years, I especially wish to thank my doctoral advisor, John F. Ward, for teaching me how to think like a physicist. I am also grateful for a number of mentors and colleagues who offered sage advice during my years at Los Alamos National Laboratory, including Joseph Figueira, Allen Hartford, and Edgar Hildebrand. Since coming to Boston University, I could not have asked for a more collegial and intellectually stimulating group of colleagues than my fellow faculty members. This is a truly supportive academic environment, and I feel privileged to be part of this enterprise. My family's support has been critical. I especially want to thank my wife Ruth for her infinite forbearance during the long writing odyssey, my son Aaron for his camaraderie and faith in me, and my daughter Erica, who cheered me on and shared advice on how to "stay the course."

SF: I would like to express my profound gratitude to a number of my teachers and mentors, from elementary school in Piazza San Salvi – Firenze to postdoctoral training at the University of Illinois at Urbana-Champaign: Anna Maria Betti, Leda Stoppato, Mara Cecchini, Marcella Cantelli, Francesco Savelli, Nello Taccetti, Tito Arecchi, Riccardo Pratesi, Marco Zoppi, Lorenzo Ulivi, and Enrico Gratton. In particular, I owe a special debt of gratitude to Enrico Gratton for guiding my early steps in the field of biomedical optics. I also thank my colleagues at the Department of Biomedical Engineering at Tufts University for making it such a pleasant, productive, stimulating, and fun work place. Lastly, and most importantly, I wish to express my heartfelt gratitude to my family: my parents, Enio and Luigina – siete veramente grandiosi, grazie di tutto!; my brother Fabio – un mito sempre presente, sei proprio un grande; my wife Tanya, my wonderful soulmate, who showed so much patience and support during this long book project; my daughter Lisa, so utterly fantastic and my infinite source of joy; and last but certainly not least, my stepson Marco, whose wit never ceases to amaze me and who is such a truly awesome boy.

Among the great experiences associated with writing this book, we are both happy to reckon the countless work lunches that we had, alternating between the areas around the campuses of Boston University and Tufts University. The discussions during these meetings were intriguing and illuminating in their exposition of the multitude of ways to describe the topics of biomedical optics (and opera!); we both learned how to envision and teach the many facets of this evolving field from new angles and perspectives.

Finally, we have thoroughly enjoyed working with the editorial staff at Cambridge University Press: Heather Brolly, Michelle Carey, Sarah Marsh, and Elizabeth Davey; and we are grateful for the patience with which they followed this book project during its long years of gestation.

IJB and SF

1 Nomenclature

Is a rose a rose? To crudely paraphrase Shakespeare, "What's in a name?" Can the same flower be described and characterized differently when viewed from different perspectives? Different scientific and technical fields tend to develop variations of the nomenclature for the same physical quantities. Within each field there is generally a degree of consistency, but interdisciplinary topics that bring together scientists, engineers, and practitioners from disparate fields often face inconsistencies in the symbols, units, and terminology utilized for quantitative analysis. The field of biomedical optics has not been immune to this problem, and publications in the field have invoked a variety of terms and symbols. In some cases, the same terms and symbols, as used in different fields, carry subtle but crucial differences in their meanings. An important example is the different meaning of the term *intensity* in physics (power per unit area), in radiometry (power per unit solid angle), and in heat transfer (power per unit area, per unit solid angle). Another example is that of the molar extinction coefficient for optical absorption of hemoglobin, which is typically defined using the logarithm to base-10 in chemistry and biology, or to base-*e* in physics, and may refer to one functional heme group or to the full molecule (four heme groups), depending on the specific physiological or biochemical characterizations. This may lead to a mismatch by as much as a factor of 9.2 (i.e., $4 \times \ln(10)$) among the numerical values reported for the molar extinction coefficient of hemoglobin according to different conventions. For a long time, the three fields that most commonly make use of and describe (and teach) the methods of quantitative optical measurements have been physics, astronomy, and electrical engineering (or its subfield, optical engineering), although chemistry and biology also make use of optical characterization techniques. Because biomedical optics is a broad interdisciplinary field, which is based on contributions from researchers in a variety of specialty areas, it is important that a common language be used to describe and characterize its key quantitative parameters.

In this chapter, we define the nomenclature used in this book for the quantities that describe the optical radiation field and its interaction with biological tissue. In cases of ambiguity across disciplines, we have adopted the notation of physics rather than radiometry, such that, for example, we define the intensity (I) as the

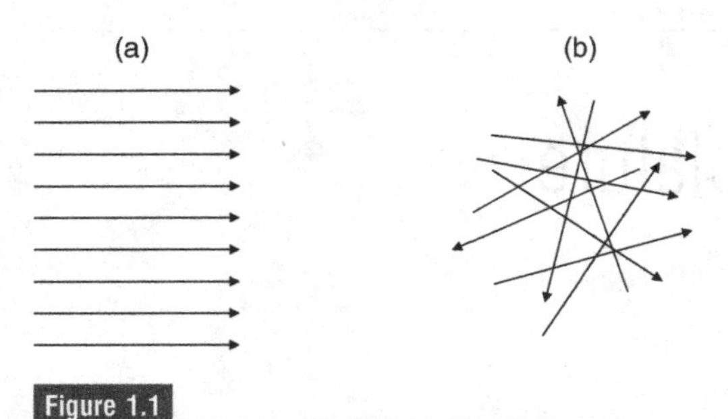

Figure 1.1

(a) A uniform, collimated optical field propagating in a vacuum or non-scattering medium. (b) Light strongly scattered in a turbid medium, featuring a net energy flow from left to right.

power per unit area incident on a surface, and we identify the power per unit solid angle emitted by a light source as radiant angular intensity ($\mathcal{J}$). In most cases, the physical dimensions of a parameter provide clear indications of its meaning, but this is not always the case: for example, intensity and fluence rate both have dimensions of power per unit area, but they represent different measures of an optical field. This is discussed below, and the definitions listed in this chapter provide a rigorous description of the main quantities and parameters used in biomedical optics.

1.1. Describing the optical radiation field and its interactions with tissue

Sometimes the units alone are not enough to clearly define a physical parameter because the units alone do not fully convey complex directional information, even when the parameters are expressed as vectors. Thus, while it is important to be cognizant of whether the quantity is a scalar or vector parameter, there may be more to the story. Thus, for vector parameters, a simple unit vector may provide information about the direction of a net flow of energy (or photons), for example, yet it may not provide all the necessary information to properly describe the flow; information about a *distribution* of directions may be needed. Consider, for example the two cases represented in Figure 1.1. Both optical fields illustrated in panels (a) and (b) of Figure 1.1 represent a net flux of optical energy from left to right, a quantity often referred to in physics in terms of the Poynting vector (named after the British physicist John Henry Poynting [1852–1914]).

Figure 1.1(a) represents the simple propagation, from left to right, of a collimated optical field in a vacuum or non-scattering medium. Figure 1.1(b) represents a field of diffusely scattered light in a turbid medium, with photons propagating in multiple directions; this field, nonetheless, has a net flow of optical energy from left to right as in Figure 1.1(a). Although these two figures represent optical fields with a net flow of energy in the same direction, the first case can be fully described by a single vector quantity, whereas the latter case would require at least two measures to properly describe the optical field: a vector quantity (similar to the first case) representing the net flux of energy, plus a scalar quantity to indicate what the total optical exposure would be, at a given point within the turbid medium, as a consequence of photons impinging from all directions.

The field of physics tends to be rigorous about such distinctions – consider, for example the clear distinctions among *speed, velocity,* and *velocity distribution* – but newly developing interdisciplinary fields, for example those related to the biological sciences, are often less rigorous about units and nomenclature, especially when reporting physical quantities. As a consequence, it is not uncommon to hear the terms *power* and *intensity* utilized interchangeably in the field of radiology for describing the emission properties of an X-ray source.

In 1996 the American Association of Physicists in Medicine, under the auspices of the American Institute of Physics, published a reference document titled "Recommended Nomenclature for Physical Quantities in Medical Applications of Light" (Hetzel et al., 1996). This document recognized the disparate and inconsistent usage in the published literature, and proposed a codification of the definitions. The International Organization for Standardization (ISO) has also been publishing standard nomenclature and definitions for quantities associated with light and electromagnetic radiation (ISO, 2008). What follows below is a listing of the principal optical parameters (and associated units) of relevance in biomedical optics, attempting to be consistent with those documents, while nonetheless invoking some deviations that we have introduced for some of the symbols for the purpose of clarity and consistency with the majority of the biomedical optics literature. For a few of the parameters describing the optical radiation field itself, we also note whether the parameter is conserved in a non-dissipative system, and explain why. (Such conservation need not be limited to a closed system. This can be important for understanding the efficiency of optical system designs and methods for detection.) The modern metric system (or International System of Units, abbreviated SI from the French *Système International d'Unités*) is mainly used in this book, although centimeters (cm), grams (g), and seconds (s) (CGS system and the related Gaussian system of units, named after the German mathematician Carl Friedrich Gauss [1777–1855]) will sometimes take preference over meters (m), kilograms (kg), and seconds (s) (MKS units, the basis for the SI system), for convenience or because the associated dimensions are closer to those of observed quantities.

The most common terms used to describe the radiant field, the flow of optical energy, and the interactions of light with tissue and biological media are listed below, including indications of the preferences utilized in this book.

1.2 Quantities describing the optical radiation field

Electric field: $\mathbf{E}(\mathbf{r}, t)$ – SI units: V/m (V: volt, named after the Italian physicist Alessandro Volta [1745–1827]).

Magnetic field: $\mathbf{H}(\mathbf{r}, t)$ – SI units: A/m (A: ampere, named after the French physicist André-Marie Ampère [1775–1836]).

Magnetic induction: $\mathbf{B}(\mathbf{r}, t)$ – SI units: T (T: tesla: named after the Serbian-American engineer Nikola Tesla [1856–1943]).

The *electric field* (**E**) and the *magnetic field* (**H**), or *magnetic induction* (**B**), vectors are the basic vector quantities that describe electromagnetic radiation. **E** and **H** are not independent as the electric and magnetic fields are linked by Maxwell's equations, named after the Scottish physicist James Clerk Maxwell [1831–1879]. The electric and magnetic field vectors are orthogonal to each other and to the direction of propagation of the optical wave. The square of the electric field's magnitude and the square of the magnetic field's magnitude are both proportional to the energy density (and the intensity) associated with the optical radiation.

Radiant energy: $Q(\Delta t)$ – SI units: J (joule, named after the English physicist James Prescott Joule [1818–1889]).

Radiant energy describes the emission or delivery of optical energy. $Q(\Delta t)$ is the energy of an optical field that has been emitted by a source over a time span $\Delta t = t_2 - t_1$, or that is delivered to, reflected from, transmitted through, or absorbed by a medium over time Δt. $Q(\Delta t)$ is related to the radiant power $P(t)$, defined below, by the relationship:

$$Q(\Delta t) = \int_{t_1}^{t_2} P(t)dt. \tag{1.1}$$

The radiant energy can be expressed in terms of the number of photons (emitted, delivered, reflected, transmitted, or absorbed over time Δt), which are the energy quanta of the electromagnetic field, each carrying an energy hf, where h is Planck's constant ($h = 6.626 \times 10^{-34}$ J s) (named after the German physicist Max Planck [1858–1947], who received the Nobel Prize in physics in 1918) and f is the frequency of light. In this text, we will predominantly use the angular frequency, $\omega = 2\pi f$, for the optical field, which leads to the photon energy being represented as $\hbar\omega$, where $\hbar \equiv h/(2\pi)$ is the reduced Planck's constant. Therefore, the radiant energy associated with $N_{\Delta t}$ photons emitted, delivered, reflected, transmitted, or absorbed over time Δt at frequency f is given by:

$$Q(\Delta t) = N_{\Delta t} hf = N_{\Delta t} \hbar\omega. \tag{1.2}$$

In the absence of dissipative effects (conversion of optical energy to other forms, such as heat) total optical energy is conserved.

Radiant power: $P(t)$ – SI units: W (watt, named after the Scottish engineer James Watt [1736–1819]).

The *radiant power*, $P(t)$, is the radiant energy emitted, delivered, reflected, transmitted, or absorbed per unit time. The *average power* over time $\Delta t = t_2 - t_1$ is:

$$\langle P \rangle_{\Delta t} = \frac{1}{\Delta t} \int_{t_1}^{t_2} P(t) dt = \frac{Q(\Delta t)}{\Delta t}, \tag{1.3}$$

while the instantaneous power at time t is:

$$P(t) = \frac{dQ}{dt}, \tag{1.4}$$

where dQ is the infinitesimal radiant energy over time dt about t. The average power associated with $N_{\Delta t}$ photons emitted, delivered, reflected, transmitted, or absorbed over time Δt at angular frequency ω is:

$$\langle P \rangle_{\Delta t} = \frac{N_{\Delta t} \hbar \omega}{\Delta t}. \tag{1.5}$$

No directional information is conveyed by the radiant power, which is thus a scalar quantity. Radiant energy flux and radiant power are synonymous, and both terms are found in the literature. We make preferential use of the terms *power* or *photon rate* over *energy flux*, following the physics tradition. In the absence of dissipation, the optical radiant power is conserved for continuous sources and can only be altered for pulsed sources if the optical pulses are compressed or expanded temporally. Moreover, the average radiant power is conserved whenever the observation time, Δt, is longer than the emission time of the source.

Radiant angular intensity: $\mathcal{J}(\hat{\Omega})$ – SI units: W/sr (sr: steradian, SI unit for solid angle, from the Greek *stereos* [solid] and the Latin *radius* [ray]).

The *radiant angular intensity* describes the power per unit solid angle emitted by a light source along a given direction $\hat{\Omega}$. In the field of radiometry, the word "intensity" is used to describe the quantity defined here. By contrast, in physics, intensity is defined as the power incident on a surface per unit area. Because we have decided to adopt the latter definition of intensity, we have added the "angular" specifier here, to indicate the power emitted per unit solid angle as the "angular intensity." If $dP(\hat{\Omega})$ is the infinitesimal radiant power emitted within the infinitesimal solid angle $d\Omega$ about $\hat{\Omega}$, then the radiant angular intensity is defined as:

$$\mathcal{J}(\hat{\Omega}) = \frac{dP(\hat{\Omega})}{d\Omega}. \tag{1.6}$$

The radiant angular intensity is not used in this book, but the reader is cautioned to take note of the actual units when the term "intensity" appears in the literature. Sometimes the term *brightness* is used for the angular intensity, although it can lead to confusion because without including the factor of area^{-1} (as in the definition of *intensity* below), the quantity is not conserved. It is important to note that even in a closed system this quantity is not conserved, because the value can be changed by simply using a lens to alter the divergence angle of the light.

Radiance: $L(\hat{\Omega})$ – SI units: W/(m^2-sr).

The *radiance* emitted by a light source (or traveling through a medium) along direction $\hat{\Omega}$ is defined as the radiant angular intensity per unit area perpendicular to the direction $\hat{\Omega}$. If $d\mathcal{J}$ is the infinitesimal radiant angular intensity transmitted through a material (or emitted by a source) over the infinitesimal solid angle $d\Omega$ about $\hat{\Omega}$ and over an infinitesimal surface area dA that is normal to the unit vector $\hat{\mathbf{n}}$, then the radiance is defined as:

$$L(\hat{\Omega}) = \frac{d\mathcal{J}}{dA\hat{\Omega} \cdot \hat{\mathbf{n}}} = \frac{d^2P}{d\Omega dA\hat{\Omega} \cdot \hat{\mathbf{n}}}. \tag{1.7}$$

There is a long tradition in the fields of astronomy and physics of the term *brightness* for this parameter, and the term *specific intensity* can also be found in the literature in place of radiance. Directional information is indicated by the unit vector along the direction considered ($\hat{\Omega}$), and the radiance may be expressed as the vector $\mathbf{L} = L\hat{\Omega}$ to indicate optical flux along direction $\hat{\Omega}$.

Radiance is often the most important parameter to use in describing the emission properties of a light source, especially for tissue spectroscopy. It is important to understand how the radiance of a source governs the amount of light that gets through an optical system, a spectrometer, etc. For example: this parameter is the only true measure that allows one to determine how much light can be coupled from a light source into an optical fiber. This is because both the divergence angle *and* the effective area of the emitting surface are invoked. Thus, lasers generally feature a much higher radiance than incoherent sources, even if incoherent sources may generate a higher optical power. As such, in the absence of dissipation (or temporal manipulation of pulsed light), radiance is conserved, because it is a fundamental property of the emitted field that cannot be altered (increased or diminished) by use of non-dissipative optical manipulations (e.g., lenses, mirrors, etc.). It is interesting to note that if one attempts to increase the power per unit area (or intensity, as defined below) of a beam of light by focusing it (say, with a lens) down to a smaller area, this is inevitably accompanied by a commensurate increase in the solid angle of the converging beam, such that the radiance remains unchanged. Texts on spectroscopy often refer to this phenomenon as "conservation of *étendue*," after the French word meaning *extent* or *scope*.

Radiant exposure: $H(\Delta t)$ – SI units: J/m^2.

The *radiant exposure* is a useful parameter to describe the optical energy delivered per unit area on a surface, and is sometimes referred to as incident energy, exposure, or absorbed energy. If A is the surface area to which a radiant energy $Q(\Delta t)$ is delivered, then the radiant exposure is defined as:

$$H(\Delta t) = \frac{Q(\Delta t)}{A}. \tag{1.8}$$

$H(\Delta t)$ is related to the intensity $I(t)$, defined below, by the relationship:

$$H(\Delta t) = \int_{t_1}^{t_2} I(t)dt. \tag{1.9}$$

Intensity: $I(t)$ – SI units: W/m^2.

The *intensity* describes the amount of optical energy delivered per unit time, per unit area on a surface. In radiometry, the term *irradiance* is used for this parameter, but here we follow the physics nomenclature that is also commonly adopted in biomedical optics. If A is the surface area to which a radiant power $P(t)$ is delivered, then the intensity is defined as:

$$I(t) = \frac{P(t)}{A}. \tag{1.10}$$

The average intensity over time $\Delta t = t_2 - t_1$ is:

$$\langle I \rangle_{\Delta t} = \frac{1}{\Delta t} \int_{t_1}^{t_2} I(t)dt = \frac{H(\Delta t)}{\Delta t}, \tag{1.11}$$

and the instantaneous intensity at time t is:

$$I(t) = \frac{dH}{dt}, \tag{1.12}$$

where dH is the infinitesimal radiant exposure over time dt about t. For a surface (actual or conceptual) normal to a unit vector $\hat{\mathbf{n}}$, the intensity resulting from all directions having positive ($I_{+\hat{\mathbf{n}}}$) or negative ($I_{-\hat{\mathbf{n}}}$) components along $\hat{\mathbf{n}}$ is given by the following angular integrals involving the radiance $L(\hat{\Omega})$:

$$I_{+\hat{\mathbf{n}}} = \int_{\hat{\Omega}\cdot\hat{\mathbf{n}}>0} L(\hat{\Omega})\hat{\Omega}\cdot\hat{\mathbf{n}}d\Omega, \tag{1.13}$$

$$I_{-\hat{\mathbf{n}}} = \int_{\hat{\Omega}\cdot\hat{\mathbf{n}}<0} L(\hat{\Omega})\hat{\Omega}\cdot(-\hat{\mathbf{n}})d\Omega. \tag{1.14}$$

Like radiant angular intensity, in a non-dissipative system the intensity can easily be altered with a lens, and thus is not conserved.

Fluence rate: $\phi(\mathbf{r}, t)$ – SI units: W/m^2.

The concept of *fluence rate* is particularly relevant in biomedical optics. It provides a measure of the optical energy per unit time, per unit area, incident from *any* direction within optically turbid media (such as most biological tissues) as opposed to the case of surface illumination (in which case the intensity is the proper parameter). Inside optically turbid media, generally, optical fields (or photons) propagate along directions covering the entire solid angle. Therefore, to obtain a measure of the power per unit area about a given point $\mathbf{r}$, it is appropriate to consider the optical energy per unit time that reaches an infinitesimal area about the point from all possible directions. Sometimes, the term *total exposure rate* is used for the fluence rate. Because the radiance $L(\hat{\Omega})$ describes the optical power per unit area propagating along direction $\hat{\Omega}$, it is natural to define the fluence rate by integrating $L(\hat{\Omega})$ over the entire solid angle (4π) as follows:

$$\phi(\mathbf{r}, t) = \int_{4\pi} L(\mathbf{r}, t, \hat{\Omega})d\Omega. \tag{1.15}$$

It is important to understand the difference between the integral in Eq. (1.15) and those in Eqs. (1.13) and (1.14). First, the integral in Eq. (1.15) is carried out over the entire solid angle 4π, whereas those in Eqs. (1.13) and (1.14) are carried out over a half-space solid angle of 2π. Second, the integrands in Eqs. (1.13) and (1.14) contain the additional factor $\hat{\Omega}\cdot\hat{\mathbf{n}}$ or $\hat{\Omega}\cdot(-\hat{\mathbf{n}})$. This means that while $I_{+\hat{\mathbf{n}}}$ and $I_{-\hat{\mathbf{n}}}$ provide a power per unit area associated with a flat surface element normal to $\hat{\mathbf{n}}$, ϕ provides a power per unit area associated with the surface of an infinitesimal sphere centered at position $\mathbf{r}$. This is illustrated in Figure 1.2.

Fluence: $\psi(\mathbf{r}, \Delta t)$ – SI units: J/m^2.

The *fluence* extends the concept of fluence rate to describe the optical energy delivered per unit area over a time interval $\Delta t = t_2 - t_1$ from *all* directions about a given point $\mathbf{r}$ within an optically turbid medium. Sometimes, the symbol H_0 is

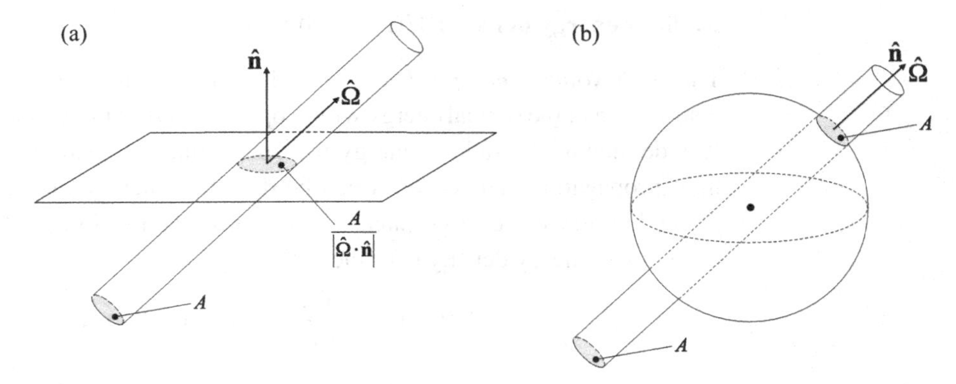

Figure 1.2

To translate the radiance L along direction $\hat{\Omega}$ (power per unit area normal to $\hat{\Omega}$, per unit solid angle) into the intensity (power per unit area) on a surface, one needs to take into account the relationship between the area of a general surface element and the area A of the corresponding surface element normal to $\hat{\Omega}$. (a) In the case of a planar surface characterized by a normal unit vector $\hat{\mathbf{n}}$, the area of the surface element is larger than A by a factor $1/|\hat{\Omega} \cdot \hat{\mathbf{n}}|$ so that the intensity on the surface over solid angle $d\Omega$ about $\hat{\Omega}$ is $L(\hat{\Omega})\hat{\Omega} \cdot \hat{\mathbf{n}} d\Omega$. (b) In the case of a sphere for which $\hat{\Omega}$ is a radial direction, the area of the surface element is equal to A so that the intensity on the surface over solid angle $d\Omega$ about $\hat{\Omega}$ is simply $L(\hat{\Omega})d\Omega$.

also used to indicate that the fluence is a radiant exposure (H) associated with light propagating along all possible directions, or radii of an infinitesimal sphere (suggested by the subscript 0) around the point considered. The fluence is given by a time integral of the fluence rate as follows:

$$\psi(\mathbf{r}, \Delta t) = \int_{t_1}^{t_2} \phi(\mathbf{r}, t)dt. \tag{1.16}$$

An alternative term for the fluence is *total exposure*.

Angular energy density: $u(\mathbf{r}, \hat{\Omega}, t)$ – SI units: J/(m^3-sr).

The *angular energy density* $u(\mathbf{r}, \hat{\Omega}, t)$ is defined as the energy per unit volume, per unit solid angle that is propagating about direction $\hat{\Omega}$, position $\mathbf{r}$, and time t. It is related to the radiance by the relationship:

$$u(\mathbf{r}, \hat{\Omega}, t) = \frac{L(\mathbf{r}, \hat{\Omega}, t)}{c_n}, \tag{1.17}$$

where c_n is the speed of light in the medium, given by c/n, with c speed of light in vacuum and n index of refraction of the medium. The angular photon density, defined as the number of photons per unit volume, per unit solid angle traveling along direction $\hat{\Omega}$ can be indicated as $u_N(\mathbf{r}, \hat{\Omega}, t)$.

Radiant energy density: $U(\mathbf{r}, t)$ – SI units: J/m^3.

This is a volume *energy density* and it is important in describing the spatial distribution of the optical energy within optically turbid media such as most tissues. It is defined as the radiant energy per unit volume, considering all directions of light propagation. Therefore, it can be defined either by the fluence rate $\phi(\mathbf{r}, t)$ divided by the speed of propagation of light in the medium, or by the integral of the angular energy density over the entire solid angle:

$$U(\mathbf{r}, t) = \frac{\phi(\mathbf{r}, t)}{c_n} = \int_{4\pi} u(\mathbf{r}, \hat{\boldsymbol{\Omega}}, t) d\Omega. \tag{1.18}$$

An important alternative, often found in the literature, is to express the energy density in terms of the photon number density, which has units of m^{-3} and lacks the information about the photon energy ($\hbar\omega$). The photon number density may be represented by the symbol $U_N(\mathbf{r}, t)$.

Net flux: $\mathbf{F}(\mathbf{r})$ – SI units: W/m^2.

We have seen above (in the definition of radiance) how an elemental flux vector can be defined as $\mathbf{L} = L\hat{\boldsymbol{\Omega}}$ to indicate optical flux along a given direction $\hat{\boldsymbol{\Omega}}$. If light propagates along different directions through a point $\mathbf{r}$, such as the case in optically turbid media, it is useful to define a *net flux* vector $\mathbf{F}(\mathbf{r})$ by integrating the elemental flux vector over the entire solid angle:

$$\mathbf{F}(\mathbf{r}) = \int_{4\pi} L(\mathbf{r}, \hat{\boldsymbol{\Omega}})\hat{\boldsymbol{\Omega}} d\Omega. \tag{1.19}$$

The magnitude of the net flux vector provides the net energy per unit time, per unit area that propagates through an elemental area at point $\mathbf{r}$, while the direction of the net flux vector provides its direction of propagation. Thus, in the limit of perfectly isotropic, diffusely scattered light, the net flux can be zero, even though the fluence rate and energy density may be finite. In this text the net flux, $\mathbf{F}(\mathbf{r})$, takes the place of the Poynting vector that is frequently used in physics texts.

1.3 Quantities describing the optical properties of tissue and the interactions between the radiation field and tissue

Index of refraction: n – dimensionless.

The *index of refraction* of a medium is the ratio of the speed of light in vacuum to the speed of light in the medium. It can vary with the frequency of light (to be discussed in Section 4.8), but the spatial dependence of n on a microscopic scale is of particular importance in biomedical optics. In fact, the discontinuities and gradients in the refractive index at the cellular and subcellular level are the major sources of light scattering in tissue, as will be discussed in Chapters 7 and 8.

Absorption coefficient: μ_a – SI units: m^{-1}.

The optical *absorption coefficient* represents the absorption probability per unit distance traveled by a photon. As a result, the absorption probability of a photon traveling over an infinitesimal distance dx is given by $\mu_a dx$. Even though the absorption probability per unit length traveled by individual photons is constant (in a homogeneous medium with uniform μ_a), the photon population decreases exponentially with the distance traveled (see the Beer-Lambert law as described in Sections 2.5 and 10.3), so that the number of absorption events per unit distance also decreases exponentially. This fact is described by the exponential absorption probability density $[f_a(x, \mu_a)]$ over distances $x \geq 0$ traveled by a photon:

$$f_a(x, \mu_a) = \mu_a e^{-\mu_a x}, \tag{1.20}$$

from which one can derive the absorption cumulative distribution function $F(L, \mu_a)$ that specifies the probability that a photon is absorbed when traveling over a distance L:

$$F(L, \mu_a) = \int_0^L \mu_a e^{-\mu_a x} dx = 1 - e^{-\mu_a L}. \tag{1.21}$$

The average distance traveled by a photon before being absorbed, called the *absorption mean free path* (mfp_a), is given by the inverse of the absorption coefficient, as it may be derived directly by calculating the expectation value of x using the absorption probability density function given in Eq. (1.20):

$$\text{mfp}_a = \int_0^\infty x \mu_a e^{-\mu_a x} dx = \frac{1}{\mu_a}. \tag{1.22}$$

Scattering coefficient: μ_s – SI units: m^{-1}.

The *scattering coefficient* represents the scattering probability per unit distance traveled by a photon, and (in the absence of absorption) it leads to the same expressions for the scattering probability density $[f_s(x, \mu_s)]$ and scattering cumulative distribution function $[F_s(L, \mu_s)]$ as in Eqs. (1.20) and (1.21), respectively, once μ_a is replaced by μ_s. The one difference is that, in the scattering case, it is the unscattered photon population, rather than the photon population itself, that decreases exponentially with the traveled distance x. The *scattering mean free path* (mfp_s), which is the average distance traveled by photons between successive scattering events, is given by the inverse of the scattering coefficient:

$$\text{mfp}_s = \int_0^\infty x\mu_s e^{-\mu_s x} dx = \frac{1}{\mu_s}. \tag{1.23}$$

Total attenuation coefficient: μ_t – SI units: m^{-1}.

The combination of absorption and scattering events leads to the introduction of a *total attenuation coefficient*:

$$\mu_t = \mu_a + \mu_s. \tag{1.24}$$

Because absorption and scattering events are mutually exclusive, one can derive the total attenuation probability density $[f_t(x, \mu_t)]$ and the associated cumulative distribution function $[F_t(L, \mu_t)]$ as in Eqs. (1.20) and (1.21), respectively, once μ_a is replaced by μ_t. The optical *mean free path* (mfp) is defined as the average distance traveled by a photon before it is *either* absorbed *or* scattered, and it is given by the inverse of the total attenuation coefficient:

$$\text{mfp} = \int_0^\infty x\mu_t e^{-\mu_t x} dx = \frac{1}{\mu_t}. \tag{1.25}$$

Single-scattering albedo: a – dimensionless.

The *single-scattering albedo* (a) is defined as:

$$a = \frac{\mu_s}{\mu_a + \mu_s} = \frac{\mu_s}{\mu_t}, \tag{1.26}$$

and it specifies the relative contributions of absorption and scattering to light attenuation. A value of a approaching 1 means that scattering dominates, whereas a value of a near 0 means that absorption dominates.

Phase function: $p(\hat{\Omega}, \hat{\Omega}')$ – SI units: sr^{-1}.

The *phase function* is the angular probability density (probability per unit solid angle) of photon scattering from direction $\hat{\Omega}$ into direction $\hat{\Omega}'$. It is defined so as to satisfy the normalization conditions:

$$\int_{4\pi} p(\hat{\Omega}, \hat{\Omega}') d\Omega = \int_{4\pi} p(\hat{\Omega}, \hat{\Omega}') d\Omega' = 1, \tag{1.27}$$

which state that the total probabilities of scattering from all possible directions into a given direction $\hat{\Omega}'$ or from a given direction $\hat{\Omega}$ into all possible directions are both equal to 1. For the case of unpolarized light and isotropic scattering particles,

$p(\hat{\Omega}, \hat{\Omega}')$ depends only on the angle between $\hat{\Omega}$ and $\hat{\Omega}'$ (the scattering angle), so that it can be written as $p(\hat{\Omega} \cdot \hat{\Omega}') = p(\cos\theta)$, where θ is the scattering angle. In this case, the normalization condition for the phase function becomes:

$$2\pi \int_0^\pi p(\cos\theta)\sin\theta d\theta = 2\pi \int_{-1}^{1} p(\cos\theta) d(\cos\theta) = 1. \tag{1.28}$$

We observe that, in this case, one can express the probability of scattering over a solid angle element $(d\Omega')$ equivalently in terms of the phase function $p(\cos\theta)$ or the product of probability densities per unit polar angle, $p_\theta(\theta)$, and per unit azimuthal angle, $p_\varphi(\varphi)$. Explicitly: $p(\cos\theta)d\Omega' = p_\theta(\theta)d\theta\, p_\varphi(\varphi)d\varphi$. Considering that scattering, in this case of unpolarized light and an isotropic particle, is equally probable over the full range (2π) of the azimuthal angle φ, i.e., $p_\varphi(\varphi) = 1/(2\pi)$, and that $d\Omega' = \sin\theta d\theta d\varphi$, one finds $p_\theta(\theta) = 2\pi\sin\theta\, p(\cos\theta)$. Consequently, in the special case of isotropic scattering, i.e., $p(\cos\theta) = 1/(4\pi)$, the scattering probability per unit polar, or scattering, angle is not uniform over θ. (In fact, $p_\theta(\theta) = \frac{1}{2}\sin\theta$.)

This result of a non-uniform probability density per unit scattering angle, $p_\theta(\theta)$, in the case of isotropic scattering may seem surprising and counterintuitive, but one can see its reason on the basis of a physical argument. Isotropic scattering implies that directional irradiation of the scattering particle with a given radiant power P_0 (units: W) results in a uniform radiant angular intensity $\mathcal{J}_0$ (units: W/sr) along any direction. Consider a sphere of radius R centered at the scattering particle; the surface element of the sphere that intersects the range $d\theta$ about a scattering angle θ is $2\pi R^2 \sin\theta d\theta$. This surface element is greatest for $\theta = \pi/2$ and gets smaller and smaller as the scattering angle approaches 0 (forward scattering) or π (backscattering). By normalizing the area of this surface element by the area of the sphere $(4\pi R^2)$ one finds a scattering probability over $d\theta$ about θ of $\frac{1}{2}\sin\theta d\theta$, in agreement with the above derivation. This is why a uniform scattering probability per unit solid angle results in a non-uniform scattering probability per unit polar angle, or scattering angle, θ.

We conclude this section on the phase function by noting that its name has nothing to do with the phases of the incident or scattered optical radiation, which are in fact irrelevant for the description of the distribution of scattering angles provided by the phase function. The terminology originates from astronomy, where the phase angle for a given planet (or the moon) is defined as the angle between the line joining the sun and the planet (the direction of illumination) and the direction joining the planet and the earth (the "scattering" direction). This astronomy angle offers a similarity to the scattering angle defined in optics once you translate the astronomical sun-planet-earth into the optical source-particle-detector. Because this astronomy angle defines the phase of the planet or the phase of the moon – the angle is ~ 0 at new moon and $\sim \pi$ at full moon – it is given the name of phase angle.

Average cosine of the scattering angle: g – dimensionless.

The *average cosine of the scattering angle* is defined as the expectation value of $\hat{\Omega} \cdot \hat{\Omega}'$ according to the scattering distribution described by the phase function:

$$g = \int_{4\pi} (\hat{\Omega} \cdot \hat{\Omega}') p(\hat{\Omega} \cdot \hat{\Omega}') d\Omega' = \langle \cos \theta \rangle. \tag{1.29}$$

It represents the average scattering angle over many scattering events. In the case of unpolarized light and isotropic scattering particles, g is independent of the direction of propagation of the incoming photons. The average cosine of the scattering angle is most commonly referred to as the *anisotropy factor*, and is also sometimes called the *asymmetry parameter*, *scattering anisotropy*, or simply *anisotropy*.

Reduced scattering coefficient: μ_s' – SI units: m^{-1}.

The *reduced scattering coefficient* (μ_s') (also referred to as *transport scattering coefficient*) involves a generalized definition of the scattering coefficient that takes into account the angular scattering distribution through the average cosine of the scattering angle:

$$\mu_s' = \mu_s(1 - \langle \cos \theta \rangle) = \mu_s(1 - g). \tag{1.30}$$

Its inverse ($1/\mu_s'$) represents the distance over which photons "lose memory" of their original direction of propagation, and it is therefore a length scale that characterizes isotropic or effectively isotropic scattering.

Total reduced attenuation coefficient: μ_t' – SI units: m^{-1}.

The combination of absorption and reduced scattering effects results in the introduction of a *total reduced attenuation coefficient*:

$$\mu_t' = \mu_a + \mu_s', \tag{1.31}$$

which is sometime referred to as *total transport coefficient*.

Reduced single-scattering albedo: a' – dimensionless.

The *reduced single-scattering albedo* (a'), which is also referred to as *transport single-scattering albedo*, is defined as:

$$a' = \frac{\mu_s'}{\mu_a + \mu_s'} = \frac{\mu_s'}{\mu_t'}, \tag{1.32}$$

and it specifies the relative contributions of absorption and effectively isotropic scattering to light attenuation. A value of a' approaching 1 means that effectively isotropic scattering dominates, whereas a value of a' near 0 means that absorption dominates.

Effective attenuation coefficient: μ_{eff} – SI units: m^{-1}.

The *effective attenuation coefficient* (μ_{eff}) is defined as follows in terms of the absorption and reduced scattering coefficients:

$$\mu_{\text{eff}} = \sqrt{3\mu_a(\mu_s' + \mu_a)}. \tag{1.33}$$

It is highly significant in biomedical optics because its inverse ($1/\mu_{\text{eff}}$) provides a length scale, in the diffusion regime, over which the optical energy attenuates in tissues. Such a length scale is referred to as the *diffusion length* (L_D) or the *optical penetration depth* (to be dealt with in more detail in Section 10.2.1).

Single-distance diffuse reflectance: $R(\rho)$ – SI units: m^{-2}.

The *single distance diffuse reflectance* [$R(\rho)$], or *local reflectance for point illumination*, represents the optical signal (normalized by the strength of the optical illumination) that is measured on the tissue surface at a distance ρ from the illumination point. In this case, the measured signal represents an optical intensity (optical power over the area of light collection) whereas the illumination signal represents an optical power (not an intensity, because in the case of point illumination what matters is the optical energy delivered per unit time, regardless of the area about the illumination point over which it is delivered). Therefore, the single-distance diffuse reflectance has units of an inverse area.

Total diffuse reflectance for point, plane wave, diffuse, or spatially modulated illumination: $\dot{R}_d$, $\bar{R}_d$, $\ddot{R}$, $\tilde{R}_d$ – dimensionless.

The case in which the optical signal resulting from surface illumination is collected from the entire tissue surface, rather than from a single point, realizes a *total diffuse reflectance* measurement. In the case of point irradiation, the illumination optical signal represents an optical power, and the total diffusely reflected signal also represents an optical power, because it results from the integration of a decaying intensity over the entire surface area of the medium. The total diffuse reflectance for point illumination, indicated with the symbol $\dot{R}_d$, is therefore a dimensionless ratio of illumination-to-collection optical power. In the case of broad-beam illumination (ideally by an infinitely broad beam), both illumination and diffusively reflected signal represent an optical intensity. Consequently, the diffuse reflectance under broad-beam illumination is a dimensionless ratio of illumination-to-collection optical intensity. In this book, we introduce different symbols to identify the diffuse reflectance for broad-beam plane-wave illumination ($\bar{R}_d$), diffuse illumination ($\ddot{R}$), or spatially modulated illumination ($\tilde{R}_d$).

Problems – answers to problems with * are on p. 645

1.1* How many photons per second are emitted by a light source that radiates light in vacuum with a power of 3 mW at a wavelength of 680 nm?

1.2 A light source is turned on at time $t = 0$ and directed onto a tissue surface. The radiant energy delivered by this light source onto the tissue surface from time 0 to time t is given by $Q(t) = Q_0\sqrt{t/t_0}$, where $Q_0 = 12\,\text{mJ}$ and $t_0 = 2\,\text{s}$. What is the source power at time $t = 8\,\text{s}$?

1.3* A point light source emits a uniform radiant angular intensity of 14.9 W/sr over a solid angle of π (symmetrically distributed around the polar axis of emission z). A large flat screen, orthogonal to z, is placed at a distance of 50 cm from the source.

(a) What is the maximum polar angle (the angle θ with the positive z axis) at which this light source radiates?

(b) What is the area of the illuminated surface on the screen?

(c) What is the total radiant power on the screen?

(d) How long does it take to deliver a radiant energy of 176 J to the screen?

1.4 Consider a case in which the maximum permissible radiant exposure of the skin is 1.1 J/cm^2 for laser illumination over a time period of 1 s. What is the minimum illumination spot size that can be used to stay within this limit if a laser emitting a power of 15.4 mW is used?

1.5* Suppose that you place a flat surface of area $A = 0.5\,\text{cm}^2$ inside a broad (consider it to be infinitely wide) collimated laser beam of uniform intensity 5 W/cm^2.

(a) How much power is intercepted by the surface if it is perpendicular to the beam?

(b) How much power is intercepted by the surface if its normal makes an angle of 30 degrees with the direction of the collimated beam?

1.6 The radiant angular intensity emitted by the sun is 3×10^{25} W/sr. The average distance between the sun and the earth is 1.5×10^{11} m.

(a) What is the radiant power incident on the head of a person walking outside in a sunny day? (Assume a surface area of 260 cm^2 for the person's head.)

(b) What is the radiant intensity incident on the person's head?

(c) What is the radiant energy delivered by the sun to the person's head in one hour?

1.7 For broadband light, it is relevant to define the radiant energy density per unit frequency (u_f) or per unit wavelength (u_λ) such that $u_f df$ and $u_\lambda d\lambda$ represent the radiant energy density over the infinitesimal frequency element df about f and the corresponding infinitesimal wavelength element $d\lambda$ about λ, respectively.

(a) Find the relationship between u_f and u_λ.

(b) Consider the case of a total radiant energy density $U = 350\,\text{mJ/cm}^3$ over the wavelength range $\lambda_1 \leq \lambda \leq \lambda_2$, with $\lambda_1 = 500\,\text{nm}$ and $\lambda_2 = 800\,\text{nm}$. Assuming that $c_n = 2.26 \times 10^{10}$ cm/s, find the total photon density U_N over the wavelength range $\lambda_1 \leq \lambda \leq \lambda_2$ for the following cases:

(b1) constant radiant energy per unit wavelength ($u_\lambda =$ constant for $\lambda_1 \leq \lambda \leq \lambda_2$);

(b2) constant radiant energy per unit frequency ($u_f =$ constant for $\lambda_1 \leq \lambda \leq \lambda_2$).

1.8* For a case of isotropic radiance $L(\hat{\Omega}) = L_0 = 8.34$ mW/(cm^2-sr), find the hemispherical intensity $I_{+\hat{\mathbf{n}}}$ along any direction $\hat{\mathbf{n}}$.

1.9 Consider the radiance $L(\hat{\Omega}) = \frac{L_0}{4\pi}(3 + 2\hat{\Omega} \cdot \hat{\mathbf{z}})$, where $L_0 = 0.6$ W/(cm^2-sr), and $\hat{\mathbf{z}}$ is the unit vector along the positive z axis. Consider also two optically transparent flat surfaces S_{xy} and S_{xz} parallel to the xy plane and xz plane, respectively, and both having an area $A = 4$ cm^2.

(a) Find the radiant power incident on S_{xy} from below (i.e., from lower to higher z) and from above (i.e., from higher to lower z).

(b) Find the radiant power incident on S_{xz} from the left (i.e., from lower to higher y) and from the right (i.e., from higher to lower y).

(c) Find the net radiant power (value and direction) flowing through S_{xy} and S_{xz}.

(See the hint in Problem 1.10.)

1.10* Find, in terms of the constant A, the fluence rate ϕ and the net flux $\mathbf{F}$ associated with the following radiances:

(a) $L = A\delta(\hat{\Omega} - \hat{\mathbf{x}})$ (collimated radiance along $\hat{\mathbf{x}}(\theta = \pi/2, \varphi = 0)$; here δ is the Dirac delta defined in Appendix 10.A);

(b) $L = \frac{A}{4\pi}$ (isotropic radiance);

(c) $L = \frac{A \sin\theta}{\pi^2}$;

(d) $L = \frac{A\theta}{2\pi^2}$.

(Hint: By using the polar angle θ [angle with the positive z axis: $0 \leq \theta \leq \pi$] and the azimuthal angle φ [angle from the positive x axis in the x-y plane: $0 \leq \varphi < 2\pi$], one can write:

$$\hat{\Omega} = \sin\theta \cos\varphi\, \hat{\mathbf{x}} + \sin\theta \sin\varphi\, \hat{\mathbf{y}} + \cos\theta\, \hat{\mathbf{z}};$$

$$d\Omega = \sin\theta d\theta d\varphi;$$

$$\delta(\hat{\Omega} - \hat{\Omega}') = \frac{1}{\sin\theta'}\delta(\theta - \theta')\delta(\varphi - \varphi').)$$

1.11* What is the probability that a photon travels a distance of at least 1 mm in a medium with $\mu_a = 0.01\ \text{cm}^{-1}$ and $\mu_s = 31\ \text{cm}^{-1}$ without experiencing any absorption or scattering events?

1.12 Find the normalization factor K for the phase function $p(\hat{\Omega}, \hat{\Omega}') = K \sin\theta$ and its anisotropy factor g.

1.13* In a given medium, the photon mean free path for scattering is 250 μm and the reduced scattering coefficient is $8.2\ \text{cm}^{-1}$. What is the mean value of the cosine of the scattering angle?

1.14 Can the reduced single-scattering albedo a' possibly be greater than the single-scattering albedo a? If so, what is the condition for $a' > a$?

1.15* What is the condition for which the diffusion length and the typical length scale for effectively isotropic scattering are the same?

References

Hetzel, F., Patterson, M., Preuss, L., and Wilson, B. (1996). *Recommended Nomenclature for Physical Quantities in Medical Applications of Light*. AAPM Report No. 57. Woodbury, NY: American Institute of Physics.

Quantities and Units: Part 7: Light, International Organization for Standardization, ISO 80000–7:2008.

Further reading

DeCusatis, C., ed. (1998). *Handbook of Applied Photometry*. New York: Optical Society of America and Springer.

Welch, A. J., and van Gemert, M. J. C., eds. (2011). *Optical-Thermal Response of Laser-Irradiated Tissue*, 2nd edn. Dordrecht: Springer, Chapter 1.

2 Overview of tissue optical properties

In Chapters 5, 6, and 7, we will examine in detail the molecular transitions and microscopic-scale structural features of tissue that are responsible for its optical properties on any scale, and that afford opportunities for diagnostic assessment. In this chapter, however, we introduce the primary constituents and structures of tissue that are responsible for its bulk optical properties, with primary attention to the wavelength range of near-UV through visible to near-IR. Of course, the heterogeneity of tissue lends complexity to the interpretation of the interactions of light with tissue, but, prior to addressing the effects of the heterogeneity, it is instructive to understand the factors that govern the averaged optical properties on a macroscopic scale.

The two optical properties that govern light transport in tissue, and that must be understood thoroughly, are absorption and scattering.

> **Absorption** is due to a variety of chromophores, compounds and fluids that absorb light at certain wavelengths and are commonly found in tissue. The word *chromophore* derives from the Greek *chroma*, for color. This would imply that a chromophore exhibits an optical absorption band in the visible range of wavelengths (400–700 nm), although common usage in biomedical optics includes the near-infrared (NIR) region as well. In mammalian tissue, the origins of optical absorption in the visible-NIR range are primarily due to broad absorption features of electronic transitions of π-electrons in large organic molecules and biomolecular constituents.
>
> **Scattering** is due to gradients and discontinuities in the optical refractive index of the various architectural microstructures and sub-cellular components of tissue.

Other intrinsic optical properties of tissue that can affect light transport include *birefringence* and *optical activity*. These are subtle effects, however, that do not generally affect bulk transport properties of tissue, but can be relevant to microscopy based on phase contrast or differential interference contrast, which will be studied in Chapter 17 (Section 17.2).

Whether absorption or scattering dominates the propagation of light in tissue depends on the specifics of the tissue optical properties, which are generally a function of wavelength, and on their spatial distribution at microscopic and

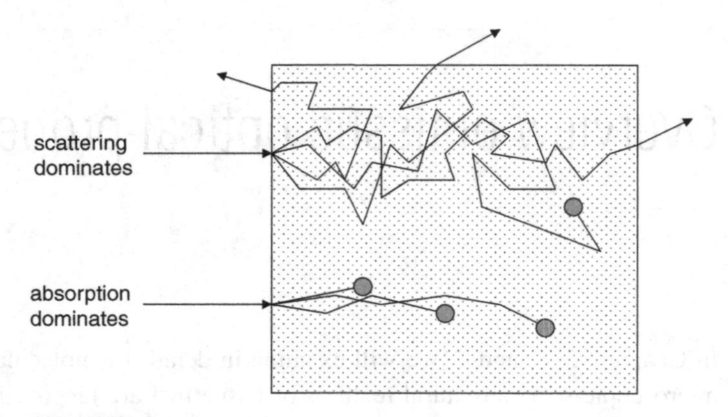

Figure 2.1

Schematic of possible photon paths in a turbid medium. For the upper beam $\mu_s > \mu_a$. For the lower beam $\mu_a > \mu_s$. The darker spots represent absorption events.

macroscopic levels. Figure 2.1 illustrates two cases, wherein either absorption or scattering dominates.

2.1 Absorption and scattering coefficients

We address the bulk optical properties, for which typical scale lengths are long compared to both the wavelength of light and the individual elements of tissue microstructure (cells, for example). The commonly used terms were introduced in Chapter 1 (Section 1.3). We repeat here the expressions for the absorption and scattering coefficients, and note that these are commonly reported in units of cm^{-1}, rather than SI units of m^{-1}, because centimeters are much closer to dimensions of physiological structures and to the scale of geometries used in optical measurements for diagnostic applications in tissue. Thus,

$$\mu_a[\text{cm}^{-1}] \equiv \text{probability of absorption per unit pathlength}$$
$$= \text{inverse of the mean-free-path for absorption}$$

and

$$\mu_s[\text{cm}^{-1}] \equiv \text{probability of scattering per unit pathlength}$$
$$= \text{inverse of the mean-free-path for scattering}$$

The absorption coefficient is related to the intrinsic properties of an absorbing species through the expression:

$$\mu_a = C\varepsilon, \tag{2.1}$$

where C[M] is the molar concentration (M = mol/L) of the absorbing compound, and $\varepsilon[\text{cm}^{-1}/\text{M}]$ is its molar extinction coefficient, i.e., its absorption per unit concentration. If there are multiple absorbing species in a medium, as is generally the case in tissue, then Eq. (2.1) can be expanded as:

$$\mu_a = \sum_i C_i \varepsilon_i, \tag{2.2}$$

with the units as in Eq. (2.1).

Similarly, one may write an expression for the scattering coefficient in terms of the combined contributions to the bulk scattering probability of individual scattering particles as

$$\mu_s = \sigma_s N, \tag{2.3}$$

where σ_s (cm^2) is the scattering cross section of an individual particle, and N(cm^{-3}) is the number density of the particles. Equation (2.3) is primarily accurate for low-to-modest concentrations of small particles, where there is minimal "shadowing" effect of particles in front of others. One can examine further the scattering cross section of a single particle, such as a sub-cellular organelle or other biological microstructure, and this is addressed later in this chapter (and in detail in Chapter 7; see Eq. (7.32)); for now, however, the scattering coefficient is considered as a bulk optical property of tissue, which makes sense at spatial scales that are long compared to the size of typical cellular tissue constituents. We also note that the scattering coefficient, in this terminology, refers only to elastic scattering: the case for which photons change direction but not energy. (Inelastic scattering, such as Raman scattering, will be covered in Chapter 6.)

In Chapter 1 we also defined the single-scattering albedo, Eq. (1.26): $a = \mu_s/(\mu_a + \mu_s)$. Intuitively, the albedo tells us how reflective (i.e., how "white") a turbid medium would appear from the outside. Note that while the absorption, scattering and total attenuation coefficients have units of inverse length, the albedo is dimensionless. Its value ranges between 0 and 1. Thus, a highly reflective object would have an albedo approaching 1, and would appear white to the naked eye (think of white clouds or milk), whereas an object with low albedo would appear to be dark (think of coal dust or India ink).

2.2 Survey of the primary chromophores in tissue

Hemoglobin At visible wavelengths, the "800-pound gorilla" of absorption in mammalian tissue is *hemoglobin.* Hemoglobin is a globular metalloprotein, located in red blood cells, that is responsible for oxygen transport. A remarkable

Figure 2.2
Hemoglobin is a globular protein with four globular subunits of the same structure. Most of the protein structure of each globule is alpha-helical, with non-helical connecting segments, and its overall structure is governed by hydrogen bonding. Each globular unit has a strongly bound heme group. (OpenStax College [http://cnx.org/content/col11496/1.6/]; converted to grayscale.)

feature of hemoglobin is its reversible oxygenation, enabling it to readily bind and later release molecular oxygen. Whether in its oxidized and reduced state (*oxy-hemoglobin* and *deoxyhemoglobin*, respectively), it is the most important native chromophore in most types of tissue. (For efficiency, these two compounds will often be referred to as HbO and Hb in this text, and the word hemoglobin will denote either form.) The component of hemoglobin where the action occurs is the *heme group* (also referred to as the *heme complex* or, simply, the *heme*), whose chemical formula is $C_{34}H_{32}FeN_4O_4$, with a molecular weight of 616.48, whereas the molecular weight of the entire hemoglobin protein is approximately 64,500. (This means that one mole of hemoglobin has a mass of 64.5 kg.) The heme, with its central iron atom, is the component that binds molecular oxygen from the blood. As depicted in Figure 2.2, there are four heme groups in a hemoglobin protein molecule, each capable of binding one molecule of oxygen.

The term *saturation*, or *oxygen saturation*, of hemoglobin refers to the percentage of the heme groups, of all the hemoglobin in a sample, that have bound oxygen. Thus, 90% saturation would mean that 9 of 10 heme groups in a blood sample have bound oxygen molecules. This percentage, when averaged over the microvessels in peripheral tissue, is sometimes abbreviated as StO_2; and in the case of arterial blood, it is often abbreviated SaO_2. A full discussion of hemoglobin saturation and its relationship to the partial pressure of dissolved oxygen will be developed in Chapter 15 (Section 15.1).

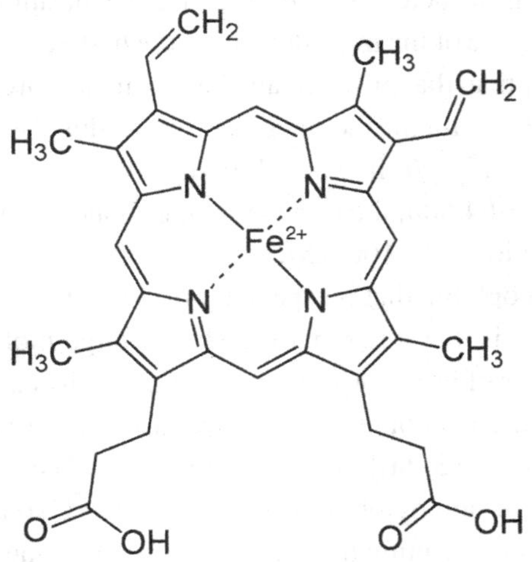

Figure 2.3

Molecular diagram of the heme group, with the central iron atom that is responsible for binding oxygen.

Figure 2.2 represents the gross molecular structure of hemoglobin, and Figure 2.3 shows the detailed molecular structure of the heme group.

So, how much hemoglobin do we have in our bodies? To start, approximately 35–40% of the volume of *red blood cells* (RBCs), or *erythrocytes*, is filled with hemoglobin. Importantly, in healthy tissue, hemoglobin exists *only* within erythrocytes, and is quickly broken down and metabolized as erythrocytes die. Thus, outside blood vessels there is essentially no hemoglobin. Next, the percentage of the volume of whole blood occupied by red blood cells, called the *hematocrit*, is typically in the range 40–46% for healthy adult males and 37–43% for healthy adult females. (These ranges can vary with age, diet, degree of physical activity and health status. We will assume an average value of 40% for both sexes combined.) The next factor is the percentage of tissue volume that is occupied by blood. This, of course, varies for different types of tissue. Blood constitutes about 6–8% of the entire body mass of a healthy adult (about 5 liters), but some of that is flowing in major blood vessels, such that, on average, about 4% of tissue volume is occupied by the perfusion of smaller vessels, mainly capillaries, arterioles and venules. This percentage is often referred to as the *blood volume fraction* or *perfusion* of the tissue. The tissues of some organs, such as liver and kidney, are more highly perfused (12–14%), commensurate with their functions, whereas adipose tissue (fat) has a lower blood volume fraction (~2%).

The final calculation, to determine the molar concentration of hemoglobin in tissue, requires an estimate of the mass density of the hydrated hemoglobin protein, which is slightly larger than that of water, and has been measured as ~1.1–1.2 g/mL (Grover, 2011). Together with the factors given above, this then leads to an average mass density of Hb of ~170 g/L in whole blood, and ~6.8 g/L in tissue; and, with a molecular weight of ~64,500, the corresponding tissue molar concentration (for 4% blood volume fraction) is ~100 μM.

The property of hemoglobin that is especially important, and relevant to the optical properties of tissue, is that it has an exceedingly large molar extinction coefficient at the visible peaks of its absorption spectrum, which is due almost entirely to the heme group. Measurement of an accurate spectrum of the extinction coefficient of hemoglobin is not a simple task, for several reasons. There are molecular variations in the structure of hemoglobin of different species, so accurate knowledge of the spectral properties of human hemoglobin requires measurements starting with human blood. Further, within humans the exact structure of hemoglobin varies with age, with fetal and infant hemoglobin differing significantly from adult hemoglobin. Moreover, hemoglobin is relatively unstable outside erythrocytes, so a non-scattering solution of hemoglobin is difficult to maintain. If hemoglobin is extracted from red blood cells, and if it is purified and lyophilized (freeze-dried), it reverts to methemoglobin (pronounced met-hemoglobin). In methemoglobin the iron is oxidized (ferric), with a 3+ charge, whereas in normal hemoglobin in living cells the iron is ferrous, in the Fe^{2+} state. Methemoglobin cannot carry oxygen, and its absorption spectrum is dramatically different from that of hemoglobin (Zijlstra et al., 1991). Additionally, even when the proper form of hemoglobin is isolated and maintained in the Fe^{2+} state, it is experimentally difficult to control a sample such that it will be either purely oxyhemoglobin or purely deoxyhemoglobin, to enable obtaining the individual spectra.

Can you really "turn blue" by holding your breath? Numerous authors have published absorption spectra for oxy- and deoxyhemoglobin, but the most rigorous of these have typically covered small segments of the full (near-UV to near-IR) optical spectrum, and/or a limited range of oxygen saturation. Scott Prahl, of the Oregon Health Sciences University, has published online a set of spectra for oxy- and deoxyhemoglobin for the full spectral range 250–1000 nm (Prahl, 1999). These highly quoted spectra are the result of a meta-analysis, wherein Dr. Prahl combined spectral segments from different published sources, in some cases deducing consensus values, to synthesize a "best guess" full spectrum for each of the redox states. A plot of the resulting "best values" spectra is shown in Figure 2.4. We hasten to note that these plotted values are for an extinction coefficient, ε_{10}, for *absorbance* (typically represented as A), which is a base-10

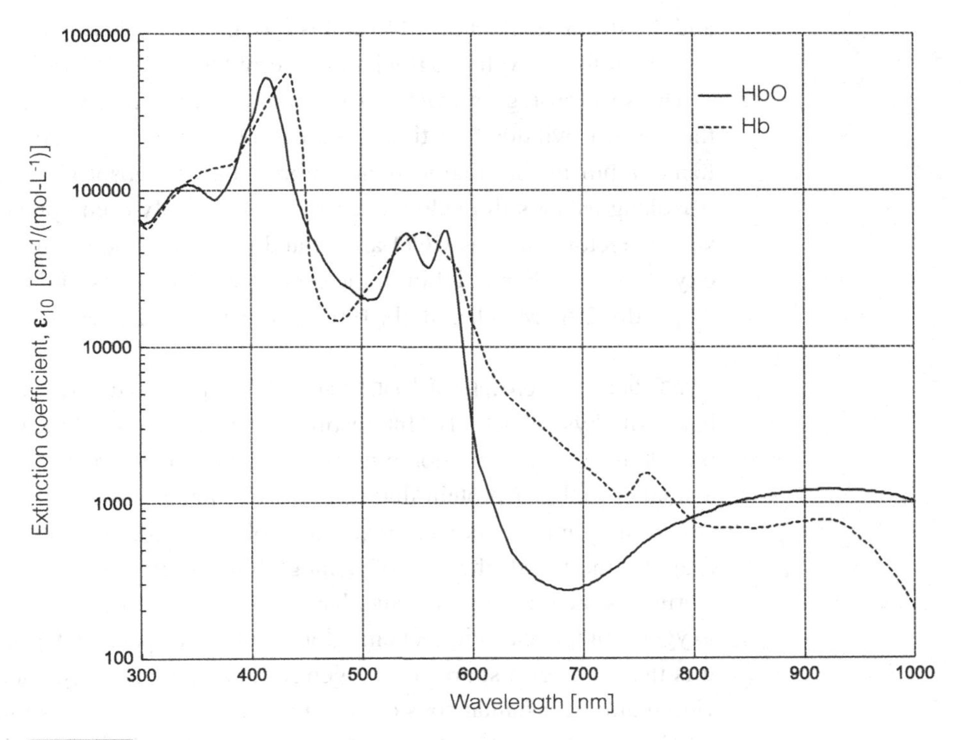

Figure 2.4

The wavelength dependence of the extinction coefficients for oxy- and deoxyhemoglobin. Note that this extinction coefficient is for base-10, ε_{10}, rather than base-e for ε, which is our convention for Eq. (2.1). Note also that the ordinate scale is logarithmic (data adapted from Prahl, 1999).

logarithmic scale, whereas the extinction coefficient, ε, related to the absorption coefficient, μ_a as used in our convention (Eq. (2.1)), is base-*e*. The conversion is, thus, $\varepsilon = \ln(10)\varepsilon_{10} \sim 2.3\varepsilon_{10}$.

From this spectral plot in Figure 2.4, the very strong absorption in the blue-green spectral region (400–600 nm) is readily evident, and one can see that blood would appear red, since the absorption in the red region (>610 nm) is much weaker. It should also be clear that in the red region there is, nonetheless, a significant difference between the extinction coefficients of oxyhemoglobin and deoxyhemoglobin, with HbO exhibiting much lower absorption than Hb. While the absorption due to Hb in the red region is weaker than in the blue-green, it is still significant, hence the dusky tone that tissue takes on when the hemoglobin is deoxygenated. The very strong absorption band at ~420 nm is often referred to as the *Soret band* (named for the Swiss chemist Jacques-Louis Soret [1827–1890]), and the strong doublet at ~440 and ~480 nm for oxyhemoglobin, or the singlet at ~460 nm for deoxyhemoglobin are referred to as the *Q bands*.

The spectral region from ~650 nm to ~1000 nm is sometimes referred to as the "optical window" in tissue. As can be seen from its spectra, absorption by hemoglobin in this region is at a minimum, yet absorption variations at these wavelengths are still readily detected by commonly used optical instrumentation, with detectors such as CCD arrays and cameras. The extinction coefficients of oxy- and deoxyhemoglobin in the optical window 650–1042 nm are tabulated in Appendix 2.A (Matcher et al., 1995; Kolyva et al., 2012).

Myoglobin (found predominantly in muscle) What powers us to engage in sudden, short bursts of physical activity? Interestingly, the main cause of the red color of skeletal muscle is not the hemoglobin in the blood vessels that perfuse muscle tissue, but a smaller globular protein that contains one heme group. Myoglobin has protein structural components similar to those of a quartile unit of hemoglobin, and is roughly one-fourth the size of hemoglobin. Myglobin, which is responsible for storing oxygen in muscle tissue, having only one heme group, has one-fourth the oxygen-binding capacity of hemoglobin but can be packed densely in muscle cells. It is the short-term source of oxygen for anaerobic activity, such as a sprint, providing almost instantaneous oxygen to muscle cells, whereas longer-term activity requires aerobic replenishment of oxygen from the blood. Myoglobin is especially abundant in the muscles of diving sea animals (such as whales, seals and penguins), which can hold their breath for a long time, with the myoglobin acting as an oxygen "bank" with greater capacity, due to the large muscle mass, than the blood supply itself. The oxygen-dissociation curve of myoglobin is described in Section 15.2.1. The molar extinction coefficient of myoglobin has a similar spectral shape to that of hemoglobin, but with about one-fourth the value (in proportion to the number of heme groups).

Bilirubin Old and dying erythrocytes are processed in the spleen, where the protein chains of hemoglobin are broken up into amino acids, and the hemes are released. The hemes are metabolized to form bilirubin, which is yellow in color (see Figure 2.5). Bilirubin is mostly processed by the liver, and an overabundance of bilirubin in the blood or in tissue can be an indication of liver disease. Such a condition can manifest as a yellow hue in the skin or schlera of the eye, typically referred to as *jaundice*, from the French word for yellow (*jaune*). Jaundice often occurs in neonates when the liver is too immature to process bilirubin fast enough. Since excess bilirubin in the body is toxic, potentially causing lifelong health problems, neonates with excess bilirubin are often placed under a "bili-light," a blue light source, since blue light helps accelerate the breakdown of bilirubin.

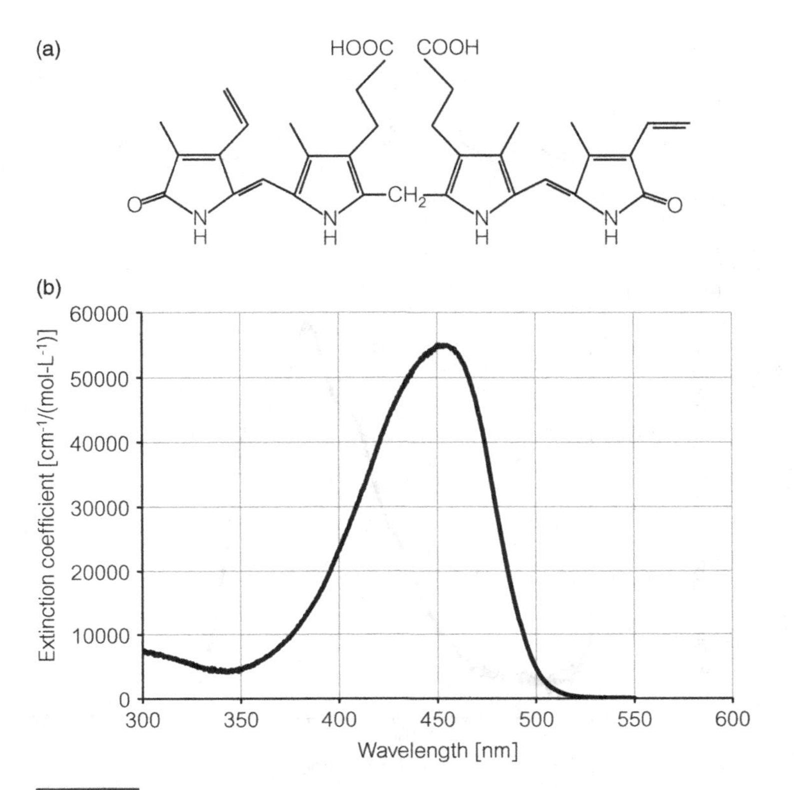

Figure 2.5

(a) The molecular structure or bilirubin, a breakdown product of the heme; (b) plot of the extinction coefficient of bilirubin, indicating strong absorption in the violet-blue spectral range (data from PhotochemCAD 2: Dixon et al., 2005).

Beta-carotene *Beta-carotene* (or β-carotene) is a strongly orange-colored compound (see Figure 2.6), found in many fruits and vegetables, and in abundance in carrots (hence, the source of its name). It is a precursor and is enzymatically converted to vitamin A in the human body. Beta-carotene is lipid soluble, is stored in the liver and is also found (in humans) in adipose (fatty) tissue, as well as arterial plaque. It is beta-carotene that gives human fat its pale-yellow color, whereas fatty tissue in pigs and cattle is more white in color.

Flavins Flavins (from the Latin for yellow, *flavus*) are proteins that serve as enzymes, which catalyze the redox processes that are elemental to the energy cycle of cells and cellular respiration, including the Krebs cycle. (The oxidation cycle in the mitochondrial matrix named after the German-born, British biochemist Hans Krebs [1900–1981].) These include flavin adenine dinucleotide (FAD) and flavin mononucleotide (FMN). While these can be detected by fluorescence, their

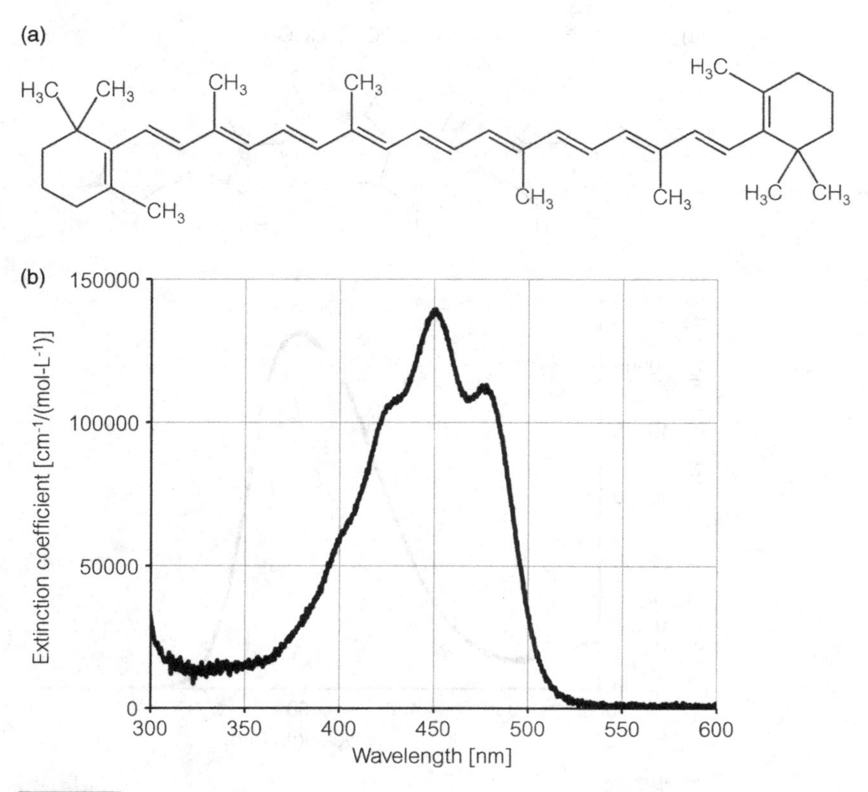

Figure 2.6

(a) The molecular structure of beta-carotene, a precursor to vitamin A found in adipose tissue and other lipids; (b) plot of the extinction coefficient of beta-carotene, indicating strong absorption from the violet to blue-green spectral range (data from PhotochemCAD 2: Dixon et al., 2005).

concentrations in tissue are typically much lower than hemoglobin, such that their contribution to the absorption coefficient of typical tissue, in the middle of the visible spectrum, is minimal, resulting in a negligible perturbation of the absorption spectrum.

Cytochrome *c* and cytochrome *c* oxidase *Cytochrome c* is a small protein incorporating a heme, which is found in mitochondria. One important role is in electron transport to cytochrome *c* oxidase (a mitochondrial trans-membrane protein), which is involved in the energy cycle role of ATP synthesis. Measurement of the redox state of cytochrome *c* oxidase, which can serve as a tool for evaluating metabolic rates in tissue, is described in Section 15.1.3.2. Cytochrome *c* exhibits absorption similar in spectral shape to that of hemoglobin. Its limited abundance, however, results in a relatively small perturbation to the tissue absorption spectrum due to hemoglobin and/or myoglobin. In cytochrome *c* oxidase, the optical absorption changes are due to the state of electron charge transfer in the copper atoms of a copper A complex.

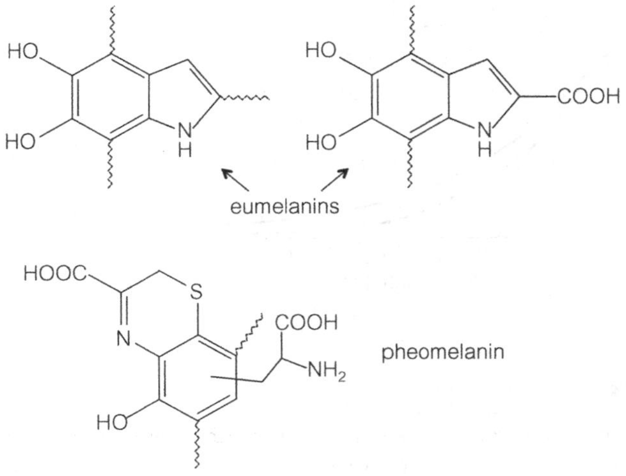

Figure 2.7

The molecular structures of the monomeric building elements of eumelanin and pheomelanin. The wavy lines indicate binding sites for polymerization.

The difference in the extinction coefficients for the oxidized state (Cyt|ox) minus the reduced state (Cyt|red) of cytochrome *c* oxidase is tabulated in Table 2.A.1 from 650 to 986 nm.

Melanin *Melanin* (from the Greek *melas*, for black or dark) is often, externally, the most obvious chromophore in humans, since it is produced by *melanocytes* that reside in the basal layer of the skin and in the eye. In the eye, the melanocytes are located in the *choroid* and in the *retinal pigmented epithelium* and, in the case of brown-eyed people, in the iris. The melanocytes store some of the melanin they produce, but also pack it into membrane-bound organelles, called *melanosomes*, which are typically 100–500 nm in diameter and can be exuded by the melanocytes. The melanosomes can then move through the extracellular fluid space, spreading into nearby tissue. In the skin, the melanosomes generally migrate toward the surface and are eventually shed along with the keratinocytes.

In humans, melanin occurs in two primary forms: *eumelanin* and *pheomelanin*. Eumelanin comes in a range from brown to almost black and is what colors brown and black hair, various shades of brown in skin, and brown eyes. Pheomelanin results in "red" hair and freckles in red-haired people. Both eumelanin and pheomelanin are polymers of small monomeric structures, and are found in a large range of degrees of polymerization and tertiary molecular conformations. The molecular structures of the monomers are shown in Figure 2.7.

The polymeric structures of melanin and the packing density of melanosomes vary significantly, and can be affected by the synthesis rate as stimulated by UV

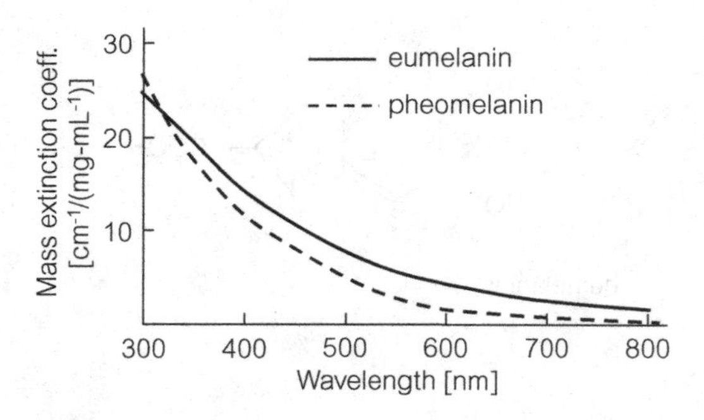

Figure 2.8

Mass extinction spectra of eumelanin and pheomelanin based on mass density (data extracted from Sarna and Swartz, 1988, and Riesz, 2007).

exposure or disease status. In any given sample there is a diverse mixture of polymer sizes. Consequently, there is no fixed value for a molecular weight, and it is thus not possible to assign specific values for a molar extinction coefficient. It is, however, possible to determine values of an extinction coefficient based on mass density, ε_m (cm^{-1}/(mg-ml^{-1})), although structural variations also generate variations in the values for this form of extinction coefficient. Figure 2.8 is a plot of the spectra of ε_m for eumelanin and pheomelanin, typical of melanin from melanosomes in the skin.

2.3 Non-chromophore absorbers

Some molecules commonly found in tissue do not fall in the class of chromophores because they do not have dipole-allowed electronic transitions in the visible-NIR range, and hence exhibit very weak optical absorption. Nonetheless, if they exist at large concentrations in tissue, as is the case for lipids and water, then the overtones of their vibrational modes can cause significant absorption in the near-infrared region. Optical measurement of their overtone bands can provide information about the local abundance of such molecules, which can lead to diagnostic information about tissue condition.

2.3.1 Lipids

Lipids constitute a class of molecules that are found in most tissues of the body, and are the primary constituents of fats, sterols (e.g., cholesterol), glycerides (e.g.,

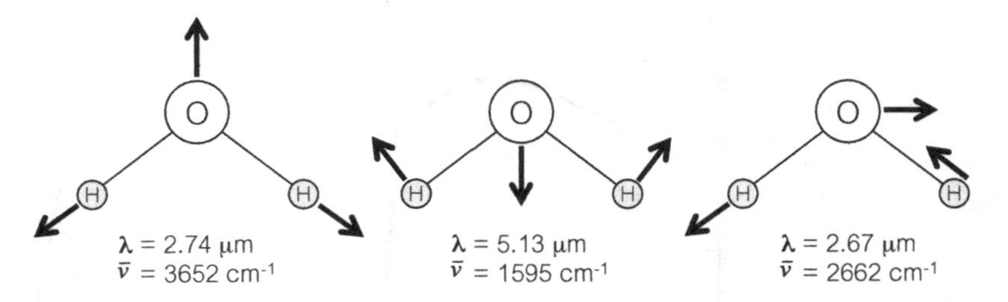

Figure 2.9

The fundamental vibrational modes of a free (gaseous) water molecule. The energies of the vibrational modes are denoted by the unit called *wavenumbers*, defined as $\bar{\nu} = 1/\lambda$. A detailed explanation of this unit is found in Section 6.1.

triglyceride), phospholipids and fat-soluble compounds (e.g., vitamins A, D and E). Lipids store energy and participate in signaling; and, importantly, the lipid bilayer forms the membrane of all cells and many organelles (whereas other organelles are separated by single-lipid layers). A representative spectrum of lipid absorption is tabulated from 650 to 1042 nm in Appendix 2.A (van Veen et al., 2004). There is an identifiable NIR absorption band at ~930 nm, which is readily distinguishable from the absorption spectra of water and hemoglobin, thus enabling determination of lipid concentrations in tissue by diffuse optical methods, such as those described in Chapters 10–14.

2.3.2 Water, water, everywhere – and vibrational modes

Every schoolchild knows that the majority constituent of the human body (and most living things) is water. Simply put, water is ubiquitous in biology. Water, like lipids, does not exhibit any electronic (dipole) transitions in the visible-NIR range and, consequently, is not regarded as a chromophore. Water does, nonetheless, exhibit strong absorption in the near-infrared, due to overtones of its primary vibrational modes, which absorb in the mid-infrared. The fundamental vibrational modes and their wavelengths/frequencies are shown in Figure 2.9. The wavelengths and *wavenumbers* of the modes are associated only with the gaseous phase of a single water molecule. The units of wavenumbers, $\bar{\nu} = 1/\lambda$, are cm^{-1}. The use of this unit instead of the more familiar definition of frequency (c/λ), when representing the energy of vibrational transitions, will be explained in detail in Chapter 6 (Section 6.1).

While gaseous water (vapor) has sharp spectral absorption features in the NIR to infrared, based on the fundamental modes and their various overtones, these are not of immediate relevance to tissue optical properties because the lines are

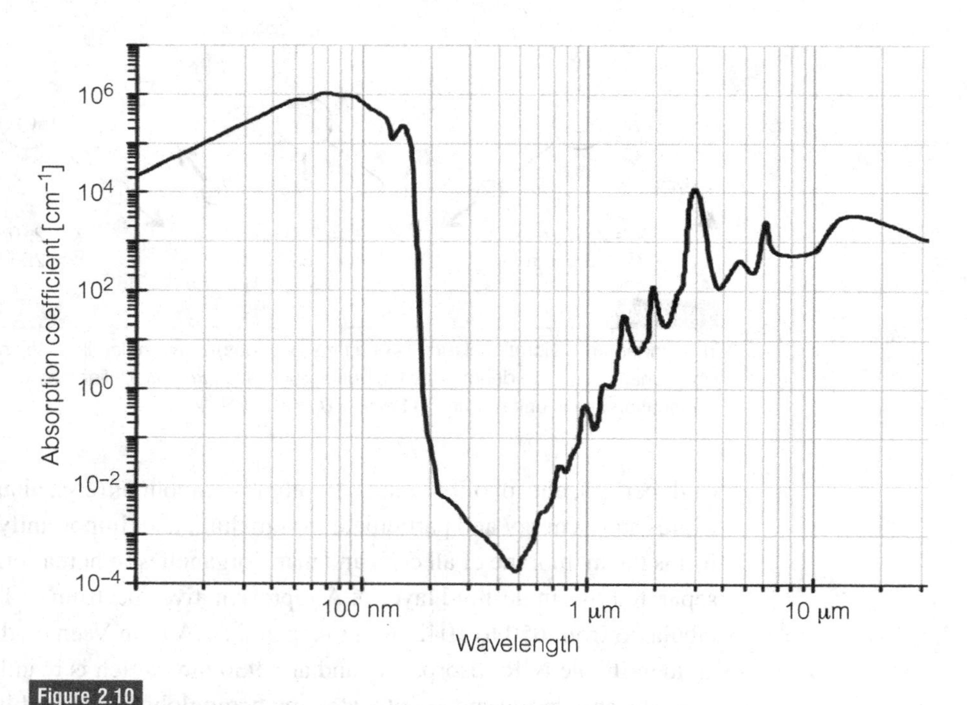

Figure 2.10

Absorption spectrum of liquid water (data adapted from Wikimedia [public domain], based on metadata compendium at http://omlc.org/spectra/water/abs/index.html).

dramatically broadened in liquid water due to collisional interactions and, especially, to the strong dipole-dipole coupling among nearest neighbors. The absorption spectrum of liquid water is shown in Figure 2.10 and tabulated in Appendix 2.A from 650 to 1042 nm (Segelstein, 1981). This is not presented as a molar extinction coefficient because water is generally the solvent, not the solute, so it is more logical to present the absorption coefficient, μ_a, of the pure liquid directly. The intense permanent-dipole field of the water molecule also exerts strong effects on the electronic transitions of compounds that are dissolved in water, such that their spectra in aqueous solution can differ significantly from that in, say, methanol.

Vibrational modes and the complexity of infrared spectra for biomolecules Inspection of the molecular structures depicted in several of the figures in this chapter (e.g., Figures 2.3, 2.5, 2.6) should make it evident that a large variety of stretching, bending and twisting modes can be manifest is such molecules, in addition to combinations of vibrational modes from neighboring atomic groups. The fundamental frequencies of almost all these vibrational modes are in the mid-infrared region (2–20 μm), and such complexity of the mid-IR spectra of organic molecules leads to the absorption spectra being referred to as the "fingerprint" spectral region

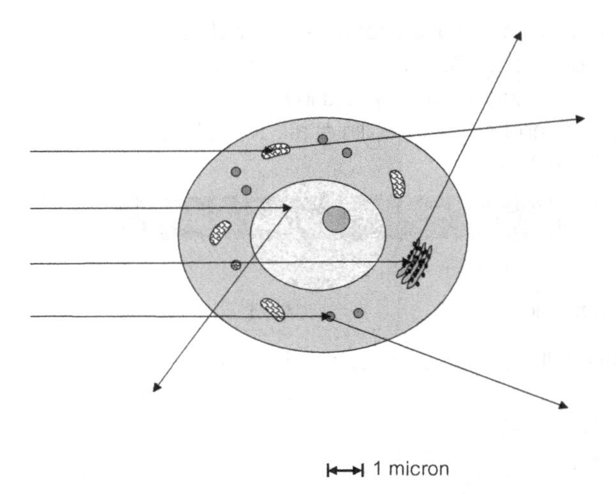

Figure 2.11

A physicist's cartoon of a cell, with rays representing random scattering from sub-cellular organelles of photons in an incident plane wave.

of biomolecules. These infrared-absorption bands, however, have little effect on the optical properties of tissue at visible-NIR wavelengths. Moreover, the fact that the absorption by water strongly dominates propagation of light in tissue at wavelengths $\gtrsim 1.3$ μm means that the vibrational absorption bands of biomolecules are generally not treated as significant factors in the visible-NIR optical properties of tissues. As we will see in Chapter 6, nonetheless, spectroscopic measurement of the vibrational modes of biomolecules can play a valuable role in tissue identification and diagnosis.

2.4 The scattering properties of tissue

2.4.1 Sources of scattering in tissue

Light will scatter off a microscopic water droplet in a cloud, whose typical size is 1–15 μm in diameter, as a consequence of the difference in refractive index between the water and surrounding air. Similarly, in tissues scattering is due to gradients and discontinuities in the index of refraction. (The electromagnetic theory of the scattering will be addressed in Chapter 7, Sections 7.2 and 7.3.) As represented in Figure 2.11, organelles, membranes and other intracellular structures have higher refractive indices than the intracellular fluid that surrounds them. Similarly, extracellular structural proteins also have higher refractive indices than the extracellular fluid that surrounds them. Factors that determine the effective

Table 2.1 Approximate values for refractive indices of tissue components at 500 nm. The question marks indicate a significant range of values in literature sources. Lipids and proteins: Bolin et al., 1989; DNA: Kwon et al., 2012.

Tissue component	Refractive index
water	~1.33
extracellular fluid	~1.34–1.35
intracellular fluid	~1.35–1.36
proteins	~1.40?
lipids	~1.45?
DNA	~1.44–1.49?
bilipid membrane	~1.46
melanin	~1.65

index of an organelle include dissolved salts and sugars, suspended amino acids, proteins, nucleic acids and DNA, as well as lipids and internal membrane structures. When local heterogeneity is on a scale that is small compared to the wavelength of light, we can often treat the index of an organelle as being homogeneous, at least for estimating first-order scattering properties.

Table 2.1 is a list of estimated refractive indices for different components of tissue, some gleaned from a variety of measurements reported in the literature.

The question marks for some components in Table 2.1 denote the fact that there is a sizable range of index values reported in the literature, and that the actual index depends on the specific tertiary structure and/or aggregation of proteins and DNA, as well as the degree of hydration. More detailed discussion of the effective indices of components and organelles is found in Chapter 7, Section 7.1.

2.4.2 The scattering cross section and its relation to the scattering coefficient

The probability of scattering by a single particle is represented by the *scattering cross section,* σ_s. The scattering cross section has units of area and can be larger or smaller than the actual physical cross section. The two are related by the scattering efficiency, Q_s, which is dimensionless.

$$\sigma_s = Q_s A_s \tag{2.4}$$

For a dielectric particle in a dielectric medium, Q_s ranges in value from 0 to ~4. Particles (organelles and structures), or *scattering centers*, in tissue are not generally spherical, so the cross section will depend on the orientation of the particle with respect to the direction of propagation of the impinging light. Nonetheless, it is often a reasonable approximation to treat scattering centers in tissue as spheres, when the effects of multiple scattering are being examined, given the averaging effects of random orientations.

We can now return to the concept of the scattering coefficient, which is a bulk property of a turbid medium, such as tissue, and relate it to the cross section for a single particle. For a medium containing multiple scattering particles, at a concentration, or number density, N (cm^{-3}), then the scattering coefficient is determined by the expression in Eq. (2.3).

In consideration of Eq. (2.4), Eq. (2.3) can also be written as

$$\mu_s = Q_s A_s N \tag{2.5}$$

Tissue is highly scattering, with typical values for μ_s of ~100 cm^{-1} for visible-NIR wavelengths; and in the wavelength range from the red and near-infrared (~640–1200 nm) the scattering coefficient is much larger than the absorption coefficient: $\mu_s \gg \mu_a$. With these parameters, the fluence rate (which includes light impinging from all directions) within a scattering medium can greatly exceed the intensity of the light impinging on the medium surface. Furthermore, diffuse reflection from the medium adds to the fluence rate outside the medium, which is therefore greater than that resulting solely from the light source. This interesting phenomenon is illustrated in Figure 2.12, which shows the fluence rate in and near a block of turbid medium, with light incident from a distant source on the left, and for the case where the absorption coefficient is much smaller than the scattering coefficient. The reader might experience this effect personally in the circumstance of climbing a mountain and being in the sun, with a layer of cloud below. It may seem intuitive that one would be more at risk of sunburn, compared to climbing on a cloudless day, because of backscattered light from below adding to the direct sunlight from above. It might be less intuitive, however, that as the climber then descends into the cloud layer, the exposure actually increases further, within the cloud, until the layer of cloud above becomes large.

2.4.3 The anisotropy factor in tissue

While the scattering cross section relates to the ability of an individual particle to scatter light, it does not tell us what the *scattering angle* is likely to be. In

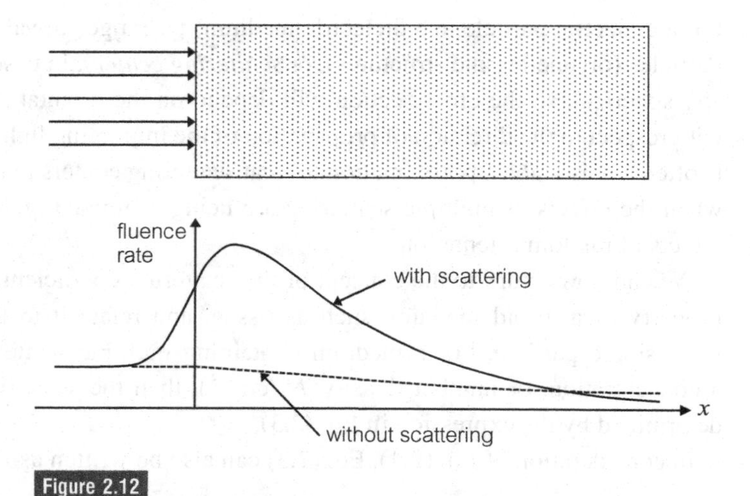

Figure 2.12

Fluence rate near and inside a turbid medium with weak absorption. The horizontal axis represents the distance, x, from the illuminated medium surface.

Chapter 1 (Section 1.3) we introduced the *scattering phase function*, $p(\hat{\Omega}, \hat{\Omega}')$, which, for unpolarized light incident upon isotropic particles, can be written as $p(\theta)[\text{sr}^{-1}]$ (without the subscript θ), representing the probability, per unit solid angle, of scattering at an angle θ *relative to the incident direction*; we noted that $p(\theta)$ is normalized such that the total scattering probability integrated over all solid angle is 1. We also defined the anisotropy factor, g, as the average cosine of the scattering angle, $\langle\cos\theta\rangle$, which can be calculated from the phase function. The phase function in detail is a property of the scattering by a single particle, and, as we will see in Chapter 7 (Section 7.2.2), it depends on the size, shape and refractive index of the particle. Thus, every different particle will have a different phase function.

Any representation of a phase function as a bulk optical property, in a heterogeneous medium (like tissue), is an expression approximated to average the properties of the multiple microscopic scattering events. That is, for describing the bulk properties of tissue, the phase function gets averaged over the many different types of structures and particles that light encounters in such a heterogeneous medium. It is in this context, of light-propagation properties of bulk media, where the *anisotropy factor*, g, is the more (and, in some circumstances, the only) appropriate parameter to represent an average phase function that itself represents the scattering by particles with effectively isotropic properties. For a medium with isotropic bulk properties (the medium is not birefringent, and the various scattering particles are randomly oriented), Eq. (1.29) determines the value of g for such an average phase function, and it can be rewritten in slightly simpler

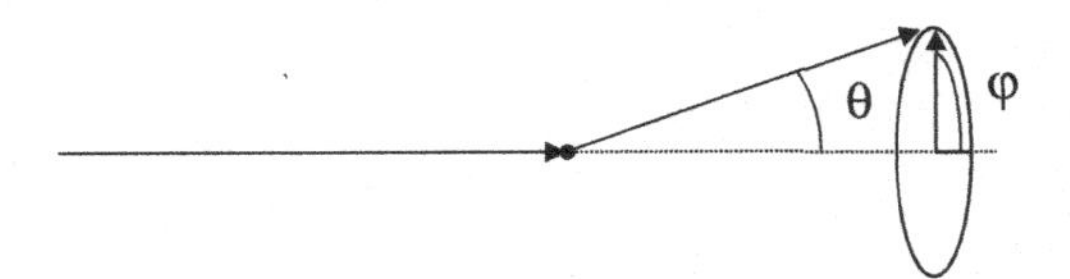

Figure 2.13

Geometry depicting the axes of the polar, θ, and azimuthal, φ, angles of scattering.

format:

$$g \equiv \langle \cos\theta \rangle = \int_{4\pi} \cos\theta p(\theta) d\Omega \tag{2.6}$$

where the integral is taken over all solid angle, $\Omega = 4\pi$.

The term *birefringence* refers to the property of some structured media that are optically anisotropic, in that they have a refractive index that is different for different polarizations and propagation directions of light. On a microscopic to mesoscopic scale (microns to millimeters) a variety of biopolymers exhibit birefringence, a salient example of which is collagen in its various forms. For most of the theoretical formalisms to be developed in this text, however, we will assume isotropic optical properties of the medium, and where the propagation parameters may be polarization-dependent, we will specifically note the polarization components, or otherwise indicate that the formalism presented is for unpolarized light.

More detailed discussions of the phase function and its mathematical representations are found in Sections 7.2.2 and 8.2.1–8.2.5. For individual particles of arbitrary shape, scattering can be a function of both the polar angle θ and the azimuthal angle, φ, which define the scattering direction $\hat{\Omega}'$ (see Figure 2.13), the direction of incident light, $\hat{\Omega}$, and its polarization angle relative to particle shape and properties, but we are limiting this discussion to unpolarized light and spherical particles with isotropic properties.

For unpolarized light and isotropic particles, there is no azimuthal dependence (on φ) and no dependence on the incident direction $\hat{\Omega}$; thus, Eq. (2.6) becomes,

$$g \equiv \langle \cos\theta \rangle = 2\pi \int_0^{\pi} \cos\theta p(\theta) \sin\theta d\theta \tag{2.7}$$

The value of g is bound between 0 and 1 for dielectric particles in a dielectric medium. (Negative values can obtain for conducting particles.) Again, for the bulk properties of tissue, g represents an average for a distribution of particle sizes, shapes and refractive indices. Small values of g, approaching 0, characterize particles much smaller than the wavelength of the light, indicating that scattering from very small particles is nearly isotropic, with equal probability in all directions

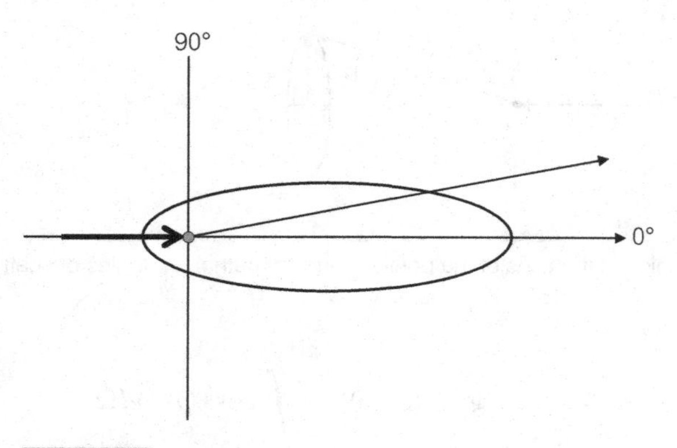

Figure 2.14

A numerically generated polar plot of an example of a phase function, $p(\theta)$, for a value of g of ~0.9.

(for unpolarized light). (Theoretically, $g = 0$ corresponds to any scattering distribution $p(\theta)$ that is symmetrical about a peak at $\theta = \pi/2$, but in biomedical optics it generally indicates isotropic scattering.) Large values of g, approaching 1, characterize particles much larger than the wavelength, representing scattering that is preferentially peaked in the forward direction. In the case of tissue, with a wide range of particle sizes, typical values of g range from ~0.75 to 0.98 (Cheong et al., 1990), with a value of 0.9 used frequently to represent average tissue properties. This corresponds to an average value for θ, $\langle\theta\rangle$, of 26°, consistent with the scattering in tissue being strongly forward-biased. Figure 2.14 shows a polar plot for a numerically constructed example of a phase function that would yield a value $g \cong 0.9$. One should note that although the average scattering angle is 26°, the *most probable* scattering angle for this example (and for any phase function representative of real tissue) is 0°.

2.4.4 The reduced scattering coefficient in tissue

Since most scattering events in tissue are in the near-forward direction, this means that it takes many scattering events before light becomes truly diffused, losing information as to the original direction of propagation. Think of this as the photon losing its "memory" of the location of its source. In common experience, when there is a thin cloud layer in the sky, and sunlight reaches the ground after only a few scattering events, one can look up and determine the approximate location of the sun, even if it cannot be clearly discerned. On the other hand, when there is heavy cloud cover one cannot even determine the sun's general location in the sky, despite the fact that much light still reaches the ground.

As defined in Chapter 1 (Eq. (1.30)), it is convenient to introduce a *reduced scattering coefficient*, μ_s', which relates to the probable distance of major changes in direction for a photon being scattered. The reduced scattering coefficient is expressed in terms of the anisotropy factor as defined in Eq. (1.30): $\mu_s' = \mu_s(1-g)$. The inverse of the reduced scattering coefficient, $1/\mu_s'$, can be thought of as representing the mean free path between effectively isotropic scattering events. This expression, Eq. (1.30), is sometimes referred to as a *similarity relation*, which is based on the condition that observed optical measurements are equivalent for any combination of g and μ_s that results in the same value for μ_s'. Such conditions hold for large source-detector separations, but not for short separations. Detailed development of phase function representations for small source-detector separations will be presented in Sections 8.2.1–8.2.5.

In the case of tissue, with $\mu_s \approx 100\ \text{cm}^{-1}$, for a value $g = 0.9$ we get $\mu_s' = (1-0.9)\,\mu_s = 10\,\text{cm}^1$. This means that the mean free path for effectively isotropic scattering is ~1 mm, as an average representative value for tissue.

2.5 The Beer-Lambert law to describe absorption and weak scattering

When a medium absorbs light but causes negligible scattering, it is easy to calculate the attenuation of light propagating through the medium. A simple exponential expression, commonly known as the *Beer-Lambert law*, relates the intensity of a collimated light source at any distance, L, along the direction of propagation to the incident intensity and the absorption coefficient:

$$I = I_0 e^{-\mu_a L} \tag{2.8}$$

The historical development of this simple relation is often overlooked (Perrin, 1948). This law is really the result of the work and observations of three scientists. The French scientist Pierre Bouguer [1698–1758] realized that the introduction of multiple equal layers of an absorbing substance results in a geometrical progression of optical energy reduction, which translates into a logarithmic relationship between the transmitted intensity and the sample thickness. Had he utilized the recently invented differential calculus – by the English physicist and mathematician Isaac Newton [1643–1727] and the German mathematician Gottfried Leibniz [1646–1716] – he could have determined something closer to the modern expression. The Swiss mathematician, physicist and philosopher Johann Lambert [1728–1777] explicitly expressed such a logarithmic relationship in a concise mathematical form, stating that the logarithm of the inverse of the light transmission ratio (which is defined as the absorbance) is proportional to the sample thickness. Finally, the German mathematician and chemist August Beer [1825–1863]

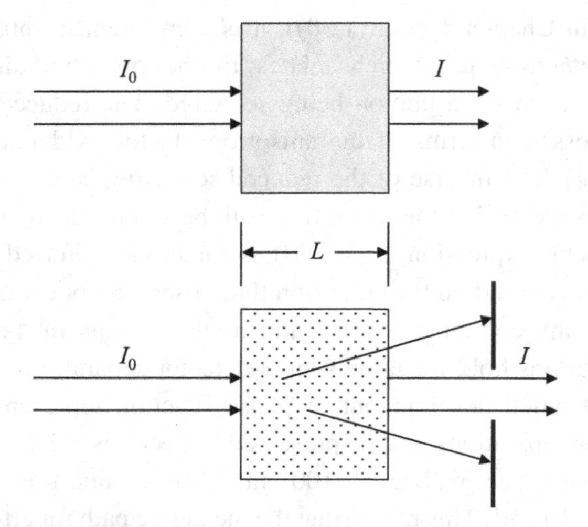

Figure 2.15

Figurative depiction of propagation of collimated light through purely absorbing (top) and purely, but weakly, scattering media (bottom). The use of an aperture allows detection of only the still-collimated, unscattered light.

determined that the proportionality factor is given by the product of the absorber concentration and the molar extinction coefficient, as expressed in Eq. (2.1), which is sometimes referred to as *Beer's law*.

Following in this train of logic, consider a slab of medium that features scattering but is non-absorbing. If the scattering is weak enough and/or if the slab thickness, L, is thin enough that there is negligible probability of multiple scattering events by any traversing photon (i.e., if $\mu_s L \ll 1$), then one could calculate the amount of unscattered light, after a distance L though the medium, with Eq. (2.9).

$$I = I_0 e^{-\mu_s L} \tag{2.9}$$

Figure 2.15 illustrates the attenuation of a collimated light beam when traversing either a purely absorbing slab, or a slab that is purely, but weakly, scattering. Using such an arrangement, wherein the incident intensity and slab thickness are known and the transmitted intensity is measured, Eqs. (2.8) and (2.9) enable determination of the absorption coefficient of a non-scattering medium (upper illustration) or the scattering coefficient of a weakly scattering but non-absorbing medium (lower illustration).

The situation becomes more complicated, however, when trying to determine the intensity of light as it propagates through a medium that is both absorbing and scattering. One might write an equation similar to (2.8) or (2.9), but using the *total attenuation coefficient*, μ_t, as defined in Eq. (1.24). This can serve as an approximation for any absorption coefficient when the slab thickness, L, is much less than $1/\mu_s$, or for cases when absorption is much stronger than scattering. For more turbid media or longer pathlengths, however, the problem is that the total pathlength of a typical photon through the medium is greater than L, due to the scattering, and the pathlength cannot be calculated analytically because of the stochastic nature of the scattering process. That is, the effective pathlength presents a greater probability of *absorption* because the path is longer than it would be in a non-scattering medium due to the more tortuous path taken by the photons; that photon pathlength, however, is affected by both the scattering properties and the absorption itself. In this case it becomes difficult to analytically separate the absorption and scattering coefficients, unless the measurement geometry and the medium properties enable validity of the diffusion approximation to the transport equation. For short source-detector separations that do not satisfy the diffusion approximation, empirically determined analytical expressions can facilitate separation of the absorption and scattering coefficients (see Sections 8.3–8.6), and results can be validated by stochastic computational methods, based on *Monte Carlo* simulation codes (see Sections 8.1.1 and 14.5.2). The extension of the Beer-Lambert law to highly scattering media is referred to as the *modified Beer-Lambert law*, and it will be introduced in Section 10.3.

Appendix 2.A: Tabulated near-infrared spectra of HbO, Hb, cytochrome *c* oxidase, water and lipids

Table 2.A.1 reports tabulated extinction or absorption spectra of five of the most relevant near-infrared absorbing species in tissue. These are the chromophores oxyhemoglobin (HbO), deoxyhemoglobin (Hb) and cytochrome *c* oxidase, and the bulk constituents water and lipids. The spectral values are reported every nanometer in the range 650–1042 nm. In all cases, the absorption coefficients are expressed in units of cm^{-1} and are defined using the logarithm to base-e, so that they exceed by a factor of $\ln(10) \cong 2.3$ corresponding coefficients defined using the logarithm to base 10. For HbO and Hb we report on the full functional unit containing four heme groups. The extinction coefficients of HbO and Hb represent the absorption coefficients per unit concentration. In the case of cytochrome *c* oxidase, we report the difference of the extinction coefficients for the oxidized state (Cyt|ox) minus the reduced state (Cyt|red). For water and lipids, we report the absorption coefficients of liquid water (concentration: ~55.5 M) and mammalian fat. The near-infrared-absorption spectra of the five components considered here are illustrated in Figure 2.A.1, where we have assumed nominal tissue concentrations of $[\text{HbO}] = 50\,\mu\text{M}$, $[\text{Hb}] = 50\,\mu\text{M}$, $\left[\text{Cyt|tot}\right] = 10\,\mu\text{M}$, $[\text{water}] = 80\%(\text{v/v})$, and $\left[\text{lipids}\right] = 50\%\ (\text{v/v})$.

Table 2.A.1 Tabulated near-infrared extinction coefficients of oxyhemoglobin (ε_{HbO}), deoxyhemoglobin (ε_{Hb}), the difference of oxidized and reduced cytochrome *c* oxidase ($\varepsilon_{Cyt|ox}$-$\varepsilon_{Cyt|red}$), and absorption coefficients of water and lipids.

λ nm	ε_{HbO} $cm^{-1}M^{-1}$	ε_{Hb} $cm^{-1}M^{-1}$	$\varepsilon_{Cyt\|ox}$- $\varepsilon_{Cyt\|red}$ $cm^{-1}M^{-1}$	$\mu_a^{(water)}$ cm^{-1}	$\mu_a^{(lipids)}$ cm^{-1}
650	890.6	8598	11350	0.003233	0.00471
651	873.1	8526	11265	0.003266	0.00472
652	856.8	8455	11180	0.003303	0.00467
653	842.3	8382	11068	0.003343	0.00464
654	829.0	8316	10956	0.003385	0.00457
655	816.1	8251	10829	0.003429	0.00448
656	804.5	8185	10703	0.003476	0.00438
657	794.4	8119	10564	0.003523	0.00428
658	785.0	8053	10426	0.003571	0.00424
659	777.2	7987	10278	0.003617	0.00422
660	770.3	7921	10131	0.003661	0.00418
661	763.8	7841	9975	0.003702	0.00414
662	758.3	7762	9818	0.003739	0.00413
663	754.1	7682	9652	0.003772	0.00414
664	750.5	7594	9486	0.003802	0.00419
665	747.0	7500	9313	0.003829	0.00428
666	744.5	7406	9140	0.003852	0.00439
667	742.9	7312	8974	0.003874	0.00447
668	741.2	7211	8808	0.003894	0.00452
669	740.3	7111	8639	0.003915	0.00453
670	739.6	7011	8471	0.003938	0.0045
671	739.4	6904	8304	0.003963	0.00452
672	739.4	6795	8137	0.003990	0.00457
673	739.9	6686	7973	0.004017	0.00462
674	739.9	6577	7809	0.004043	0.00463
675	739.2	6466	7648	0.004068	0.00457

Table 2.A.1 (*cont.*)

λ nm	ε_{HbO} $cm^{-1}M^{-1}$	ε_{Hb} $cm^{-1}M^{-1}$	$\varepsilon_{Cyt\|ox}$ − $\varepsilon_{Cyt\|red}$ $cm^{-1}M^{-1}$	$\mu_a^{(water)}$ cm^{-1}	$\mu_a^{(lipids)}$ cm^{-1}
676	738.5	6355	7488	0.004091	0.00446
677	738.0	6244	7330	0.004115	0.00433
678	737.8	6135	7172	0.004141	0.00422
679	736.9	6027	7035	0.004172	0.00413
680	735.3	5919	6899	0.004209	0.00406
681	733.4	5812	6790	0.004255	0.00398
682	732.3	5706	6681	0.004311	0.00387
683	730.4	5601	6589	0.004372	0.00375
684	728.1	5495	6497	0.004436	0.0037
685	726.7	5395	6413	0.004499	0.00369
686	725.4	5296	6329	0.004559	0.00366
687	722.8	5196	6257	0.004615	0.00361
688	721.7	5102	6185	0.004673	0.00354
689	719.6	5012	6123	0.004737	0.00348
690	718.9	4922	6061	0.004813	0.00342
691	718.7	4833	6007	0.004905	0.00338
692	718.7	4755	5952	0.005010	0.00335
693	718.9	4677	5907	0.005124	0.00331
694	719.4	4598	5863	0.005241	0.00328
695	720.5	4527	5810	0.005357	0.00324
696	722.8	4459	5758	0.005470	0.00321
697	724.7	4391	5710	0.005586	0.00324
698	728.4	4326	5663	0.005715	0.00327
699	732.3	4266	5611	0.005867	0.00325
700	736.4	4206	5560	0.006052	0.00323
701	741.0	4146	5507	0.006274	0.0032
702	746.5	4092	5453	0.006528	0.00319

(*cont.*)

Table 2.A.1 (*cont.*)

λ nm	ε_{HbO} $cm^{-1}M^{-1}$	ε_{Hb} $cm^{-1}M^{-1}$	$\varepsilon_{Cyt\|ox}$ − $\varepsilon_{Cyt\|red}$ $cm^{-1}M^{-1}$	$\mu_a^{(water)}$ cm^{-1}	$\mu_a^{(lipids)}$ cm^{-1}
703	752.5	4039	5395	0.006805	0.00322
704	758.7	3987	5337	0.007099	0.00326
705	764.9	3936	5284	0.007403	0.00323
706	771.6	3887	5231	0.007711	0.00318
707	779.5	3839	5167	0.008023	0.00313
708	786.8	3790	5103	0.008339	0.0031
709	794.2	3745	5039	0.008658	0.00311
710	802.2	3699	4975	0.008980	0.00315
711	811.2	3654	4917	0.009308	0.0032
712	820.4	3609	4860	0.009644	0.00321
713	829.2	3564	4793	0.009992	0.0032
714	837.7	3519	4726	0.01036	0.00321
715	847.1	3474	4663	0.01074	0.00326
716	857.3	3432	4599	0.01114	0.00334
717	866.7	3389	4543	0.01157	0.00343
718	876.4	3346	4487	0.01204	0.00351
719	887.2	3307	4437	0.01253	0.00358
720	897.5	3268	4386	0.01306	0.00367
721	907.7	3229	4346	0.01362	0.00376
722	918.0	3194	4305	0.01423	0.00383
723	928.9	3161	4277	0.01488	0.0039
724	939.9	3129	4248	0.01556	0.00401
725	951.4	3098	4226	0.01629	0.00415
726	962.5	3074	4204	0.01706	0.0043
727	973.7	3050	4185	0.01784	0.00447
728	985.5	3026	4166	0.01864	0.00463
729	997.2	3011	4157	0.01942	0.0048

Table 2.A.1 (*cont.*)

λ nm	ε_{HbO} $cm^{-1}M^{-1}$	ε_{Hb} $cm^{-1}M^{-1}$	$\varepsilon_{Cyt\|ox}$ − $\varepsilon_{Cyt\|red}$ $cm^{-1}M^{-1}$	$\mu_a^{(water)}$ cm^{-1}	$\mu_a^{(lipids)}$ cm^{-1}
730	1009	2999	4148	0.02019	0.00499
731	1021	2987	4146	0.02092	0.00516
732	1033	2980	4144	0.02161	0.00535
733	1045	2981	4143	0.02224	0.00552
734	1057	2982	4142	0.02281	0.00564
735	1070	2983	4146	0.02330	0.00576
736	1082	2999	4150	0.02370	0.00589
737	1094	3015	4154	0.02405	0.006
738	1107	3030	4158	0.02435	0.00616
739	1120	3056	4168	0.02462	0.00636
740	1133	3087	4178	0.02487	0.00654
741	1145	3118	4193	0.02512	0.0068
742	1158	3153	4207	0.02534	0.00707
743	1171	3197	4221	0.02555	0.00735
744	1185	3241	4235	0.02572	0.00768
745	1198	3284	4247	0.02585	0.00801
746	1211	3338	4259	0.02595	0.00837
747	1224	3392	4269	0.02601	0.00874
748	1238	3446	4278	0.02606	0.00908
749	1251	3502	4290	0.02609	0.00941
750	1265	3558	4301	0.02613	0.00972
751	1279	3615	4320	0.02617	0.01001
752	1292	3669	4339	0.02622	0.0103
753	1306	3716	4355	0.02626	0.01059
754	1319	3763	4370	0.02628	0.01089
755	1333	3810	4378	0.02629	0.0112
756	1348	3832	4386	0.02628	0.01155

(*cont.*)

Table 2.A.1 (*cont.*)

λ nm	ε_{HbO} cm^{-1}M^{-1}	ε_{Hb} cm^{-1}M^{-1}	$\varepsilon_{Cyt\|ox}$ - $\varepsilon_{Cyt\|red}$ cm^{-1}M^{-1}	$\mu_a^{(water)}$ cm^{-1}	$\mu_a^{(lipids)}$ cm^{-1}
757	1361	3852	4394	0.02625	0.0119
758	1375	3872	4401	0.02621	0.01227
759	1388	3871	4408	0.02616	0.01258
760	1403	3855	4415	0.02612	0.01278
761	1417	3838	4421	0.02610	0.01289
762	1432	3813	4428	0.02607	0.0129
763	1446	3762	4439	0.02604	0.01273
764	1461	3711	4451	0.02597	0.01249
765	1475	3660	4464	0.02586	0.01206
766	1489	3587	4477	0.02569	0.01153
767	1504	3513	4488	0.02547	0.01089
768	1518	3438	4500	0.02522	0.01021
769	1533	3359	4512	0.02497	0.00948
770	1547	3275	4525	0.02476	0.00871
771	1562	3193	4536	0.02458	0.00793
772	1577	3111	4547	0.02441	0.00719
773	1591	3031	4560	0.02425	0.00653
774	1605	2952	4573	0.02407	0.00596
775	1620	2873	4593	0.02388	0.00553
776	1635	2803	4612	0.02366	0.00515
777	1650	2733	4640	0.02342	0.00481
778	1664	2663	4667	0.02317	0.00451
779	1680	2601	4693	0.02291	0.00426
780	1694	2544	4719	0.02266	0.00409
781	1708	2487	4745	0.02240	0.00391
782	1723	2433	4771	0.02216	0.00378
783	1738	2388	4800	0.02192	0.00366

Table 2.A.1 (*cont.*)

λ nm	ε_{HbO} $cm^{-1}M^{-1}$	ε_{Hb} $cm^{-1}M^{-1}$	$\varepsilon_{Cyt\|ox}-\varepsilon_{Cyt\|red}$ $cm^{-1}M^{-1}$	$\mu_a^{(water)}$ cm^{-1}	$\mu_a^{(lipids)}$ cm^{-1}
784	1752	2342	4829	0.02170	0.0036
785	1768	2296	4853	0.02147	0.00357
786	1782	2261	4876	0.02126	0.00355
787	1797	2226	4897	0.02105	0.00353
788	1813	2190	4917	0.02085	0.00349
789	1828	2160	4934	0.02067	0.00346
790	1843	2133	4951	0.02049	0.00346
791	1858	2105	4966	0.02033	0.00347
792	1872	2080	4981	0.02018	0.00347
793	1887	2058	5010	0.02004	0.00347
794	1902	2036	5038	0.01993	0.00352
795	1918	2015	5068	0.01984	0.00359
796	1932	1997	5097	0.01978	0.00369
797	1947	1980	5129	0.01973	0.00379
798	1962	1962	5160	0.01969	0.00387
799	1977	1948	5184	0.01966	0.00393
800	1992	1933	5207	0.01963	0.00403
801	2006	1919	5218	0.01961	0.00413
802	2021	1907	5229	0.01960	0.00424
803	2036	1896	5240	0.01962	0.00434
804	2051	1885	5251	0.01967	0.00442
805	2066	1875	5265	0.01976	0.00454
806	2081	1867	5278	0.01991	0.00467
807	2095	1859	5292	0.02012	0.00482
808	2110	1851	5307	0.02036	0.00497
809	2124	1845	5321	0.02063	0.00511
810	2139	1839	5334	0.02091	0.00527

(*cont.*)

Table 2.A.1 (*cont.*)

λ nm	ε_{HbO} $cm^{-1}M^{-1}$	ε_{Hb} $cm^{-1}M^{-1}$	$\varepsilon_{Cyt\|ox}$ − $\varepsilon_{Cyt\|red}$ $cm^{-1}M^{-1}$	$\mu_a^{(water)}$ cm^{-1}	$\mu_a^{(lipids)}$ cm^{-1}
811	2153	1833	5343	0.02120	0.00547
812	2168	1828	5353	0.02150	0.00567
813	2182	1824	5359	0.02181	0.00583
814	2196	1820	5366	0.02212	0.00594
815	2211	1816	5372	0.02243	0.00603
816	2225	1813	5378	0.02275	0.00615
817	2239	1810	5384	0.02310	0.00633
818	2253	1807	5390	0.02351	0.00654
819	2267	1805	5396	0.02399	0.00674
820	2281	1803	5403	0.02457	0.00691
821	2296	1802	5405	0.02526	0.00708
822	2309	1801	5406	0.02605	0.00724
823	2323	1799	5403	0.02687	0.0074
824	2337	1798	5399	0.02766	0.00758
825	2351	1798	5393	0.02837	0.00775
826	2365	1797	5387	0.02900	0.00786
827	2379	1797	5380	0.02956	0.00796
828	2393	1797	5372	0.03007	0.00799
829	2405	1796	5364	0.03052	0.00802
830	2419	1796	5356	0.03095	0.00803
831	2432	1796	5348	0.03135	0.00803
832	2445	1796	5341	0.03177	0.00801
833	2458	1796	5335	0.03222	0.00794
834	2472	1797	5329	0.03273	0.00785
835	2486	1797	5325	0.03332	0.00775
836	2498	1797	5321	0.03403	0.00764
837	2511	1797	5319	0.03483	0.00757

Table 2.A.1 (*cont.*)

λ nm	ε_{HbO} $cm^{-1}M^{-1}$	ε_{Hb} $cm^{-1}M^{-1}$	$\varepsilon_{Cyt\|ox^-}$ $\varepsilon_{Cyt\|red}$ $cm^{-1}M^{-1}$	$\mu_a^{(water)}$ cm^{-1}	$\mu_a^{(lipids)}$ cm^{-1}
838	2523	1798	5317	0.03567	0.00751
839	2536	1798	5315	0.03645	0.00739
840	2549	1799	5313	0.03713	0.00726
841	2562	1799	5309	0.03768	0.00715
842	2574	1800	5305	0.03816	0.00703
843	2586	1801	5299	0.03863	0.00692
844	2598	1802	5293	0.03915	0.00685
845	2610	1803	5288	0.03977	0.00678
846	2622	1804	5283	0.04055	0.00669
847	2634	1806	5279	0.04144	0.00661
848	2646	1807	5276	0.04235	0.00651
849	2658	1808	5274	0.04319	0.00643
850	2669	1810	5273	0.04388	0.00637
851	2681	1812	5267	0.04443	0.00633
852	2692	1814	5261	0.04489	0.00632
853	2703	1816	5250	0.04532	0.00631
854	2714	1819	5239	0.04576	0.00633
855	2726	1821	5222	0.04628	0.00639
856	2736	1824	5206	0.04690	0.00647
857	2747	1827	5186	0.04760	0.00656
858	2758	1830	5166	0.04831	0.00662
859	2768	1833	5150	0.04898	0.00666
860	2779	1836	5134	0.04955	0.00674
861	2789	1840	5124	0.05003	0.0068
862	2798	1844	5114	0.05044	0.0069
863	2808	1847	5103	0.05081	0.00705
864	2819	1852	5094	0.05116	0.00724

(*cont.*)

Table 2.A.1 (*cont.*)

λ nm	ε_{HbO} $cm^{-1}M^{-1}$	ε_{Hb} $cm^{-1}M^{-1}$	$\varepsilon_{Cyt\|ox}$- $\varepsilon_{Cyt\|red}$ $cm^{-1}M^{-1}$	$\mu_a^{(water)}$ cm^{-1}	$\mu_a^{(lipids)}$ cm^{-1}
865	2828	1856	5080	0.05151	0.00751
866	2837	1860	5067	0.05189	0.00786
867	2847	1865	5051	0.05228	0.00822
868	2856	1870	5035	0.05270	0.0086
869	2866	1875	5012	0.05314	0.00899
870	2875	1880	4989	0.05359	0.00938
871	2884	1885	4956	0.05407	0.00979
872	2893	1891	4923	0.05457	0.01022
873	2901	1896	4893	0.05508	0.01074
874	2910	1902	4864	0.05560	0.01133
875	2919	1907	4841	0.05611	0.01198
876	2926	1913	4818	0.05661	0.01264
877	2934	1919	4799	0.05709	0.01345
878	2942	1925	4780	0.05754	0.0143
879	2950	1931	4761	0.05794	0.01527
880	2957	1936	4744	0.05830	0.01636
881	2965	1942	4726	0.05863	0.01752
882	2971	1948	4709	0.05895	0.01867
883	2978	1954	4687	0.05928	0.01992
884	2986	1960	4665	0.05964	0.02121
885	2993	1966	4642	0.06007	0.02262
886	2999	1972	4619	0.06056	0.02414
887	3006	1978	4595	0.06109	0.02581
888	3012	1984	4571	0.06163	0.02754
889	3017	1990	4541	0.06213	0.02932
890	3024	1996	4511	0.06256	0.03116
891	3031	2001	4481	0.06293	0.03305

Table 2.A.1 (*cont.*)

λ nm	ε_{HbO} $cm^{-1}M^{-1}$	ε_{Hb} $cm^{-1}M^{-1}$	$\varepsilon_{Cyt\|ox}$ – $\varepsilon_{Cyt\|red}$ $cm^{-1}M^{-1}$	$\mu_a^{(water)}$ cm^{-1}	$\mu_a^{(lipids)}$ cm^{-1}
892	3036	2007	4452	0.06328	0.03491
893	3041	2012	4421	0.06366	0.03674
894	3047	2018	4389	0.06409	0.03833
895	3053	2023	4355	0.06462	0.03983
896	3058	2027	4320	0.06528	0.04121
897	3063	2032	4286	0.06605	0.04256
898	3067	2036	4252	0.06684	0.04384
899	3071	2040	4223	0.06759	0.04506
900	3077	2044	4193	0.06822	0.04633
901	3081	2048	4170	0.06872	0.04766
902	3085	2051	4146	0.06916	0.04915
903	3089	2054	4128	0.06961	0.05078
904	3093	2057	4110	0.07015	0.05267
905	3097	2059	4094	0.07084	0.05472
906	3100	2061	4078	0.07176	0.05721
907	3103	2062	4063	0.07297	0.05994
908	3107	2063	4048	0.07453	0.06316
909	3109	2064	4034	0.07651	0.06663
910	3111	2064	4019	0.07898	0.07046
911	3114	2064	3997	0.08192	0.07438
912	3117	2062	3974	0.08517	0.07837
913	3120	2060	3952	0.08851	0.08234
914	3122	2058	3929	0.09174	0.08621
915	3124	2056	3906	0.09470	0.08994
916	3126	2053	3884	0.09750	0.09354
917	3127	2050	3852	0.1004	0.09683
918	3128	2045	3821	0.1035	0.1

(*cont.*)

Table 2.A.1 (*cont.*)

λ nm	ε_{HbO} $cm^{-1}M^{-1}$	ε_{Hb} $cm^{-1}M^{-1}$	$\varepsilon_{Cyt\|ox}$ - $\varepsilon_{Cyt\|red}$ $cm^{-1}M^{-1}$	$\mu_a^{(water)}$ cm^{-1}	$\mu_a^{(lipids)}$ cm^{-1}
919	3127	2041	3788	0.1071	0.1032
920	3128	2036	3756	0.1114	0.1065
921	3130	2030	3723	0.1166	0.1099
922	3131	2023	3690	0.1228	0.1133
923	3131	2017	3652	0.1299	0.1167
924	3131	2009	3613	0.1379	0.12
925	3131	2001	3567	0.1468	0.1231
926	3132	1992	3520	0.1563	0.126
927	3131	1983	3477	0.1662	0.1282
928	3131	1973	3434	0.1759	0.1299
929	3131	1963	3401	0.1850	0.1307
930	3130	1952	3367	0.1934	0.131
931	3129	1941	3333	0.2012	0.1299
932	3128	1929	3300	0.2087	0.1283
933	3127	1916	3260	0.2165	0.1253
934	3125	1903	3220	0.2247	0.1218
935	3124	1889	3177	0.2339	0.1175
936	3121	1875	3134	0.2443	0.1126
937	3119	1860	3091	0.2559	0.107
938	3118	1845	3048	0.2683	0.1008
939	3115	1829	3005	0.2811	0.09433
940	3112	1813	2961	0.2939	0.08754
941	3110	1796	2929	0.3063	0.08088
942	3108	1779	2897	0.3182	0.07446
943	3105	1761	2871	0.3293	0.06822
944	3100	1744	2845	0.3394	0.0625
945	3097	1725	2822	0.3483	0.05704

Table 2.A.1 (*cont.*)

λ nm	ε_{HbO} $cm^{-1}M^{-1}$	ε_{Hb} $cm^{-1}M^{-1}$	$\varepsilon_{Cyt\vert ox}$ − $\varepsilon_{Cyt\vert red}$ $cm^{-1}M^{-1}$	$\mu_a^{(water)}$ cm^{-1}	$\mu_a^{(lipids)}$ cm^{-1}
946	3095	1706	2798	0.3562	0.05234
947	3091	1687	2766	0.3634	0.04802
948	3087	1667	2733	0.3701	0.04445
949	3083	1647	2693	0.3767	0.04109
950	3079	1627	2653	0.3834	0.03799
951	3074	1607	2613	0.3905	0.03519
952	3069	1587	2572	0.3980	0.03274
953	3065	1568	2534	0.4056	0.03046
954	3060	1546	2496	0.4130	0.02852
955	3053	1523	2474	0.4198	0.02666
956	3048	1500	2451	0.4257	0.02504
957	3043	1478	2426	0.4308	0.02358
958	3038	1456	2401	0.4350	0.02242
959	3032	1434	2384	0.4386	0.02133
960	3026	1412	2367	0.4415	0.02047
961	3021	1390	2361	0.4438	0.01975
962	3015	1368	2355	0.4457	0.01917
963	3009	1345	2353	0.4472	0.0186
964	3002	1322	2351	0.4484	0.01808
965	2997	1300	2322	0.4495	0.01753
966	2991	1278	2293	0.4505	0.01693
967	2985	1255	2257	0.4515	0.01639
968	2976	1233	2221	0.4524	0.01589
969	2970	1211	2183	0.4531	0.01554
970	2962	1189	2145	0.4534	0.01528
971	2955	1167	2117	0.4533	0.0151
972	2948	1145	2090	0.4527	0.01491

(*cont.*)

Table 2.A.1 (*cont.*)

λ nm	ε_{HbO} cm^{-1}M^{-1}	ε_{Hb} cm^{-1}M^{-1}	$\varepsilon_{Cyt\|ox}$ - $\varepsilon_{Cyt\|red}$ cm^{-1}M^{-1}	$\mu_a^{(water)}$ cm^{-1}	$\mu_a^{(lipids)}$ cm^{-1}
973	2940	1123	2071	0.4516	0.01463
974	2932	1102	2052	0.4502	0.01431
975	2922	1080	2040	0.4485	0.01413
976	2913	1058	2029	0.4467	0.01408
977	2906	1037	2020	0.4446	0.01414
978	2898	1016	2010	0.4423	0.01421
979	2889	995.2	1996	0.4398	0.01423
980	2881	974.4	1981	0.4371	0.01417
981	2871	954.4	1958	0.4342	0.01407
982	2861	934.8	1934	0.4313	0.01406
983	2853	915.0	1909	0.4285	0.0142
984	2843	895.9	1883	0.4260	0.01441
985	2833	877.1	1863	0.4239	0.01481
986	2822	858.4	1843	0.4222	0.01524
987	2813	839.8		0.4205	0.01564
988	2802	821.1		0.4188	0.01607
989	2793	802.7		0.4169	0.01652
990	2782	784.7		0.4147	0.01716
991	2772	767.5		0.4120	0.01793
992	2761	750.2		0.4087	0.01891
993	2748	733.6		0.4049	0.01994
994	2738	717.5		0.4010	0.02087
995	2727	701.4		0.3969	0.02178
996	2714	685.5		0.3929	0.02272
997	2701	669.7		0.3890	0.02366
998	2692	654.0		0.3851	0.02464
999	2679	639.0		0.3811	0.0256

Table 2.A.1 (*cont.*)

λ nm	ε_{HbO} $cm^{-1}M^{-1}$	ε_{Hb} $cm^{-1}M^{-1}$	$\varepsilon_{Cyt\|ox}$ − $\varepsilon_{Cyt\|red}$ $cm^{-1}M^{-1}$	$\mu_a^{(water)}$ cm^{-1}	$\mu_a^{(lipids)}$ cm^{-1}
1000	2665	624.8		0.3770	0.02654
1001	2655	610.6		0.3726	0.02751
1002	2642	596.8		0.3680	0.0287
1003	2629	583.2		0.3632	0.02996
1004	2616	569.9		0.3583	0.03136
1005	2602	557.0		0.3534	0.03282
1006	2591	544.3		0.3486	0.0343
1007	2579	531.7		0.3438	0.03579
1008	2564	519.7		0.3392	0.03735
1009	2551	508.2		0.3348	0.03894
1010	2535	496.4		0.3306	0.04057
1011	2521	485.6		0.3267	0.04222
1012	2508	474.8		0.3230	0.04388
1013	2492	464.0		0.3194	0.04554
1014	2477	453.8		0.3159	0.04722
1015	2465	443.9		0.3124	0.04877
1016	2452	433.8		0.3089	0.05024
1017	2436	424.6		0.3053	0.05152
1018	2418	415.4		0.3017	0.05279
1019	2406	405.9		0.2978	0.05386
1020	2390	397.4		0.2938	0.05481
1021	2374	388.7		0.2895	0.05561
1022	2358	380.2		0.2849	0.05631
1023	2342	372.6		0.2800	0.05683
1024	2325	365.0		0.2750	0.05734
1025	2309	357.6		0.2698	0.0579
1026	2296	350.5		0.2646	0.05857

(*cont.*)

Table 2.A.1 (*cont.*)

λ nm	ε_{HbO} cm^{-1}M^{-1}	ε_{Hb} cm^{-1}M^{-1}	$\varepsilon_{Cyt\|ox}$ − $\varepsilon_{Cyt\|red}$ cm^{-1}M^{-1}	$\mu_a^{(water)}$ cm^{-1}	$\mu_a^{(lipids)}$ cm^{-1}
1027	2277	343.5		0.2594	0.05948
1028	2265	336.4		0.2542	0.0605
1029	2242	329.7		0.2491	0.06164
1030	2225	322.8		0.2441	0.06274
1031	2210	316.4		0.2394	0.06389
1032	2191	310.2		0.2349	0.06503
1033	2168	303.9		0.2306	0.0661
1034	2147	298.2		0.2264	0.06717
1035	2134	292.9		0.2224	0.06799
1036	2118	287.8		0.2186	0.06871
1037	2100	282.5		0.2148	0.06926
1038	2083	277.5		0.2112	0.0697
1039	2059	272.4		0.2077	0.07004
1040	2044	267.3		0.2042	0.07029
1041	2022	262.7		0.2008	0.07018
1042	1997	257.9		0.1974	0.06998

The sources of the reported spectra are as follows:

Oxyhemoglobin and deoxyhemoglobin (up to 1000 nm), and cytochrome *c* oxidase: Kolyva et al., 2012.

Oxyhemoglobin and deoxyhemoglobin (1000–1042 nm): Matcher et al., 1995.

Water: Segelstein, 1981. We have performed a spline interpolation of the data points originally reported about every 5 nm. The data from Segelstein (1981) report the complex refractive index $\tilde{n} = n + ik$ where the real part n is related to the phase velocity of the light wave and the imaginary part k is related to the optical attenuation. By recalling that the electric field of a plane wave propagating along the z axis can be written:

$$\mathbf{E}(z,t) = \mathbf{E}_0 e^{i\left(\frac{2\pi}{\lambda}\tilde{n}z - \omega t\right)} = \mathbf{E}_0 e^{-\frac{2\pi}{\lambda}kz} e^{i\left(\frac{2\pi}{\lambda}nz - \omega t\right)}, \tag{2.A.1}$$

and by recalling that the optical intensity is proportional to the amplitude square

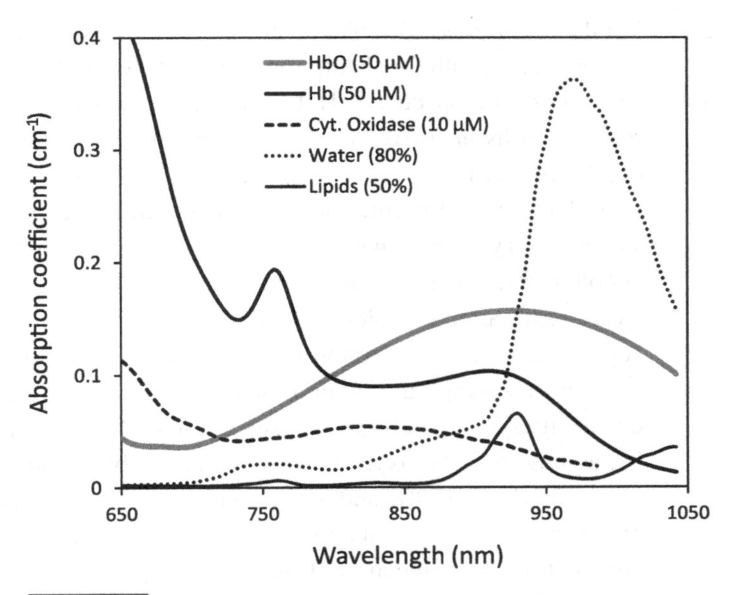

Figure 2.A.1

Near-infrared-absorption spectra of oxyhemoglobin (HbO), deoxyhemoglobin (Hb), cytochrome *c* oxidase (Cyt. oxidase), water, and lipids for the nominal tissue concentration values of 50 μM (for HbO and Hb), 10 μM (for cytochrome *c* oxidase), 80% by volume (for water), and 50% by volume (for lipids).

of the electric field, the relationship between the absorption coefficient and k is as follows:

$$\mu_a = \frac{4\pi}{\lambda}k. \tag{2.A.2}$$

Lipids: van Veen et al., 2004. This spectrum was measured on oil extracted and purified from pig lard.

Problems – answers to problems with * are on p. 645

2.1* In a person with active sickle-cell anemia, the concentration of hemoglobin in their whole blood is ~70 g/L. What is the effective molar concentration of hemoglobin in the average tissue in that case?

2.2 Based on an assumed tissue molar concentration for hemoglobin of 100 μM, and for the case of a hemoglobin oxygen saturation of 85% averaged over the capillary bed, estimate the absorption coefficient, μ_a, of tissue due to total hemoglobin at a wavelength of 700 nm. (You can use values of the extinction coefficients from Table 2.A.1.) What would be the mean free path for absorption?

2.3* For the conditions of Problem 2.2, what would be the absorption coefficient and mean free path for absorption at 420 nm (using Figure 2.4)?

2.4 The absorption spectra of HbO and Hb are equal at ~800 nm, which is referred to as an *isosbestic point*, a wavelength at which the extinction coefficients of the two forms of hemoglobin (oxy- and deoxy-) are equal, such that the total absorption by hemoglobin is independent of the redox stoichiometry. Explain why it is useful to make spectroscopic measurements at both 800 nm and another wavelength, say 650 nm, when determining the oxygen saturation of a blood sample.

2.5 A *pulse oximeter* (see Section 15.1.2) is a device that attempts to determine the oxygen saturation of arterial blood, by measuring the intensity of light that diffuses across peripheral tissue like a fingertip or earlobe. For simplicity, assume that the oxygen saturation of the blood on the arterial side of the capillaries is 100%, and for the venus side is 60%. Also assume that the blood volume fractions of both sides of the capillaries are equal. If the blood volume fraction of the arterial side fluctuates by ±5% from its mean value as a result of blood-pressure variations with the pulse (and the blood volume fraction of the venus side is constant), what will be the percent size of the measured fluctuations detected in diffusely transmitted light at 800 nm, if the average effective photon pathlength is 2 cm?

2.6* A turbid medium used for a tissue phantom is made of an aqueous suspension of scattering microparticles with an average particle diameter of 1.5 μm.

(a) If the value of the scattering efficiency for the particles is 0.5, what will be the density of the particles (in number of particles per cm^3) needed to achieve a scattering coefficient, μ_s, of 50 cm^{-1}?

(b) What would need to be the effective anisotropy factor, g, of the particles to result in a reduced scattering coefficient, μ_s', of 10 cm^{-1}?

References

Bolin, F. P., Preuss, L. E., Taylor, R. C., and Ference, R. F. (1989). Refractive index of some mammalian tissues using fiber optic cladding method. *Applied Optics*, 28, 2297–2303.

Cheong, W. F., Prahl, S. A., and Welch, A. J. (1990). A review of the optical properties of biological tissues. *IEEE Journal of Quantum Electronics*, 26, 2166–2185.

Dixon, J. M., Taniguchi, M., and Lindsey, J. S. (2005). PhotochemCAD 2: a refined program with accompanying spectra databases for photochemical calculations. *Photochemistry and Photobiology*, 81, 212–213.

Grover, W. H., Bryan, A. K., Diez-Silva, M., et al. (2011). Measuring single-cell density. *Proceedings of the National Academy of Sciences USA*, 108, 10992–10996.

Kolyva, C., Tachtsidis, I., Ghosh, A., et al. (2012). Systematic investigation of changes in oxidized cerebral cytochrome c oxidase concentration during frontal lobe activation in healthy adults. *Biomedical Optics Express*, 3, 2550–2566.

Kwon, Y.-W., Choi, D. H., and Jin, J.-I. (2012). Optical, electro-optic and optoelectronic properties of natural and chemically modified DNAs. *Polymer Journal*, 44, 1191–1208.

Matcher, S. J., Elwell, C. E., Cooper, C. E., Cope, M., and Delpy, D. T. (1995). Performance comparison of several published tissue near-infrared spectroscopy algorithms. *Analytical Biochemistry*, 227, 54–68.

Perrin, F. H. (1948). Whose absorption law? *Journal of the Optical Society of America*, 38, 72–74.

Prahl, S. (1999). http://omlc.org/spectra/hemoglobin/.

Riesz, J. (2007). The spectroscopic properties of melanin. PhD thesis, The University of Queensland, Australia.

Sarna, T., and Swartz, H. M. (1988). The physical properties of melanins. In *The Pigmentary System*, ed. J. J. Nordlund et al. Oxford: Oxford University Press.

Segelstein, D. J. (1981). The complex refractive index of water. MSc thesis, University of Missouri-Kansas City.

van Veen, R. L. P., Sterenborg, H. J. C. M., Pifferi, A., Torricelli, A., and Cubeddu, R. (2004). Determination of VIS-NIR absorption coefficients of mammalian fat, with time- and spatially resolved diffuse reflectance and transmission spectroscopy. *Optical Society of America Biomedical Topical Meeting*. Miami Beach, FL, April.

Zijlstra, W. G., Buursma, A., and Meeuwsen-van der Roest, W. P. (1991). Absorption spectra of human fetal and adult oxyhemoglobin, de-oxyhemoglobin, carboxyhemoglobin, and methemoglobin. *Clinical Chemistry*, 37, 1633–1638.

Further reading

Jacques, S. L. (2013). Optical properties of biological tissues: a review. *Physics in Medicine and Biology*, 58, R37–R61.

3 Introduction to biomedical statistics for diagnostic applications

Many of the concepts and technologies of biomedical optics, as introduced in this text, are intended as novel, and often minimally invasive, diagnostic methods that are capable of measuring or sensing a variety of physiological parameters. To facilitate understanding the value of novel diagnostics, we introduce here some of the statistical concepts that are most commonly used to assess the reliability of a diagnostic test. As such, we are limiting discussion to a narrow range of the statistical issues in medicine. Among a substantial number of other classes of statistical analysis are: verification that one treatment is superior to another (or to placebo); determining dosage equivalence of two treatments; estimating the mean value of a physiological parameter in a population; measuring the prevalence of a disease in a sample population (for example, to determine whether the disease prevalence is growing); estimating the prevalence of a condition in the general population based on measurements in a sample of the population. All of these require determinations of statistical significance, often invoking measures such as paired and unpaired "*t*-tests" and "*p*-values," and each type of assessment has different criteria, but those considerations are outside the scope of this chapter.

When developing a new diagnostic method or medical test it is *always* necessary to determine how well the method works by statistical methods. Why? Let us pose a very simple, non-medical problem (and not related to diagnosis). Imagine you have a friend who lives in a small town that recently installed its first traffic light at a central intersection, which was previously controlled by four-way stop signs. After two years the mayor announces that the accident rate at that intersection was lower in 2013 than it had been in 2012. Consequently, the mayor's office is asking the Town Council to raise taxes so that they can install five more traffic lights at other intersections. At a Town Council meeting a concerned citizen asks for the numbers: how many accidents have there been in those two years? The mayor, who wants to take credit for the improvement, is quick to respond that there were 42 accidents in 2012 and "only" 37 in 2013, a 12% drop. He claims that this is "significant" and that he expects the improvement to continue. Economic times are tough, however, so installation of more lights is delayed a couple of years. Then, a year later the police records show that there were 43 accidents for 2014 at the same intersection, and people are looking for someone to blame. You are visiting your friend, and you read about this controversy in the local paper. More importantly, you are a well-trained engineer, and you immediately realize what the problem is. (If not, keep reading).

3.1 The importance of proper statistics

"There are three kinds of lies: lies, damned lies, and statistics." This statement, attributed variously to Mark Twain (the great American author Samuel Langhorne Clemens [1835–1910]) or Benjamin Disraeli (British statesman and novelist [1804–1881]), suggests that one should be cautious about how numbers are used or interpreted. Statistics can be used properly or improperly. Even when used properly, however, statistical analysis cannot tell us the "truth," but it can tell us, quantitatively, what is the confidence level that the results represent the truth. So, when visiting your friend, you tell him that the accidents are intrinsically random events, which do not occur at prescribed rates, and as such are subject to the variations of counting statistics, best considered in terms of probabilities. You explain, that if the yearly accident rate, averaged over several years, had been, say, $n = 42$, then the likely annual variation, due purely to statistical fluctuations, would be $\pm\sqrt{n}$, or about ± 6.5, for about two-thirds of the time periods considered. That is, one would expect the count to be within the range 35.5–48.5 two-thirds of the times, and outside that range one-third of the times that yearly counts are taken. Thus, the one-year drop from 42 accidents to 37, and the subsequent rise from 37 to 43, were probably due to statistical fluctuations, and those changes could not be regarded as *statistically significant*. (This type of statistical variation is sometimes called, not surprisingly, the "square-root-of-n rule" for counting statistics.) You add, nonetheless, that had there been a drop of double that amount ($2 \times 6.5 = 13$), say from 42 to 29, then there would be a 95% confidence level that this drop was not due to chance.

It is comforting, therefore, that if a company approaches the FDA, seeking approval for a new instrument or method to diagnose a disease, then, in addition to wanting proof of safety, the FDA will require proof of *efficacy*. How well does it work? With what degree of statistical confidence? Is that degree of accuracy good enough? What are the consequences of error?

An interesting and instructive example was provided by the University Group Diabetes Program, an extensive multi-center clinical trial performed in the 1960s to determine the efficacy of various treatment options for type-2 diabetes. The investigators found themselves in a dilemma when they discovered a higher cardiac death rate for the patients treated with tolbutamide, a commonly prescribed anti-diabetic drug at that time. What were they supposed to do? Were their results statistically significant? Once they convinced themselves that the answer to this question was "yes", they felt that they had no choice but to announce their findings and stop the tolbutamide portion of the study. This resulted in a vitriolic public debate in the general and scientific press, involving pharmaceutical companies, statisticians, and patient advocacy groups. Statisticians argued both sides of the case, sometimes in a strong and vituperative tone, showing how statistical analysis, no matter how careful the study design, often leaves room for interpretation. The interested reader will find several references on this case in the further reading section at the end of this chapter. These describe the original study, some representative reports in favor and against its findings about tolbutamide, and some more recent perspectives that offer a dispassionate view of the entire controversy. A careful and critical study of this case can serve as a lesson about the value of statistical analysis, with important practical implications about its power and pitfalls, strengths and limitations.

The example above relates to measurement of the incidence rate of a side effect. Again, the material in this chapter is limited to statistical methods as applied to diagnostic tests, which are more common for methods of biomedical optics, whereas suggested sources listed under further reading address a broader range of topics in biostatistics.

3.2 Assessing the efficacy of dichotomous diagnostic tests

The statistics of any new diagnostic test are established by comparing its performance to that of a "*gold standard*." The gold standard for any test (or treatment) is the one that is currently regarded as being the most reliable by the medical profession. Of course, no current standard is 100% accurate, so the gold standard might have no more luster than tarnished silver, but it is still the standard that must be used. In some cases the most reliable ultimate determination of a patient's condition is the long-term outcome. Regulatory bodies (e.g., the FDA), however, do not often have the luxury of waiting for long-term outcomes to decide whether to approve a new test or treatment, if it can provide significant patient benefits. This is why we often hear that a drug or method that has been in use for a number of years is later removed from the market, or has restrictions imposed on its use, after finding negative effects that were not evident in initial testing.

So, how can the efficacy of a new test be assessed? We start by defining terms to be used in a discussion of the statistics concepts for such a test, which, for our discussion, is one that has been developed to determine whether a person has a specific disease condition. Additionally, we restrict the analysis to the case of a *dichotomous decision*, with only two possible test outcomes: the condition for which the test is performed is, or is not, present. A diagnostic test can also be *continuous* (for example, measurement of serum cholesterol) or have multiple sequential outcomes (for example, a test that determines risk of a certain disease as low, medium or high). In fact, however, a dichotomous test is often derived from a continuous laboratory measurement: for example, a blood test that measures the concentration of certain proteins in the blood to determine whether a patient has suffered a heart attack. (Such proteins are released when heart-muscle cells die.) This becomes a *dichotomous test* based on an "established" threshold for the measured concentrations, but that threshold is a variable that affects the statistics. Nonetheless, sometimes a test is *intrinsically dichotomous*, as in a test for the presence or absence of a gene mutation. For a dichotomous test, the two test outcomes are defined as:

Test positive (T^+): A positive test result, which *suggests* the presence of the disease.
Test negative (T^-): A negative test result, which *suggests* absence of the disease.

These definitions lead to the following more-specific terms for a dichotomous test:

True positive (TP) ≡ A positive test in a case where the disease is present according to the gold standard (a correct positive test result).
False negative (FN) ≡ A negative test in a case where the disease is present according to the gold standard (incorrect negative test result). A false negative is also referred to as a *type-II error*.
True negative (TN) ≡ A negative test in a case where the disease is absent according to the gold standard (correct negative test result).
False positive (FP) ≡ A positive test in a case where the disease is absent according to the gold standard (incorrect positive test result). A false positive is also referred to as a *type-I error*.

Figure 3.1 illustrates how the threshold setting can affect the result for the case of a dichotomous test that is determined by measurement of a continuous variable. This illustration plots what could be, for example, the probability distribution of concentrations in blood of the proteins related to muscle cell death, for a broad population of post-attack patients. It is assumed that the same distribution for healthy subjects is shifted toward lower protein concentrations. Those to the right

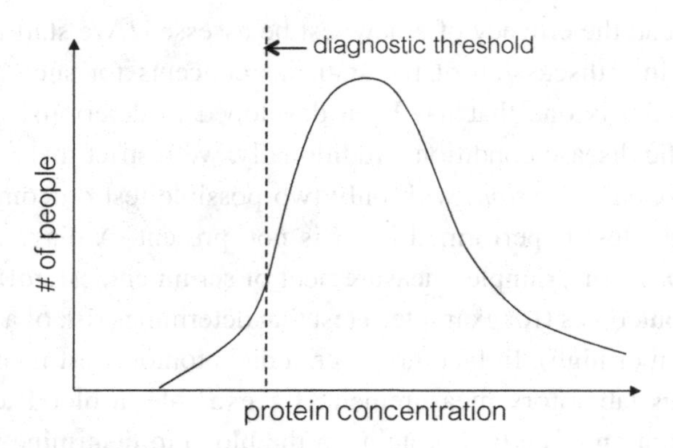

Figure 3.1

An example of setting a threshold for a measured value of a continuous test variable for determination of a dichotomous diagnostic decision. What is represented here is the distribution of the biomarker protein levels for people who have suffered a heart attack. (The distribution for healthy people would, of course, be at lower protein concentrations than this spread.)

of the threshold are TP test results, and those to the left are FN test results. (Assessment of dual distributions that invoke both the diseased and "normal" subjects will be addressed further below.)

We examine the test results for a study performed on a total number n of test subjects, a subset of whom, n^+, have the disease (D^+), the remainder, n^-, not having the disease (D^-), where $n = n^+ + n^-$. Let us, for a moment, assume that the gold standard is perfect, thus making it equivalent to absolute truth. In the expressions that follow, we abbreviate "number of true positives" as TP, and so on. Thus, we can then write a "Truth Table" as follows:

		Gold standard assessment of disease		
		Present (D^+)	Absent (D^-)	
	Test-positive (T^+)	TP	FP	
Test	Test-negative (T^-)	FN	TN	
	Total	n^+	n^-	$= n$

noting that

$$n^+ = \text{TP} + \text{FN},$$
$$n^- = \text{TN} + \text{FP},$$
$$T^+ = \text{TP} + \text{FP}, \text{ and}$$
$$T^- = \text{TN} + \text{FN}.$$

3.2.1 Sensitivity, specificity, positive and negative predictive values

There are four primary measures of the performance (efficacy) of a test. We start by defining them conceptually, and introduce them in the language of Bayesian probability notation (named after the British theologian and mathematician Thomas Bayes [1701?–1761]) for conditional probability, $P(A|B)$, the probability of event A, given that event B has occurred:

sensitivity $(Se) \equiv P(T^+|D^+)$: the probability that the test yields a positive result (T^+), given that the subject has the disease (D^+)

specificity $(Sp) \equiv P(T^-|D^-)$: the probability that the test yields a negative result (T^-), given that the disease is absent in the subject (D^-)

positive predictive value (PPV) $\equiv P(D^+|\, T^+)$: the probability that the subject actually has the disease (D^+), given that the test result is positive (T^+)

negative predictive value (NPV) $\equiv P(D^-|T^-)$: the probability that the subject is disease-free (D^-), given that the test result is negative (T^-)

But it is also important to know the *prevalence* of the disease in the test population, the fraction of the test population that is D^+: in other words, the probability that any randomly selected individual in the test population has the disease (sometimes called the *sample proportion*). This factor is important, and is a critical factor in the predictive values, and also in the test *accuracy*, as we shall see. Whereas sensitivity and specificity are intrinsic to the test, and independent of prevalence, predictive values are not because they depend on the prevalence of the tested condition. This also means that it is important to observe that the prevalence of the disease in the tested population is independent of the diagnostic test performance.

In a *retrospective study* of patients with known gold-standard diagnosis, the prevalence of the disease in the tested population is set by the investigator. For example, when assessing a diagnostic study for breast cancer, it may be appropriate to select a study group consisting of the same number of healthy women and women with cancer. In a *prospective* screening study based on randomized patient recruitment, however, the prevalence of the disease reflects the prevalence in the investigated population; for example, the prevalence of skin cancer is significantly different in fair-skinned vs. dark-skinned people. The prevalence of the disease in the tested population can be written as:

$$P(D^+) \equiv \frac{n^+}{n} = \frac{\text{TP} + \text{FN}}{n}. \tag{3.1}$$

The reader should be aware that the disease prevalence is sometimes confused with disease *incidence* (sometimes called *occurrence rate*), which refers to the proportionate *rate* of occurrence of *new* disease in a given population: that is, the number of new cases per unit population *per unit time*. According to this definition, the incidence multiplied by a short time period yields the probability that a (randomly chosen) and currently disease-free individual will develop the disease within that specific time period (most commonly quoted for one year).

Analogous to Eq. (3.1), the expression

$$P(D^-) \equiv \frac{n^-}{n} = \frac{\text{TN} + \text{FP}}{n} = 1 - P(D^+) \tag{3.2}$$

can be used to represent the fraction of the test population that is currently disease-free (again, assuming a dichotomous decision). It follows, simply, that $P(D^+) + P(D^-) = 1$.

In this chapter, we will abbreviate the disease prevalence (and absence) values, $P(D^+)$ and $P(D^-)$, as P^+ and P^-, respectively.

In writing expressions for Se and Sp we utilize the definition of conditional probabilities, $P(A|\,B) = P(A, B)/P(B)$, where $P(A, B)$ is the *joint probability*, i.e., the probability of two events both happening. For example, one could calculate the joint probability that a person is infected with flu virus *and* has been exposed to infected people, which are not completely independent events. (For truly independent events, the joint probability is simply the product of the two independent probabilities.) Following the definition of conditional probability, the test sensitivity and specificity can be expressed as

$$Se = P(T^+|D^+) = \frac{P(T^+, D^+)}{P^+} = \frac{\text{TP}/n}{n^+/n} = \frac{\text{TP}}{\text{TP} + \text{FN}} \tag{3.3}$$

and

$$Sp = P(T^-|D^-) = \frac{P(T^-, D^-)}{P^-} = \frac{\text{TN}/n}{n^-/n} = \frac{\text{TN}}{\text{TN} + \text{FP}}. \tag{3.4}$$

The possible range for either Se or Sp is 0 to 1.0 (or 0–100%, as a percentage). If the test outcome were based on simply flipping a coin for each patient, both Se and Sp would average (over many coin flips) at 1/2, or 50%. If the test outcome were based on rolling a die many times and assigning a positive test outcome for a roll of 1 and a negative outcome for any roll from 2 to 6, then Se would average 1/6 and Sp would average 5/6. In fact, for any fixed probability p of a test being positive (i.e., a test whose outcome is independent of the patient examined, definitely not a good test to apply for diagnostics!), the sensitivity would be equal to p and the specificity would be equal to $1 - p$. Sensitivity and specificity can each take any value between 0 and 1, depending on the specific test design. Even more important, *sensitivity by itself is a meaningless representation of the efficacy of a test if not quoted together with specificity* (and *vice versa*). For example, if a test device is broken, always yielding a positive result (regardless of the truth), the test sensitivity would be 100% ($Se = 1$), because the test output would be positive for all diseased individuals. This would, of course, be useless information because the specificity would be 0.

Even though disease prevalence (or absence rate) appears explicitly in the denominator of the probability expressions in Eqs. (3.3) and (3.4), it is also invoked in the joint probability expression in the numerator, and therefore cancels. Consequently, *Se and Sp can be determined individually without knowledge of the prevalence*. In fact, *Se* can be determined by running the test on a cohort of subjects who are *all* diseased, and *Sp* can be determined by testing a group composed entirely of normal subjects. Moreover, *Se* and *Sp*, once established by such a measure, would each be the same (within statistical variation) for any test population, regardless of the disease prevalence. This should make a researcher appropriately wary of using only *Se* or *Sp* (or, even, both) to evaluate the efficacy of a test, because results can be potentially misleading, as will be shown below and in Section 3.3.

The positive- and negative-predictive values can be written with more explicit formulas by following *Bayes's rule* as:

$$\mathrm{PPV} = P(D^+|T^+) = \frac{P(T^+|D^+)P^+}{P(T^+)} = \frac{P(T^+|D^+)P^+}{P(T^+|D^+)P^+ + P(T^+|D^-)P^-} = \frac{\mathrm{TP}}{\mathrm{TP} + \mathrm{FP}} \tag{3.5}$$

and

$$\mathrm{NPV} = P(D^-|T^-) = \frac{P(T^-|D^-)P^-}{P(T^-)} = \frac{P(T^-|D^-)P^-}{P(T^-|D^-)P^- + P(T^-|D^+)P^+} = \frac{\mathrm{TN}}{\mathrm{TN} + \mathrm{FN}} \tag{3.6}$$

In contrast to *Se* and *Sp*, PPV and NPV are inextricably connected to disease prevalence, such that any predictive values determined for a specific test population would *only* apply to a new test group if the prevalence in that population is the same. Again, caution in the interpretation of statistical results is advised, because predictive values can also lead to misunderstanding. For a low-prevalence disease, in particular, it is easy for the NPV to be exceptionally high, possibly instilling false confidence in the efficacy of the test. Consider, for example, a test for a disease that occurs in only 1 of 10,000 subjects in the test population. Then, even if the sensitivity and specificity are each only 0.5 (as would be achieved by a coin toss), the NPV would be 0.9999 (99.99%). Of course, the PPV would be very poor.

On the other hand even excellent values for *Se* and *Sp* may not be good enough when one considers the PPV of a low-prevalence disease. Consider again the case where the prevalence of the condition in the test population is low (say, 1 in 10,000 subjects). Now, this time assume that both *Se* and *Sp* are very high, at

0.99 (99%). Note that this means that the false positive and false negative rates will be "only" 1%. In terms of predictive value, on the other hand, this means that the positive predictive value, PPV, will be only $\sim$ 1% ($\frac{1}{1+100}$). Thus, for every person correctly diagnosed with the condition (1 in 10,000), 100 people will be incorrectly designated as positive (1% $\times$ 10, 000 = 100).

Could this be harmful? That depends on the consequence of a false positive. To exemplify the situation more explicitly, imagine a large company with 10,000 employees. The company president suspects that one employee is disloyal and is providing trade secrets to a competitor. The president wants to root out that one employee, and prosecute him/her aggressively. Finding the suspect will be accomplished by subjecting all employees to a lie-detector test, and the company that provides the test services guarantees the test "accuracy" (Se and Sp) to be 99%, which is unlikely, but we'll allow the assumption for the sake of the illustration. The president thinks this means the test will be fair to all because any individual honest employee will be found innocent, with 99% probability. The scientists in the company, fortunately, understand the statistics and are quick to inform the president that the lives of approximately 100 honest employees will be harmed irreparably by being falsely accused and prosecuted. This is clearly a case where the PPV value must be exceptionally high to minimize harm to innocent people. Perhaps other methods need to be utilized to find the "spy." It is not surprising, consequently, that lie-detector tests are not permitted as evidence in courts of law in the USA (and one wonders why such tests are used for investigative purposes by various government agencies).

It is interesting to note that if the sensitivity and specificity of a test are known from a study performed on any sample population (with any disease prevalence), or from selected diseased and disease-free cohorts, the PPV can be determined for a different population as long as the disease prevalence is known for that new set. This observation derives from Bayes's theorem, and can be seen from Eq. (3.5), which can be rewritten as:

$$\text{PPV} = \frac{Se \times P^+}{Se \times P^+ + (1 - Sp)(P^-)}. \tag{3.7}$$

A similar expression for NPV of a test can also be determined from any measure of Se and Sp, if the prevalence in a new population is known.

3.3 Test accuracy

When discussing a diagnostic test, the term *accuracy* is often used to indicate, in a general sense, how good both Se and Sp are, and in some contexts also including the predictive values. Sometimes it is desired, however, to express the overall accuracy of a test, ACC, as a single number. This is frequently expressed as the overall fraction (or percentage) of correct test results:

$$\mathrm{ACC} \equiv \frac{\mathrm{TP}+\mathrm{TN}}{\mathrm{TP}+\mathrm{FN}+\mathrm{TN}+\mathrm{FP}} = \frac{\mathrm{TP}+\mathrm{TN}}{n} = Se \times P^{+} + Sp \times P^{-} \quad (3.8)$$

As can be seen from Eq. (3.8), and in contrast to Se and Sp themselves, the value for ACC depends directly on the disease prevalence, because it is a prevalence-weighted average of Se and Sp. *However, if the disease prevalence is very high (or very low), this definition can lead to misleadingly high values of accuracy.* Let's recall the broken test device, but this time we imagine that its output is always negative, even for diseased individuals. If the disease prevalence is very low, say 1/10,000, then the overall accuracy will be almost 100%, even though the sensitivity is zero (TP = 0), because $\mathrm{TN} \approx n$. (A similar situation would arise, regardless of the specificity, if almost all of the test population were diseased.) To avoid the overstated accuracy for an imbalanced dataset, a *"balanced" accuracy"* is sometimes quoted

$$\text{balanced accuracy} \equiv \mathrm{ACC}_{bal} = \frac{Se + Sp}{2}. \quad (3.9)$$

With this definition, Se and Sp are equally important, and the measure of accuracy is more meaningful.

3.4 Dichotomous decisions based on a continuous variable

As mentioned in Section 3.2, frequently a dichotomous test is based on a specific threshold value for a continuously measured parameter. To understand how the chosen threshold affects the accuracy of a test on a population of subjects that invoke both diseased and non-diseased individuals, consider a test that measures blood glucose concentration for the purpose of determining whether a person is a potential diabetic (which can impact decisions about follow-up tests that may be more expensive or invasive). This is essentially the risk-assessment of a disease based on the measurement of a symptom.

If one were to conduct a screening test on a large population base, one might be dealing with a cohort of subjects for which the distribution of blood glucose levels is shown in Figure 3.2, which plots two distributions of blood glucose readings, separated into the populations of subjects who are truly "normal" and those who are undiagnosed or untreated diabetics. (It is presumed that these distributions can be known by other means, or that the prevalence in the test population is known.) If a decision about the subject's risk is to be based on this test, then decisions for values to the left of the gray overlap area would always correctly be negative (although there would be many false positives), and decisions for subjects to the right of

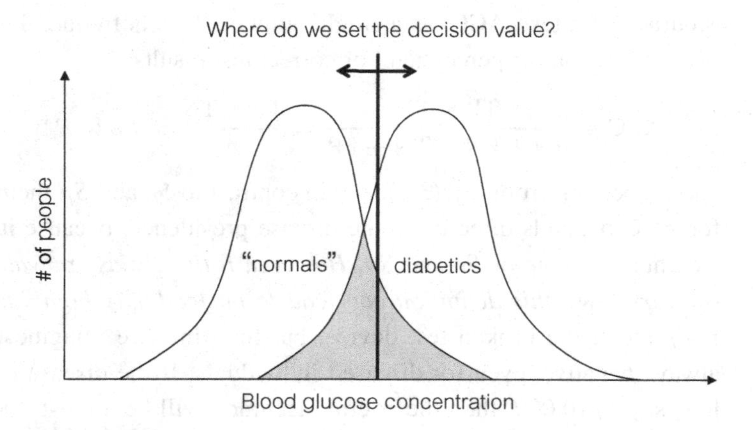

Figure 3.2

Blood glucose concentrations for populations of "normal" subjects (left distribution curve) and untreated diabetics (right distribution curve). (This graph is imaginary, for purposes of illustration, and should not be construed as representing medical data.)

the overlap would always be correct as positive (although there would be many false negatives). Any diagnostic decision made within the overlap area, however, would entail a variable probability of both false positives and false negatives, with the relative probabilities determined by the specific glucose value chosen for the decision threshold. This should not be surprising. "Normal" subjects in the gray zone might have just finished consuming a piece of cake with a sweet soda, whereas a diabetic in that zone may have just completed a morning exercise after skipping breakfast.

This type of situation with a diagnostic test is common. The choice for the threshold value of the test parameter to be used as the cutoff point may require factoring the consequences of false positives or false negatives, and the patient's individual status. Often the consequence of a "borderline" test result is additional testing, to either confirm or reject the initial result. Again, in the example above, it would be easy to achieve high sensitivity, at the expense of lower specificity, by placing the decision boundary near the left end of the gray overlap area. The inverse would occur if the decision boundary were located near the right-hand end of the overlap area.

3.4.1 The ROC curve

A graphical method for displaying the accuracy statistics of a test, which is commonly used in medical and biomedical engineering literature, is the *receiver operating characteristic curve*, or the *ROC curve*, for short. This type of plot provides

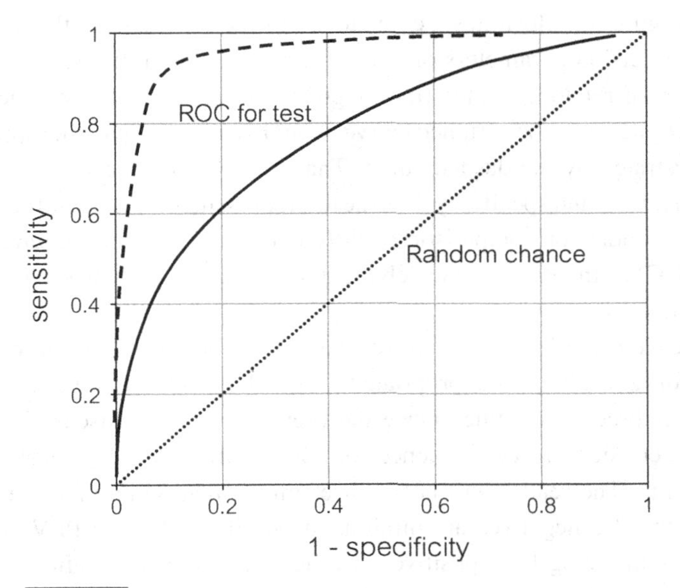

Figure 3.3

An example of ROC curves for two different diagnostic tests. The test represented by the dashed curve exhibits higher accuracy than the one represented by the solid curve. The straight dotted line represents what the curve would look like for random chance, i.e., when there is a fixed probability that a test is positive, independent of the patient or sample being examined.

a good visualization of the relative effects of a variable threshold for a test parameter on the sensitivity versus the specificity, and may be more instructive than the expressions of Eqs. (3.8) or (3.9). As exemplified in Figure 3.3, an ROC graph plots Se against $(1 - Sp)$, as a function of the diagnostic parameter threshold. In the example above, the test for diabetes, the diagnostic parameter would be the chosen threshold value of blood glucose concentration for a positive or negative test diagnosis. Another way of saying this is that the ROC curve is a plot of the "hit" rate versus the false-alarm rate, as the threshold for the diagnostic parameter is varied.

The origins of the ROC curve come from the early days of radar and radio receivers. Electrical engineers wanted a way to display the rate of detecting a target (e.g., an airplane) compared to the noise-generated "signals" as an instrumental parameter was varied: say, the gain of the amplifier in the radar receiver. If the gain is too high, noise generates false blips on the radar screen, whereas real targets could be missed if the gain is set too low. The ROC graphical method of data display has been widely adopted by the biomedical research community, although, often, too much emphasis is placed on Se and Sp as a measure of test performance, as has been discussed above.

Figure 3.3 illustrates a comparison of the ROC curves that result from varying the decision parameters of two different tests with different effective accuracies, or from random chance (the diagonal straight dotted line). Ideally, for a highly accurate test, the graphed curve would reach far into the upper left corner, as illustrated by the dashed curve. That test would achieve a better combination of sensitivity and specificity than the test represented by the solid curve. A quantitative assessment for comparison of different tests would be the *area under the curve* (AUC) of the ROC, for which the dashed curve clearly has a higher value than the solid curve.

Given the definitions above and the various potential circumstances for misleading results, it is appropriate to ask: which is more important: a test's accuracy or its predictive value? Once the prevalence of the disease is known, one must also consider the consequences of false positives or false negatives among the test results. The reader will note, for example, that *Se* and NPV are both optimized when false negatives are minimized. Similarly, *Sp* and PPV are both optimized by minimizing false positives. To illustrate the considerations (in addition to the scenarios discussed above), another example may be found in the administration of tests that screen for HIV in blood donated at a blood bank. In the United States, the prevalence is generally low, although it can be surprisingly high among people who donate blood at blood banks that pay for the blood. There must be minimal risk of infecting recipients of transfusions, especially those with large exposure, such as hemophiliacs. In this case, one may think that it is most important to have a high NPV, such that if the test says the donor is disease-free, the probability of being disease-free is very high. However, unless the parameters of the test were developed using a population with higher prevalence, the NPV value for a low-prevalence population may be misleadingly optimistic. In that case, the *Se* value may be more important, as long as the consequence of a number of false positives is acceptable. Since an initial positive test for HIV will always be followed up by additional tests to confirm the result, the main consequence of false positives is that some blood donors will be nervous until the additional tests disprove the initial FP result.

3.5 Required sample size and reliability of a test

When planning a study to assess the accuracy of a diagnostic test, one is always confronted with the issue of how large the sample size needs to be in order to establish a desired level of statistical confidence about the determined values of the sensitivity and specificity or the predictive values for a known prevalence. Let's say we believe the sensitivity of a test to be 90%, but we want to be sure that it is no less than, say, 85%. This would correspond to a maximum acceptable error of ~5% of the determined sensitivity. Thus, we want the sample size to be large enough to provide the desired statistical confidence, but not larger than necessary, if for no other reason than to limit the cost or time. Determination of the minimal sample size is often referred to as "powering a study" (not to be confused with the frequently used term, "power of a test," which is generally synonymous with sensitivity).

No single formula can be presented to calculate a needed sample size because the calculation is different for each specific study objective or biomedical question (Suresh and Chandrashekara, 2012; Li and Fine, 2004; Flahaulta et al., 2005). Moreover, the calculation depends on a set of assumptions that must be made, which cannot be known with certainty, leading to additional factors that are sometimes called "design effect" (Fosgate, 2009). For example, one may need to make an initial assumption about what the test sensitivity is expected to be. In short, the necessary sample size depends on the study design and its intended outcome. There are too many variations to cover in this section – after all, biostatisticians have jobs for good reason! – so we provide a few examples, again specific to diagnostic tests, and direct the reader to the suggested sources in the further reading list. Given the many types of studies, and the various sets of study parameters, reference books of calculated tables are published to aid researchers in their study designs (e.g., Machin et al., 2008).

For assessing the sample size needed for determining the precision of the performance parameters of a diagnostic test, it is useful to introduce expressions for the *standard error* (or standard deviation of the mean of the sampled distribution, if the sample is representative of the population) of each of the four measures. If we have an estimate for the sensitivity (Eq. (3.3)), then the standard error for that estimate is

$$\mathrm{SE}_{Se} = \sqrt{Se(1 - Se)}/\sqrt{n^+}. \tag{3.10}$$

Similarly, for the other performance measures:

$$\mathrm{SE}_{Sp} = \sqrt{Sp(1 - Sp)}/\sqrt{n^-}; \tag{3.11}$$

$$\mathrm{SE}_{\mathrm{PPV}} = \sqrt{\mathrm{PPV}(1 - \mathrm{PPV})}/\sqrt{T^+}; \tag{3.12}$$

$$\mathrm{SE}_{\mathrm{NPV}} = \sqrt{\mathrm{NPV}(1 - \mathrm{NPV})}/\sqrt{T^-}. \tag{3.13}$$

For any of these parameters, then, the statistical range of the estimated performance factor can be calculated (for a normal distribution) as

$$\mathrm{CI} = F_{est} \pm Z_{1-\alpha/2} \times \mathrm{SE}, \tag{3.14}$$

where CI is the *confidence interval*, spanning from the lower confidence limit to the upper confidence limit, and F_{est} stands for the estimated value of *Se*, *Sp*, PPV, or NPV. The factor $Z_{1-\alpha/2}$, which can be found in tables in most statistics textbooks, is the $(1 - \alpha/2)$ quantile of a standard normal variable *Z*, which is defined by $P(Z \leq Z_{1-\alpha/2}) = 1 - \alpha/2$. The factor α represents the type-I error rate, and $\alpha/2$ is the probability associated to values that fall outside the confidence interval on each side of it (so that α is the total probability associated with values outside the confidence interval). The factor $Z_{1-\alpha/2}$ has a value of 1.96 for a 95% confidence level of the interval limits (i.e., for $\alpha = 0.05$). In the context of testing for a single condition, the presence of a disease, the null hypothesis would state that the disease is absent in a test subject. If the null hypothesis is true, but is rejected (the test outcome is positive for disease), then the type-I error is what we call a false positive.

Thus, the *precision* of each performance estimate is the *width of its confidence interval*, δ, or (for 95% confidence):

$$\delta = Z_{1-\alpha/2} \times \text{SE} = 1.96 \times \text{SE}. \tag{3.15}$$

(For clarity about the terminology, we note that the precision, or what is called the width, δ, is actually the half-width of the full range of the CI, which is $2 \times \delta$.)

3.5.1 Example 1: testing sensitivity and specificity with pre-diagnosed sets

The sensitivity and specificity of a test are not dependent on disease prevalence, but the availability of a necessary sample size for evaluation of a test is. Nonetheless, sometimes a new diagnostic method can be tested on a controlled population, all of whose members are already known to be either disease-positive or disease-free by prior reliable diagnosis. With these preselected sample sets, the sensitivity of this dichotomous test can be assessed using a set that is composed of only known-positive subjects, and the specificity can be assessed with only "controls" (known disease-free subjects), thereby bypassing the issue of prevalence. But how many subjects, n_{Se}, in this selected group are needed to establish the test sensitivity? To plan the sample sizes in advance, the researcher must still have estimates of the expected test sensitivity and specificity, and must choose the desired precision in the determination of each of those values.

Let's say we expect the sensitivity of a test to be ~0.80, and we wish to verify that sensitivity value with a precision of ±0.1 at the 95% confidence level, such that the confidence interval will be 0.70 to 0.90, covering the range of the estimated sensitivity $\pm 1.96 \times \text{SE}_{Se}$. Thus, we want the standard error to be no larger than 0.1 / 1.96, or ~0.05. For this type of controlled-sample study, in which the effective prevalence within the test sample for *Se* is 1.0, we can directly apply Eq. (3.10), yielding a needed sample size of $n^+ = 64$. Since all of the subjects in this set are disease-positive, we have the needed sample size for this measure, $n_{Se} = n^+ = 64$. The calculation for the sample size for *Sp* is analogous, and the result is the same if the expected specificity is also 0.8; thus, equal numbers of disease-positive and disease-negative subjects are needed for this determination, regardless of the disease prevalence in the population. Of course, if the initial estimated values for the *Se* and *Sp* of this test are significantly off target, then a second iteration may be required to achieve the desired precision after the initial study.

Note that this simplified approach cannot be used for determining the predictive values, as they depend intrinsically on the disease prevalence, by definition.

3.5.2 Example 2: determining *Se* and *Sp* for a population sample with an expected prevalence

A common testing condition is for a new test to be compared alongside an existing "gold standard" test when screening a random sample of the general population, rather than a controlled population sample. Typically, the prevalence of disease-positive subjects will not be the same as the prevalence of disease-free individuals. That is, there is not, typically, a 50/50 split in the general population. Thus, for a specific desired precision (with 95% confidence), the sample numbers needed to estimate sensitivity and specificity (or PPV and NPV) will be different. Commonly, the larger of the two is chosen as the sample size for both variables.

An initial assumption here for random screening samples is that the sampling method is such that the expected disease prevalence in the population sample is approximately the same as in the general population or in the intended population where the test will be applied. (Alternatively, it can be assumed that the relative prevalence values are known.) Then, in order to achieve the desired precision for the calculated values of the test performance, three parameters must be specified by the researcher to enable determination of the required sample size (Buderer, 1996): (1) the desired precision (or 95% confidence interval); (2) the disease prevalence in the test population; and (3) an initial estimate of the performance of the test for the sensitivity and specificity (or PPV and NPV).

Again, using Eqs. (3.10) and (3.15), the number of disease-positive subjects needed for determining the sensitivity (assuming we want a 95% confidence in the precision, δ) is calculated as

$$\mathrm{TP} + \mathrm{FN} = n^+ = Z^2_{1-\alpha/2} \frac{Se(1 - Se)}{\delta^2}. \tag{3.16}$$

With the known prevalence, P^+, of the disease in the test population, this leads (for a 95% confidence level of the desired precision, δ, to the number of test subjects, n_{Se}, needed to establish the sensitivity:

$$n_{Se} = n^+/P^+ = (1.96)^2 \frac{Se(1 - Se)}{P^+ \delta^2}. \quad (3.17)$$

If we apply the same values as in Section 3.5.1 for the desired precision (0.1) and the expected sensitivity (0.8), but now we assume a disease prevalence, P^+, of 10% (or 0.1), suddenly the required sample size has mushroomed to 640. Clearly the study will be less expensive when applied to a controlled test population (although that luxury would not exist for assessing PPV or NPV).

Equations analogous to Eqs. (3.16) and (3.17) can provide an estimate for the needed sample size, n_{Sp}, for establishing the specificity with the desired precision. Finally, the researcher chooses the randomly selected sample number for the study, n, to be the larger of n_{Se} and n_{Sp}.

3.5.3 Example 3: measuring the prevalence of a disease in a population

As the final example, which can be addressed with similar formalism, we consider the epidemiological question of using a diagnostic test to determine the prevalence of a disease in a target population. This type of situation might arise, for example, if a government health agency wants to know whether the prevalence of a disease is higher or lower (compared to a historical prevalence) as a result of, say, environmental factors or government policies. This is typically a two-population mean-value type of study, comparing, for example, a disease prevalence before and after implementation of a prevention method, or comparing parallel populations, one in which the measures have been implemented, with the other as a control population. The statistics of such two-set studies follow similar rules of probability (Fosgate, 2009), but entail more complex formalism than is addressed in this chapter, especially if the diagnostic test itself is imperfect.

We simplify to measuring the current disease prevalence in a random sampling of the general population, while assuming that we know with high precision what the prevalence is in the historical data or in the comparison population, and also assuming that the *Se* and *Sp* of the test are both near perfect (close to 1.0). Thus, since we start with an initial estimate of the prevalence, our assumption of near-perfect *Se* and *Sp* also leads to the expectation of good predictive values of the test. So, we want to know how large the randomly sampled set must be to enable determination of a change in prevalence (compared to a known prior value) that is adequately representative of the larger population.

Analogous to the prior examples, for a sample size n (the number of screened subjects from the population), the standard error of the disease prevalence (or proportion), is given by

$$\mathrm{SE}_{P+} = \sqrt{P^+(1-P^+)}/\sqrt{n}, \tag{3.18}$$

and if we want a precision δ, with a specific level of confidence, the needed sample size is

$$n = Z^2_{1-\alpha/2}\frac{P^+(1-P^+)}{\delta^2}, \tag{3.19}$$

where, again, $Z_{1-\alpha/2}$ represents the $(1-\alpha/2)$ quantile of a standard normal variable Z, which has a value of 1.96 for a 95% confidence interval (of the chosen precision).

Two things are interesting to note:

1. For a given precision, the required sample size is largest when the prevalence is 0.5. For a small prevalence, the sample size becomes smaller, for the same precision. However, for small prevalence the critical factor then becomes the desired precision.
2. It must be noted that the precision is an absolute value, not a relative value. That is, δ is not a percentage of the measured proportion, but an acceptable range of that proportion.

Thus, if the desired precision is as good as 1/1000, that may seem like high precision, but it can represent a 100% change (up or down) in a measured proportion whose comparison prevalence value is also 1/1000. Consequently, the issue of precision can be the limiting factor in the statistical significance of a study, if the prevailing prevalence is small.

Cases of this limitation have arisen frequently in population health studies. For example, in a town of, say, 10,000 people living near a nuclear research center, one might ask whether the prevalence of 1.5×10^{-3} (15 cases for the population of 10,000) for leukemia constitutes a significant increase over the prevalence in the general US population, which is approximately 1/1000. Can one say with 95% statistical confidence that this difference is significant? Examination of Eq. (3.19) tells us that we would need to count the cases in a population sample of ~16,000 in order to determine (with 95% confidence) that the prevalence of 0.015 is significantly higher than 0.01 of the US population. Since this is more than the entire population of the town, nothing can be said with high statistical confidence about the disease prevalence.

3.6 Prospective vs. retrospective studies

All of the discussion thus far in this chapter has dealt with diagnostic tests that measure a specific parameter that is assumed to be a direct measure of a disease condition. Sometimes, however, a new diagnostic technology that is being developed measures a multivariate combination of factors that are assumed to be associated with a disease, with a range of correlation strengths. For example, automated image analysis algorithms may look at a variety of factors in a digital radiography image, to establish correlations of image features with disease risk for, say, breast cancer. Similarly, for some of the optical methods to be presented in later chapters, a measured optical spectrum in tissue (e.g., a broadband diffuse-reflectance spectrum) may comprise several optical spectral features relating to various physiological parameters, which either singly or in combination provide probable indications of a pathological state. The challenge is to identify the set of spectral features (and their changes from those of a disease-free state) that provide the highest degree of correlation with disease risk. In such cases a diagnostic algorithm must first be "trained" in a *retrospective study* to recognize features that correlate with disease, and this is done by statistical methods of identifying the spectral features that best correlate with the test condition. This type of problem invokes the statistical tools of pattern recognition, sometimes referred to as methods or tools for *classification analysis* or *machine learning*. (For a review of these methods, see the suggested sources under further reading.)

For the initial training one presents the classification tool with a set of known cases whose diagnosis has been established by a "gold standard." The classification tool examines a multiplicity of features (say, spectral features) and determines those that best correlate with the pre-determined diagnosis, enabling the creation of an algorithm that will perform well when the diagnostic method is tested on this set. This is referred to as the "training" set, and in effect the training process proceeds by numerical methods in which the sensitivity and specificity are checked iteratively against a large multi-dimensional array of candidate features, until the set of features is found that yields the best values for Se and Sp. Of course, this newly established diagnostic algorithm will perform well *retrospectively* on the data set that was used to train it. This does *not* mean, however, that the test, employing this algorithm, will perform equally well on a new, "naïve" set of cases. If fact, unless the training set is very large, a *prospective study* on a naïve set will almost always yield worse performance, especially if the new set is not sampled from the same population base as was used for the training set.

The message here is that a new diagnostic method, after being "trained," must always be tested prospectively on a new data set, with sample sizes and selection methods as discussed in the sections above. Often, unfortunately, the number of subjects available for study of such a new diagnostic method is limited by issues of cost. The researcher can take measurements on a specific total number of subjects, who are presumably a random subset of the population for which the new test method is intended. The question is then: what is the best way to divide up this group into subsets for training and prospective testing? This question itself is a subject of current research in statistical methods of classification (Rodriguez-Diaz et al., 2011). The answer depends on a number of factors, including the number of features of the signal being classified, the relative weightings of those features, and the population distribution widths of those features, etc. (Dobbin and Simon, 2011). A facile, but not uncommon, conclusion is to have more sample cases than needed, and split them 50/50.

Problems – answers to problems with * are on p. 645

3.1* A new noninvasive diagnostic test for skin cancer (using optical spectroscopy) is being studied. The test is intended to assist the doctor seeing patients who present at the dermatologist's office with a suspected skin lesion. The test has a simple binary output: either positive for cancer or negative for benign lesions. The study cohort included 220 patients, each with one suspected skin lesion. When measured by the new optical test, 98 of the lesions were designated as cancerous. When surgical biopsies (tissue specimens) were taken from all of the lesions, and were assessed by a pathologist (the "gold" standard), 109 lesions were found to be cancer, and 93 of those real cancers were among those designated as cancer by the new test.

(a) What are the numbers for true positives (TP), false negatives (FN), true negatives (TN), and false positives (FP)?

(b) Calculate the values of the sensitivity (Se) and specificity (Sp) for the new test.

3.2 Write an expression for NPV in terms of the Se, Sp, and the disease prevalence, P^+.

3.3* For the numbers in Problem 3.1, calculate the values of PPV and NPV for:

(a) a study population with a disease prevalence of 50%;

(b) a study population with a disease prevalence of 10%;

(c) a study population with a disease prevalence of 1%;

(d) what conclusions can you draw about the dependence of PPV and NPV on prevalence from the above three answers?

3.4 Provide examples for situations (not covered in this book) in which the PPV or NPV might be more important than the *Se* and *Sp*, respectively. Explain why.

3.5* Using the ROC curves shown in Figure 3.3:

(a) If the cutoff parameter for the dashed curve is chosen to yield a sensitivity of 90% or better, what is the best specificity that can be obtained?

(b) What would be the best sensitivity for the solid curve if a specificity of >80% was desired?

3.6 Consider prostate cancer, for which a symptom is an elevated level of prostate-specific antigen (PSA), determined with a blood test. In a population of men over 60, suppose that, when using a threshold blood concentration of PSA of >4 ng/ml, the test yields elevated levels for 80% of men with prostate cancer and 20% of men who are disease-free. What are the PPV and NPV values for this test, assuming that 20% of men over 60 actually harbor prostate cancer?

3.7* If the expected sensitivity of a new test is 0.9, and the prevalence of the disease is 5%, calculate the sample size needed if you want to verify the sensitivity to within ±0.05 for a 95% confidence interval.

3.8 The national average for the prevalence of a specific disease is 1/300. You wish to know the prevalence in your community with a precision of 30% at a confidence level 95%. What is the sample size needed to determine the prevalence in your community?

References

Buderer, N. M. (1996). Statistical methodology: I. Incorporating the prevalence of disease into the sample size calculation for sensitivity and specificity. *Academic Emergency Medicine*, 3, 895–900.

Dobbin, K. K., and Simon, R. M. (2011). Optimally splitting cases for training and testing high dimensional classifiers. *BMC Medical Genomics*, 4, 2–8.

Flahaulta, A., Cadilhaca, M., and Thomas, G. (2005). Sample size calculation should be performed for design accuracy in diagnostic test studies. *Journal of Clinical Epidemiology*, 58, 859–862.

Fosgate, G. T. (2009). Practical sample size calculations for surveillance and diagnostic investigations. *Journal of Veterinary Diagnostic Investigation*, 2, 3–14.

Li, J., and Fine, J. (2004). On sample size for sensitivity and specificity in prospective diagnostic accuracy studies. *Statistics in Medicine*, 23, 2537–2550.

Machin, D., Campbell, M. J., Tan, S. B., and Tan, S. H. (2008). *Sample Size Tables for Clinical Studies*, 3rd edn. London: BMJ Books.

Rodriguez-Diaz, E., Castanon, D., Singh, S. K., and Bigio, I. J. (2011). Spectral classifier design with ensemble classifiers and misclassification-rejection: application to elastic-scattering spectroscopy for detection of colonic neoplasia. *Journal of Biomedical Optics*, 16, 067009–06715.

Suresh, K. P., and Chandrashekara, S. (2012). Sample size estimation and power analysis for clinical research studies. *Journal of Human Reproductive Sciences*, 5(1), 7–13.

Further reading

General statistics and biostatistics

Gray, J. A. M. (2001). *Evidence-Based Healthcare*, 2nd edn. Edinburgh: Churchill Livingstone.

Rosner, B. (2011). *Fundamentals of Biostatistics*, 7th edn. Boston, MA: Brooks/Cole Cengage Learning Press.

Spiegel, M. R., Schiller, J. J., and Srinivasan, R. A. (2009). *Schaum's Outlines, Probability and Statistics*, 3rd edn. New York: McGraw Hill.

Woolson, R. F. and Clarke, W. R. (2002). *Statistical Methods for the Analysis of Biomedical Data*, 2nd edn. (John Wiley, New York).

Analysis of diagnostic tests

Shapiro, D. E. (1999). The interpretation of diagnostic tests. *Statistical Methods in Medical Research*, 8, 113–134.

Statistical pattern recognition

Jain, A. K., Duin, R. P. W., and Mao, J. (2000). Statistical pattern recognition: a review. *IEEE Transactions on Pattern Analysis and Machine Intelligence*, 22, 4–37.

Webb, A. R., and Copsey, K. D. (2011). *Statistical Pattern Recognition*, 3rd edn. John Wiley & Sons.

The University Group Diabetes Program

Bradley, R. F., Dolger, H., Forsham, P. H., and Seltzer, H. (1975). Settling the UGDP controversy? *Journal of the American Medical Association*, 232, 813–817.

Cornfield, J. (1971). The University Group Diabetes Program: a further statistical analysis of the mortality findings. *Journal of the American Medical Association*, 217, 1676–1687.

Feinstein, A. R. (1971). Clinical biostatistics-VIII: An analytical appraisal of the University Group Diabetes Program (UGDP) study. *Clinical Pharmacology and Therapeutics*, 12, 167–191.

Klimt, C. R., Knatterud, G. L., Meinert, C. L., and Prout, T. E. (1970). A study of the effects of hypoglycemic agents on vascular complications in patients with adult-onset diabetes: I. Design, methods and baseline results. *Diabetes*, 19, 747–783.

Klimt, C. R., Knatterud, G. L., Meinert, C. L., and Prout, T.E. (1970). A study of the effects of hypoglycemic agents on vascular complications in patients with adult-onset diabetes: II. Mortality results. *Diabetes*, 19, 789–830.

Leibel, B. (1971). An analysis of the University Group Diabetes Study Program: Data results and conclusions. *Canadian Medical Association Journal*, 105, 292–294.

Schor, S. (1971). The University Group Diabetes Program: a statistician looks at the mortality results. *Journal of the American Medical Association*, 217, 1671–1675.

Schwartz, T. B., and Meinert, C. L. (2004). The UGDP Controversy: Thirty-four years of contentious ambiguity laid to rest. *Perspectives in Biology and Medicine*, 47, 564–574.

4 General concepts of tissue spectroscopy and instrumentation

The word *spectroscopy* comes from the Latin-Greek hybrid, wherein *spectrum* is from the Latin word for "an appearance" (specter or apparition = a ghost!), and *skopein* is the Greek word for seeing or observing. Why does spectroscopy mean observing an appearance or a ghost? One explanation is based on the first systematic observation (among the many contributions to modern mathematics and science made by English physicist and mathematician Sir Isaac Newton [1642–1727]) of how the propagation of sunlight through a prism results in the "appearance" of a band of different colors. A second explanation may be that when one uses light to study an object, the observations are based on the effects of that object on light, so that one does not directly view the object, but one see its appearance, or a "ghost" of it. Finally, one can think of a spectral decomposition of sunlight as being filled with ghosts: the set of dark lines first observed by the English scientist William Wollaston [1766–1828] and the German physicist Joseph Fraunhofer [1787–1826], and the bands of invisible "colors" beyond the red on one side and beyond the violet on the other. The word spectrum has evolved in English to mean a range of ordered values or colors. For our purposes *optical spectroscopy* may be loosely defined as *the study of atoms, molecules and particles by means of their wavelength-dependent interactions with light*. Relevant to the methods discussed in this chapter *light* is broadly defined to encompass the range from the ultraviolet to mid-infrared wavelengths of the electromagnetic spectrum, or about 200 nm to 15 μm, which are transmitted through air with only moderate attenuation (see Figure 4.1). Over this range, the components used for optical instrumentation are most conveniently explained by appealing to the wave-like nature of light.

When it comes to scattering interactions and the diffusion of light in turbid media such as tissue, however, it is often more convenient to treat light as particles, or *photons*, rather than waves. Scattering can be *elastic*, in which the photon changes direction but not energy, or *inelastic*, in which the photon can both exchange energy and change its direction of propagation. Scattering phenomena are often described in probabilistic terms, by specifying the likelihood of a photon being scattered and the probable angular distribution of the scattering.

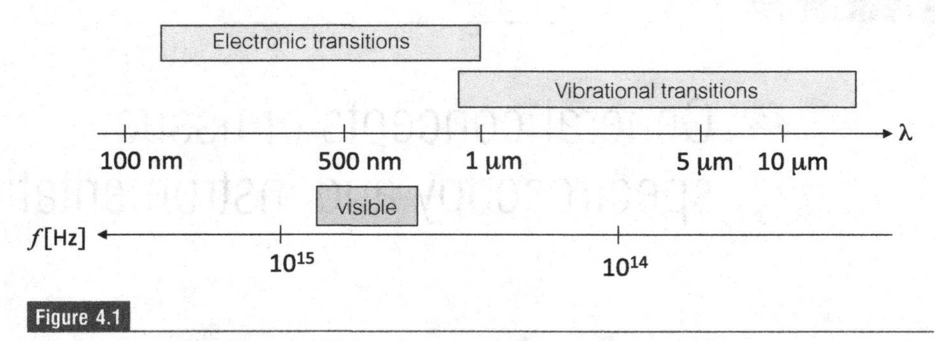

Figure 4.1

The range of the electromagnetic spectrum that is relevant to optical spectroscopy.

Most of the categories of tissue spectroscopy that are introduced in this chapter will be covered in greater depth in Chapters 5–8.

As represented in Figure 4.1, the visible portion of the electromagnetic spectrum, 400–700 nm, covers only a small fraction of the range we are referring to as "optical." Optical spectroscopy has proven to be of great value for a variety of biomedical diagnostic applications, and this chapter provides a broad introduction to the topic, with emphasis on how optical spectroscopy can be employed to reveal information about tissue and biological systems. Given that the wavelength (or photon energy) of light determines how it interacts with matter, the properties studied with optical spectroscopy can include the energy levels of atoms and molecules, fundamental processes in molecules, and, importantly, the structural properties and molecular constituents of tissues. Processes of interest include absorption, emission and scattering, both elastic and inelastic. These interactions can serve as probes to obtain information about tissue noninvasively, and such information can lead to the identification and quantification of tissue constituents, and the sensing of cellular and sub-cellular structural properties, all relating to tissue condition or pathology.

4.1 Atomic spectroscopy

Unlike molecules, which can have vibrational and rotational energy transitions, the energy states of atoms (other than the translational, or kinetic, energy) are purely electronic. A useful way to represent atomic energy levels is to represent them as a column of horizontal lines where the vertical coordinate represents energy. The energy associated with electrons bound to the atom is, by convention, negative, indicating that it is necessary to increase the energy of a bound electron (i.e., transfer positive energy from an external source) to raise the electron to

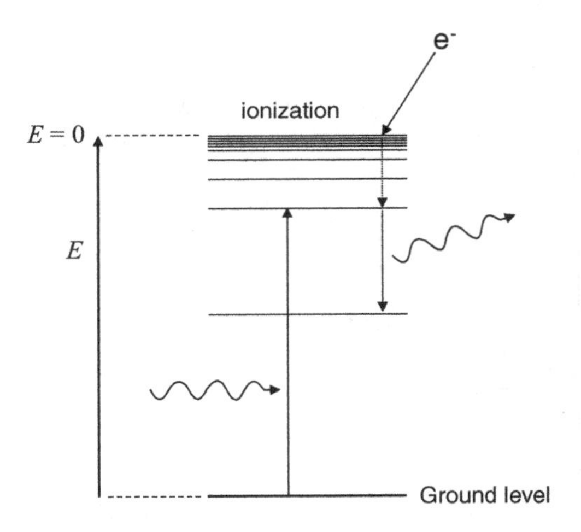

Figure 4.2

Energy-level diagram of atomic energy levels. Upper levels can be populated by various mechanisms of excitation, including absorption of photons from the ground state. Relevant to laser-induced breakdown spectroscopy (LIBS), if the atom is already ionized, the upper levels can be populated following recombination of a free electron with the ion. Thus, emission from upper levels can occur as electrons combine with ionized atoms.

the ionization level of zero energy, where the now "free" electron can be ejected from the atom. For atomic species, *energy-level diagrams* are often limited to representing single-electron (singlet, or *S*) states. Such a diagram of atomic singlet states is depicted in Figure 4.2. Excitation from a lower to a higher energy level can occur by a variety of mechanisms, including absorption of a photon that has an energy matching the energy difference between the two levels. Similarly, a primary mechanism for decay from excited states is by photon emission, where the energy of the emitted photon is equal to the energy difference between the initial and the final atomic energy levels. Thus, spectroscopic methods can be used to identify atoms by their characteristic optical absorption or emission frequencies.

Since the constituents of tissue are almost entirely molecular, there is little in the way of optical atomic spectroscopy that has led to biomedical applications. In other fields of science, atomic spectroscopy is generally associated with elemental analysis or fundamental research. Nonetheless, there is one form of atomic spectroscopy that, although infrequently exploited to date for measurements in living systems, has the potential to prove useful in diagnosing conditions such as heavy-metal poisoning. This method is called *laser-induced breakdown spectroscopy*, or LIBS.

As the name implies, and as depicted in Figure 4.3, a short-pulse (typically, nanoseconds), high-power laser beam is focused to a small spot, such that the

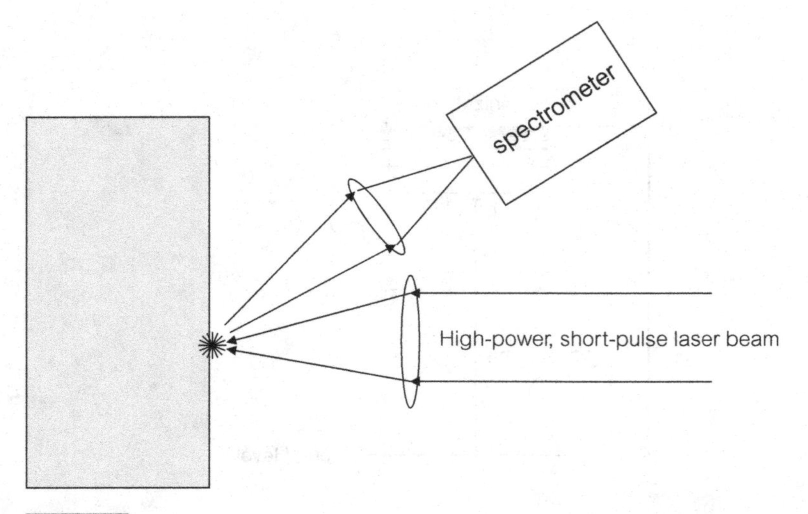

Figure 4.3

Schematic of a LIBS measurement. The beam of a short-pulse, high-peak-power laser is focused onto the surface to be examined, generating a "spark" of light from the optically induced "breakdown" or ionization of the surface material. The recombination light is collected by a lens and imaged to the entrance slit of a spectrometer for spectral analysis.

resulting strong optical electric field at the focus generates electrons with energies that far exceed the ionization potential of atoms and molecules, causing *ablation* ("vaporization") and ionization of the surface matter. The result is a small plume of *plasma*, which is manifested as a breakdown "spark." In this context, the plasma is a medium with an overall approximately neutral charge, consisting of unbound charged particles: negative electrons and positive ions. The strong electrical interactions among these charged particles account for the unique properties of plasma, which is sometimes referred to as a "fourth state" of matter (in addition to the solid, liquid, and gaseous states). Reliable breakdown at the surface of a solid (or liquid) typically occurs when the instantaneous optical power density, or intensity, exceeds ~ 1 GW/cm^2. When focused to a spot size of ≤ 1 mm diameter, this intensity is readily achieved with laser pulse durations of ≤ 10 ns and pulse energies of ≥ 100 mJ (parameters that are common for Q-switched solid-state lasers).

At such intensity values, the powerful optical electric field strength ($> 10^6$ V/cm) of the laser light causes molecules to be dissociated into their elemental components, all of which are initially ionized in the spark, a consequence of the effective temperatures in the plasma plume, which can reach 100,000 K. As the plasma cools, electrons recombine with the ionized atoms, and the resulting highly excited atoms then emit light as they decay from high-lying states towards lower levels, on the way to the ground energy level. The recombination light can be readily detected and analyzed by a spectrometer, and the detected sharp spectral lines conveniently

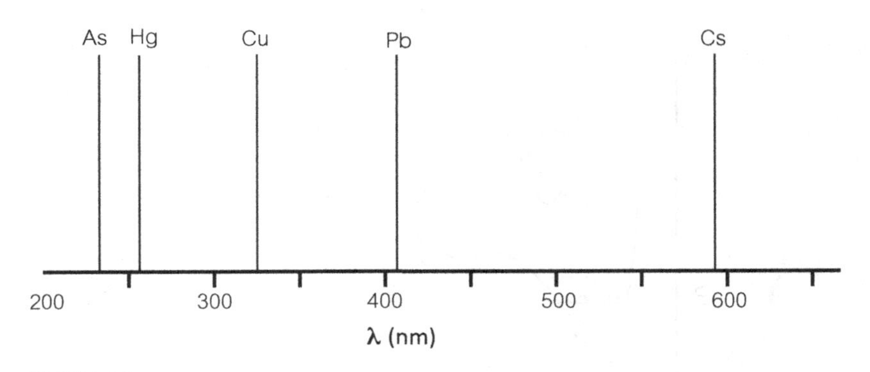

Figure 4.4

Atomic emission lines of common and potentially toxic elements that can affect human health.

represent the specific atomic elements (because their energies are associated with the energy levels of the corresponding atoms), and are easily identified. Figure 4.4 shows a sampling of the wavelengths of some atomic lines seen in the recombination light of a few elemental species that are sometimes suspected in heavy-metal (plus arsenic) poisoning.

LIBS has been applied for sensing of heavy-metal poisoning in measurements applied to hair and fingernails samples (Hosseinimakarem and Tavassoli, 2011). Hair can retain contaminants long after exposure, although contaminants can also be picked up directly, without ingestion. That type of application has been used primarily for forensic purposes, and is most useful for identifying past exposures, over weeks to months. As a potential medical diagnostic application relevant to near-term exposures, on the other hand, we speculate that skin that has been superficially cleaned (to remove simple surface contamination) will reveal toxic agents shortly after ingestion or inhalation. The heavy-metal elements are most likely to be found as salts (HgCl, for example), but the elemental dissociation occurs along with charge dissociation in the LIBS spark. For such an application, the LIBS laser beam would be focused on the skin surface. This type of test would be less invasive, and potentially more sensitive, than more conventional tissue or blood sampling followed by chemical assay; and with LIBS the laser pulse would be less invasive, causing no more discomfort than a small pin-prick.

4.2 Molecular spectroscopy: from diatomic molecules to biological molecules

The complexity of spectroscopic analysis increases dramatically in going from atomic species to molecules, especially biomolecules, with their large molecular

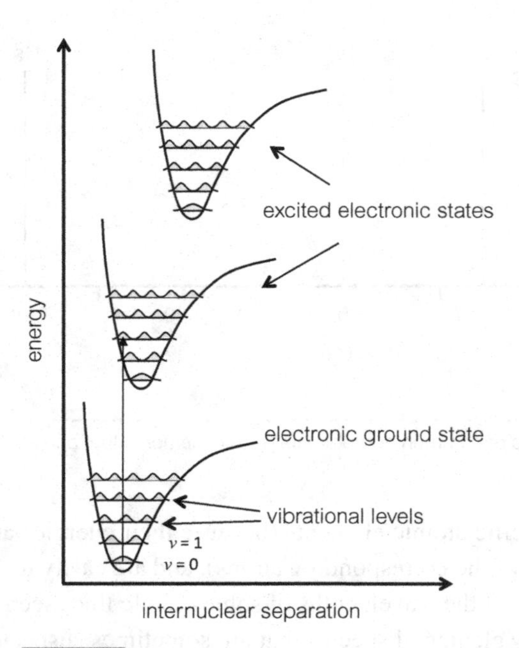

Figure 4.5

Energy diagram for a diatomic molecule. Excited electronic potential energy wells typically have larger mean internuclear separations than the ground electronic state. Within the potential well of each electronic state a series of vibrational sublevels is populated. The up arrow illustrates that, for this case, an optical absorption event from the ground vibrational level is more likely for photon energies that coincide with a v_0-to-v_2 transition, because this transition would not require much change of nuclear positions.

weights of the order of 10^2–10^6 daltons. The *dalton* (symbol: Da) is a *unified atomic mass unit* equal to 1 / 12 the mass of an atom of carbon-12, and can be approximated by the total number of nucleons (protons and neutrons) in the molecule. (The unit is named after John Dalton, English physicist and chemist [1766–1844]). It is the complexity of molecular spectra, however, that provides rich and valuable information, and the opportunity for identifying tissues and disease conditions.

As illustrated in Figure 4.5, molecular electronic energy states are no longer "flat." Rather, the energy varies with internuclear separation, and within each electronic potential energy well are a sequence of vibrational levels. Note that the lowest (or "ground") vibrational level is above the lowest point of the electronic energy well. This reminds us that a molecular vibrational mode, or any quantum mechanical harmonic oscillator, has a non-zero energy for its ground state. The probability distribution for internuclear separation of a given vibrational state is determined by quantum mechanical rules and qualitatively looks like the "oscillations" shown in Figure 4.5, indicating that for each vibrational level there are specific ranges of internuclear separation where the molecule is more likely to be found. These probability distributions are determined from the squares of the quantum mechanical *wave functions* of the nuclear pair.

Up- and down-transitions between electronic states, corresponding to absorption or emission of photons (respectively), may also invoke changes in the vibrational levels. The transition probability (which is directly related to the so-called absorption or emission *cross section*) between a specific vibrational level in the ground electronic state and a specific vibrational level in an excited electronic state is higher when the maximum internuclear-separation probabilities of the two states overlap. This results from the fact that electronic transitions are fast, whereas changes in internuclear separation require motion of the more massive nuclei, and are thereby slower. Consequently, electronic transitions have higher probability if they do not require much change of the positions of the nuclei. It should also be noted that the mean nuclear position of the potential well of an excited state is not found, in general, at the same internuclear separation as for the ground electronic state; consequently, the most probable internuclear separation of the $v = 0$ vibrational level of the ground state may best overlap with, say, the $v = 0$ level of the excited state. The result, in that case, would be a larger cross section (probability) for a $v_0 - v_2$ transition than a $v_0 - v_0$ transition.

More formally, these concepts are derived from the *Franck-Condon principle* (named after German physicist, and Nobel Prize winner, James Franck [1882–1964] and American physicist Edward Uhler Condon [1902–1974]), which explains that an electronic transition is most likely to occur with minimal change in the positions of the nuclei in the molecule. In the language of quantum mechanics, the formulation of this principle from quantum mechanical rules is that the probability of a vibrational transition is proportional to the square of the integral of overlap between the wave functions of the two vibro-electronic states of the transition.

The discussion above relates to diatomic molecules. Electronic and vibrational spectra of polyatomic molecules, our main interest for spectroscopy of biological tissues, are even more complicated, and the complexities of the transition rules and energy states are left to specialized texts on the topic. In this text we will confine ourselves to a more qualitative description, with relevance to biological molecules. In polyatomic molecules, electrons in given levels are typically shared among the atoms of a specific covalent bond, or over a range of covalent bonds. The photoactive electrons of larger molecules are classed as Σ- and π-electrons, loosely analogous to electrons in *s*- and *p*-orbitals of atoms. In these larger molecules, π-electrons typically roam the backbone of the molecular structure, whereas Σ-electrons are generally confined to a specific covalent bond. The absorption and emission spectra in the near-UV to near-IR range (~300–900 nm) are due, predominantly, to electronic transitions of the π-electrons. Given the enormous variety of potential interactions seen by these electrons, the spectral absorption and emission bands are typically broad and devoid of sharp spectral features. (See, for example, Figures 2.4, 2.5, 2.6.)

Some of the features of macromolecular (and biomolecular) structure that dominate the properties of UV-visible to NIR spectra are the following:

- π-electrons are not associated with specific covalent bonds and roam the molecular spine; these are responsible for most of the UV/visible/NIR absorption and fluorescence.
- The transitions of more-localized electrons (often, Σ-electron transitions) are responsible for hard-UV ($\leq$250 nm) absorption.
- A large variety of vibrational modes, with energies in the mid-IR, characterizes more detailed structural features of molecular sub-components, covalently bound groups of atoms, and is characteristic of those groups.
- Tertiary structure (i.e., the three-dimensional structure of a large molecule), which is predominantly governed by hydrogen bonding, as in the case of protein folding, and hydrogen bonds themselves perturb the vibrational modes of sub-molecular groups, inducing frequency shifts; such details of the vibrational spectroscopy can aid in the identification of the specific parent molecules.

The various components of molecular energy, in descending strengths, are listed diagrammatically here:

$$\left.\begin{aligned} E = E_{\text{electronic}} \\ + E_{\text{vibrational}} \end{aligned}\right\} \text{We care about these}$$

$$+ E_{\text{rotational}}$$

$$+ E_{\text{spin coupling}}$$

$$+ E_{\text{translational}}$$

Since tissues are condensed matter (liquid or solid), rotational modes are not relevant; and the spin-coupling and translational energy contributions are too small to affect the measurable frequencies (or wavelengths) in optical spectroscopy.

Optical wavelengths associated with molecular transitions

Electronic transitions: 200–1000 nm

Vibrational transitions: 900–25,000 nm (0.9–5 μm)

the near IR is defined by the range ~0.7–2 μm

the mid-IR is defined by the range ~2–25 μm

Even though the fine-structure effects of electron spin coupling are not seen in biomolecular spectroscopy from tissues, the effects of the quantum-mechanical rules of electron spin on optical spectroscopy are important, because they have a significant influence on the strengths of spectral features.

4.3 Electronic transitions – the issue of electron spin

One of the quantum mechanical properties of electrons, referred to as *spin*, is not unlike the spin that imparts angular momentum in a classic child's spinning top. Electrons, being small things, however, live in a quantum mechanical world, and the angular momentum associated with electron spin corresponds with a *spin quantum number* ($s = 1/2$); and its projection relative to a coordinate system, to be measured along a chosen direction (say, the positive z axis), can take on only two values: $\pm 1/2\hbar$, where $\hbar = h/2\pi$ is the reduced Planck's constant. These two values are described by a *secondary spin quantum number* (m_s), which can assume values ranging from $-s$ to $+s$ by steps of 1, so that m_s can only assume values of $-1/2$ or $+1/2$ in the case of the electron. Physicists tend to call these "up spin" (for $m_s = +1/2$) and "down spin" (for $m_s = -1/2$). (An electron's charge and spin also result in its *spin magnetic moment*.)

The importance of electron spin, and why we need to account for it, is that spin strongly affects the probability of an electronic transition and, hence, the likelihood of optical absorption or emission, and the consequent spectra of a molecule. Just as classic conservation laws – to wit, conservation of angular momentum – make it difficult to flip a spinning top with its gyroscopic forces, the total quantized spin of electronic states resists change. The electrons in molecules are generally *paired*, two to an orbital, and these paired electron states are typically the "optically active" energy states. Nonetheless, electrons in an orbital can also be *unpaired*. These terms, paired and unpaired, relate to how the spins of the two electrons are coupled, and the energy states are denoted as *singlet states* or *triplet states*, respectively. A pictorial representation with up- and down-arrows can aid in understanding the distinction.

Singlet states In this case the electrons are *paired*, with opposite spins, such that the total spin of the pair is $s_T = 0$, and the associated secondary spin quantum number, $m_{s,}$ can only be 0.

$$\uparrow\downarrow \Rightarrow \text{total spin, } s_T = 0, \quad \text{and } m_s = 0.$$

Triplet states In this case the electrons are *unpaired*, such that each electron can have either up- or down-spin, and the total spin of the two electrons is $s_T = 1$, so that the associated secondary spin quantum number m_s can take on three values: $+1, 0$ or -1.

$$\begin{array}{ccccc} \uparrow + \uparrow & \text{or} & \uparrow + \downarrow & \text{or} & \downarrow + \downarrow \\ s_T = 1 & & s_T = 1 & & s_T = 1 \\ m_s = 1 & & m_s = 0 & & m_s = -1 \end{array}$$

It may seem inconsistent, and even paradoxical, that the combination of spins up and down results in a total spin of 0 in the singlet state and a total spin of 1 in the triplet state. This result is justified by the fact that the case in which the two electrons have opposite spins is really a superposition, or combination of two states, a first state in which the first electron has spin up and the second electron has spin down, and a second state in which the first electron has spin down and the second electron has spin up. The combination of these two states is different in the singlet state and in the triplet state, and this difference accounts for the different values of the total spin in the two cases.

The electronic ground state of most molecules is a singlet state. During an electronic transition (with the exception of collision-induced transitions and non-radiative coupling, as discussed in Sections 5.1 and 5.8), a photon is either absorbed or emitted. Although photons do have and exchange angular momentum (quantized as $\pm\hbar$), they do not exchange spin in the same sense as an electron. Consequently, an *"allowed" transition*, corresponding to the absorption of a photon by the ground state, will result in an excited singlet state, since a change in electron spin is, technically, a "forbidden" process. (We remind the reader that a "forbidden" process in the world of quantum mechanics, which deals with probabilities, is one that is unlikely to happen, not one that "absolutely can't happen.") It is also possible for an excitation to move one of the electrons of a pair into an empty orbital, thereby releasing them from necessarily having their spins paired; but most optically induced transitions in biomolecules are singlet-to-singlet transitions.

4.3.1 Fluorescence and phosphorescence

Once a molecule has been excited to an upper singlet state, it may decay directly back to the ground state by emitting a photon as *fluorescence*. (In biomedical optics the term "autofluorescence" is sometimes used to denote fluorescence emission that originates from molecules that are present natively in the tissue.) As depicted in Figure 4.6, the fluorescence emission usually follows thermal relaxation of an upper vibrational level within the excited electronic state. Thus, the emitted light is almost always at a longer wavelength (lower energy) than that of the

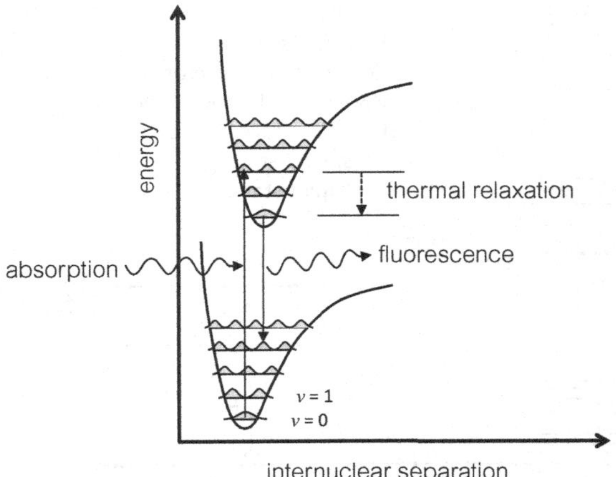

Figure 4.6

Absorption and emission (fluorescence) by allowed transitions in a molecular system are depicted by the up- and down-arrows. The energy of the fluorescence photon is almost always lower than that of the excitation photon, as a consequence of thermal relaxation to lower vibrational levels within the excited electronic state, and is also affected by the Franck-Condon factors that govern transition probabilities.

absorbed photon. Since the thermal relaxation invokes motion of the nuclei, there is a delay between absorption and emission events, which is described in terms of the *fluorescence lifetime*. For biological molecules the fluorescence lifetime (for π-electron transitions) is commonly a few nanoseconds, but a variety of factors can perturb the system and result in shorter or longer lifetimes, factors to be discussed in greater detail in Sections 5.6 and 5.8.

For the case of most simple diatomic molecules or for atoms, it is interesting to note that the emission linewidth (which correlates to the range of values ΔE for the energy difference between the excited and ground energy levels) and the *excited-state lifetime*, τ, are inversely related and can be expressed through the Heisenberg uncertainty principle, named after the German physicist Werner Heisenberg [1901–1976] (Nobel Prize for physics in 1932):

$$\Delta E \times \tau \gtrsim \frac{\hbar}{2} \tag{4.1}$$

That is, ΔE is directly associated with the residence time in the excited state, such that the linewidth is determined by the lifetime of the excited state. This relation is analogous to the classic relation between two Fourier conjugate variables, such as time and angular frequency, for which the product of the variances is ≥ 1. In more complex polyatomic and biomolecules, however, ΔE is determined more

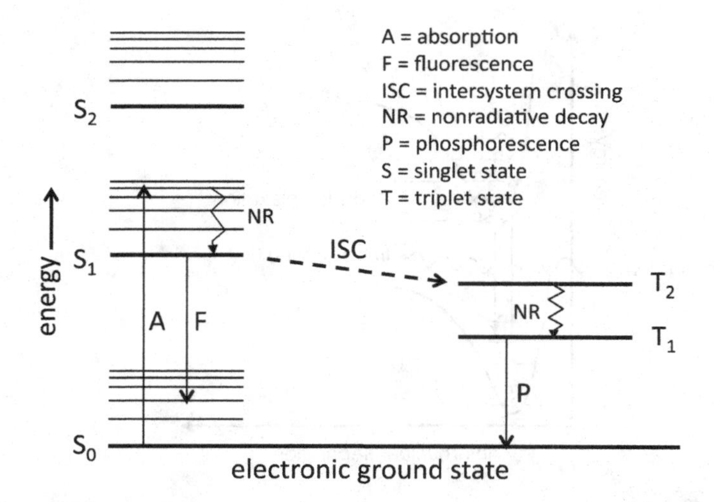

Figure 4.7

Jablonski diagram representing molecular intersystem crossing from an excited singlet state to a triplet state. Population accumulates in the triplet state and the emission resulting from the slow "forbidden" transition to the ground singlet state is called phosphorescence.

by the complexity and structure of the vibrational levels that perturb the energy of the electronic state, resulting in much broader linewidths without commensurately shorter lifetimes. This comes about because of the diverse and varied perturbations of the electronic states, induced by the numerous and interconnected vibrational modes.

Phosphorescence is the term used to denote a slow (of the order of milliseconds) or very slow (seconds, minutes or hours) radiative decay of a long-lived excited state, from which emission is weak but lasts a long time. Such a slow decay will happen when a molecule finds itself in an excited state from which there is no "allowed" transition to the ground state. This can be the result of an *intersystem crossing* (ISC) from an excited singlet state to its corresponding excited triplet state. Collisions or thermal coupling often aid in intersystem crossing. In a manner similar to the energy-level diagram for atoms (Figure 4.2) we illustrate the molecular states with a simplified energy-level diagram, Figure 4.7, with flat energy levels, and in which electron-spin states are separated horizontally in different columns. Such a representation is often referred to as a *Jablonski diagram* (named after the Polish physicist, Aleksander Jablonski [1898–1980]).

Once a molecule is in the excited triplet state, de-excitation to the ground singlet state is "forbidden," which means that it takes a long time. Consequently, the molecule "hangs" in the triplet state, with an effective lifetime, commonly in organic molecules, of milliseconds to seconds. Bioluminescence, such as that from phytoplankton in sea water, or the light from fireflies, is often a form of phosphorescence, wherein a chemical reaction leaves a molecule in its excited triplet state. It should be noted that the lowest, or ground, energy state of larger molecules is almost always a singlet state, such that molecular triplet states only exist at excited levels.

Fluorescence spectroscopy is treated in greater detail in Chapter 5.

4.4 Vibrational transitions: IR absorption spectroscopy and Raman spectroscopy

During a transition between vibrational levels, motion of nuclei is invoked, as depicted by the specific positional probability distributions for each vibrational level, which derive from the wave functions of the sub-molecular group (Figures 4.5 and 4.6). Nonetheless, the cross sections for *direct* photon absorption or emission at the infrared wavelengths can still be large. As listed in Section 4.2, the wavelengths associated with differences in vibrational levels typically range from the NIR to mid-IR spectral regions. The precise wavelengths (energies) associated with vibrational transitions characterize the specific structures of the sub-molecular groups, and hence provide identifying information about the parent molecule.

There are two types of spectroscopy commonly employed to measure the energies of vibrational transitions, *infrared-absorption spectroscopy* and *Raman spectroscopy*. Individual vibrational modes may lend themselves more readily to measurement by one or the other of these types of spectroscopy. Detailed development of the principles underlying these two types of spectroscopy will be covered in Chapter 6. Here we briefly define them:

Infrared-absorption spectroscopy: measurement of direct absorption of infrared-wavelength light, resulting in an increase in the energy level of a vibrational mode. (See Figure 4.8(a).) *Infrared-active* modes are those for which the permanent dipole moment of the molecular bond changes with internuclear separation. These transitions exhibit large cross sections (strong probability) for both absorption and stimulated emission; and, in the language of molecular physics, these modes are called *dipole transitions*. IR-absorption spectroscopy, however, has practical impediments for in vivo spectroscopy, because, as described in Section 2.3.2, the strong broadband absorption of infrared light by water, which is ever-present in tissue, strongly attenuates light used to measure IR absorption.

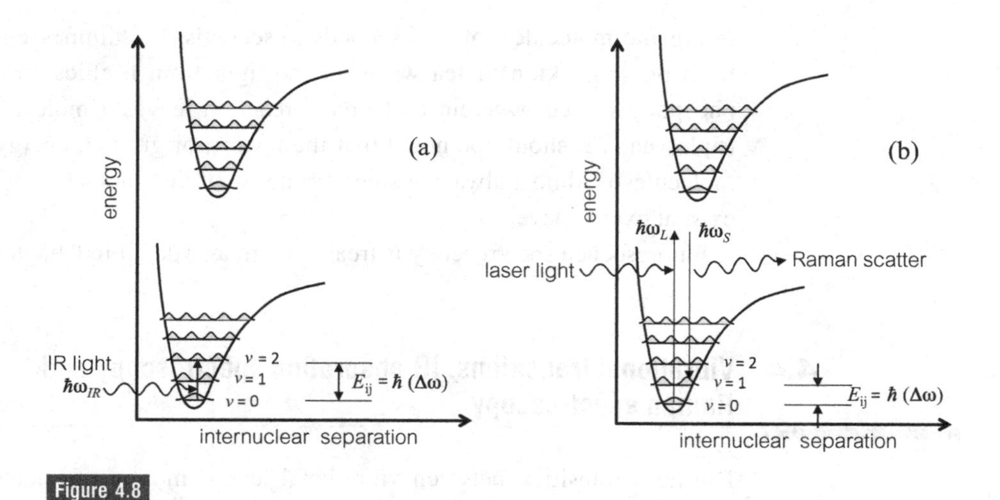

Figure 4.8

Vibrational spectroscopy by direct IR absorption (a), or by Raman scattering (b). The subscripts *L* and *S* refer to laser excitation light and Stokes emission, respectively.

Raman scattering spectroscopy: measurement of photons that result from an excitation to a *"virtual" electronic state*, followed by instantaneous decay to the ground electronic state, but at a higher vibrational level. (See Figure 4.8(b).) A virtual, i.e., not actually existing, energy level may be occupied for only a very short time because of the exceedingly short lifetimes associated with the large energy spread of the virtual energy level, as described by the Heisenberg principle (see Eq. (4.1)). This resonance lifetime is typically of the order of 1 femtosecond, much shorter than fluorescence lifetimes (typically >1 nm). Raman-active modes are those for which there is a significant change in the *polarizability* associated with a change in internuclear separation. This interaction mechanism is commonly referred to as *Raman scattering*, an example of inelastic scattering, in which the scattered photon has different (typically lower) energy than the incident photon. The term "scattering" is used because there is no true transition to an upper state, as no upper state exists for the energy of the excitation photon. Both the excitation and scattered wavelengths are, typically, in the near-UV to near-IR range, and the difference in photon energies is a direct measure of the energy difference between the initial and final vibrational levels. Raman scattering is a "weak" (low-probability) process, with a cross section that is several orders of magnitude smaller than that for IR absorption, generally requiring the high intensity of a laser beam for the excitation source to produce enough scattered signal. Some bonds, however, do not have a permanent dipole moment, and hence are not dipole active, such that the only way to study them is by Raman scattering. (In physics-speak, a Raman scattering event can be

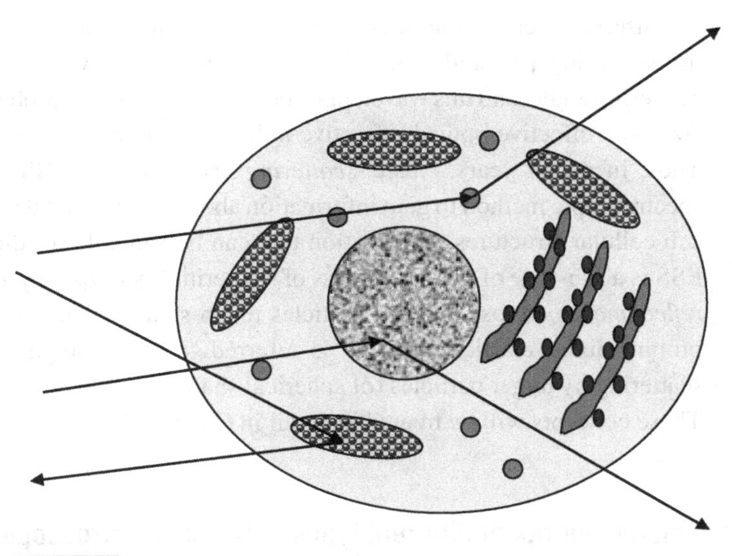

Figure 4.9

Light scatters from gradients or discontinuities in the refractive index. The scattering probability, its wavelength dependence, and its angular distribution depend on the size and relative refractive index of the scattering center.

referred to as an interaction with the molecular group's *quadrupole moment*.) Moreover, since the excitation and scattered photons for Raman scattering are, in practice, commonly in the visible range, this type of spectroscopy is not hampered by absorption from water, which is weak in the visible spectral range.

It is possible, and common in biomolecules, for the vibrational mode of a bond or molecular group to be both dipole-active and Raman-active. The choice of spectroscopy method is then determined by the specific experimental circumstances and exigencies. The spectroscopy of vibrational modes will be treated in greater depth in Chapter 6.

4.5 Scattering by particles: elastic-scattering spectroscopy

The interactions of photons with atoms and molecules described in the preceding sections all invoke exchange of energy between the photon and the molecule, by absorption, emission or inelastic scattering. As illustrated (simplistically!) in Figure 4.9, similar to Figure 2.11, however, photons can also scatter elastically without energy exchange (hence, no change in wavelength), from interactions with small particles, including particles as small as molecules. The reason this can lead to a useful spectroscopic measurement is that although no energy is exchanged, the scattering probability itself generally does vary with wavelength and with scattering angle, and this wavelength-dependent scattering probability relates to the sizes and effective optical refractive indices of the particles or scattering centers. Thus, in recent years, *elastic-scattering spectroscopy* (ESS) has emerged as a spectroscopic method to gain information about the size and density of cellular and sub-cellular structures, information that can be valuable for diagnostic purposes. ESS is a sub-type of a broader class of scattering spectroscopy, often called *diffuse reflectance spectroscopy*. For particles much smaller than the wavelength of the probing light, elastic scattering is referred to as Rayleigh scattering, whereas scattering by larger particles (of spherical shape) is often described by Mie theory. These concepts will be treated in detail in Chapter 7.

4.6 Relative merits of different types of tissue spectroscopy

We summarize here some of the factors that govern the utility and practicality of different types of spectroscopy for assessment of tissue condition or detection of disease. Each of these will be developed in detail in Chapters 5–8, including all the terms not fully defined here, but we provide this "high-altitude" overview for the purposes of comparison.

4.6.1 Fluorescence measurements and spectroscopy

Advantages

- Can be used for diagnostic imaging of a tissue surface with a large field-of-view (at a few wavelengths). Detecting or imaging fluorescence is, fundamentally, a dark-background technique, which facilitates wide-field spectral imaging with good *signal-to-noise ratio* (SNR). Hence, this method offers good potential for clinical applications.
- Imaging can be further enhanced with exogenous drugs. Fluorescence techniques are not limited to intrinsic (native) fluorophores. Many dyes and fluorophores have been developed that target specific biomolecules in cells and tissue, offering large excitation cross sections and high quantum efficiency as biomarkers.
- It is possible to detect fluorophores whose absorption bands are found among other strong absorbers. Given the dark-background nature of fluorescence detection, even if a fluorophore emits at a wavelength in the absorption band of a

local chromophore (often, hemoglobin), if the interfering absorber itself is not fluorescent (e.g., again, hemoglobin) the attenuated biomarker fluorescence can still often be detected.

Disadvantages

- The signal can be affected by the local electrochemical environment and patient condition (e.g., blood pH, oxygenation, etc.). The effects of a number of potential confounding factors limit the utility of fluorescence measurements for quantitative or absolute measurements of concentrations of target species.
- Autofluorescence signals are weak, but using an exogenous drug is medically invasive (requiring drug approval for use in humans). Intrinsic fluorophores are generally found in low abundance naturally, and often have small cross sections for the excitation transitions and/or low fluorescence quantum efficiencies.

4.6.2 Infrared-absorption spectroscopy

Advantages

- Sharp spectral features of vibrational transitions facilitate separation and identification of sources. Infrared-absorption spectra are highly detailed, and it is often possible to identify the specific molecular bonds associated with specific features for this "fingerprint" type of spectroscopy.
- IR absorption bands are strong. The large cross sections of dipole-allowed IR transitions facilitate experimental methods that exhibit good SNR.
- There is no interference from fluorescence. Unlike the UV-visible-NIR spectral range, there is no fluorescence emission in the infrared to confound spectral measurements.

Disadvantages

- It is difficult to avoid water absorption in tissue. Except in the case of measurements on thin sections of tissue, the broad and strong optical absorption by water in the mid-IR obscures absorption bands of less-abundant molecular species in hydrated samples.
- It is difficult to implement fiber transmission for endoscopic use. Materials commonly used for optical fibers do not transmit light in the mid-IR, and materials that can serve that purpose are either fragile, toxic or expensive.
- Instrumentation for fast parallel spectroscopy in the IR is expensive and limited in performance. Most mid-IR spectrometers utilize wavelength-sweeping

dispersive components (see Section 4.7.2), or scanned interferometry with a single broadband detector and Fourier-transform analysis, methods that slow the generation of a recorded spectrum.

4.6.3 Raman spectroscopy

Advantages

- Raman spectra have sharp spectral features: easy to separate and identify sources. Like IR-absorption spectra, Raman spectra are highly detailed, often enabling identification of specific molecular bonds.
- Flexibility with excitation wavelength and detection ranges: illumination and detection in the broad visible range avoids absorption due to water and facilitates measurement with conventional optics.
- Raman spectroscopy can be used to detect compounds that do not fluoresce. Molecules can be studied that do not exhibit fluorescent bands, or even absorption bands.

Disadvantages

- Raman scattering efficiency is very weak, requiring intense/long excitation and relatively expensive instrumentation. It is generally challenging to implement Raman spectroscopy studies in vivo with adequate SNR. Raman measurements are also, typically, confounded by a strong background of autofluorescence emission, which is generally significantly stronger.
- Resonance enhancement Raman scattering can enhance signal strength, but requires strong UV illumination, which carries the risk of genetic damage.

4.6.4 Elastic-scattering spectroscopy and diffuse reflectance spectroscopy

Advantages

- Strong signal: allows use of lower-cost components. Elastically scattered light is generally orders of magnitude stronger than either fluorescence or Raman scattering.
- Scattering spectroscopy is sensitive to both tissue structure and biochemistry. While elastic-scattering spectroscopy is generally aimed at extracting micro-architectural properties of cells and tissues, it is nonetheless sensitive to strong absorbing species, which can also yield diagnostic information.

- Diffuse reflectance spectroscopy is sensitive to important chromophores that are not fluorescent (e.g., hemoglobin).

Disadvantages

- Scattering spectroscopy is not directly applicable for imaging. ESS and certain other variations of diffuse reflectance spectroscopy rely on specific geometries of illumination source and detection points to facilitate quantitative extraction of information about tissue. To extract such information over larger tissue surfaces requires specialized scanning schemes.
- Scattering spectroscopy may not be sensitive to subtle biochemical changes preceding structural changes. Often micro-structural changes are preceded by local biochemical or metabolic changes, which generally cannot be sensed by ESS.

4.7 Introduction to general principles of spectroscopic measurements and instrumentation

4.7.1 What is light?

In Chapters 1 and 2 we have used the terms "photon" and "light" interchangeably, albeit with predominant use of photon for most definitions of light or for description of interactions of light with matter. For centuries physicists struggled with (and fought over!) the question: is light a particle or a wave? This challenge of wrestling with the growing body of experimental evidence had been billed as a contest between the "corpuscular" and wave theories of light, and evolved into the "duality question." Brilliant experiments were devised to demonstrate that light behaves one way or the other. In the early part of the twentieth century, however, consensus developed about the dual nature of light: that it behaves with properties of *both* particles and waves. (It is now understood that duality applies to all waves or particles at the quantum level, including particles with rest mass, such as electrons and protons, and that both their wave-like and particle-like natures can be demonstrated experimentally.) Regarding the various properties exhibited by light, some, like diffraction, interference and polarization, are most easily explained in terms of waves; on the other hand, the photoelectric effect and momentum exchange are more conveniently explained in terms of the particle nature of light. In this text we will use, and denote, either approach as is most appropriate for the phenomenon being described.

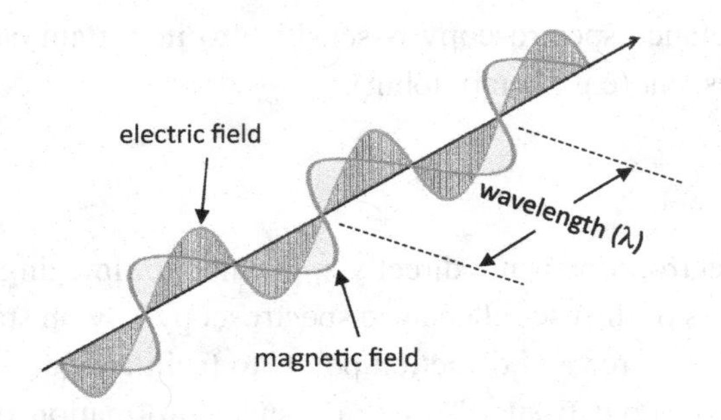

Figure 4.10

A cartoon depiction of a propagating light wave, with orthogonal electric and magnetic fields.

As is the case for any electromagnetic wave, light is a transverse wave, in which the electric and magnetic fields are orthogonal to each other and to the direction of propagation of the field. For most interactions of light with matter, and for all circumstances with which we will be concerned, the effects of the optical magnetic field are too small to be significant, and we can focus on interactions of the electric field.

Classically, any wave is characterized by its frequency and amplitude (or power), and the amplitude has no specific restrictions. In the quantum world, however, when the power of a light source (or any electromagnetic wave) is reduced to very low levels, we get into a regime wherein an optical detector will no longer detect a continuous wave. The detected optical energy will then come in small quantized bits, each representing the energy of a single *photon*. This unit of energy, the photon, is the minimum energy that can represent light at that frequency, f (or angular frequency $\omega = 2\pi f$, or wavelength, λ).

$$E = hf = \hbar\omega = \frac{hc}{\lambda}, \tag{4.2}$$

where h, again, is Planck's constant ($h = 6.63 \times 10^{-34}$ J-s) and $\hbar = h/2\pi$ is the reduced Planck's constant.

Fractional photons cannot exist, so the next larger amount of energy that could be measured during the integration time of a detector would correspond to two photons, or $2\hbar\omega$. In general, for a steady-state light source, if an average of N photons (or the energy equivalent to N photons) are detected during the integration time of the detector, the random measurement-to-measurement statistical fluctuations in the number sensed will be $\sim\sqrt{N}$. These variations are a consequence of *counting statistics*, as discussed in Chapter 3, and in the case of measurements with optical detectors such variations are referred to as the "*shot noise*" of the measurement; that is, photon counting follows a Poisson statistical distribution

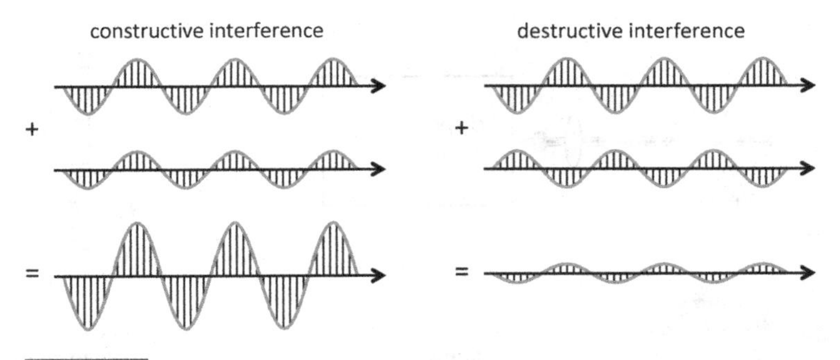

Figure 4.11

The interference between two optical waves is determined by the relative phase between the waves.

(named for the French mathematician Siméon Poisson [1781–1840]). (See Section 13.3.6.) When the number of photons detected during the integration time of the detector reaches the point where $\sqrt{N} \ll N$, things tend to smooth out and behave like the classical regime, and the principal limits of precision are determined by other sources of measurement noise.

We represent the optical electric field (for propagation in the $+z$ direction) as the vector quantity

$$\mathbf{E}(z, t) = \hat{\mathbf{p}} E_0 \cos(kz - \omega t) \tag{4.3}$$

or its equivalent, but mathematically more convenient, exponential notation:

$$\mathbf{E}(z, t) = \hat{\mathbf{p}} E_0 e^{i(kz - \omega t)} \tag{4.4}$$

where $\hat{\mathbf{p}}$ is the polarization unit vector, $k = 2\pi/\lambda$ is the *wavenumber*, and E_0 is the electric-field amplitude. (Note: later, in Chapter 6, the plural "wavenumbers" is defined as $1/\lambda$.) The wave period, $T = 2\pi/\omega$, is the time it takes for one full wave to pass a stationary point. (It is important to remember that only the real part of the RHS of Eq. (4.4) is physical.) In later chapters we will often use a vector symbol, $\mathbf{E_0}$, as a combined representation of the information about the optical polarization direction (and, sometimes, also the spatial distribution of the field) and the electric-field amplitude.

With this language, the concepts of constructive and destructive interference, at least in the simple case of monochromatic light, are easy to illustrate pictorially, as shown in Figure 4.11. The property of interference is key to the function of commonly used diffraction gratings and other dispersion elements in spectroscopy instrumentation, as well as interferometric methods used in various types of microscopy, to be discussed in Chapter 17.

Throughout all of the discussion here and in following chapters regarding instrumentation, it must be remembered that an optical detector does not measure the

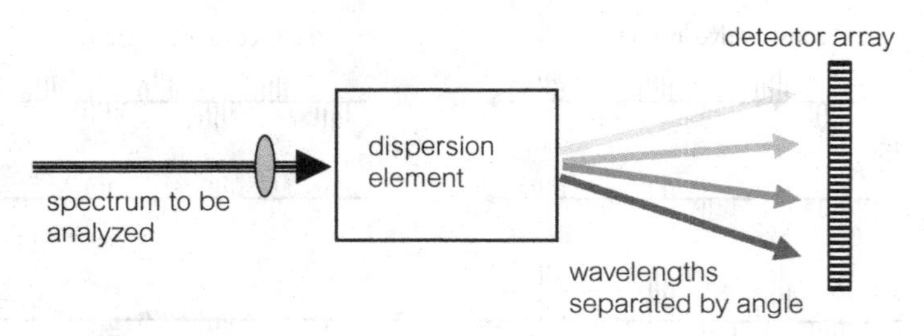

Figure 4.12

The basic elements of a spectrometer consist of optical components to shape and direct the light to a dispersive element, which separates the different wavelengths, typically in an angular spread. The spatially dispersed spectrum can then be read by an array of detector elements.

electric-field amplitude of light, but the associated intensity, $I(t)$, which is proportional to the square of the electric-field amplitude: $I \propto E_0^2$.

We also briefly introduce the concepts of spatial and temporal coherence of light, which will be developed more completely in Chapters 17 and 18. An optical field is said to have good *spatial coherence* or *transverse coherence* if the phase of the optical wavefront is close to constant across a plane (for collimated light) or a spherical surface (for converging or diverging light) that is normal to the direction of propagation. The optical field is said to have good *temporal coherence* if the phase measured at one instant in time can be correlated with, or predicted from, the phase at an earlier or later time, when measured at the same location. (Temporal coherence is equivalent to *longitudinal coherence*, which is a measure of the phase correlation along the direction of propagation for an instant in time.)

4.7.2 Anatomy of a spectrometer

The main elements of a spectroscopy measurement system are: (1) a light source and optical elements to direct light to and from the sample to be studied; (2) a means of collecting the light from the sample and directing it to a dispersive element; (3) a dispersive element that separates the different wavelengths of the light; and (4) a detection system to record the relative strengths of light at a range of wavelengths. Elements 3 and 4, along with some components of element 2, are what typically make up a *spectrometer*, as represented in Figure 4.12.

Photographic film played the role of the detector "array" in early spectrometers, prior to solid-state electronics. The next step, historically, was when highly sensitive electrical photodetectors were developed (e.g., photomultiplier tubes). A spectrum could be measured by placing a narrow exit slit after the dispersing

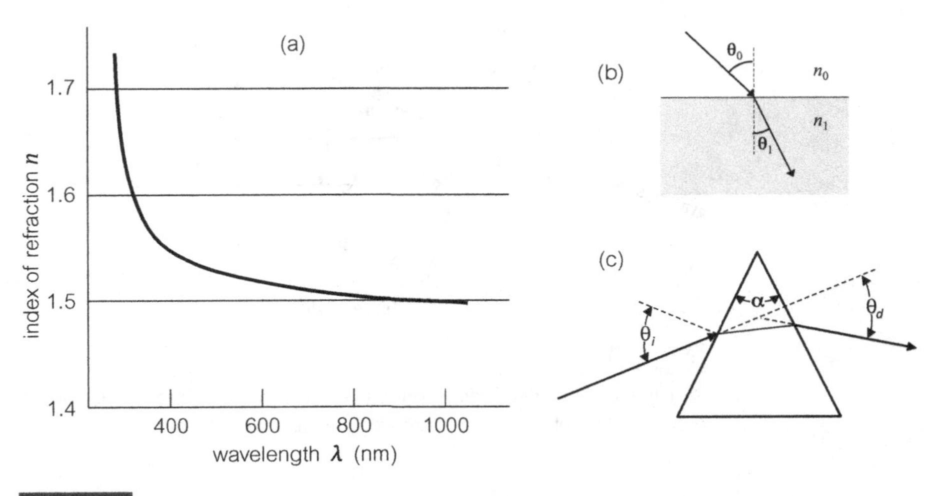

Figure 4.13

(a) The index dispersion curve for borosilicate glass; (b) the geometry of Snell's law; (c) wavelength-dependent refraction of a ray of light by a glass prism.

element, enabling only a narrow range of wavelengths to reach the detector. The system optical components were rotated or moved so as to scan the spectrally dispersed light in front of the slit, such that the temporal signal from the detector corresponded to the spectrum. Such instruments are called "scanning monochromators." In most modern spectrometers, however, the dispersed wavelengths of the spectrum are detected simultaneously (in parallel) by a detector array (typically a CCD or photodiode array), which consists of a large number of small detector elements that can be read individually and quickly.

4.7.2.1 *Dispersive element: prisms*

The dispersive element of a spectrometer is key to its performance, and it helps to understand how it works. In the earliest spectrometers, hailing back to the time of Newton, the dispersive element was a glass prism. (The detector in those days was the observer's eye – a remarkably sensitive detector for visible light!) A prism separates the different wavelengths of light by angle of refraction as a consequence of the *dispersion* (wavelength dependence) of the optical refractive index of glass: $n = n(\lambda)$. Figure 4.13(a) plots the refractive index of borosilicate glass as a function of wavelength. As can be seen, the refractive index is higher for shorter wavelengths, resulting in stronger refraction. The refraction at each interface is governed by Snell's law (named after the Dutch mathematician/physicist Willebrord Snellius [1580–1626]):

$$n_0 \sin\theta_0 = n_1 \sin\theta_1, \tag{4.5}$$

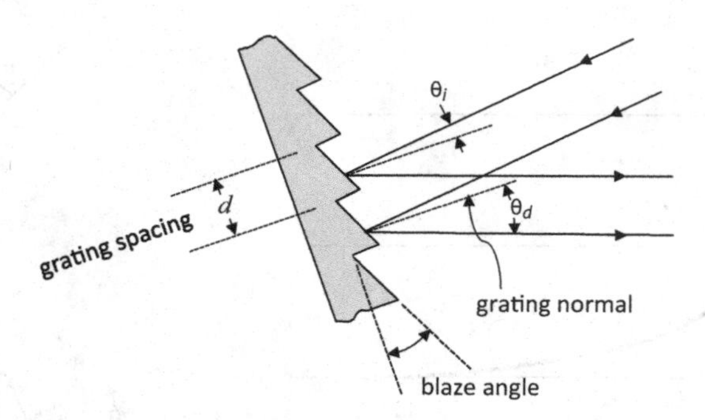

Figure 4.14

Structure of a reflective diffraction grating. The groove "blaze" angle can be chosen to maximize the grating optical efficiency for a preferred range of wavelengths.

where n_0 and n_1 are the refractive indices in media 0 and 1, respectively, and θ_0 and θ_1 are the corresponding ray propagation angles measured with respect to the surface normal. The relation is depicted in Figure 4.13(b).

By applying Eq. (4.5) at each interface of a prism, the deviation angle, θ_d, of an exiting ray, with respect to the incoming ray, is given by the "*prism equation*":

$$\theta_d = \theta_i + \sin^{-1}\left[(\sin\alpha)\sqrt{n_1^2 - \sin^2\theta_i} - \sin\theta_i\cos\alpha\right] - \alpha \tag{4.6}$$

where θ_i is the angle of incidence with respect to the normal to the entrance surface of the prism, α is the angle of the prism apex, and $n_1 = n(\lambda)$ is the prism glass refractive index at the specific wavelength. The refractive index of air, n_0, is approximated to be ≈ 1.0.

4.7.2.2 *Dispersive element: diffraction gratings*

Today, most modern spectrometers use a reflective diffraction grating as the dispersive element. Diffraction gratings can be customized to provide the required *free-spectral range* (the wavelength range that will reach the detector array) with a high degree of angular dispersion. Figure 4.14 illustrates the structure and elements of a reflective diffraction grating.

The angle for maximum diffraction of light at wavelength λ is given by the "*grating equation,*" which follows from the concepts of interference discussed above.

$$d(\sin\theta_i + \sin\theta_d) = m\lambda \tag{4.7}$$

where d is the groove spacing of the grating (see Figure 4.14), θ_i and θ_d are the angles (with respect to the normal of the grating substrate) of the incident and

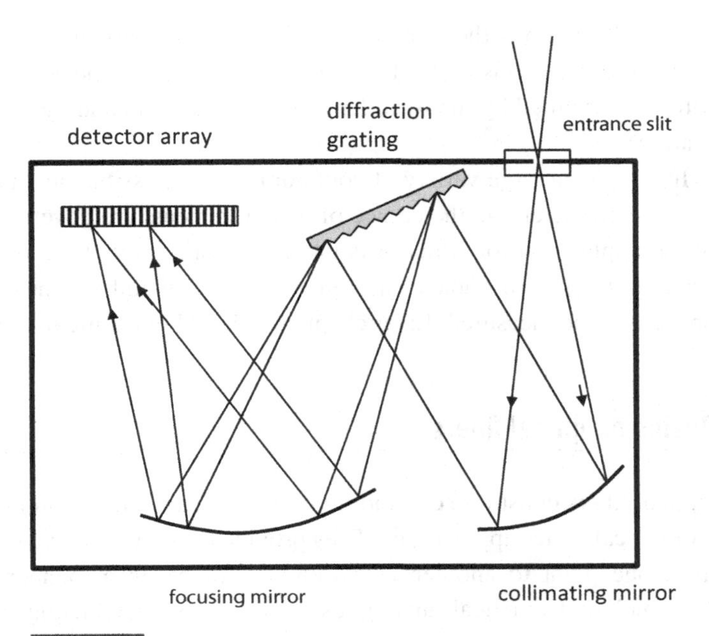

Figure 4.15

Common configuration for a grating-based spectrometer.

diffracted rays, respectively, and m is an integer. For any specific wavelength, preferential diffraction occurs at an angle for which the pathlength difference between rays reflected from different grooves is an integer multiple of the wavelength. The differential dispersion, $d\theta/d\lambda$, is greater for smaller groove spacing, d, with the practical limitation that $d \geq \lambda/2$. Reflective gratings can be fabricated with curved surfaces, so as to focus light while also angularly separating wavelengths, which can be useful for spectrometers with various configurations.

Diffraction gratings can also be of a transmission type, wherein fine structure in the thickness of a sheet of transparent material (e.g., glass) induces a periodic phase shift in the wavefront of light that is transmitted, causing wavelength-dependent angular diffraction in a manner analogous to the reflective grating. The grating equation (4.7) applies for transmission gratings as well, with the RHS representing the relative phase shift of light transmitted through thick vs. thin segments. In practice, the use of reflective gratings predominates, due to lower cost and flexibility with surface shapes.

4.7.2.3 *Exemplary configuration of a grating spectrometer*

Figure 4.15 pictorially illustrates a common layout of components in a grating spectrometer. Parameters that affect the spectral resolution of this type of design include the width of the entrance slit, the groove density of the grating, the spacing of the detector pixels in the detector array, and the distances between components (often determined by the focal lengths of the mirrors and the physical size of the grating).

In practice a large variety of configurations is possible, and the specific design of a spectrometer and the choice of its components are governed by the specifics of the application for which it is intended. Especially, the choice of light source used for the spectroscopic measurement depends mainly on the biological or tissue property to be measured. Later chapters will address some of these variables.

4.8 Basics of optical fibers

Optical fibers constitute commonly found optical elements in instrumentation for biomedical optics applications. They provide the convenience of transporting light from one point to another with minimal losses, in a manner that is uniquely "flexible" and practical, analogous to the use of conducting wires to transport electrical energy.

Optical fibers make use of the phenomenon of *total internal reflection* (TIR) to confine light within the core of the fiber and transport it with low loss, limited by absorption and scattering within the core material of the fiber. To understand TIR, we recall Snell's law, and write it for a light ray propagating within the core of an optical fiber with refractive index n_1 that is surrounded by a *cladding* with a lower refractive index, n_2:

$$n_1 \sin \theta_1 = n_2 \sin \theta_2, \tag{4.8}$$

where the angles are specified with respect to the normal to the interface between the core and the cladding. Depicted in Figure 4.16, as θ_1 is increased, a *critical angle*, θ_c, is reached, given by the expression

$$\theta_c = \sin^{-1}(n_2/n_1), \tag{4.9}$$

for which the refracted ray is parallel to the interface. For incident angles larger than θ_c (depicted as θ_3) the ray is reflected with zero loss – no light propagates out of the core – under the condition of TIR.

For incident angles below the critical angle, the relative proportion of light that is reflected vs. refracted depends on the polarization state of the light ray and on the index difference. Above the critical angle, both polarizations (p- and s-polarization) are totally reflected. Here we use a common representation for optical field polarization directions, defined as being parallel (p-polarization) to the plane formed by the incident and reflected or refracted rays, and normal (s-polarization)

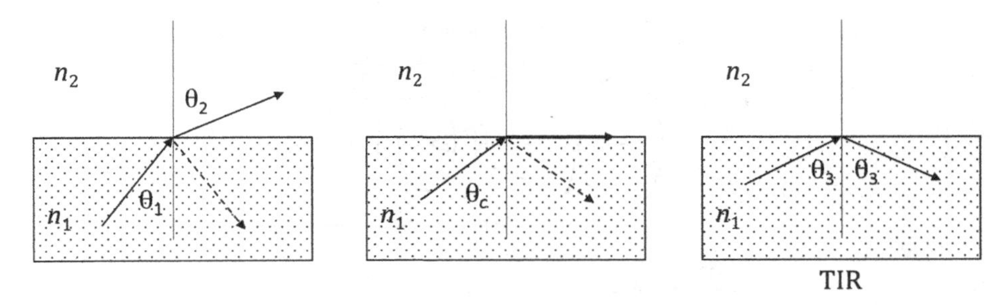

Figure 4.16

Light incident is on an interface from a core of higher refractive index to a surrounding medium of lower index. Left: below the critical angle the ray is partially coupled out and refracted, and partially reflected. Center: at the critical angle, the refracted ray is parallel to the surface, with effectively zero amplitude. Right: above the critical angle, all light is reflected.

to that plane. For efficient propagation of light confined in an optical fiber, both the core and the cladding must exhibit very low absorption at the propagating wavelengths.

Commonly, as in Section 1.3, the refractive index is defined as the ratio of the speed of light in vacuum, c, to that in the medium, c_i: $n_i = c/c_i$. A more complete definition of the refractive index is as a complex variable that depends on the optical frequency:

$$N(\omega) = n(\omega) + in'(\omega) \tag{4.10}$$

where the real part, n, is the familiar index of refraction, and the imaginary part, n', relates to the absorption coefficient of the medium. When optical absorption is weak, the absorption coefficient, μ_a, is linearly proportional to n'. For low-loss propagation in an optical fiber, it is important that both the core and cladding exhibit a very low absorption coefficient. (As the reader may well imagine, when optical fibers are used for long-distance telecommunications, achieving minimal losses due to absorption is critical.) The reason why the cladding must also have low absorption is that, even though there is no *propagating* refracted ray outside the core, there is an *evanescent field* that penetrates into the cladding from the core. The decay of this field is an exponential function of the distance from the interface, with a decay constant of the order of the wavelength. Thus, within a distance of a few wavelengths, the evanescent field becomes negligible. Nonetheless, this evanescent field oscillates at the same optical frequency as the propagating light, and if the evanescent field experiences absorption in the cladding medium, the absorption will lead to transfer of optical power from the core to the cladding, as a loss mechanism for the guided wave.

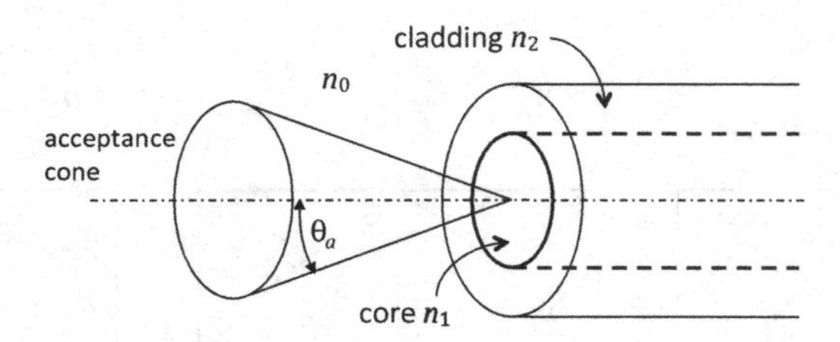

Figure 4.17

The acceptance cone of an optical fiber. Rays impinging at an angle $>\theta_a$ will exceed the critical angle θ_c inside the core, and will experience high losses.

4.8.1 Numerical aperture of a fiber

One of the important parameters describing an optical fiber is the maximum range of angles for rays that can be confined by TIR. It is convenient to define the acceptance cone angle of an optical fiber with a dimensionless parameter, the sine of the half-angle, called the *numerical aperture*, NA, which is determined by the fiber optical indices:

$$\mathrm{NA} \equiv \sin\theta_a = \frac{1}{n_0}\left[n_1^2 - n_2^2\right]^{1/2}, \tag{4.11}$$

where n_0 is the refractive index of the medium. Equation (4.11) can be derived from Eqs. (4.8) and (4.9). In the vast majority of the literature, and in the product listings of fiber manufacturers, the *NA* is commonly specified for the fiber in air, where $n_0 \cong 1$.

$$\mathrm{NA_{air}} \cong \left[n_1^2 - n_2^2\right]^{1/2}. \tag{4.12}$$

The geometry is depicted in Figure 4.17.

The thickness of the cladding layer must be enough to significantly exceed the extent of the evanescent field (several wavelengths), but in practice is much thicker for reasons of mechanical stability of the fiber, especially in the case of single-mode fibers (see below). We also note that Eq. (4.12) is approximately correct only for multimode fibers, whereas in single-mode fibers the internal propagating mode structure also affects the NA.

4.8.2 Basic fiber types

There are two fundamental classes of optical fibers and three basic types of configurations for fabrication, although there are multiple variations on those three themes. The fundamental classes are *multimode* and *single-mode*. Here the term

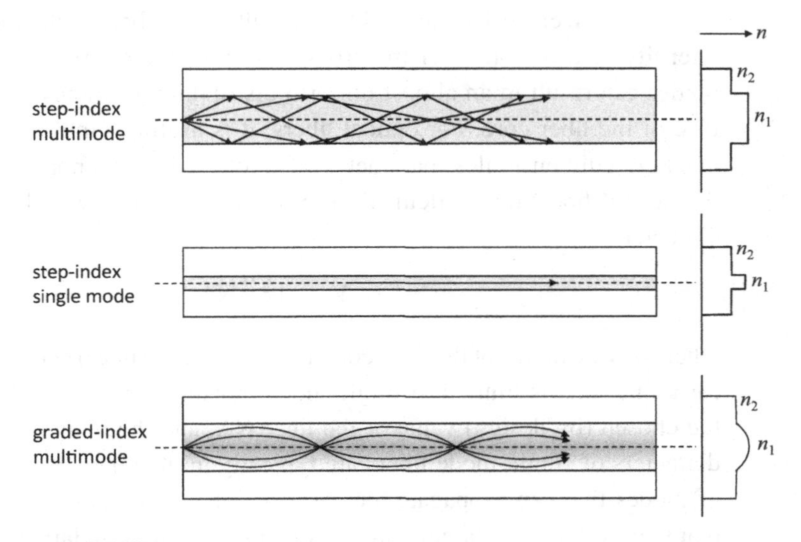

Figure 4.18

Three common structural configurations for optical fibers. Top and center: multimode and single-mode fibers with a stepped change in refractive index between core and cladding. Bottom: a multimode fiber with a graded index in the core.

"mode" refers to the solution of the electromagnetic wave equation in a waveguide, and reminds us that an optical fiber is an electromagnetic waveguide. The basic types of structural configurations for optical fibers are illustrated in Figure 4.18 and include step-index (single mode, multimode) and graded-index (multimode), referring to the changes in index between core and cladding materials.

4.8.3 Single-mode vs. multimode fibers

Electromagnetic waves propagate through waveguides as a set of modes, determined by the boundary conditions (continuous tangential component of the electric field and normal component of the magnetic field at the boundary) for the transverse electromagnetic (TEM) waves. Similarly, light propagates through optical waveguides as a set of modes. For planar waveguides, a mode of the waveguide is the sum of multiple reflections of a ray at a specific angle to the surface. Each mode travels along the axis of the waveguide with a group velocity determined by its internal angle. The case for cylindrical waveguides (optical fibers) is approximately the same. The lowest-order mode travels paraxially, has a planar wavefront (normal to the fiber axis) and has an intensity distribution that is approximately Gaussian. Higher-order modes are also solutions to the wave equation in the fiber, and correspond to effectively higher angles of incidence to the surface of the core, limited by the critical angle for TIR. In multimode fibers each mode has a different intensity distribution (over the cross section of the core), and the sum of many modes can result in an almost uniform distribution of intensity across the output face of the fiber core. For optical fibers, it is useful to introduce the so-called *V number*, a dimensionless parameter (also referred to as a normalized frequency in the optical fiber) that is defined in terms of the optical wavelength and the core diameter:

$$V = 2\pi \frac{a}{\lambda} \mathrm{NA}, \tag{4.13}$$

where a is the radius of the fiber core. In a cylindrical (fiber) geometry, the condition for a fiber that is limited to single-mode transmission is $V \leq 2.4$. Depending on the chosen (by design) value of the fiber NA, and the intended wavelength, core diameters of single-mode fibers are typically in the range 3–10 μm. The number of modes that can propagate increases as integers for increasing core diameters that yield values of V larger than the single-mode limit. In larger multimode fibers, the number of sustainable modes, M, can be approximated as:

$$M \approx \frac{4}{\pi^2} V^2. \tag{4.14}$$

One advantage of multimode fibers, especially those with larger values of NA, is that the larger core and acceptance cone angle enable coupling of more light from incoherent light sources, such as LEDs, xenon arc lamps or incandescent bulbs. Another advantage is the possibility of coupling light from high-power lasers, which, albeit coherent, might damage the core of a small single-mode fiber if the laser beam were focused down to that size. Consequently, multimode fibers are typically used for directing high-power laser light in medical applications like tissue ablation and thermotherapy (to be addressed in Section 19.3). Common core sizes of multimode fibers are in the range 50–400 μm, or even larger (often limited by the mechanical stiffness for larger core sizes).

In addition to the advantage of freedom from mode dispersion (see below), light propagated through a single-mode fiber retains spatial (transverse) coherence by virtue of having a uniform phase front of the lowest-order mode. This feature is important for applications like *confocal microscopy* and *optical coherence tomography* (OCT), which will be described in Chapters 17 and 18, respectively.

4.8.4 Graded-index fibers

One consequence of transmitting multiple modes in a larger fiber is that different modes travel at different group velocities (related to propagating at different

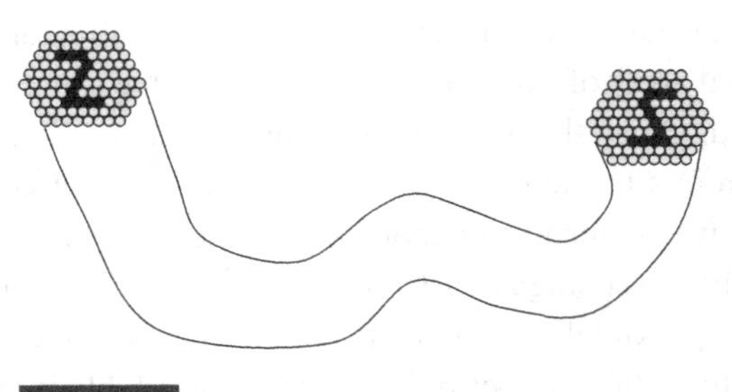

Figure 4.19

An ordered fiber bundle can serve as a flexible image relay system for medical endoscopy.

angles), which leads to temporal spreading of short optical pulses or loss of phase information for modulated waves. This phenomenon is called *mode dispersion*. Mode dispersion is the reason why fiberoptic telecommunications require information to be carried by a single optical mode, even if the fiber is capable of multimode transmission. A partial solution to the problem of mode dispersion can be achieved with graded-index multimode fibers. As illustrated in the bottom panel of Figure 4.18, the index of refraction of the core is not constant, but is peaked at its center and tapers down to the value of the index of the cladding at the interface. The lowest-order mode propagates paraxially, as in the case with a step-index fiber, but higher modes now follow serpentine paths. Although the serpentine paths are longer than the straight paraxial path, the paraxial ray experiences a higher refractive index over most of its path, compared with the average index experienced by higher-mode rays, whose paths include regions of lower index. Hence the paraxial ray travels at a lower speed, and the result is that different modes arrive at the exit face at about the same time, having traveled approximately equal *optical* distances.

4.8.5 Imaging fiber bundles

Aside from microscopes and familiar instruments used by optometrists, one of the more widely recognized biomedical optical instruments is the flexible imaging *endoscope*. Endoscopes have become indispensible medical tools for viewing internal spaces within the body, especially in hollow organs. Endoscopes are used by several specialties of medicine, and are ubiquitous in gastroenterology. Early endoscopes were rigid tubes with series of small lenses to relay an image from, say, a patient's stomach to the physician's eye, but a real impact on medical practice did not occur until the advent of flexible, steerable endoscopes.

Flexible endoscopy was first enabled by the development of "coherent" fiber bundles in the late 1950s. As illustrated in Figure 4.19, such a bundle consists of a large number of small optical fibers, in close-pack configuration, and assembled so that the ordered array of fibers at the object end is the same as (or mirror-image of) the image end. Given the more rigorous use of the word "coherent" in optical physics, which refers to the deterministic temporal or spatial phase information of a light wave, we prefer to use more appropriate terminology to represent an imaging fiber bundle as being an *ordered* bundle. A small lens at the distal end and an eyepiece at the proximal end complete the image transfer to the physician's eye. In an ordered bundle, the number of image pixels available is equal to the number of independent fibers. With a bundle diameter limit of ~5 mm when used for, say, colonoscopy (endoscopy of the colon), early flexible endoscopes had about 10,000 resolution elements, with later improved versions exceeding 50,000 fibers.

In recent years, for many applications, video endoscopes have become more common, wherein the imaging fiber bundle has been replaced by a very small CCD camera incorporating a micro lens (similar to the cameras found in modern "smart" phones), at the distal end of the endoscope, and with wires conveying the image electronically to a video screen that replaces the eyepiece for viewing. In these versions, the illumination light is still conveyed to the end of the endoscope by optical fibers, thus keeping the heat source outside the body. Nonetheless, some applications, such as endoscopic confocal microscopy and very small-diameter endoscopes, still invoke the use of ordered fiber bundles.

4.8.6 Photonic-crystal fibers (PCF)

A special class of optical fibers has been developed in recent years, called *photonic-crystal fibers* (PCF), which confine propagating modes in the core based on the filtering effects of transverse periodic micro-structural modifications (photonic crystal structures), rather than by total internal reflection at a refractive index gradient. An electron micrograph of one such fiberoptic structure is shown in Figure 4.20. The periodic, wavelength-size structure leads to "photonic band gaps," in a manner similar to the effect of the crystalline structure leading to electronic band gaps in semiconductor materials. The photonic band gaps can block, reflect or transmit specific wavelengths and angles, analogous to X-ray Bragg diffraction by crystals (Yablonovitch, 1987; John, 1987). In nature, the wavelength-selective reflectance and attenuation of photonic crystals results in, among other things, the colors of butterfly wings and peacock feathers. For optical fibers, the transverse periodic structure, with discontinuous changes in refractive index, results in propagation parameters for the core that are highly nonlinear.

In the type of PCF shown in Figure 4.20 (with a solid core), the array of holes acts to filter propagation modes, such that the lowest-order mode is trapped and propagates with low losses, whereas higher-order modes escape.

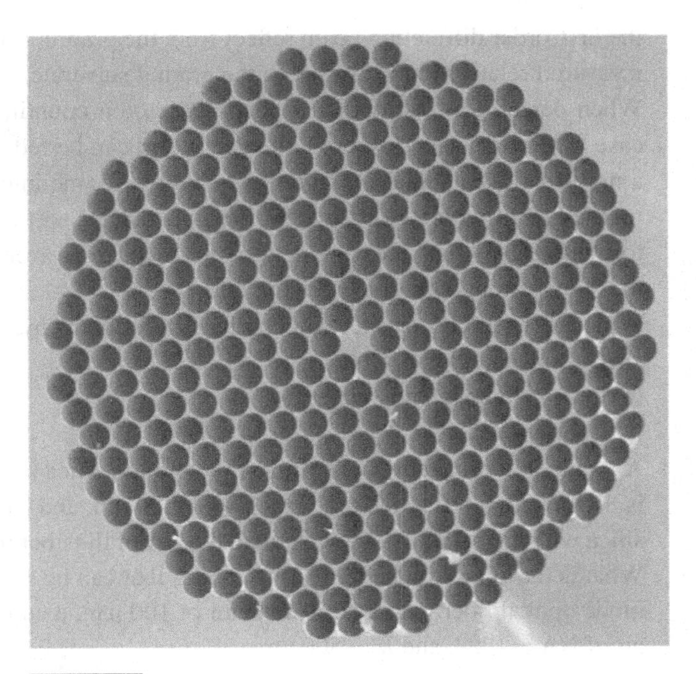

Figure 4.20

Scanning electron-microscope image of end of a photonic-crystal fiber (full diameter not shown). The central propagation core is ~5 μm diameter, whereas the hollow channels are ~4 μm diameter. (Courtesy US Naval Research Laboratory [public domain].)

At the time of this writing, the most prominent application for PCF in biomedical optics is for the efficient generation of *supercontinuum* light for spectroscopic applications. When ultrashort laser pulses are injected into the propagation core of a PCF, the high peak powers and channeled confinement lead to dramatic spectral broadening by a high multiplicity of nonlinear optical effects (Wadsworth et al., 2002). Since all wavelengths are still limited to propagation in their lowest-order transverse mode, the result is a "white" laser beam with excellent spatial (transverse) coherence. The cost of such sources has been dropping recently, and they are ideal for a variety of spectroscopic and imaging applications.

Problems – answers to problems with * are on p. 645

4.1 Derive an expression for the differential dispersion of a diffraction grating, $d\theta/d\lambda$, in terms of the grating spacing, d, the wavelength, λ, and the diffraction order, m.

4.2* If a collimated beam of light at 500 nm impinges, at normal incidence, on a diffraction grating whose groove density is 400 lines/mm, at what angle does the first-order diffraction beam reflect from the grating surface. (Angles are measured relative to the normal to the grating substrate.)

4.3* When detecting weak optical signals by photon counting (as is often the case for Raman spectroscopy), the shot noise can be reduced by averaging a number of sequential measurements with the same integration time. If a measurement with a given integration time detects an average of 400 photons, how many such measurements would need to be averaged to determine the signal strength with a precision of $\pm 1\%$?

4.4 What are the effects on spectrometer resolution, and why, when the entrance slit is made smaller:
  (a) for an incoherent source?
  (b) for a laser source?

4.5* An optical fiber is fabricated with a core of pure silica (fused quartz, which is SiO_2) with a refractive index of 1.46 at 500 nm, and a cladding of doped silica with an index of 1.44. Calculate the NA of the fiber in air and in water.

4.6 What is the approximate number of modes that can be sustained in a multimode optical fiber with a core diameter of 100 μm, a numerical aperture in air of NA = 0.22, and $\lambda = 500$ nm.

References

Hosseinimakarem, Z., and Tavassoli, S. H. (2011). Analysis of human nails by laser-induced breakdown spectroscopy. *Journal of Biomedical Optics*, 16, 057002.

John, S. (1987). Strong localization of photons in certain disordered dielectric superlattices. *Physical Review Letters*, 58, 2486–2489.

Wadsworth, W. J., Ortigosa-Blanch, A., Knight, J. C., et al. (2002). Supercontinuum generation in photonic crystal fibers and optical fiber tapers: a novel light source. *Journal of the Optical Society of America*, B19, 2148–2155.

Yablonovitch, E. (1987). Inhibited spontaneous emission in solid-state physics and electronics. *Physical Review Letters*, 58, 2059–2062.

Further reading

General optics relevant to spectroscopy

Hecht, E. (2002). *Optics*, 4th edn. Boston, MA: Adisson-Wesley.

Saleh, B., and Teich, M. (2007). *Fundamentals of Photonics*, 2nd edn. Hoboken, NJ: Wiley Interscience.

Optical spectroscopy basics and instrumentation

Tkachenko, N. V. (2006). *Optical Spectroscopy: Methods and Instrumentations*, Amsterdam: Elsevier Science.

Basics of optical fibers

Buck, J. A. (2004). *Fundamentals of Optical Fibers*, 2nd edn. Hoboken, NJ: Wiley Interscience.

Laser-induced breakdown spectroscopy

Cremers, D. A., and Redziemski, L. J. (2013). *Handbook of Laser-Induced Breakdown Spectroscopy*, 2nd edn. Hoboken, NJ: Wiley.

General molecular spectroscopy

Steinfeld, J. I. (2005), *Molecules and Radiation: An Introduction to Modern Molecular Spectroscopy*, 2nd edn. Mineola, NY: Dover Publications.

Photonic-crystal fibers

Russell, P. (2003). Review: photonic crystal fibers. *Science*, 299, 358–362.

5 Autofluorescence spectroscopy and reporter fluorescence

When people are in a nightclub with "black light" illumination, one can often observe various articles of clothing that "glow in the dark." What is happening is that the "black light" is illuminating at near-ultraviolet wavelengths (a broad range around 350 nm), and various dyes in the clothing absorb UV light and remit light (fluoresce) at longer (visible) wavelengths. (This includes dyes from laundry detergent intended to make your clothes "brighter than white"!) Your eyes do not see most of the UV wavelengths, and the nightclub room is otherwise rather dark, so strongly fluorescent dyes, which are often called *fluorophores*, stand out dramatically. What you cannot see is that people's skin is also fluorescent, albeit much more weakly.

A variety of *endogenous* (naturally occurring in the body) molecules in tissue exhibit some degree of fluorescence. By spectroscopically measuring, or by imaging, the fluorescence emission from tissue, one can obtain information about the constituents that are present in the tissue, which can serve as valuable diagnostic information. Alternatively, an *exogenous* (not naturally occurring in the body) fluorophore can be formulated, or conjugated with targeting moieties, such that it will bind preferentially to certain tissue types, or to cells with a specific disease state. In that case, if the fluorophore is administered, it can serve as a tissue-selective "optical reporter." Of course, for human use such a compound would be treated as a new drug, requiring FDA approval, following extensive testing to demonstrate safety and efficacy.

One of the major advantages of fluorescence measurements is that they are essentially "background-free" measurements, meaning that there is typically no fluorescence emission unless the molecules are illuminated with shorter-wavelength excitation light, which is readily filtered out at detection. This dark-background type of measurement enables sensing of fluorophores at exceedingly small concentrations, as small as femtomolar, or even at a single-molecule level. Although such measurements are always more complicated in living tissue, a variety of sensing and diagnostic applications in biomedical optics are based on fluorescence detection or imaging.

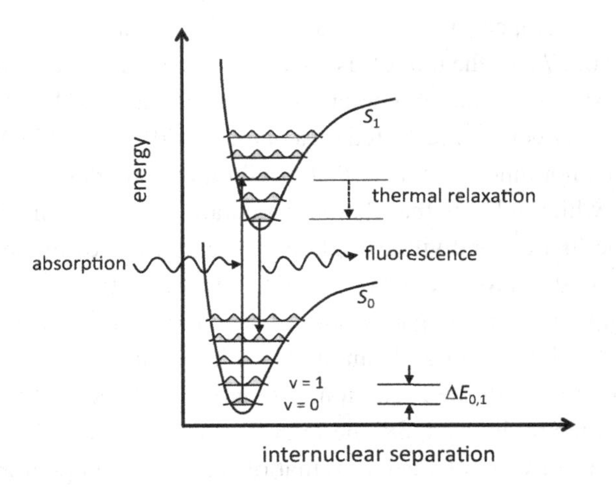

Figure 5.1

Energy-level diagram indicating absorption occurring predominantly from the lowest vibrational level of the ground electronic state, S_0, to higher vibrational levels of the S_1 excited electronic state, as a consequence of the Franck-Condon factors affecting transition probabilities.

5.1 Fluorescence emission relevant to measurements in biological systems

Why is the wavelength of fluorescence emission from biomolecules almost always longer than that of the excitation light? A dominant factor governing the emission wavelength can be understood from examination of Figure 5.1 (revised from Figure 4.6), as well as Figures 4.5–4.8, which indicates optical absorption occurring from the lowest vibrational level of the ground electronic state, because that is the state occupied by most molecules under normal conditions. That is, although, under typical conditions, some molecules are found in excited vibrational modes at room temperature (~293 K) or body temperature (~310 K), most are in the lowest vibrational level. The relative level-occupancy numbers can be calculated under conditions of thermal equilibrium, for which the relative statistical occupancy rates of vibrational levels, i and j, are given by the Boltzmann distribution (first explained by Austrian physicist Ludwig Boltzmann [1844–1906]):

$$\frac{N_j}{N_i} = e^{-(\Delta E_{i,j}/k_B t)}, \tag{5.1}$$

where $\Delta E_{i,j}$ is the energy difference between levels j and i ($\Delta E_{i,j} = E_j - E_i$), and k_B is Boltzmann's constant: $k_B = 1.38 \times 10^{-23}$ J/K. (Note: $\Delta E_{i,j}$, the energy difference between levels, should not be confused with ΔE, as defined in Section 4.3.1, to represent the linewidth of the transition.)

Thus, body temperature can be associated with an energy, $k_B T$, of $\sim 4.28 \times 10^{-21}$J ≈ 0.0267 eV (the unit eV is the *electron volt*, a non-SI unit of energy often used in physics to denote photon energies, and defined as the energy acquired by an electron after being accelerated by a potential difference of 1 V). This energy is less than, but not much less than, that of vibrational modes common in biological molecules, which tells us that there is generally small, but not negligible, occupancy in the first excited vibrational energy levels. For example, some important vibrational modes have energies in the $\sim$0.1 eV range (above the "ground" level), which would result in occupancy rates of approximately $\sim$2% for the first excited vibrational level, but negligible amounts for levels higher than that.

Under conditions of the excitation originating in the lowest vibrational level of the ground electronic state, S_0, the largest possible energy for a photon emitted from the S_1 level would be equal to that of the excitation photon. What happens, however, is that most molecules undergo internal relaxation (by thermal mechanisms), which occurs on a time scale that is short compared to the fluorescence emission lifetime. As we have seen in Chapter 4, fluorescence emission lifetimes, for S_1–S_0 transitions in large biomolecules, are typically of the order of a few nanoseconds, easily enough time for the internal relaxation processes, which happen on a time scale of picoseconds. Consequently, almost all emission starts from the lowest vibrational level of the excited electronic state, S_1. Moreover, due to the same Franck-Condon factors that govern the quantum-mechanical probabilities for absorption (see Section 4.2), the emission from the S_1 state is generally to an excited vibrational level of S_0, one that results in minimal change of internuclear separation during the transition. We wish to point out that, for applications in biomedical optics, fluorescence measurements that involve electronic states higher than S_1 are generally not invoked, because absorption (or emission) between those higher states and the S_0 state is at wavelengths in the "hard-UV" (<250 nm), which do not transmit in tissue or water. (Also discussed in Section 4.3.1, emission from a triplet state, a "forbidden" transition, is much slower, and is referred to as phosphorescence.)

The end result of these considerations is that absorption probabilities and emission probabilities are such that the fluorescently emitted light from biological molecules is (essentially) always at a longer wavelength (lower energy) than that of the excitation light. Figure 5.2 illustrates a common relationship between absorption and emission spectra for large biomolecules and organic dyes. Plots that are qualitatively like this are commonly found in the literature and in product information from companies that sell fluorophore dyes for biomedical research.

The reader may note that the absorption and emission spectral shapes in Figure 5.2 are represented as being approximately mirror images of each other (reflected in wavelength). This occurs for some (but not all) fluorophores of

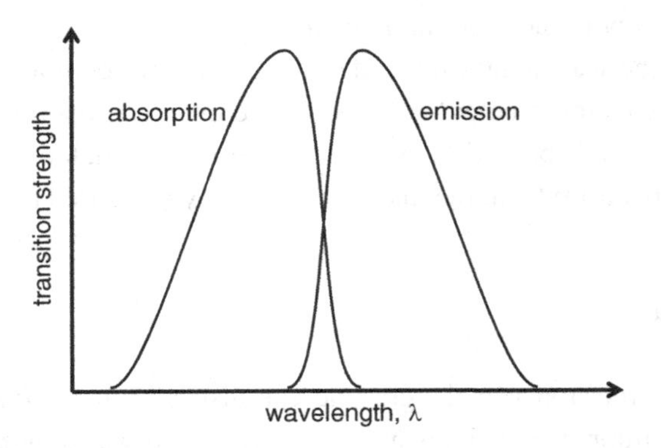

Figure 5.2

Plot of possible absorption and emission wavelengths for a model organic dye molecule or biomolecule.

interest in biomedical studies, and is a consequence of two factors: (1) the spacing of vibrational levels is approximately the same for the ground (S_0) and excited (S_1) electronic states (i.e., vibrational modes are only slightly perturbed by electronic state occupation); and (2) both absorption and emission transitions originate predominantly from the ground vibrational level of their respective electronic states. Thus, absorption and emission processes sample the same array of vibrational transition probabilities, but in opposite directions.

The two spectra in Figure 5.2 have significant overlap, and one may wonder whether the overlap zone indicates that emission wavelengths can be shorter than the excitation (absorption) wavelength, which would be inconsistent with the above discussion. Don't be fooled! What is not readily apparent from this type of plot is that for each excitation wavelength, there is a range of emission wavelengths that are possible, specific to that excitation. Conversely, for a fluorescence photon emitted in the "blue" (shorter-wavelength) end of the emission spectrum, the excitation photon responsible for that emission must have been at a shorter wavelength in the blue end of the absorption spectrum. The traces in Figure 5.2 represent the full absorption spectrum, and a superposition of the range of possible emission spectra for all excitation wavelengths. More rigorously, for each excitation wavelength there is a corresponding fluorescence spectrum, relating to the wavelength-dependent emission for that excitation photon energy. In general, however, the shape of the fluorescence spectrum is essentially insensitive to the specific excitation wavelength because, as discussed above, fluorescence emission originates almost exclusively from the lowest vibrational level of the S_1 state.

In practice, in real tissue, fluorescence spectroscopy is subject to distortions induced by the scattering and absorption properties of the tissue. The tissue properties affect both the spectral intensity of excitation light that reaches the fluorophores, and the wavelength-dependent losses of emitted fluorescence traversing tissue before reaching a detector. If the native optical properties of the tissue are known, a degree of spectral "correction" can be calculated to retrieve a native fluorescence spectrum from the one that is actually measured (Müller et al., 2001).

5.1.1 Quantum yield

For a specific emission wavelength one can also plot an *excitation spectrum* by measuring the fluorescence intensity at that emission wavelength as a function of the excitation wavelength. The fluorescence quantum yield, Φ_F, of a specific pair of electronic energy levels is defined as the ratio of the number of emitted fluorescence photons to the number of absorbed photons, regardless of the emission wavelength. (Sometimes the term *quantum efficiency* is used, but that term is more commonly associated with the conversion ratio between number of electrons generated and number of photons absorbed in a photodetector.) The quantum yield represents the percentage of excited molecules that relax to the ground state by radiative decay associated with fluorescence emission. By introducing the rate constants for radiative decay, k_r, associated with photon emission and for non-radiative decay, k_{nr}, reflecting thermal relaxation processes, the *fluorescence quantum yield* is defined as the ratio of the radiative decay rate to the total decay rate:

$$\Phi_F = \frac{k_r}{k_r + k_{nr}} = \frac{\#\, photons\ emitted}{\#\, photons\ absorbed}. \tag{5.2}$$

It is common to find in the literature, or in the data sheets from manufacturers of fluorescent dyes, a single fluorescence spectrum without specification of the excitation wavelength that was used to generate that spectrum. Experimenters need to be cognizant of the assumption that the excitation wavelength for such plots must be shorter than any of the plotted emission wavelength range.

5.1.2 Excitation-emission matrix

An excitation-emission matrix (EEM) graph of fluorescence emission strengths can serve as a more complete display of information about the relation between excitation and emission wavelengths. An EEM can be plotted as a quasi "three-dimensional" graph of excitation-vs-emission wavelengths, a "contour map" with the third dimension being represented by height, color or shade, and indicative of the relative emission probabilities in complex media. Given that there are generally

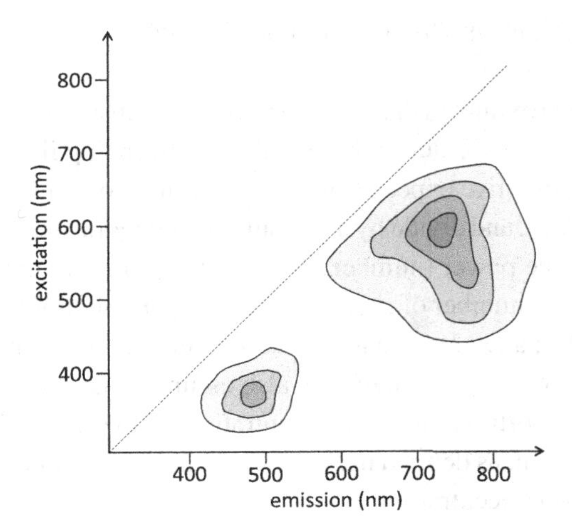

Figure 5.3

A fluorescence excitation-emission matrix (EEM) for an imagined biological sample with two fluorescent species. Within the shaded areas, darker shades indicate stronger fluorescence emission (for the same excitation intensity). The contour lines represent equal values of quantum yield for the specific excitation and emission wavelengths. To minimize interference by strong elastic scattering of the excitation light, detection is typically blocked by variable filters for emission wavelengths near the equal excitation wavelength: i.e., near the main diagonal that defines equal excitation and emission wavelengths.

several fluorophores in tissue, EEMs can be especially useful for added specificity in tissue diagnostic applications of fluorescence spectroscopy (Zuluaga et al., 1999; Brewer et al., 2001). A representation of such a plot can be seen in Figure 5.3, which plots an EEM for an imagined biological sample that contains two distinguishable fluorescent species, with emission peaks at 480 and 720 nm. The reader will note that all of the emission is found on the "red" (long-wavelength) side of the diagonal that denotes equal excitation and emission wavelengths. Using this graph, if one were to plot the fluorescence intensity along a horizontal line located at a specific excitation wavelength, the result would be the emission spectrum for that excitation wavelength. Similarly, a vertical line at a specific emission wavelength would yield the excitation spectrum for that fluorescence wavelength.

It should also be noted that strong absorption by electronic transitions does not always lead to fluorescence. To wit, the "tissue" represented in Figure 5.3 might exhibit strong absorption at ~440 nm, where the fluorescence intensity is nonetheless low. As mentioned previously, an important specific chromophore in tissues is hemoglobin, which has very strong absorption bands in the visible range, but is not measurably fluorescent.

5.1.3 Fluorescence intensity vs. fluorophore concentration

It is sometimes approximated that the fluorescence intensity is linearly proportional to the concentration of the fluorophore in the sample being illuminated. The logic to this seemingly intuitive expectation follows from two assumptions: (1) one can reconfigure Eq. (5.2), and divide by a measurement integration time, to simply state that the fluorescence power (number of emitted photons per unit time) is equal to the absorbed power (number of absorbed photons per unit time) times the quantum yield: $P_F = P_A \Phi_F$; and (2) for a given illumination intensity, if the fluorophore's absorption coefficient, μ_a, is small, one also assumes that the absorbed power per unit volume is proportional to the concentration of the fluorophore, because the absorption coefficient, as defined in Eq. (2.1), is the product of the molar extinction coefficient and the concentration.

In practice, the relation between fluorescence intensity and fluorophore concentration is more complicated. While, in general, the intensity will depend monotonically on the concentration, the relationship is rarely linear. Here are some of the factors that confound the determination of a fluorophore concentration (even relative concentration) from measurements of its fluorescence intensity:

- The molar extinction coefficients of some fluorophores of interest in biomedicine exhibit strong dependence on concentration. (See an example in Section 5.4.1.)
- The relation between excitation intensity and rate of photons absorbed is generally nonlinear, due to saturation of the absorber and/or bleaching effects; and the saturation depends on concentration.
- In turbid media (like tissue) the photon pathlength, for both illumination light and emitted fluorescence, depends on the details of the medium and on the wavelengths of excitation and emission light. Thus, the distribution of absorbed fluence will change with fluorophore concentration, which will affect the rate of photons absorbed.
- Factors that affect the quantum efficiency, such as pH and electronegativity of the medium, can themselves be affected by the fluorophore concentration.
- Other absorbers in tissue, compounded by wavelength-dependent scattering effects, will distort the expected absorption and emission spectra (Müller et al., 2001).
- If the emission wavelength measured is within the overlap with the absorption band of the fluorophore itself, reabsorption by the fluorophore itself occurs, perturbing the relation to concentration.

5.2 Fluorescence lifetime

The average time required for a molecule in the excited state S_1 to relax to the ground state, by an allowed dipole transition, is defined as the *fluorescence lifetime*, τ_F. The ideal case, with an absence of any thermal or other non-radiative relaxation processes, corresponds to an intrinsic lifetime, τ_0, that is exclusively associated with the *radiative decay rate* and is given by:

$$\tau_0 = \frac{1}{k_r}. \tag{5.3}$$

The observed fluorescence lifetime, however, must take into account all non-radiative decay processes, i.e., absorption not followed by fluorescence, such that:

$$\tau_F = \frac{1}{k_r + k_{nr}}. \tag{5.4}$$

A comparison of Eqs. (5.2)–(5.4) leads to the following relationship between fluorescence quantum yield and lifetimes:

$$\Phi_F = \frac{\tau_F}{\tau_0}. \tag{5.5}$$

While the radiative decay rate, k_r (and, consequently, τ_0), is an intrinsic quantum-mechanical property of the vibro-electronic molecular energy levels, it can nonetheless be affected by the molecular environment, in particular by factors that affect the transition dipole moment and/or local electric (optical) field. These include the pH, dielectric constant of the medium, and the electronegativity, or conductivity, of the medium (typically related to ions from dissolved salts), as well as water that may be hydrogen-bonded with the fluorophores. On the other hand, the non-radiative decay rate, k_{nr} (hence, τ_F), is typically influenced by kinetic interactions between the molecule of interest and solute molecules, which can quench the excited state: i.e., factors such as the temperature and viscosity of the medium, among others. Fluorescent probes can therefore play the role of "timers" to measure dynamic and local environment effects that influence the fluorescence lifetime.

One can examine what happens following an instantaneous (much shorter than the excited state lifetime) pulse of excitation light, which results in an initial population, N_0, of molecules in the excited electronic state. (More specifics about methods for measuring fluorescence lifetime will be discussed in Section 5.6.) From the definition of the decay rates k_r and k_{nr}, the change in the number of excited molecules (dN) over an infinitesimal time interval dt is given by:

$$dN = -(k_r + k_{nr})Ndt, \tag{5.6}$$

which yields the following time dependence for the number of molecules in the excited state, $N(t)$, following the excitation pulse:

$$N(t) = N_0 e^{-(k_r + k_{nr})t} = N_0 e^{-t/\tau_F}, \tag{5.7}$$

and its time derivative is:

$$\frac{dN}{dt} = -\frac{N_0}{\tau_F} e^{-t/\tau_F}. \tag{5.8}$$

Continuing to address the case of a single fluorophore, with a mono-exponential decay rate, we note that the detected intensity of the fluorescence emission is proportional to the product of the fluorescence quantum yield and the magnitude of the time-derivative of the excited-state population, $I_F \propto \Phi_F \left|\frac{dN}{dt}\right|$, such that:

$$I_F(t) \propto C_D \Phi_F \frac{N_0}{\tau_F} e^{-t/\tau_F} \tag{5.9}$$

where we have introduced a "detection constant," C_D, a factor that accounts for the geometric efficiency of detection of the total fluorescence emission; this factor can sometimes be approximated (for unpolarized fluorescence measurements) as the solid angle subtended by the detector, divided by 4π.

In the presence of multiple fluorophores, the single exponential expression in Eq. (5.7) becomes a summation of exponential terms:

$$N(t) = \sum_i N_{0i} e^{-t/\tau_{Fi}}, \tag{5.10}$$

where N_{0i} and τ_{Fi} indicate the initial number of excited molecules and the lifetime, respectively, for the i-th fluorophore.

5.3 Endogenous fluorescent molecules

Spectroscopic assessment of endogenous (intrinsic) fluorophores in tissue has proven to be a promising tool among optical methods for minimally invasive detection of cancer (Müller et al., 2003). This has motivated investigation of the diagnostic potential of a range of naturally occurring fluorophores. Table 5.1 lists some of the more commonly encountered and naturally occurring fluorophores in human tissues, selected, in particular, for those likely to be found in greater abundance in tissue. Amino acids are the building-block molecular groups of proteins, but their fluorescence is in the UV, and their emissions overlap with strong absorption by abundant constituents in tissue, requiring careful experimental technique to separate species spectroscopically. Nonetheless, research groups have demonstrated fluorescence imaging of, for example, tryptophan, as a biomarker for

Table 5.1 Some absorption and emission band centers for native fluorophores found in tissue. From Ramanujam (2000).

Endogenous fluorophores	Excitation maxima (nm)	Emission maxima (nm)
Amino acids		
Tryptophan	280	350
Tyrosine	275	300
Phenylalanine	260	280
Structural proteins		
Collagen	325	400, 405
Elastin	290, 325	340, 400
Enzymes and coenzymes		
FAD, flavins	450	535
NADH	290, 340–350	440, 460
NADPH	336	464
Lipids		
Phospholipids	436	540, 560
Lipofuscin	340–395	540, 430–460
Ceroid	340–395	430–460, 540
Porphyrins	400–450	630, 690

neoplasia (Banerjee et al., 2012). The other compounds in the list emit fluorescence in the visible range, and are readily detected in fluorescence spectroscopy or imaging.

The structural proteins, collagen and elastin, especially, are abundant in normal tissue, as they constitute the bulk of extracellular matrix and connective tissues. They exhibit characteristic fluorescence in the near-UV to visible-violet spectral range. A number or research groups have noted that tumors exhibit reduced fluorescence from collagen and elastin, when compared with healthy tissue, because of the loss of extracellular matrix consistent with the structurally disorganized nature of tumors (e.g., Georgakoudi et al., 2002). Fluorescence from structural proteins can also be reduced as a consequence of increased blood perfusion in malignant and pre-malignant tissues, which will result in increased absorption by hemoglobin of the fluorescence from collagen or elastin.

As an example of potential clinical applications, studies of fluorescence (measured at an epithelial surface) as a marker for cancer have indicated that both mechanisms, loss of extracellular matrix and increased perfusion, may be involved in the apparent reduction in collagen fluorescence in patients at risk of cervical cancer (Chang et al., 2005; Thekkek and Richards-Kortum, 2008). Those studies utilize both diffuse reflectance spectroscopy (see Chapter 8) and fluorescence spectroscopy to complement each other and yield statistically significant results.

The metabolic enzymes and coenzymes are important potential diagnostic indicators, because they are intimately involved in the energy cycle of cells. The processes and dynamics of living systems impose a continual demand for energy. Endothermic processes can include mechanical work (muscles), transport of signals (nerve impulses) or of molecules and ions (trans-membrane transport), and biochemical synthesis of macromolecules (proteins, DNA, etc.). The role of coenzymes, like flavin adenine dinucleotide (FAD) and nicotinamide adenine dinucleotide (NAD), is to facilitate oxidation of fuels (carbohydrates and sugars) and store the free energy in ATP (adenosine triphosphate) by hydrolysis, and for enabling ATP to transfer its energy by phosphorylation of other compounds that use the energy for physiological functions. Fluorescence detection/imaging can be used to monitor rates of metabolic activity because the redox states of these coenzymes affect their fluorescence. Namely, the reduced state, NADH, is fluorescent, while the oxidized state (NAD) is not; and, conversely, FAD is fluorescent, whereas its reduced form, $FADH_2$, is not. Consequently, significant research effort has been expended to explore the potential for imaging the fluorescence from these compounds, especially NADH (or the decrease of NADH fluorescence), as an indicator of, for example, cancer risk in epithelial tissues. Such research has included a number of clinical studies (e.g., Skala et al., 2007; Mayevsky and Chance, 2007).

The molar extinction coefficients of the structural proteins and enzymes/coenzymes are strong: in the range 10^3–10^4 cm^{-1}/M, and quantum yields are of the order of 0.1 (Dawson et al., 1989). Thus, it is often possible to measure fluorescent signals from these species with good signal-to-noise ratios in tissue. Compared with emission from those classes, fluorescence from lipids is generally weak, although measurable fluorescence from beta-carotene in adipose tissue (in humans) can constitute indirect evidence of lipid concentrations. There are a number of other endogenous compounds that are fluorescent, including several of the vitamins and other essential compounds, but these are found in much smaller concentrations, and generally do not play roles in optical imaging or sensing in vivo.

As mentioned above, some compounds that absorb visible light may not fluoresce. Melanin, for example, has a very large extinction coefficient, but does not fluoresce. The "800-pound gorilla" of visible-wavelength optical absorption in our bodies is hemoglobin, which, unlike melanin, is abundant in all tissues that are perfused with blood vessels, i.e., all but a very few places (e.g., the cornea of the eye) in the body. As already mentioned, hemoglobin has very strong absorption bands at visible wavelengths (see Chapter 2), but exhibits negligible fluorescence. Nonetheless, although hemoglobin itself is not fluorescent, some of the precursors in the *heme cycle* (the biochemical cycle by which cells produce hemoglobin), such as porphyrins, and some of the breakdown products of metabolized hemoglobin, such as biliverdin and bilirubin, are indeed fluorescent. Thus, sensing the florescence of porphyrins or heme metabolites can provide diagnostic information about the tissue or its host organism, and inform on the production and/or breakdown of hemoglobin.

5.4 Exogenous "reporter" fluorophores

When using endogenous fluorophores as optical biomarkers, one must deal with the limitations of what is found naturally in the organism of interest. An administered foreign compound, on the other hand, can be tailored with properties designed to facilitate the optical sensing or imaging application of interest. With the methods of modern synthetic chemistry, an exogenous fluorophore can be endowed with almost all the properties needed for the application. Chief among the variables that are typically desirable are:

1. a strong molar extinction coefficient at a wavelength range that does not overlap the absorption bands of strong naturally occurring chromophores (e.g., hemoglobin);
2. a wavelength range for fluorescence emission that also evades strong chromophores;
3. a high value for the fluorescence quantum yield;
4. photostability: the compound should be robust to strong illumination at the excitation wavelengths;
5. biochemical properties that enable targeting specificity for the biomolecule, cell marker, or tissue of interest, with high contrast compared to neighboring tissues, generally a more challenging task.

Again, when dealing with optical reporters for proposed use in humans, there is the exigency of no or minimal toxicity. That requirement can often be at odds with the molecular design features that lead to good performance for the five desired properties listed above.

5.4.1 Exogenous fluorophores in common clinical use: fluorescein and indocyanine green

Exogenous fluorophores can serve as biomarkers for a variety of tissues, physiological functions and disease states. Two FDA-approved fluorophores that exhibit

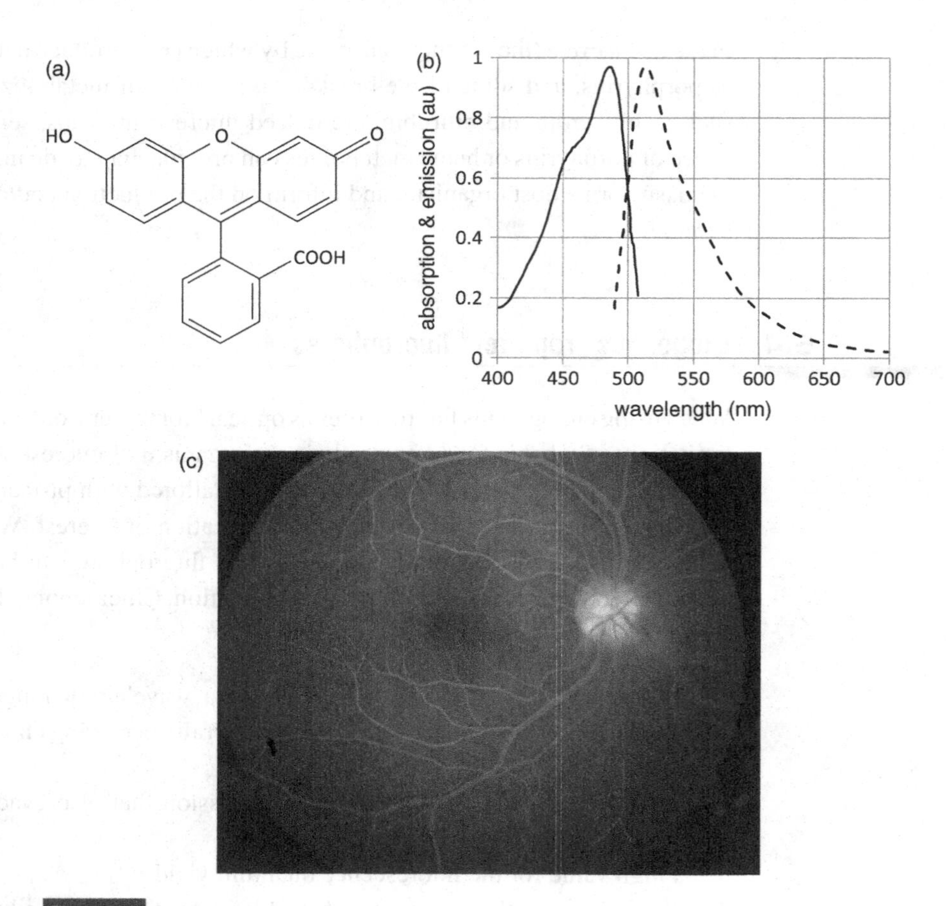

Figure 5.4

(a) Molecular structure of fluorescein. (b) Excitation (left) and emission (right) spectra of aqueous solution of fluorescein (data extracted from Auger et al., 2011), which look qualitatively like those in Figure 5.2, with the excitation peak around 485 nm and the emission peak (for excitation at 488 nm) at ~515 nm. (c) Image of retinal vasculature with fluorescein.(Wikimedia Commons [public domain].)

minimal toxicity, when properly administered in humans, and have been in clinical use for a long time are *fluorescein* and *indocyanine green*.

Fluorescein (Figure 5.4(a)) absorbs violet-blue light and fluoresces in the green spectral range. Despite the fact that both its absorption and emission bands overlap the strong absorption bands of hemoglobin (Figure 5.4(b)), fluorescein has found valuable clinical applications, especially in ophthalmology. Since the 1960s one important ophthalmic application has been mesoscopic imaging of the vasculature of the retina, variously called *ophthalmic photography*, *fundus imaging*, or *fluorescein angiography*. An example is shown in Figure 5.4(c). Such images

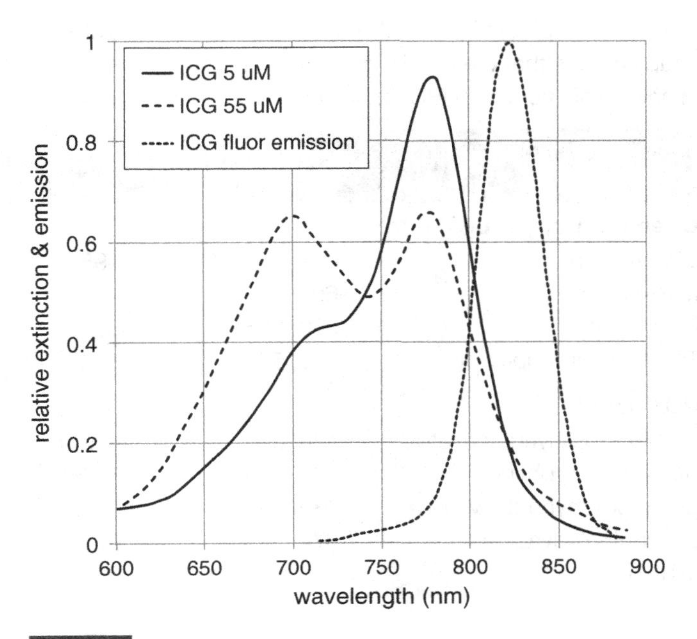

Figure 5.5

Relative extinction coefficients of ICG at two different concentrations in distilled water. The extinction coefficient also changes dramatically when ICG is bound to proteins, such as serum blood albumin. Also shown is the fluorescence emission spectrum for ICG in water. (There are only subtle changes in the fluorescence spectrum with concentration.)

are valuable tools for diagnosing and guiding treatment of various retinal disorders, such as diabetic retinopathy and age-related macular degeneration, which are clearly manifested as major changes of the vascular pattern. Additional ophthalmic applications include topical application of fluorescein to the cornea (and viewing the corneal fluorescence in a darkened room) for finding damage sites on the corneal surface. In these applications, as well as for studying microscopic imaging of vasculature in other tissues, the imaging is not hampered by absorption from hemoglobin, because there is minimal intervening perfused tissue between the imaged vessels and the camera optics.

Indocyanine green (ICG) has also been extensively applied for clinical sensing and imaging. A significant advantage of ICG is that its absorption and emission spectra are centered in the near-infrared (NIR), in the so-called "optical window" of tissue, where absorption due to hemoglobin is much lower than in the visible and ultraviolet. (See Figure 2.A.1.) The absorption peak of ICG in water is typically at ~780 nm, and its fluorescence emission peak is typically at ~820 nm. (See Figure 5.5.) Fluorophores that work in the NIR spectral range can be used as biomarkers for sub-surface imaging and sensing at depths of ~1 cm or deeper.

Table 5.2 Sampling of chemotherapy drugs and photodynamic therapy agents that are readily measured or imaged optically (ALA, aminolevulinic acid; mTHPC, *m*-tetrahydroxyphenylchlorin).

	Absorption maxima (nm)	Emission maxima (nm)
Chemotherapy drugs		
Doxorubicin	479, 496, 529	595
Mitoxantrone	608, 671	680
Mitomycin C	360, 555	600
Taxol – fluorophore tagged	500, 570	610
PDT agents		
Hematoporphyrin derivative	430, 630	625, 700
ALA/protoporphyrin IX	430, 635	620, 634
Benzoporphyrin derivative	430, 690	690, 720
Zinc phthalocyanine	674	679
mTHPC	420, 520, 615	652

The applications of some of these diffuse optics methods will be presented in Chapter 15. For ICG, however, one of the confounding factors in a variety of applications is that the absorption spectrum changes when ICG is bound to proteins in blood (e.g., serum albumin); additionally, as shown in Figure 5.5, the magnitude of the molar extinction coefficient (as well as the spectral shape of the absorption) of ICG changes nonlinearly with concentration, such that the relation between fluorescence intensity and concentration is highly nonlinear. This confounding property is a consequence of intermolecular quenching, whereby fluorophore molecules that are in close proximity can quench each other's emission by non-radiative coupling mechanisms. (Similar concentration effects also occur for some native fluorophores, including NADH.)

In response to these shortcomings of ICG, several companies have been developing other NIR fluorophores for animal studies and pre-clinical research, and it is hoped that some of these will gain FDA approval in the future.

5.4.2 Drugs that are fluorescent

There are also classes of approved chemotherapy agents that are fluorescent, and their biodistributions can therefore be studied by fluorescence imaging. Similarly, many photodynamic therapy (PDT) agents, by design, absorb and emit in the NIR spectral range. Table 5.2 provides a list of some of the more common chemotherapy drugs and PDT agents that can be tracked optically.

5.4.3 Fluorescent biomolecular probes

Over the past two decades a large industry has developed, devoted to the synthesis of fluorescent dyes that can be used to label different biomolecular components of cells and tissues. These fluorescent probes are generally designed to be photostable and to localize, often by hydrogen bonding, within or next to specific analytes, without covalently bonding to them, thus minimally altering the inherent biochemistry of the target biomolecule. Some are designed to exhibit enhanced fluorescence when adhering to hydrophobic regions of proteins. Others are photostable variations of more common dyes, such as coumarins, fluorescein, rhodamine and Texas Red. These "Alexa-Fluor" series of fluorophore dyes have been developed by commercial companies (e.g., Invitrogen, Inc.).

The three principal classes of biomolecules that are targeted with fluorescent biomarkers are proteins, membrane structures and nucleic acids. Protein labeling includes dyes that target a huge variety of bioactive proteins (e.g., enzymes, hormones, contractile and motile proteins, etc.) and structural proteins (e.g., collagen, elastin, etc.). Labels for membranes, which are lipid bilayers, tend to be much smaller structures, often fatty-acid analogs or phospholipids. The human genome program was launched successfully, in part, because of the development of fluorophores specific to DNA bases, with high enough quantum yield to enable single base-pair detection. Additionally, a range of fluorescent probes have been developed for reporting chemical conditions in cells and tissues, such as pH, calcium- and potassium-ion concentrations.

To date, almost all of these molecular probes are limited to laboratory and pre-clinical research, as none of them, as of the time of this publication, have been approved for use in humans, although efforts are under way to seek approval for a few dyes that may prove highly valuable for specific clinical diagnostic applications.

5.4.4 Nanoparticles as fluorescent biomarkers: quantum dots

Over the past decade there has been a great surge of interest in nanoparticles as optical reporters. The greatest amount of development effort on fluorescent particles has been directed to a class of particles that are called *quantum dots*. Quantum dots (QDs) are nanoscale (typically ≤ 10 nm) crystals of semiconductor compounds, whose electrical properties exhibit quantum-mechanical behavior due to "quantum confinement" of the excitons (the coupled electron-hole excitation modes of the semiconductor) to the three-dimensional spatial limits of the particle. In other words, and consequently, the effective semiconductor band-gap is coupled

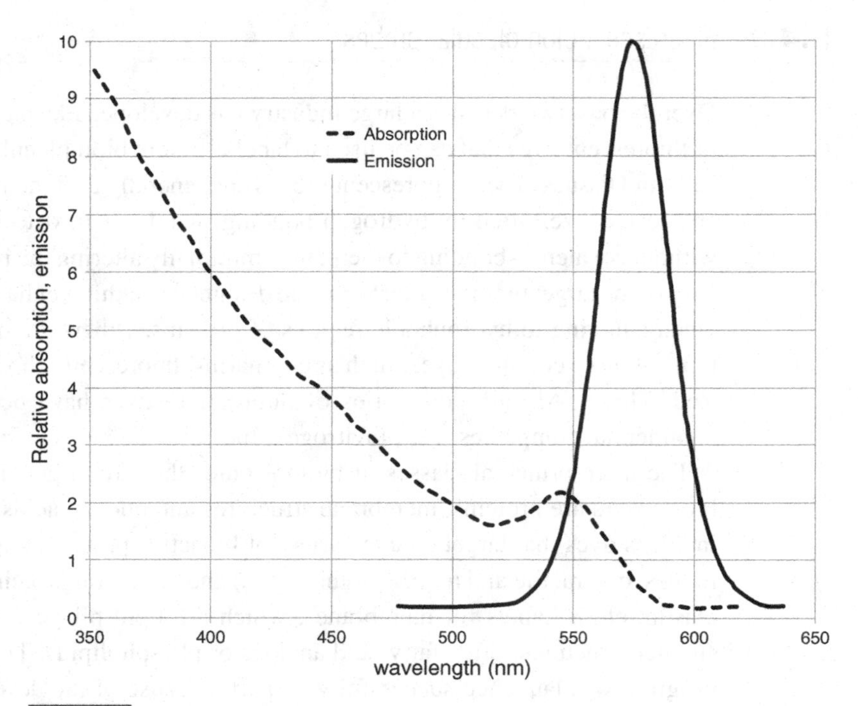

Figure 5.6

Absorption (dashed) and emission (solid) spectra of a quantum dot whose peak emission was tuned to 570 nm. Data extracted from Dennis and Bao, 2008.

to the dimensions and shape of the particle, which can be controlled during synthesis (Murray et al., 2000). The optical absorption cross section per particle can be exceptionally large, leading to values for the effective molar extinction coefficients easily exceeding 10^6 cm^{-1}/M. These values are ~10-times greater than the highest extinction coefficients that can typically be achieved with organic dyes. The strong absorption by QDs is coupled with very high fluorescence quantum yields, commonly ≥80%, again significantly better than values found in organic dyes. Moreover, quantum dots can be tuned to the wavelengths that best suit the application, since both the absorption and emission frequencies are tunable by controlling the nanocrystal dimensions during synthesis (as well as the materials). Figure 5.6 shows the absorption and emission spectra of a QD that has been tuned to an emission wavelength of 570 nm. As can be seen, there is broad absorption in the visible to near-UV spectral range, while emitting at a controlled, narrow range in the visible.

It is no surprise that a great flurry of scientific activity has ensued as a consequence of such a remarkable combination of optical properties. Interest in the potential applications of QDs for biomedical research took a big jump when it was demonstrated that QDs could be made water-soluble (Bruchez et al., 1998). Quantum dots have been shown to be powerful tools in biomedical studies such as imaging of cancer biomarkers, cell trafficking, biodistribution of drug delivery, cell-protein binding, etc. Quantum dots are also especially well suited for use in studies in molecular dynamics like protein-protein binding and enzymatic activity, when used as a component of Förster resonance energy transfer (FRET), which is described in Section 5.8 below.

With researchers looking forward to applications of QDs in humans, the question of toxicity has become a major issue. The earliest and, to date, the most developed quantum dots have been based on the semiconductor cadmium selenide (CdSe) (Ekimov and Onushchenko, 1981). Large quantities of CdSe QDs can be manufactured in colloidal suspension, and the cost has been dropping. Cadmium, however, is a highly toxic heavy metal, and its use in consumer products is restricted in many countries. Quantum dots based on CdSe are still highly useful for in vitro studies with cells (Dahan et al., 2003; Tokumasu et al., 2005; Howarth et al., 2008), and for in vivo studies with small animals (Åkerman et al., 2002; Ballou et al., 2004). To overcome such limitations for biomedical applications, recent research has aimed at developing quantum dots based on less toxic semiconductors (Subramaniam et al., 2012). These, unfortunately, are less optically efficient than CdSe. Another approach has been to design CdSe quantum dots that are prophylactically coated with polymers (Pelley et al., 2009). An additional concern, for any nanoparticle that is eventually aimed at human use, is the rate and mechanism of excretion (Choi et al., 2007). Future research will determine whether toxicity and elimination rates can be in the range that will allow for applications of QDs in humans.

5.4.5 Molecular beacon probes based on oligonucleotide hybridization

Finally, a class of fluorescent reporter probes, for use in laboratory studies and commonly referred to as *molecular beacon probes*, has been developed that is based on oligonucleotide hybridization, i.e., a specific nucleic acid sequence (single-stranded DNA) that can bind to a matching target DNA nucleotide sequence and become fluorescent when bound (Tyagi and Kramer, 1996). The probe is typically 25 nucleotides long, and has a fluorophore bound on one end and a quenching molecule (which stops fluorescence) on the other end of the sequence. Prior to binding of the probe to the target DNA sequence, the probe is folded into a hairpin shape, such that the fluorophore and quencher molecules are adjacent to each other, and no fluorescence can occur. The DNA in the sample under study is denatured thermally, to generate single-stranded DNA, and the probe is introduced, which can then bind to the target sequence, thus unfolding and separating the quencher from the fluorophore. The fluorophore can then emit if illuminated with its excitation wavelength. A more detailed explanation of the interactions between the fluorophore and quencher will be provided in Section 5.8.

5.5 Instrumentation for fluorescence sensing and imaging

The general concept of florescence sensing or imaging is simple: the sample is illuminated with light whose wavelength is in the excitation band of the fluorophore, and fluorescently emitted light is measured or imaged. The main concern is that scattered light of the excitation source must be filtered out before the detector, and background ambient light must also be suppressed. In the case of endogenous fluorophores in tissue, the diffusely backscattered light of the excitation source is typically >100-times stronger than any fluorescence emission from the tissue. In the case of detection through a spectrometer, even though the dispersive element (commonly the diffraction grating) can separate the excitation wavelength from emission wavelengths, in practice if the excitation light is strong, some of it will reflect off surfaces in the spectrometer and "fog" the detector array. (A common challenge in spectrometer design is to minimize stray light.) Wavelength-specific optical filters, which selectively transmit certain wavelengths and reflect others, can be used to block most of the scattered excitation light prior to the detector. If the excitation source is narrowband, as with lasers, the scattered excitation light can be blocked with *long-pass filters* or *notch filters*, which transmit or reflect a narrow band of wavelengths. Ambient light, at the same wavelengths as the fluorescence emission, may still be able to pass through any filters that allow the fluorescence light to reach the detector, and this can also lead to excessive background signal. For clinical applications with, for example, the bright lights in a surgical suite or the illumination light of an endoscope, the ambient light can easily be much brighter than the fluorescence. Consequently, spectroscopy instrumentation must be engineered to suppress the background light. This can often accomplished with temporally gated illumination and/or detection.

5.5.1 Instrumentation for point measurements

Figure 5.7 shows a generalized schematic of the common elements for measuring the fluorescence spectrum on a single point at the surface of a tissue. The large majority of cancers are epithelial in origin (called *carcinomas*) and, as such, start in the superficial layer of most organs. Many of these are readily accessible, as in hollow organs via endoscopes or catheters (e.g., in the gastrointestinal tract, or

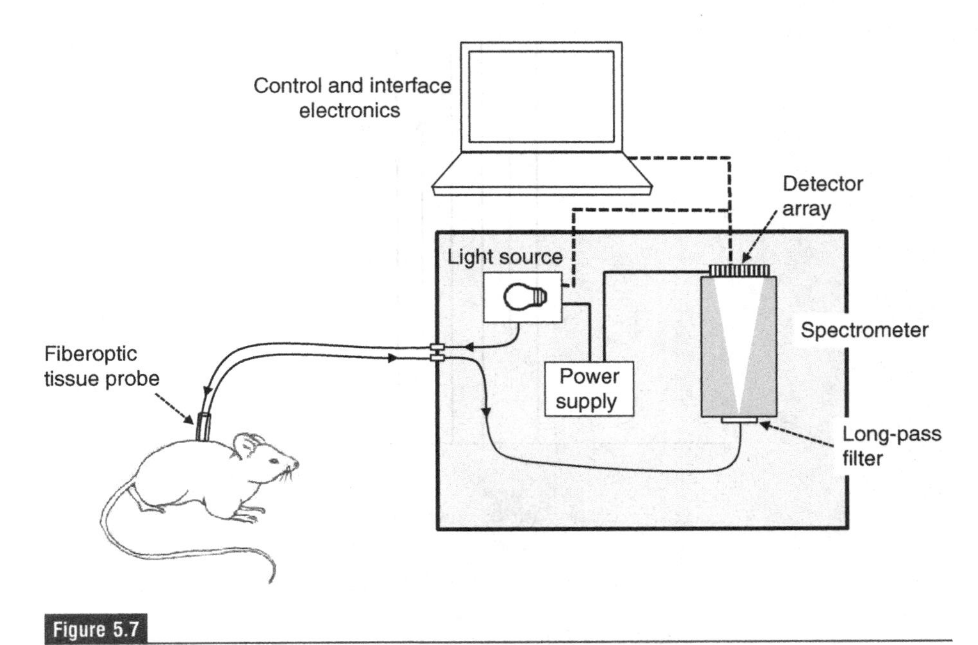

Figure 5.7

Generalized schematic of typical instrumentation for point measurements of tissue fluorescence, using optical fibers for all light transmission.

bladder), or directly accessible (oral cavity, cervix, skin). Consequently, optical fibers often serve as the most convenient method for directing the light in both directions. In the configuration depicted in Figure 5.7 one optical fiber conveys excitation light to the tissue surface, and an adjacent fiber collects some of the fluorescence that diffuses in the tissue near the "source" fiber. This configuration simplifies optical coupling in both directions.

The two-fiber configuration, however, does not optimize efficiency for collecting the diffusely emitted fluorescence light from a tissue surface. The strongest fluorescence will be emitted from the tissue volume experiencing the highest fluence rate of excitation light. When an illuminating fiber is in contact with the surface of a turbid medium, this volume of maximum excitation fluence is, not surprisingly, immediately under the fiber tip. (See Figure 5.8.) Thus, it is more efficient to collect the scattered fluorescence with the same optical fiber that was used for illumination. In such a configuration the excitation light and the collected fluorescence can be separated with a *dichroic beamsplitter.* A typical experimental arrangement of optical elements for a single-fiber probe is depicted in Figure 5.9. A properly designed dichroic beamsplitter (typically a glass or silica substrate with a multilayer dielectric coating) will be able to transmit almost 100% of the excitation wavelength, while reflecting at, say, an angle of 45°, almost 100% of the

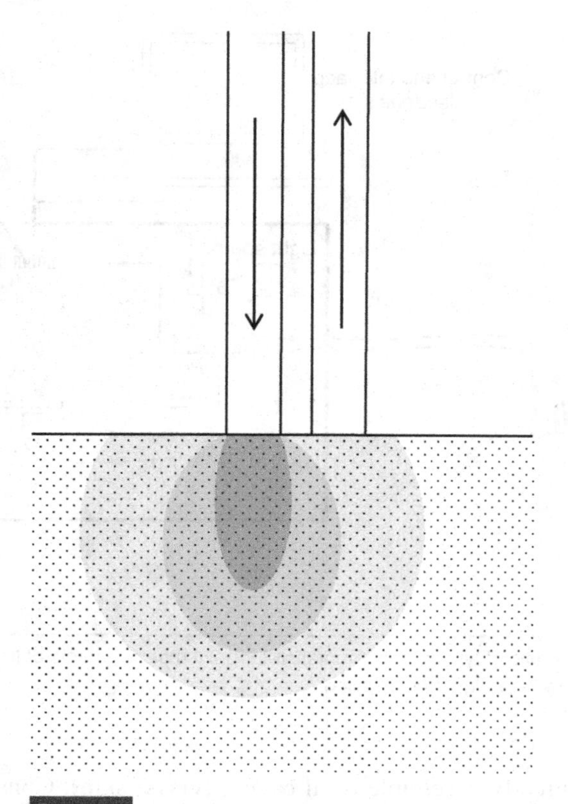

Figure 5.8

The relative distribution of fluorescent fluence rate mimics the distribution of fluence rate from the excitation light source. While a nearby detection fiber can collect emitted fluorescence, the surface area for the highest probability of collecting fluorescence is the same surface where the illuminating fiber itself is located, hence the benefit of a single-fiber configuration as depicted in Figure 5.9.

longer fluorescence wavelengths. In both two-fiber and single-fiber configurations a long-pass filter at the entrance of the spectrometer helps to reject the strong excitation light that is elastically scattered and inevitably collected by the detection fiber.

5.5.2 Instrumentation for fluorescence imaging

One of the advantages of fluorescence measurements is that they can be readily adapted for imaging the distribution of fluorescence from a large field of tissue surface. This approach can be valuable if one wants to determine the biodistribution of a compound. Figure 5.10(a) is a schematic representation of a simple layout for

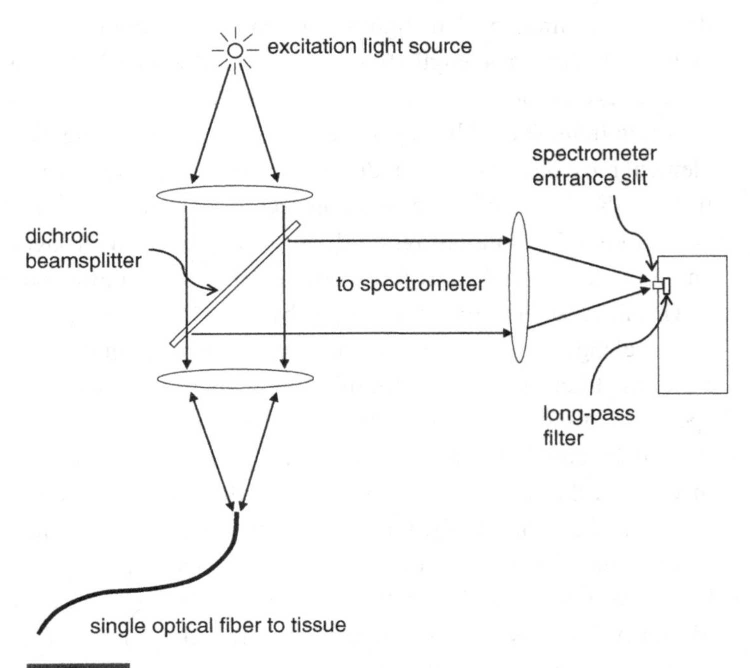

Figure 5.9

Given the wavelength difference between excitation and fluorescence emission, it is often possible to implement measurements with a single fiber, which optimizes the collection efficiency from the surface of a tissue.

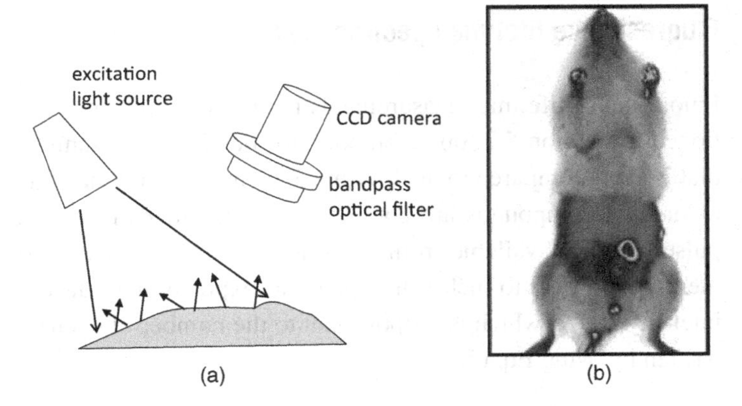

Figure 5.10

(a) Schematic of the simple geometry for imaging the fluorescence emission from sources near the surface of tissue. In vivo imaging of small animals is commonly effected with this type of system. (b) An example of an in vivo fluorescence image of a mouse, exhibiting strong signal from the location of an orthotopic prostate tumor that has been tagged by a fluorophore with emission peak at 800 nm (courtesy Licor, Inc.).

fluorescence imaging. The light source can be a laser or incoherent source of light at the excitation wavelength. If it is a broadband source, bandpass filters can be used to block excitation light that would encroach on the band of emission wavelengths, resulting in unwanted background "noise" at the detecting camera. The detection element must have its own bandpass filter, so that the shorter-wavelength excitation light is also blocked, and only fluorescence emission from the tissue is recorded. Backscattered excitation light will generally be much stronger than fluorescence emission, and would constitute a strong background noise source.

The fluorescence image acquired by the CCD camera will represent a band of wavelengths, which can be narrow or broad, depending on the design of the bandpass filter, but will not, in this simple format, provide information about the spectrum of the fluorescence. If the center wavelength of a bandpass filter before the camera can be tuned, it then becomes possible to record a *hyperspectral image* in which a fluorescence spectrum is recorded for each pixel (or group of pixels) in the image. Commonly, the filter is sequenced through the wavelength range, and a camera image is recorded for each collection wavelength. The result is a large "*data cube*" with a recorded fluorescence spectrum for every image pixel. Similarly, it is possible to generate a hyperspectral image of excitation spectra by sequencing through wavelengths of the excitation source while recording at a fixed wavelength. With fast-frame-rate cameras and the reduced cost of flash memory, hyperspectral imaging is no longer the *tour-de-force* it was a few years ago.

5.6 Fluorescence lifetime spectroscopy

Fluorescence lifetime measurements require time-resolved approaches. As mentioned in Section 5.2, one method is to illuminate the sample with a light pulse that is short compared to the lifetime to be measured. Since fluorescence lifetimes of organic compounds are commonly of the order of nanoseconds, picosecond pulses, readily available from mode-locked lasers, generally satisfy this requirement. In response to such a short pulse, an expression for the detected fluorescence intensity, $I_F(t)$, which is proportional to the number of excited molecules, can be written by using Eq. (5.9):

$$I_F(t) = I_{F0} e^{-t/\tau_F}, \tag{5.11}$$

where I_{F0} incorporates all proportionality factors, including $1/\tau_F$ and the geometric detection factor, from Eq. (5.9). Time-resolved detection of the fluorescence intensity allows for the determination of the fluorescence lifetime as the exponential time constant of the measured intensity decay, which is the mean time associated with the measured fluorescence intensity decay:

$$\tau_F = \frac{\int_0^\infty t I_F(t)dt}{\int_0^\infty I_F(t)dt}. \tag{5.12}$$

This method can also be referred to as the *time-domain method.*

Another method for lifetime measurements is based on illuminating the sample with light that is intensity-modulated at an angular frequency ω, which is generally called the *frequency-domain method.* The angular modulation frequency should be such that $\omega\tau_F \sim 1$, so that the delay associated with fluorescence emission translates into a measurable phase shift between the modulated illumination and the modulated fluorescence emission. Since τ_F is often in the range 1–10 nanoseconds, typical modulation frequencies, $\omega/(2\pi)$, are ~100 MHz. In the frequency-domain approach, one measures a sinusoidal modulation of fluorescence intensity, $I_F(\omega)$, which is characterized by an average value, $I_{F,\mathrm{DC}}$, a modulation amplitude, $I_{F,\mathrm{AC}}(\omega)$, and a phase, θ_F:

$$I_F(\omega) = I_{F,\mathrm{DC}} + I_{F,\mathrm{AC}}(\omega)\sin\left[\omega t + \theta_F(\omega)\right]. \tag{5.13}$$

A fluorescence modulation depth factor, $m_F(\omega)$, is defined as the ratio of the amplitude of the fluorescence intensity oscillation to its average value:

$$m_F(\omega) = \frac{I_{F,\mathrm{AC}}(\omega)}{I_{F,\mathrm{DC}}}. \tag{5.14}$$

The *frequency-domain method* is based on measuring the phase and the modulation of the fluorescence intensity and is, consequently, also referred to as the *phase-modulation method.* If one assumes that the modulation depth of the excitation light source is 100%, then it is possible to express the fluorescence lifetime in terms of either the phase or the detected modulation depth of the fluorescence, and subscripts θ and m are introduced to specify the two cases:

$$\tau_{F,\theta} = \frac{\tan\left[\theta_F(\omega)\right]}{\omega}, \tag{5.15}$$

$$\tau_{F,m} = \frac{\sqrt{1 - m_F^2}}{\omega m_F}. \tag{5.16}$$

5.6.1 Measurement of multiple fluorophores

The ability of fluorescence lifetime measurements to detect and characterize more than one fluorophore in the same tissue areas is a major advantage over

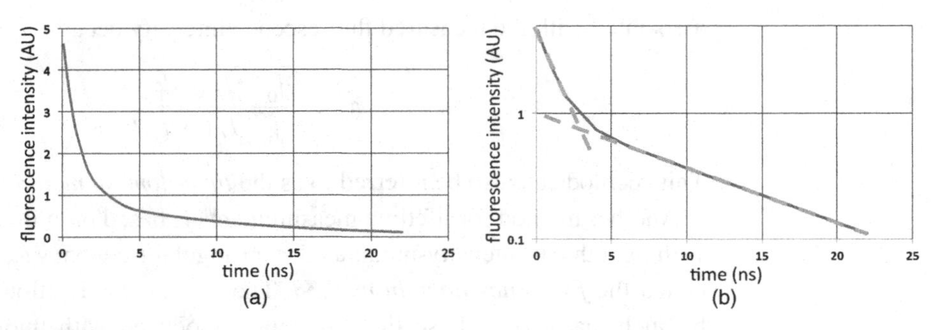

Figure 5.11

Two plots of detected fluorescence intensity for the case of a dual-exponential decay: (a) linear plot; (b) semi-log plot. The decay constants for the two components are 1.0 ns and 10 ns, with the concentration (or quantum efficiency) of the shorter-decay species being 4-times larger than that of the longer-decay species. The intensity is in arbitrary units. The decay rates shown are typical of many organic fluorophores.

continuous-wave fluorescence measurements. If the fluorescence emission bands are sufficiently separated, then detecting the individual fluorophores can be accomplished by simple spectral filtering (or dispersing) at the detector. On the other hand, if the fluorescence wavelengths of different fluorophores overlap, they cannot be separated spectrally by simple methods. Nonetheless, in such cases it may still be possible to separately detect the fluorophores if they exhibit sufficiently different values of their excited-state lifetimes. Alternatively, a given fluorophore may exhibit a different emission lifetime when bound to a target molecule, and valuable information may be obtained by determining the relative amounts of bound vs. unbound fluorophore.

In the case of multiple fluorophores, the time-domain Eq. (5.11) can be replaced with a summation of single exponential terms associated with each fluorophore, and Eq. (5.12) only yields an apparent fluorescence lifetime. In the frequency domain, the two expressions for the fluorescence lifetime, Eqs. (5.15) and (5.16), which coincide for a single-exponential decay, yield different values for multiple fluorophores. More sophisticated methods must be introduced to measure multiple lifetimes, and these typically involve multi-exponential fitting procedures in the time domain, or measurements at multiple modulation frequencies in the frequency domain.

In Figure 5.11, a dual exponential decay of fluorescence power is plotted for times following excitation by a short pulse of excitation light. (The fluorescence intensity is in arbitrary units.) The plots exemplify a case of two excited-state lifetimes, differing by a factor 10. The plot on the left is linear, but with a factor 10 difference in decay rates it is already evident by inspection that there is more than one rate. When plotted in semi-log mode, however, the two components stand out and the early and late times can be fit to different straight lines.

5.6.2 Example applications of fluorescence lifetime measurements

Fluorescence lifetime measurements can be used to monitor the binding of fluorescent molecules, both endogenous and exogenous, to proteins and other biomolecules. For example, a very important compound found in all cells, and a key facilitator of cellular metabolism, is the coenzyme nicotinamide adenine dinucleotide (NAD). As mentioned in Section 5.3, its reduced form, NADH, is fluorescent, and the fluorescence lifetime is strongly dependent on whether the molecule is bound to proteins. Whereas the lifetime is 0.4 ns for NADH in aqueous solution, when bound to dehydrogenases (enzymes that participate in cellular redox reactions) the fluorescence lifetime can be longer than 8 ns. Thus, measurement of the NADH fluorescence lifetime has proven to be a valuable method for studying cellular metabolic functions (Jameson et al., 1989; Lakowicz et al., 1992).

Similarly, ethidium bromide (EtBr) is a fluorescent dye that is commonly used as a non-specific tag for DNA (or other nucleic acids), because it readily intercalates (locates itself between helical or planar layers) into the DNA structure. The bright orange emission of EtBr facilitates microscopic imaging and gel electrophoresis measurements of DNA. Higher-quality imaging is facilitated by taking advantage of the fluorescence lifetime contrast, given that unbound EtBr exhibits a fluorescence lifetime of 1.7 ns, whereas DNA-intercalated EtBr has a lifetime of $\sim$23 ns (Heller and Greenstock, 1994). Thus, by simply delaying the imaging of the sample by $\sim$5 ns following a short-pulse excitation, even small amounts of dye-tagged DNA will stand out against a background of free dye.

5.7 Polarization and anisotropy of fluorescence emission

For the geometry depicted in Figure 5.8, with diffusely scattered excitation light and emitted fluorescence, the collected light will carry no molecular directional information. Similarly for the geometries depicted in Figure 5.9 (even without the coupling fiber) and Figure 5.10, molecules suspended in a fluid will be randomly oriented; in that case, if the excitation light is unpolarized the fluorescence emission will be isotropic, again losing any molecular directional information. For an individual molecule, however, the probability of excitation (and, hence, consequent fluorescence emission) is determined by the degree of alignment of the molecular dipole moment with the optical electric field of the excitation light. With linearly polarized light, the probability of absorption is proportional to $\cos^2\theta$, where θ is the angle between the optical electric field vector and the molecular dipole moment. Moreover, the emission of that dipole will be polarized (maximally for detection in a plane normal to the axis of the radiating dipole) and exhibit directional variation.

Referring again to Figures 5.9 and 5.10, the same experimental configurations can be modified with the addition of linear polarizers to the light source and detection arms of each setup. In such an experiment one can define the polarization of the fluorescence emission as:

$$p_f = \frac{I_\parallel - I_\perp}{I_\parallel + I_\perp} \tag{5.17}$$

for which the reference frame, for defining the parallel and perpendicular designations of the polarization of emitted fluorescence intensities ($I_\parallel$ and $I_\perp$, respectively), is the axis of the linear polarization of the excitation light, regardless of the propagation direction, and is therefore a measure that varies with the specific measurement geometry. As will be discussed in more detail in Section 7.2.1, the emission pattern of a radiating dipole depends on the polarization component that is measured. For the discussion here, we are considering only the degree of polarization of the detected fluorescence relative to a linearly polarized excitation source.

A number of papers in the literature cite the fluorescence anisotropy, A_f, which is defined slightly differently as:

$$A_f = \frac{I_\parallel - I_\perp}{I_\parallel + 2I_\perp}. \tag{5.18}$$

By these definitions, given that either p_f or A_f can be defined in terms of the other, the information content for either is equivalent, and choice of usage is somewhat arbitrary, generally determined by prior convention in different strings of publications.

Even if fluorescent molecular dipoles are randomly oriented, as in a fluid, emitted fluorescence will exhibit a degree of polarization if the excitation is polarized. As illustrated in Figure 5.12, in a sample with randomly oriented molecular dipoles, linearly polarized excitation light will preferentially excite those dipoles that are oriented closer to the parallel with the excitation electric field direction. Each of those molecules will then radiate in a pattern determined by the axis of its transition dipole, which, initially following excitation, will be preferentially oriented parallel to the excitation polarization.

We remind the reader that for measurements of fluorescence from a fluorophore in a turbid medium, scattering will degrade the degree of polarization for both the

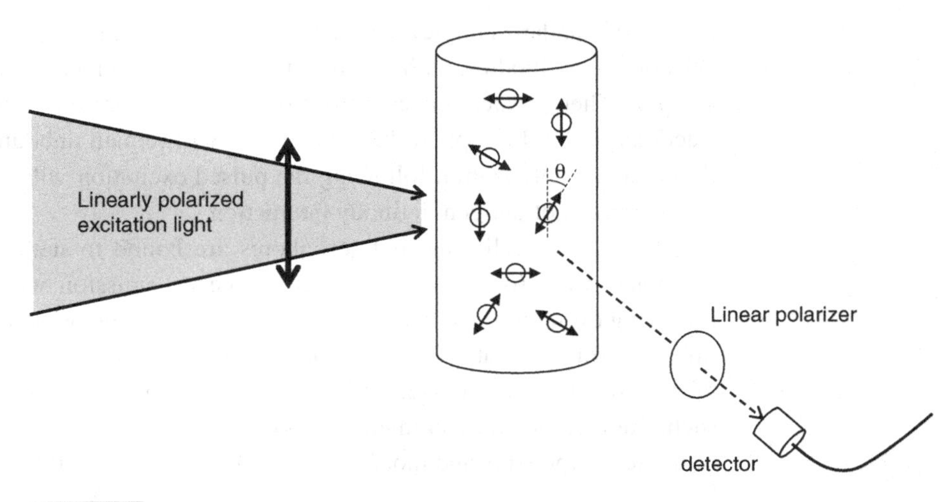

Figure 5.12

Randomly oriented molecules with an absorbing dipole are illuminated by a linearly polarized (vertical, in this illustration) excitation optical field. Those molecules with their dipoles oriented closer to vertical will be more strongly excited.

excitation and emitted light, limiting the information content related to molecular orientation.

5.7.1 Combining polarization and time-dependent fluorescence measurements

By combining a polarization measurement with a time-dependent ("lifetime") measurement of fluorescence, i.e., by measuring the temporal decay of fluorescence polarization following a pulsed excitation, useful information can be gleaned about the interactions between the fluorescent molecules and the medium in which they are suspended. Although a short pulse of linearly polarized excitation light will interact most strongly with molecules whose absorption dipoles are closer to parallel with the excitation polarization, if the fluorophores can rotate during the fluorescence lifetime then the degree of polarization of the emission will decrease with time following the excitation (in addition to the temporal decay of the overall emission intensity). However, if the fluorophores are bound to larger molecules, whose motion and rotation is much slower than that of the free fluorophores, then the degree of polarization will persist longer during the emission lifetime.

This phenomenon is the basis for, among other diagnostic tests, well-established clinical laboratory immunoassays of human serum or blood (Jolley et al., 1981). A fluorophore-labeled analyte specific to binding sites on antibodies is added to a serum sample, at a molar concentration that has been determined to be optimum by a series of standard sample studies. The labeled analyte then competes, for binding sites on the antibodies, with unlabeled analyte (the antigen) that is present in the sample. When greater concentrations of antigen are present in the sample, larger fractions of the fluorophore-labeled analyte will remain unbound. The degree of fluorescence polarization following the pulsed excitation will then be a measure of the amount of antigen-antibody interaction.

It follows logically that if fluorophores are bound to static tissue structures, the polarization (or anisotropy) of fluorescence emission will remain constant following excitation, but may be a strong function of the orientation of the excitation polarization relative to the tissue structural geometry. (The accuracy of such measurements is, of course, limited for measurements in a turbid medium.) As such, fluorescence measurements can elucidate structure and structural dynamics of tissue components and motile proteins (Forkey et al., 2003).

5.8 Sensing molecular dynamics with fluorescent biomarkers: FRET

When the energy of an optical electronic transition in one molecule is close to that of another molecule, the transitions are said to be "resonant" with each other. If, in addition to being energetically resonant, two molecular entities are in close enough proximity that their dipole moments can interact directly, energy can be transferred from one to the other without the emission and absorption of a photon. This process, which is fundamentally a set of variations of the implementation of quenching mentioned for molecular beacons in Section 5.4.5, is called *Förster resonance energy transfer* (FRET) (named after German scientist Theodor Förster [1910–1974], who first described the process of non-radiative coupling between two chromophores [Förster, 1948]). Since this phenomenon is usually based on two fluorescent chromophores, and the efficiency of energy transfer is typically obtained from fluorescence measurements, the "F" in FRET is often taken to indicate fluorescence. It is important to note, however, that fluorescence is not involved in the process of energy transfer between the two chromophores. The chromophores are not limited to molecules: nanoparticles that fluoresce or absorb light can also be participants in FRET. What is important is that the FRET process refers to the non-optical transfer of energy between two resonant entities, when they are close enough for dipole-dipole coupling, generally a few nanometers.

In essence, FRET happens by direct dipole-dipole resonant coupling between a *donor* and an *acceptor*, wherein energy is transferred resonantly from the donor in its excited state to the acceptor in its ground state. The efficiency of direct energy transfer is proportional to the inverse sixth power of the distance between the donor and acceptor, which means that the process is extremely sensitive to that distance. When a fluorophore is optically excited, the relative probability, P_t, that it will transfer its energy to a nearby acceptor, rather than emit a photon, can be described by the following proportionality:

$$P_t \propto \frac{\mathcal{J}\kappa^2}{R^6}, \tag{5.19}$$

where $\mathcal{J}$ is the overlap integral of the emission spectrum of the donor and the absorption spectrum of the acceptor, κ^2 is a geometric factor representing the relative orientation of the two dipoles, and R is the donor-acceptor distance. While the relative orientation of the two dipole moments is an important factor, the FRET efficiency is dominated by the donor-acceptor distance. The *Förster distance*, R_F, is defined as the distance between donor and acceptor for which the FRET efficiency is 50%, meaning that half of the donor energy is transferred by non-radiative mechanisms, and half is emitted as fluorescence photons.

The value of FRET in studying biomolecular processes is that the method can be employed to enable monitoring of enzymatic activity and, more generally, mechanisms of molecular binding or the moving apart of biomolecular structures. Typically, at least one of the entities is a fluorescent protein, which, for example, may be used as a marker for a binding site on a cellular surface. Among other applications, FRET can be used to quantify the density of binding sites on the membrane surface of a cell. High density would correspond to closer average distance between nearest sites. FRET can be used to monitor the distance between donors and acceptors in several ways:

1. The donor is excited optically and emits fluorescence, and the fluorescence is significantly quenched when an acceptor binds to the donor. The binding event is sensed as a drop in fluorescence power.
2. Conversely, the donor fluorescence is initially quenched by the acceptor, and donor fluorescence increases when the entities are cleaved, say by protease activity (or, in the case of molecular beacons, by oligonucleotide hybridization). The cleavage is sensed as an increase in measured donor fluorescence.
3. The acceptor is excited by FRET from the donor, and emits fluorescence only when the two are bound. The acceptor fluorescence is at a different (longer) wavelength than the donor fluorescence. This option is illustrated pictorially in Figure 5.13. In this case, one is initially measuring fluorescence at the emission wavelength of the donor, and binding is indicated by a shift in the fluorescence power to the longer wavelength of the acceptor emission.
4. The ratio of detected fluorescence wavelengths (donor emission relative to acceptor emission) can indicate the percentage of binding of either species. In general, there is a continuous range of the ratio between donor emission and acceptor emission, depending on the binding strength, and the relative concentrations of the two species. Measurement of the ratio of emission at the two wavelengths can be used as an indicator of, say, enzymatic activity.

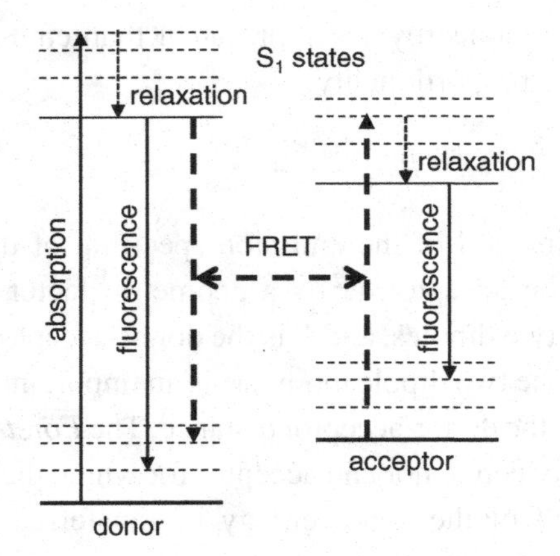

Figure 5.13

Energy-level diagram of FRET interaction between donor and acceptor species, depicting the interaction option 3 described in the text. The sequence of energy-transition events, generally, proceeds from left to right, except that the FRET exchange can occur faster than fluorescence emission from the donor species when the separation distance is small enough.

Quantum dots, described in Section 5.4.4, can serve as ideal donors for FRET applications. The strong blue-UV absorption spectrum for excitation of QDs enables the use of an excitation wavelength that does not overlap the absorption spectrum of the acceptor molecule, thus minimizing unwanted background caused by direct excitation of the acceptor. Moreover the QD can still be tuned so that its emission overlaps ideally with the excitation spectrum of the target acceptor (Dennis and Bao, 2008). As QDs are being developed with "coatings" to minimize toxicity, while also being functionalized for targeting/binding of cellular membranes, proteins, etc., FRET is emerging as one of the more powerful tools in cellular/molecular biology.

Problems – answers to problems with * are on p. 646

5.1* One of the most important vibrational modes in biological molecules is the C=O stretch mode of the amide group in proteins. (See Chapter 6.) The precise energy of this mode varies with other factors of molecular structure, but is generally in the ~0.16 eV range. Relative to the ground level, how many amide molecular groups are likely to be found with a vibrational excited amide-I mode at body temperature (37°C)?

5.2* The temperature dependence of the non-radiative decay rate of a specific fluorescent molecule is such that non-radiative decay is twice as fast at body temperature (37°C) as it is at room temperature (20°C). When the fluorescence of a solution of the dye is measured in the lab at 20°C, the lifetime is determined to be 3 ns; when the same dye solution is injected in a mouse, the fluorescence lifetime is measured to be 2 ns. What is the intrinsic fluorescence lifetime of the electronic transition for that dye?

5.3 For the values given in Problem 5.2, what is the quantum yield of the fluorophore at the two temperatures?

5.4 An experimenter applies ethidium bromide to a cell culture to facilitate imaging of the DNA. After equilibration, in the resulting sample 50% of the EtBr is intercalated in DNA and 50% is in free solution. Following a short-pulse excitation of UV light, how long a delay is required until the background fluorescence (from the free EtBr) is less than 1% of the total fluorescence intensity? (Use the lifetimes stated in Section 5.6.2.)

5.5* A small volume of randomly oriented fluorescent molecules is illuminated with unpolarized light through the objective lens of a microscope. The numerical aperture of the lens is 1.0. What is the maximum percentage of the emitted fluorescence that can be collected by the same lens?

5.6 The excitation light source that is illuminating a sample for a fluorescence lifetime measurement is intensity modulated at 100 MHz.

(a) The phase lag of the measured fluorescence emission is found to be $\pi/4$. What is the effective fluorescence lifetime of the fluorophore?

(b) For a different fluorophore the modulation depth of the emission is measured to be 10%. What is the effective fluorescence lifetime of this fluorophore?

5.7* When proteins "dock" with drug molecules, the distance between the two can be affected by the level of hydration of the protein molecule, since water can act as a buffer between the two molecules. If the Förster distance for a functionalized quantum dot and a fluorescent protein is 2 nm, at what distance will the FRET efficiency be 90%?

References

Åkerman, M. E., Chan, W. C., Laakkonen, P., Bhatia, S. N., and Ruoslahti, E. (2002). Nanocrystal targeting *in vivo*. *Proceedings of the National Academy of Sciences USA*, 99, 12617–12621.

Auger, A., Samuel, J., Poncelet, O., and Raccurt, O. (2011). A comparative study of non-covalent encapsulation methods for organic dyes into silica nanoparticles. *Nanoscale Research Letters*, 6, 328.

Ballou, B., Lagerholm, B. C., Ernst, L. A., Bruchez, M. P., and Waggoner, A. S. (2004). Noninvasive imaging of quantum dots in mice. *Bioconjugate Chemistry*, 15, 79–86.

Banerjee, B., Renkoski, T., Graves, L. R., et al. (2012). Tryptophan autofluorescence imaging of neoplasms of the human colon. *Journal of Biomedical Optics*, 17, 016003.

Brewer, M., Utzinger, U., Silva, E., et al. (2001). Fluorescence spectroscopy for in vivo characterization of ovarian tissue. *Lasers in Surgery and Medicine*, 29, 128–135.

Bruchez, M., Moronne, M., Gin, P., Weiss, S., and Alivisatos, A. P. (1998). Semiconductor nanocrystals as fluorescent biological labels. *Science*, 281, 2013–2016.

Chang, S. K., Mirabal, Y. N., Atkinson, E. N., et al. (2005). Combined reflectance and fluorescence spectroscopy for in vivo detection of cervical pre-cancer. *Journal of Biomedical Optics*, 10(2), 024031.

Choi, H. S., Liu, W., Misra, P., et al. (2007). "Renal clearance of quantum dots." *Nature Biotechnology*, 25(10), 1165–70.

Dahan, M., Lévi, S., Luccardini, C., et al. (2003). Diffusion dynamics of glycine receptors revealed by single-quantum dot tracking. *Science*, 302(5644), 442–5.

Dawson, R. B., Elliott, D. C., Elliott, W. H., and Jones, K. M. (1989). *Data for Biochemical Research*, 3rd edn. Oxford: Clarendon Press.

Dennis, A. M., and Bao, G. (2008). Quantum dot-fluorescent protein pairs as novel fluorescence resonance energy transfer probes. *Nano Letters*, 8, 1439–1445.

Ekimov, A. I., and Onushchenko, A. A. (1981). Quantum size effect in three-dimensional microscopic semiconductor crystals. *JETP Letters*, 34, 345–349.

Forkey, J. N., Quinlan, M. E., Shaw, M. A., Corrie, J. E., and Goldman, Y. E. (2003). Three-dimensional structural dynamics of myosin V by single-molecule fluorescence polarization. *Nature*, 422(6930), 399–404.

Förster, T. (1948). Zwischenmolekulare Energiewanderung und Fluoreszenz [Intermolecular energy migration and fluorescence]. *Annalen der Physik* (in German), 437, 55–75.

Georgakoudi, I., Jacobson, B. C., Mueller, M. G., et al. (2002). NAD(P)H and collagen as quantitative fluorescent biomarkers of epithelial precancerous changes. *Cancer Research*, 62, 682–687.

Heller, D. P., and Greenstock, C. L. (1994). Fluorescence lifetime analysis of DNA intercalated ethidium bromide and quenching by free dye. *Biophysical Chemistry*, 50, 305–312.

Howarth, M., Liu, W., Puthenveetil, S., et al. (2008). Monovalent, reduced-size quantum dots for imaging receptors on living cells. *Nature Methods*, 5, 397–399.

Jameson, D. M., Thomas, V., and Zhou, D. M. (1989). Time-resolved fluorescence studies on NADH bound to mitochondrial malate dehydrogenase. *Biochimica et Biophysica Acta*, 994, 187–190.

Jolley, M. E., Stroupe, S. D., Wang, C. H. J., et al. (1981). Fluorescence polarization immunoassay I: Monitoring aminoglycoside antibiotics in serum and plasma. *Clinical Chemistry*, 27(7), 1990–1997.

Lakowicz, J. R., Szmacinski, H., Nowaczyk, K., and Johnson, M. L. (1992). Fluorescence lifetime imaging of free and protein-bound NADH. *Proceedings of the National Academy of Sciences USA*, 89(4), 1271–1275.

Mayevsky, A., and Chance, B. (2007). Oxidation–reduction states of NADH *in vivo*: From animals to clinical use. *Mitochondrion*, 7(5), 330–339.

Müller, M. G., Georgakoudi, I., Zhang, Q., Wu, J., and Feld, M. S. (2001). Intrinsic fluorescence spectroscopy in turbid media: Disentangling effects of scattering and absorption. *Applied Optics*, 40, 4633–4646.

Müller, M. G., Valdez, T., Georgakoudi, I., et al. (2003). Spectroscopic detection and evaluation of morphologic and biochemical changes in early human oral carcinoma. *Cancer*, 97, 1681–1692.

Murray, C. B., Kagan, C. R., and Bawendi, M. G. (2000). Synthesis and characterization of monodisperse nanocrystals and close-packed nanocrystal assemblies. *Annual Review of Materials Research*, 30(1), 545–610.

Pelley, J. L., Daar, A. S., and Saner, M. A. (2009). State of academic knowledge on toxicity and biological fate of quantum dots. *Toxicological Sciences*, 112(2), 276–296.

Ramanujam, N. (2000). Fluorescence spectroscopy in vivo. In *Encyclopedia of Analytical Chemistry*, ed. R. A. Meyers. Hoboken, NJ: John Wiley, pp. 20–56.

Skala, M. C., Riching, K. M., Gendron-Fitzpatrick, A., et al. (2007). *In vivo* multiphoton microscopy of NADH and FAD redox states, fluorescence lifetimes, and cellular morphology in precancerous epithelia. *Proceedings of the National Academy of Sciences USA*, 104, 19494–19499.

Subramaniam, P., Lee, S. J., Shah, S., Patel, S., Starovoytov, V., and Lee, K. B. (2012). Generation of a library of non-toxic quantum dots for cellular imaging and siRNA delivery. *Advanced Materials*, 24(29), 4014–4019.

Thekkek, N., and Richards-Kortum, R. (2008). Optical imaging for cervical cancer detection: solutions for a continuing global problem. *Nature Reviews Cancer*, 8, 725–731.

Tokumasu, F., Fairhurst, R. M., Ostera, G. R., et al. (2005). Band 3 modifications in Plasmodium falciparum-infected AA and CC erythrocytes assayed by autocorrelation analysis using quantum dots. *Journal of Cell Science*, 118, 1091–1098.

Tyagi, S., and Kramer F. R. (1996). Molecular beacons: probes that fluoresce upon hybridization. *Nature Biotechnology*, 14, 303–308.

Zuluaga, A. F., Utzinger, U., Durkin, A., et al. (1999). Fluorescence excitation emission matrices of human tissue: a system for in vivo measurement and method of data analysis. *Applied Spectroscopy*, 53, 302–311.

Further reading

Fluorescence spectroscopy

Lakowicz, J. R. (2006). *Principles of Fluorescence Spectroscopy*, 3rd edn. Springer: New York.

Ramanujam, N. (2000). Fluorescence spectroscopy in vivo. In *Encyclopedia of Analytical Chemistry*, ed. R. A. Meyers. Hoboken, NJ: John Wiley, pp. 20–56.

Wagnieres, G. A., Star, W. M., and Wilson, B. C. (1998). Review: *in vivo* fluorescence spectroscopy and imaging for oncological applications. *Photochemistry and Photobiology*, 68(5), 603–632.

Fluorescent probes

Johnson, I., and Spence, M. T. Z., eds. (2010). *The Molecular Probes Handbook – A Guide to Fluorescent Probes and Labeling Technologies*, 11th edn. Carlsbad, CA: Life Technologies.

Molecular beacon probes (oligonucleotide hybridization probes)

Goel, G., Kumar, A., Puniya, A. K., Chen, W., and Singh, K. (2005). A review – Molecular beacon: a multitask probe. *Journal of Applied Microbiology*, 99, 435–442.

Fluorescence lifetime imaging

Berezin, M. Y., and Achilefu, S. (2010). Fluorescence lifetime measurements and biological imaging. *Chemical Reviews*, 110, 2641–2684.

Digman, M. A., Caiolfa, V. R., Zamai, M., and Gratton, E. (2008). The phasor approach to fluorescence lifetime imaging analysis. *Biophysical Journal*, 94, L14-L16.

Lakowicz, J., Szmacinski, H., Nowaczyk, K., Berndt, K. W., and Johnson, M. (1992). Fluorescence lifetime imaging. *Analytical Biochemistry*, 202(2), 316–330.

FRET

Clegg, R. M. (2007). The history of FRET: from conception through the labors of birth. In *Reviews in Fluorescence*, ed. C. D. Geddes and J. R. Lakowicz. Springer, New York, pp. 1–45.

Medintz, I., and Hildebrandt, N., eds. (2014). *FRET-Förster resonance energy transfer: From theory to applications.* Wiley-VCH, Weinheim.

Helms, V. (2008). Fluorescence resonance energy transfer. In *Principles of Computational Cell Biology.* Weinheim: Wiley-VCH, p. 202.

Wu, P. and Brand, L. (1994). Resonance energy transfer: methods and applications. *Analytical Biochemistry*, 218, 1–13.

FRET and quantum dots

Medintz, I. L., Clapp, A. R., Brunel, F. M., et al. (2006). Proteolytic activity monitored by fluorescence resonance energy transfer through quantum-dot-peptide conjugates. *Nature Materials*, 5(7), 581–589.

6 Raman and infrared spectroscopy of vibrational modes

In Chapter 4 we introduced the concepts of IR-absorption spectroscopy and Raman spectroscopy in general terms. Both are methods that enable study of the vibrational modes of molecules. Here we explore these quantitatively. While IR-absorption spectroscopy is frequently used in the fields of chemistry and biochemistry, Raman spectroscopy has proven more practical for biomedical applications, especially for in vivo measurements, thanks to the absence of interference from water absorption in the visible range, and to the convenience of using fiberoptic components, which are generally not available for the mid-IR wavelengths associated with the vibrational modes that generate IR-absorption bands. (Raman scattering can also result from the interaction of light with the rotational modes of a molecule. This type of scattering, however, is not addressed here because molecular rotation is only significant in the gaseous state, whereas with tissue we are dealing with condensed matter.)

The Raman effect was discovered in 1928 by Indian physicist Sir Chandrasekhara Venkata Raman [1888–1970]. Raman's early experiments (Raman and Krishnan, 1928), conducted long before the development of lasers, utilized sunlight transmitted through a violet spectral filter as a narrowband light source. This narrowband light was directed through a transparent material being studied and then to a spectrograph, where it featured additional spectral lines, slightly red-shifted from the illumination wavelength. Raman presented his explanation of the phenomenon at a scientific conference in Bangalore, India, and received the Nobel Prize in physics, remarkably a mere two years later, in 1930.

6.1 Vibrational modes in biological molecules

The number of possible different vibrational modes in a biological molecule is truly large. Each of N atoms in a molecule has three degrees of freedom, which results in $3N$ possible variations of motion. The number can be reduced slightly for biomolecules in liquid or solid media (condensed matter), where we can ignore translational and rotational degrees of freedom, leading to $3N-6$ possible internal

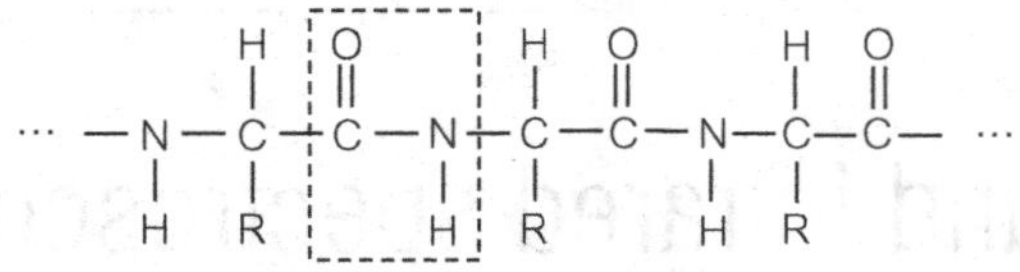

Figure 6.1

Amino acids are linked by peptide (amide) groups, one of which is denoted in the dashed box, to form the backbone of most structural proteins. The "R" designates residues, which are complexes (such as methyl groups, CH_3, and side chains) that define the individual amino acids.

degrees of freedom, or vibrational modes. This translates to a lot of detailed spectral structure for a biological molecule, mostly in the near-IR to IR range, which is why the infrared or Raman spectra of biomolecules are sometimes referred to as spectral "fingerprints."

A typical protein molecule may have more than 20,000 normal modes of vibration, with many overlapping in frequency. Consequently, one might fear that attempts to decipher vibrational spectra would be futile. Nonetheless, despite the large number of possible assignments for spectral features relating to vibrational modes in a biomolecule, in practice some modes are more prominent spectroscopically than others, especially in important polymeric molecules (e.g., proteins, DNA), because specific sub-groups of these biopolymers are repeated many times. For example, the *peptide group*, seen highlighted in Figure 6.1, is a core group that links amino acids, building chains of polypeptides, which are the type of polymers that form the backbones of proteins. Consequently, the vibrational modes of peptide groups (called the *amide modes*) are frequently strong spectral features, given that hundreds of identical bonds in a single molecule can contribute to the signal.

We note that the most common unit used to specify the energy or frequency of a vibrational transition is *wavenumbers* (cm^{-1}), most commonly as the plural. The wavenumbers, sometimes denoted by the symbol $\bar{\nu}$, is defined as the inverse of the vacuum wavelength (λ) of the electromagnetic wave (i.e., $\bar{\nu} = 1/\lambda$). Its name indicates that it represents the number of wavelengths per unit length along the direction of propagation of the electromagnetic wave. There are two versions of the term wavenumber(s) in common use, differing by the factor 2π. The version utilized here, $\bar{\nu}$, is sometimes referred to as the "spectroscopy version" of wavenumbers, and is generally pluralized. In Chapter 4, the term wavenumber (singular) was also introduced, referring to the number of *radians* of phase per unit length (in the direction of propagation), represented by the symbol k, where $k = 2\pi/\lambda$. This "physics version" of wavenumber is also commonly used, especially in the more general formalisms of electromagnetic waves, but may be more precisely called *angular wavenumber*.

It should be noted that with Raman scattering there is no actual electromagnetic wave propagating at the vibrational frequency, but wavenumbers is used to denote the energy difference between the two states of the vibrational transition or, equivalently, the energy difference between an incident photon and a Raman-scattered photon. Thus, for Raman scattering, where the incident light has wavelength λ_0, and the scattered light has wavelength λ_s, the frequency (which is proportional to the energy) of the vibrational transition is expressed in terms of the wavenumbers shift (often simply called the *Raman shift*), which is the difference of the wavenumbers between the incident light and scattered light:

$$\Delta\bar{\nu} = \left(\frac{1}{\lambda_0} - \frac{1}{\lambda_s}\right), \tag{6.1}$$

for which one expresses the values of λ_0 and λ_s in cm. This nomenclature has persisted over the years, despite the fact that the wavenumbers does not directly have units of energy, but is proportional to the energy (E) through the product of Planck's constant, h, times the speed of light in vacuum, c.

6.1.1 Frequencies of some biomolecular vibrational modes

Most of the normal modes of biomolecules are couplings of various individual stretch motions and bending motions of a small group of neighboring atoms, but sometimes a single stretch mode dominates a group, albeit with minor perturbations induced by the coupled motions of other nearby atoms. In such cases, measuring the frequency of the mode precisely can provide an indication of the perturbing bonds and atomic groups, facilitating the identification of the parent molecule. For example, the C=O stretch mode (the well-studied *amide-I mode*) of the peptide group has a fundamental frequency of $\sim$1650 cm^{-1}. But in specific proteins that frequency can vary from 1640 to 1690 cm^{-1}, depending on factors such as water binding, neighboring groups, and the tertiary structure of the molecule: i.e., its folding and global shape, which are often governed by weaker hydrogen-binding forces.

Table 6.1 provides a selected list of some of the more commonly studied vibrational modes that are measured by Raman spectroscopy for biomolecular analysis. These are listed in order of increasing wavenumbers (vibrational transition energy). An alternative way to list common biomolecular modes would be as separate lists of Raman shifts for different molecular species types (e.g., proteins, lipids, nucleic acids, etc.). If one were conducting Raman spectroscopy in predominantly one species or another, then grouping a table by molecular type (e.g., proteins or lipids) would offer easier identification of correlations, facilitating assignments.

Table 6.1 Identifications of Raman-active molecular bands in key biomolecular constituents of tissue: proteins, nucleic acids and lipids. Some bands can be associated with specific molecular bonds or vibrational/bending modes of bond groups, and are noted, whereas others are more simply associated with the parent molecule because they are due to a complex combination of stretching, bending and torsional motions of various coupled atomic groups. It is also clear from the table that many of the spectral feature bands of different molecules overlap directly. Sources include public domain, own work and Schrader (1989) (some lipids and proteins).

Raman shift (cm^{-1})	Likely assignment
653–673	guanine
781–790	cytosine, uracil
842–855	tyrosine, lipid
887–905	CH_2 rock (lipid), structural protein
930–939	C–C stretch (protein backbone)
978	phospholipid
1004	protein, phenyl-ring breathing
1062	DNA/RNA, in-plane ring mode, phospholipid, C–C stretch
1084	protein, C–N of protein backbone, phospholipid, O–P–O, C–C
1055–1069	C–C stretch (lipid)
1090	DNA/RNA, O–P–O, C–N of protein backbone
1162–1184	aromatic amino acids, C–C stretch
1185–1211	aromatic amino acids
1212–1290	amide III, nucleic acid O–P–O
1291–1324	CH_2 twist (protein, lipid), nucleic acids
1325–1424	CH_2 twist and bend (protein, lipid), CH_3 deformation (nucleic acids)
1435–1460	CH_2 deformation (protein, lipid)
1562–1573	guanine, adenine
1609	aromatic amino acid (protein)
1652–1690	amide I (C=O stretch), lipid (C=C)
1744	C=O (lipid)
2723	C–CH_3
2850–2936	CH_2 and CH_3 symmetric, CH_2 asymmetric (lipid), CH_3 stretch (protein), CH (lipid, protein)
2940–2960	asymmetric CH_3 (lipid and protein)
2979	CH stretch (protein)

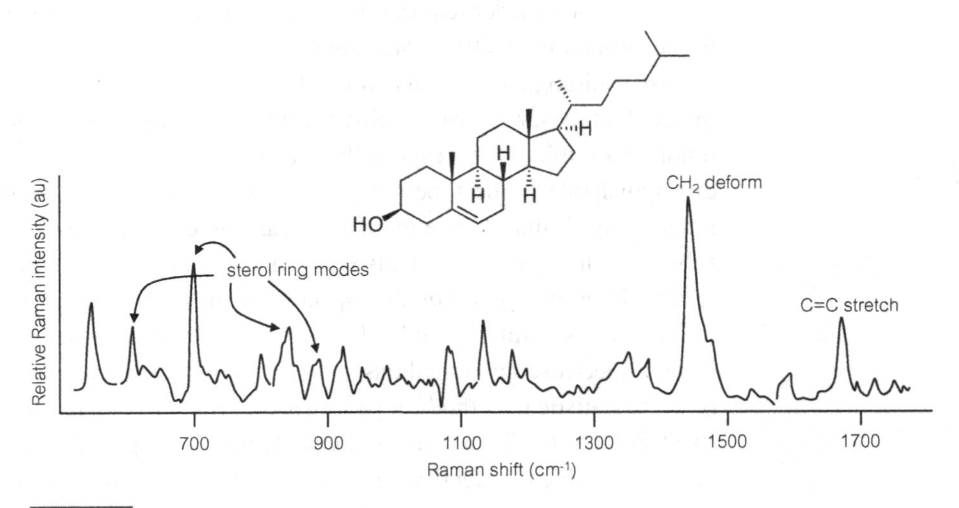

Figure 6.2

Molecular structure of cholesterol and the Raman spectrum of the monohydrate crystal. Spectral data extracted from Hawi et al. (1997) and Weinmann et al. (1998).

6.1.2 The complexity of biological molecules and feature assignment

In samples of pure small molecules, or combinations of very simple compounds, it is indeed possible to make robust assignments of specific Raman spectral features to molecular modes. But even small biomolecules invoke fairly complex spectra. Cholesterol, for example, is a lipid molecule that is synthesized in most mammalian cells, and serves many essential functions in animal physiology, including the role as a structural component of all mammalian cell membranes. As biomolecules go, and as shown in Figure 6.2, the basic form of cholesterol is simple. It is a relatively small molecule, $C_{27}H_{46}O$, with three sterol rings (benzene-like, six-sided), one aromatic ring (five-sided) and a molecular mass of "only" 386.6. (The reader is reminded that the molecular mass is commonly indicated with units of daltons (Da), which represent the mass of the molecule relative to 1 / 12 the mass of ^{12}C, as defined in Section 4.2. It is important to observe that the molecular mass also represents the mass in grams of one mole of that species, so that one may express the molecular mass in units of g/mol.)

The Raman spectrum shown in the figure is for the monohydrate of cholesterol in purified crystalline form. Some of the features can indeed be associated with specific vibrational/bending modes of bond groups. Nonetheless, the spectrum is not simple, and clean assignment of many of the individual spectral features, especially the broader features, is challenging when the cholesterol is measured in tissue (Römer et al., 1998; Weinmann et al., 1998).

In real biological systems, generally, there are always numerous contributing molecular types, with overlapping spectra, rendering rigorous assignment difficult, if not impossible. This is especially true in living systems. Raman spectra of readily distinguishable tissue types (e.g., adipose tissue, skeletal muscle and skin) are remarkably similar. While there are measurable differences, those differences are subtle, requiring measurements with good signal-to-noise ratios for reproducible results. Even with good-quality spectra, unambiguous assignment of the spectral differences is rarely feasible. For this reason, many research groups studying Raman spectroscopy in real tissue, to distinguish, say, cancer from normal tissue, resort to statistical methods of pattern recognition to correlate tissue condition with spectral features. Such statistical correlation of unspecified spectral features can be more reliable for identifying tissue type than attempting to employ model-based identification of the specific molecular bonds associated with the spectral features (e.g., Hanlon et al., 2000; Haka et al., 2009; Draux et al., 2009).

One way to narrow the range of contributing species to a Raman spectrum (of an ex vivo tissue sample) is to perform the spectroscopy through a high-resolution microscope, in which spectra can be recorded from a specific region of a cell, and then to perform *chemometric analysis*. The chemometric analysis constitutes fitting the measured spectrum with a weighted combination of spectra for a small number of potential constituents (e.g., Römer et al., 1998; Kunapareddy et al., 2008). As an example, Raman spectra for primary constituents in mammalian cell nuclei are shown in Figure 6.3.

Kunapareddy (2007) and Kunapareddy et al. (2008) took this approach to identify changes in the proportions of the major constituents as a means to determine differences between live cells and necrotic cells shortly after cell death. Figure 6.4 shows the Raman spectra for the two cell states. One can see that the spectra exhibit all the same features, but with variations in the relative intensities. Over measurements in many cells, the authors found statistically reliable changes in the relative proportions of the constituents. Even with these carefully controlled experimental conditions, nonetheless, only a small number of strong features can be individually attributed to specific molecular groups. Most evident in this case are the protein C=O stretch (amide-I band) overlapping the C–C stretch from lipids at $\sim$1660 cm^{-1}, the CH_2 deformation band from lipids and, possibly, the C–C stretch of the protein backbone at $\sim$935 cm^{-1}.

We remind the reader that spectra of purified biomolecular species, such as those shown in Figures 6.2 and 6.3, can be compared to known tabulations of such features, listed by molecular species. Compendia of such tables are published (e.g., Schrader, 1989) and can serve as resources relevant for the study of

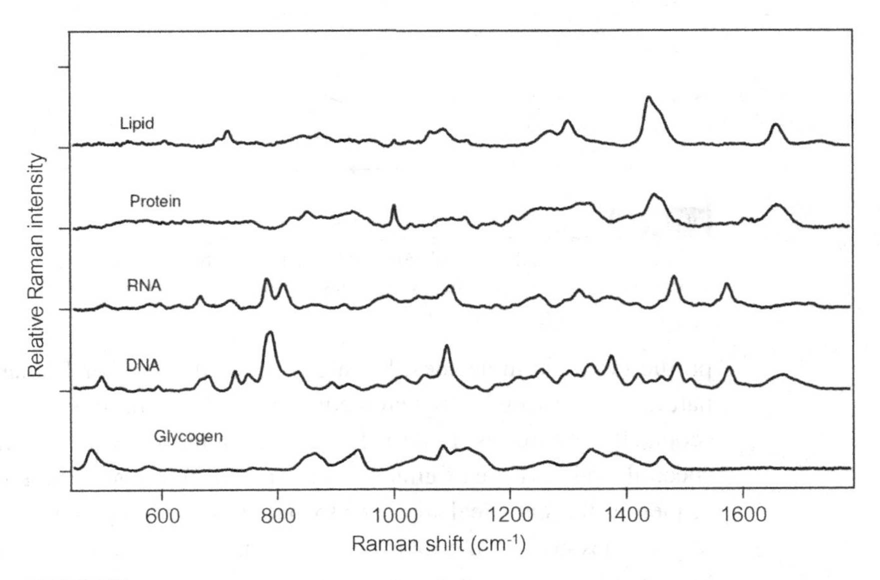

Figure 6.3

Raman spectra of isolated molecular constituents of cell nuclei. Relative intensities are normalized to equal mass concentrations. Data extracted from Kunapareddy (2007).

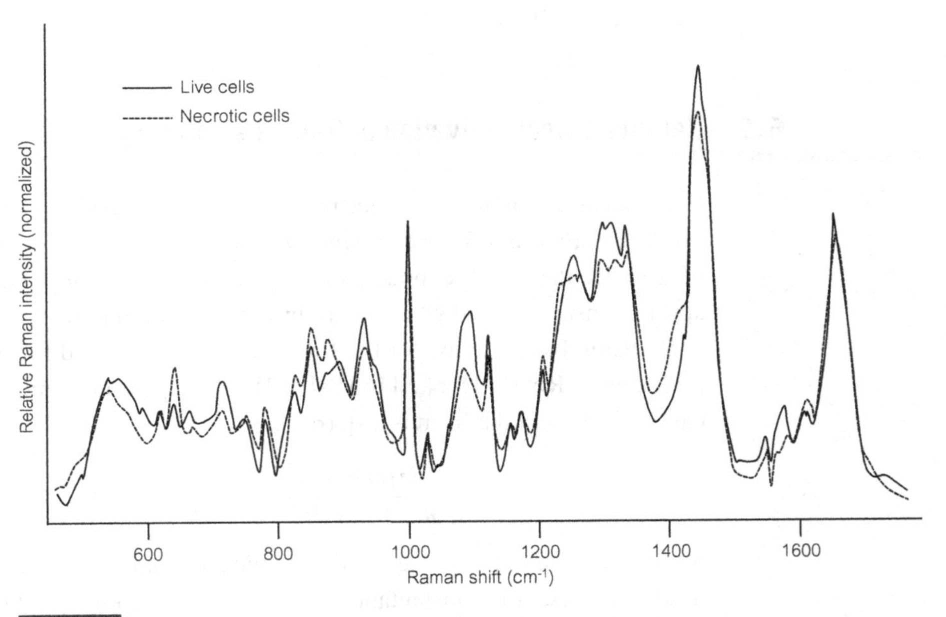

Figure 6.4

Raman spectra of live cells and necrotic cells (within 24 hours) of the human melanoma cell line MEL-28. Data adapted from Kunapareddy (2007).

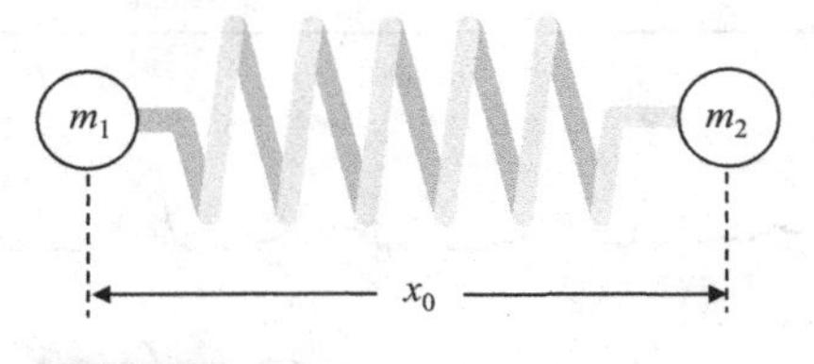

Figure 6.5

The vibration of a diatomic molecule is represented as two masses connected by a spring, of spring constant K. The distance x_0 denotes the separation of the two masses at equilibrium.

purified organic molecules, but are of limited value for Raman spectroscopy of naturally occurring biological media, even when singling out components microscopically. Again, as Table 6.1 indicates, almost all of the other (and weaker) spectral features have origins in at least two different molecular constituents. Consequently, the statistical approach to matching the full pattern of individual molecular species and adding their contributions yields the most reliable assignments. Further, many research groups have used statistical pattern recognition methods to correlate the complex Raman spectra of biological tissue with diagnosis by gold standard (e.g., histopathology), thus training diagnostic algorithms in a manner that ignores any assignment of detailed features (e.g., Haka et al., 2005, 2009; Kanter et al., 2009).

6.2 Semi-classical derivation of Raman scattering

Consider a diatomic molecule, represented as a classical vibrational harmonic oscillator (Figure 6.5), comprising two masses (m_1 and m_2), representing atoms of a molecular bond, connected by a spring, representing the binding force, of spring-constant K, and whose equilibrium separation is x_0.

By using Hooke's law for the force of a spring (named for the English natural philosopher Robert Hooke [1635–1703]), the equation of motion for this system can be written in the form of Eq. (6.2):

$$\frac{m_1 m_2}{m_1 + m_2}\frac{d^2 q}{dt^2} = \mathcal{M}\frac{d^2 q}{dt^2} = -Kq \tag{6.2}$$

where $\mathcal{M}$ is the reduced mass of the atomic pair, and $q = x - x_0$ is an abbreviation used to represent the instantaneous displacement from equilibrium. The solution to Eq. (6.2) is that of the classical harmonic oscillator:

$$q = q_m \cos(\omega_m t) \tag{6.3}$$

where q_m is the amplitude of nuclear displacement from equilibrium, and the mechanical vibration angular frequency of the bond, ω_m, is a function of the reduced mass and the spring constant (bond strength):

$$\omega_m = \sqrt{\frac{K}{\mathcal{M}}} \tag{6.4}$$

Putting this in the context of molecular vibrations, the vibrational frequency of a bond is proportional to the square root of the *bond force constant*, K (sometimes called the *bond strength*) and inversely proportional to the square root of the reduced mass of the participating atoms. The SI units of the bond strength, K, are N/m = kg/s^2. The reduced mass can be easily calculated using the atomic masses expressed in atomic mass units, Daltons (1 Da = 1/12 the mass of ^{12}C), and the conversion to SI units is given by 1 Da = 1.66×10^{-27} kg.

Keeping in mind that vibration of a diatomic molecule is a simplification of the mechanical motions of the atoms associated with a specific bond in a biomolecule, we continue with this representation, and now examine the interaction of a light wave with a molecular bond. The optical electrical field will exert oppositely directed forces on the nuclei and electrons associated with a bond, so as to induce a dipole moment, in addition to exerting a force on any permanent dipole moment that may be associated with the bond. (In fact, this can also happen with an isolated atom.) The induced dipole moment, or *polarization*, **p**, is proportional to the applied electric field, **E**, through the *polarizability*, α, of the bond.

$$\mathbf{p} = \alpha \mathbf{E} \tag{6.5}$$

For the moment, this representation assumes that the polarizability is isotropic, such that α is a scalar quantity, and the induced dipole moment is aligned with the applied electric field. Examining only the time-dependence of the optical electric field (i.e., incorporating the position dependence into the field amplitude, $\mathbf{E}_0$), we write a simplified version for the representation of the optical electric field, earlier given as Eq. (4.3):

$$\mathbf{E}(t) = \mathbf{E}_0 \cos(\omega_0 t), \tag{6.6}$$

where $\mathbf{E}_0$ is the spatially dependent optical electric-field amplitude, and ω_0 is the optical angular frequency.

Next, we note that the polarizability, α, is a function of the nuclear separation, x, hence a function of q, and can be written as an expansion (using the small-amplitude approximation):

$$\alpha = \alpha_0 + q\left(\frac{\partial \alpha}{\partial q}\right)_{q=0} + \cdots \tag{6.7}$$

where α_0 is the polarizability at the equilibrium separation. Combining Eqs. (6.3), (6.6) and (6.7) into Eq. (6.5), we get

$$\mathbf{p} = \left[\alpha_0 + \left(\frac{\partial \alpha}{\partial q} \right)_{q=0} q_m \cos(\omega_m t) \right] \mathbf{E}_0 \cos(\omega_o t) \tag{6.8}$$

The trigonometric identity, $\cos(A)\cos(B) = \frac{1}{2}[\cos(A+B) + \cos(A-B)]$, enables us to write Eq. (6.8) as

$$\mathbf{p} = \alpha_0 \mathbf{E}_0 \cos(\omega_o t) + \frac{1}{2} \mathbf{E}_0 q_m \left(\frac{\partial \alpha}{\partial q} \right)_{q=0} \{ \cos[(\omega_o + \omega_m)t] + \cos[(\omega_o - \omega_m)t] \} \tag{6.9}$$

Given that an oscillating dipole radiates, this simplified classical treatment of the interaction between an optical wave and a molecular bond yields three types of scattered waves:

- First term: Rayleigh scattering, unshifted in frequency, named after British scientist John William Strutt, Lord Rayleigh [1842–1919].
- Second term: *anti-Stokes* Raman scattering, in which the scattered light has a higher frequency, equal to the sum of the optical frequency and the vibrational frequency.
- Third term: *Stokes* Raman scattering (named after the Irish-British physicist Sir George G. Stokes [1819–1903]), in which the scattered light has a lower frequency, equal to the difference between the optical frequency and the vibrational frequency.

We will see in Section 6.2.2 that the strength of anti-Stokes scattering is much smaller than that of Stokes scattering at typical temperatures. It is also seen (in Figure 6.9) that a condition for Rayleigh scattering is that the polarizability of the bond at equilibrium must be non-zero; and for Raman scattering the polarizability must change as a result of vibrational motion (i.e., as a consequence of a change in internuclear separation):

$$\left(\frac{\partial \alpha}{\partial q} \right)_{q=0} \neq 0 \tag{6.10}$$

This differs from the condition for direct optical absorption, which requires that a bond have a permanent dipole moment; i.e., it must be "*dipole-active*" (and also that the derivative of the dipole moment with respect to internuclear separation be non-zero). No permanent dipole moment is required for Raman scattering.

The three processes derived in Eq. (6.9) are illustrated schematically in Figure 6.6.

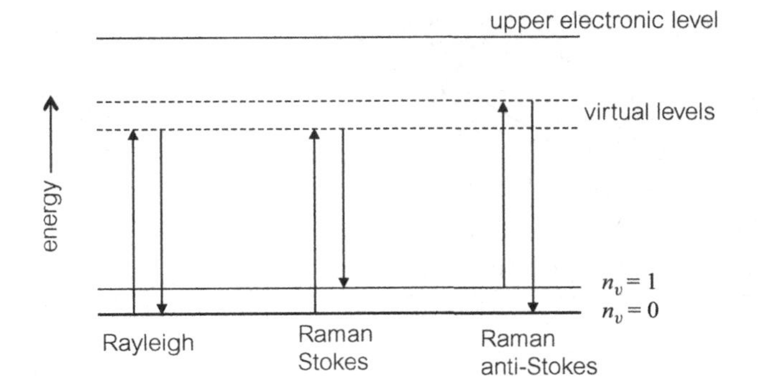

Figure 6.6

Jablonski diagram of molecular energy levels, with arrows indicating virtual transitions resulting in optical scattering from the electronic ground state.

6.2.1 Wavelength dependence of the scattered field from molecules

We wish to understand the wavelength-dependence of the optical scattering process. For this purpose, we continue with the simplified examination of a single bond, represented as an isolated diatomic molecule in an externally applied optical field; and the bond has either a permanent dipole moment, or a dipole that is induced by the optical field. We make two key approximations, which are physically good representations for the parameters of interest: that the size (separation of atoms) of the bond (dipole), x_0, is much smaller than the wavelength, λ, of the applied optical field, and that the location for detecting the scattered field is at a distance, r, from the dipole that is much greater than the wavelength: the *"far-field"* condition. Thus, the conditions are summarized as $x_0 \ll \lambda \ll r$. In this case, the radiating field of the resulting dipole oscillation behaves as a spherical wave, but with an angular dependence for the electric field amplitude (of the scattered wave) that is determined by the observation angle (θ) with respect to the dipole orientation. Figure 6.7 depicts the geometry.

It is appropriate to introduce spherical coordinates, given that the scattering bond looks like an oscillating point-dipole source in the far-field, with a radiation field varying with the viewing angle relative to the dipole direction. The radial coordinate (r) is the distance from the dipole to the observation point, the polar angle (θ) is the angle between the dipole orientation and the direction of observation, and the azimuth angle (φ) is the angle between a reference direction and the projection of the direction of observation onto the plane orthogonal to the dipole. By considering a harmonic time-dependence ($e^{i\omega t}$), the radial, polar, and azimuthal components

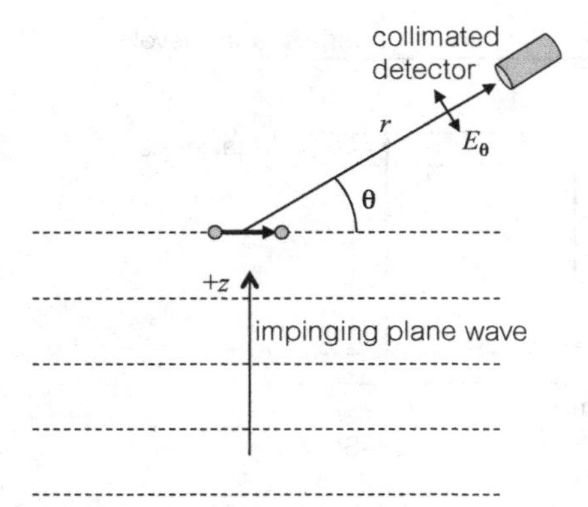

Figure 6.7

Geometry for observation of the scattered field of an induced dipole in an externally applied field propagating in the $+z$ direction. The dipole orientation is depicted in the x direction, parallel with the applied optical electric field. The detector (located at a distance, r, far from the source of field disturbance) is collimated and sees only the scattered field. The electric field amplitude E_θ denotes the component of the radiating field that is normal to the radial coordinate r.

of the electric field associated with the far-field radiation of a dipole of magnitude p are:

$$E_r \approx 0, \tag{6.11}$$

$$E_\theta\,(r, \theta, t) \approx pk^2 \sin\theta \,\frac{e^{-i(kr-\omega t)}}{r}, \tag{6.12}$$

$$E_\varphi = 0, \tag{6.13}$$

indicating that the far-field dipole radiation is a spherical wave whose electric (and magnetic) field is transverse to the direction of propagation (see, for example, Heald and Marion, 2012). (The radial component of the electric field decreases as $1/r^2$, whereas the polar component decreases as $1/r$.) The electric field of Eqs. (6.11)–(6.13) is expressed here in Gaussian units, which is an unrationalized cgs unit system (Wangsness, 1986). In Eq. (6.12), $k = \omega/c = 2\pi/\lambda$, indicating that the electric field amplitude is proportional to λ^{-2}. Because one detects intensity, which is proportional to the square of the electric field, the scattered power is proportional to λ^{-4}.

As we see in this derivation, *the $1/\lambda^4$ dependence of scattered power characterizes Raman scattering as well as Rayleigh scattering.* For Rayleigh scattering, this dependence holds for any polarizable particle that is small compared to the optical wavelength.

Motivated to explain the blue sky, Lord Rayleigh, himself, provided an elegant determination of the λ^{-4} dependence for light scattered by small particles, based purely on dimensional analysis, without any equations (Rayleigh, 1871). His reasoning can be roughly summarized (and translated to the terminology used here) as follows:

> The objective is to examine the wavelength dependence of the ratio, i, of the amplitude of the scattered light field to the amplitude of the incident light field, given that the scattered field is proportional to the incident field. This ratio can be a function of
>
> (1) the particle polarizability α, which is proportional to the particle volume, hence a^3, where a is the particle radius;
> (2) the distance, r, of the observation point from the disturbance;
> (3) the wavelength, λ;
> (4) the speed of light, c;
> (5) the electric permittivity of the medium with and without the disturbance, ϵ, ϵ'.
>
> The first three depend only on length, the fourth on length and time, and both versions of the fifth relate to the mass density of the medium. All other factors in the ratio (e.g., angle) are dimensionless. Since the ratio sought is itself independent of mass density, the permittivity can only appear equally in the numerator and denominator, so its dimensions can be ignored. Of the other four factors, only c incorporates the dimension of time. Since the ratio of scattered-to-incident light is constant (over times long compared to the optical period), c "cannot occur in its expression." This leaves α, r and λ. We know that i is directly proportional to α and inversely to r, so dimensionally, it must be proportional to $\alpha/(\lambda^2 r)$, since the ratio must be dimensionless overall. Thus, i is proportional to $1/\lambda^2$. Finally, since the ratio of amplitudes varies as $1/\lambda^2$, it follows that the ratio of intensities (what we detect) varies as $1/\lambda^4$.

The spatial distribution of scattered intensity from a small particle will be examined in Section 7.2.1.

Given the λ^{-4} wavelength dependence of Raman scattering, one might think that Raman spectroscopy is optimized by using illumination wavelengths that are short (as in the UV-blue portion of the optical spectrum). With biological samples such as tissue, however, illumination by short wavelengths induces fluorescence emission that is several orders of magnitude greater than the Raman scattering signal, rendering poor signal-to-noise ratios for the Raman measurements. As listed in Chapter 5 (Table 5.1), the excitation bands for the primary sources of native fluorescence in tissue are at wavelengths generally shorter than 550 nm. Consequently, Raman spectroscopy of biological tissues, especially for measurements performed in vivo, are typically performed with illumination wavelengths in the red-to-NIR spectral range (600–800 nm), for which native fluorescence is much weaker, by a factor that is larger than that for the wavelength-dependent reduction suffered by the Raman signal.

6.2.2 A little bit less simplification

The classical treatment of a mechanical vibration represented by Eqs. (6.2)–(6.4) ignores the quantum mechanical nature of the vibrational states in a molecule. In a classical system, the potential energy of the vibration would have a value $V = \frac{1}{2}Kq^2$, incorrectly suggesting that the amount of vibrational energy is continuously variable, and with a minimum value of zero. At a molecular scale, however, the "harmonic-oscillator problem" requires a quantum-mechanical treatment, which leads to a quantized set of energy states, with energies given by:

$$E_v = \hbar\omega_m\left(n_v + \frac{1}{2}\right) = \hbar\sqrt{\frac{K}{\mathcal{M}}}\left(n_v + \frac{1}{2}\right), \tag{6.14}$$

where $\hbar$ is Planck's constant (h) divided by 2π, the reduced Planck's constant, and n_v is the *vibrational quantum number*, which can take on values 0, 1, 2.... (This symbol E_v for the vibrational energy, a scalar, should not be confused with the symbol used for the optical electric field, discussed in the previous section.) Thus, the ground state of a quantum-mechanical harmonic oscillator has a minimum energy of $\frac{1}{2}\hbar\omega_m$, and the excited states are equally spaced above that. The observant reader will note that the non-zero ground-state energy of vibrational modes was depicted in Figures 4.5, 4.6, 4.8 and 5.1, and the distributed wave-functions were also depicted. Intuitively, we may understand the non-zero energy of a quantum-mechanical vibrational ground state (i.e., the variance of the quantum-mechanical eigenstates) through the uncertainty principle. If one tries to define the equilibrium position of the oscillator very precisely, then the uncertainty principle ($\Delta x \Delta p \geq \hbar/2$) tells us that the standard deviation of the potential energy (related to momentum by the constant c, the speed of light) must be larger; and, conversely, if one tries to define the oscillator's energy level (momentum) precisely, then its wave function spreads out, exploring higher energies in the potential-energy well.

Also, a purely classical approach, as represented by Eq. (6.9), fails to properly predict the ratio of anti-Stokes to Stokes scattering, for which quantum theory is needed to determine the occupation ratio of the vibrational energy states, which can be determined from the Boltzmann distribution. As explained in Section 5.1, and represented mathematically by Eq. (5.1), at typical temperatures the population of excited vibrational modes is much lower than those of the ground states for those modes. Consequently, at room temperature or body temperature, Stokes scattering is typically much stronger than anti-Stokes scattering. Similarly, for a bond with a permanent dipole moment, the probability of absorption relative to that of stimulated emission is determined by the same Boltzmann distribution, given that the absorption and stimulated-emission cross sections are equal, as was famously enunciated by Albert Einstein in 1917, the inimitable German-born physicist [1879–1955], who received the Nobel Prize in physics in 1921 (see, for example, *Modern Physics*, by Randy Harris, under further reading).

Moreover, the isotropic representation of the polarizability, as assumed in Eq. (6.5), is oversimplified. In general this is not true, since the polarizability of most molecules, and essentially all biomolecules (or molecular groups), is anisotropic, and the induced dipole moment, in general, does not point in precisely the same direction as the applied optical electric field. Thus, Eq. (6.5) can be written, more appropriately, as a tensor relation:

$$\mathbf{p} = [\alpha]\mathbf{E}, \tag{6.15}$$

where the polarizability is expressed as a tensor rather than as a scalar:

$$[\alpha] = \begin{pmatrix} \alpha_{xx} & \alpha_{xy} & \alpha_{xz} \\ \alpha_{yx} & \alpha_{yy} & \alpha_{yz} \\ \alpha_{zx} & \alpha_{zy} & \alpha_{zz} \end{pmatrix}, \tag{6.16}$$

resulting in the Cartesian components of **p** expressed in terms of the polarizability tensor elements:

$$\begin{aligned} p_x &= \alpha_{xx}E_x + \alpha_{xy}E_y + \alpha_{xz}E_z \\ p_y &= \alpha_{yx}E_x + \alpha_{yy}E_y + \alpha_{yz}E_z \\ p_z &= \alpha_{zx}E_x + \alpha_{zy}E_y + \alpha_{zz}E_z \end{aligned} \tag{6.17}$$

It should be noted that the polarizability tensor is symmetric, such that $\alpha_{ij} = \alpha_{ji}$. In the case of isotropic polarizability, $[\alpha]$ becomes a diagonal matrix, with equal diagonal elements and zero values for the off-diagonal elements, thus reverting to Eq. (6.5).

6.3 IR-absorption spectroscopy

If a molecular bond that exhibits a permanent dipole moment is oscillating, the dipole will radiate. Similarly, the dipole can interact directly with an external electric field and absorb energy at an optical frequency that is resonant with its vibrational frequency. This interaction can be strong. For optical fields that are resonant with the vibrational frequency of the bond (hence, the dipole oscillation frequency) the effective cross section for the interaction can be much larger than the true geometric dimensions of the bond or group of atoms participating in the normal mode. This means that the bond is able to draw in electromagnetic energy from a much larger spatial region (of the optical field) than the atomic group's own geometric extent.

If two equal and opposite charges, $+Q$ and $-Q$, are separated by a distance, x_0, the pair produces a *permanent dipole moment*, $\mathbf{p}_0$, whose strength is given by the product of the charge and the separation:

$$\mathbf{p}_0 = Q\mathbf{x}_0 \tag{6.18}$$

where the charge separation, or *displacement vector*, $\mathbf{x}_0$, points from the negative charge to the positive charge. The unit most commonly used for the dipole moment of a bond or molecule is the Debye (D), a cgs unit (named after the Dutch-American physicist Peter Debye [1884–1966], who received the Nobel Prize for chemistry in 1936). The relationship between the Debye and the SI unit for the electric dipole moment (C-m) is $1\ \text{D} = 3.33 \times 10^{-30}$ C-m. (We note that the smallest prefix in SI units is *yocto* $= 10^{-24}$; thus the Debye is a more convenient unit for quantifying the dipole moments of molecular bonds.) The values for the dipole moments of common covalent molecular bonds, and for most heteronuclear diatomic molecules, is in the range 0.1–5 D. As examples, the dipole moment of the water molecule (as a gas) is ~1.85 D, and for CO it is ~0.11 D.

For an isolated charge pair (one positive charge and one negative charge), the dipole moment approaches infinity as the charge separation becomes infinitely large. In the case of a molecular bond, however, a dipole moment exists when one nucleus is more positive than another, and the electron cloud distributes unevenly around them, resulting in an *effective* charge separation. In this case, the dipole moment of the bond is not determined by the simple linear relation of Eq. (6.18), because the electron cloud eventually reverts to the distributions around the individual atoms as the bond breaks. A qualitative representation of the dipole moment of a diatomic molecule, as a function of internuclear separation, is shown in Figure 6.8. In this example the derivative of the dipole moment with respect to internuclear separation at equilibrium is negative.

The magnitude of the dipole moment of a bond leads to another term often used to describe the probability of an absorption event: the *oscillator strength* of the transition (or state), which is found by integrating the absorption probability as a function of frequency.

The energy of the interaction between an electric field and a dipole, U_{int}, is the dot-product of the two vectors:

$$U_{\text{int}} = -\mathbf{E} \cdot \mathbf{p} \tag{6.19}$$

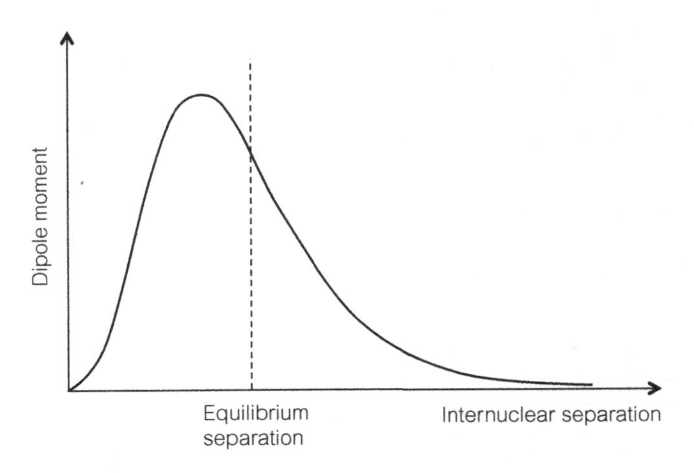

Figure 6.8

Dependence on internuclear separation for the permanent dipole moment of a heteronuclear diatomic molecule. The dependence for a bond in a larger molecule is qualitatively similar. The equilibrium position may exhibit a positive derivative, negative derivative (as illustrated here) or zero derivative. (If the derivative of the dipole moment is zero at equilibrium, the bond is IR-inactive.)

However, the amount of energy that an optical field can lose to (or gain from) a vibrational bond that absorbs (or emits) energy is quantized, and the quantum of energy exchanged is given by Eq. (6.14). The interaction energy, as expressed in Eq. (6.19), relates to the probability that the energy exchange, either absorption or stimulated emission, will happen, which is linearly proportional to the applied optical field amplitude and the dipole moment of the bond. The determination of the actual transition rate from a ground state to an excited vibrational state entails quantum-mechanical calculations of the wave-function overlap, as discussed in Sections 4.2 and 5.1.

6.4 Is it IR-active, Raman-active or both?

The details of the charge distribution of a bond at equilibrium determine whether it will exhibit IR absorption, Raman scattering or both. We illustrate, at least qualitatively, the relevant effects of charge distribution in molecular bonds, by examining simple diatomic and triatomic molecules, presented as surrogates for individual bonds in a larger molecule. Even though the related vibrational properties in macromolecules are more complicated, these simpler molecular properties are indicative of the trends to be expected for specific bonds in larger molecules.

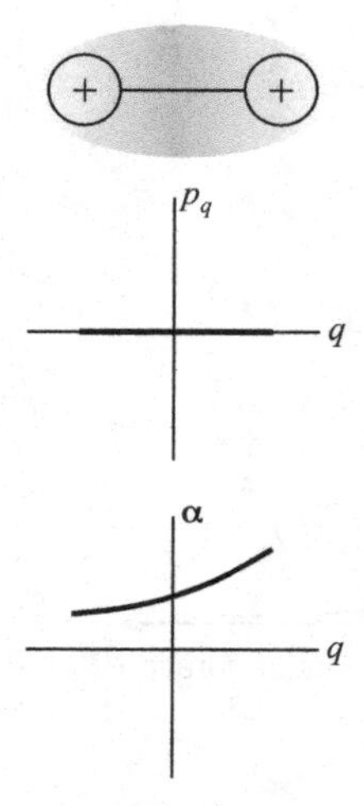

Figure 6.9

A homonuclear diatomic molecule exhibits no dipole moment, regardless of internuclear separation, but the polarizability is non-zero and its derivative is non-zero at equilibrium. In this illustration p_q represents the amplitude of the dipole moment, which is associated with any variation in q. For simple linear molecules like this, and those in Figures 6.10 and 6.11, any dipole moment will be aligned with the molecular axis, along which q is measured.

6.4.1 Homonuclear diatomic molecules

When both atoms of a diatomic molecule are identical, simple symmetry arguments dictate that there can be no permanent dipole moment, and changes in internuclear distance will not create a dipole, as the electron cloud is symmetrically distributed around the identical positively charged nuclei. The polarizability, however, is non-zero, and it increases with internuclear separation, because it is easier for an external electric field to distort the electron cloud if the cloud is more widely distributed. These dependences are illustrated in Figure 6.9. This molecule (or a bond of similar structure) is Raman-active, but is not IR-active, meaning that there is no direct absorption of electromagnetic radiation at the vibrational frequency.

6.4.2 Heteronuclear diatomic molecules

With nuclei of different charges, the molecule (or bond) has a permanent dipole moment, which varies with internuclear separation (and has a non-zero derivative at equilibrium); therefore, this bond will have a direct dipole-interaction – it is IR-active – with an externally applied field, and this interaction will be strong

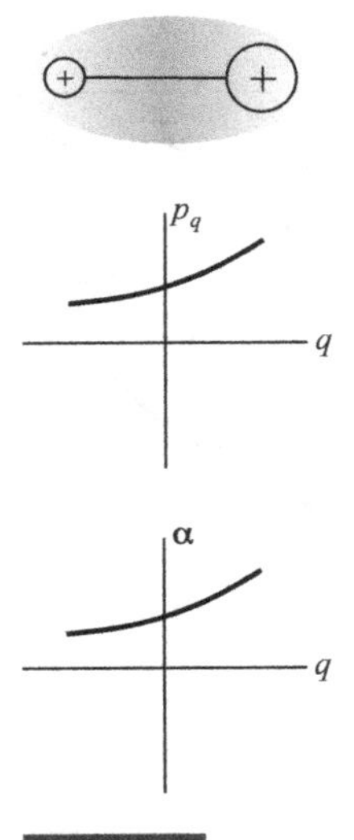

Figure 6.10

A heteronuclear diatomic molecule has a permanent dipole moment, and also exhibits polarizability that varies with internuclear separation and has a non-zero derivative at equilibrium.

when the frequency of the external field is resonant with that of the vibrational oscillation. Moreover, the polarizability will also vary with internuclear separation (and have non-zero derivative at equilibrium), so the bond is also Raman-active. This type of molecule (bond) can be studied with either IR-absorption spectroscopy or Raman spectroscopy.

6.4.3 Linear triatomic molecules

A classic example of this type is CO_2, as depicted in Figure 6.11, for which there are four possible vibrational modes: the symmetric-stretch mode, the asymmetric-stretch mode, and two bending modes (in orthogonal planes, not depicted here), which are degenerate in energy, appearing as one frequency band in spectroscopy. The symmetric stretch mode behaves similarly to the homonuclear diatomic molecule, exhibiting Raman activity but not IR activity. Although the molecule lacks a permanent dipole moment at equilibrium, the asymmetric-stretch vibration, nonetheless, does result in a dipole that oscillates with nuclear motion, and the derivative of the dipole moment is non-zero at equilibrium. On the other hand, although the polarizability is non-zero at all internuclear distances, the derivative of the polarizability at equilibrium, for this mode, is zero. Consequently, the asymmetric vibration mode of CO_2 is dipole-active but Raman-inactive. For this

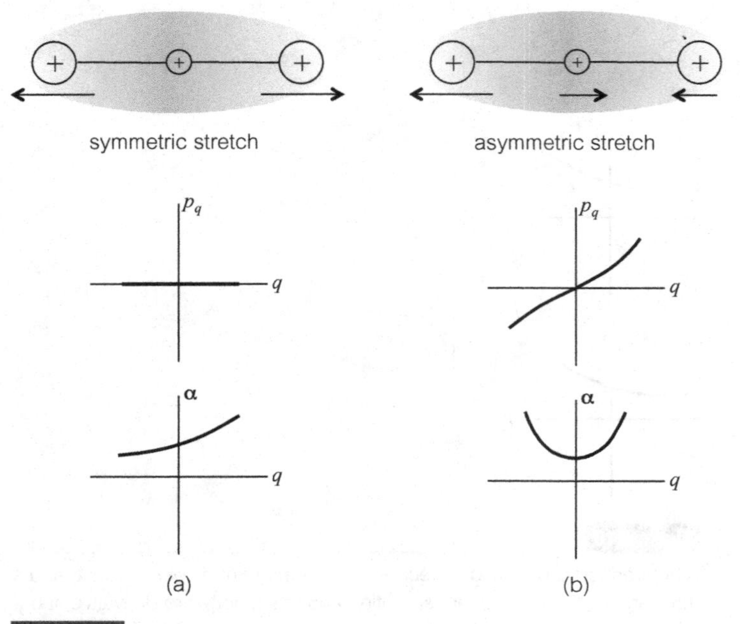

Figure 6.11

The symmetric (a) and asymmetric (b) stretch modes of a linear triatomic molecule. Note that although there is no dipole moment at equilibrium, the derivative of the dipole moment at equilibrium is non-zero for the asymmetric mode, and that mode is IR-active.

molecular configuration, the parameter q represents the displacement of the central atom relative to the midpoint between the two outer atoms and, for the geometry as depicted in Figure 6.11, is positive when the central atom moves towards the atom on the right.

For the bending mode of CO_2, not depicted here, there is no change in polarizability, but there is a non-zero derivative of the dipole moment, so the bending mode is IR-active but Raman-inactive.

6.4.4 Nonlinear triatomic molecules

The classic, and most important example of a nonlinear triatomic molecule in the study of biological systems, is water (Figure 6.12). The angle between the HO bonds is 104.5°, and, due to the fact that it is "bent," the water molecule has a strong permanent dipole moment of 1.85 D in the gas state, and an even stronger dipole moment of ~2.9 D in the liquid state (Gubskaya and Kusalik, 2002).

When water is in its liquid state, the dipole moments of individual molecules induce attractive forces, such that groups of molecules "clump" together as a

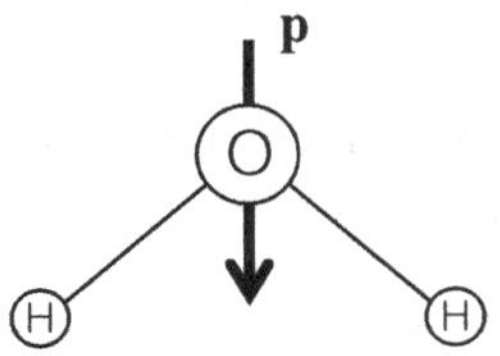

p

O

H H

Figure 6.12

The water molecule has a strong permanent dipole moment, whose direction is indicated by the heavy arrow.

result of *hydrogen bonding*. An important spectroscopic consequence of hydrogen bonding in liquid water is that the strong IR-active bands (the stretch and bending modes) are dramatically broadened as compared to those of gaseous water. As previously mentioned in Section 4.4, the strong broadband absorption of water, beginning around 1.3 μm and extending beyond the microwave band, gets in the way of IR-absorption spectroscopy of biomolecules that are typically in a natural aqueous environment when examined in tissue. For example, the absorption coefficient of water in the spectral range 3–10 μm, where the wavelengths of many important vibrational modes of biomolecules are found, is in the range 10^2–10^4 cm^{-1}. This is the reason why almost all spectroscopy of vibrational modes in viable biosystems, especially in vivo, is performed by Raman spectroscopy, for which laser illumination sources are typically in the visible-NIR wavelength range. (As mentioned in Section 6.2.1, excitation wavelengths in the NIR are often preferred over visible wavelengths, to avoid interference from fluorescence, while still being in the low-absorption range of water.)

6.5 Isotopic shifts

Even though the spectral features of some bonds are enhanced by repeated subunits of a molecule, the number of vibrational modes in biomolecules is still large, and many modes overlap in frequency, resulting in a challenge to assign a measured spectral feature to a specific bond or normal mode. Nonetheless, biochemists are often highly adept at substituting different isotopes of specific atoms in specific complexes of a larger molecule. This offers a mechanism to identify specific bonds that are contributing to a spectrum, by examining the spectral feature that shifts as a consequence of the isotopic substitution. Equation (6.4) represents a classical vibrational oscillator, but the effect of an altered reduced mass is also accurately represented for the energy associated with a molecular bond at the quantum-mechanical level. In other words, also at the quantum-mechanical level, the frequency of the mode is proportional to the inverse square root of the reduced mass: $\omega_m \propto 1/\sqrt{\mathcal{M}}$.

Of course, isotopic substitution is not relevant to applications for clinical diagnostics; nonetheless, it is a powerful tool for laboratory measurements at the research and pre-clinical stages.

6.6 Enhancements of Raman scattering

There are three methods that have generated significant interest for enhancing the signal in Raman scattering measurements. These are all proven and useful methods for laboratory measurements on cells and biomolecules, whereas practicality for clinical applications has been limited by the exigencies of in vivo conditions. The methods, reasons for their limitations and some concepts that help to overcome the limitations are briefly reviewed here. Options for further reading are provided at the end of the chapter.

6.6.1 Resonance-Raman scattering (RRS)

When the frequency of an externally applied optical field (typically from a laser) is close to that of an electronic transition (a *resonance*) of a molecule, but not close enough to induce significant direct absorption, the vibrational modes associated with that particular electronic transition exhibit a greatly increased Raman scattering intensity. In effect, the Raman cross section, which relates directly to the polarizability of the vibrational modes, is increased. The enhancement in scattered intensity for RRS compared with non-resonant Raman scattering can be several orders of magnitude, which leads to improved signal-to-noise ratio for experimental measurements.

There are, nonetheless, a number of practical challenges for RRS measurements for biomedical applications. For one, only the vibrational modes associated with the atoms involved in the specific electronic transition are enhanced, and the enhancement can be two to three orders of magnitude, masking any Raman scattering at close frequencies but from other parts of a molecule. This can be beneficial if one is interested in those specific vibrational modes, and the strong scattering from those modes also helps to uniquely identify them as being associated with that specific set of atoms. If, on the other hand, the intent is to identify a molecule that is of unknown structure, then it is often desirable to study a broader range of vibrational modes from different regions of the molecule. Another problem with RRS that is especially problematic for studies of biomolecules in tissue is that the electronic resonances of most proteins and nucleic acids, as well as lipids and membrane structures, overlap in the UV spectral region. Thus, when illuminating a biological (say, tissue) sample with UV light, some molecules may be near resonance but others will be at resonance, strongly absorbing the UV light, and then emitting strong fluorescence. That strong fluorescence generally overwhelms any Raman scattering, resulting in unacceptable signal-to-background levels. Consequently, as mentioned above, most current research activity for developing in vivo Raman diagnostic spectroscopy invokes excitation sources at near-IR wavelengths, far from any electronic resonances in biomolecules.

6.6.2 Surface-enhanced Raman scattering (SERS)

In lieu of invoking resonance enhancement from electronic transitions, in effect enlarging the Raman cross section, Raman scattering efficiency can be also enhanced by increasing the effective strength of the optical field in the vicinity of the molecules of interest. But this is not as simple as "turning up" the light source. If the power density of the light source is increased significantly, as when focusing a laser beam, this can lead to unwanted thermal damage to the sample. With SERS the light source illuminates the surface of a microscopically rough conductor (on a nanometer scale) (Jeanmaire and van Duyne, 1977), and the optical electric field in the vicinity of the small sharp points on the metal surface is dramatically amplified due to local field effects of superficial plasmons, which are excitations of the surface conduction electrons (Polman and Atwater, 2005). The degree of enhancement can be several orders of magnitude, but only for those molecules that are in contact with or very close to the conductor surface.

One of the questions frequently raised for application of SERS to biomedical diagnostics is how to get the molecules of interest close enough to the conducting surface, and in a practical manner for biomedical measurements with a directed laser beam. An interesting approach that has been introduced recently for introducing a nano-scale rough surface locally in tissue is the use of gold nanoparticles, which can generate *localized surface plasmons* (Zeng et al., 2011). Yuan et al. (2012) have synthesized star-shaped gold nanoparticles ("nanostars") that were specially coated to make them biocompatible. These biocompatible particles can be injected into sub-dermal tissue sites, which enables the nanoparticles to be optically interrogated while in direct contact with the *extracellular fluid* (ECF), where molecules in the ECF can then be studied.

6.6.3 Coherent anti-Stokes Raman spectroscopy (CARS)

CARS is a nonlinear optical method for generating strong anti-Stokes Raman scattering by simultaneously illuminating with two laser beams whose frequencies

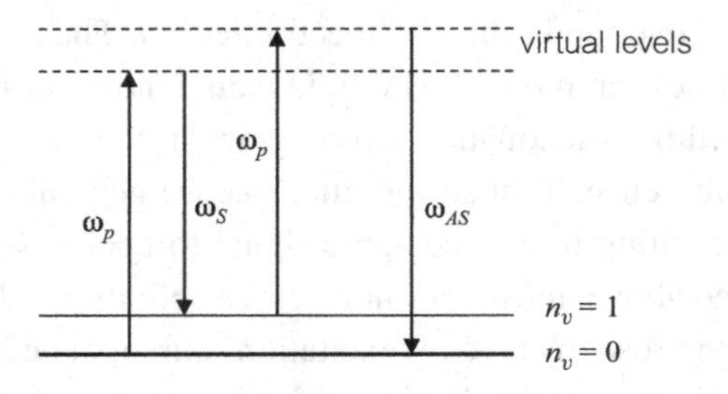

Figure 6.13

Energy-level diagram for the nonlinear optical process of CARS.

are separated by the vibrational mode being studied. In this method, the optical electric fields interact with the molecular bond through its third-order hyperpolarizability. Using the notation of this text, the most general expression relating the i-th component of the induced dipole moment to the applied electric field can be written as

$$p_i = p_{0i} + \alpha_{ij} E_j + \alpha_{ijk} E_j E_k + \alpha_{ijkl} E_j E_k E_l + \cdots \tag{6.20}$$

where the first term on the right-hand side is the i-th component of the permanent dipole moment, $\mathbf{p}_0$, the second term is the tensor relation from Eq. (6.17), and the following terms represent the nonlinear polarizabilities of increasing order. In Eq. (6.20), the summation over repeated indices is implied. In physics texts, the second-order polarizability α_{ijk}, for example, is often represented as $\chi^{(2)}$ or $\alpha^{(2)}$, with the tensor subscripts understood, and similarly for higher-order terms. In the case of CARS, the relevant term is the third-order polarizability, α_{ijkl}, for which there are three applied electric fields. (It should be noted that each nonlinear polarizability term can represent different nonlinear processes, depending on the frequencies of the applied fields and the process symmetries. For example, the third-order polarizability can also represent third-harmonic generation, when the three applied fields derive from the same laser field.)

In Figure 6.13, the three applied fields are represented as the two "pump fields," at frequency ω_p, and the "Stokes field," ω_s (sometimes called the "probe field"), which are typically provided by two applied laser fields, with the two pump waves derived from the same laser beam.

When the difference in frequency between the applied pump field and Stokes fields is resonant with the energy difference between the two vibrational levels, there is a strongly (resonantly) enhanced field at the anti-Stokes frequency, ω_{AS}, which is a coherent field, whose direction and phase are determined by the applied fields. In the method of CARS spectroscopy, to probe vibrational resonance frequencies of bonds, the wavelength (frequency) of one of the two applied laser fields is tuned, and the emitted power of anti-Stokes emission is recorded as a function of the variable difference between the pump and Stokes frequencies. Since the anti-Stokes field can be strong and is directional, it can be easy to achieve good signal-to-noise ratios in experiments. The resulting anti-Stokes intensity (for a thin sample, such as monolayer or cells) is proportional to the nonlinear polarizability and the applied optical intensities as follows:

$$I_{AS} \propto \left|\alpha^{(3)}\right|^2 I_p^2 I_S \tag{6.21}$$

Given that CARS is a coherent nonlinear process, for thick samples ($\gtrsim 0.1$ mm for water) the signal intensity also depends on factors that govern the wavevector mismatch between the applied fields and the propagating anti-Stokes wave. The phase mismatch results from the dispersion in the speed of light in the sample medium, such that the induced polarization wave in a macroscopic medium and the anti-Stokes wave that it generates travel at different velocities and eventually get out of phase with each other.

These issues of wavevector mismatch are minimized when CARS (or any non-linear optical process) is mediated through a microscope objective lens with a large numerical aperture, because most of the signal is generated in the short distance of the focal beamwaist. Consequently, the main application of CARS for biomedical research has been for microscopic imaging with the pump and Stokes beams tuned to a specific vibrational resonance, enabling imaging of the abundance of a specific molecular species in living cells, without any exogenous labeling. An excellent review of CARS for biomedical microscopy by Evans and Xie (2008) is listed under further reading.

6.7 Instrumentation for IR-absorption and Raman spectroscopy

6.7.1 Fourier-transform IR (FTIR) spectroscopy

The analytical technique predominantly used for infrared spectroscopy of organic compounds is called Fourier transform infrared (FTIR) spectroscopy. FTIR spectroscopy, most commonly, invokes transmission measurements of a thin (often dehydrated) sample with a broadband IR light source. The detection is based not on dispersion of the wavelengths, but on scanning an interferometer, generating a temporal Fourier transform (named for the French mathematician and physicist, Joseph Fourier [1768–1830]) of the spectrum that is detected by a broadband cooled IR sensor. Given that the methods of IR spectroscopy are rarely used in vivo or with tissue samples in their natural (hydrated) state, the applications of IR spectroscopy in biological science are limited and generally fall outside the scope of biomedical optics as defined for this text. The applications are generally limited to laboratory analysis of prepared (generally dehydrated) samples. The

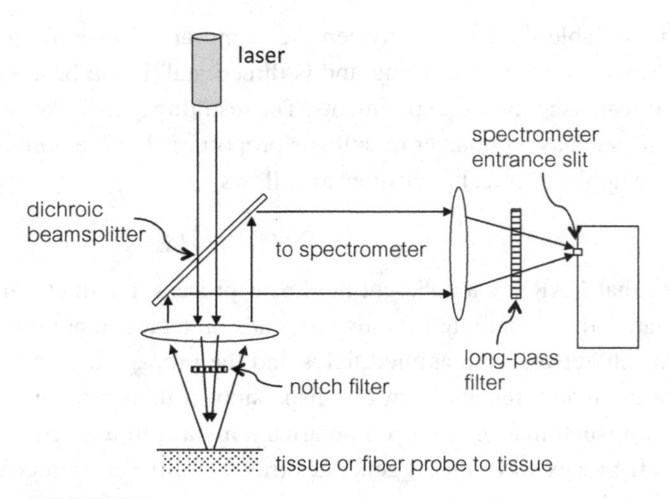

Figure 6.14

A schematic of a Raman spectroscopy setup, indicating some of the key optical components that help to enhance the signal-to-noise ratio for the weak Raman signal in the presence of strong background from other sources.

interested reader is referred to a reference manual by Griffiths and de Hasseth (2007), and an informative compilation of a broader range of current methods edited by Diem et al. (2008), both listed under further reading. There is, however, a significant amount of interest and activity in biomedical applications of Raman spectroscopy, with potential clinical applications becoming practical as a result of recent advances in optical technologies.

6.7.2 Basic instrumentation for Raman spectroscopy of biological samples

In Section 5.5 the primary elements of instrumentation for fluorescence spectroscopy were introduced, with a representative configuration shown in Figure 5.9. Fundamentally, the instrumentation for Raman spectroscopy of tissue follows the same themes. Light at an excitation wavelength illuminates the tissue, and that wavelength must be blocked from the detection arm because the light that we want to measure, which is at a longer wavelength, is weak compared to backscattered illumination light. This wavelength-selectivity in the detection is more important for Raman spectroscopy than for fluorescence measurements, because the Raman scattered light is several orders of magnitude weaker than typical fluorescence signals, and more than a million times weaker than the illuminating field. Figures 6.14 and 6.15 identify some of the components that are used to enhance the discrimination of Raman signal from other sources of confounding signal. The illumination source is almost always a laser, given the need for an intense illuminating field

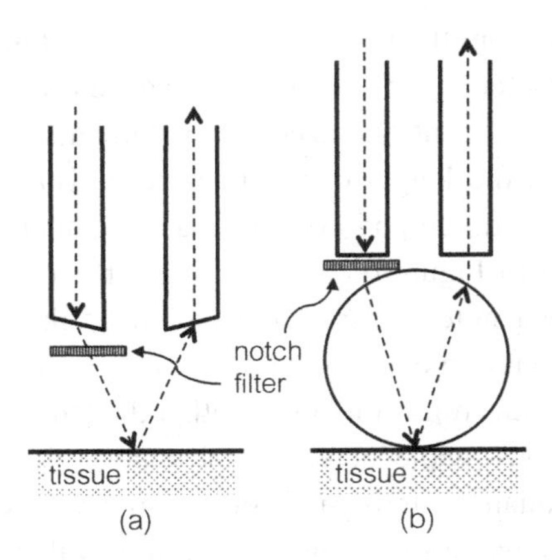

Figure 6.15

Two possible fiber probe configurations that enable overlap of the illumination and detection spots on the tissue, and can incorporate a notch filter on the illumination side: (a) the facets of the optical fibers are polished at an angle, such that the refraction angle leads to overlap of the optical cones; (b) a small ball lens both focuses and refracts the light cones of the illumination and collection fibers, leading to improved collection efficiency of the Raman scattered light.

strength at a narrow linewidth. Some of the other important components that are typically incorporated include: (1) a notch filter in the illumination arm; this is an optical filter with a narrow bandbpass (at the laser wavelength) and strong rejection of other wavelengths, which blocks spontaneous emission lines and other optical "noise" originating with the laser, at wavelengths other than the main laser line; (2) a dichroic beamsplitter to preferentially direct red-shifted light to the spectrometer; and (3) a long-pass filter before the entrance to the spectrometer, to further block backscattered laser light. One might think that the spectral dispersion in the spectrometer should accomplish the separation of the laser light from the Raman-shifted light on the detector array, but backscattered laser light can be so strong (compared to the weak Raman signal) that even a small percentage reflecting off optical components can "fog" the detector.

6.7.3 Instrumentation invoking fiberoptic probes

In a number of applications, a fiberoptic probe is desired for more convenient access to the tissue site where the measurement is to be made. (In Figure 6.14, an optical fiber would have one end at the focal point of the laser beam, and the notch filter would be relocated to the other end of the fiber, the sample end.) In this case, unlike the case of fluorescence spectroscopy (Figure 5.9), the fiber probe generally has two or more fibers, because the fiber used to convey the illumination light from the laser must be separate from the fiber(s) used to collect Raman-scattered light and convey it to the spectrometer. The reason is that the material of the illumination fiber itself can generate significant fluorescence or Raman signal, given the high laser field intensity within the length of the fiber core. That confounding light would backscatter from the tissue and constitute a large source of noise for the collection arm. As shown in Figure 6.15, a multiple-fiber configuration enables placement of an additional notch filter at the output of the illuminating fiber, blocking any additional emission coming from the fiber itself.

In addition to optimizing the rejection of light other than the Raman signal itself, various optical configurations enable increasing the solid angle for collection of the Raman scattered light. Raman scattering is strongest in the backward and forward directions. (The geometric pattern for Raman scattering follows that of Rayleigh scattering, which will be treated in Chapter 7, Section 7.2.1.) One of the simpler schemes is to surround the illumination fiber with a ring of collection fibers, which convey light to the spectrometer slit, where the fibers are rearranged into a line parallel with the slit (Hanlon et al., 2000).

6.7.4 Data treatment for Raman spectra of biological media

Once a raw Raman spectrum has been acquired, a few steps of data treatment are generally needed to generate a meaningful spectrum. These are as follows:

- Given that the Raman signal is generally weak, spectral noise is typically the result of photon-counting statistics (shot noise). The small pixel sizes of commonly used CCD array detectors also result in a degree of pixel-to-pixel statistical variation. Thus a first step of data treatment often constitutes a degree of computational smoothing, while being cautious not to remove small sharp spectral features with real information content.
- With real biological samples, especially tissues, and even when the illumination wavelength is in the near-infrared, there is always a background of fluorescence that is generally much stronger than the Raman signal itself. In the NIR, this fluorescence background is commonly monotonic (decreasing) with wavelength, so it can be fit quite accurately with a fourth- or fifth-order polynomial. Thus, the second step is to perform a "baseline fit" and subtract the fluorescence background, leaving the Raman spectrum.
- A third step is often to check the wavelength calibration of the system by measuring a spectrum of a material that has well-known and sharp spectral features that cover the range of Raman shift that is of interest. Polystyrene is often used for this purpose for spectral studies of biological samples.

6.8 Examples of pre-clinical and clinical applications of Raman spectroscopy

As discussed above, while IR-absorption spectroscopy can be useful for laboratory measurements of molecular samples, the majority of published research related to vibrational spectroscopy for biomedical applications invokes Raman spectroscopy, in its various forms. For clinical translational applications most studies have focused on diagnostic detection of tissue disease, with a heavy emphasis on detection of cancer. The two cases described here are exemplary of a number of published reports.

Haka et al. (2009) conducted a prospective test on freshly excised breast tissue samples. In prior work, the group had performed a retrospective "training" study to identify Raman spectral signatures that could be used to distinguish reliably between normal, benign (nonmalignant disease, such as fibroadenoma) and breast cancer (Haka et al., 2005). In that study the Raman signatures related to adipose tissue (fat = lipids) and collagen (a structural protein) were found to be the most reliable spectral features distinguishing between cancer and other tissue states, when retrospective correlations with histopathology were assessed. Referring to spectral bands listed in Table 6.1, differences (between breast cancer and benign tissues) can be seen in their published spectra for the relative band intensities of protein bands at 1270 and 1660 cm^{-1}, and in the strong lipid band at 1440 cm^{-1}. In their prospective study, the fitting parameters previously derived were tested on a new set of ex vivo tissue samples from a new set of patients, and the instrumentation was similar to a combination of what is illustrated in Figures 6.14 and 6.15(b). By using an illumination wavelength of 830 nm, background fluorescence was reduced, but detector integration times of 10–30 seconds were required for a good signal-to-noise ratio at each tissue spot. Practical in vivo clinical application generally requires faster measurement times, pointing towards further development.

In distinguishing breast cancer from all benign conditions, the classification scheme that was employed yielded sensitivity and specificity (SE and SP as defined in Section 3.2) of 83% and 93%, respectively. For the proposed clinical application of assessing resection margins during surgery, a high negative-predictive value (NPV) is important, and in this case an NPV of 99% was achieved. Nonetheless, given that the incidence of cancer samples in this dataset was small (6 of 129 measured sites), as discussed in Chapter 3 one must be mindful that achieving a high NPV can be deceptively easy with a low occurrence rate. In this case the high NPV was countered by a PPV of 34%.

In another study, Kanter et al. (2009) performed in vivo measurements on 90 patients to assess the potential for Raman spectroscopy to aid in the detection of early cervical cancer during colposcopy, a diagnostic procedure for the visual examination of the cervix. Again, the optical illumination and collected signal were mediated by a fiberoptic probe, placed in gentle contact with the tissue surface, incorporating spectral filters to minimize interference from fluorescence or backscattering of the illumination light. In this study the illumination wavelength was 785 nm, and a detector integration time of 5 seconds was found to be sufficient for an adequate signal-to-noise ratio. This shorter integration time approaches a condition that is clinically friendly. Nonetheless, as is common for non-resonant Raman spectroscopy of tissue, room lights and colposcope illumination were turned off during Raman measurements to minimize background signal, although the researchers were able to leave indirect, low-level incandescent lights on. For distinguishing different pathology conditions of the cervix, the most robust spectral differences were reported for Raman lines at approximately 1270, 1324, 1450 and 1655 cm^{-1}. As seen in Table 6.1, these bands can be tentatively assigned to the amide-III, the CH_2 twist, the CH_2 bend and deformation, and the amide-I bands, respectively. The amide bands are from protein backbones, and the CH_2 modes are from either proteins or lipids.

In that study, rather than a binomial classification (cancer vs. benign), and driven by clinical needs, a multi-class data analysis was developed with the intent to distinguish between four tissue classes: high-grade dysplasia (pre-cancer), low-grade dysplasia (early pre-cancer), metaplasia (benign cell growth), and normal ectocervix. Classification accuracy varied among class pairings. For example, the researchers reported sensitivity and specificity values, $Se = 92\%$ and $Sp = 96\%$, for distinguishing high-grade dysplasia from normal ectocervix, and lower scores for distinguishing low-grade dysplasia from metaplasia or normal tissue. The statistical strength of these classification values, however, is limited by spreading the available data among four classes. Additionally, this was a retrospective study, in which the dataset used for training the statistical algorithm also served as the testing set (in this case, performed as a leave-one-patient-out cross-validation). Consequently, there is a built-in statistical bias. Nonetheless, the results were encouraging, and do indicate that Raman spectroscopy has the potential to provide real-time diagnostic information in clinical procedures.

Problems – answers to problems with * are on p. 646

6.1 Units and dimensions: write the conversions that lead to the answers.

(a) A variety of units are often used to define the energy or wavelength of a photon emitted or absorbed by a molecule. If the energy of a photon is listed as 1 eV (electron-volt), what is the vacuum wavelength of its electromagnetic wave?

(b) What is the wavenumber (cm^{-1}) value for a vibrational transition that emits a photon with an energy of 0.1 eV. What is the wavelength (nm) of the emitted wave?

6.2* The C=O stretch vibration has a wavenumber value of ~1650 cm^{-1} in many biological molecules. What is the approximate value of the bond force constant, K?

6.3 Identify the normal modes of the following simple molecules, and determine whether each mode is IR-active, Raman-active or both: NO (nitric oxide), N_2O (nitrous oxide), CH_4 (methane). Draw simple diagrams of the molecules in their equilibrium states, with arrows indicating motions for the normal modes. Hint: look up the equilibrium shapes of these molecules.

6.4 In Table 6.1 there is a large gap in the listing of Raman shifts, between ~1750 and 2800 cm^{-1}. Moreover the Raman spectra at those larger wavenumber values tend to be very broad and devoid of fine spectral features. Can you explain why?

6.5* In studying the Raman spectra of a protein sample, a researcher desires to determine whether the protein was expressed by a cancer cell, or came from a different source. By growing the cells in a $C^{18}O_2$ environment, all of the oxygen atoms of the resulting protein peptide groups consist of the ^{18}O isotope. Assuming that the primary contribution to the amide-I mode is the C=O stretch vibration, and that the peak of the Raman line for the normal-isotope version is at 1650 cm^{-1}, what will be the approximate wavelength peak of the shifted isotopic variant?

6.6* In another study (similar to Problem 6.5) where the researchers wish to identify a specific CH– bond by its stretch mode, there is the option to label with ^{13}C or with D (deuterium). Which isotope substitution will yield a larger change in the vibrational frequency? Why?

References

Diem, M., Griffiths, P., and Chalmers, J. M., eds. (2008). *Vibrational Spectroscopy for Medical Diagnosis*. Hoboken, NJ: Wiley Interscience.

Draux, F., Jeannesson, P., Beljebbar, A., et al. (2009). Raman spectral imaging of single living cancer cells: a preliminary study. *Analyst*, 134, 542–548.

Gubskaya, A. V., and Kusalik, P. G. (2002). The total molecular dipole moment for liquid water. *Journal of Chemical Physics*, 117, 5290.

Haka, A. S., Shafer-Peltier, K. E., Fitzmaurice, M., et al. (2005). Diagnosing breast cancer by using Raman spectroscopy. *Proceedings of the National Academy of Sciences USA*, 102, 12371–12376.

Haka, A. S., Volynskaya, Z., Gardecki, J. A., et al. (2009). Diganosing breast cancer using Raman spectroscopy: prospective analysis. *Journal of Biomedical Optics*, 14(5), 054923.

Hanlon, E. B., Manoharan, R., Koo, T. W., et al. (2000). Topical review: prospects for in vivo Raman spectroscopy. *Physics in Medicine and Biology,* 45, R1-R59.

Hawi, S. R., Nithipatikom, K., Wohlfeil, E. R., Adar, F., and Campbell, W. B. (1997). Raman microspectroscopy of intracellular cholesterol crystals in cultured bovine coronary artery endothelial cells. *Journal of Lipid Research*, 38, 1591–1597.

Heald, M. A., and Marion, J. B. (2012). *Classical Electromagnetic Radiation*, 3rd edn. New York: Dover.

Jeanmaire, D. L., and van Duyne, R. P. (1977). Surface Raman electrochemistry Part I. Heterocyclic, aromatic and aliphatic amines adsorbed on the anodized silver electrode. *Journal of Electroanalytical Chemistry*, 84, 1–20.

Kanter, E. M., Majumder, S., Vargis, E., et al. (2009). Multi-class discrimination of cervical precancers using Raman spectroscopy. *Journal of Raman Spectroscopy*, 40, 205–211.

Kunapareddy, N. (2007). Raman spectroscopy for the study of cell death. Doctoral dissertation, Boston University.

Kunapareddy, N., Freyer, J. P., and Mourant, J. R. (2008). Raman spectroscopic characterization of necrotic cell death. *Journal of Biomedical Optics,* 13, 054002.

Polman, A., and Atwater, H. A. (2005). Plasmonics: optics at the nanoscale. *Materials Today*, 8, 56.

Raman, C. V., and Krishnan, K. S. (1928). A new type of secondary radiation. *Nature*, 121, 501–502.

Rayleigh, Lord. (1871). On the light from the sky, its polarization and colour. *Philosophical Magazine*, 41, 107–120, 274–279. Reprinted: (1964). *Scientific Papers by Lord Rayleigh, Vol I: 1869–1881*. New York: Dover, Ch. 8, pp. 90–91.

Römer, T. J., Brennan, J. F., Schut Tkker, T. C., et al. (1998). Raman spectroscopy for quantifying cholesterol in intact coronary artery wall. *Atherosclerosis*, 141, 117–124.

Schrader, B. (1989). *Raman/Infrared Atlas of Organic Compounds*. Weinheim: VCH.

Wangsness, R. K. (1986). *Electromagnetic Fields*, 2nd edn. New York: John Wiley.

Weinmann, P., Jouan, M., Dao, N. Q., et al. (1998). Quantitative analysis of cholesterol and cholesteryl esters in human atherosclerotic plaques using near-infrared Raman spectroscopy. *Atherosclerosis*, 140, 81–88.

Yuan, H., Khoury, C. G., Hwang, H., et al. (2012). Gold nanostars: surfactant-free synthesis, 3D modeling, and two-photon photoluminescence imaging. *Nanotechnology*, 23, 075102.

Zeng, S., Yong, K. T., Roy, I., et al. (2011). A review on functionalized gold nanoparticles for biosensing applications. *Plasmonics*, 6, 491–506.

Further reading

General vibrational spectroscopy

Chalmers, J. M., and Griffiths, P. R., eds. (2002). *Handbook of Vibrational Spectroscopy*. New York: John Wiley.

Diem, M., Griffiths, P., and Chalmers, J. M., eds. (2008). *Vibrational Spectroscopy for Medical Diagnosis*. Hoboken, NJ: Wiley Interscience.

Fourier transform infrared spectroscopy

Griffiths, P., and de Hasseth, J. A. (2007). *Fourier Transform Infrared Spectrometry*, 2nd edn. Hoboken, NJ: Wiley-Blackwell.

Biomolecular structure

Branden, C., and Tooze, J. (1999). *Introduction to Protein Structure*. New York: Garland.

Raman spectroscopy in biomedical research

Diem, M., Griffiths, P., and Chalmers, J. M., eds. (2008). *Vibrational Spectroscopy for Medical Diagnosis*. Hoboken, NJ: Wiley Interscience.

Evans, C. L., and Xie, X. S. (2008). Coherent anti-Stokes Raman scattering microscopy: chemical imaging for biology and medicine. *Annual Review of Analytical Chemistry*, 1, 883–909.

Muskovits, M. (2006). Surface-enhanced Raman spectroscopy: a brief perspective. In *Surface-Enhanced Raman Scattering – Physics and Applications*, ed. K. Kneipp, M. Moskovits and H. Kneipp. New York: Springer, pp. 1–18.

Tu, A. T. (1982). *Raman Spectroscopy in Biology: Principles and Applications*. Hoboken, NJ: John Wiley. Enhanced Raman spectroscopy.

Quantum theory of transitions

Harris, R. (2007). *Modern Physics*, 2nd edn. Reading, MA: Addison-Wesley.

7 Elastic and quasi-elastic scattering from cells and small structures

In Chapter 5 we reviewed methods of fluorescence spectroscopy and fluorescence measurements related to bioscience applications that enable monitoring of cellular metabolism, drug distribution, enzymatic processes, etc. In Chapter 6 we covered methods for spectroscopy of vibrational modes, which can yield diagnostic information by identification of the molecular "fingerprints" of biomolecules. These methods are powerful, and they predominantly provide information on the biochemistry of cells and tissue. Nonetheless, when considering potential clinical diagnostic applications, one might ask, "How does a pathologist diagnose cancer, when looking at a *histology* slide under a microscope?" After all, the pathologist is using an optical detector/camera (the human eye) to examine light coming from a sample of tissue through an optical instrument (the microscope). The term histology (from Greek: *histos* = tissue; *logia* = science or study) refers to the study of the microscopic structure of cells and tissue. Thus, the traditional practice of *histopathology* (the microscopic study of diseased tissue, usually for the purpose of diagnosis), which has been around for more than 120 years, may be regarded as one of the earliest techniques of biomedical optics to be put into practice.

In the common practice of histopathology, a small tissue sample is surgically cut from the organ where disease is suspected in a patient, a procedure called *biopsy*, and the sample is examined microscopically as part of the clinical diagnosis process. The biopsy sample is first fixed in formalin, which dehydrates it. It is then embedded in paraffin (for mechanical stability), and a very thin (typically 3–10 μm) slice is cut with a microtome and placed on a microscope slide. The thin slice is typically stained, most commonly with *hematoxylin* and *eosin stains* (when combined, called *H&E stain*), which stain for DNA (hence, nuclei) and other dense structures, mostly proteins. It is interesting to note that without such staining, very little cellular structure is visible in conventional bright-field microscopy, due to lack of contrast.

Getting back to the question about what a pathologist looks for when reading a slide and determining a diagnosis for disease (say, cancer), most of the time the pathologist is not performing biochemical assays or conducting genetic analysis: most of the time the pathologist is visually examining the cellular architecture, shapes and structures. Are the nuclei enlarged? What is the ratio of nuclear volume to cell volume? Is the nuclear staining especially dense (*hyperchromatic*), indicating excessive DNA? Are cell shapes and clustering patterns varied and chaotic? Is there a breakdown of the extracellular matrix? Under high-resolution microscopy: are the numbers and sizes of mitochondria different? Is the distribution of *chromatin* (the structural complex of DNA and proteins that form chromosome structures) in the nucleus granular or condensed?

All of these structural and architectural features can be expected to affect the way cells and tissue scatter light elastically. When we speak of *elastic scattering*, we are using the language of physics, in which the term "elastic" indicates a scattering event in which the photon does not exchange (gain or lose) energy (although momentum can be exchanged, since the photon may change its direction of propagation). This is to be contrasted with Raman scattering (see Chapter 6), which is an example of *inelastic scattering*, in which the scattered photon has, typically, a lower energy, and hence a longer wavelength than the incident photon. Another type of inelastic scattering is *Brillouin scattering* (named after French physicist Léon Nicolas Brillouin [1889–1969]), where photons lose or gain energy by interactions with acoustic waves (phonons) in a medium. (Brillouin scattering will be mentioned again in Section 16.2.1 on acousto-optic techniques in biomedical optics.) Scattering measurements that are sensitive to the *motion* of particles include *quasi-elastic Rayleigh scattering* (QERS), where the light frequency is broadened slightly because of random Doppler shifts caused by, say, Brownian motion of the scattering particles. With coherent light, such motion also induces fluctuations in the scattering signal intensity due to varying interference effects of light scattered from multiple particles. (Doppler shifts and coherent interference effects can be seen most easily when the light source has a narrow linewidth, as from a laser.) Other terms often used for QERS include *quasi-elastic light scattering* (QELS), *photon correlation spectroscopy* (PCS), and *dynamic light scattering* (DLS), and we will adopt the latter term here.

In Chapters 1 and 2 we introduced the concepts and formalism for describing the bulk scattering properties of tissue, and wrote the scattering coefficient in terms of the scattering cross section of individual particles (Eqs. (2.3), (2.4) and (2.5)). In this chapter we will examine in more detail the factors that determine the scattering cross section responsible for the scattering of light from individual particles, and how that scattering relates to physiological properties on a microscopic scale.

7.1 Sources of light scattering in biological systems

We re-present Figures 2.11 and 4.9 with a pair of physicist's cartoons (Figures 7.1 and 7.2) in an attempt to illustrate, in an oversimplified way, some structural

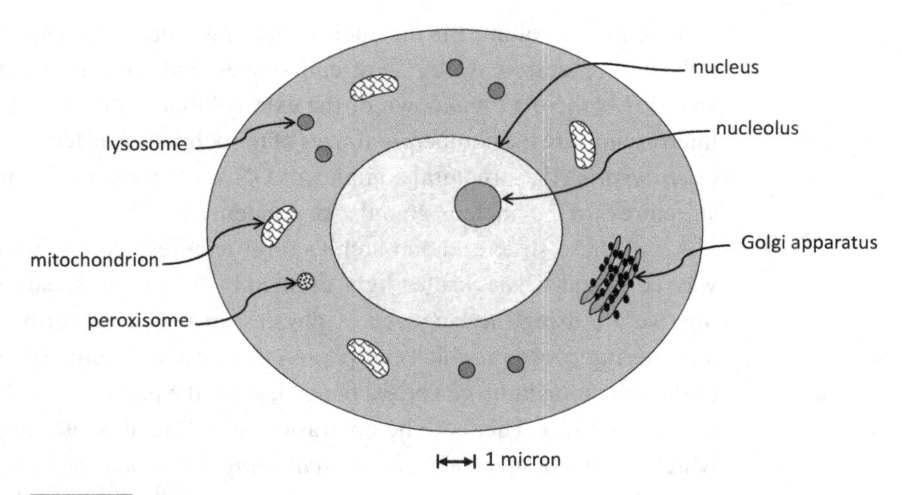

Figure 7.1

A physicist's cartoon of the primary components of a mammalian cell that are responsible for most light scattering from cells. Organelles are depicted in simplified (and generalized) shapes, and at their approximate sizes, roughly to scale.

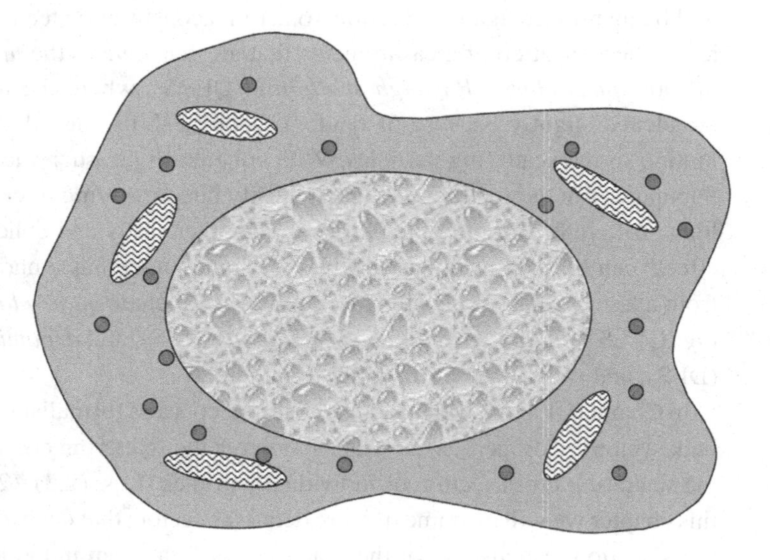

Figure 7.2

A qualitative representation of a cancer cell, indicating such altered features as enlarged nucleus, changed shapes of mitochondria, and granularity (condensation) of the DNA in the nucleus.

differences between a "normal" cell and a cancer cell. Light scatters from gradients and discontinuities of the refractive index, and in cells those index gradients arise from the fact that the effective refractive indices of organelles and other cellular structures are higher than the index of the cytoplasm.

In Table 2.1 we listed some approximate values for the refractive indices of various "building materials" and sub-structures found in tissue. Not depicted in Figure 7.2 are ultrastructural changes, such as breakdown of the cytoskeleton. (See further description of cellular structures below.) The indices listed in Table 2.2 with question marks (?) indicate that there is a large range of values to be found in the literature. The refractive index is related directly to the dielectric constant, and, consequently, to the polarizability of a medium. Proteins and DNA, in living organisms, always have some "bound" water, and even lipids are never found in tissue without some water. The effective index consequently depends on the fraction of the volume occupied by the solids, the remainder, generally in the range 60–80%, assumed to be water with various concentrations of dissolved salts and sugars. The primary exception to this is adipose tissue (fat), in which the percentage of water is generally less than ~15%. For lipids, for example, these index values are consistent with the known refractive indices of pure oils (~1.44–1.47).

The mass densities of the solid constituents, hence the specific index values, depend also on the molecular tertiary structure (folding, compaction, etc.), especially for globular proteins; and in the case of DNA the index depends on the degree of condensation of the chromatin. We remind the reader that although the refractive index of the lipid bilayer (also called the bilipid membrane) is relatively high, the mammalian cell membrane is very thin (reported variously in the range 3–9 nm), much less than the wavelength of visible light, and the refractive indices of the fluids on either side (extra- and intracellular fluids) are almost equal, due to the cellular machinery that maintains osmotic balance. Under these conditions, the membrane (hence, the whole cell in the presence of extracellular fluid) induces very little scattering of light. Consequently, organelles in cells and extracellular matrix proteins are the structures that are primarily responsible for light scattering in soft tissues. In situations where the index of the extracellular fluid changes significantly (say, due to a person ingesting a large quantity of glucose), the cell's mechanisms for responding to osmotic pressure will quickly balance concentrations across the membrane, maintaining a very small differential in concentrations of solutes. Nonetheless, such rebalancing does not occur across the boundaries of organelles inside the cells, and the index mismatch between organelles and the cytoplasm can indeed change. The result is that an increase of the refractive index of the extracellular (and intracellular) fluid will generally result in a decrease in scattering within tissue (Maier et al., 1994).

We provide here a (non-exhaustive) list of the typical structures in mammalian cells that are likely to contribute to the total scattering probability because they exhibit gradients in refractive index. These are listed in descending order of approximate size.

Nucleus: enclosed in its own bilipid membrane, the nucleus is a spheroid of 3–8 μm in diameter, although internal features, including the nucleolus and condensations of DNA, constitute smaller structures. During relatively static phases in the cell cycle of healthy cells, the DNA is distributed throughout the nucleus, which then appears as a relatively isotropic structure to an optical wave. In diseased (e.g., cancer) cells, or in dynamic states such as apoptosis, the DNA may be condensed and granular, presenting effectively smaller structures despite the overall enlargement of the nucleus, or generally denser due to *polyploidy*. *Ploidy* refers to the number of sets of chromosomes in the nucleus, which can be abnormally increased in disease states. Normal cells that are not undergoing cell division, or *mitosis*, are *diploid*, having two copies of the genome. Some notable exceptions are red blood cells that do not have nuclei (*nulliploid*), or sperm and egg cells whose nuclei have a single set of chromosomes (*haploid*).

Mitochondria: these energy generators of the cell (producing much of the cell's ATP, adenosine triphosphate) are often in a shape that resembles a kidney bean or as prolate spheroids, with a major axis length of 1–2 μm and diameter of 0.3–0.8 μm. The mitochondrion is enclosed in a bilipid membrane and has an internal ultrastructure of an undulating, folded membrane (forming "*crystae*") lined with a variety of nanoscale structures and containing mitochondrial DNA. The number of mitochondria in the cytoplasm can vary from a few to thousands.

Golgi apparatus: this organelle (named after the Italian scientist Camillo Golgi [1843–1926]) is a stack of several membrane-enclosed flat structures called *cisternae*, with a total stack thickness around ~200 nm and transverse dimension of up to 2 μm. An important role of the Golgi *apparati* is to package proteins that are destined for secretion outside the cell.

Lysosomes: these organelles are the recycling agents of the cell, digesting various compounds and cellular debris. Their sizes are in the range 0.2–0.9 μm, and they can be variously shaped, often spherical. Their numbers also vary by cell type.

Peroxisomes: smaller in number and less dense than lysosomes, peroxisomes are similar in size and predominantly spherical. Their major role is the breakdown of very long-chain fatty acids.

Structures smaller than 100 nm: there are a number of nanoscale organelles and structural elements in the cell that are important for its function and that help maintain spatial relationships among organelles. These include the *endoplasmic reticulum*, composed of tubules and sheets of membrane with dimensions in the range 30–100 nm, which (together with ribosomes) are involved in protein synthesis or lipid metabolism; *ribosomes*, the civil engineers that build proteins according to directions from *messenger-RNA*, and are 25–30 nm in diameter; the *cytoskeleton*, which is a lattice of protein filaments and *microtubules* that are around 25 nm in diameter, and help provide global mechanical structure for the cell, as well as guiding intracellular transport of molecules and particles.

It is interesting to note that, for dimensions greater than about 100 nm, the distribution of particle sizes in a mammalian cell is **not** continuous, but punctate. In a given cell type, the diameter of the nucleus is typically found to be within a narrow (±10%) range, or narrower during plateau phases of cell growth. A common nuclear size in, say, a squamous epithelial cell will often be around 5–8 μm, depending on the tissue type. There is nothing larger within the cell, and the next available smaller organelle size corresponds to that of the mitochondrion, whose major axis is generally <2 μm, with a diameter of <1 μm. There is nothing with a size between those of nuclei and mitochondria. Moving down from there, the next smaller particle size range is that of the lysosomes, with a gap in between them and the mitochondria. It is important to keep the size distribution in mind when trying to extract particle sizes by inverse solutions of theoretical scattering formalisms.

Non-cellular tissue structure: aside from the cellular sources of scattering, there is also, of course, the remaining bulk of tissue, which is predominantly composed of the structural proteins of the *extracellular matrix* (ECM), most importantly the fibrous proteins *collagen* and *elastin*. The individual fibrils are as small as 1.5 nm, but are typically cross-linked to form larger fibers. Depending on the tissue type, the ECM can contribute a significant percentage of the scattering strength of the tissue, but it generally behaves as structures that are smaller than 100 nm, consequently contributing to the *Rayleigh-scattering* element of the spectrum (to be discussed below in Section 7.2.1).

7.2 Scattering by a single particle: coordinates and formalism

If one is thinking in terms of the particle nature of light, it is easy to picture photons scattering off particles that have a discontinuously higher (or lower) index of refraction than their surrounding medium (see Figure 7.1), not unlike photons scattering from water droplets in clouds, or from fine air bubbles in water. While that approach is convenient for discussing the bulk scattering properties of tissue, if we want to understand the scattering processes that enable calculation of the scattering cross section for an individual particle, it is most convenient to revert to the wave-like description of light.

There are two primary functional relations that are needed to describe the scattering process for individual particles: the wavelength dependence of the total scattering probability (over all angles), and the angular probability distribution

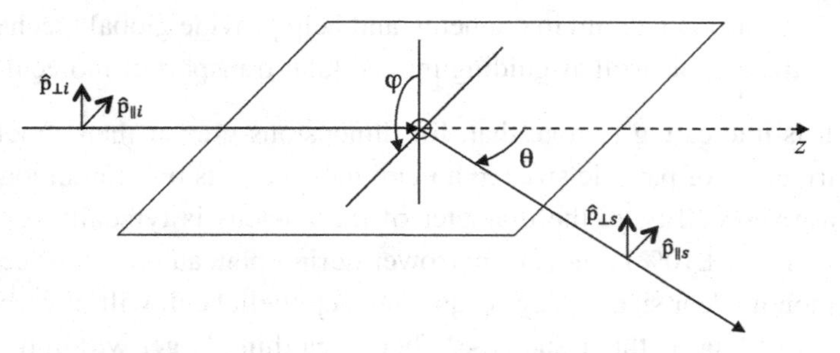

Figure 7.3

The coordinate system for describing the scattered field from a single particle is defined in the context of the plane of the scattering event, which is determined by the direction of the incident plane wave and the location of the detector measuring the scattered field. The unit vectors $\hat{\mathbf{p}}_{||i}$, $\hat{\mathbf{p}}_{||s}$, $\hat{\mathbf{p}}_{\perp i}$, $\hat{\mathbf{p}}_{\perp s}$ denote the directions of the electric field (polarization) components that are parallel or normal to the scattering plane for the incident and scattered waves.

for scattering at a specific wavelength. Recalling Eq. (2.4), which defines the scattering cross section as $\sigma_s = Q_s A_s$, where Q_s is the scattering efficiency and A_s is the physical cross section of the particle, we note that even though Q_s is dimensionless, it is a function of wavelength, particle size, and the indices of refraction of the particle and the medium. Similarly, the angular probability distribution for single-particle scattering given by the phase function, $p(\theta, \varphi)$, is also a function of those parameters. Thus, it is reasonable to expect that changes in the sizes and densities (hence refractive indices) of cellular organelles exhibited by diseased tissue will lead to changes in Q_s and $p(\theta, \varphi)$ that may be detectable with optical measurements. (For the discussions in this chapter, given the assumption of azimuthal symmetry in scattering, the phase function will be presented as $p(\theta)$.)

It is instructive to calculate the microscopic scattering properties of individual particles from first principles of electromagnetic theory. Although cellular organelles and other structures in tissue are of varied shapes and orientations, much can be learned by approximating them as uniform spheres to facilitate theoretical and computational determination of the scattered field. We start by establishing a coordinate system (Figure 7.3) defined by the direction of propagation of the impinging wave and by a scattering direction given by the polar angle θ relative to the incident direction and by the azimuthal angle φ relative to a reference direction orthogonal to the incident direction. The scattering direction is determined by the chosen location of a detector, and the resulting *scattering plane* is the basis for defining the polarization of incident (i) and scattered (s) rays as being parallel to the plane (*p-polarization*) or normal to the plane (*s-polarization*), denoted by subscripts $||$ and $\perp$, respectively, in the mathematical formalism.

For mathematical convenience, we use the exponential notation for the spatial and temporal dependence of the optical electric field from Eq. (4.4), in this case for a wave propagating in the $+z$ direction. (Again, we suppress the complex conjugate, and assume that physical properties are represented by the real part of complex exponential expressions.)

$$\mathbf{E} = (E_{\|}\hat{\mathbf{p}}_{\|} + E_{\perp}\hat{\mathbf{p}}_{\perp})e^{i(kz-\omega t)}, \tag{7.1}$$

where $E_{\|}$ and $E_{\perp}$ are the complex amplitudes of the components of the electric field parallel and normal, respectively, to the scattering plane, k is the angular wavenumber $(2\pi/\lambda)$, and ω is the angular frequency. The optical wave is *linearly polarized* when the two components $E_{\|}$ and $E_{\perp}$ are in phase; it is *circularly polarized* when $E_{\|}$ and $E_{\perp}$ have the same magnitude and are out of phase by $\pi/2$; it is *elliptically polarized* when $E_{\|}$ and $E_{\perp}$ have a phase difference other than 0 or $\pi/2$ (or they have a phase difference of $\pi/2$ and unequal magnitudes), and it is *unpolarized* when the direction of the electric field varies unpredictably in the plane transverse to the direction of propagation. A general expression for the scattered field ($\mathbf{E}_s$) at a distance, r (that is large compared to both the particle size and the optical wavelength), includes contributions from the incident plane wave, and the reradiated spherical wave from the induced polarization in the particle. This can be written in terms of the incident field ($\mathbf{E}_i$) by introducing a *scattering matrix* whose elements are indicated with S_i ($i = 1,2,3,4$):

$$\begin{bmatrix} E_{\|s} \\ E_{\perp s} \end{bmatrix} = \frac{e^{-ik(r-z)}}{ikr} \begin{bmatrix} S_2 & S_3 \\ S_4 & S_1 \end{bmatrix} \begin{bmatrix} E_{\|i} \\ E_{\perp i} \end{bmatrix}, \tag{7.2}$$

where the factor in front of the matrix contains a term representing the incident plane wave (e^{ikz}), and a term representing the scattered spherical wave $[e^{-ikr}/(ikr)]$. (The time dependence cancels on both sides of the equation.) The matrix elements provide the angular dependence of the scattering probability for each polarization direction. If we approximate the particle as being isotropic and spherical, then scattering will be azimuthally symmetric. (Symmetry arguments obviate any variation relative to the azimuthal angle, φ, about the z axis, whereas the transverse optical field direction of the incident wave does experience variation about the polar angle, θ.) Therefore, $S_3 = S_4 = 0$. (That is, for such symmetry and isotropy there can be no off-diagonal elements that would mix polarizations.) Consequently, in that case of azimuthal symmetry, $S_1 = S_1(\theta)$ and $S_2 = S_2(\theta)$.

As a result, we have

$$E_{\|s} = \frac{e^{-ik(r-z)}}{ikr} S_2(\theta) E_{\|i} \tag{7.3}$$

and

$$E_{\perp s} = \frac{e^{-ik(r-z)}}{ikr} S_1(\theta) E_{\perp i} \tag{7.4}$$

We also note that a detector does not measure electric field amplitude: it measures optical intensity, which is proportional to the square of the field amplitude. So, for an intensity in the far field resulting from an isotropic scatterer, the scattering matrix relation can be written as:

$$\begin{bmatrix} I_{\parallel s} \\ I_{\perp s} \end{bmatrix} = \frac{1}{k^2 r^2} \begin{bmatrix} |S_2|^2 & 0 \\ 0 & |S_1|^2 \end{bmatrix} \begin{bmatrix} I_{\parallel i} \\ I_{\perp i} \end{bmatrix}, \tag{7.5}$$

which leads to the expressions:

$$I_{\parallel s} = \frac{1}{k^2 r^2} |S_2|^2(\theta) I_{\parallel i}, \tag{7.6}$$

$$I_{\perp s} = \frac{1}{k^2 r^2} |S_1|^2(\theta) I_{\perp i}. \tag{7.7}$$

For unpolarized light, $I_{\parallel i} = I_{\perp i} = I_i/2$ and $I_{\parallel s} = I_{\perp s} = I_s/2$, so that Eqs. (7.6) and (7.7) yield the following relationship between the scattered and incident intensities:

$$I_s = \frac{1}{k^2 r^2} \left[\frac{|S_1|^2(\theta) + |S_2|^2(\theta)}{2} \right] I_i = \frac{1}{k^2 r^2} S_{11}(\theta) I_i, \tag{7.8}$$

where $S_{11}(\theta)$ denotes the total scattering probability at angle θ:

$$S_{11}(\theta) = \frac{|S_2|^2(\theta) + |S_1|^2(\theta)}{2}. \tag{7.9}$$

The scattering cross section (σ_s) can be defined as the total scattered power (i.e., the scattered intensity integrated over the surface of a sphere of radius r centered at the particle), divided by the incident intensity, resulting in the following expression:

$$\sigma_s = \frac{\int_0^{2\pi} d\varphi \int_0^{\pi} I_s r^2 \sin\theta d\theta}{I_i} = \frac{2\pi}{k^2} \int_0^{\pi} S_{11}(\theta) \sin\theta d\theta. \tag{7.10}$$

By comparing Eq. (7.10) with the normalization condition for the scattering phase function (Eq. (1.28)), one can see that $S_{11}(\theta) = p(\theta) k^2 \sigma_s$ (where k is the optical wavenumber), which provides the relationship between $S_{11}(\theta)$ and $p(\theta)$.

7.2.1 Rayleigh scattering

A semi-classical derivation of Rayleigh scattering, based on the interaction between an optical wave and the dipole moment of a molecular bond, was described

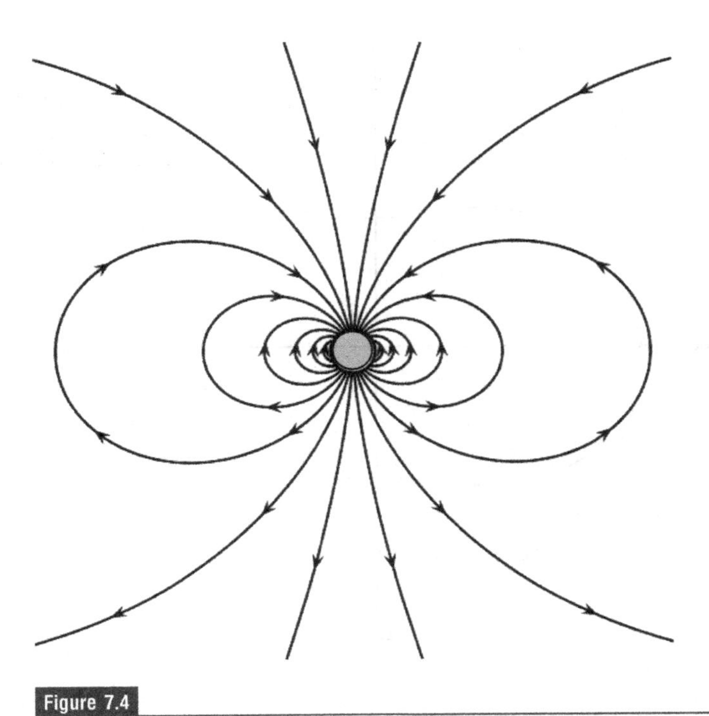

Figure 7.4

Field lines of a radiating dipole, excited by a p-polarized optical field propagating horizontally.

in Chapter 6. In the limit that the size of a particle is much smaller than the wavelength of light, one can treat the optical electric field in the particle as being uniform. In that circumstance, the particle behaves as a driven dipole, radiating in all directions with the dipole-field distribution pattern that is illustrated in Figure 7.4.

In this small particle limit, and with isotropic polarizability, α, of the particle, the scattering amplitude matrix can be written as:

$$\begin{pmatrix} S_2 & S_3 \\ S_4 & S_1 \end{pmatrix} = ik^3\alpha \begin{pmatrix} \cos\theta & 0 \\ 0 & 1 \end{pmatrix}, \tag{7.11}$$

which can be derived from a combination of Eqs. (6.12) and (6.15), noting the change in coordinate designations. The $\cos\theta$ factor of the matrix element S_2 and the angle-independent factor (unity) of S_1 can be understood intuitively by examining the electric-field component directions of the incident and scattered waves in Figure 7.3. The p-polarized components (of the incident wave and the scattered field) overlap as $\cos\theta$, whereas the overlap of the s-polarized components is unity, regardless of scattering angle. Remembering that we measure intensity,

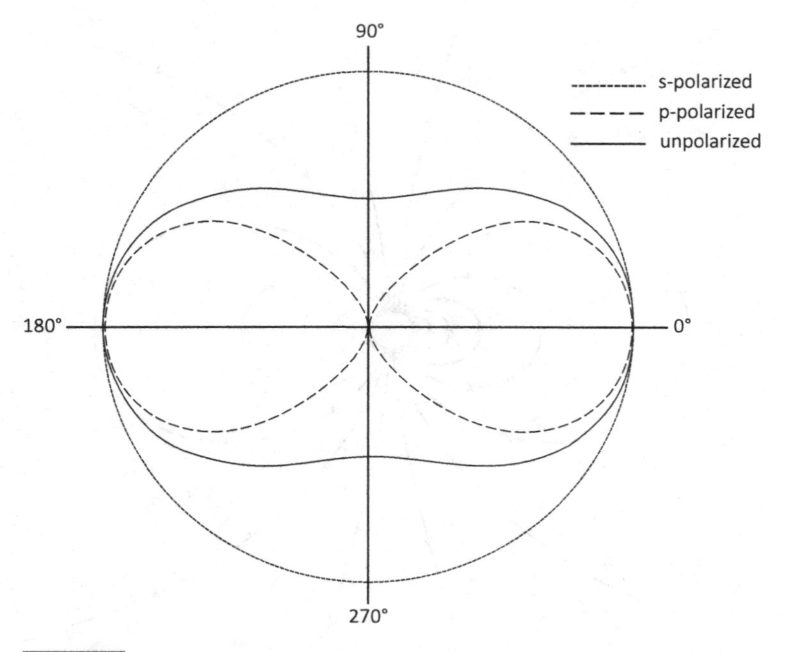

Figure 7.5

Polar plot of the relative θ-dependent intensity of scattered light from a small particle with isotropic polarizability, for incident light that is s-polarized, p-polarized and unpolarized light.

which is proportional to the square of the field amplitude, as written in Eq. (7.5), it is instructive to generate a polar plot of the scattered intensity as a function of θ, for incident light that is p-polarized light, s-polarized light and unpolarized (the sum of p- and s-polarized components), as illustrated in Figure 7.5.

From Eqs. (7.8) and (7.11), we can also deduce that the scattered intensity of unpolarized light relates to the incident intensity (in cgs-Gaussian units) as:

$$I_s = I_i(2\pi)^4 \frac{(1+\cos^2\theta)}{2r^2}\frac{\alpha^2}{\lambda^4} \tag{7.12}$$

where r is the distance from the particle, and α is the polarizability (total electric susceptibility) of the particle. As defined in Chapter 6 (Eq. (6.5)), the polarizability is the proportionality factor between the induced dipole moment and the applied electric field, and it has cgs units of length cubed. (We use cgs units, wherein the permittivity of free space is unity, in this instance because of the simplicity compared to the SI units of polarizability: C-m^2-V^{-1}.) Given that α is proportional to the volume of the particle, the scattering cross section for Rayleigh scattering depends on the sixth power of the particle size (as specified by a linear dimension a) and the negative-fourth power of the wavelength:

$$\sigma_s \propto \frac{a^6}{\lambda^4} \tag{7.13}$$

The negative-fourth power dependence on the wavelength, discussed in Section 6.2.1, leads to the classical explanation of the blue sky as a consequence of preferential scattering of the shorter wavelengths of the sunlight by atmospheric molecules. Given that many structures in tissue are smaller than 50 nm, Rayleigh scattering is almost always a significant component of light scattered from tissue.

7.2.1.1 *Polarized Rayleigh scattering*

In addition to explaining the blue color of the sky, the principles of Rayleigh scattering also reveal that the blue skylight is partially polarized. As can be deduced from Figure 7.5, the maximum polarization ratio is observed when viewing the sky at 90° to the direction of the sunlight. This effect is often used by photographers to deepen (darken) the apparent blue shade of the sky in outdoor photography, by properly orienting the transmission axis of a linear polarizing filter on the camera lens. (The effect can also be observed when wearing polarizing sunglasses on a clear day, and tilting one's head to cross the axis of the polarization of the lens filter to that of the skylight.)

7.2.2 Mie theory

As mentioned in Section 7.2.1, in the Rayleigh approximation (for particles whose diameter is much smaller than the wavelength of the incident field), the applied field can be treated as spatially invariant within the particle, thereby serving as a uniform, oscillating driving field for the induced oscillating dipole moment. For particles that are too large for the Rayleigh scattering approximation to hold, we seek more general solutions to the electromagnetic wave equation. Since many organelles and structures in tissue are of dimensions that are comparable to or larger than the wavelengths of visible light, this regime is important for understanding light scattering by tissue. *Mie theory* or *Mie scattering* is the most commonly used moniker for the exact solution to Maxwell's equations that describes the scattering of a monochromatic electromagnetic wave by a homogeneous dielectric sphere (of arbitrary size) with an isotropic refractive index in an isotropic dielectric medium. The attribution refers to the German physicist Gustav Mie [1869–1957] for his 1908 publication in the journal *Annalen der Physik*, although the principles of

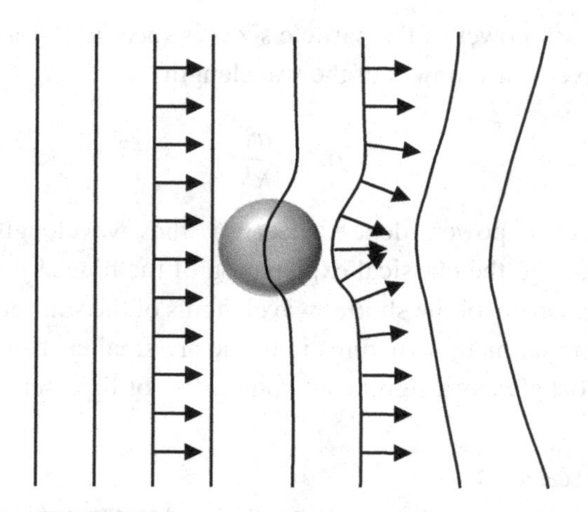

Figure 7.6

A plane wave propagating from left to right becomes locally distorted after interacting with a sphere whose diameter is of the order of the wavelength of the light. The distorted wavefront translates into scattered light propagating in different directions in the far field. In Mie theory, the scattered wave is derived from a combination of the external field and the spherically radiating internal field, obeying appropriate boundary conditions. (Not represented here is a back-reflected set of waves that also contribute to the scattered field.)

the solution were explained earlier by the Danish mathematician/physicist Ludvig Lorenz [1829–1891] in 1890 (in the Danish journal *Det Kongelige Danske Videnskabernes Selskabs Skrifter*) and published posthumously in 1898 in French. Consequently, the method is also sometimes referred to as *Lorenz-Mie theory*.

The representation of scattering particles in tissue (e.g., most organelles) as being uniform spheres is an idealized simplification, of course. Nonetheless, many particles are in fact spheroids that behave (optically) approximately as spheres of an equivalent dimension. Moreover, even non-spheroidal shapes, when oriented randomly, can exhibit *averaged* scattering properties similar to spheres of some representative size. Consequently, when many particles are involved in the measurement, which is commonly the case in biomedical optics, valuable semi-quantitative information is available from theoretical representation of the particles as spheres.

To help gain an intuitive sense of what is happening, we can imagine the distortion of the wavefront of the incident electromagnetic plane wave after interacting with a dielectric sphere (say, in air), as depicted in Figure 7.6. The change in direction of the local Poynting vector for regions of the wavefront in the near field will translate to scattered light in the far field that is traveling at a range of angles from the original direction of propagation.

The Mie theory approach is to consider the three fields, incident, internal and scattered, and solve Maxwell's equations for the scattered field, subject to the requirement that the tangential components of the electric and magnetic fields be continuous at the spherical boundary of the particle. (It turns out that the normal component of the magnetic field is also continuous, whereas the normal component of the electric field is not continuous at a boundary between media with different indices of refraction.) The electromagnetic *normal modes* of the sphere are expressed as series expansions of vector spherical harmonics, and the far field is a superposition of those normal modes, with relative contributions determined by coefficients that depend on the particle properties. (In the case of isotropic particles, with no azimuthal dependence, cylindrical functions [various types of Bessel functions [after the German astronomer Friedrich Wilhelm Bessel [1784–1846]] and their inherent relation to Legendre functions [after the French mathematician Adrien-Marie Legendre [1752–1833]] can be invoked, as explained above and detailed below.) The key to the Mie theory approach is the determination of those expansion coefficients. Our condensed discussion outlines the highlights of the formalism of C.F. Bohren and D.R. Huffman (B&H) in their now-classic book: *Absorption and scattering of light by small particles* (Wiley-Interscience, New York, 1983). The representation here is limited to an isotropic dielectric particle of refractive index n_i in a dielectric medium of index n_0.

The first step is to write the incident plane wave as an expansion of vector spherical harmonics, and this is the most arduous part of the formalism, because a plane wave does not naturally lend itself to expansion in spherical harmonics. We do know that this can be done, nonetheless, because any arbitrary field shape can be represented by a series of functions that constitute a complete orthonormal set for the appropriate number of dimensions. (This is analogous to representing an arbitrary one-dimensional waveform with a set of sine waves, which can be determined from the Fourier analysis of the waveform.) For such a representation of a plane wave, the ambitious reader is directed to B&H, where the derivation is guided by the desire to perform the expansions with specific polynomials and functions (e.g., spherical Bessel and Hankel functions) that lend themselves readily to computational calculation by series expansions. Amusingly, B&H themselves remark: "the reader who has painstakingly followed the derivation [of the expansion of a plane wave in vector spherical harmonics], and thereby acquired virtue through suffering, may derive comfort from the knowledge that it is relatively clear sailing from here on."

The utility of expanding the incident plane wave in a set of vector spherical harmonics is that such representation lends itself to combining with the internal field and the scattered field when meeting the boundary conditions (on the spherical surface of the particle) and determining which normal modes of the spherical particle can be excited by the incident field. The essence of the approach is to construct vector spherical harmonics that will satisfy the vector wave equation from functions that satisfy the *scalar* wave equation. Vector spherical harmonics are generated by taking the curl of $\mathbf{r}\psi$, where $\mathbf{r}$ is the radius vector of a spherical polar coordinate system, and ψ is any function that is a solution to the homogeneous scalar wave equation (i.e., $\nabla^2\psi + k^2\psi = 0$) in spherical polar coordinates, referred to as the *generating function*. The specific choices of generating functions, the mathematical identities and the notations employed, as well as the unit systems chosen, vary among a number of authors over the years. An interesting historical review of the various formalisms of Mie theory has been provided by Logan (1965). In the case of B&H, the generating function ψ is of the form:

$$\psi_{en} = \cos\phi P_n^1(\cos\theta) z_n(kr) \tag{7.14}$$

and

$$\psi_{on} = \sin\phi P_n^1(\cos\theta) z_n(kr) \tag{7.15}$$

where the subscripts e and o refer to the even and odd elements of the function expansions, respectively, and z_n can be any of the spherical Bessel functions (j_n, y_n), or the spherical Hankel functions $(h_n^{(1)}, h_n^{(2)})$ (named after the German mathematician Hermann Hankel [1839–1873]). Then, in the B&H notation, one vector field ($\mathbf{M}$) is defined as follows:

$$\mathbf{M} = \nabla \times (\mathbf{r}\psi), \tag{7.16}$$

such that $\mathbf{M}$ is a solution to the vector wave equation in *spherical coordinates*; and it turns out that $\mathbf{M}$ is tangential to any sphere $|\mathbf{r}| = \text{constant}$. (That is, $\mathbf{r} \cdot \mathbf{M} = 0$ due to the condition of spherical symmetry.)

A second vector field $\mathbf{N}$ can be defined in terms of $\mathbf{M}$ as follows:

$$\mathbf{N} = \frac{\nabla \times \mathbf{M}}{k} \tag{7.17}$$

Again, since ψ is a solution to the scalar wave equation in spherical coordinates, $\mathbf{N}$ is also a solution to the vector wave equation (i.e., $\nabla^2\mathbf{M} + k^2\mathbf{M} = \nabla^2\mathbf{N} + k^2\mathbf{N} = 0$). Furthermore, each has zero divergence ($\nabla \cdot \mathbf{M} = \nabla \cdot \mathbf{N} = 0$) because the divergence of the curl of a vector field is always zero.

With this formalism, the result for the incident plane wave is of the form:

$$\mathbf{E}_i = E_0 \sum_{n=1}^{\infty} i^n \frac{2n+1}{n(n+1)} \left(\mathbf{M}_{on}^{(j)} - i\mathbf{N}_{en}^{(j)}\right), \tag{7.18}$$

where the subscripts (on) and (en) indicate vector fields defined from ψ_{en} and ψ_{on}, respectively, and the superscript (j) refers to the specific use of the spherical Bessel function j_n in Eqs. (7.14) and (7.15).

The second step of the Mie-theory approach is to write the expressions for the internal field and the scattered field, which are of the same form as Eq. (7.18), but with the addition of the expansion coefficients (a_n, b_n, c_n, d_n) for the vector fields **M** and **N**:

$$\mathbf{E}_{int} = E_0 \sum_{n=1}^{\infty} i^n \frac{2n+1}{n(n+1)} \left(c_n \mathbf{M}_{on}^{(j)} - i d_n \mathbf{N}_{en}^{(j)}\right) \tag{7.19}$$

$$\mathbf{E}_s = E_0 \sum_{n=1}^{\infty} i^n \frac{2n+1}{n(n+1)} \left(-b_n \mathbf{M}_{on}^{(h)} + i a_n \mathbf{N}_{en}^{(h)}\right) \tag{7.20}$$

where the superscript (h) refers to the specific use of the spherical Hankel function h_n in Eqs. (7.14) and (7.15). We note that E_0 in Eqs. (7.19) and (7.20) is the electric-field amplitude of the *incident* wave. (i.e., the amplitudes of the internal and scattered fields are proportional to the amplitude of the incident wave). Expressions of similar form are generated for the magnetic fields, $\mathbf{H}_i$, $\mathbf{H}_{int}$ and $\mathbf{H}_s$, of the incident, internal and scattered waves, respectively. For the scattered electric field, $\mathbf{E}_s$, which is our primary interest, the expansion coefficients a_n and b_n are the key to determining the scattering matrix elements, S_1 and S_2.

The third step is to write the boundary conditions (of continuous tangential field components) at the surface of the particle:

$$(\mathbf{E}_i + \mathbf{E}_s - \mathbf{E}_{int}) \times \hat{\mathbf{r}} = (\mathbf{H}_i + \mathbf{H}_s - \mathbf{H}_{int}) \times \hat{\mathbf{r}} = 0 \tag{7.21}$$

where $\hat{\mathbf{r}}$ is the unit vector normal to the particle surface (which is spherical and centered on the origin). These cross products lead to four independent (scalar) equations in four unknowns (for each value of n): a_n, b_n, c_n and d_n. In addition to the boundary conditions, the form of the expansions for the internal fields and the scattered fields is determined by the form of the expansion of the incident field and the orthogonality of the vector harmonics.

With elastic scattering, the energy of the photons (hence the frequency of the light) is unperturbed; nonetheless, the speed of light, and hence the wavelength of light, depends on the index of refraction of the medium in which it is propagating. Noting also that the refractive indices of the particle and external medium are different, two main parameters are introduced to characterize a scattering particle (and to simplify the notation of the solutions of the four equations). The first is the unitless *size parameter*, x:

$$x = \frac{2\pi n_0 a}{\lambda}, \tag{7.22}$$

where a is the particle radius; and the second is the *relative refractive index*, m:

$$m = \frac{n_1}{n_0}, \tag{7.23}$$

where n_1 and n_0 are the refractive indices internal to the particle and of the surrounding medium, respectively. (This use of m should not be confused with the azimuthal expansion index in Eqs. (7.14) and (7.15).) It is convenient and intuitive to use the unitless size parameter to represent the dimensions of particles, because the wavelength of the light is the size of the "ruler" that is being used to take a measure of the particle. This formalism is, therefore, equally valid for assessing the scattering by a golf-ball-sized sphere with microwaves, or by a basketball-sized sphere with radio waves (as long as the observer is far enough away from the interaction site to be in the far field).

The resulting expansion coefficients for the scattered wave, after significant algebraic manipulation, are:

$$a_n = \frac{m\psi_n(mx)\psi_n'(x) - \psi_n(x)\psi_n'(mx)}{m\psi_n(mx)\zeta_n'(x) - \zeta_n(x)\psi_n'(mx)} \tag{7.24}$$

$$b_n = \frac{\psi_n(mx)\psi_n'(x) - m\psi_n(x)\psi_n'(mx)}{\psi_n(mx)\zeta_n'(x) - m\zeta_n(x)\psi_n'(mx)}, \tag{7.25}$$

where ψ_n and ζ_n are the Riccati-Bessel functions (Jacopo Riccati [1676–1754], Italian mathematician), which are defined in terms of spherical Bessel and Hankel functions, respectively, ($\psi_n(x) = x j_n(x)$ and $\zeta_n(x) = xh_n^{(1)}(x)$), and the prime denotes the first derivative with respect to the argument. Similar expressions are obtained for the expansion coefficients of the internal wave.

We note, after this condensed derivation, that the choice of Bessel (or Hankel) functions (which are convenient for cases of cylindrical symmetry) for the expansion coefficients is consistent with the conditions of an isotropic particle medium, leading to azimuthal symmetry in the scattered fields. It is also gratifying to note that when the relative refractive index m approaches 1, the scattering coefficients a_n and b_n both approach zero. (It should not be surprising that when there is no index mismatch, there is no scattering!)

By combining Eqs. (7.3) and (7.4) with the boundary-matching equations that emerge from Eq. (7.21), we can now write the scattering matrix elements in terms of the expansion coefficients.

$$S_1(\theta) = \sum_{n=1}^{\infty} \frac{2n+1}{n(n+1)} (a_n \pi_n(\cos\theta) + b_n \tau_n(\cos\theta)) \tag{7.26}$$

and

$$S_2(\theta) = \sum_{n=1}^{\infty} \frac{2n+1}{n(n+1)} (b_n \pi_n(\cos\theta) + a_n \tau_n(\cos\theta)) \tag{7.27}$$

where π_n and τ_n are functions of the scattering angle:

$$\pi_n(\cos\theta) = \frac{1}{\sin\theta} P_n^1(\cos\theta) \tag{7.28}$$

$$\tau_n(\cos\theta) = \frac{d}{d\theta} P_n^1(\cos\theta) \tag{7.29}$$

and P_n^1 is the *associated Legendre function* of degree n and order 1. The angle-dependent functions have $\cos\theta$ as their argument as a result of the use of that form of the Lengendre polynomials, as in Eqs. (7.14) and (7.15), and the assumed spherical particle.

When properly normalized, S_{11} (see Eq. (7.10)) then can lead to the angular phase function, $p(\theta)$, of the scattering particle (for unpolarized light). Specifically, we recall that $p(\theta) = S_{11}/(\sigma_s k^2)$, where k is the optical wavenumber. The dimensionless scattering efficiency, Q_s, can also be calculated from S_{11} as follows:

$$Q_s = \frac{1}{x^2} \int_0^{\pi} 2S_{11}(\theta) \sin\theta d\theta \tag{7.30}$$

If angular distribution information is not needed, Q_s for scattering over all angles can be calculated more simply from:

$$Q_s = \frac{2}{x^2} \sum_{n=1}^{\infty} (2n+1)(|a_n|^2 + |b_n|^2) \tag{7.31}$$

As expressed in Eq. (2.4), the scattering cross section of a particle, σ_s, is then defined as the product of the scattering efficiency, Q_s, and the actual geometrical cross-sectional area of the particle, $A_s = \pi a^2$:

$$\sigma_s = Q_s A_s \tag{7.32}$$

which then enables determination of the bulk scattering coefficient, μ_s, using Eq. (2.5). (This also gets us back to Eq. (7.10).)

It is instructive to write the expansion of Q_s for the special case of no absorption (i.e., when $m = n_1/n_0$ is real), and for all angles (ignoring angular dependence). In this approximation, the first three terms for Q_s are expressed as

$$\begin{aligned} Q_s = & \frac{8x^4}{3}\left(\frac{m^2-1}{m^2+2}\right)^2 \\ & \times \left[1 + \frac{6}{5}x^2\left(\frac{m^2-2}{m^2+2}\right) + x^4\left\{\frac{3}{175}\left(\frac{m^6+41m^4-284m^2+284}{(m^2+2)^2}\right)\right.\right. \\ & \left.\left. + \frac{1}{900}\left(\frac{m^2+2}{2m^2+3}\right)^2 [15 + (2m^2+3)^2]\right\} + \cdots\right] \end{aligned} \tag{7.33}$$

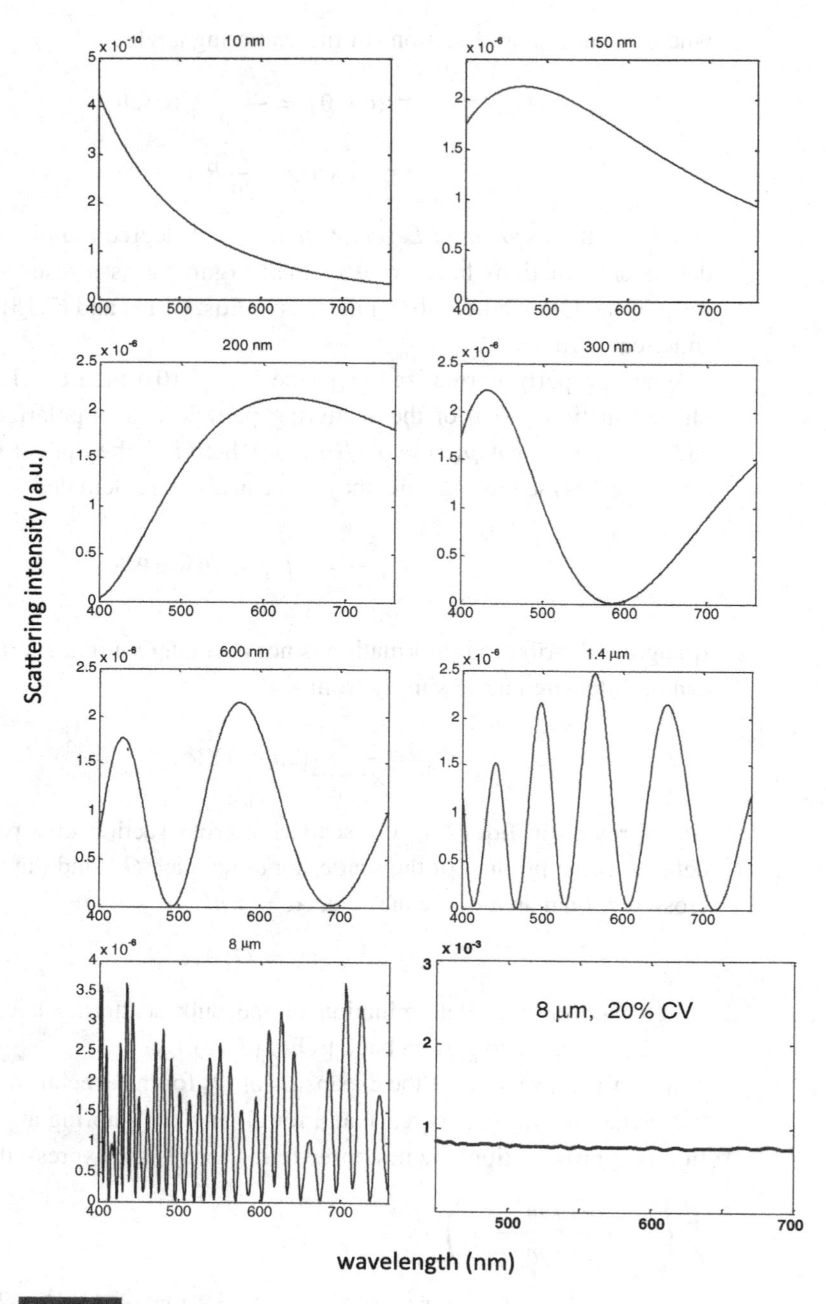

Figure 7.7

Plots of relative scattered intensities for a range of particle diameters (indicated in each panel) with $m = 1.03$. The lower-right plot is the average scattering intensity for particles with an average diameter of 8 μm, and with a distribution of sizes with a coefficient of variation (CV), i.e., the ratio of standard deviation to the mean, of 20%. The ordinate axes of the single-size particle plots are in arbitrary units, but are scaled relative to each other, illustrating the fact that larger particles exhibit larger scattering cross sections.

The first term inside the large square brackets is the number 1, followed by a term with the second power of the size parameter, x^2, and the third term with the factor x^4. (Subsequent terms have increasing even powers of x.) In the limit of particles that are much smaller than the wavelength, the size parameter $x \ll 1$, and terms after the first, 1, can be ignored. It is encouraging that this small particle limit of Mie theory leads to the correct representation of Rayleigh scattering, as can be seen by combining the first term of Eq. (7.33) into Eq. (7.32). For larger particle sizes, however, the subsequent terms in the expansion become important, and the scattering cross section is more complex, and also more interesting than the simple λ^{-4} dependence of Rayleigh scattering. That increased complexity is also more informative about particle features including size and index.

In Figure 7.7 we plot the wavelength-dependence of relative scattered intensities for a number of particle sizes that are typical of organelles and small structures in tissue. For all of these plots the relative refractive index chosen was $m = 1.03$ (a reasonable value for some cellular organelles), and the measured scattering angle was 180°. The upper-left plot, for 10-nm particles, exhibits the expected λ^{-4} wavelength dependence for Rayleigh scattering. The next six plots exhibit the increasing spectral complexity for larger particles. All but the lower-right plot are for single particles, or sparse groups of particles with *monodisperse* size distributions: i.e., all particles of equal size. The high-frequency spectral oscillations for the 8-μm-diameter particle (sized like a nucleus) would wash out for scattering from particles that are not all of the same size. The lower-right plot has been added to show the effect of a 20% coefficient of variation in the particle size for the 8-μm particles, as might be expected in real tissue. The wavelength dependence of scattering for large particles (with a variation coefficient >5%) approaches an approximate power law that is proportional to $\lambda^{-0.4}$.

The scattering phase function can be plotted in two ways, each of which is illustrative. Figure 7.8 shows the phase function (for the full range of θ: 0–180 degrees) for three different sizes of particle, with $m = 1.17$ (for polystyrene spheres in water), at a wavelength of 500 nm, as well as the Henyey-Greenstein (HG) phase function at $g = 0.85$ for comparison. (The HG phase function, which will be discussed in greater detail in Chapter 8, is often used as a mathematically simple approximation for the scattering phase function (Henyey and Greenstein, 1941). It is named after the American astrophysicists Louis Henyey [1910–1970] and Jesse Greenstein [1909–2002].) As in the case of the scattered intensity, if there is a significant variance in the particle size of larger particles, then scattering from multiple particles will wash out the fine oscillation in $p(\theta)$, but larger features will remain. It should be noted that the ordinate in this plot is logarithmic, and that the scattering is highly forward-biased for the two larger particle sizes.

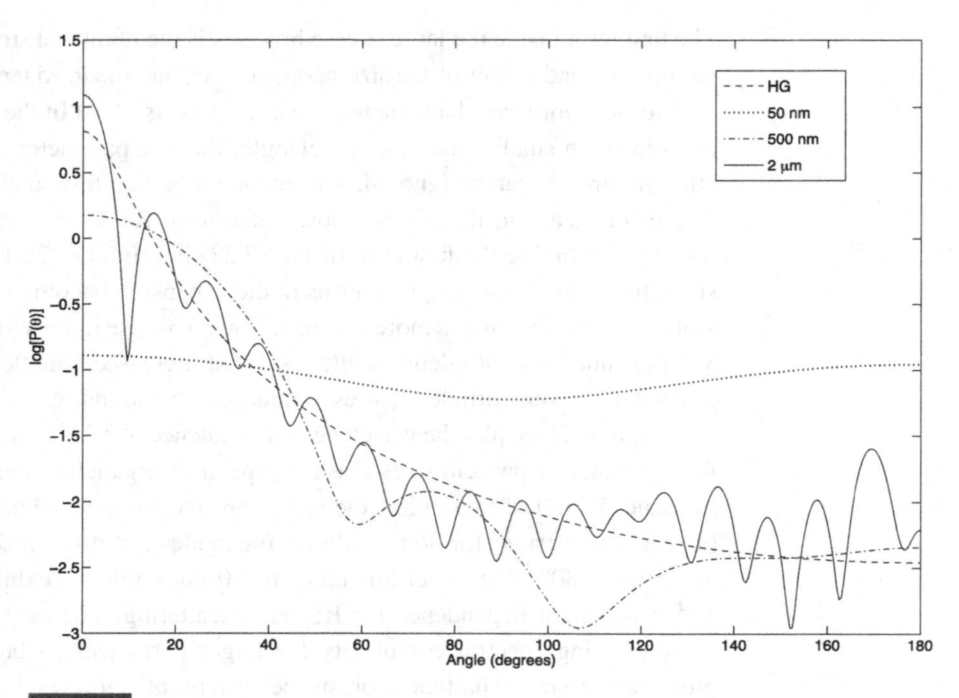

Figure 7.8

The Mie-theory scattering phase functions at $\lambda = 500$ nm for three particle diameters, compared with the Henyey-Greenstein phase function for $g = 0.85$. For the Mie-theory calculations, the particle material was polystyrene ($n \cong 1.56$), and the medium was water ($n = 1.33$). This combination is often used for tissue phantoms intended for testing instrumentation performance and data analysis methods.

Another instructive way to plot the phase function is as a polar plot, comparable to Figure 7.5, but with logarithmic radial axis. This can be plotted in two dimensions, because, for the assumptions made for the Mie-theory formalism, the scattering is azimuthally symmetric. Figure 7.9 shows a polar plot for the scattering phase function of a particle with $m = 1.04$, and diameter 500 nm. The scattering is strongly forward-biased, but there are oscillations and variations over larger angles, including direct backscattering.

To help illustrate the useful information content for different parameter ranges, we plot in Figure 7.10 the relative scattered intensity as a function of both wavelength and angle. (This can be extracted from calculations of the total scattering probability $S_{11}(\theta)$.) The reader will notice that for scattering in the near-forward direction (angles $<\sim 20°$), the wavelength dependence is featureless, whereas at large angles the scattering gains significant complexity. This is consistent with the experience of several research groups (Mourant et al., 1996; Canpolat and

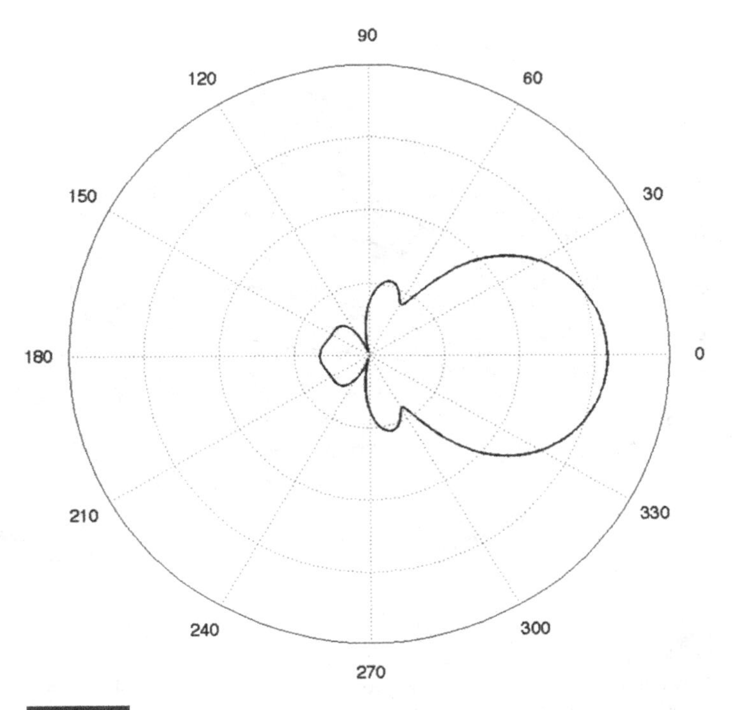

Figure 7.9

Polar plot of the scattering phase function at a wavelength of 500 nm for a 500-nm diameter particle with $m = 1.04$. The radial axis is logarithmic.

Mourant, 2001; Amelink et al., 2003) who have found that measurements of the scattering spectrum at large scattering angles ("backscattering") is more informative of changes in cellular structure than the more intense forward scattering. This illustrates how guidance from theoretical modeling of scattering can help design experimental geometries that enhance sensitivity to diagnostic information.

7.3 Exact and numerical calculations of scattering by single particles of arbitrary shape

Analytical methods like Mie theory limit us to idealized spherical approximations of biological micromorphology. It is sometimes valuable, however, to be able to calculate scattering by other shapes. If the shape of a particle is such that its boundaries coincide with a coordinate system in which the wave equation is separable (such that the methods of separation-of-variables can be applied), then methods similar to Mie theory can be applied. Mie-like solutions have been

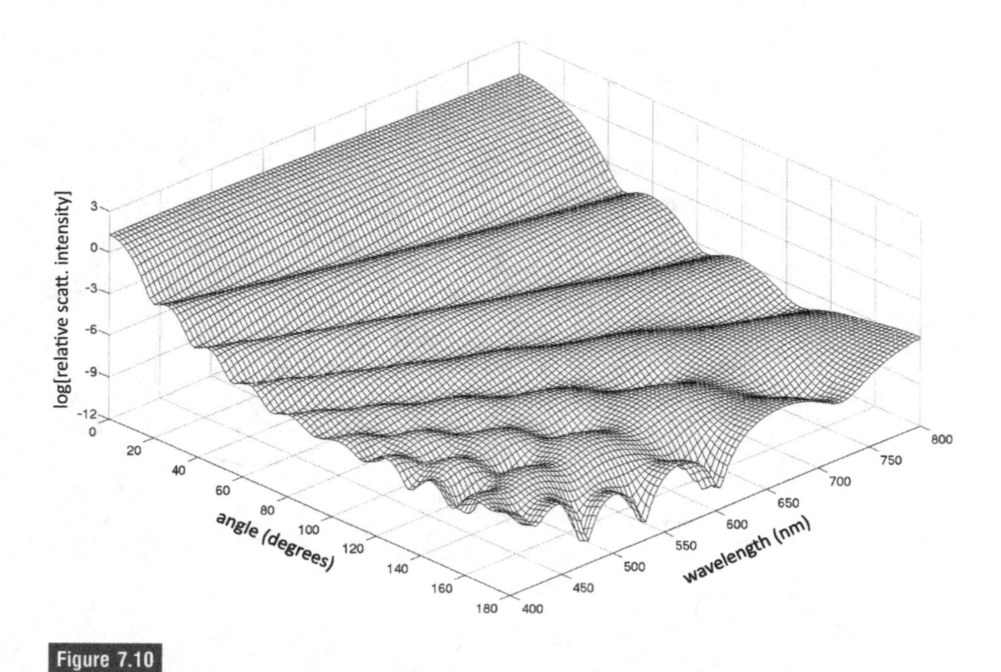

Figure 7.10

A plot of relative scattered intensity (vertical axis) as a function of both wavelength and angle for a 1-μm polystyrene particle in water.

published for long cylinders (fully derived in B&H), and for oblate and prolate spheroids (Asano and Yamamoto, 1975). (Oblate and prolate spheroids have a polar axis that is shorter or longer, respectively, than the equatorial diameter.) For biological particles, for which the relative refractive index is generally <1.05, comfort can also be drawn from Rayleigh-Gans theory (Richard Gans [1880–1954] was a German physicist). The Rayleigh-Gans approximation refers to "soft" particles for which $m - 1$ is small (i.e., $m \approx 1$), and leads to scattering matrix elements for which particles of arbitrary shape can be treated as spheres of an equivalent radius. (See, for example, B&H Chapter 6.)

Nonetheless, one may sometimes need more detailed information about particles of arbitrary shape and index. One rigorous numerical approach for calculating scattering by a particle of arbitrary shape is based on the *finite-difference time-domain* (FDTD) method. Maxwell's equations teach that the time derivative of the electric field (**E**-field) is dependent on the change in the spatial distribution of the magnetic field (**H**-field) through its *curl*. In the absence of free currents, the microscopic version of the equation can be written as:

$$\nabla \times \mathbf{H} = \varepsilon \frac{\partial \mathbf{E}}{\partial t}, \tag{7.34}$$

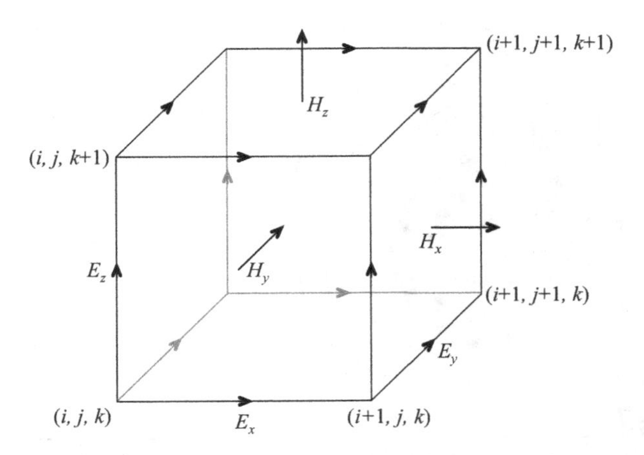

Figure 7.11

The voxel cell for FDTD computations as formulated by Yee (1966), showing the Cartesian components of the electric and magnetic fields.

where ε is the medium permittivity. In the FDTD method, the computational domain includes the volume of the scattering particle and some of its surrounding medium, and is divided into a grid of typically rectangular unit cells, or *voxels*, in a Cartesian-coordinate array. In one convenient formalism (Yee, 1966) the **E**-field is specified at the edges of each voxel and the magnetic **H**-field is specified normal to each face. (See Figure 7.11.) The optical field source is represented as a plane wave, and the computation is carried out in a time-step manner for every point in the grid. For each point, the time-iterated value of the **E**-field is dependent on the prior value of the **E**-field and the numerical curl of the local distribution of the **H**-field in space around it. Similarly, the **H**-field is stepped in time for each point, with each iteration depending on the prior value of the **H**-field and the curl of the surrounding **E**-field (calculated numerically).

One way to picture the process of an FDTD numerical computation is with each grid element (voxel) becoming an individual radiating element, and the resultant field is the sum of all the fields from the individual radiators (Figure 7.12).

Since the FDTD method is a time-domain calculation, if the light source emits a short pulse, a broad range of frequencies are provided by the same simulation. This allows calculation for a range of optical wavelengths with one simulation (Drezek et al., 2000). A disadvantage of the FDTD method is that the grid, which covers the entire computational domain, must have a voxel spacing that is fine enough to cover the smallest spatial gradients of structure that are of interest and must also be much smaller than the shortest optical wavelength of interest. This can lead to extremely large grid domains for computation, and very long computing times.

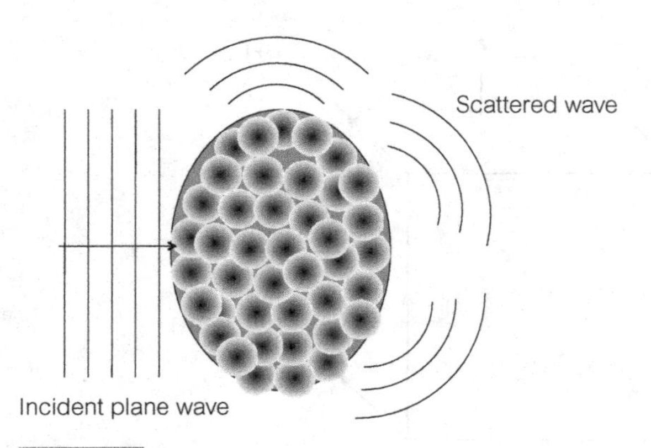

Figure 7.12

In an FDTD computation, each of the voxels (see Figure 7.11) in the volume of the particle, with its own local properties, can serve as a radiation source for a spherical wave radiating from that voxel. The fields from those individual radiators can then be summed at an observation point.

(For modeling the structure of a cell, the structural features dominate this issue, since they can be much smaller than the optical wavelength.) Another issue with the FDTD method is that it is essentially a near-field calculation, whereas measurements and detection of light scattered from tissue are almost always performed in the far field, at a distance large compared to the structure domain and the wavelength. Determining the resultant far field can require extensive postprocessing for transposition of near-field to far-field conditions. It is reassuring to note that Dunn et al. (1997) have shown that the FDTD method applied to a spherical particle yields the same result as Mie theory, when the near-field to far-field conversion is applied.

Other numerical methods, including the *T-matrix* method, have been proposed. Although these have not been widely applied in biomedical optics, they can offer insight on scattering from non-spherical shapes. The T-matrix approach is a computational method for calculating scattering by non-spherical particles, including particles with a non-homogeneous refractive index. In this method, as with Mie theory, the incident and scattered electric fields are expanded in a series of vector spherical harmonics (Mishchenko et al., 1996). Although the method can be applied to particles of arbitrary shape, it is well suited for particles that have an axis of rotational symmetry (axisymmetric particles) (Nilsson et al., 1998; Quirantes, 2005), as in the case of prolate or oblate spheroids. Invoking the linearity of Maxwell's equations, the scattered and incident field coefficients are related by a transition matrix (T-matrix). By choosing the "natural" reference frame (e.g., the z axis being the axis of rotational symmetry), the matrix elements can be averaged for all directions of incident and scattered fields, which is equivalent to averaging over particle orientations, and can be helpful for simulating scattering by cells in tissue (Giacomelli et al., 2010).

7.4 Single-scattering methods to study cell properties: instrumentation and applications

Chapters 9–14 will develop the theory and methods of analysis of scattering and absorption in the diffusion regime, with large source-detector separations, for which photons have scattered many times before they are collected. Chapter 8 will address the regime where the source and detector are closer to each other, and collected photons have scattered a few times, for which the analysis of diffuse reflectance entails empirical formalisms and numerical simulations. (Under those conditions, generally, one cannot invoke the methods of the diffusion approximation to the transport equation.) In this section we address some of the methods that enable measurement of photons that have scattered only once, permitting the benefit of analytical description by Mie theory and other methods discussed above. Measurements of single scattering from individual cells enable such applications as differentiation of cell types or monitoring of internal cellular processes and dynamics. The measurements are of two parameters: the angular dependence of scattering probability at a given wavelength, or the wavelength dependence of scattering at a given angle, or range of angles.

7.4.1 Measuring the scattering phase function: methods and devices

As demonstrated in Figures 7.8, 7.9 and 7.10, the scattering phase function exhibits informative complexity for particles that are larger than the Rayleigh-scattering size range ($\gtrsim$50 nm). The experimentally simplest method for measuring the angular dependence (the phase function) of scattering from living cells is to do just that (!), as implemented with an optical *goniometer* (from the Greek: *gonia* = angle; and *metron* = measure). A more precise term for a device intended to measure scattered light at different angles would be a *polar nephelometer* (Hansen and Evans, 1980) (from the Greek, *nephos*, for cloud). Two common types of design of polar nephelometers can be differentiated: the most common type, as depicted in Figure 7.13, uses a single detector adjustable to different angles, and the second one measures simultaneously the light scattered with many detectors arrayed at a range of angles around the test space. The first type offers controllable angular resolution

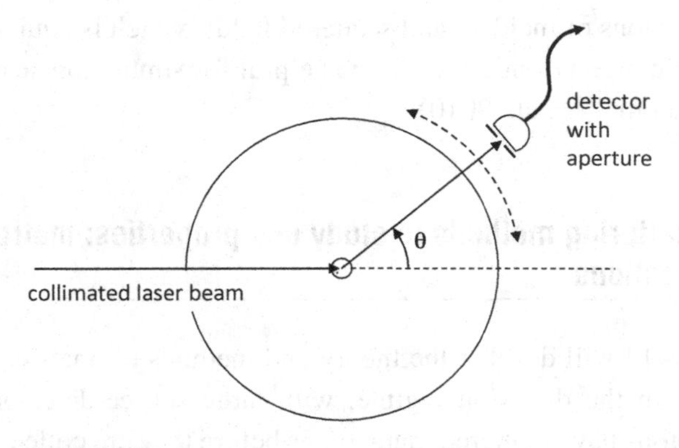

Figure 7.13

The simplest, and most common, implementation of a polar nephelometer for measuring the scattering phase function of cells and organelles. The cells are in fluid suspension, at a low enough concentration to ensure single scattering of the measured light.

but is slow, whereas the second type has limited angular resolution but performs the measurement almost instantaneously. In the more common approach, cells in suspension are placed in a cylindrical vial located at the axis of a protractor-like rotating assembly (Figure 7.13). A collimated laser beam illuminates the sample at right angles to the rotation axis, and a detector (typically, a photomultiplier or avalanche photodiode) is rotated incrementally around the sample. The parallel-detection version of this system simply has multiple detectors arrayed around the arc, often coupled by fiber optics to separate pixel groups on a common CCD detector.

When a polar nephelometer is used for particle sizing (by inversion of scattering data using Mie theory), the angular resolution is particularly important, as the angular frequency of the oscillations in the scattering phase function increases with particle size (Figure 7.8). Also, when characterizing a polydisperse range of particle sizes, as with living cells, the measurement time is crucial for the correct determination of the phase function, as the motion of cells should be small with respect to the time it takes to scan the angular field of measurement. New designs of polar nephelometers have been demonstrated to achieve both rapid measurement and high angular resolution, albeit with a more complex optical system (Castagner and Bigio, 2007).

As an example of the application of angular scattering measurements to reveal properties of cells, Ramachandran et al. (2007) measured the phase functions of tumorigenic and non-tumorigenic cell lines of the same basic cell type, and correlated the differences in the measured phase function to alterations in cellular microarchitecture, which they modeled with consistent size distributions.

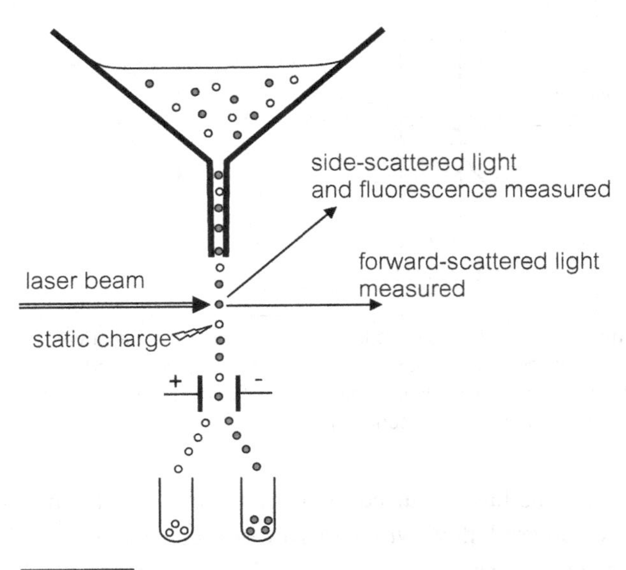

Figure 7.14

Key elements of a typical flow cytometer include hydrodynamic flow structures to generate a stream of single cells; an illuminating laser, or set of lasers at different wavelengths; various detectors to sense fluorescence from labels, or scattering at side and forward angles. Individual cells or cell droplets can carry static charge to facilitate sorting by a controlled electric field following selection by the system of detectors.

7.4.2 Flow cytometry

One of the earliest applications of a method that assesses a scattering property of cells is *flow cytometry*. This general term can refer to any method of rapid analysis of individual cells in a high-throughput manner, by flowing the cells "single file" in a small laminar stream past a detection/sensing zone. The earliest versions measured cell volume or density based on electrical impedance or capacitance. These were introduced by the American engineer Wallace H. Coulter [1913–1998] in 1953, and modern versions are still sold as "Coulter counters." Flow cytometers based on optical measurements of cells were developed in the 1960s and 1970s, originally based on counting cells that have been selectively tagged with a fluorophore. More recently, flow cytometers have included the capability of comparing side-scattering intensity (at ~90°) to near-forward scattering, and have added the capability of sorting the cells after measurement. As illustrated in Figure 7.14, a laser beam interrogates the stream of cells, one cell at time, and optical detectors at various angles sense emitted fluorescence from fluorophore tags, or measure the side-to-forward scattering ratio, which provides information about the relative sizes of cells or their internal constituents. As the engineering

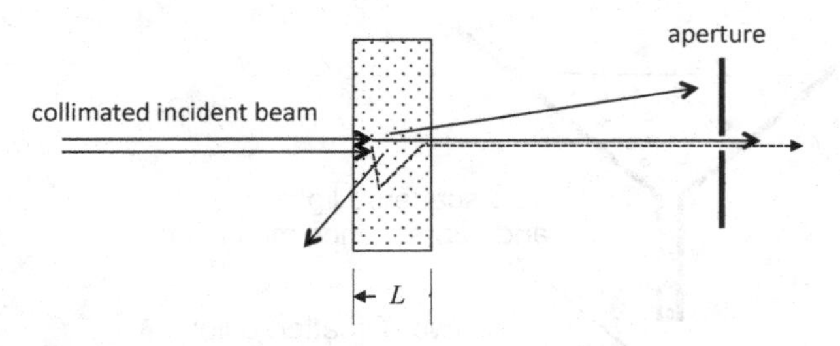

Figure 7.15

A simple measurement to determine the scattering coefficient can invoke detection of light passing through a collimating aperture at a distance from the scattering medium, and comparison of that with the incident light. Any multiply scattered light that passes through the aperture, as exemplified by the dashed-line photon path, contributes a confounding signal.

of these systems has advanced, and the variety of available fluorescent cellular labels has increased, flow cytometers are now capable of assessing a large range of cellular properties and can sort thousands of cells per second. (See, for example, Shapiro, 2003.)

7.4.3 Measurements of single scattering events in tissue in a transmission geometry

Equation (7.33) is an expression for the total scattering efficiency (over all angles), Q_s, of a single spherical particle of known size and relative refractive index. Equation (7.32) expresses the scattering cross section in terms of Q_s, and we also recall the expression for the macroscopic scattering coefficient for bulk tissue, under low number-density conditions, from Eq. (2.5):

$$\mu_s = \sigma_s N = Q_s A_s N, \tag{7.35}$$

where N is the number density of particles. We also noted that, when absorption is negligible, the Beer-Lambert law can be used to approximate the loss of unscattered light through a thin slab of scattering medium with thickness L:

$$I(\lambda) = I_0(\lambda) e^{-\mu_s L} \tag{7.36}$$

But Eq. (7.36) is only relevant for a transmission geometry (or for a configuration with known geometric pathlength) where a collimated beam passes through a thin (or weakly scattering) slab, in which only single scattering events are likely, as was depicted in Figure 2.15. This is an approximation because of the assumption that there are no multiple scattering events that could result in a photon scattering more than once, and still end up collinear with the original beam, as depicted by the dashed-line photon path in Figure 7.15. Nonetheless, if the probability of multiple scattering is very small, then measurement of the wavelength dependence of the ratio of transmitted to incident light can yield the wavelength dependence of μ_s (through Eq. (7.36)); and if the number density of particles, N, is known, then direct information about the particles can be gleaned from Mie theory.

Thus, this approach is to learn about the particles, under conditions where light has scattered only once, by measuring the unscattered light and making deductions about the scattered light from that. It is also possible to measure the wavelength dependence for an integral of most of the scattered light from a transmission geometry as in Figure 7.15, by use of an *integrating sphere*, which averages the light propagating in all directions, and presents the average to a detector.

7.4.4 Measurement of singly scattered photons in a backscattering geometry

As illustrated in Figure 7.10, the wavelength dependence of backscattered light is typically more informative than that of near-forward scattered light. For measurements in a backscattering geometry from bulk tissue, however, especially with a two-fiber configuration, as in Figures 5.7 and 5.8, most of the collected light has been scattered more than once, and Beer's law cannot be applied, unless the average photon pathlength for the collected photons is somehow otherwise known. (A special case will be introduced in Section 8.6.1, in which the mean pathlength of multiply scattered photons can be determined.) Here we consider experimental conditions that facilitate selective detection of singly scattered photons, for which Mie theory can be used to understand the relation between scattered wavelength dependence and cellular microscopic structural properties.

7.4.4.1 *Scattering from cell monolayers*

Microscopy of cells is generally facilitated by growing a monolayer of cells on a culture plate. Such a geometry is also convenient for measurement of single backscattering events from the cells, as all backscattered light will be scattered only once, given the vanishingly small probability of scattering more than once within a single cell. By use of an optical geometry that restricts the detected light to a narrow range of backscattering angles, a spectral measurement of the scattered light over a broad range of wavelengths (often called *elastic-scattering spectroscopy*) enables direct correlations with Mie theory. As an example, the geometry illustrated in Figure 7.16, using two small optical fibers, provides a convenient geometry for this type of measurement. Although the cone angles of the optical fibers ($\sim 26°$ for fibers with a numerical aperture, NA, of 0.22) enables illumination of a multi-cell area of the cultured cell monolayer, enhancing the collected signal, the range of angles between any illumination ray and its collection ray for a single cell is small. For two adjacent fibers with cores of 100 μm (center-to-center separation ~ 120 μm), whose ends are at a distance of

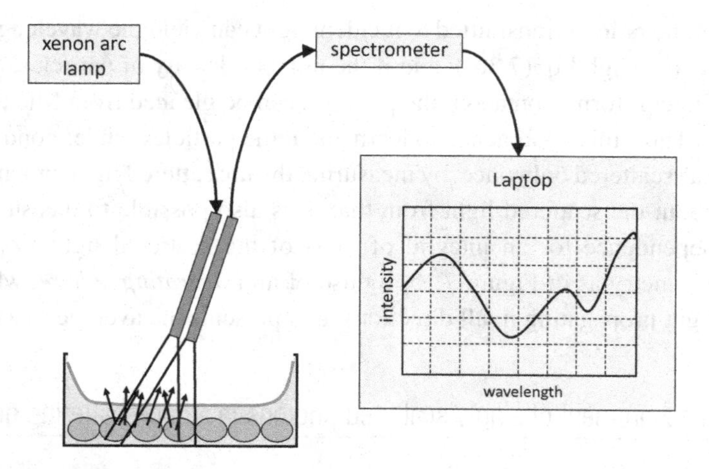

Figure 7.16

Diagram of the optical geometry for measurement of the spectrum of light backscatterd at almost 180° from a monolayer of cells on a culture plate.

~6–7 mm from the cell monolayer, the average backscattering angle is 179°, and all possible angles between an illumination ray and the associated collection ray of the collected backscattered light are within ±0.8° of that angle (formed between the centers of the fiber cores and any cell in the monolayer).

This geometry was utilized by Mulvey et al. (2009, 2011a) in a series of studies to monitor *apoptosis* in cultures of cells treated with various chemotherapeutic agents. In these studies the spectrum was measured from 400 to 800 nm of backscattered light at 179.2 ± 0.6° from cultures of epithelial cells, including cancer cell lines, as a function of time following the administration of chemotherapy agents. Apoptosis, sometimes called "cell suicide," is a type of programmed cell death, and is generally the earliest observable response to treatment, if the cells indeed respond. In the reported studies, changes in the backscattered spectrum could be reliably measured at times (10–20 minutes) that were considerably earlier than the earliest time points at which established assays for apoptosis (e.g., measurement of caspase cascade activity) can typically detect changes. The authors attributed the early response to a combination of ultrastructural changes of mitochondria and other organelle structures at nano-scales that are too small to be seen microscopically, and to alterations in the relative refractive index between cytoplasm and organelles (Mulvey et al., 2011b).

7.4.4.2 *Selection of singly backscattered light from bulk tissue*

For backscattering measurements from bulk tissue, however, most of the backscattered light has scattered more than once (see Figure 5.8), imposing limitations to analysis by Mie theory or other exact theories of single scattering. Some of the research efforts to develop clinical applications of elastic scattering spectroscopy, notably for the detection of pre-cancer or cancer, have focused on techniques to enhance the detection of singly scattered photons over multiply scattered photons. The more common methods have been based on polarization-sensitive measurement of the backscattered light (Backman et al., 1999; Qiu et al., 2010). In this approach, the illumination source is polarized, and measurements are made of components of the backscattered light, I_s, that are parallel ($I_s^{\|}$) and perpendicular ($I_s^{\perp}$) to the polarization of the illumination. These methods take advantage of the fact that singly backscattered light retains the polarization of the illumination, whereas diffusely scattered light has become unpolarized after several scattering events. This leads to the following simple assumptions:

(i) that backscattered light, I_s, is composed of both singly scattered light, $I_{s,1}$, and multiply scattered light, $I_{s,m}$:

$$I_s = I_s^{\|} + I_s^{\perp} = I_{s,1} + I_{s,m}; \tag{7.37}$$

(ii) that all of the singly scattered light is polarized parallel to the illumination light:

$$I_{s,1} = I_{s,1}^{\|}; \tag{7.38}$$

(iii) that multiply scattered light is composed of equal parts of light polarized parallel and perpendicular to the polarization of the illumination light:

$$I_{s,m} = I_{s,m}^{\|} + I_{s,m}^{\perp} \tag{7.39}$$

and

$$I_{s,m}^{\|} = I_{s,m}^{\perp} \tag{7.40}$$

It also follows that all scattered light that is polarized perpendicular to the illumination is multiply scattered light. The scattered light intensity can then be expressed as:

$$I_s = I_{s,1}^{\|} + I_{s,m}^{\|} + I_{s,m}^{\perp} \tag{7.41}$$

and it follows that:

$$I_{s,1} = I_s^{\|} - I_{s,m}^{\|} = I_s^{\|} - I_s^{\perp}. \tag{7.42}$$

Thus, the spectrum of singly scattered light can be obtained from the difference of two polarized measurements of the backscattered light: one with the polarizing filter parallel to the illumination polarization, and the other with the polarizing filter perpendicular to the illumination polarization. In practice, if only a small percentage of backscattered light is singly scattered, then application of Eq. (7.42) entails subtracting measurements of two large signals that are close to each other, such that the resulting difference will be noisy.

Another approach for selectively detecting singly scattered photons in the presence of diffusely backscattered light is to perform both illumination and collection of the light through a single, small-diameter fiber that is in contact with the tissue surface (Canpolat and Mourant, 2001; Amelink et al., 2003). This method enhances the proportion of singly scattered light that is collected, but does collect some amount of multiply scattered diffuse light as well. The relative proportion of singly-to-multiply scattered light that is collected depends on the fiber diameter and numerical aperture, and on the scattering properties, including the phase function, of the cellular components of the tissue surface (epithelium). If the fiber diameter is smaller than the mean free path for scattering ($1/\mu_s$), and if the cone angle (numerical aperture) for illumination/collection is narrow, then most of the photons that have scattered two or more times will have traveled out of the most likely collection zone in the tissue, whereas photons that scatter directly back, following one scattering event, are more likely to be collected. (The relative contribution of the multiply scattered light can be further reduced by a factor of 2 by placement of a linear polarizer between the fiber and the tissue.)

These optical geometries are predominantly sensitive to the cells only at the surface of the tissue, which is ideally suited for detecting signs of early cancer or pre-cancer transformations for many cancer types. This follows from the fact that most cancers, the type called *carcinomas*, originate in the epithelial (superficial) layer of the organ, which is typically only 100–300 μm thick. In applying the above approaches for clinical applications, many of the implementations have sought to provide diagnostic information based on a calculation of the sizes of the nuclei of the epithelial cells, derived from an inverse fitting of Mie theory to the spectral oscillations. *Dysplastic* (premalignant) and cancerous cells often have nuclei that are enlarged and, often, denser than nuclei in normal cells. For example, Qiu et al. (2010) applied this principle to the diagnosis of premalignant conditions of the esophagus, in which the nuclei of dysplastic cells are significantly enlarged. They generated size distributions of nuclei and related those to risk of dysplasia. When compared with histopathology from correlated biopsy samples, sensitivity and specificity values were obtained that were comparable with the inter-observer agreement rate among pathologists.

7.4.4.3 *Angle-resolved low-coherence interferometry (aLCI)*

Another method for determining the large-angle range of the phase function for single-scattering has been developed by Wax and coworkers (2001). The method, dubbed *angle-resolved low-coherence interferometry* (aLCI), combines *coherence gating* of backscattered light, in a manner similar to depth sectioning by optical coherence tomography (see Chapter 18), with angle-dependent measurements. (Coherence gating is a method of timing the detection of light that is based on detecting the phase delay of a low-coherence spectrum, i.e., which has a short coherence length, to obtain the time of propagation.) In the method of aLCI, photons that have scattered only once, and at a specific depth within a tissue sample, are selectively detected by the coherence gating, based on interference from a reference beam. In addition to the coherence gating, the method invokes detection of backscattering at a range of angles (in one implementation, ~140–170°) with a CCD camera, or with an array of collection fibers located to detect at different angles. The angular dependence enables determination of nuclear size distribution, by fitting to the phase function from Mie theory, and the depth selectivity enables sensing of the epithelial cells that are most likely to exhibit disease-induced changes. In a recent study, Terry et al. (2011) deployed a fiberoptic version of aLCI through an endoscope in a clinical study that demonstrated detection of dysplasia (pre-cancer) in the esophagus.

In summary, the portent of noninvasive optical sensing of premalignant and malignant conditions in epithelial tissues, opening the door for noninvasive instant diagnosis, is of great potential for clinical applications under development in the area of optical diagnostics.

7.5 Scattering from particles in motion

All of the discussion above has dealt with scattering from particles that are effectively stationary for the duration of the measurement, and the coherence of the light source is not important (with the exception of aLCI, for which the light source must have low coherence). The measurements give information on the sizes and densities (refractive indices) of the constituent particles. If, on the other hand, the light source is temporally coherent, it is possible to extract information about the *motion* of scattering particles. Extraction of information about particle motion can be accomplished by measurement of the subtle wavelength shifts of the scattered light, or of the temporal variations in the detected optical intensity.

7.5.1 Doppler shifts and laser Doppler velocimetry

The perceived change in pitch of the horn from a car or train, as the vehicle approaches (or recedes from) a stationary listener, is generally referred to as the *Doppler shift* (named after the Austrian physicist Christian Doppler [1803–1853]). The principles are the same for sound or light (as long as the relative speed of the source and detector is small compared with the speed of the wave in the medium). As depicted in Figure 7.17, the detected frequency increases (wavelength

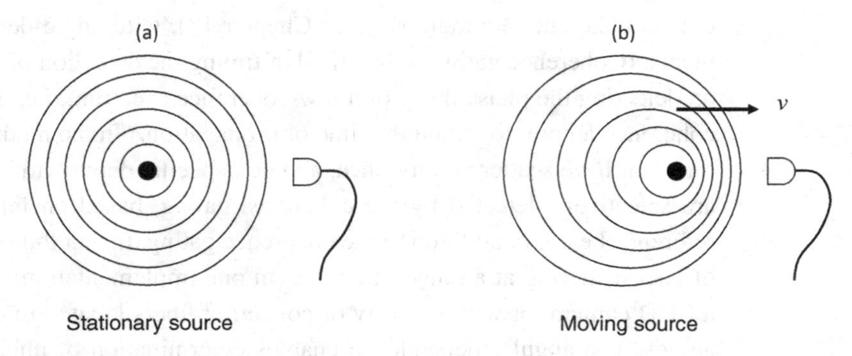

Figure 7.17

When source and detector are stationary relative to each other (a), the detector senses the frequency as emitted by the source; if the source and detector are approaching each other (b), the phasefronts are compressed, resulting in an increase in the frequency that is sensed by the detector.

shortens) when the source approaches the detector, and this is often referred to as a "*blue shift*" of the detected wave. Conversely, a receding source results in a *red-shifted* signal at the detector. (The reader may recall that astrophysicists commonly estimate the distance to stars from the amount of red shift in the spectrum of known atomic emission lines. This is based on the assumption that everything started at one point during the "big bang," and that the distance now of any extra-galactic source from earth is proportional to the receding speed between the two.) Although we preferentially use the angular frequency, ω, for electromagnetic waves in most of this text, for the discussion below we revert to frequency, $f = \omega/2\pi$, to be consistent with the notation that is used pervasively in the literature for formalisms of Doppler shift calculations.

The Doppler-shifted frequency perceived by the detector, f', in the case of a source moving toward (+) or away (–) from it, can be written as:

$$f' = f\left(1 \pm \frac{v}{c}\right) \tag{7.43}$$

where f is the frequency emitted by the source, v is the speed of the source relative to the detector, and c is the propagation speed of the wave (in this case light) in the medium. In the more general case where there is a non-zero angle, θ, between the velocity vector of the source relative to the detector ($\mathbf{v} = \mathbf{v}_S - \mathbf{v}_D$) and the position vector of the detector relative to the source ($\mathbf{r} = \mathbf{r}_D - \mathbf{r}_S$), Eq. (7.43) is generalized to:

$$f' = f\left(1 + \frac{v\cos\theta}{c}\right), \tag{7.44}$$

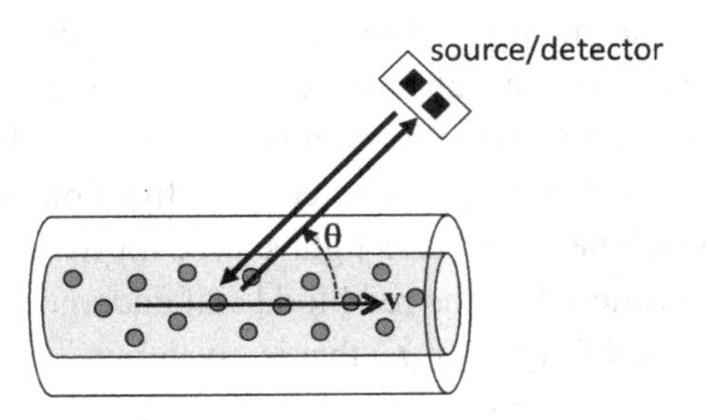

Figure 7.18

A Doppler velocimeter is used to measure the velocity of blood cells flowing in a blood vessel. The light from the source in the instrument is backscattered by flowing cells, which can then be treated as moving sources of light whose frequency is already Doppler-shifted (once) from that of the source. The backscattered light that is detected (by the detector adjacent to the source) is then perceived by the detector to have a frequency that has been Doppler-shifted twice, with respect to that of the source.

where the positive or negative sign of the Doppler shift is now accounted for by the $\cos\theta$ term, which is positive if the component of $\mathbf{v}$ along $\mathbf{r}$ is positive (source and detector approaching each other) and negative otherwise. We observe that, as long as $v \ll c$, Eq. (7.44) describes equally well the cases of a moving source and a moving detector, so that what really counts is their relative speed ($\mathbf{v} = \mathbf{v}_S - \mathbf{v}_D$) as indicated above.

This principle has been applied, to great advantage, for the measurement of blood flow in vivo, in the method generally called *laser Doppler velocimetry* (LDV) or *laser Doppler flowmetry* (LDF), wherein the velocity of blood flow in vessels near the tissue surface is measured. In the schematic depicted in Figure 7.18, a narrowband (temporally coherent) and collimated laser beam is aimed at blood vessel(s) near the surface of the tissue, and light is scattered back to a detector adjacent to the source, mainly by red blood cells, which are strong scatterers. For such a measurement, the *coherence length* of the laser beam, $L_c \cong 0.44\lambda^2/\Delta\lambda$, must be long compared to the round-trip distance between the source/detector and the tissue being interrogated. (A more detailed treatment of the concepts of optical coherence will be provided in Section 18.1.) The factor 0.44 obtains precisely for a laser beam with a Gaussian bandwidth, and is approximate for any beam that is close to collimated. (Both the coherence length and the optical wavelength are specific to the refractive index of the medium in which the light is propagating.) If an optical fiber is used to collect light from tissue, it must be a single-mode fiber to preserve optical phase information over the round-trip distance. Moreover, we restrict the discussion to measurements of flow near the surface, because unambiguous determination about blood flow and optimum signal-to-noise ratio are obtained for singly-backscattered photons.

In this configuration, a moving red blood cell (RBC) conceptually acts as a moving detector and perceives a light frequency, f', that is Doppler-shifted per Eq. (7.44) and that is reemitted in the form of scattered light. (Red blood cells are responsible for the large majority of scattered light from blood.) Because of the motion of the RBC, which behaves like a light source for its scattered light, the f' frequency of the light scattered by the red blood cell undergoes a second Doppler shift resulting in a detected frequency f'' that is given by:

$$f'' = f'\left(1 + \frac{v\cos\theta}{c}\right) = f\left(1 + \frac{v\cos\theta}{c}\right)^2. \tag{7.45}$$

For $v \ll c$, it can be shown that:

$$f'' \cong f\left(1 + \frac{2v\cos\theta}{c}\right). \tag{7.46}$$

It would, of course, be quite difficult to measure the optical frequency directly, but it is easy to perform *heterodyne* detection, whereby a sample of the original laser beam is detected together with the backscattering light, such that the detector signal oscillates at the beat-frequency between the two, $\Delta f = |f'' - f|$. Remembering that the laser wavelength is related to its frequency as $\lambda = c/f$, one can derive an expression for the velocity of the red blood cells:

$$v = \Delta f \frac{\lambda}{2\cos\theta} \tag{7.47}$$

LDF measurements can be made through the skin to sense superficial vessels, or on exposed tissues during surgery. Faster blood flow can mean that an artery is constricted at the site of measurement, whereas slower blood flow can mean that there is a restriction "upstream" from the point of measurement. If a measurement is made from the capillary bed of a tissue, where flow can be in many directions within the volume being interrogated, the information about general blood perfusion is extracted from a broadening of the backscattered spectrum rather than a simple shift, and temporal correlations, as described in the next section, can then be utilized. An excellent review of methods and applications of Laser Doppler flowmetry is provided by Rajan et al. (2009).

7.5.2 Dynamic light scattering

Doppler measurements are "snapshot" determinations of flow, over a time that is short compared to the time scale of variations in the flow. Aside from blood flow, however, cells and other smaller (and larger) structures in tissue are constantly in motion, due to a variety of causes, including Brownian motion, diffusion and physiological kinetics. These types of quasi-random motion and flow will cause temporal variations in the scattered optical field. If the light source is continuous and coherent, then, even in the single-scattering regime, a temporal measurement of individual particles can provide information about motion similar to the information from an ensemble-average measurement over many particles. In short, the temporal fluctuations in the scattered optical field can be assessed to extract information about the medium itself through the dynamics of the particles in the medium. In the discussion that follows, the reader should note that the terms g_1 and g_2, as used here, represent different parameters than the moments of the phase function as will later be discussed in Chapter 8 (Section 8.2.2).

The normalized temporal autocorrelation function of the scattered field is defined as:

$$g_1(\tau) = \frac{\langle E(t)E^*(t+\tau)\rangle}{\langle E(t)E^*(t)\rangle} \tag{7.48}$$

Here τ is the autocorrelation delay time, the asterisk (*) indicates the complex conjugate, and the purpose of the denominator is to normalize $g_1(\tau)$ to 1 at $\tau = 0$, whereas $g_1(\tau)$ tends to zero for long delay times. The theory of DLS expresses the field autocorrelation function, $g_1(\tau)$, in terms of the motion of scattering particles, as described by their mean squared displacement, $\langle \Delta r^2(\tau)\rangle$, over time, τ.

Since intensity is the parameter that is measured experimentally, the normalized temporal intensity autocorrelation function is the relevant quantity that is accessible to measurement:

$$g_2(\tau) = \frac{\langle I(t)I(t+\tau)\rangle}{\langle I(t)^2\rangle}. \tag{7.49}$$

The denominator in Eq. (7.49) normalizes $g_2(\tau)$ to 1 for $\tau \to \infty$. The temporal intensity autocorrelation function is determined directly by recording the temporal fluctuations of the detected intensity over a period of time that is long compared with the longest fluctuation period. The connection between the measured $g_2(\tau)$ and the mean-square displacement of the scatterers (which characterizes particle movement) appearing in $g_1(\tau)$ is realized by the Siegert relation (named after the German-American physicist Arnold Siegert [1911–1995]), which holds in a case of a Gaussian distribution of scattered field amplitudes:

$$g_2(\tau) = 1 + \beta|g_1(\tau)|^2. \tag{7.50}$$

The factor β in Eq. (7.50), which ideally takes a value of 1, is a constant instrumental response factor, which depends on the details of the optical elements of the measurement system and the solid angle of the detector, which determines the number of speckles intercepted. If fewer speckles are detected, signal-to-background is better, because this results in larger intensity fluctuations and a more peaked autocorrelation function. Consequently, single-mode optical fibers are ideally suited for light collection in dynamic light scattering measurements.

There are various types of instrumentation for DLS, often used in physics (and biology) to determine particle sizes (and medium viscosity) by measuring the autocorrelation function of scattered light and extracting a temporal decay rate that gives information on the particle size distribution and/or the fluid properties. We remind the reader that in biomedical optics we are more frequently interested in "static" (elastic) light scattering as a means of learning about cellular and tissue micro-architecture, and in this chapter we have focused on single scattering events. Nonetheless, an extension of the single-scattering technique of DLS to multiple scattering conditions, termed *diffusing wave spectroscopy* (DWS) or *diffuse correlation spectroscopy* (DCS), will be introduced in Chapters 8 (Section 8.9.1) and 9 (Section 9.9) for multiple scattering and diffuse scattering, respectively. Additionally, the discussion here was limited to the description of scattering by dielectric (rather than conducting) particles in a dielectric medium, which is most representative of tissue.

Although the techniques of temporal autocorrelation methods are applicable for single-scattering circumstances, laser Doppler flow methods are more commonly used in biomedical optics to characterize the motion of scattering particles in single-scattering conditions. By contrast, temporal autocorrelation methods are employed for DCS in the multiple-scattering and diffusion regimes. However, it is important to observe that the spectral shift associated with the Doppler effect and the temporal fluctuations associated with the intensity autocorrelation function are just different manifestations of the same phenomenon. In fact, the temporal autocorrelation function (which reflects the intensity fluctuations associated with varying interference effects) and the frequency spectrum (which represents the spectral broadening associated with Doppler shifts) are a Fourier transform pair, so that they are mathematically correlated. The temporal autocorrelation techniques used in DLS and the spectrally resolved techniques used in laser Doppler flowmetry do differ in their experimental approach and methods of data analysis, but they are conceptually analogous.

Problems – answers to problems with * are on p. 646

7.1 Refer to the series-expansion expression for the total scattering efficiency, Q_s, which is derived from Mie theory (Eq. (7.33)). This expression includes the first three terms of the expansion from the scattering matrix elements.

(a) For wavelength $\lambda = 600\,\text{nm}$, medium index, $n_0 = 1.35$, index ratio $n_1/n_0 = 1.06$, and particle radius $a = 0.02\ \mu\text{m}$, calculate and list the values of the three terms.

(b) If accuracy of 1% is the goal for part (a), will the number of terms needed for convergence be <3, 3 or >3? Explain.

(c) Now repeat step (a), but this time using a particle radius of 3 μm.

(d) Again, if accuracy of 1% is the goal, will the number of terms needed for convergence be <3, 3 or >3? Explain.

(e) For $\lambda = 600\,\text{nm}$ and $a = 0.02\ \mu\text{m}$, compare the values of Q_s (sum of the three terms) for two different values of index ratio: $m = n_1/n_0 = 1.06$ and $m = 1.02$. Which yields a larger scattering efficiency? Why?

7.2* As explained in Section 7.2.1, light from the blue sky is partially polarized. Describe the scattering plane, in the case of visual observation of the sky, according to its definition in Section 7.2. Assume the sun to be to the left or right with respect to the viewing direction to the sky. With respect to the scattering plane, in which direction is the preferential polarization oriented? (i.e., normal or parallel). Explain your reasoning.

7.3* Show that a result equivalent to that of Eq. (7.42), to determine the singly scattered spectrum, can also be accomplished by two measurements, only one of which invokes a polarizer: the backscattered light without a polarizing filter, and the backscattered light through a polarizing filter that is perpendicular to the illumination polarization (calibrated for the transmission efficiency of the polarizer).

7.4* A laser Doppler flowmeter (LDF) is being used to measure the velocity of blood flow in one specific vein near the surface of the arm. The laser source ($\lambda = 800$ nm) and the detector of the LDF instrument are adjacent to each other, and the beam is aimed at the vein with an angle of 45° to the flow direction. The beat frequency (between the backscattered light and the original laser light) measured by the instrument is 600 kHz. What is the speed (in cm/s) of the blood flowing in the vein?

7.5 Show that Eq. (7.46) can be derived from Eq. (7.45) for the condition that $v \ll c$.

7.6 Derive Eq. (7.47).

References

Amelink, A., Bard, M. P. L., Burgers, S. A., and Sterenborg, H. J. C. M. (2003). Single-scattering spectroscopy for the endoscopic analysis of particle size in superficial layers of turbid media. *Applied Optics*, 42, 4095–4101.

Asano, S., and Yamamoto, G. (1975). Light scattering by a spheroidal particle. *Applied Optics*, 14, 29–49.

Backman, V., Gurjar, R., Badizadegan, K., et al. (1999). Polarized light scattering spectroscopy for quantitative measurement of epithelial cellular structures in situ. *IEEE Journal of Selected Topics in Quantum Electronics*, 5, 1019–1027.

Bohren, C. F., and Huffman, D. R. (1983). *Absorption and Scattering of Light by Small Particles*. Wiley: New York.

Canpolat, M., and Mourant, J. R. (2001). Particle size analysis of turbid media with a single optical fiber in contact with the medium to deliver and detect white light. *Applied Optics*, 40, 3792–3799.

Castagner, J.-L., and Bigio, I. J. (2007). Particle sizing with a fast polar nephelometer. *Applied Optics*, 46, 527–532.

Drezek, R., Dunn, A. K., and Richards-Kortum, R. R. (2000). A pulsed finite-difference time-domain (FDTD) method for calculating light scattering from biological cells over broad wavelength ranges. *Optics Express*, 6(7), 147–157.

Dunn, A. K., Smithpeter, C. L., Welch, A. J., and Richards-Kortum, R. R. (1997). Finite-difference time-domain simulation of light scattering from single cells. *Journal of Biomedical Optics*, 2(3), 262–266.

Giacomelli, M. G., Chalut, K. J., Ostrander, J. H., and Wax, A. (2010). Review of the application of T-matrix calculations for determining the structure of cell nuclei with angle-resolved light scattering measurements. *IEEE Journal of Selected Topics in Quantum Electronics*, 16, 900–908.

Hansen, M. Z., and Evans, W. H. (1980). Polar nephelometer for atmospheric particulate studies. *Applied Optics*, 19, 3389–3395.

Henyey, L. G., and Greenstein, J. L. (1941). Diffuse radiation in the galaxy. *Astrophysics Journal*, 93, 70–83.

Maier, S. J., Walker, S. A., Fantini, S., Franceschini, M. A., and Gratton, E. (1994). Possible correlation between blood glucose concentration and the reduced scattering coefficient of tissues in the near infrared. *Optics Letters*, 19(24), 2062–2064.

Mie, G. (1908). Beiträge zur Optik trüber Medien, speziell kolloidaler Metallösungen. *Annalen der Physik*, 330(3), 377–445.

Mishchenko, M. I., Travis, L. D., and Mackowski, D. W. (1996). T-matrix computations of light scattering by nonspherical particles: a review. *Journal of Quantitative Spectroscopy and Radiative Transfer*, 55, 535–575.

Mourant, J. R., Boyer, J., Hielscher, A., and Bigio, I. J. (1996). Influence of the scattering phase function on light transport measurements in turbid media performed with small source-detector separations. *Optics Letters*, 21, 546–548.

Mulvey, C. S., Sherwood, C. A., and Bigio, I. J. (November, 2009). Wavelength-dependent backscattering measurements for quantitative real-time monitoring of apoptosis in living cells. *Journal of Biomedical Optics*, 14(6), 064013.

Mulvey, C. S., Zhang, K., Liu, W. H. B., Waxman, D. J., and Bigio, I. J. (2011a). Wavelength-dependent backscattering measurements for quantitative monitoring of apoptosis – Part I: early and late spectral changes are indicative of the presence of apoptosis in cell cultures. *Journal of Biomedical Optics*, 16(11).

Mulvey, C. S., Zhang, K., Liu, W. H. B., Waxman, D. J., and Bigio, I. J. (2011b). Wavelength-dependent backscattering measurements for quantitative monitoring of apoptosis – Part II: Early spectral changes during apoptosis are linked to apoptotic volume decrease. *Journal of Biomedical Optics*, 16(11).

Nilsson, A. M. K., Alsholm, P., Karlsson, A., and Andersson-Engels, S. (1998). T-matrix computations of light scattering by red blood cells. *Applied Optics*, 37(13), 2735–2748.

Qiu, L., Pleskow, D. K., Chuttani, R., et al. (2010). Multispectral scanning during endoscopy guides biopsy of dysplasia in Barrett's esophagus. *Nature Medicine*, 16(5), 603.

Quirantes, A. (2005). A T-matrix method and computer code for randomly oriented, axially symmetric coated scatterers. *Journal of Quantitative Spectroscopy and Radiative Transfer*, 92, 373–381.

Rajan, V., Varghese, B., van Leeuwen, T. G., and Steenbergen, W. (2009). Review of methodological developments in laser Doppler flowmetry. *Lasers in Medical Science*, 24, 269–283.

Ramachandran, J., Powers, T. M., Carpenter, S., et al. (2007). Light scattering and microarchitectural differences between tumorigenic and non-tumorigenic cell models of tissue. *Optics Express*, 15, 4039.

Shapiro, H. M. (2003). *Practical Flow Cytometry*, 4th edn. New York: John Wiley.

Terry, N. G., Zhu, Y., Rinehart, M. T., et al. (2011). Detection of dysplasia in Barrett's esophagus with in vivo depth-resolved nuclear morphology measurements. *Gastroenterology*, 140, 42–50.

Wax, A., Yang, C., Backman, V., et al. (2001). Cellular organization and substructure measured using angle-resolved low-coherence interferometry. *Biophysical Journal*, 82, 2256–2264.

Yee, K. (1966). Numerical solution of initial boundary value problems involving Maxwell's equations in isotropic media. *IEEE Transactions on Antennas and Propagation*, 14(3), 302–307.

Further reading

Tissue and cellular architecture

Marieb, E. N. (2012). *Human Anatomy and Physiology*, 9th edn. New York: Pearson.

Historical review of scattering by small spheres

Logan, N. A. (1965). Survey of some early studies of the scattering of plane waves by a sphere. *Proceedings of the IEEE*, 53, 773–785.

Mie theory and related theories for scattering by small particles

Bohren, C. F., and Huffman, D. R. (1983). *Absorption and Scattering of Light by Small Particles*. New York: Wiley.

van der Hulst, H. C. (1981). *Light Scattering by Small Particles*. New York: Dover.

General survey of topics in scattering of light by biological structures

Wax, A., and Backman, V., eds. (2010). *Biomedical Applications of Light Scattering*. New York: McGraw Hill.

Finite-difference time-domain computational method

Kunz, K. S., and Luebbers, R. J. (1993). *The Finite Difference Time Domain Method for Electromagnetics*. Boca Raton, FL: CRC Press.

Taflove, A., and Hagness, S. C. (2005). *Computational Electrodynamics: The Finite-Difference Time-Domain Method*, 3rd edn. Artech House: Norwood, MA.

Yee, K. (1966). Numerical solution of initial boundary value problems involving Maxwell's equations in isotropic media. *IEEE Transactions on Antennas and Propagation*, 14(3), 302–307.

Flow cytometry

Shapiro, H. M. (2003). *Practical Flow Cytometry*, 4th edn. New York: John Wiley.

Laser Doppler

Briers, J. D. (2001). Laser Doppler, speckle and related techniques for blood perfusion mapping and imaging. *Physiological Measurement*, 22, R35-R66.

Rajan, V., Varghese, B., van Leeuwen, T. G., and Steenbergen, W. (2009). Review of methodological developments in laser Doppler flowmetry. *Lasers in Medical Science*, 24, 269–283.

Dynamic light scattering

Berne, B. J., and Pecora, R. (2000). *Dynamic Light Scattering: With Applications to Chemistry, Biology, and Physics*. Mineola, NY: Dover.

8 Diffuse reflectance spectroscopy at small source-detector separations

Chapter 7 introduced theoretical methods for calculating single scattering events of optical fields from individual particles, and illustrated examples of applications for measurement of individual cell properties by backscattering spectroscopy or by measurement of the scattering phase function. Chapters 9–12 will cover the concepts and theoretical framework for measurement of bulk tissue optical properties in the diffusion regime, where source-detector distances are large compared to $1/\mu_s'$ (in most tissues, $\gtrsim 0.5$ cm), to enable application of the *diffusion approximation* to the radiative transport equation. Chapter 9, especially, rigorously addresses the parameter space for which the diffusion approximation to the transport equation is valid. In this chapter we cover the ground in between: the measurement of light that has scattered a few times, but not so many times that the light can be considered diffuse. The formalisms presented here, nonetheless, can be extended into the beginning of the diffusion regime, and may be advantageous where the P_1 approximation holds (to be presented in Section 9.4). For the typical optical properties of tissue, this generally corresponds to distances less than 0.5 cm. Moreover, whereas the diffusion approximation also requires that the absorption coefficient be much lower than the reduced scattering coefficient, in the formalisms for short source-detector distances that are presented in this chapter that restriction is lifted. Consequently, the wavelengths typically used for short source-detector distances may span from the near-UV through the near infrared (NIR). In contrast, for applications in real tissue, the useful wavelength range for the diffusion regime is limited to the red-NIR wavelengths, where absorption by hemoglobin is minimal.

Given these considerations, it may seem confusing that the terms *diffuse reflectance* and *diffuse reflectance spectroscopy* (DRS) are commonly used to denote measurements of reflectance at any distance from the light source, including short distances. Such broad application of those terms may be partially justified by the fact that "diffuse" reflectance can be intended simply to distinguish between reflectance from a diffusing medium and single backscattering events or specular reflection (as from a mirror). DRS encompasses terms that more specifically relate to variations of the optical geometry and/or methods of the spectral analysis. These include *elastic scattering spectroscopy* (Bigio and Mourant, 1997), *light scattering spectroscopy* (Backman et al., 2001), *differential pathlength spectroscopy* (Amelink et al., 2004), and *optical pharmacokinetics* (Mourant et al., 1999) among others.

8.1 Practical optical geometries

In this chapter our discussions will treat the tissue to be interrogated as a semi-infinite medium, with light source and detector, commonly mediated by optical fibers, applied at the tissue surface. Within those broad constraints, a variety of optical geometries can be employed for DRS measurements, ranging from single-fiber through multiple-fiber configurations and imaging schemes, and spanning a range of incidence angles and numerical apertures for source(s) and detector(s). In describing the basics of the formalisms for extraction of tissue property information, we will limit most of our discussion to the simple two-fiber configuration, as illustrated in Figure 8.1, which represents the most commonly used geometry for measurements of diffuse reflectance at short source-detector separations. The principles, nonetheless, apply to the broader range of geometric configurations and multiple-fiber probes.

In the two-fiber geometry the medium is treated as semi-infinite in extent, with source and detector fibers in contact with the tissue surface. Details of the fibers depend on the specific intent of the measurement method being employed, but most geometries can be regarded as a single source and single collector, conceptually (and mathematically) for purposes of theoretical treatment, regardless of the actual illumination and collection geometries. For example, a single fiber can serve as both source and detector (at zero separation), or a ring of illumination fibers equidistantly surrounding one collection fiber can increase the collected signal and still behave, conceptually, as a two-fiber configuration. Unless otherwise noted, the geometry illustrated in Figure 8.1 is the basis for the formalisms presented in this chapter. Also, all discussions in this chapter assume steady-state conditions, wherein any temporal variations of the light source are slow compared with the photon transit times between source and detector.

8.1.1 Monte Carlo simulations: where have the collected photons been?

A valid question to ask is: "Where have the collected photons been before they reached the detection fiber?" For the semi-infinite geometry of Figure 8.1, the stochastic spatial distribution of paths (often referred to as *visitation histories*) for the photons that actually reach the detector is frequently referred to in the literature

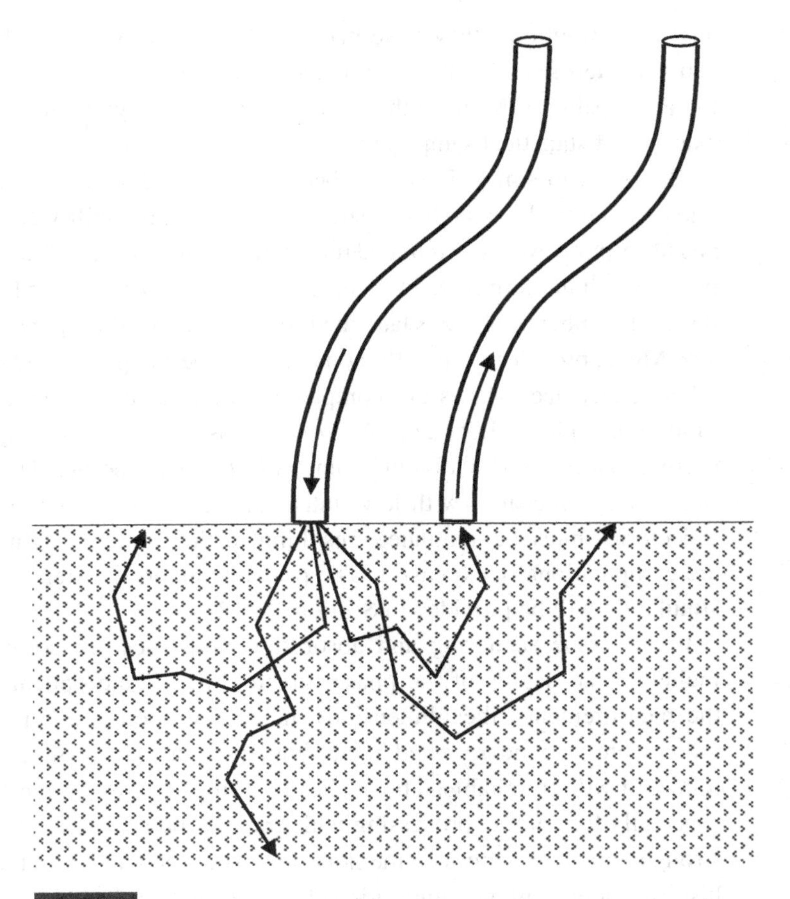

Figure 8.1

Optical geometry most commonly used for measurement of diffuse reflectance at short source-detector separations.

as the "*photon banana.*" In general, for short source-detector separations, it is not possible to analytically determine the spatial distribution of the photon banana, because of the lack of an analytical representation derived from first principles. (As will be seen in Section 14.1, such determination *is* possible under conditions of the diffusion approximation, given the availability of an analytical Green's function for the diffusion equation.)

What is both possible and convenient, for the small-separation regime, is to computationally determine the photon visitation histories of collected photons using *Monte Carlo simulations*. Originally developed to model computationally the random diffusion of neutrons in the fuel material of nuclear reactors, Monte Carlo (MC) simulations are a class of computational techniques used to simulate physical processes that are stochastic in nature and not easily amenable to analytical solutions; they have been used extensively in a wide range of applications (Metropolis, 1987). The name of the method pays homage to the city in the Principality of Monaco that is famous for casino gambling: the quintessential example of statistical sampling.

Monte Carlo simulations have been widely used to model the propagation of photons in turbid media, like tissue, based on the probability profiles of scattering and absorption events. Simulated reflectance values are calculated as the fraction of photons that are terminated (collected) at a particular detector location relative to the total number of photons launched from the source during the entire simulation. The MC approach is generally considered to be the gold standard against which other reflectance models are compared. Unlike analytical models, there are no limitations to its validity, and MC simulations can therefore be easily modified to accommodate any variation in geometry or tissue properties. However, to achieve satisfactory precision (with low statistical variance), a large number of photons trajectories must be simulated, resulting in the limitation that MC simulations are computationally expensive. The principal algorithm steps for a Monte Carlo simulation will be provided in Section 14.5.2.

If the tissue scattering and absorption properties are correctly represented, including the details of the phase function, and if the optical specifications of the measurement (fiber dimensions, numerical apertures, separation, refractive indices, etc.) are incorporated in the simulation, then the resulting stochastic determination will be reliable. The challenge of this type of Monte Carlo simulation, especially for parallel processing with graphics processing units, is that there is a strong demand on processor memory due to the need to record the voxel visitation histories of each photon until it is either collected or abandoned.

A plot of a Monte Carlo simulation depicting voxel density of visitation histories of collected photons is shown in Figure 8.2. This case is for two fibers with 200-μm core diameters, at a center-to-center separation of ~280 μm. Both fibers have the same numerical aperture (0.22), and the source fiber is modeled as emitting photons (within the NA) at a uniform radiance across its surface. As may be expected, stochastically the probability of a photon succeeding in reaching the collection fiber (at an angle within its numerical aperture) is higher for shorter distances between the injection point and collection point of the photon. This occurs for the photons that emerge from an emission location on the source fiber that is closer to the collection fiber, and more photons are collected at the area of the collector surface that is closer to the source fiber. This short-distance bias is displayed as being darker for the higher densities of voxel visitations that are near surface at the face of each fiber, in volumes that are closer to the opposing fiber, representing the shorter source-detector separations. It can also be seen that some of the collected photons visited voxels at the surface near one of the fiber faces, and, rather than leaving the medium and being lost, reflected off the surface back

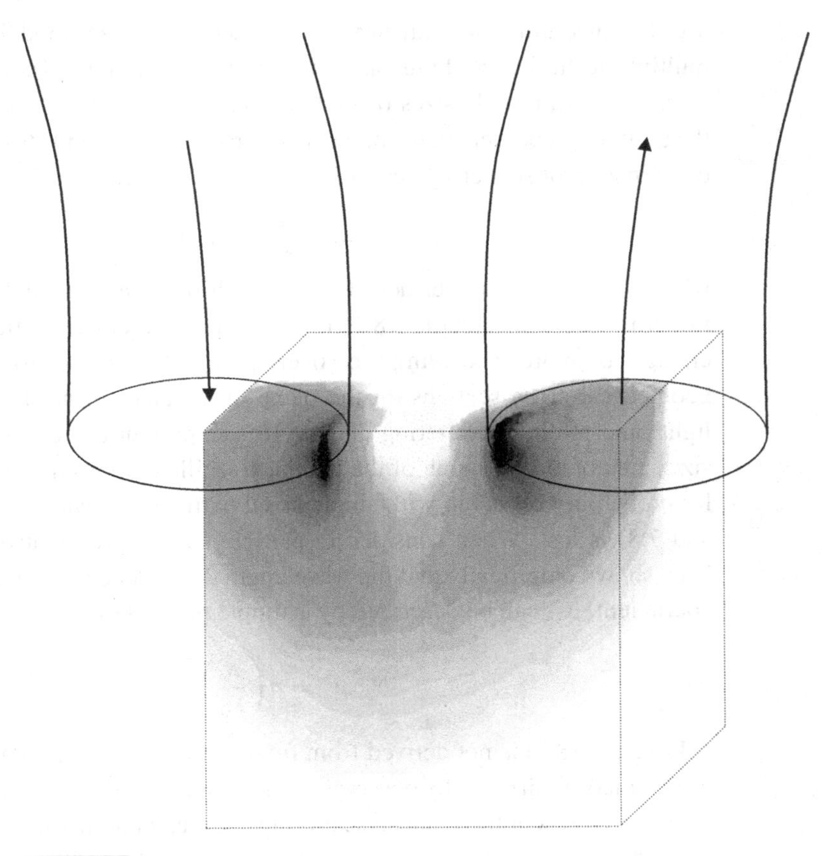

Figure 8.2

A cube of tissue volume, showing results of a Monte Carlo simulation of the density of voxel visitation histories of photons that are collected in the geometry of a semi-infinite medium. The positions of the source and detector fibers are shown with respect to the cube. Each fiber has a 200-μm core diameter and an NA of 0.22, and the center-to-center separation is ~280 μm. The medium is non-absorbing, with a scattering coefficient $\mu_s = 100$ cm^{-1} and an anisotropy factor $g = 0.9$.

into the medium, eventually making it into the collection fiber (We note that if a similar depiction were to be made for source and detector fibers separated by a large enough distance to meet the conditions of the diffusion regime, the collection voxel densities at the fiber faces would be essentially uniform.)

8.2 Representing tissue scattering properties in the partially diffuse regime

Methods like Mie theory (see Section 7.2.2) or finite-difference time domain calculations (see Section 7.3) can yield detailed information about the scattering efficiency and phase function of an individual particle. If one were to examine a turbid medium where all particles are the same size, modeling the effects of multiple scattering could invoke simplifying assumptions. Tissues are composed, of course, of multiple sizes of particles and, since scattering is a linear process, the scattering coefficient, μ_s, of a collection of particles and structures is the linear combination of scattering coefficients for each particle size:

$$\mu_s = \sum_i \mathcal{N}_{s,i}\sigma_{s,i} \tag{8.1}$$

where $\mathcal{N}_{s,i}$ is the number density of the i-th particle type, and $\sigma_{s,i}$ is its scattering cross section. Equation (8.1) holds for particle concentrations that are small enough to ignore shadowing (i.e., overlapping of the projections of the particles' geometrical cross sections on the plane orthogonal to the direction of incident light) and multiple scattering effects. It is important to note that when multiple sizes are combined, much of the complex oscillatory spectral structure and angular probability observed with single-sized particles (as was shown in Figures 7.7 and 7.8) is lost. When considering particle sizes representative of tissue, it has been shown empirically that the wavelength dependence of the reduced scattering coefficient, μ_s', can be described by a simple power law:

$$\mu_s' = a\left(\frac{\lambda}{\lambda_0}\right)^{-b} \tag{8.2}$$

Equation (8.2) is not derived from fundamental principles, but is an expression determined empirically by observation. The incorporation of λ_0 allows for the variable a to represent the reduced scattering coefficient at that reference wavelength, λ_0. When the distribution is skewed towards small particle sizes, the exponent b approaches the value 4 (for Rayleigh scattering), and as the size distribution increases to particle sizes that are larger than the wavelength, b approaches 0.37 (Mourant et al., 1997; Graaff et al., 1992). Thus, the value of b can provide an indication about the mean size of the particles in the sample, if the distribution is not broad. However, as reported by Mourant et al. (1997), there is not a robust relationship that directly equates b to an average particle size, meaning that the value of b can only provide intuition into the relative size distribution.

For size distributions where both small and large particle sizes are represented, it has been proposed to model the wavelength dependence of the reduced scattering coefficient as a combination of Eq. (8.2) plus a term representing the Rayleigh limiting case (Lau et al., 2009; Saidi et al., 1995):

$$\mu_s' = a\left(\frac{\lambda}{\lambda_0}\right)^{-b} + c\left(\frac{\lambda}{\lambda_0}\right)^{-4} \tag{8.3}$$

This enables the exponent b to provide a better representation of the effective size distribution specific to the larger particles.

8.2.1 Importance of the phase function for short source-detector separations – phase functions for multiple scattering

When propagating in a diffusing medium, a photon scatters, on average, once per mean free path ($1/\mu_s$). The *diffusion regime* is generally assumed to be reached when the source-detector separation is greater than a few (two to four) times the *reduced* mean free path ($1/\mu_s'$). From the definitions of μ_s, μ_s', and the anisotropy factor g (see Section 1.3), the mean number of scattering events experienced by a photon over a distance of the reduced mean free path is $\mu_s/\mu_s' = 1/(1-g)$. If, for the sake of the current discussion, we assume that a distance twice the reduced mean free path is sufficient to establish the diffusion regime, then the minimum number of scattering events (for photons reaching the detector) for the diffusion regime is in the $2/(1-g)$ range. In a tissue with $g = 0.9$, this translates to a minimum of 20 scattering events before the radiance appears to be approximately isotropic. Conversely, when the source-detector separation is small enough (in the case of $g = 0.9$) that the average photon has scattered fewer than 20 times before reaching the detection point, directional probabilities for the individual scattering events become important for determining how many photons reach the detector.

As introduced earlier, the phase function, $p(\hat{\Omega}, \hat{\Omega}')$, of a particle specifies the probability distribution, per unit solid angle, of scattering from direction $\hat{\Omega}$ into direction $\hat{\Omega}'$. Recalling Eq. (1.27), the phase function is normalized to 1 over all solid angles:

$$\int_{4\pi} p(\hat{\Omega}, \hat{\Omega}')d\Omega = 1 \tag{8.4}$$

where light is described as scattering from the direction denoted by the unit vector $\hat{\Omega}$ into the direction $\hat{\Omega}'$. For a medium with multiple sizes of scatterers, each particle will have its own phase function, and the average phase function for scattering is directly related to the distribution of scatterer sizes present in the sample. The resulting function varies with wavelength (and relative refractive index). Larger particles are characterized by more forward scattering, yielding a larger anisotropy factor, g, whereas smaller particles produce more isotropic scattering, resulting in a smaller g value.

Before describing several phase functions commonly used in tissue optics, it is important to clarify that the phase function represents the scattering probability per unit solid angle. As described in Section 1.3, in the case of unpolarized light and isotropic particles, the phase function only depends on the scattering angle θ, i.e., the polar angle between $\hat{\Omega}$ and $\hat{\Omega}'$, and we indicate it as $p(\cos\theta)$ or $p(\theta)$. Alternatively, one can introduce a scattering probability distribution per unit scattering angle (rather than per unit solid angle), $p_\theta(\theta)$, so that the normalization condition for $p_\theta(\theta)$ is $\int_0^\pi p_\theta(\theta)d\theta = 1$. Recalling the discussion following Eq. (1.28), $p_\theta(\theta) = 2\pi \sin\theta\, p(\cos\theta)$. The probability density for scattering angles, $p_\theta(\theta)$, becomes important when simulating photon transport in Monte Carlo simulations, and its distinction from the phase function $p(\cos\theta)$, or $p(\theta)$, is important and must be properly taken into account (Binzoni et al., 2006).

8.2.2 Higher moments of the phase function

Given that a phase function for isotropic particles and unpolarized incident light has azimuthal symmetry, it only depends on $\hat{\Omega}\cdot\hat{\Omega}' = \cos\theta$ and can be expanded in a series of Legendre polynomials, $P_n(\cos\theta)$, for further characterization:

$$p(\cos\theta) = \frac{1}{4\pi}\sum_n (2n+1)\, g_n P_n(\cos\theta), \tag{8.5}$$

where the coefficients g_n are the n-th-order Legendre moments defined by:

$$g_n = 2\pi \int_0^\pi P_n(\cos\theta)p(\cos\theta)\sin\theta d\theta. \tag{8.6}$$

The first Legendre polynomial ($P_1(\cos\theta) = \cos\theta$) gives us the first-order moment, g_1, which is equivalent to the anisotropy factor, g, introduced earlier (Eq. (1.29)):

$$g_1 = g = 2\pi \int_0^\pi \cos\theta\, p(\cos\theta)\sin\theta d\theta = \langle\cos\theta\rangle \tag{8.7}$$

Similarly, using the second Legendre polynomial, $\frac{1}{2}(3\cos^2\theta - 1)$, the second moment, g_2, can be calculated as:

$$g_2 = 2\pi \int_0^\pi \left[\frac{1}{2}(3\cos^2\theta - 1)\right] p(\cos\theta)\sin\theta d\theta \tag{8.8}$$

From these two moments, a new similarity parameter can be defined (Bevilacqua and Depeursinge, 1999):

$$\gamma = \frac{(1-g_2)}{(1-g_1)} \tag{8.9}$$

Under the conditions considered in this chapter (i.e., when scattered light is detected after experiencing only a few scattering events), this additional parameter helps to more accurately describe the scattering properties of a medium, in addition to μ_s'. As described in Chapter 2 (Section 2.2.4), the similarity relationship

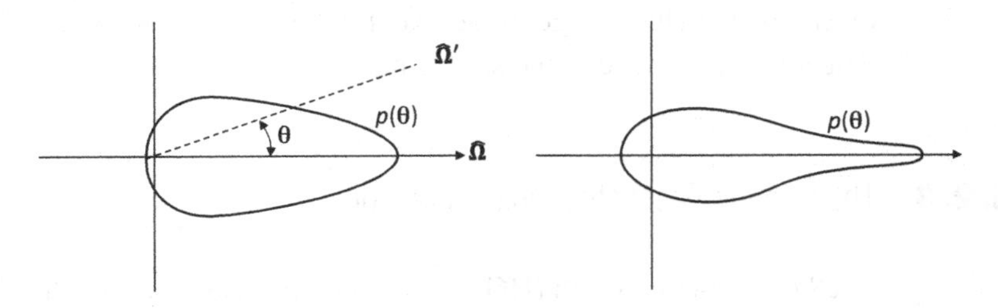

Figure 8.3

Polar plots (light incident from the left) for two hypothetical phase functions with the same value for the anisotropy factor, $g \approx 0.9$. The function on the left ($g_2 \approx 0.7$, $\gamma \approx 3$) has low probability for scattering angles larger than 90°, whereas the function on the right ($g_2 \approx 0.9$, $\gamma \approx 1$), despite also being strongly forward biased, does have significant scattering probability at large angles.

(Eq. (1.30)) for the reduced scattering coefficient, μ_s', is based on the condition that observed optical measurements are equivalent for any combination of g and μ_s that result in the same μ_s'. *This is valid only for the diffuse regime*, however, and when photons have only undergone a few scattering events before being detected, that relationship does not hold. Under such conditions, observed optical measurements will be equivalent for any combination of g, μ_s, **and** g_2 that results in the same μ_s' **and** γ. This relationship was confirmed empirically by Bevilacqua and Depeursinge (1999). From a physical perspective, the parameter γ is representative of (and inversely related to) the relative contribution of large-angle (near backward) scattering in the phase function. That is, for a given value of $g = g_1$, a larger g_2 (hence, smaller γ) leads to a larger relative proportion of scattering at large angles.

To illustrate the issue of the relative contribution of large-angle scattering events to the phase function, Figure 8.3 shows polar plots of two imagined phase functions, each of which could represent an anisotropy factor, $g = g_1$, of ~0.9, but with different values of g_2. These plots are not quantitatively computed from specific functions, but are estimates drawn to illustrate the effect of higher-order moments on the phase function. The phase function on the left (representative of values for $g_2 \approx 0.7$, $\gamma \approx 3$) exhibits little scattering probability (per unit solid angle) for angles >90°, whereas the phase function on the right (representative of values for $g_2 \approx 0.9$, $\gamma \approx 1$) models scattering at angles up to 180°. (The plots are symmetric about $\theta = 0°$, consistent with the assumption of azimuthal symmetry for scattering by isotropic particles.) Examination of the optical geometry in Figure 8.1 (and the simulation in Figure 8.2) may provide an intuitive understanding of why a phase function with greater large-angle scattering contribution (e.g., the one on the right in Figure 8.3) can lead to increased reflectance near the source. In fact, if an illumination photon does not "turn around" within a small number of scattering events that include large-angle events, the probability of the photon reaching a detector that is close to the source rapidly diminishes.

8.2.3 The Henyey-Greenstein phase function

The Henyey-Greenstein (HG) phase function (Henyey and Greenstein, 1941) was originally developed in the field of astrophysics to describe diffuse scattering of light by dust in interstellar space, but has gained widespread use in the biomedical optics community as a computationally simple approximation of tissue scattering. It is defined by the equation:

$$p_{\mathrm{HG}}(\cos\theta) = \frac{1}{4\pi}\frac{1-g^2}{\left(1+g^2-2g\cos\theta\right)^{3/2}}, \tag{8.10}$$

where $g = g_1 = \langle\cos\theta\rangle$ is bound between 0 and 1, for dielectric particles in a dielectric medium. (Negative values are possible for conducting [metallic] particles, which are not relevant for representing tissue.) As with all appropriately constructed phase functions, its integral over all solid angle is unity.

By selecting a given g_1 value of the Henyey-Greenstein phase function ($g_{1,\mathrm{HG}}$), all other Legendre moments ($g_{n,\mathrm{HG}}$) are automatically fixed: $g_{n,\mathrm{HG}} = g^n_{1,\mathrm{HG}}$; and thus the possible values of γ range from 1.0 to 2.0, as constrained because $g_1 \leq 1.0$.

8.2.4 The modified Henyey-Greenstein phase function

From the few experimental studies on true effective phase functions in tissue, it has been observed that the HG phase function underestimates high-angle backward scattering (Mourant et al., 1996; Canpolat and Mourant, 2000; Bevilacqua and Depeursinge, 1999). To address this drawback, and motivated by the attraction of a simplified analytical expression for the phase function, a modified version of the HG function was proposed that combines the standard HG function with an added large-angle scattering component that contributes the desired high-angle scattering (Jacques et al., 1987). Known simply as the modified Henyey-Greenstein (MHG) phase function, it is defined as:

$$p_{\mathrm{MHG}}(\cos\theta) = \beta\, p_{\mathrm{HG}}(\cos\theta) + (1-\beta)\frac{3}{4\pi}\cos^2\theta \tag{8.11}$$

where β is the fractional contribution of the standard HG phase function, and conversely $(1-\beta)$ is the fractional contribution of the large-angle component $(0 \leq \beta \leq 1)$.

The first and second moments are easily calculated and are given by the following equations (Bevilacqua and Depeursinge, 1999):

$$g_{1,\mathrm{MHG}} = \beta g_{1,\mathrm{HG}} \tag{8.12}$$

and

$$g_{2,\mathrm{MHG}} = \beta g_{1,\mathrm{HG}}^2 + \frac{2}{5}(1-\beta) = \beta g_{2,\mathrm{HG}} + \frac{2}{5}(1-\beta) \tag{8.13}$$

where g_{HG} is the anisotropy factor used to construct the HG phase function contribution using Eq. (8.10). With the added flexibility of the MHG phase function, both g and γ can be controlled individually. It should be noted, however, that values of γ for the MHG phase function are still limited to the range 1.0–2.0 due to mathematical constraints of the function, whereas values greater than 2.0 are indeed representative of some tissues that are highly forward-scattering (e.g., skeletal muscle). In such cases, Mie theory is more appropriate for determining the phase function than approximations such as the MHG phase function.

8.2.5 Mie-theory phase function for multiple scattering

The most rigorous estimation of the phase function in tissue can be calculated from Mie theory by describing tissue as a distribution of discretely sized spheres. By combining the individual scattering distributions for each sphere size, an average phase function can be constructed to represent the bulk medium. The key here is in selecting the appropriate sizes and distribution to best approximate tissue. One approach is to model the size distribution using a fractal description, based on observations that refractive index spatial variations measured by phase-contrast microscopy scale according to a power law (Wang, 2000; Sharma and Banerjee, 2005). The number, N_i, of particles with diameter d_i contained within a reference spherical volume of diameter d_0 is then defined as:

$$N_i = \left(\frac{d_0}{d_i}\right)^{\alpha}, \tag{8.14}$$

where α is the fractal volume dimension, which relates to the ratio of small versus large particles. A large α value corresponds to a greater ratio of small to large particles. (Note that N_i is not a number density, but simply the effective number of particles in the volume associated with d_0.)

The sphere sizes span a range that is representative of structures found in cells and extracellular matrix, which were described in Section 7.1. Each of these organelles and structures has a unique index of refraction. For the purpose of the fractal model, however, Schmitt and Kumar (1998) proposed modeling the index variations by a statistically equivalent volume of discrete particles having the same index but different sizes. Referring to Table 2.1, the refractive index of the intra- and extra-cellular fluids is taken as $n_f = 1.35$, and the average particle index is calculated from the indices of collagen/elastin fibers, nuclei and organelles ($n_p = 1.39$–1.45). The range is dependent on the collagen/elastin content, which is specific to the tissue type. For each discrete particle size, d_i, Mie theory is used to calculate the scattering cross section, $\sigma_s(d_i)$, and the anisotropy factor, $g(d_i)$, assuming n_f and n_p to be the medium and particle indices of refraction, respectively. Then, using the assumption that waves scattered by individual particles add linearly, the bulk scattering coefficient and anisotropy factor are defined as:

$$\mu_s = A \sum_{i=1}^{m} \left(\frac{d_0}{d_i} \right)^{\alpha} \sigma_s\,(d_i) \tag{8.15}$$

$$g = \frac{\sum_{i=1}^{m} d_i^{-\alpha} \sigma_s(d_i) g(d_i)}{\sum_{i=1}^{m} d_i^{-\alpha} \sigma_s(d_i)} \tag{8.16}$$

where m is the total number of particle sizes, and A is a constant factor relating to the total sphere volume. Such representation has been shown to provide good predictability of tissue optical properties. First introduced by Gélébart et al. (1996), this approach has been adopted by a number of research groups (e.g., Sharma and Banerjee, 2005). These equations are simply linear combinations, weighted by the scattering cross section and number density of the i-th particle sizes. The combined phase function, $p(\cos\theta)$, has often been reported in the same form as Eq. (8.16), with $p_i\,(\cos\theta)$ (i.e., the phase function for particles with diameter d_i) replacing $g\,(d_i)$.

The reader should note, however, that constructing the equivalent phase function in this manner, and directly calculating its anisotropy factor, g, from Eq. (8.7), does *not* yield a result equivalent to the g value that is calculated from Eq. (8.16). Instead, a linear combination based on the scattering amplitude, $S_{11}(\theta)$ (see Eq. (7.9)), scaled by the number density of spheres is likely to offer a more accurate solution:

$$p(\theta) = \frac{\sum_{i=1}^{m} d_i^{-\alpha} S_{11}(\theta, d_i)}{\sum_{i=1}^{m} d_i^{-\alpha}} \tag{8.17}$$

This follows from the fact that the scattered radiant angular intensity is already dependent on particle size, eliminating the need to scale with respect to the scattering cross section, σ_s. There is no discussion in the literature as to why the two calculation methods are not equivalent, but, given that Eq. (8.17) is a more intuitive solution, it is provided here as the more reliable approximation of the equivalent phase function. Calculation of g_1 and g_2 can then be performed using Eqs. (8.7) and (8.8). An important distinction of a Mie-theory-based phase function from

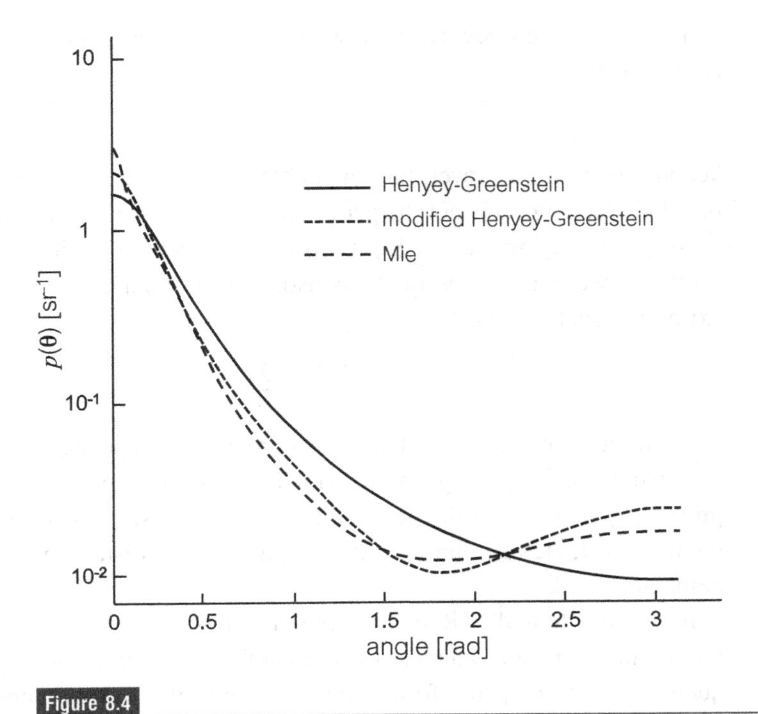

Figure 8.4

Comparison of Henyey-Greenstein, modified Henyey-Greenstein and Mie-theory-based phase functions, all with $g = 0.7$, but with different γ values: HG $\gamma = 1.7$; MHG and Mie $\gamma = 1.4$. (Adapted from Calabro and Bigio, 2014.)

the HG and MHG functions is that values of $\gamma > 2$ are possible (and appropriate for some tissue types), enabling representation of media that are strongly forward scattering.

For purposes of comparison, Figure 8.4 illustrates constructed functions for the HG, MHG and Mie phase function descriptions. All three functions are for $g = 0.7$. However, the HG function in Figure 8.4 has $\gamma = 1.7$, while the MHG and Mie functions in Figure 8.4 have values $\gamma = 1.4$, which result in the differences in high-angle scattering among the HG, MHG and Mie functions that are evident in the traces. Note also that despite having the same γ value, the MHG and Mie phase functions in Figure 8.4 are not identical. Differences between the two result from, and can be observed in, variations in the higher-order Legendre moments (g_n).

8.3 Representing tissue absorption properties in the partially diffuse regime

The details of light absorption also need to be examined for this regime. The Beer-Lambert law, which was introduced in Chapter 2 for a homogeneous medium

(Eq. (2.8)), is reproduced here with the added designation of its wavelength dependence:

$$I(\lambda) = I_0(\lambda)e^{-\mu_a(\lambda)L}, \tag{8.18}$$

keeping in mind, however, that the photon pathlength, L, in a turbid medium (to be addressed below) is not the geometrical distance between source and detector.

As previously written in Eq. (2.2) the absorption coefficient, $\mu_a(\lambda)$, depends on the concentrations, c_i, of the constituent chromophores, and their individual extinction coefficients, $\varepsilon_i(\lambda)$.

$$\mu_a(\lambda) = \sum_i c_i \varepsilon_i(\lambda) \tag{8.19}$$

Concentration, c, is in units of moles per unit volume, and the extinction coefficient, ε, is in units of $(\text{length})^2$ per mole, which gives μ_a units of inverse length. As a bulk property, μ_a represents the probability of photon absorption event per unit optical pathlength or, viewed alternatively, $1/\mu_a$ is the mean distance photons travel before being absorbed.

In the visible and NIR spectral regions, absorption due to oxy- and deoxyhemoglobin is typically much stronger than that of other absorbing species. Consequently, we develop the formalism assuming only those components, although addition of others (e.g., water, beta carotene, etc.) is straightforward using Eq. (8.19). With this assumption, the total absorption coefficient *in tissue* can be written as:

$$\mu_a(\lambda) = f_1[f_2\mu_{a,\text{HbO}}(\lambda) + (1 - f_2)\,\mu_{a,\text{Hb}}(\lambda)] + \text{other} \tag{8.20}$$

(with "other" being ignored for now in Eq. (8.20)) where $\mu_{a,\text{HbO}}$ and $\mu_{a,\text{Hb}}$ are the absorption coefficients of oxygenated and deoxygenated hemoglobin, respectively, when at a concentration representative of hemoglobin in whole blood ($\sim$2.3 mM); f_1 is the blood volume fraction in tissue; and f_2 is the oxygen saturation of hemoglobin, defined as:

$$f_2 \equiv \text{SO}_2 = c_{\text{HbO}}/(c_{\text{HbO}} + c_{\text{Hb}}). \tag{8.21}$$

8.3.1 Correction for the vessel-packing effect

When considering bulk tissue properties like μ_a, the distribution of chromophores is assumed to be homogeneous. In reality, however, tissue is generally heterogeneous, and this is especially true for the distribution of hemoglobin, which becomes an important consideration when considering small source-detector separations. This is a consequence of the fact that blood, hence hemoglobin, is confined within blood vessels. It has been observed that the effective absorption coefficient of hemoglobin in tissue is *not* equivalent to the absorption coefficient of a medium with the same total concentration of blood homogeneously distributed throughout the volume. This is sometimes referred to as the "*vessel-packing effect*," and has been examined by a number of authors who have heuristically developed scaling relationships to account for the decrease in effective absorption coefficient as a function of average blood vessel radius, r. The effect is stronger for larger absorption coefficients and larger blood-vessel radii. The most commonly applied expression for a correction factor ($C_{\text{corr}}(r, \lambda)$) that accounts for blood absorption and the size of blood vessels was developed by Svaasand et al. (1995):

$$C_{\text{corr}}(r, \lambda) = \frac{1 - e^{-2\mu_{a,\text{bl}}(\lambda) r}}{2\mu_{a,\text{bl}}(\lambda) r}, \tag{8.22}$$

where $\mu_{a,\text{bl}}(\lambda)$ is the absorption coefficient of whole blood:

$$\mu_{a,\text{bl}}(\lambda) = c_{\text{Hb,bl}}\left[f_2 \varepsilon_{\text{HbO}}(\lambda) + (1 - f_2)\varepsilon_{\text{Hb}}\right] \tag{8.23}$$

where $c_{\text{Hb,bl}}$ is the concentration of hemoglobin in whole blood (~2.3 mM). The effective tissue absorption coefficient is then:

$$\mu_a(\lambda) = f_1 C_{\text{corr}}(r, \lambda)\, \mu_{a,\text{bl}}(\lambda). \tag{8.24}$$

Since the correction factor is a function of both wavelength and average blood-vessel radius, the scaling will be different at each wavelength, resulting in a change in the spectral shape of the effect of hemoglobin absorption on DRS spectra. As an example, the Soret absorption band (around 420 nm) of hemoglobin (see Section 2.2) has an absorption coefficient in whole blood of nearly 3000 cm^{-1}. If a capillary radius of 5 μm is assumed, the correction factor is 0.32, resulting in an *effective* μ_a (under these conditions) that is one-third of what would be expected if the blood were homogeneously distributed in the tissue. In contrast, at a wavelength of 650 nm, the absorption coefficient of whole blood is approximately 5 cm^{-1}, which results in a correction factor of 0.99. Thus, for diffuse reflectance measurements that are performed in the NIR, the consequences of the vessel-packing effect are negligible. The effect of the correction factor can be readily seen in the UV-visible DRS spectrum of tissue, as exemplified in Figure 8.5. The measured effective depth of the absorption feature due to the Soret band at ~415 nm is only ~2× the depth seen for the Q-bands at 540 and 580 nm, even though the extinction coefficient at 415 nm is almost 10× that at 540 or 580 nm. (See Figure 2.4.)

8.4 Empirical models for the diffuse reflectance

Empirically derived models combine the advantages of both diffusion theory (a closed form functional equation) and Monte Carlo simulations (larger valid

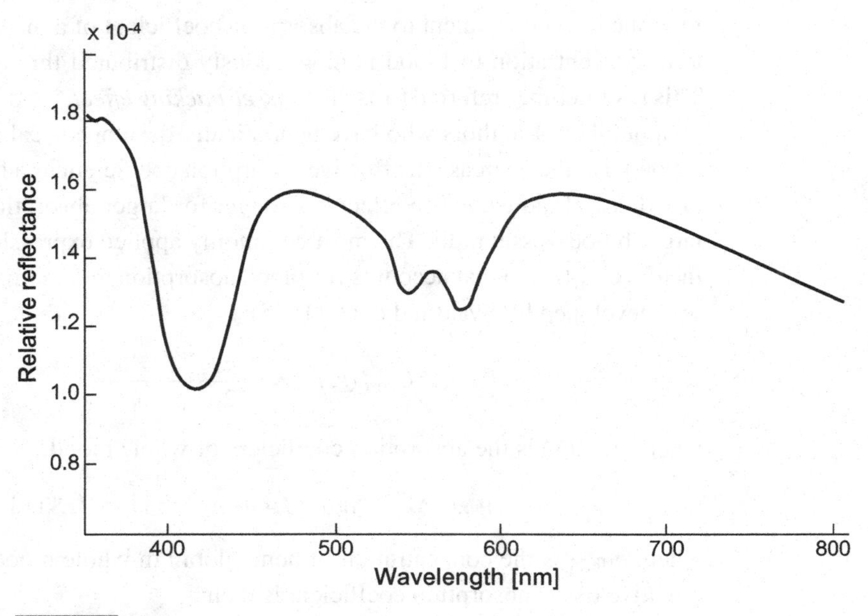

Figure 8.5
An example of a diffuse reflectance spectrum (elastic-scattering spectrum) for an in vivo measurement on colon mucosa with the optical geometry of Figure 8.1. The dips at approximately 415, 540 and 580 nm are due to absorption by (predominantly) oxyhemoglobin, whereas the remainder of the spectral shape is due to the wavelength dependence of scattering from this tissue.

parameter space). The motivation for the development of empirical models has come from a need to describe reflectance at small source-detector separation distances and/or large values of absorption coefficient. For a number of biomedical applications, small source-detector separations are necessary: for example, when the fiber probes must be designed to fit inside the working channel of an endoscope or laparoscopic tool, or for highly localized "point" measurements as in cases where sensitivity to only a shallow tissue depth is desired. All *carcinomas*, which account for 85% of cancers, begin in the *epithelial* (surface) layer of an organ. Detecting early cancer or pre-malignant conditions, often referred to as *carcinoma in situ*, or *dysplasia* or *neoplasia*, means detecting the disease while it is still confined to the epithelium, when it is most easily treated. Given that the thickness of the epithelial layer is typically in the range 100–300 μm for most hollow organs, this means that one would like to detect changes in the optical properties of a thin near-surface layer, with minimal background/masking signal from sub-epithelial tissues.

Considering the geometry depicted in Figure 8.1, we define the absolute, steady-state *reflectance*, on the surface, at a distance ρ from the illumination source, as the local intensity (I_{coll}) at the collector divided by the total illumination power:

$$R(\lambda, \rho) = \frac{I_{coll}(\lambda, \rho)}{P_{illum}(\lambda)} \tag{8.25}$$

(Note, defined as such, $R(\lambda, \rho)$ has units of inverse area, and the radiant power of illumination results from integrating the intensity over the surface area of the illumination optical fiber.) When reported for a range of wavelengths, measured reflectance produces a spectrum, like the spectrum illustrated as an example in Figure 8.5.

In general, at short distances, stronger scattering leads to an increase in measured reflectance, and stronger absorption leads to a decrease in reflectance. As can be seen from the representative spectrum of Figure 8.5, the overall spectral shape for wavelengths longer than ~650 nm, where absorption is much weaker than scattering, exhibits an approximate power-law dependence, as described by the μ_s' relationship in Eq. (8.2). The reflectance is broadly attenuated around 415 nm and 560 nm by strong absorption bands of hemoglobin, and the spectral shape of those bands is indicative of the oxygen saturation level of the hemoglobin. The typical DRS spectrum of Figure 8.5 qualitatively suggests how a measured spectrum may then be used to extract information about the density and size of cellular components through scattering, as well as the volume fraction and oxygen saturation of blood from absorption.

For practical applications, it is commonly difficult to reliably measure the absolute illumination power, due to lamp intensity fluctuations, loss of signal through the optical probe and variations in the optical contact efficiency at the tissue surface. Consequently, reflectance is more commonly reported as a relative wavelength-dependence of detected intensity to illumination power. Normalization of the spectrum can be achieved via several approaches: by normalizing to the total area under the curve (AUC), to the value at a specific wavelength, or relative to a measurement from an optical calibration phantom of known reflectance. All three conserve the shape of the spectrum, but in the last approach the overall amplitude (of one tissue measurement compared to another) is also conserved. This allows for retention of information concerning the overall scattering intensity of the sample, which itself can provide important diagnostic information, in addition to the spectral shape.

Empirical reflectance models are relatively recent advances in the field. Several have been modified from diffusion theory (Zonios et al., 1999; Hull and Foster, 2001; Hayakawa et al., 2004; Vitkin et al., 2011), whereas others are novel developments based on heuristic intuition of scattering and absorption interactions in tissue (Zonios and Dimou, 2006; Amelink et al., 2004; Reif et al., 2007).

8.4.1 A representative empirical model

Important note before proceeding: Having explained in Section 8.2.2 that the similarity relationship for μ_s' is not valid for sub-diffusive distances, we will nonetheless proceed (as we have done with Eqs. (8.2) and (8.3), and as a variety of authors have done) to use μ_s' in the development of the empirical models described here. At short distances, μ_s' does not represent the "inverse of the mean free path for isotropic scattering events," but it can still serve as a mathematical representation of the combination of g and μ_s through the product $(1-g)\,\mu_s$. In fact, if g is specified within a narrow range for a collection of scattering particles (say, ~0.9 for tissue), it is possible to identify a meaningful g value for a single representative particle, as obtained from its phase function. Moreover, the value of μ_s' that is determined by these methods at short distances would be the applicable to measurements over larger distances in the same medium, wherein *only* μ_s' would matter (and not μ_s or g individually).

We present an adaptation of the approach that has been described by Reif et al. (2007), as exemplary of a class of empirical models to relate the diffuse reflectance at short distances to the tissue optical properties. The model relates the optical coefficients of a material to *relative reflectance* rather than absolute reflectance. Relative reflectance (which is dimensionless) is defined as the ratio of absolute reflectance from the tissue, $R_{\text{tiss}}^{\text{abs}}(\lambda)$, to the absolute reflectance from a calibration phantom, $R_{\text{cal}}^{\text{abs}}(\lambda_0)$, whose optical properties are known at a given wavelength, λ_0, as given by:

$$R_{\text{tiss}}^{\text{rel}}(\lambda) = \frac{R_{\text{tiss}}^{\text{abs}}(\lambda)}{R_{\text{cal}}^{\text{abs}}(\lambda_0)} \tag{8.26}$$

The choice of calibration wavelength, λ_0, is arbitrary, as long as the same one is consistently used. When using the same calibration phantom, relative reflectance measurements are then comparable between systems and over time.

The structure of the empirical model is built around the framework of the Beer-Lambert law, where the reflectance from a sample with negligible absorption is designated as R_0, analogous to I_0 in Eq. (8.18). Tissue reflectance is then defined by exponentially scaling R_0, based on the absorption coefficient, μ_a, and the average photon pathlength, $\langle L\rangle$, for the collected photons, expressed as:

$$R_{\text{tiss}}^{\text{rel}} = R_0 e^{-\mu_a \langle L\rangle} \tag{8.27}$$

We observe that Eq. (8.27) is only an approximation since it ignores the variability of $\langle L\rangle$ for absorption coefficients in the range 0–μ_a. A strictly correct version of Eq. (8.27) is obtained by replacing $\langle L\rangle$ with the mean value of $\langle L\rangle$ over the absorption range 0–μ_a, which may be indicated as $\langle \overline{L} \rangle$ (Sassaroli and Fantini, 2004), leading to the so-called *modified Beer-Lambert law* (see Section 10.3). The differential form of the modified Beer-Lambert law is given in Eq. (10.14). Another way to modify Eq. (8.27) to yield a rigorously correct equation is to consider a single photon time-of-flight, t, from the source to the detector, which corresponds to a single photon pathlength $c_n t$, where c_n is the speed of light in tissue. This approach leads to the so-called *microscopic Beer-Lambert law*, which is given in Eq. (9.69).

Remembering that the use of μ_s' requires an assumed value of g, scattering-only (zero-absorption) reflectance, R_0, is modeled to be a function of μ_s', and $\langle L \rangle$ is modeled to be a function of μ_s' and μ_a. These optical coefficient values are assumed to be homogeneous in the sample, while incorporating the vessel-packing correction factor as needed. The empirical (observed) relationship between R_0 and μ_s' is modeled to be linear when $\mu_a = 0$.

$$R_0 = s\mu_s' + s_0; \tag{8.28}$$

and the average pathlength of collected photons, $\langle L \rangle$, is modeled by the empirical relation:

$$\langle L \rangle = l_1 \left(\frac{\mu_a^* \mu_s'^*}{\mu_a \mu_s'} \right)^{l_2}, \tag{8.29}$$

where s, s_0, l_1 and l_2 are fitting coefficients that are unique for a given fiber probe geometry, and are determined with the calibration phantom, whereas μ_a^* and $\mu_s'^*$ are constant reference values of tissue absorption and reduced scattering coefficients, which can be chosen to have values that are in the range of many tissues, such as 0.1 cm^{-1} and 10 cm^{-1}, respectively. Since the units of $\langle L \rangle$ are length, the units of l_1 must also be length. (The coefficient s has units of length, and s_0 is dimensionless.) Together with Eqs. (8.28) and (8.29), then, Eq. (8.27) provides a closed-form relationship between measured relative reflectance and the optical coefficient values, μ_a and μ_s':

$$R_{\text{tiss}}^{\text{rel}} = \left(s\mu_s' + s_0\right) \exp\left[-\mu_a l_1 \left(\frac{\mu_a^* \mu_s'^*}{\mu_a \mu_s'} \right)^{l_2} \right] \tag{8.30}$$

For measurements made in the visible range, in blood-perfused tissue, Eq. (8.24) can be used to replace μ_a with a corrected absorption coefficient. It should be noted that l_1 is monotonically related to the source-detector separation, ρ.

Experimental measurements on phantoms of known optical properties are plotted in Figure 8.6 as exemplary demonstrations of how to establish the fitting coefficients for Eq. (8.30). The data in Figure 8.6 were taken with the optical

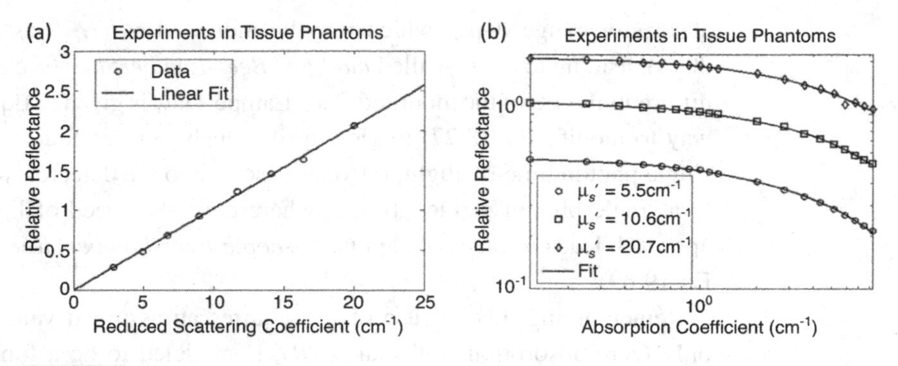

Figure 8.6

Reflectance relationships: (a) relative reflectance versus reduced scattering coefficient in the absence of absorption; (b) relative reflectance versus absorption coefficient for three values of reduced scattering coefficient. The values of μ_s' in these plots are for an assumed $g = 0.9$, and could be replaced by their equivalent values of μ_s. (Reproduced, with permission, from Reif et al., 2007.)

geometry depicted in Figure 8.1, using two 200-μm core fibers with center-to-center separation of 250 μm.

With these measurements, the values of s, s_0, l_1 and l_2 can be experimentally determined in a two-part procedure. First, the relationship between μ_s' and relative reflectance in the absence of absorption is evaluated to estimate s and s_0, as shown in Figure 8.6(a), based on fitting reflectance data to Eq. (8.28); then, using those parameters as known values for the test medium, reflectance data that included absorption are then fit to Eq. (8.29) for each value of μ_s', using μ_a as the independent variable, as shown in Figure 8.6(b). The resulting fitting values of l_1 and l_2 for each μ_s' data set are averaged to determine final model values. Alternatively, if a nonlinear least-squares fit is applied in two dimensions, i.e., fitting reflectance data using both μ_a and μ_s' as independent variables, optimized fitting values can be determined simultaneously for the entire data set. Again, the resulting fit parameters are unique to the probe specifications and the particular calibration phantom. Nonetheless, once the parameters are determined for those components, Eq. (8.30) can be used for the "*inverse problem*": extracting the optical properties of an unknown medium.

We conclude this section by stressing again that at source-detector separations that are $\lesssim 1/\mu_s'$, the diffuse reflectance depends on μ_a, μ_s, and on the details of the scattering phase function (including the anisotropy factor g). However, for cases where a collection of scattering particles has g values that are distributed over a narrow range about a known value, one can scale μ_s by $(1 - g)$, to represent scattering properties with μ_s', as done in Figure 8.6. This approach would not be applicable to collections of particles with a broad distribution of g values, and it also does not give insight into the individual dependence of the diffuse reflectance on μ_s and g. However, it does provide meaningful information on the reduced scattering coefficient, which is the optical parameter that describes the bulk scattering properties of highly scattering media such as most biological tissues.

8.4.2 Lookup tables

In contrast to using mathematical formalisms to estimate reflectance based on optical coefficient values, a simpler "brute-force" approach is to store a table with known reflectance and coefficient value combinations *for an assumed value of g typical of tissue*, and use it to 'look up' a reflectance value based on μ_a and μ_s' inputs. A given table would be specific to a given set of optical and geometric parameters of the measurement. Because a table with a continuous distribution of coefficient values is not possible, reflectance values for μ_a and μ_s' combinations that are not explicitly listed are estimated using standard interpolation techniques. The data that populate the table are generally experimental, either from Monte Carlo simulations or from phantom experiments (Rajaram et al., 2012; Nichols et al., 2012; Erickson et al., 2012; Bouchard et al., 2010).

8.5 The inverse problem: extracting tissue properties

The primary motivation for building formalisms that describe reflectance as a function of optical coefficient values is **not** for use in forward-direction calculations (estimating reflectance given values for μ_a and μ_s'), but rather for use in solving the *inverse problem*: determining μ_a and μ_s' from measured reflectance, when a good estimate of g can be assumed. The fundamental challenge is that there is no unique combination of coefficient values corresponding to one reflectance value; a minimum of two distinct reflectance measurements are required. This can be achieved by taking measurements either at multiple source-detector separations or at multiple wavelengths. Each of these approaches, in practical experimental methods, imposes different requirements for assumptions about the phase function (to be discussed in the following two sections). We also note that more reliable results are generally obtained when the number of data points (i.e., distinct reflectance measurements) is larger. This produces an over-determined system, which can be solved using a least-squares fitting approach. One of the most common least-squares minimization routines is known as the Levenberg-Marquardt

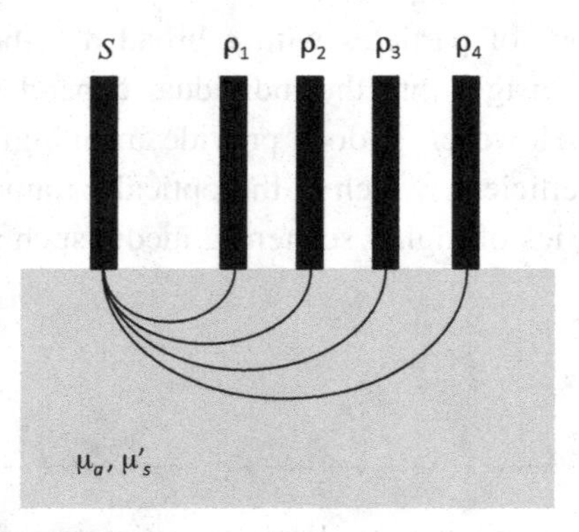

Figure 8.7

Radially dependent diffuse reflectance measurements. S is the source, and ρ_1–ρ_4 are the separate detectors at those distances. As explained in the text, if ρ_1 is $\geq\sim 0.7$ mm, g does not need to be separately specified.

method, named after the American statisticians Kenneth Levenberg [1919–1973] and Donald Marquardt [1929–1997]. (It is readily applied using the fitting toolbox in Matlab^{TM}, Mathworks, Natick, MA.)

8.5.1 Varying the source-detector separation

At a specific wavelength, with the steady-state, semi-infinite models discussed here, the only controllable variable is the source-detector separation distance, ρ (and the assumed value of g for tissue), since μ_a and μ_s' are generally unknown. (As will be seen in Section 10.4.4, the same applies for the steady-state solution of the diffusion equation, in the diffusion regime.) The inverse solution method has traditionally been implemented by making reflectance measurements at multiple distances, ρ, and then fitting those data to the model to estimate the values of μ_a and μ_s'. This is performed for each wavelength independently. It should be noted that while this approach is most commonly used with diffusion theory, for other models of quantities that are functions of measurement distance the same principles would apply (Andree et al., 2010). The measurement geometry is depicted in Figure 8.7. In practice, it is experimentally easiest to perform such measurements at ρ values $\geq\sim 0.7$ mm, and at those distances (for scattering parameters typical of tissue) higher moments of the phase function have small effect on the reflectance, so approximations of diffusion theory are applicable (Bevilacqua and Depeursinge, 1999), and it is not necessary to assume a value for g in order to extract μ_s'. The same consequence applies to fitting empirical models to reflectance spectra.

In practice, for accurate extraction of the optical coefficients, curves for reflectance vs. separation, $R(\rho)$, must be sufficiently unique. The result is consistently a plot of an exponential decay, with the only variant being the exponent of the decay rate. It is important to note that the model fitting is based only on the normalized shape of the curve, because, as mentioned before, the absolute reflectance amplitude is difficult to determine for typical experimental conditions. Even if one of the two parameters, μ_a or μ_s', is known a priori, distinction between the curves can still be experimentally challenging, especially with large values of the absorption and reduced scattering coefficients. When fitting for both μ_a and μ_s' simultaneously, the uncertainty in the extracted coefficients is magnified, leading to a degree of "cross-talk" between the two determined optical coefficients (Bargo et al., 2005). Also, as depicted in Figure 8.7, the measurements at different distances do not interrogate the same volumes, which can lead to inaccuracies in layered or heterogeneous tissues.

8.5.2 Measurements at multiple wavelengths

Given the challenges of the distance-dependent inverse model approach, many researchers have followed the alternative approach of making multiple reflectance measurements as a function of wavelength. Steady-state diffuse reflectance measurements are commonly performed over a wide range of wavelengths in parallel (measured by a spectrometer with an array detector), enabling data acquisition with a single (broadband) measurement with a single detection fiber. Although the models described in this chapter are not inherent functions of wavelength, they invoke the wavelength-dependent functions of and μ_a and μ_s' (μ_s with an assumed value of g typical of tissue). In tissue, this involves using the spectral μ_s' and μ_a relationships defined by Eqs. (8.2) and (8.24), respectively, in any of the forward models.

We hasten to note that measurements of wavelength dependence can readily be performed at very short source-detector separations, including, essentially, zero values for separation, as would pertain to measurements made with a single fiber (Amelink et al., 2004). Under these conditions, the higher moments of the phase function do indeed affect the reflectance, and even an assumed value for g is inadequate, as knowledge of γ would also be required for empirical formalisms such as described in Section 8.4.1. Nonetheless, a number of groups have found success with such empirical formalisms (Zonios and Dimou, 2006; Bevilacqua et al., 1999; Amelink et al., 2004; Reif et al., 2007), and we submit that this is probably because the use of several fitting parameters for the spectral reflectance data essentially sets the value of γ relative to that of the calibration phantom that is used.

With measurements at a broad range of wavelengths, it is possible to implement additional fitting parameters to reveal more specific information about the tissue physiology, while actually reducing the total number of fitting parameters. Thus:

$$R = f(\mu_s', \mu_a), \tag{8.31}$$

where:

$$\mu_s' = f(\lambda, a, b), \tag{8.32}$$

and:

$$\mu_a = f(\lambda, f_1, f_2, r) \tag{8.33}$$

leading to:

$$R = f\,(\lambda, a, b, f_1, f_2, r)\,. \tag{8.34}$$

This reduces the number of unknowns from two at *each* wavelength (μ_s' and μ_a) to a total of five over all wavelengths. These five unknown fitting parameters include a, the scattering coefficient at a normalization wavelength λ_0, and b, the scattering "power-law exponent" related to the average scatterer size, and f_1, f_2 and r, representing the blood volume fraction (%), the blood oxygen saturation (%) and the average blood vessel radius, respectively. Thus, each measured spectrum can be characterized uniquely by five optical and physiological parameters. The example reflectance spectrum in Figure 8.5 illustrates how wavelength-dependent measurements have more distinct features that are useful to the fitting routine when uniquely calculating the fitting coefficients.

This method, unfortunately, also comes with some drawbacks. First, as described for the representative empirical model, the reflectance curves must be normalized to a measurement from a calibration phantom with known optical properties. Also, the wavelength-dependent relationships for μ_a and μ_s' are only based on assumptions about the composition of the tissue. It is always possible that such assumptions are incomplete or incorrect, which could alter the fitting results. *In general, the more that is known about the tissue, the more reliable will be the parameters extracted by the inverse solution.*

8.6 Special distances for diffuse reflectance measurements: "isosbestic" points

In spectroscopy an *isosbestic point* (from the Greek *iso* = equal, and *sbestos* = extinguished) refers to a wavelength at which the molar absorption (extinction) coefficients for two species of a molecule are equal. For example, as evident in

Figure 2.4, the extinction coefficients for oxyhemoglobin and deoxyhemoglobin are equal at ~800 nm, such that the absorption coefficient at that wavelength due to hemoglobin in tissue will be independent of the hemoglobin redox stoichiometry. In a similar spirit, we describe optical measurement geometries for which the value of one measured parameter (reflectance) is insensitive to the value of a specific tissue optical property. These determinations are based on heuristic arguments, and have been verified computationally and experimentally.

8.6.1 Source-detector distances for which the average photon pathlength is insensitive to the tissue reduced scattering coefficient

Sometimes it is desirable to be able to measure the absorption coefficient of a turbid medium (e.g., tissue) without needing to worry about how variations in the reduced scattering coefficient can affect the measurement. If the absorption coefficient of the tissue includes the contribution of an administered drug, one can write a variant of the Beer-Lambert law (see Eq. (8.27)) to address the measurement of reflectance for a given source-detector separation (ρ) in the semi-infinite geometry, and to incorporate the time dependence of a dynamic drug concentration:

$$R(\lambda, \rho, t) = R_0 e^{-\mu_a(\lambda, t)\langle L \rangle}. \tag{8.35}$$

We recall that Eq. (8.35) is an approximation, as pointed out in the paragraph following Eq. (8.27). If we "invert" Eq. (8.35), write it for two different time points ($t = t_0, t_1$), and subtract the two results, we can derive an expression for the change in the absorption coefficient between the two time points:

$$\mu_a(t_1) - \mu_a(t_0) \equiv \Delta\mu_a = -\frac{1}{\langle L \rangle}\ln\left[\frac{R(t_1)}{R(t_0)}\right], \tag{8.36}$$

where the wavelength dependence is assumed. If the absorption change $\Delta\mu_a$ is small, so that $R(t_1) \approx R(t_0)$, then Eq. (8.36) coincides with Eq. (10.14), which expresses the modified Beer-Lambert law. (After replacing the diffuse reflectance R with the fluence rate ϕ, and considering that $\ln(1 + x) \approx x$ for $x \ll 1$.) If, for the moment, we make the simplifying assumption that the concentrations of native chromophores (e.g., Hb and HbO) are constant, then the change in the absorption coefficient comes only from the administration of a chromophoric drug. If we further assume that the drug concentration in tissue is zero at time t_0, prior to administration, then the right-hand side of Eq. (8.36) defines the absorption coefficient due to drug at time t_1. Remembering that the absorption coefficient is the product of the concentration and the extinction coefficient, which is known independently, since the reflectance values at the two time points are measured quantities, the only remaining unknown needed to determine the absolute drug concentration is the average photon pathlength, $\langle L \rangle$. The challenge is that $\langle L \rangle$ in this geometry generally depends on the scattering properties of the medium.

It has been shown that the value of $\langle L \rangle$ can be determined for a specific optical geometry, independent of scattering properties. The underlying idea is that, when light delivery and collection are performed with optical fibers in contact with the tissue surface (a semi-infinite medium), there is an optimum separation between the source and the detector fibers, ρ, for which the average photon pathlength is independent of μ_s' (or g, as will be explained below) over the range of scattering parameters that are relevant to biological tissues. The following heuristic argument has been validated by Monte Carlo simulations and by physical experiments with controlled scattering media (Mourant et al., 1997, 1999):

> The idea arises from examination of the pathlengths of collected photons (in the semi-infinite geometry) for the extremes of very small or large source-detector separations. In general, for very small fiber separations ($\rho \ll 1/\mu_s'$), the average pathlength of the collected photons is longer for less-scattering media because the photons need to undergo a certain number of high-angle scattering events to reverse their direction of travel and reach the collection fiber (Mourant et al., 1996). In a weakly scattering medium, most collected photons will have traveled deep into the tissue before experiencing a turnaround scattering event, whereas a strongly scattering medium will backscatter light close to the source. On the other hand, at very large fiber separations it is intuitively obvious that the pathlength is longer for the more highly scattering media, for reasons similar to the case of light transmission through a slab. Therefore, there must be some crossover regime of source-detector separation for which the average pathlengths are insensitive to moderate variations in the scattering properties.

For the range of values of μ_s' that are plausible for tissues (5–15 cm^{-1}), the optimum value for ρ is $\sim$1.7 mm, and the sensitivity of $\langle L \rangle$ to the value of μ_s' remains low for the range of $\rho = 1.5$–2.3 mm. This result was initially obtained with Monte Carlo simulations that incorporated details of the phase function, but given that the ρ values obtained are $\gtrsim 1/\mu_s'$, it is valid to assume insensitivity to g (or higher moments of the phase function) for a large range of $g \gtrsim 0.5$, as well as insensitivity to μ_s'. Thus, for a specific fiber probe with source-detector separation in that range, $\langle L \rangle$ can be determined by measuring the reflectance of a phantom with known absorption coefficient and with scattering anywhere in the range typical of tissue, and that same value of $\langle L \rangle$ can then be applied to measurements in tissue with unknown absorption coefficient, enabling extraction of drug concentrations, as described in Section 8.8.2.

So, what is a typical resulting value of $\langle L \rangle$? Using Monte Carlo simulations, as well as phantom measurements, Mourant et al. (1997) determined an effective pathlength of $\sim$1 cm, for two fibers of 200-μm core diameter and numerical aperture 0.22, at a separation of $\rho = 1.7$ mm (center-to-center). This effective pathlength was found to be essentially independent of μ_s' over the range 5–15 cm^{-1}. (This consequence is not incompatible with Eq. (8.29), which is only valid for source-detector separations that are $\ll 1/\mu_s'$.) For a pathlength of $\sim$1 cm, measurements of relatively weak absorption coefficients, using Eq. (8.36), become possible with acceptable signal-to-noise ratios.

It should be noted, nonetheless, that while $\langle L \rangle$ is insensitive to μ_s' in this regime, it is still affected by the total absorption coefficient itself (due to all absorbing constituents), if absorption is comparable to or greater than scattering (i.e., effective pathlengths are shorter in turbid media with stronger absorption). This effect is significant at visible wavelengths in tissue, where there is strong absorption due to hemoglobin. As described by Mourant et al. (1999), the effect can be accounted for by mathematical calibration of a specific probe configuration by measurements in phantoms with known absorption.

8.6.2 Source-detector separations for which the reflectance is insensitive to the tissue reduced scattering coefficient

In a similar study, Kumar and Schmitt (1997) identified a source-detector separation for which the *reflectance* is insensitive to variations in the reduced scattering coefficient. The optical geometry considered was the same as for the case above: one illumination fiber and one detection fiber, in contact with the surface of a semi-infinite medium, with the fiber axes at normal incidence to the medium surface.

The logic of the consideration of a special source-detector separation is similar to that in Section 8.6.1, and the value of g (or γ) need not be assumed, given the value of ρ obtained with Monte Carlo simulations. When ρ is large, fewer photons reach the detector (i.e., the reflectance or surface fluence is lower) for larger values of μ_s', which can be seen from solutions to the steady-state diffusion equation (see Eq. (10.23)) and Figure 10.5), or from heuristic arguments. In fact, when the source-detector separation is large, the reflectance at the detector is equally sensitive to changes in scattering or absorption. When the source-detector separation is very small, on the other hand, the reflectance becomes much less sensitive to absorption (due to the short photon pathlengths), whereas the sensitivity of reflectance to scattering changes sign, such that stronger scattering increases the probability of light scattering back to the nearby detector. Kumar and Schmitt reported a set of Monte Carlo simulations incorporating the Henyey-Greenstein phase function, as well as a set of phantom measurements, that identified a value of $\rho \approx 3$ mm as the optimum distance for which the surface fluence (or reflectance) is least sensitive to variations in reduced scattering coefficient over ranges relevant to tissues.

It should be noted that the logic behind, and the validity of, the "special" distances (i.e., for the insensitivity to a range of values for μ_s' typical of tissue) holds for turbid media with large range of anisotropy values $\gtrsim$0.5. (Monte Carlo simulations have not been performed for g values outside the broadest range associated with tissue.)

8.6.3 Source-detector separation for which the reflectance is insensitive to the phase function parameter γ

At shorter source-detector separations there is another special value that may offer benefits for reflectance measurements. It has been shown that, in the diffusion regime, the reflectance (at a given ρ) is insensitive to the value of the anisotropy factor, $g = g_1$, as long as μ_s' is constant, whereas, in general, the reflectance in the non-diffuse regime *does* depend on g_1 and g_2 (hence, γ), even if μ_s' is constant (Bevilacqua and Depeursinge, 1999; Calabro and Bigio, 2014). That is, the reflectance does depend on the proportion of scattering events that are at large angles, as well as the anisotropy factor, which is an overall average. A new observation for short distances is the existence of an "isosbestic" point, where reflectance is not dependent on γ (Calabro and Bigio, 2014). This was found to occur at a *dimensionless scattering* value of approximately 0.7, where dimensionless scattering is defined as the product $\rho\mu_s'$. (Thus, for example, at a typical tissue value of $\mu_s' = 10$ cm^{-1}, a dimensionless scattering value of 0.7 corresponds to a separation $\rho = 0.7$ mm.) While reflectance is known to be insensitive to phase function in the diffusion regime (at distances ρ such that $\rho\mu_s' > \sim 4$), where the optical radiance is approximately isotropic, it is an unexpected revelation that there is an additional singular dimensionless scattering distance, well below the diffusion regime, at which insensitivity to the detailed shape of the phase function exists.

For small dimensionless scattering values, a high γ corresponds to fewer high-angle scattering events and, hence, to lower reflectance. As dimensionless scattering increases, however, a point is reached at which reflectance is insensitive to the value of γ. And at yet larger dimensionless scattering values, a larger value of γ results in a larger reflectance. Intuitively, this is because, in tissues with low γ, photons with more backward scattering and less forward scattering are more likely to remain close to the source. With high γ, more forward scattering and fewer backscattering events allow a photon to travel farther from the source before experiencing the larger-angle scattering events that will direct it back towards the surface for collection. At large scattering and source-detector separations this

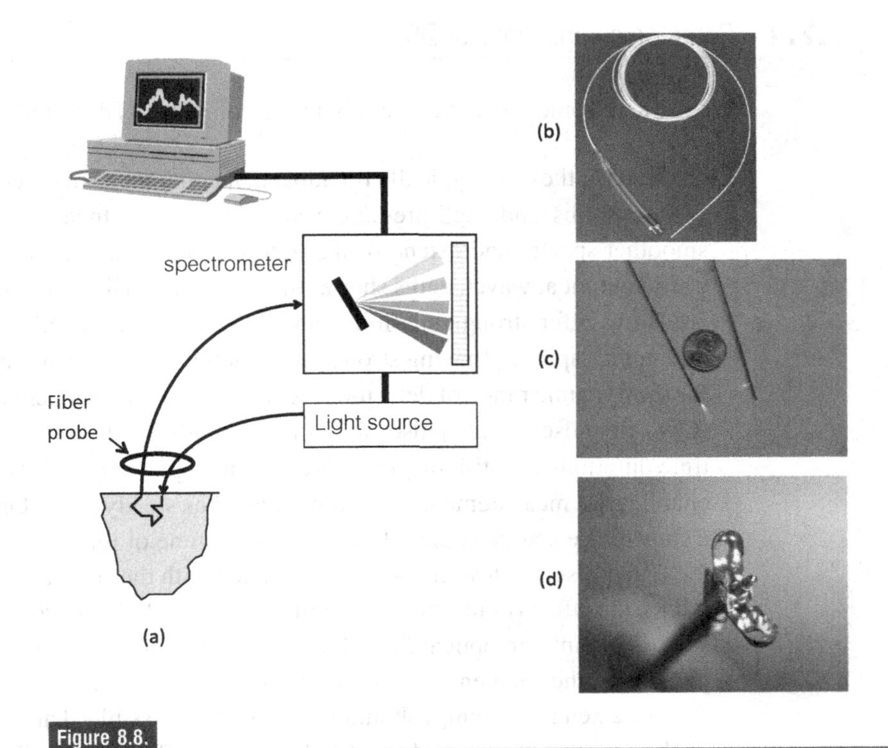

Figure 8.8.

(a) Schematic of the primary elements of a compact system for DRS measurements; (b) a flexible fiberoptic probe for use through endoscopes; (c) transdermal needles incorporating fiberoptic probes; (d) an integrated biopsy forceps tool incorporating a fiberoptic probe.

means that the proportion of photons collected at larger source-detector separations to photons collected at smaller separations is higher for large γ, leading to higher reflectance.

8.7 Typical instrumentation for diffuse reflectance spectroscopy

One of the advantages of steady-state diffuse reflectance spectroscopy is that the instrumentation can often be relatively simple and inexpensive. Moreover, these systems are, typically, readily transferable to a clinical setting. As depicted schematically in the left panel of Figure 8.8, a representative DRS system houses a broadband light source, a spectrometer with a linear CCD array detector and an interface computer. Some examples of fiberoptic probes that have been used clinically are also shown. As can be seen, thanks to the small source-detector separations, fiberoptic probes can be integrated into tools that are used endoscopically or laparoscopically, and can also be incorporated in small needles.

8.7.1 Common components for DRS

We list here some more specifics about the components depicted in Figure 8.8:

Light source: these are typically broadband, incoherent sources, commonly incandescent bulbs and high-pressure xenon arc lamps. Incandescent bulbs have smoother spectra and extend further into the NIR range (past 1.5 μm) but lack useful output at wavelengths shorter than ~420 nm. High-pressure xenon short-arc lamps offer strong emission from UV to NIR, but the NIR spectrum above 800 nm is "spikey" (having strong line structure), which can present challenges for the dynamic range of detectors. Xenon lamps can be either continuous-wave (CW) or pulsed, with pulse durations typically in the range 3–30 μs. Such time durations are still long compared to any light transit times in the medium, enabling all measurements to still be treated as steady-state. One advantage of a short-pulse source is that, if the integration time of the detector is matched to the light pulse, the low duty-cycle associated with the short integration time can reduce the effects of interference from background light. To couple as much light as possible into an optical fiber, for use as the illumination source at the tissue, the size of the filament (for an incandescent bulb) or the size of the discharge arc (for a xenon arc lamp) should be as small as possible. This is a consequence of the fact that these incoherent light sources effectively radiate into all (4π) solid angle, whereas the optical fiber has a limited cone angle (determined by its numerical aperture – see Section 4.8.1) of light acceptance. Thus, more light can be coupled into the fiber if the source has higher *radiance* (W/(m^2-sr)), for a given radiant power, which means that the dimensions of the source should be as small as possible. Supercontinuum sources (see Section 4.8.6), based on nonlinear spectral broadening of ultrashort laser pulses, can also provide high-radiance "white" light, but these are not commonly used for CW spectroscopy due to their high cost (and short pulse lengths). They can be valuable, however, for time-domain techniques, to be described in Chapter 11.

Spectrometer: in the near-UV to NIR wavelength range, the spectral features to be seen in light that is elastically scattered from tissue are generally broad, such that there would be no benefit to detection with spectral resolution narrower than ~3–4 nm. Referring to Section 4.7.2 and Figure 4.15, this relaxed specification for the resolution allows for use of a larger width of the entrance slit, since the detected bandwidth of monochromatic light at the detector scales linearly with the slit width. This results from the fact that the entrance slit is effectively imaged (albeit with wavelength dispersion) onto the detector array by the optical elements of the spectrometer. A larger slit width, in turn, leads to a stronger signal, because more light is imaged onto the detector. Another beneficial consequence of the modest spectral resolution requirement is that the overall dimensions of the spectrometer can be quite small, and some of the commercially available spectrometers often used for this type of application are as small as a pack of cigarettes.

Linear detector array: there are two common classes of linear detector arrays that are used for compact spectrometers: charge-coupled device (CCD) arrays and photodiode (PD) arrays. The total length of a line of sensor elements in an array is typically 1.5–3 cm, and the number of sensor pixels is generally 1024–4048 for CCD arrays and 256–512 for PD arrays. Some of the factors that are considered for determining the choice of detector are the desired wavelength range, desired responsivity vs. sensitivity, range of possible integration times, dynamic range, availability of element cooling, etc. Additional discussion about optical detectors is presented in Section 13.3.

8.7.2 Calibration of system response

The most common method to calibrate system spectral response for a system intended for spectroscopy, as in DRS, is to record a reference spectrum, $R_{ref}(\lambda)$, using the complete system (including the specific fiberoptic probe) by shining the light source on a reference material whose diffuse reflectivity is known for the entire wavelength range of interest. Commercially available diffuse reflectivity standards (e.g., Spectralon™ [Labsphere, Inc.] or Fluorilon™ [Avian Technologies, Inc.]) exhibit high (>98%) spectrally flat diffuse reflectivity from the near-UV through the NIR. Thus, the ratio of a tissue spectrum, $R_{tiss}(\lambda)$, to the reference spectrum, $R_{ref}(\lambda)$, will constitute a spectrally accurate measurement of the tissue reflectance, corrected for any variation in spectral response of the system. It should be noted that this reference spectrum is not to be confused with the phantom calibration measurement used in Eq. (8.26). (One does not need to know anything about the intrinsic optical properties of the reference material, only that it yields spectrally flat diffuse reflectance.) We also note that, if the illumination spectrum in the denominator of Eq. (8.26) is also divided by a reference spectrum (if, for example, the system response to the illumination source were recorded), each component would then be spectrally accurate, but the reference spectrum would simply cancel out. In practice, the reference spectrum is often incorporated for real-time display of measured tissue spectra, to enable real-time observation of the true spectrum, rather than post-measurement correction of spectra.

8.7.3 Tissue "phantoms" for validation

There is currently no gold standard approach for measuring the optical properties of a small volume of tissue in vivo. Therefore, the reflectance models and extraction techniques cannot be validated directly from biological samples in vivo. Instead, it is common to use materials of known properties that mimic the optical properties of tissue, called *phantoms*, for experimentally validating that a technique extracts the correct values over a range of wavelengths. The use of phantoms has the added benefit of enabling direct control of the properties of the samples. The fabrication and characterization of such phantoms is an active area of development, helping to meet the need for standardized approaches, as evidenced by a growing number of publications on the topic (Bouchard et al., 2010; Di Ninni et al., 2011; Martelli and Zaccanti, 2007; Michels et al., 2008).

In developing phantoms to test diffuse reflectance spectroscopy techniques, two primary components are necessary: a source of scattering, and an absorbing constituent. These can then be combined either with water to produce liquid phantoms or with silicone, epoxy resins or various gels to produce solid phantoms (Pogue and Patterson, 2006). Liquid phantoms are easier to produce and offer a refractive index that is more closely matched to that of tissue, but solid phantoms are more stable in the long term. Therefore, liquid phantoms are well suited for validation experiments in a laboratory environment, whereas solid phantoms may be preferable for use as calibration standards when collecting clinical measurements at various times and locations.

Traditional scattering compounds include titanium dioxide powder, polystyrene microspheres and lipid emulsions (such as Intralipid™ and Liposyn™). Titanium dioxide (TiO_2) is inexpensive and stable, but when mixed in water has a tendency to settle, requiring that the phantoms be stirred regularly. This problem is avoided when TiO_2 is used in solid resin phantoms. Further, due to TiO_2 having a refractive index of 2.6 and an anisotropy value g close to 0.5, the scattering properties of TiO_2 particles are not an optimal representation of biological tissue. Polystyrene spheres are an expensive option, but offer the greatest control of optical properties, since the scattering and phase function details can be calculated directly from Mie theory based on the diameter and refractive index of the microspheres used. Their refractive index of 1.56 is still higher than that of scatterers in tissue, but it is a better match than that of titanium dioxide. Fat emulsions like Intralipid provide the best approximation to tissue scattering properties, since they are composed of lipid vesicles, with a refractive index (1.45–1.47) similar to structures in tissue. The primary clinical use of Intralipid is as a nutrient supplement for intravenous feeding. Thus, although Intralipid is stable long-term without affecting its properties if stored unopened, the effective shelf-life is only a few weeks after diluting for phantoms (and exposure to air). Depending on the fat concentration, anisotropy values range from ~0.6 to 0.9. Recently, effort has been made to better characterize and standardize the use of Intralipid, making it an attractive choice among scattering materials currently available (Di Ninni et al., 2011).

Absorption is more easily incorporated into phantoms than scattering. A frequent and effective choice is often food dyes. Historically, India ink (black) dye has been widely used, especially for studies in the NIR, but since the sizes of its absorbing particles are large enough to introduce a non-negligible amount of additional scattering, it is now in disfavor. To more closely mimic tissue conditions, it is also possible to use hemoglobin, which can be purchased in freeze-dried powdered form and can be added to any liquid phantom; however, it must first be converted from met-hemoglobin, and the phantom shelf-life is further reduced. The extinction coefficient, ε, of any non-scattering chromophore is easily measured in a spectrophotometer, so that the absorption coefficient is readily known when a measured concentration is added to a scattering medium.

8.8 Examples of clinical and pre-clinical applications of DRS with incoherent light

For the regime of small source-detector separations there are two general classes of measurement sought when using the methods of diffuse-reflectance spectroscopy: those that seek to maximize information about the microscopic structural properties of tissue through determination of the scattering properties, and those that seek to measure absorption coefficients (while minimizing the effects of scattering variations) to provide information on the biochemistry of the tissue. The general features of the instrumentation described above are common to both classes.

8.8.1 Elastic-scattering spectroscopy for detection of dysplasia

Diffuse reflectance spectroscopy, when applied to study the scattering properties of tissue (at the smallest source-detector separations), has demonstrated high potential for practical applications in recent years, having been applied in clinical studies with the goal of diagnosing disease in vivo (Bigio and Bown, 2004; Brown et al., 2009) in a wide range of organs such as brain (Bevilacqua et al., 1999; Canpolat et al., 2009), cervix (Weber et al., 2008), skin (Salomatina et al., 2006), colon (Rodriguez-Diaz et al., 2011) and breast (Palmer et al., 2006), among others. As an illustrative example, we summarize work reported by Rodriguez-Diaz et al. (2011) on statistical methods for classification of DRS spectra, as applied to a large-scale clinical study for detection of pre-cancerous conditions (*dysplasia* or *metaplasia*) during colonoscopy (optical examination of the colon through an endoscope). The clinical study in that report demonstrated the sensing of tissue changes associated with pre-cancer, in the optical scattering properties of the colon epithelial wall, the DRS method referred to as elastic-scattering spectroscopy (ESS).

In the United States, colorectal cancer (CRC) is the second leading cause of cancer and cancer death for both sexes (behind lung cancer), with nearly 150,000 new cases and 50,000 deaths annually in recent years. Current standards of clinical practice in the USA aim at early detection of cancer and pre-cancer by screening the entire population for CRC beginning at age 50. Of the various medical modalities for screening that have been implemented, colonoscopy has emerged as the most effective method, because a high-quality examination permits simultaneous detection and removal of precancerous polyps (small growths on the wall of the colon), thus detecting risk and preventing CRC progression. The effectiveness of colonoscopic cancer prevention hinges on the complete removal of all polyps detected during a white-light endoscopy. An optical method of sensing disease can improve the efficiency of endoscopy, and can reduce the cost burden of surgical removal and histopathology analysis of every polyp that is found.

A two-fiber sensing probe was integrated into the endoscopic biopsy forceps used for tissue removal (illustrated in Figure 8.8(d)), which enabled the ESS measurements to be coregistered precisely with the physical biopsy. The integrated "optical biopsy" forceps incorporated two 200-μm core fibers, each with a numerical aperture of 0.22 in air, at a center-to-center separation of $\sim$250 μm. The spectra from the ESS measurements were recorded for the range 300–800 nm. The clinical dataset consisted of ESS measurements from 297 polyps from 134 patients. The statistical methods concentrated on binary classifiers, so the samples were grouped into two clinically relevant classes, non-neoplastic (benign) samples and neoplastic (pre-cancerous) samples. Approximately 1/3 of the polyps were deemed neoplastic by histopathology. Figure 8.9 shows some typical ESS spectra (displayed for 350–700 nm) of colon polyps, measured in vivo during colonoscopy procedures.

The tissue spectra were classified with statistical methods based on multi-parameter pattern recognition to determine the spectral features that correlated best with the pathology reports. Half of the dataset was used to train a diagnostic algorithm by this method, and then the algorithm was frozen and tested prospectively on the other half of the data, which had not been used in the training. The sensitivity, specificity, positive- and negative-predictive values were in the 90+% range, which is comparable with the inter-observer agreement rate among pathologists. Since histopathology is the "gold standard" against which new diagnostic methods are compared, the accuracy of this method, and others like it, is on a par with current practice. Long-term outcome studies would be required to determine whether the optical methods may actually be more accurate than the subjective reading of histology slides by pathologists.

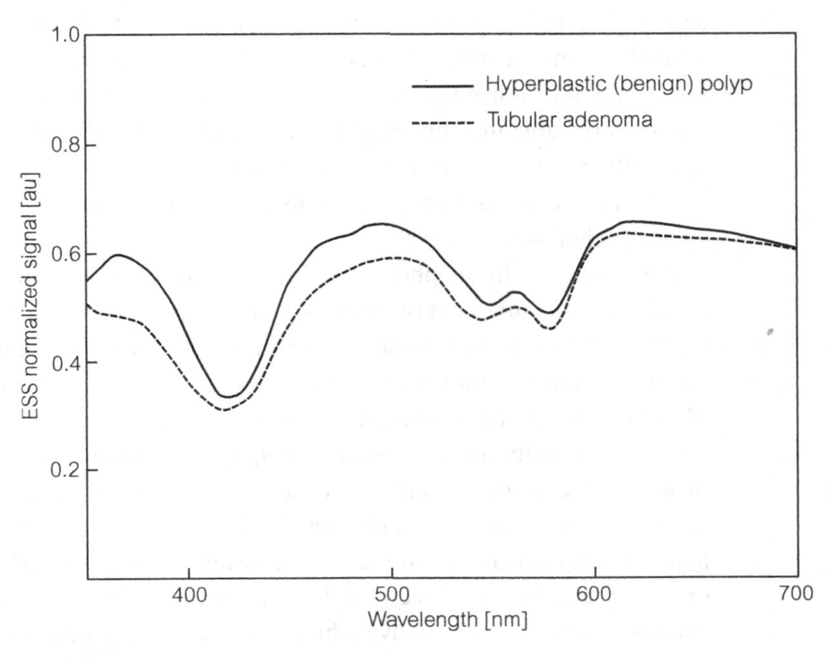

Figure 8.9

Typical ESS spectra of colon polyps taken during colonoscopy procedures. The fiber probe was incorporated into the biopsy forceps, and comprised two 200-μm-core fibers (NA = 0.22) separated by 250 μm center-to-center. The two spectra have been normalized to the same value at 700 nm. The tubular adenoma is considered to be potentially pre-malignant, and must be excised to reduce cancer risk. The hyperplastic polyp is benign and conveys no significant cancer risk.

8.8.2 Measurement of drug concentrations in tissue

The ability to noninvasively measure the concentrations of drugs in tissues could provide a variety of benefits for pharmacology research and, ultimately, clinical applications. There are several advantages of noninvasive site-specific measurements to determine concentrations of drugs used in chemotherapy and photodynamic therapy (PDT; see Section 19.3.1). Clinically, the therapeutic benefit of anti-cancer drugs is a function of the site-specific concentration-time profile in tumor tissue, whereas most side effects are a consequence of systemic concentrations. Therefore, the ability to track the location (*biodistribution*) and time-history (*pharmacokinetics*) of drug concentrations in tissue noninvasively would be advantageous for the development of new chemotherapy drugs and for monitoring treatment (Ding and Wu, 2012). The concentrations and kinetics of drugs at specific locations in the body are generally difficult to determine, given only the administered dosage or blood serum measurements. Other than new optical methods and costly imaging methods, minimally invasive methods with site-specificity are limited, the most commonly reported method being *microdialysis*, which involves implantation of a needle into the tissue to sample the extracellular fluid through a semi-permeable membrane. Despite its inherent invasiveness and slow response (typically 5–30 minutes), microdialysis has been developed in recent years as an *in situ* method of measurement for localized pharmacokinetics of chemotherapy agents (Schmidt et al., 2008).

If a drug is a chromophore, with a strong absorption band in the visible-NIR range, then measurement of its contribution to the absorption coefficient in tissue can lead to determination of the drug concentration. A number of chemotherapy agents (and all photodynamic therapy agents, by design) are chromophores. Moreover, for animal studies, it is common to conjugate experimental drugs with fluorescent tags for imaging, and fluorophores are always chromophores. In the diffusion regime, analytical methods enable separation of the absorption and reduced scattering coefficients, as will be detailed in Chapters 10–12. However, site specificity of drug sensing is limited for measurements in the diffusion regime due to the large required source-detector separation, ρ. For short distances, although analytical solutions are not available, empirical methods can be employed, taking advantage of the special value for ρ explained in Section 8.6.1. Given that the effective pathlength, $\langle L \rangle$, can be ≥ 1 cm, this intermediate regime still enables measurement of small concentrations that are relevant to therapeutic dosages.

In practice, reflectance, $R(t)$, is measured before and at various times after drug administration, and Eq. (8.36) is solved for the sum of all absorbing constituents:

$$\Delta\mu_a = \Delta\mu_{a,\mathrm{Hb}} + \Delta\mu_{a,\mathrm{HbO}} + \mu_{a,\mathrm{drug}}(t_1) = -\ln\left[\frac{R(t_1)}{R(t_0)}\right] \Big/ \langle L \rangle, \quad (8.37)$$

where $\mu_{a,\mathrm{drug}}(t_1)$ is the absolute concentration of the drug at time t_1, rather than the change in concentration, simply because the drug concentration is zero prior to administration. This expression represents an absorption spectrum as a function of wavelength. Since the spectra of the extinction coefficients of oxy- and deoxyhemoglobin, and of the drug, are known, the three constituents can be fit to the spectrum represented by the right-hand side of Eq. (8.37) using standard chemometric methods, or by various computational fitting "toolkits" available in common scientific programming environments.

In an early pre-clinical study, the concentrations of two chromophoric chemotherapy agents, doxorubicin and mitoxantrone, were tracked in dorsal subcutaneous tumors grown in immune-suppressed mice (Mourant et al., 1999). Measurements in the xenograft tumors were taken through the skin, which was deemed to constitute a minimal perturbation, due to the fact that photon transit through the thin skin layer (~250 μm) was a small percentage of the full pathlength through the tumors. The optically measured drug concentrations agreed well with chemical assays performed on tumors extracted from mice that were sacrificed at specific time-points. One key advantage of the optical method is that a full pharmacokinetic curve can be generated from a single animal, given that the DRS measurement take less than 1 second. This is compared with hundreds of animals required by conventional methods of pharmacokinetics to cover a range of time points and to average for animal-to-animal variations.

More recent studies have included monitoring of the time-dependent biodistribution of photosensitizer drugs (Austwick et al., 2011) and assessment of the disruption of the blood-brain barrier to facilitate delivery of drugs to brain tumors (Ergin et al., 2012).

8.9 Quasi-coherent variants of diffuse reflectance spectroscopy

In this section we introduce the concepts relating to useful phenomena in the multiple-scattering regime that can be measured when the light source is coherent, either temporally coherent (*diffuse correlation spectroscopy* and *speckle contrast imaging*) or spatially coherent (*low-coherence enhanced backscattering*).

8.9.1 Diffuse correlation spectroscopy in the short source-detector distance regime

In Section 7.5.2 we described the slight spectral broadening that is imposed on singly scattered monochromatic light, when the scattering centers are moving randomly. With multiply scattered light, as is more common in measurements from tissue, the effect is also measurable as long as the coherence length, L_c, of the laser light is longer than the total pathlength from source to detector, including the multi-scattered path in the turbid medium. Figure 8.10 depicts two possible paths for photons from the point of entry of the laser beam into the tissue, to a specific location for detection (at a distance ρ on the surface from the source), corresponding to two different times.

As depicted, each of the scattering center locations is in a slightly different location for the later photon path, due to random motion of the particles. Of course, there are many more possible paths, engaging different scattering particles, and each path will vary in time if the particles are in motion. Figure 8.11 shows an imagined temporal history of the detected intensity at the chosen output point.

The temporal variations result from the phase differences between different photon paths, resulting in temporally dependent interferometric effects of constructive and destructive interference at the point of detection. In the later photon path,

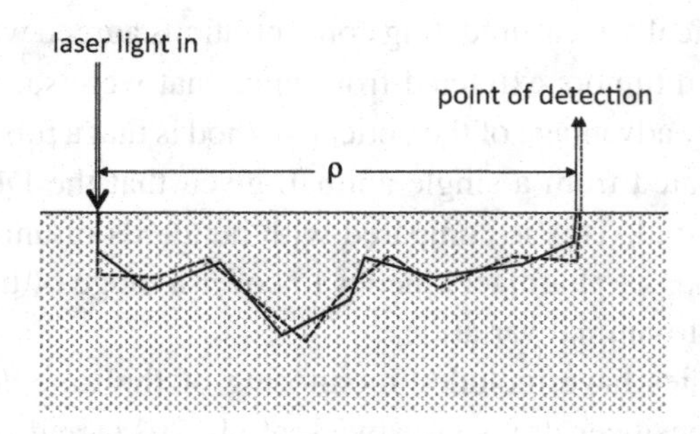

Figure 8.10

Laser light entering at one point can experience scattering multiple times from moving particles, resulting in different possible paths (two depicted) to reach the detection point at two different time points, separated by time-difference τ.

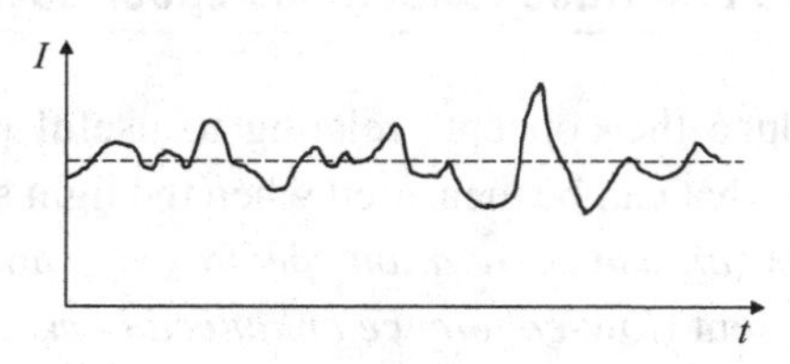

Figure 8.11

The measured intensity at a given detection point, some distance from the input, will vary temporally about a mean value.

each of the scattering centers may have moved randomly, and consequently can contribute randomly to the accumulation of the phase shift, thus inducing a faster decay of the temporal autocorrelation function. That is, the fields from individual photon paths are uncorrelated, such that the phase differences of the emerging photons can vary with the motion of the individual scattering centers. From the definition expressed by Eq. (7.48), one can derive the following expression for the normalized temporal autocorrelation function of the scattered field for *one* photon path (see, for example, Yu et al., 2010):

$$g_1(\tau)_{\text{one}} = e^{i\omega\tau} e^{-\frac{1}{3}k^2\mathcal{N}(1-g)\langle\Delta r^2(\tau)\rangle}, \tag{8.38}$$

where $\mathcal{N}$ is the number of scattering events along the path, k is the optical wavenumber ($2\pi/\lambda$), ω is the optical angular frequency, g is the anisotropy factor (either averaged for the different particles or assumed to be the same for all scatterers), and $\langle\Delta r^2(\tau)\rangle$ is the mean-square displacement of the particles over time τ. (Please note: this use of the term g_1 should not be confused with its use to represent the first moment of the phase function (the anisotropy factor), as discussed in Section 8.2.)

The number of scattering events along the photon path, $\mathcal{N}$, is (on average) given by the total pathlength, L, divided by the photon mean free path, $1/\mu_s$. Consequently,

$$\mathcal{N}(1-g) = L\mu_s', \tag{8.39}$$

which indicates how the anisotropy factor affects the decay rate of the autocorrelation function. Of course, the relevant field autocorrelation function should include the contributions of all photon paths. In a strongly scattering medium, which is a reasonable assumption for tissue, it is possible to represent the combination of all photon paths with an integral over all possible pathlengths, L:

$$g_1(\tau) = e^{i\omega t} \int_{\rho}^{\infty} p(L) e^{-\frac{1}{3}k^2 L\mu_s' \langle \Delta r^2(\tau) \rangle} dL \tag{8.40}$$

where $p(L)$ represents the probability density distribution of possible photon pathlengths. The pathlength distribution associated with $p(L)$ depends on the tissue properties and on the source-detector separation, ρ. The distribution can be determined with Monte Carlo simulations, or experimentally by, for example, time-resolved measurements using ultrashort laser pulses (see Chapter 11). In practice, the detectors for diffuse correlation measurements are often fast, photon-counting avalanche photodiodes, and photon counts are compared for sequential delays of a measurement time period.

Demonstrations of practical applications of localized DCS have focused on the determination of tissue viability by assessing microvascular blood flow. Good examples of recent clinical studies are found in de Mel et al. (2013) and Shang et al. (2013).

8.9.2 Laser speckle contrast imaging

Just as interferometric effects with randomly moving particles can yield temporal variations in the measured intensity at a single point on the tissue surface, the light reaching different *spatial* locations will exhibit random intensity variations that result from the interferometric effects of the various photon paths leading to the specific surface location, generating a spatially random surface intensity. If a scattering medium is illuminated with laser light and imaged with a camera (or the eye!), the image exhibits a spatially random intensity distribution, which is referred to as a *speckle pattern*. This effect is almost ubiquitous when observing scattered coherent light, regardless of particle motion, and the variations exhibit higher contrast if all scattering centers are static. The reader can easily observe speckle by spreading the light of a laser pointer with a lens and illuminating a diffusely reflecting surface (such as a wall painted with flat paint); the effect can be enhanced by purposely defocusing one's eye. It should be noted, however, that the detected photons have traveled on paths that include the distance from the emitting surface to the detector, such that the observed speckles are not actually located on the illuminated wall itself, but on the retina of the observer's eye (or on a camera's detector). This is the same for imaging speckles of light emerging from a tissue surface, and the speckle spacing will depend on the details of the imaging optics.

Speckle constitutes a common problem in ultrasound imaging, and is also generally regarded as a troublesome source of noise in optical imaging with coherent light, especially when laser light sources are used in microscopy, as in the cases, for example, of confocal microscopy and optical coherence tomography (see Chapters 17 and 18). Much effort has been expended to reduce the effects of speckle in microscopy (e.g., Schmitt, 1997; Glazowski and Rajadhyaksha, 2012). If the scattering particles are in motion, however, then the spatial pattern of the speckle will randomly vary in time; and if the integration time of an imaging camera is longer than the temporal variations of the speckles, then the spatial pattern of the speckle will blur: i.e., the contrast of the speckles will be reduced. This effect can be used to advantage for imaging of superficial blood flow in vessels near the tissue surface, and imaging of this reduction in contrast is the basis of the method called *speckle contrast imaging*.

In its simplest implementation, speckle contrast imaging is realized by imaging the light scattered from a tissue surface that is illuminated by a laser source. The camera integration time is set longer than the temporal variation time of speckles in areas of blood vessels with flowing blood. The simple experimental configuration is similar to that illustrated in Figure 5.10, while taking care to minimize specular reflections from the surface. The camera image is analyzed computationally by defining a small area, or "window," of pixels that is large enough to encompass several imaged speckles, but small enough to allow acceptable spatial resolution of the final image. A quantitative measure of the speckle contrast in that window is determined, and the analysis is repeated while scanning the window across the image to generate a new image wherein a color code or gray level represents the local value of the speckle contrast.

The speckle contrast, K_{spec}, is generally defined (Boas and Dunn, 2010) as the ratio of the standard deviation of the speckle intensity, σ_{spec}, divided by the average intensity, $\langle I \rangle$, both within the window:

$$K_{\text{spec}} = \frac{\sigma_{\text{spec}}}{\langle I \rangle}. \tag{8.41}$$

The value of K_{spec} can range from 0 to 1, where 0 obtains if the speckles are completely blurred, and 1 obtains in a static situation of no motion, hence no

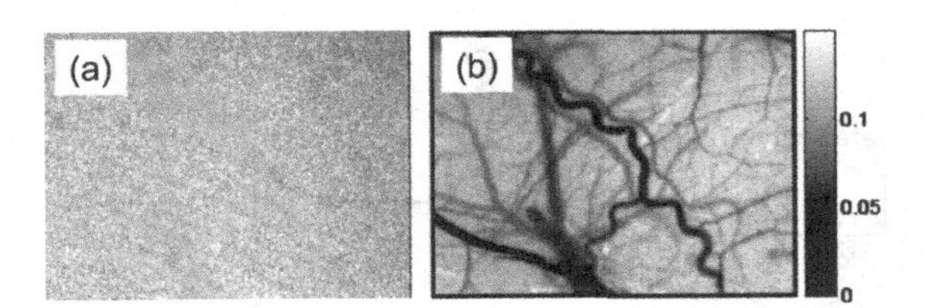

Figure 8.12

(a) The grainy image of the surface of an exposed rat brain illuminated with laser light; (b) the superficial blood vessels are clearly revealed when the image is processed such that areas of lower speckle contrast are presented as darker shades. (Copyright Boas and Dunn, 2010, by permission *Journal of Biomedical Optics*.)

blurring. The contrast is a function of the integration time, and it is possible to extract blood flow velocity information from the relationship between the integration time and the contrast. The formalisms relating speckle contrast to integration time and to the size of the analysis window are developed and reviewed in Boas and Dunn (2010). An example of an image of superficial blood vessels with and without speckle contrast analysis is shown in Figure 8.12. The vascular detail that is revealed is striking.

8.9.3 Low-coherence enhanced backscattering

A hybrid of scattering spectroscopy and coherence effects has recently been developed by Backman and co-workers (Kim et al., 2006). In this method, which they call *low-coherence enhanced backscattering* (LEBS), the temporal (longitudinal) coherence of the light source can be short (consistent with a broad spectral bandwidth), whereas the transverse coherence is required to be good, consistent with a flat phase front of collimated light, as for a plane wave. As illustrated in Figure 8.13(a), for any given photon that emerges after scattering a few times, there is a "time-reversed" photon path in precisely the reverse direction, for which a backward photon emerges from the starting point of the "forward" photon. The two emerging photons started in phase at the points of entry, and travelled exactly the same path, albeit in opposite directions, emerging in phase at the exit points. If the backscattered fields are observed at 180°, the two fields will exhibit constructive interference, because the total distances travelled to the detector will be identical. As the angle of observation (measured with respect to 180° backscattering), θ, changes from 0°, the constructive phase relationship is lost, and interference variations with angle become too fine to observe.

Under optimum conditions the degree of coherence enhancement at $\theta = 0°$ can be as much as a factor of two (the ratio of adding two waves coherently

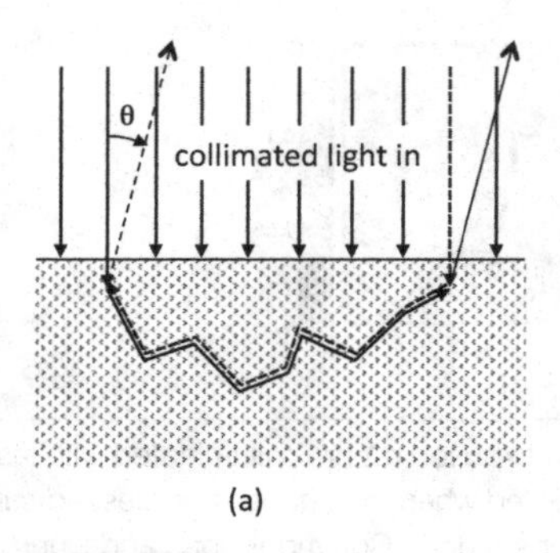

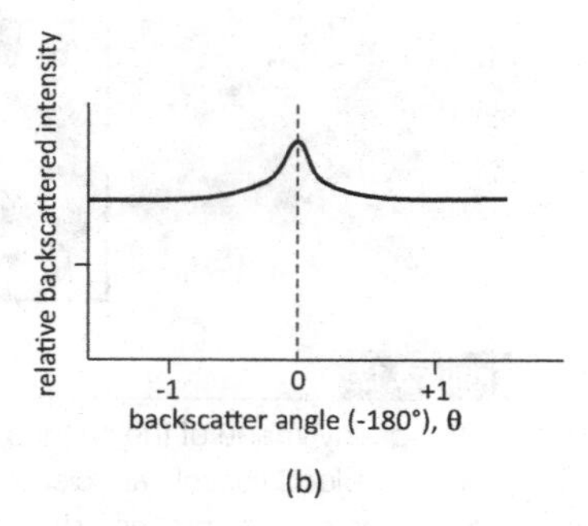

Figure 8.13.

Low-coherence enhanced backscattering (LEBS) measurement: (a) collimated light with a uniform wavefront illuminates the tissue surface. For each potential photon pathway, there is an identical pathway for photons in the reverse direction. The angle of observation, θ (relative to the 180° backscattering direction), will determine the relative phase of the emitted photons at the detector; (b) the constructive interference enhancement of backscattering is measurable over a small range of angles in the backscattering direction.

vs. incoherently), but in practice the measured enhancement in tissues is several percent. It has been shown (Turzhitsky et al., 2011) that it is possible to model the LEBS process to predict the average penetration depth of photons collected at a selected angle. This feature is beneficial for assessing epithelial diseases (e.g., pre-cancerous changes) where selective sensitivity to the superficial layer helps to determine the cellular properties of the specific tissue layer being investigated.

Problems – answers to problems with ∗ are on p. 646

8.1* What are the units of the coefficients a and c in Eq. (8.3)?

8.2 Show that the Henyey-Greenstein phase function, $p_{HG}(\cos\theta)$ (Eq. (8.10)) satisfies the normalization requirement, as specified by Eq. (8.4).

8.3 Show that the modified Henyey-Greenstein phase function, $p_{MHG}(\cos\theta)$ (Eq. (8.11)) also satisfies the normalization requirement.

8.4 The Figure Problem 8.4 shows a comparison of an accurate Mie-theory phase function, $p(\theta)$, for a specific spherical particle size and index ratio, with the Henyey-Greenstein approximation for the same value of g (the anisotropy factor).

(a) For typical tissue scattering parameters (for example, $\mu_s' = 10\ \text{cm}^{-1}$) explain why the Henyey-Greenstein phase function is good for source-detector separations $\rho > 1$ cm.

(b) Explain why a more accurate phase function, such as that from Mie theory, is necessary when ρ is small (say, <1 mm).

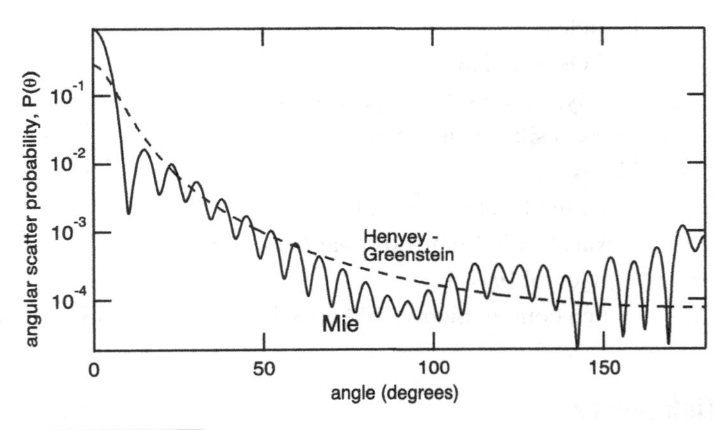

Figure Problem 8.4

8.5 Using any convenient programming environment, construct polar plots of $p_{HG}(\theta)$ for $g = 0.2, 0.5, 0.8$ and 0.95.

8.6* (a) Estimate the size of the "vessel packing effect" at 420 nm, as determined by Eq. (8.22), for tissue incorporating mostly arterioles of 30 μm diameter, carrying mostly oxygenated blood.

(b) What would be the factor at 550 nm?

(c) How would the resulting ratio of those two spectral features compare when measured in live tissue to an equal concentration of hemoglobin uniformly distributed in a scattering medium?

8.7* Equation (8.35) is an approximation based on a variant of the Beer-Lambert law. Assume that μ_a, and consequently the measured reflectance, R, are changing with time because of changes in the drug concentration (or hemoglobin concentration) in the tissue being measured. Derive the expression for the change in absorption coefficient: i.e., derive Eq. (8.36) from Eq. (8.35).

8.8* Frequently in the literature, values for reflectance are plotted against dimensionless scattering, which is defined by the product $\rho\mu_s'$. What is the advantage of this type of plot?

8.9* Recall the development of the Mie theory for scattering in Chapter 7. Also, assume a representative refractive index of 1.41 for organelles, and 1.35 for surrounding fluid. Which combination of particles and fluids might best be used in developing a tissue phantom that mimics the scattering properties of tissue? Explain why. (It may be assumed that the particle sizes can be chosen to mimic organelle sizes.)

Particles:
TiO_2 ($n = 2.6$)
polystyrene spheres ($n = 1.56$)
fused silica spheres ($n = 1.45$)
Fluids:
distilled water ($n = 1.33$)
water with dissolved sugar ($n = 1.39$)
glycerol ($n = 1.47$)
Which combination would constitute the second-best alternative?

References

Amelink, A., Sterenborg, H. J. C. M., Bard, M. P. L., and Burgers, S. A. (2004). In vivo measurement of the local optical properties of tissue by use of differential path-length spectroscopy. *Optics Letters*, 29, 1087–1089.

Andree, S., Reble, C., Helfmann, J., Gersonde, I., and Illing, G. (2010). Evaluation of a novel noncontact spectrally and spatially resolved reflectance setup with continuously variable source-detector separation using silicone phantoms. *Journal of Biomedical Optics*, 15, 067009.

Austwick, M. R., Woodhams, J., Chalau, V., et al. (2011). Optical measurement of photosensitizer concentrations in vivo. *Journal of Innovative Optical Health Sciences*, 4, 97–111.

Backman, V., Gopal, V., Kalashnikov, M., et al. (2001). Measuring cellular structure at submicrometer scale with light scattering spectroscopy. *IEEE Journal of Selected Topics in Quantum Electronics*, 7, 887–893.

Bargo, P. R., Prahl, S. A., Goodell, T. T., et al. (2005). In vivo determination of optical properties of normal and tumor tissue with white light reflectance and an empirical light transport model during endoscopy. *Journal of Biomedical Optics*, 10, 034018.

Bevilacqua, F., and Depeursinge, C. (1999). Monte Carlo study of diffuse reflectance at source – detector separations close to one transport mean free path. *Journal of the Optical Society of America A*, 16, 2935–2945.

Bevilacqua, F., Piguet, D., Marquet, P., et al. (1999). In vivo local determination of tissue optical properties: applications to human brain. *Applied Optics*, 38, 4939–4950.

Bigio, I. J., and Mourant, J. R. (1997). Ultraviolet and visible spectroscopies for tissue diagnostics: fluorescence spectroscopy and elastic-scattering spectroscopy. *Physics in Medicine and Biology*, 42, 803–814.

Bigio, I. J., and Bown, S. G. (2004). Spectroscopic sensing of cancer and cancer chemotherapy, current status of translational research. *Cancer Biology and Therapy*, 3(3), 259–267.

Binzoni, T., Leung, T., Gandjbakhche, A. H., Rufenacht, D., and Delpy, D. T. (2006). The use of the Henyey–Greenstein phase function in Monte Carlo simulations in biomedical optics. *Physics in Medicine and Biology*, 51, N313–N322.

Boas, D. A., and Dunn, A. K. (2010). Review article: Laser speckle contrast imaging in biomedical optics. *Journal of Biomedical Optics*, 15(1), 011109.

Bouchard, J. P., Veilleux, I., Jedidi, R., et al. (2010). Reference optical phantoms for diffuse optical spectroscopy. Part 1 – Error analysis of a time resolved transmittance characterization method. *Optics Express*, 18, 11495–11507.

Brown, J. Q., Vishwanath, K., Palmer, G. M., and Ramanujam, N. (2009). Advances in quantitative UV-visible spectroscopy for clinical and pre-clinical application in cancer. *Current Opinion in Biotech*, 20(1), 119–131.

Calabro, K., and Bigio, I. J. (2014). Influence of the phase function in generalized diffuse reflectance models: review of current formalisms and novel observations. *Journal of Biomedical Optics*, 19, 075005.

Canpolat, M., and Mourant, J. R. (2000). High-angle scattering events strongly affect light collection in clinically relevant measurement geometries for light transport through tissue. *Physics in Medicine and Biology*, 45, 1127–1140.

Canpolat, M., Akyuz, M., Gokhan, G. A., Gurer, E. I., and Tuncer, R. (2009). Intra-operative brain tumor detection using elastic light single-scattering spectroscopy: a feasibility study. *Journal of Biomedical Optics*, 14, 054021.

Ding, H., and Wu, F. (2012). Image guided biodistribution of drugs and drug delivery. *Theranostics*, 2, 1037–1039.

Ergin, A., Wang, M., Zhang, J. Y., et al. (2012). The feasibility of real-time in-vivo optical detection of blood brain barrier disruption with indocyanine green. *Journal of Neurooncology*, 106(3), 551–560.

Erickson, T. A., Mazhar, A., Cuccia, D., Durkin, A. J., and Tunnell, J. W. (2012). Lookup-table method for imaging optical properties with structured illumination beyond the diffusion theory regime. *Journal of Biomedical Optics*, 15, 036013.

Gélébart, B., Tinet, E., Tualle, J. M., and Avrillier, S. (1996). Phase function simulation in tissue phantoms: a fractal approach. *Pure and Applied Optics*, 5, 377–388.

Glazowski, C., and Rajadhyaksha, M. (2012). Optimal detection pinhole for lowering speckle noise while maintaining adequate optical sectioning in confocal reflectance microscopes. *Journal of Biomedical Optics*, 17(8), 085001.

Graaff, R., Aarnoose, J. G., Zijp, J. R., et al. (1992). Reduced light-scattering properties for mixtures of spherical particles: a simple approximation derived from Mie calculations. *Applied Optics*, 31, 1370–1376.

Hayakawa, C. K., Hill, B. Y., You, J. S., et al. (2004). Use of the delta-P1 approximation for recovery of optical absorption, scattering, and asymmetry coefficients in turbid media. *Applied Optics*, 43, 4677–4684.

Henyey, L. G., and Greenstein, J. L. (1941). Diffuse radiation in the galaxy. *Astrophysics Journal*, 93, 70–83.

Hull, E. L., and Foster, T. H. (2001). Steady-state reflectance spectroscopy in the P3 approximation. *Journal of the Optical Society of America A*, 18, 584–599.

Kim, Y. L., Turzhitsky, V. M., Liu, Y., et al. (2006). Low-coherence enhanced backscattering: review of principles and applications for colon cancer screening. *Journal of Biomedical Optics*, 11, 041125.

Kumar, G., and Schmitt, J. M. (1997). Optimal probe geometry for near-infrared spectroscopy of biological tissue. *Applied Optics*, 36, 2286–93.

Jacques, S. L., Alter, C. A., and Prahl, S. A. (1987). Angular dependence of HeNe laser light scattering by human dermis. *Lasers in the Life Sciences*, 1, 309–333.

Lau, C., Sćepanović, O., Mirkovic, J., et al. (2009). Re-evaluation of model-based light-scattering spectroscopy for tissue spectroscopy. *Journal of Biomedical Optics*, 14, 024031.

Martelli, F., and Zaccanti, G. (2007). Calibration of scattering and absorption properties of a liquid diffusive medium at NIR wavelengths . CW method. *Optics Express*, 15, 486–500.

de Mel, A., Birchall, M. A., and Seifalian, A. M. (2013). Assessment of tissue viability with blood flow measurements. *Angiology*, 64(6), 409–10.

Metropolis, N. (1987). The beginning of the Monte Carlo Method. *Los Alamos Science*, Special Issue, 125–130.

Michels, R., Foschum, F., and Kienle, A. (2008). Optical properties of fat emulsions. *Optics Express*, 16, 5907–5925.

Mourant, J. R., Boyer, J., Hielscher, A. H., and Bigio, I. J. (1996). Influence of the scattering phase function on light transport measurements in turbid media performed with small source–detector separations. *Optics Letters*, 21, 546–548.

Mourant, J. R., Bigio, I. J., Jack, D., Johnson, T. M., and Miller, H. D. (1997). Measuring absorbance in small volumes of highly-scattering media: source-detector separations for which pathlengths do not depend on scattering properties. *Applied Optics*, 36, pp. 5655–5661.

Mourant, J. R., Johnson, T. M., Los, G., and Bigio, I. J. (1999). Noninvasive measurement of chemotherapy drug concentrations in tissue: preliminary demonstrations of in vivo measurements. *Journal of Physics in Medicine and Biology*, 44, 1397–1417.

Nichols, B. S., Rajaram, N., and Tunnell, J. W. (2012). Performance of a lookup table-based approach for measuring tissue optical properties with diffuse optical spectroscopy diffuse optical spectroscopy. *Journal of Biomedical Optics*, 17, 057001.

Di Ninni, P., Martelli, F., and Zaccanti, G. (2011). Intralipid: towards a diffusive reference standard for optical tissue phantoms. *Physics in Medicine and Biology*, 56, N21–N28.

Palmer, G. M., Zhu, C., Breslin, T., et al. (2006). Monte Carlo-based inverse model for calculating tissue optical properties. Part II: Application to breast cancer diagnosis. *Applied Optics*, 45, 1072–1078.

Pogue, B. W., and Patterson, M. S. (2006). Review of tissue simulating phantoms for optical spectroscopy, imaging and dosimetry. *Journal of Biomedical Optics*, 11, 041102.

Rajaram, N., Nguyen, T. H., and Tunnell, J. W. (2012). Lookup table-based inverse model for determining optical properties of turbid media. *Journal of Biomedical Optics*, 13, 050501.

Reif, R., A'Amar, O., and Bigio, I. J. (2007). Analytical model of light reflectance for extraction of the optical properties in small volumes of turbid media. *Applied Optics*, 46, 7317–7328.

Rodriguez-Diaz, E., Castanon, D. A., Singh, S. K., and Bigio, I. J. (2011). Spectral classifier design with ensemble classifiers and misclassification-rejection: application to elastic-scattering spectroscopy for detection of colonic neoplasia. *Journal of Biomedical Optics*, 16, 067009.

Saidi, I. S., Jacques, S. L., and Tittel, F. K. (1995). Mie and Rayleigh modeling of visible-light scattering in neonatal skin. *Applied Optics*, 34, 7410–7418.

Salomatina, E., Jiang, B., Novak, J., and Yaroslavsky, A. N. (2006). Optical properties of normal and cancerous human skin in the visible and near-infrared spectral range. *Journal of Biomedical Optics*, 11, 064026.

Sassaroli, A., and Fantini, S. (2004). Comment on the modified Beer-Lambert law for scattering media. *Physics in Medicine and Biology*, 49, N255–N257.

Schmidt, S., Banks, R., Kumar, V., Rand, K. H., and Derendorf, H. (2008). Clinical microdialysis in skin and soft tissues: an update. *Journal of Clinical Pharmacology*, 48(3), 352–364.

Schmitt, J. M. (1997). Array detection for speckle reduction in optical coherence microscopy. *Physics in Medicine and Biology*, 42, 1427.

Schmitt, J. M., and Kumar, G. (1998). Optical scattering properties of soft tissue: a discrete particle model. *Applied Optics*, 37, 2788–2797.

Shang, Y., Gurley, K., and Yu, G. (2013). Diffuse correlation spectroscopy (DCS) for assessment of tissue blood flow in skeletal muscle: recent progress. *Anatomy & Physiology*, 3(2), 128.

Sharma, S. K., and Banerjee, S. (2003). Role of approximate phase functions in Monte Carlo simulation of light propagation in tissues. *Journal of Optics A: Pure and Applied Optics*, 5, 294–302.

Sharma, S. K., and Banerjee, S. (2005). Volume concentration and size dependence of diffuse reflectance in a fractal soft tissue model. *Medical Physics*, 32, 1767–1774.

Svaasand, L. O., Fiskerstrand, E. J., Kopstad, G., et al. (1995). Therapeutic response during pulsed laser treatment of port-wine stains: dependence on vessel diameter and depth in dermis. *Laser in Medical Science*, 10, 235–243.

Turzhitsky, V., Mutyal, N. N., Radosevich, A. J., and Backman, V. (2011). Multiple scattering model for the penetration depth of low-coherence enhanced backscattering. *Journal of Biomedical Optics*, 16(9), 097006.

Vitkin, E., Turzhitsky, V., Qiu, L., et al. (2011). Photon diffusion near the point-of-entry in anisotropically scattering turbid media. *Nature Communications*, 2, 587, 1599.

Wang, R. K. (2000). Modelling optical properties of soft tissue by fractal distribution of scatterers. *Journal of Modern Optics*, 47, 103–120.

Weber, C. R., Schwarz, R. A., Atkinson, E. N., et al. (2008). Model-based analysis of reflectance and fluorescence spectra for in vivo detection of cervical dysplasia and cancer. *Journal of Biomedical Optics*, 13, 064016.

Yu, G., Durduran, T., Zhou, C., Cheng, R., and Yodh, A. G. (2010). Near-infrared diffuse correlation spectroscopy for assessment of tissue blood flow. In *Handbook of Biomedical Optics*, ed. D.A. Boas, C. Pitris and N. Ramanujam. Oxford: Taylor & Francis.

Zonios, G., Perelman, L. T., Backman, V., et al. (1999). Diffuse reflectance spectroscopy of human adenomatous colon polyps in vivo. *Applied Optics*, 38, 6628–6637.

Zonios, G., and Dimou, A. (2006). Modeling diffuse reflectance from semi-infinite turbid media: application to the study of skin optical properties. *Optics Express*, 14, 8661–8674.

Further reading

Empirical models for diffuse reflectance spectroscopy at short distances

Calabro, K. W., and Bigio, I. J. (2014). Influence of the phase function in generalized diffuse reflectance models: review of current formalisms and novel observations. *Journal of Biomedical Optics*, 19, 075005.

Analytical models for diffuse reflectance at short distances

Vitkin, E., Turzhitsky, V., Qiu, L., et al. (2011). Photon diffusion near the point-of-entry in anisotropically scattering turbid media. *Nature Communications*, 2, 587–595.

Tissue phantoms

Pogue, B. W., and Patterson, M. S. (2006). Review of tissue simulating phantoms for optical spectroscopy, imaging and dosimetry. *Journal of Biomedical Optics*, 11, 041102.

Diffuse correlation spectroscopy

Yu, G., Durduran, T., Zhou, C., Cheng, R., and Yodh, A. G. (2010). Near-infrared diffuse correlation spectroscopy for assessment of tissue blood flow. In *Handbook of Biomedical Optics*, ed. D. A. Boas, C. Pitris and N. Ramanujam. Oxford: Taylor & Francis Group.

Speckle contrast imaging

Boas, D. A., and Dunn, A. K. (2010). Review article: Laser speckle contrast imaging in biomedical optics. *Journal of Biomedical Optics*, 15(1), 011109.

9 Transport theory and the diffusion equation

Biological tissues are strongly heterogeneous media, with spatial inhomogeneities occurring on length scales ranging from nanometers (structural proteins), to hundreds of nanometers (cellular organelles), to tens of microns (cells), all the way to millimeters and centimeters (at tissue and organ levels). These inhomogeneities are associated with spatial gradients and discontinuities in physical properties that play key roles in determining the optical properties of tissues, which govern how light propagates within tissues. For example, a discontinuity in the refractive index at a plane or interface causes reflection and refraction of light, whereas, as discussed in Chapters 7 and 8, localized optical inhomogeneities associated with particles result in light scattering, with a wavelength dependence and an angular distribution that depend on the shapes and refractive indices of particles and their sizes relative to the wavelength of light. Further, as described in Chapter 2, a number of molecules found in tissues (e.g., aromatic amino acids, melanin, bilirubin, rhodopsin, hemoglobin, cytochrome *c*, water, etc.) absorb light with spectral extinction coefficients that determine their wavelength-dependent absorption efficiency. Whereas Chapter 7 developed the theories that describe scattering from single particles, and Chapter 8 described treatment of light transport between a source and detector separated by only a few optical mean free paths, this chapter addresses the diffuse field, for which source and detector are separated by a distance large compared to the mean free path for scattering.

When considering the transport of light in bulk turbid media, it is frequently easier, conceptually, to invoke the particle representation rather than the wave nature of light. (See Section 4.7.1.) In the particle-like description of light propagation as a flow of photons, one may describe the effects of spatial heterogeneities and absorbing molecules in terms of deflections and annihilations, respectively, of traveling photons. This means that the tissue can be seen as a collection of scattering and absorbing centers that interact with photons randomly but according to quantitative rules that specify the probability of interaction and the way in which the tissue constituents deflect or absorb the incident photons. This is conceptually no different than a flow of metal balls in a pinball machine with posts (scattering centers) and holes (absorbing centers). This conceptual framework for the quantitative description of the spatio-temporal distribution of a collection of particles has been used to model a number of phenomena, such as the diffusion of molecules in air, the spread of a pollutant in ground water, the motion of neutrons in a reactor core, and heat transfer.

The starting point for such a model is a balance relationship known as the *Boltzmann transport equation* (named after the same Austrian physicist who formulated the thermodynamic occupancy distribution of energy levels; see Eq. (5.1)). The Boltzmann transport equation, which is frequently referred to as the *radiative transfer equation*, accounts for all possible sources of temporal changes in the number of particles per volume element of the phase space (i.e., per unit volume and per unit solid angle of direction of propagation). This chapter introduces the Boltzmann transport equation and its diffusion approximation, which is commonly used in biomedical optics, also covering boundary conditions required to treat bounded scattering media or interfaces between different scattering media.

9.1 The Boltzmann transport equation

The Boltzmann transport equation (BTE) is not derived from electromagnetic theory or any other fundamental theoretical framework. It is a heuristic representation of the transport of particles in diffusive media, rather than an exact analytical solution, and is based on logical assumptions on the balance of optical energy or photon number (in the case of light). The balance relationship is obtained by considering the light quanta, or photons, as particles flowing at a constant speed within a medium, where they can be absorbed (by chromophores), deflected (by scattering particles or spatial gradients in the refractive index), or emitted (by light sources). Here, we need to consider the density of photons in phase space, which includes spatial and momentum coordinates, to keep track of both the number density of photons as well as their energy and the angular distribution of their directions of propagation.

Under the assumptions that the speed of light in the medium, c_n, and the angular frequency, ω, of each photon remain constant in the medium, the magnitude of the photon momentum ($\hbar\omega/c_n$) is also a constant, and the phase space reduces to a five-dimensional space (three spatial coordinates for position plus two angular coordinates for the direction of propagation). The three spatial coordinates define the position vector $\mathbf{r}$, whereas the two angular coordinates, namely the polar angle θ (angle with the positive z axis: $0 \leq \theta \leq \pi$) and the azimuthal angle φ (angle with the positive x axis on the x-y plane: $0 \leq \phi < 2\pi$), define the directional unit vector $\hat{\mathbf{\Omega}} = \sin\theta\cos\varphi\,\hat{\mathbf{x}} + \sin\theta\sin\varphi\,\hat{\mathbf{y}} + \cos\theta\,\hat{\mathbf{z}}$ along the photon direction of propagation (where $\hat{\mathbf{x}}$, $\hat{\mathbf{y}}$ and $\hat{\mathbf{z}}$ are the unitary vectors along the positive x, y, and z axes, respectively). The volume element in this five-dimensional space is represented by the shaded volume in Figure 9.1, which also shows the angles θ,

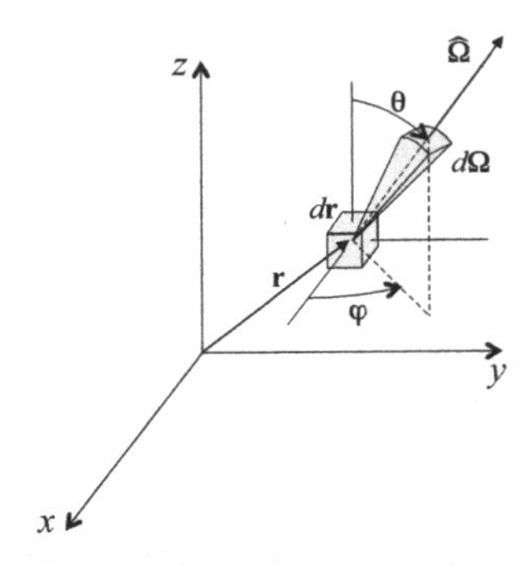

Figure 9.1

Representation of the spatial ($d\mathbf{r}$) and directional ($d\Omega$) differential elements that define the angular energy density (radiant energy per unit volume at position $\mathbf{r}$, per unit solid angle at direction of propagation $\hat{\Omega}$). The polar and azimuthal angles θ and φ that define the direction $\hat{\Omega}$ are shown.

φ, the position vector $\mathbf{r}$, the directional vector $\hat{\Omega}$, and the differential elements $d\mathbf{r}$ and $d\Omega$.

The instantaneous angular energy density in this five-dimensional space is indicated with $u(\mathbf{r}, \hat{\Omega}, t)$, and represents the radiant energy per unit volume, per unit solid angle about position $\mathbf{r}$ and propagation direction $\hat{\Omega}$ at time t. The SI units of $u(\mathbf{r}, \hat{\Omega}, t)$ are J/(m^3-sr). Of course, when considering light at a given angular frequency, ω, the angular energy density may be equivalently expressed in terms of an angular photon density, given by $u(r, \hat{\Omega}, t)/(\hbar\omega)$, which represents the number of photons per unit volume and unit solid angle that are traveling along direction $\hat{\Omega}$ at position $\mathbf{r}$ and time t. The Boltzmann transport equation denotes the overall rate of change of this angular energy density as the sum of three contributions, one from photon diffusion (driven by a spatial gradient in the angular energy density), one from photon collisions (with either absorption or scattering centers), and one from photon sources:

$$\frac{\partial u(\mathbf{r}, \hat{\Omega}, t)}{\partial t} = \left(\frac{\partial u(\mathbf{r}, \hat{\Omega}, t)}{\partial t}\right)_{\text{Diff}} + \left(\frac{\partial u(\mathbf{r}, \hat{\Omega}, t)}{\partial t}\right)_{\text{Coll}} + \left(\frac{\partial u(\mathbf{r}, \hat{\Omega}, t)}{\partial t}\right)_{\text{Sources}}. \tag{9.1}$$

The contributions from diffusion, collisions, and photon sources can be written explicitly as follows:

$$\frac{\partial u(\mathbf{r}, \hat{\Omega}, t)}{\partial t} = \underbrace{-c_n\hat{\Omega}\cdot\nabla u(\mathbf{r}, \hat{\Omega}, t)}_{\text{Diffusion}}\underbrace{-c_n(\mu_a+\mu_s)u(\mathbf{r}, \hat{\Omega}, t)+c_n\mu_s\int_{4\pi} u(\mathbf{r}, \hat{\Omega}', t)p(\hat{\Omega}', \hat{\Omega})d\Omega'}_{\text{Collisions}} \underbrace{+q(\mathbf{r}, \hat{\Omega}, t)}_{\text{Sources}}, \tag{9.2}$$

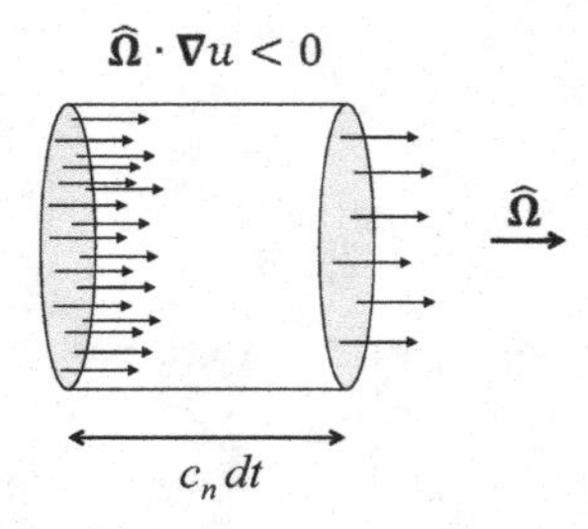

Figure 9.2

Pictorial representation of the meaning of the directional derivative $\hat{\Omega} \cdot \nabla u$, and how its negative value corresponds to a net gain of photons inside the infinitesimal volume element (there are more photons entering than exiting the volume element). For a differential time dt, the linear dimension of the volume element along the direction of propagation $\hat{\Omega}$ is $c_n dt$.

where the medium is characterized by an absorption coefficient μ_a (SI units of m^{-1}), a scattering coefficient μ_s (SI units of m^{-1}), and a scattering phase function $p(\hat{\Omega}', \hat{\Omega})$, which is the angular probability density of scattering from direction $\hat{\Omega}'$ into direction $\hat{\Omega}$ already introduced and discussed in Chapters 1, 2, 7, and 8. (Note: in Section 1.3 the phase function was defined as $p(\hat{\Omega}, \hat{\Omega}')$, with the prime indicating the final direction. Here $\hat{\Omega}'$ represents the initial photon directions. The convention to remember is that the symbol in the argument of p before the comma represents the initial direction.) The source term in Eq. (9.2) is $q(\mathbf{r}, \hat{\Omega}, t)$ (SI units of $W/(m^3\text{-sr})$). By recalling the proportionality (through the speed of light in the medium) between the angular energy density, $u(\mathbf{r}, \hat{\Omega}, \mathbf{t})$, and the radiance, $L(\mathbf{r}, \hat{\Omega}, \mathbf{t})$ (see Eq. (1.17)), we can see that the Boltzmann transport equation describes the optical radiation field in terms of its radiance.

The first term on the right-hand side of Eq. (9.2) corresponds to the diffusion contribution to the rate of change of the angular energy density u. It represents the net gain or loss of photons (per unit time) that are at position $\mathbf{r}$, time t, and travel along direction $\hat{\Omega}$, as a result of the flow of photons. For a given direction of propagation $\hat{\Omega}$, there will be a gain of photons if there are more photons entering the volume element $d\mathbf{r}$ than exiting it, and a loss of photons otherwise. The gradient of u, ∇u, is a vector perpendicular to surfaces of constant angular energy density (and pointing in the direction of the greatest rate of increase in the angular energy density), and the scalar product $\hat{\Omega} \cdot \nabla u$ represents the spatial rate of change of u along $\hat{\Omega}$ (and thus is a directional derivative). Therefore, $\hat{\Omega} \cdot \nabla u$ represents the difference between the energy per unit volume that exits and enters an infinitesimal volume element along direction $\hat{\Omega}$, as illustrated in Figure 9.2. If this difference is negative, as is the case in Figure 9.2, it represents a gain of photons in the volume element, and hence the negative sign in the first term of Eq. (9.2). The factor

c_n represents the distance traveled by light per unit time, and translates a spatial derivative into a temporal rate of change of the angular energy density.

The second term on the right-hand side of Eq. (9.2) represents the loss of photons at **r**, t, and $\hat{\Omega}$ as a result of absorption and scattering out of the direction $\hat{\Omega}$. It is the first of the two terms that describe the effect of collisions on the angular energy density u. Because this term represents a loss of photons, it has a negative sign. The fact that absorption results in a loss of photons is obvious. Scattering also results in a loss of photons from the five-dimensional volume element, because scattering changes the direction of propagation of the photon (except for the case of strictly forward scattering) and, hence, deflects the photon out of the differential solid angle element $d\Omega$ about $\hat{\Omega}$. The form of this term provides insight into the meaning of the absorption and scattering coefficients, μ_a and μ_s, respectively. In fact, since c_n is the distance traveled by a photon per unit time, the fact that $c_n(\mu_a + \mu_s)u/(\hbar\omega)$ gives the number of photons (per unit volume, per unit solid angle) experiencing absorption or scattering events per unit time indicates that μ_a and μ_s represent a probability density for absorption and scattering events, respectively (as also introduced in Chapter 1). We note that we are dealing with an infinitesimal distance traveled by photons, namely $c_n dt$, so that the interaction probability is proportional to such distance, where the factor of proportionality (i.e., $(\mu_a + \mu_s)$) represents the probability density of interaction over this infinitesimal distance. The additive feature of the two probabilities results from the mutual exclusivity of absorption and scattering events.

The third term on the right-hand side of Eq. (9.2) is the second of the two terms that include the effect of scattering. It has a positive sign because it represents the gain of photons at **r**, t, and $\hat{\Omega}$ due to scattering occurring at position **r** from any direction of propagation $\hat{\Omega}'$ (hence the integral over the entire solid angle 4π in $d\Omega'$) into the direction of interest $\hat{\Omega}$. Figure 9.3 illustrates the three possible processes associated with photon interactions (absorption and scattering), namely the loss of photons traveling in a given direction as a result of absorption or scattering into a different direction, and the gain of photons traveling in a given direction as a result of scattering from a different direction.

Finally, **the fourth term** on the right-hand side of Eq. (9.2) represents the gain of photons, hence the positive sign, due to light sources. The source term $q(\mathbf{r}, \hat{\Omega}, t)$ represents the energy emitted by a light source per unit time, per unit volume, per unit solid angle about time t, position **r**, and direction $\hat{\Omega}$. If there is no light source within the small volume about **r**, then this term is zero. (In the absence of fluorescence, this is the case that is commonly invoked for examining the photon flux at any distance from the source.) From a conceptual point of view, the source term may also correspond to a photon sink, in which case the term is negative and corresponds to a loss of photons. For example, this is the case when boundary

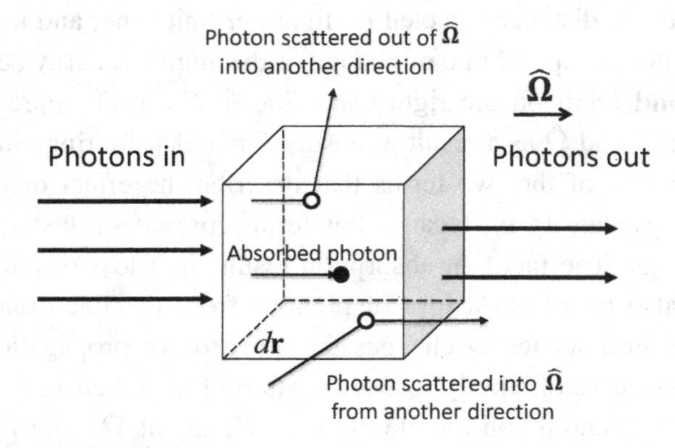

Figure 9.3.
Illustration of the effects of photon collision (absorption and scattering) on the angular energy density about position **r** (volume element $d\mathbf{r}$) and direction $\hat{\Omega}$.

conditions are modeled with the method of source images, where positive and negative photon sources are combined to satisfy specific boundary conditions (as discussed in Section 9.6.2).

9.2 Expansion of the Boltzmann transport equation into spherical harmonics

While the Boltzmann transport equation of Eq. (9.2) provides a rigorous description of photon transport in absorbing and scattering media, it is a complicated integro-differential equation, for which an analytical solution is difficult to derive, even for the simplest cases of a homogeneous medium in an infinite (Paasschens, 1997; Cai et al., 2000), semi-infinite (Liemert and Kienle, 2013), or slab (Machida et al., 2010) geometry. Besides the analytical complexity, one should appreciate a more fundamental difficulty in using the BTE to model light propagation in a realistic biological tissue. This follows from the complex spatial and temporal dependence of the absorption coefficient, scattering coefficient, and scattering phase function, as well as a need to apply boundary conditions associated with irregular tissue interfaces.

One way to simplify the transport equation is to expand the angular dependence of the angular energy density, the source term, and the scattering phase function into a proper set of basis functions. A typical choice for the basis functions is that of spherical harmonics, $Y_l^m(\theta, \varphi)$, a complete set of orthonormal functions whose angular dependence becomes more pronounced at higher values of the indices l (taking values of 0, 1, 2, ...) and m (taking integer values between $-l$ and $+l$). For definitions and properties of spherical harmonics and related topics, as presented

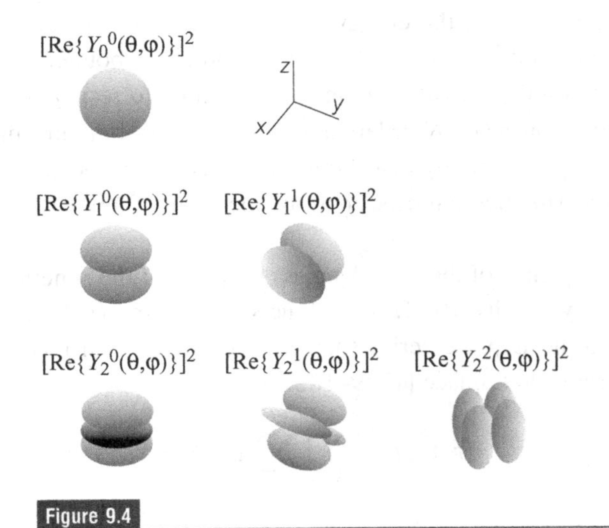

Figure 9.4

Graphical representation of the angular dependence of the real part squared of the spherical harmonics of orders $l = 0$, $l = 1$, and $l = 2$.

in the remainder of this chapter, see for example, Wyld (1994) or Arfken and Weber (2001). Because the set of two angular coordinates (θ, φ) fully determines the direction $\hat{\Omega}$ (and vice versa), we choose to indicate the spherical harmonics as $Y_l^m(\hat{\Omega})$.

The spherical harmonics for $l = 0$, $l = 1$, and $l = 2$ are listed in Appendix 9.A. The lowest-order spherical harmonic, Y_0^0, is spherically symmetric (independent of θ and φ). The Y_1^0 spherical harmonic shows a relatively smooth angular dependence with one zero as a function of θ (at $\theta = \pi/2$). The $Y_1^{\pm 1}$ spherical harmonics have two zeros as a function of θ (at $\theta = 0, \pi$) and two zeros as a function of φ (at $\varphi = \pi/2, 3\pi/2$ for the real part). The $l = 2$ spherical harmonics show a stronger angular dependence. Specifically, the Y_2^0 spherical harmonic has two zeros as a function of θ (at $\theta = \cos^{-1}(\pm\sqrt{1/3})$); the $Y_2^{\pm 1}$ spherical harmonics have three zeros as a function of θ (at $\theta = 0, \pi/2, \pi$) and two zeros as a function of φ (at $\varphi = \pi/2, 3\pi/2$ for the real part); and the $Y_2^{\pm 2}$ spherical harmonics have two zeros as a function of θ (at $\theta = 0, \pi$) and four zeros as a function of φ (at $\varphi = \pi/4, 3\pi/4, 5\pi/4, 7\pi/4$ for the real part). This is illustrated in Figure 9.4, which shows the angular dependence of the real part squared of the spherical harmonics of orders $l = 0$, $l = 1$, and $l = 2$.

The fact that the angular dependence becomes stronger for higher-order spherical harmonics is relevant because light in the red and near-infrared spectral region (which is the region of interest for the diffusion approximation, to be derived below) is much more likely to be scattered than absorbed in most biological tissues, resulting in angular energy density distributions that are nearly isotropic, at least for locations far from light sources and tissue boundaries. As a result, the lowest orders of a spherical-harmonics expansion (corresponding to the so-called P_N approximation when N orders are retained) are often adequate to accurately describe the optical energy distribution in tissues. A truncation of the expansion in spherical harmonics can be seen as an operation of low-pass angular frequency filtering.

As a consequence of the completeness property of the spherical harmonics, the angular energy density $u(\mathbf{r}, \hat{\boldsymbol{\Omega}}, t)$ and the source term $q(\mathbf{r}, \hat{\boldsymbol{\Omega}}, t)$ can be expanded according to the Laplace series (named after the French mathematician Pierre-Simon, marquis de Laplace [1749–1827]):

$$u(\mathbf{r}, \hat{\boldsymbol{\Omega}}, t) = \sum_{l=0}^{\infty} \sum_{m=-l}^{l} u_{lm}(\mathbf{r}, t) Y_l^m(\hat{\boldsymbol{\Omega}}), \tag{9.3}$$

$$q(\mathbf{r}, \hat{\boldsymbol{\Omega}}, t) = \sum_{l=0}^{\infty} \sum_{m=-l}^{l} q_{lm}(\mathbf{r}, t) Y_l^m(\hat{\boldsymbol{\Omega}}), \tag{9.4}$$

where $u_{lm}(\mathbf{r}, t)$ and $q_{lm}(\mathbf{r}, t)$ are coefficients independent of θ and φ. Because the angular energy density $u(\mathbf{r}, \hat{\boldsymbol{\Omega}}, t)$ and the source term $q(\mathbf{r}, \hat{\boldsymbol{\Omega}}, t)$ are real, the imaginary parts of the Laplace series in Eqs. (9.3) and (9.4) must cancel, which requires that $u_{l,-m} = (-1)^m u_{l,m}^*$ and $q_{l,-m} = (-1)^m q_{l,m}^*$ by considering that $Y_l^{-m}(\hat{\boldsymbol{\Omega}}) = (-1)^m Y_l^{m*}(\hat{\boldsymbol{\Omega}})$. (Here, the superscript * indicates the complex conjugate.)

The scattering phase function $p(\hat{\boldsymbol{\Omega}}', \hat{\boldsymbol{\Omega}})$ can also be expanded into spherical harmonics under the assumption that it does not depend on $\hat{\boldsymbol{\Omega}}$ and $\hat{\boldsymbol{\Omega}}'$ separately, but only on the angle between them (the scattering angle) or its cosine, given by the dot product $\hat{\boldsymbol{\Omega}}' \cdot \hat{\boldsymbol{\Omega}}$. This is the case of scattering of unpolarized light from isotropic particles or, in the case of multiple scattering, from a set of randomly oriented particles in an isotropic medium. The scattering angle is usually indicated with θ (as in Chapters 1, 2, 7, and 8) but here, since we have introduced the angles (θ, φ) and (θ', φ') to define directions $\hat{\boldsymbol{\Omega}}$ and $\hat{\boldsymbol{\Omega}}'$, respectively, we denote the scattering angle with Θ, so that $\cos\Theta = \hat{\boldsymbol{\Omega}}' \cdot \hat{\boldsymbol{\Omega}}$. Because $\cos\Theta$ (the argument of the scattering phase function for isotropic particles) assumes values in the interval $[-1,1]$, one can express $p(\hat{\boldsymbol{\Omega}}' \cdot \hat{\boldsymbol{\Omega}})$ using the Legendre series representation for functions that are continuous and have a continuous derivative in the interval $[-1,1]$:

$$p(\hat{\boldsymbol{\Omega}}' \cdot \hat{\boldsymbol{\Omega}}) = \sum_{l=0}^{\infty} \frac{2l+1}{4\pi} p_l P_l(\hat{\boldsymbol{\Omega}}' \cdot \hat{\boldsymbol{\Omega}}), \tag{9.5}$$

where P_l is the Legendre polynomial of order l, and $p_l = 2\pi \times \int_{-1}^{1} p(\cos\Theta) P_l(\cos\Theta) d(\cos\Theta)$. We observe here that for a dielectric particle in a dielectric medium (the case of interest in biomedical optics), the range of

Θ is limited to the range 0–1. Negative values (preferential backscattering) can obtain for conducting particles. Equation (9.5) is the same as Eq. (8.5), with slightly different notation. The Legendre polynomials can be written in terms of the spherical harmonics by virtue of the addition theorem, which states that $P_l(\hat{\Omega}' \cdot \hat{\Omega}) = \frac{4\pi}{2l+1} \sum_{m=-l}^{l} Y_l^{m*}(\hat{\Omega}')Y_l^m(\hat{\Omega})$, so that:

$$p(\hat{\Omega}' \cdot \hat{\Omega}) = \sum_{l=0}^{\infty} \sum_{m=-l}^{l} p_l Y_l^{m*}(\hat{\Omega}')Y_l^m(\hat{\Omega}). \tag{9.6}$$

Equation (9.6) expresses the scattering phase function in terms of spherical harmonics. By using the expressions of Eqs. (9.3), (9.4), and (9.6) for $u(\mathbf{r}, \hat{\Omega}, t)$, $q(\mathbf{r}, \hat{\Omega}, t)$, and $p(\hat{\Omega}' \cdot \hat{\Omega})$, respectively, the Boltzmann transport equation (Eq. (9.2)) becomes:

$$\sum_{l=0}^{\infty} \sum_{m=-l}^{l} \left\{ \left[\frac{\partial}{\partial t} + c_n \hat{\Omega} \cdot \nabla + c_n(\mu_a + \mu_s) \right] u_{lm}(\mathbf{r}, t)Y_l^m(\hat{\Omega}) - q_{lm}(\mathbf{r}, t)Y_l^m(\hat{\Omega}) + \right.$$

$$\left. - c_n \mu_s \int_{4\pi} u_{lm}(\mathbf{r}, t)Y_l^m(\hat{\Omega}') \sum_{l'=0}^{\infty} \sum_{m'=-l'}^{l'} p_{l'} Y_{l'}^{m'*}(\hat{\Omega}')Y_{l'}^{m'}(\hat{\Omega})d\Omega' \right\} = 0 \tag{9.7}$$

By considering the orthonormality property of the spherical harmonics, namely $\int_{4\pi} Y_l^{m*}(\hat{\Omega})Y_{l'}^{m'}(\hat{\Omega})d\Omega = \delta_{ll'}\delta_{mm'}$, one can evaluate the integral over $d\Omega'$ in Eq. (9.7), and the expression of the Boltzmann transport equation in terms of a series of spherical harmonics becomes:

$$\sum_{l=0}^{\infty} \sum_{m=-l}^{l} \left\{ \left[\frac{\partial}{\partial t} + c_n \hat{\Omega} \cdot \nabla + c_n[\mu_s(1 - p_l) + \mu_a] \right] u_{lm}(\mathbf{r}, t) \right.$$

$$\left. - q_{lm}(\mathbf{r}, t) \right\} Y_l^m(\hat{\Omega}) = 0 \tag{9.8}$$

Equation (9.8) contains $u_{lm}(\mathbf{r}, t)$ and $q_{lm}(\mathbf{r}, t)$ (the expansion coefficients of the angular energy density and the source term in spherical harmonics) and p_l (the coefficients, to within a multiplicative factor, of the Legendre polynomial expansion of the scattering phase function). The goal is to find a set of equations that establishes relationships among all of these coefficients. To this aim, we consider a specific pair of values (L, M) for the spherical harmonics indices (l, m), multiply Eq. (9.8) by $Y_L^{M*}(\hat{\Omega})$, and integrate over the entire solid angle:

$$\frac{\partial}{\partial t} u_{LM}(\mathbf{r}, t) + c_n[\mu_s(1 - p_L) + \mu_a]u_{LM}(\mathbf{r}, t)$$

$$+ c_n \sum_{l=0}^{\infty} \sum_{m=-l}^{l} \int_{4\pi} \hat{\Omega} \cdot \nabla u_{lm}(\mathbf{r}, t)Y_l^m(\hat{\Omega})Y_L^{M*}(\hat{\Omega})d\Omega = q_{LM}(\mathbf{r}, t) \tag{9.9}$$

The summations over the indices (l, m) have disappeared in the terms that feature a multiplicative factor (or a time derivative) of $u_{lm}(\mathbf{r}, t)$ and $q_{lm}(\mathbf{r}, t)$, since only the coefficients of indices (L, M) are retained as a result of the orthonormal properties of the spherical harmonics. In the case of the term containing the spatial gradient of $u_{lm}(\mathbf{r}, t)$, instead, the summations over l and m remain, as well as the integral over $d\Omega$. The integral over $d\Omega$ can be computed by using the expressions of the x, y, and z components of the vector $Y_L^M(\hat{\mathbf{\Omega}})\hat{\mathbf{\Omega}}$ in terms of spherical harmonics, which are given below as obtained from the recurrence relations for the associated Legendre functions $P_l^m(x)$ (which are defined below in Eq. (9.15)):

$$\begin{aligned} Y_L^M(\hat{\mathbf{\Omega}})\Omega_x &= Y_L^M(\hat{\mathbf{\Omega}}) \sin\theta \cos\varphi \\ &= -\frac{1}{2}\left[\frac{(L+M+1)(L+M+2)}{(2L+1)(2L+3)}\right]^{1/2} Y_{L+1}^{M+1}(\hat{\mathbf{\Omega}}) \\ &\quad + \frac{1}{2}\left[\frac{(L-M)(L-M-1)}{(2L-1)(2L+1)}\right]^{1/2} Y_{L-1}^{M+1}(\hat{\mathbf{\Omega}}) \\ &\quad + \frac{1}{2}\left[\frac{(L-M+1)(L-M+2)}{(2L+1)(2L+3)}\right]^{1/2} Y_{L+1}^{M-1}(\hat{\mathbf{\Omega}}) \\ &\quad - \frac{1}{2}\left[\frac{(L+M)(L+M-1)}{(2L-1)(2L+1)}\right]^{1/2} Y_{L-1}^{M-1}(\hat{\mathbf{\Omega}}), \end{aligned} \tag{9.10}$$

$$\begin{aligned} Y_L^M(\hat{\mathbf{\Omega}})\Omega_y &= Y_L^M(\hat{\mathbf{\Omega}}) \sin\theta \sin\varphi \\ &= -\frac{1}{2i}\left[\frac{(L+M+1)(L+M+2)}{(2L+1)(2L+3)}\right]^{1/2} Y_{L+1}^{M+1}(\hat{\mathbf{\Omega}}) \\ &\quad + \frac{1}{2i}\left[\frac{(L-M)(L-M-1)}{(2L-1)(2L+1)}\right]^{1/2} Y_{L-1}^{M+1}(\hat{\mathbf{\Omega}}) \\ &\quad - \frac{1}{2i}\left[\frac{(L-M+1)(L-M+2)}{(2L+1)(2L+3)}\right]^{1/2} Y_{L+1}^{M-1}(\hat{\mathbf{\Omega}}) \\ &\quad + \frac{1}{2i}\left[\frac{(L+M)(L+M-1)}{(2L-1)(2L+1)}\right]^{1/2} Y_{L-1}^{M-1}(\hat{\mathbf{\Omega}}), \end{aligned} \tag{9.11}$$

$$\begin{aligned} Y_L^M(\hat{\mathbf{\Omega}})\Omega_z &= Y_L^M(\hat{\mathbf{\Omega}}) \cos\theta \\ &= \left[\frac{(L-M+1)(L+M+1)}{(2L+1)(2L+3)}\right]^{1/2} Y_{L+1}^{M}(\hat{\mathbf{\Omega}}) + \left[\frac{(L-M)(L+M)}{(2L-1)(2L+1)}\right]^{1/2} Y_{L-1}^{M}(\hat{\mathbf{\Omega}}). \end{aligned} \tag{9.12}$$

Using Eqs. (9.10)–(9.12), it is now possible to calculate the integral in Eq. (9.9) by taking into account the orthonormality properties of spherical harmonics:

$$\frac{\partial}{\partial t}u_{LM}(\mathbf{r},t) + c_n\left[\mu_s\left(1-p_L\right)+\mu_a\right]u_{LM}(\mathbf{r},t)$$
$$+\frac{1}{2}\left[\frac{(L-M+1)(L-M+2)}{(2L+1)(2L+3)}\right]^{1/2}\left(\frac{\partial}{\partial x}-i\frac{\partial}{\partial y}\right)c_n u_{L+1}^{M-1}(\mathbf{r},t)$$
$$-\frac{1}{2}\left[\frac{(L+M)(L+M-1)}{(2L+1)(2L-1)}\right]^{1/2}\left(\frac{\partial}{\partial x}-i\frac{\partial}{\partial y}\right)c_n u_{L-1}^{M-1}(\mathbf{r},t)$$
$$-\frac{1}{2}\left[\frac{(L+M+2)(L+M+1)}{(2L+1)(2L+3)}\right]^{1/2}\left(\frac{\partial}{\partial x}+i\frac{\partial}{\partial y}\right)c_n u_{L+1}^{M+1}(\mathbf{r},t)$$
$$+\frac{1}{2}\left[\frac{(L-M-1)(L-M)}{(2L+1)(2L-1)}\right]^{1/2}\left(\frac{\partial}{\partial x}+i\frac{\partial}{\partial y}\right)c_n u_{L-1}^{M+1}(\mathbf{r},t)$$
$$+\left[\frac{(L+M+1)(L-M+1)}{(2L+1)(2L+3)}\right]^{1/2}\frac{\partial}{\partial z}c_n u_{L+1}^{M}(\mathbf{r},t)$$
$$+\left[\frac{(L-M)(L+M)}{(2L-1)(2L+1)}\right]^{1/2}\frac{\partial}{\partial z}c_n u_{L-1}^{M}(\mathbf{r},t) = q_{LM}(\mathbf{r},t). \quad (9.13)$$

In Eq. (9.13), for clarity, we have used notation with subscript and superscript (u_L^M) when the coefficients have indices $L \pm 1$ or $M \pm 1$. Equation (9.13) shows that the coefficients u_{LM} of the angular energy density are related to the coefficients of the source term (q_{lm}) that have the same indices $l = L$, $m = M$, and to the coefficients of the angular energy density (u_{lm}) that have nearest-neighbor indices $l = L \pm 1$, $m = M \pm 1$.

9.3 The P_N approximation

The expansion of the angular energy density, the scattering phase function, and the source term into spherical harmonics has led to writing the Boltzmann transport equation as an infinite set of equations with indices L (ranging from 0 to ∞) and M (ranging from $-L$ to L) (Eq. (9.13)). Retention of the terms with $L \leq N$ leads to the so-called P_N *approximation*, which is represented by a set of $\sum_{l=0}^{N}(2l+1) = (N+1)^2$ coupled first-order partial differential equations. The name P_N for this approximation comes from the fact that the highest-order term of the truncated Laplace series contains $Y_N^M(\hat{\Omega})$, which can be written in terms of the associated Legendre functions $P_N^M(x)$, which in turn can be expressed in terms of the Legendre polynomials $P_N(x)$:

$$Y_N^M(\hat{\Omega}) = (-1)^M\left[\frac{(2N+1)}{4\pi}\frac{(N-M)!}{(N+M)!}\right]^{1/2}P_N^M(\cos\theta)e^{iM\varphi}, \quad (9.14)$$

$$P_N^M(x) = (1-x^2)^{M/2}\frac{d^M}{dx^M}P_N(x). \quad (9.15)$$

Only odd-order P_N approximations (P_1, P_3, P_5, etc.) are typically considered. There are at least two reasons for this. One reason has to do with the fact that there is a linear relationship between the even moments of the angular photon density, so that the P_{2n} approximation at any even-order $2n$ does not feature more linearly independent moments than the preceding odd-order P_{2n-1} approximation. As a result, the P_{2n} approximation is usually not more accurate than the P_{2n-1} approximation (Gelbard, 1968). A second reason has to do with boundary conditions associated with discontinuities in the medium properties. While for any odd-order P_{2n+1} approximation it is possible to impose continuity of all moments of the angular photon density across discontinuities in the medium properties, this is not possible for even-order P_{2n} approximations (Davison, 1958). This can be seen as a result of the linear relationship, with coefficients dependent on the medium absorption and scattering coefficients, among even-order moments, which prevents continuity across interfaces for all moments (Gelbard, 1968).

9.4 The P_1 approximation

The P_1 approximation, where $L \leq 1$, is a set of four coupled differential equations that is often used to describe photon migration in tissues. The four equations, which correspond to index pairs (L, M) of (0, 0), (1, −1), (1, 0) and (1, 1) in Eq. (9.13), are the following:

$$\frac{\partial}{\partial t}u_{0,0}(\mathbf{r},t) + c_n[\mu_s(1-p_0)+\mu_a]u_{0,0}(\mathbf{r},t) + \frac{1}{2}\sqrt{\frac{2}{3}}\left(\frac{\partial}{\partial x} - i\frac{\partial}{\partial y}\right)c_n u_{1,-1}(\mathbf{r},t)$$
$$-\frac{1}{2}\sqrt{\frac{2}{3}}\left(\frac{\partial}{\partial x} + i\frac{\partial}{\partial y}\right)c_n u_{1,1}(\mathbf{r},t) + \sqrt{\frac{1}{3}}\frac{\partial}{\partial z}c_n u_{1,0}(\mathbf{r},t) = q_{0,0}(\mathbf{r},t), \tag{9.16}$$

$$\frac{\partial}{\partial t}u_{1,-1}(\mathbf{r},t) + c_n[\mu_s(1-p_1)+\mu_a]u_{1,-1}(\mathbf{r},t) + \frac{1}{2}\sqrt{\frac{2}{3}}\left(\frac{\partial}{\partial x} + i\frac{\partial}{\partial y}\right)c_n u_{0,0}(\mathbf{r},t)$$
$$= q_{1,-1}(\mathbf{r},t), \tag{9.17}$$

$$\frac{\partial}{\partial t}u_{1,0}(\mathbf{r},t) + c_n[\mu_s(1-p_1)+\mu_a]u_{1,0}(\mathbf{r},t) + \sqrt{\frac{1}{3}}\frac{\partial}{\partial z}c_n u_{0,0}(\mathbf{r},t) = q_{1,0}(\mathbf{r},t), \tag{9.18}$$

$$\frac{\partial}{\partial t}u_{1,1}(\mathbf{r},t) + c_n[\mu_s(1-p_1)+\mu_a]u_{1,1}(\mathbf{r},t) - \frac{1}{2}\sqrt{\frac{2}{3}}\left(\frac{\partial}{\partial x} - i\frac{\partial}{\partial y}\right)c_n u_{0,0}(\mathbf{r},t)$$
$$= q_{1,1}(\mathbf{r},t). \tag{9.19}$$

The coefficients $u_{0,0}(\mathbf{r}, t)$ and $u_{1,M}(\mathbf{r}, t)$ are directly associated with the energy density $U(\mathbf{r}, t)$ and the net flux vector $\mathbf{F}(\mathbf{r}, t)$, respectively. In fact:

$$U(\mathbf{r}, t) = \int_{4\pi} u(\mathbf{r}, \hat{\boldsymbol{\Omega}}, t) d\Omega = \sum_{l=0}^{\infty} \sum_{m=-l}^{l} u_{lm}(\mathbf{r}, t) \int_{4\pi} Y_l^m(\hat{\boldsymbol{\Omega}}) d\Omega = \sqrt{4\pi} u_{0,0}(\mathbf{r}, t), \tag{9.20}$$

(since $\int_{4\pi} Y_l^m(\hat{\boldsymbol{\Omega}}) d\Omega = 0$ for $(l, m) \neq (0, 0)$ and $Y_0^0(\hat{\boldsymbol{\Omega}}) = 1/\sqrt{4\pi}$), and:

$$\begin{aligned}
\mathbf{F}(\mathbf{r}, t) &= \int_{4\pi} c_n u(\mathbf{r}, \hat{\boldsymbol{\Omega}}, t) \hat{\boldsymbol{\Omega}} d\Omega \\
&= \sum_{l=0}^{\infty} \sum_{m=-l}^{l} c_n u_{lm}(\mathbf{r}, t) \int_{4\pi} (\sin\theta\cos\varphi\hat{\mathbf{x}} + \sin\theta\sin\varphi\hat{\mathbf{y}} + \cos\theta\hat{\mathbf{z}}) Y_l^m(\hat{\boldsymbol{\Omega}}) d\Omega \\
&= \sum_{l=0}^{\infty} \sum_{m=-l}^{l} c_n u_{lm}(\mathbf{r}, t) \\
&\quad \times \int_{4\pi} \sqrt{\frac{4\pi}{3}} \left\{ \begin{array}{l} \sqrt{\frac{1}{2}} \left[-Y_1^{1*}(\hat{\boldsymbol{\Omega}}) + Y_1^{-1*}(\hat{\boldsymbol{\Omega}}) \right] \hat{\mathbf{x}} \\ + \sqrt{\frac{1}{2}} \frac{1}{i} \left[Y_1^{1*}(\hat{\boldsymbol{\Omega}}) + Y_1^{-1*}(\hat{\boldsymbol{\Omega}}) \right] \hat{\mathbf{y}} + Y_1^0(\hat{\boldsymbol{\Omega}}) \hat{\mathbf{z}} \end{array} \right\} Y_l^m(\hat{\boldsymbol{\Omega}}) d\Omega \\
&= \sqrt{\frac{4\pi}{3}} c_n \left[\sqrt{\frac{1}{2}} (-u_{1,1}(\mathbf{r}, t) + u_{1,-1}(\mathbf{r}, t)) \hat{\mathbf{x}} \right. \\
&\quad \left. - i \sqrt{\frac{1}{2}} (u_{1,1}(\mathbf{r}, t) + u_{1,-1}(\mathbf{r}, t)) \hat{\mathbf{y}} + u_{1,0}(\mathbf{r}, t) \hat{\mathbf{z}} \right].
\end{aligned} \tag{9.21}$$

The four scalar Eqs. (9.16)–(9.19) that make up the P_1 approximation can be consolidated into a set of two equations (one scalar equation and one vector equation) expressed in terms of the energy density U and the net flux vector $\mathbf{F}$:

$$\frac{\partial}{\partial t} U(\mathbf{r}, t) + c_n [\mu_s (1 - p_0) + \mu_a] U(\mathbf{r}, t) + \nabla \cdot \mathbf{F}(\mathbf{r}, t) = \sqrt{4\pi} q_{0,0}(\mathbf{r}, t), \tag{9.22}$$

$$\begin{aligned}
&\frac{1}{c_n} \frac{\partial}{\partial t} \mathbf{F}(\mathbf{r}, t) + [\mu_s (1 - p_1) + \mu_a] \mathbf{F}(\mathbf{r}, t) + \frac{1}{3} c_n \nabla U(\mathbf{r}, t) \\
&= \sqrt{\frac{4\pi}{3}} \left[\sqrt{\frac{1}{2}} (q_{1,-1}(\mathbf{r}, t) - q_{1,1}(\mathbf{r}, t)) \hat{\mathbf{x}} \right. \\
&\quad \left. - i \sqrt{\frac{1}{2}} (q_{1,-1}(\mathbf{r}, t) + q_{1,1}(\mathbf{r}, t)) \hat{\mathbf{y}} + q_{1,0}(\mathbf{r}, t) \hat{\mathbf{z}} \right].
\end{aligned} \tag{9.23}$$

The vector equation (Eq. (9.23)) is obtained by combining Eqs. (9.17)–(9.19) as follows: $\sqrt{2\pi/3}$ [Eq. (9.17)–Eq. (9.19)] $\hat{\mathbf{x}} - i\sqrt{2\pi/3}$ [Eq. (9.17) + Eq. (9.19)] $\hat{\mathbf{y}} + \sqrt{4\pi/3}$ [Eq. (9.18)] $\hat{\mathbf{z}}$. It is worth noting here that the expression for $\mathbf{F}$ in Eq. (9.21) and all terms in Eqs. (9.22) and (9.23) are real. In fact, all imaginary components cancel by considering that $-Y_1^{1*} + Y_1^{-1*}$ is real, $Y_1^{1*} + Y_1^{-1*}$ is purely imaginary, and that $u_{1,-1} = -u_{1,1}^*$ and $q_{1,-1} = -q_{1,1}^*$, as discussed in conjunction with Eqs. (9.3) and (9.4). If we now use Eqs. (9.20) and (9.21) to write the P_1 approximation for the radiance ($L_{P_1}(\mathbf{r}, \hat{\mathbf{\Omega}}, t)$), we obtain an expression that reveals the physical meaning of the P_1 approximation:

$$\begin{aligned} L_{P_1}(\mathbf{r}, \hat{\mathbf{\Omega}}, t) &= c_n \sum_{l=0}^{1} \sum_{m=-l}^{l} u_{lm}(\mathbf{r}, t) Y_l^m(\hat{\mathbf{\Omega}}) \\ &= c_n u_{00}(\mathbf{r}, t) Y_0^0(\hat{\mathbf{\Omega}}) + c_n \sum_{m=-1}^{1} u_{1m}(\mathbf{r}, t) Y_1^m(\hat{\mathbf{\Omega}}) \\ &= \frac{1}{4\pi} c_n U(\mathbf{r}, t) + \frac{3}{4\pi} \mathbf{F}(\mathbf{r}, t) \cdot \hat{\mathbf{\Omega}} \end{aligned} \tag{9.24}$$

where we have used $u_{00} = U/\sqrt{4\pi}$ (from Eq. (9.20)), $Y_0^0(\hat{\mathbf{\Omega}}) = 1/\sqrt{4\pi}$, and $\mathbf{F} \cdot \hat{\mathbf{\Omega}} = (4\pi/3) c_n \sum_{m=-1}^{1} u_{1m} Y_1^m(\hat{\mathbf{\Omega}})$ (as can be shown using the expression for $\hat{\mathbf{\Omega}}$ in terms of spherical harmonics given in Eq. (9.21)). From Eq. (9.24), one can see that the radiance contains two terms in the P_1 approximation. The $l = 0$ term is isotropic, and the $l = 1$ term along direction $\hat{\mathbf{\Omega}}$ is proportional to the cosine of the angle between $\hat{\mathbf{\Omega}}$ and the net flux vector, $\mathbf{F}$. The P_1 approximation for the radiance in Eq. (9.24), which by definition ignores all terms other than $l = 0$ or 1, is a good approximation for the actual radiance $L(\mathbf{r}, \hat{\mathbf{\Omega}}, t)$, if the $l = 1$ term is small compared to the $l = 0$ term, which means that the radiance needs to be nearly isotropic. This is the reason that the P_1 approximation fails in the vicinity of sources or boundaries, where such near-isotropy condition is typically violated.

As a point of reference, the magnitude of the $l = 1$ term ($3|\mathbf{F}|/(4\pi)$) deeply inside strongly scattering tissues is of the order of 10–15% of the isotropic term ($c_n U/(4\pi)$). Figure 9.5 shows the angular dependence of the P_1 radiance for anisotropy values (i.e., the ratio between the $l = 1$ and $l = 0$ terms of the P_1 radiance) of 0, 12% (typical for regions deep inside strongly scattering tissues), 55%, and 150%. A 55% anisotropy is representative of a typical tissue-air boundary (see Section 9.6.2), whereas a 150% anisotropy is representative of a refractive-index-matched boundary between tissue and a non-scattering medium (see Section 9.6.3).

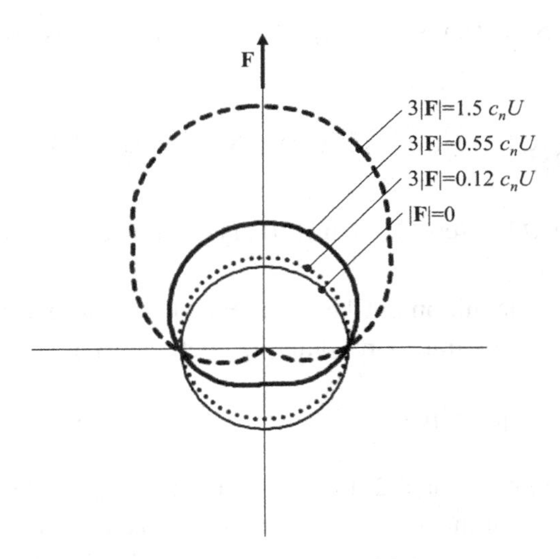

Figure 9.5

Cross section of the P_1 radiance ($L_{P_1}(\mathbf{r}, \hat{\boldsymbol{\Omega}}, t)$) along the plane containing $\mathbf{F}$ and $\hat{\boldsymbol{\Omega}}$. The three cases of 12%, 55%, and 150% anisotropy (which are typical of P_1 radiance deep in strongly scattering tissue, at a typical tissue-air interface, and at a refractive-index-matched interface with a non-scattering medium, respectively) are compared with the isotropic distribution ($|\mathbf{F}| = 0$).

We now recall the definition of the coefficients p_l (see Eq. (9.5)), from which we can determine p_0 and p_1:

$$p_0 = 2\pi \int_{-1}^{1} p(\cos\Theta) P_0(\cos\Theta) d(\cos\Theta) = 2\pi \int_{-1}^{1} p(\cos\Theta) d(\cos\Theta) = 1, \quad (9.25)$$

$$\begin{aligned} p_1 &= 2\pi \int_{-1}^{1} p(\cos\Theta) P_1(\cos\Theta) d(\cos\Theta) \\ &= 2\pi \int_{-1}^{1} p(\cos\Theta) \cos\Theta d(\cos\Theta) = \langle \cos\Theta \rangle = g, \end{aligned} \quad (9.26)$$

where in Eq. (9.25) we have used the fact that the scattering phase function is normalized according to the condition $\int_{4\pi} p(\hat{\boldsymbol{\Omega}}' \cdot \hat{\boldsymbol{\Omega}}) d\Omega' = 1$, which is equivalent to $2\pi \int_{-1}^{1} p(\cos\Theta) d(\cos\Theta) = 1$ (as introduced in Chapter 1, Eq. (1.28)). Therefore p_0 is 1 and p_1 is the average cosine of the scattering angle Θ ($< \cos\Theta >$), which is the *anisotropy factor* g previously defined in Chapter 1 (see Eq. (1.29)) and Chapter 8 (see Eq. (8.7)). The source terms in Eqs. (9.22) and (9.23) are formally a monopole term (spherically symmetric) and a dipole term, and we indicate them with $S_0(\mathbf{r}, t)$ and $\mathbf{S}_1(\mathbf{r}, t)$, respectively. The final expressions for the P_1 equations are:

$$\frac{\partial}{\partial t}U(\mathbf{r}, t) + c_n\mu_a U(\mathbf{r}, t) + \nabla \cdot \mathbf{F}(\mathbf{r}, t) = S_0(\mathbf{r}, t), \tag{9.27}$$

$$\frac{1}{c_n}\frac{\partial}{\partial t}\mathbf{F}(\mathbf{r}, t) + [\mu_s(1 - \langle\cos\Theta\rangle) + \mu_a]\mathbf{F}(\mathbf{r}, t) + \frac{1}{3}c_n\nabla U(\mathbf{r}, t) = \mathbf{S}_1(\mathbf{r}, t). \tag{9.28}$$

According to the definition in Chapter 1 (Eq. (1.30)), we introduce the *reduced scattering coefficient* (or transport scattering coefficient) μ_s':

$$\mu_s' = \mu_s(1- < \cos\Theta >) = \mu_s(1 - g). \tag{9.29}$$

As stated in Chapters 1 and 2, the reduced scattering coefficient can be interpreted as the inverse of the average distance over which photons lose memory of their initial direction of propagation, or the inverse of the average distance between effectively isotropic scattering events. In the case of isotropic scattering, $g = 0$ and $\mu_s' = \mu_s$, whereas in the case of purely forward scattering, $g = 1$ and $\mu_s' = 0$. We note that in the case of perfectly forward scattering, the medium behaves as though there is no scattering. (Even though μ_s can remain finite, it has little meaning in the context of the term "scattering," and light propagation is then better analyzed as undergoing simple phase shifts induced by regions of different refractive index.)

By using the definition of the reduced scattering coefficient, Eq. (9.28) can be written as follows:

$$\begin{aligned}\mathbf{F}(\mathbf{r}, t) &= -\frac{1}{c_n(\mu_s' + \mu_a)}\frac{\partial}{\partial t}\mathbf{F}(\mathbf{r}, t) - \frac{c_n}{3(\mu_s' + \mu_a)}\nabla U(\mathbf{r}, t) + \frac{1}{(\mu_s' + \mu_a)}\mathbf{S}_1(\mathbf{r}, t)\\ &= -\frac{3D}{c_n^2}\frac{\partial}{\partial t}\mathbf{F}(\mathbf{r}, t) - D\nabla U(\mathbf{r}, t) + \frac{3D}{c_n}\mathbf{S}_1(\mathbf{r}, t)\end{aligned} \tag{9.30}$$

where we have introduced the *diffusion coefficient* $D = c_n/[3(\mu_s' + \mu_a)]$. By using this expression for $\mathbf{F}(\mathbf{r}, t)$ in Eq. (9.27), we obtain:

$$\begin{aligned}&\frac{\partial}{\partial t}U(\mathbf{r}, t) + c_n\mu_a U(\mathbf{r}, t) - \frac{3D}{c_n^2}\frac{\partial}{\partial t}\nabla \cdot \mathbf{F}(\mathbf{r}, t)\\ &\quad - \nabla \cdot [D\nabla U(\mathbf{r}, t)] + \frac{3}{c_n}\nabla \cdot [D\mathbf{S}_1(\mathbf{r}, t)] = S_0(\mathbf{r}, t)\end{aligned} \tag{9.31}$$

Now, one can substitute the expression for $\nabla \cdot \mathbf{F}(\mathbf{r}, t)$ given by Eq. (9.27) (namely, $\nabla \cdot \mathbf{F}(\mathbf{r}, t) = S_0(\mathbf{r}, t) - \frac{\partial}{\partial t}U(\mathbf{r}, t) - c_n\mu_a U(\mathbf{r}, t)$) into Eq. (9.31) and reduce the P_1 approximation to a single equation for the energy density:

$$\frac{3D}{c_n^2}\frac{\partial^2 U(\mathbf{r},t)}{\partial t^2} + \left(1 + \frac{3D}{c_n}\mu_a\right)\frac{\partial U(\mathbf{r},t)}{\partial t} + c_n\mu_a U(\mathbf{r},t) - \nabla\cdot[D\nabla U(\mathbf{r},t)]$$
$$= \frac{3D}{c_n^2}\frac{\partial S_0(\mathbf{r},t)}{\partial t} + S_0(\mathbf{r},t) - \frac{3}{c_n}\nabla\cdot[D\mathbf{S}_1(\mathbf{r},t)] \tag{9.32}$$

In the case of a spatially independent diffusion coefficient (optically homogeneous medium), D becomes a multiplicative factor in the divergence operator, so that Eq. (9.32) becomes:

$$\frac{3D}{c_n^2}\frac{\partial^2 U(\mathbf{r},t)}{\partial t^2} + \left(1 + \frac{3D}{c_n}\mu_a\right)\frac{\partial U(\mathbf{r},t)}{\partial t} + c_n\mu_a U(\mathbf{r},t) - D\nabla^2 U(\mathbf{r},t)$$
$$= \frac{3D}{c_n^2}\frac{\partial S_0(\mathbf{r},t)}{\partial t} + S_0(\mathbf{r},t) - \frac{3D}{c_n}\nabla\cdot\mathbf{S}_1(\mathbf{r},t) \tag{9.33}$$

which has the form of the so-called *telegraph equation* used to describe the propagation of electrical signals in a transmission line.

Because the P_1 approximation is based on a truncation of the spherical-harmonics expansion of terms with $l > 1$, it is accurate only if the angular energy density does not feature high angular frequency components. This is a good approximation after photons have experienced multiple scattering events in their propagation within tissue. The additional requirements are: (a) that the absorption coefficient μ_a be much smaller than the scattering coefficient μ_s (so that photons, on average, are not absorbed before they are scattered multiple times); and (b) that the angular photon density is modeled at far distances (compared to the characteristic scattering length $1/\mu_s$) from light sources and tissue boundaries. Furthermore, we recall the assumption that the scattering phase function must only depend on the cosine of the scattering angle rather than the individual directions of the incoming and scattered photons, so that it can be written as $p(\hat{\Omega}'\cdot\hat{\Omega})$. Strictly speaking, this requires the scattering particles to be isotropic, but this assumption is also representative of the averaging effect of multiple scattering from randomly oriented particles in an isotropic medium.

9.5 The diffusion equation

The transport equation, in its quasi-isotropic P_1 approximation of Eq. (9.32), can be further simplified to the *diffusion equation* by introducing a few additional conditions. These *diffusion approximation* conditions are often satisfied in the optical study of biological tissues at a macroscopic scale, for which the diffusion equation is a powerful analytical tool. The additional conditions leading to the diffusion approximation are the following:

1. **Strong scattering.** This condition is meant in the sense that photons typically undergo many effectively isotropic scattering events before being absorbed, so that $\mu_s' \gg \mu_a$ and the reduced single-scattering albedo $a' \approx 1$. Accordingly, the diffusion coefficient $D = c_n/[3(\mu_s' + \mu_a)] \cong c_n/(3\mu_s')$, so that it is essentially independent of absorption, as also argued on a theoretical basis (Furutsu and Yamada, 1994). The issue of whether D should or should not depend on μ_a has been considered in the literature and is further explored in Appendix 9.B. Furthermore, under the strong scattering condition, $3D\mu_a/c_n \cong \mu_a/\mu_s' \ll 1$, and the second term on the left-hand side of Eq. (9.32) reduces to $\partial U(\mathbf{r}, t)/\partial t$.
2. **Isotropic source.** The source emission is isotropic, i.e., $\mathbf{S}_1(\mathbf{r}, t) = 0$ and the light sources are entirely described by the isotropic term $S_0(\mathbf{r}, t)$. Therefore, the last term on the right-hand side of Eq. (9.32) is not considered.
3. **Slow optical signals.** Temporal changes in the optical emission rate by the light source are slow compared to the characteristic transit time of photons in the medium. In other words, the time scale for changes in $U(\mathbf{r}, t)$ and $S_0(\mathbf{r}, t)$ is much longer than the characteristic time of absorption and effectively isotropic scattering events given by $1/[c_n(\mu_s' + \mu_a)] = 3D/c_n^2$. Because the time scale of changes in U and S_0 can be estimated by the inverse of their time derivative, this requirement of slow optical signals translates into the formal condition $|\partial/\partial t| \ll c_n^2/(3D)$, so that:

$$\left|\frac{\partial^2 U(\mathbf{r}, t)}{\partial t^2}\right| \ll \frac{c_n^2}{3D}\left|\frac{\partial U(\mathbf{r}, t)}{\partial t}\right|, \tag{9.34}$$

$$\left|\frac{\partial S_0(\mathbf{r}, t)}{\partial t}\right| \ll \frac{c_n^2}{3D}|S_0(\mathbf{r}, t)|. \tag{9.35}$$

As a result, the first terms on the left-hand side and right-hand side of Eq. (9.32) can be ignored.

Under these conditions, the P_1 equation (Eq. (9.32)) reduces to the diffusion equation:

$$\frac{\partial U(\mathbf{r}, t)}{\partial t} = \nabla \cdot [D\nabla U(\mathbf{r}, t)] - c_n\mu_a U(\mathbf{r}, t) + S_0(\mathbf{r}, t), \tag{9.36}$$

and the net flux $\mathbf{F}(\mathbf{r}, t)$ is related to the energy density $U(\mathbf{r}, t)$ by Fick's law (named after the German physiologist Adolf Eugen Fick [1829–1901]):

$$\mathbf{F}(\mathbf{r}, t) = -D\nabla U(\mathbf{r}, t), \tag{9.37}$$

as can be derived from Eq. (9.30) after taking into account that $\frac{3D}{c_n^2}\left|\frac{\partial \mathbf{F}(\mathbf{r},t)}{\partial t}\right| \ll |\mathbf{F}(\mathbf{r},t)|$ and $\mathbf{S}_1(\mathbf{r},t)=0$.

In the case of a spatially uniform D, Eq. (9.36) reduces to the diffusion equation for macroscopically homogeneous media:

$$\frac{\partial U(\mathbf{r},t)}{\partial t} = D\nabla^2 U(\mathbf{r},t) - c_n\mu_a U(\mathbf{r},t) + S_0(\mathbf{r},t). \tag{9.38}$$

The diffusion equation expressed in terms of the energy density, U, in Eqs. (9.36) and (9.38) can be equivalently written in terms of the fluence rate, ϕ, by recalling that $\phi(\mathbf{r},t) = c_n U(\mathbf{r},t)$ (see Eq. (1.18)).

The diffusion-approximation condition that characteristic times of variation (τ) of the energy density be much longer than the mean time between collisions (absorption or effectively isotropic scattering) is plotted in Figure 9.6(a). The frequency-domain equivalent condition (namely, that characteristic frequencies of variation must be much smaller than $1/(2\pi\tau)$), is plotted in Figure 9.6(b). Finally, the condition $\mu_a \ll \mu_s'$ for the validity of the diffusion equation is plotted in Figure 9.6(c), together with the less restrictive condition $\mu_a \ll \mu_s$ for the P_1 approximation. In Figure 9.6(c), we have assumed a value for g of 0.9, which is representative of many biological tissues (so that $\mu_s = 10\mu_s'$).

9.6 Boundary conditions

9.6.1 Infinite medium

In the case of an infinite medium, the boundary conditions require that the optical energy density vanishes at large distances from light sources:

$$\lim_{|\mathbf{r}|\to\infty} u(\mathbf{r},\hat{\mathbf{\Omega}},t) = \lim_{|\mathbf{r}|\to\infty} U(\mathbf{r},t) = 0. \tag{9.39}$$

From a practical point of view, the ideal case of an infinite medium is fulfilled when a tissue extends over distances (measured from the light source) that are much greater than the typical attenuation length for the optical energy density. Such an attenuation length is given by the inverse of the effective attenuation coefficient, $1/\sqrt{3\mu_a(\mu_s'+\mu_a)}$, which in biological tissues is typically 0.5–1 cm or less in the near-infrared spectral region. This means that tissues extending several centimeters (in all directions) from embedded light sources can often be treated as effectively infinite media.

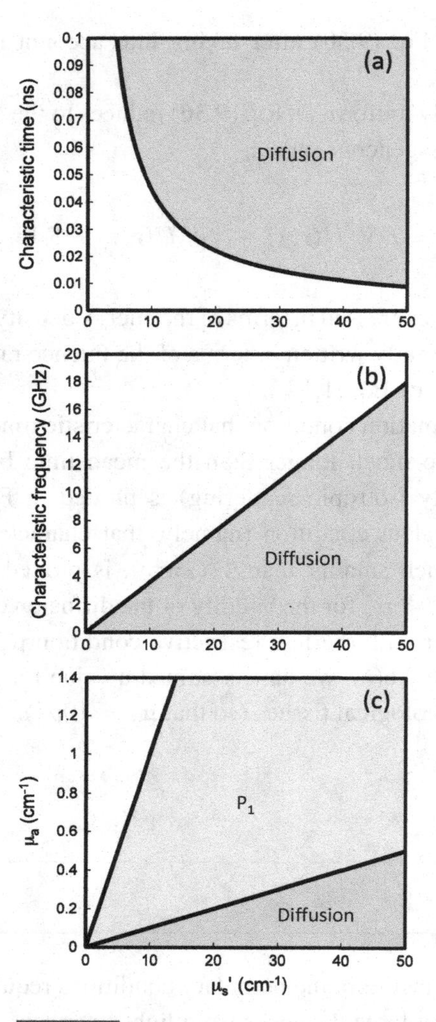

Figure 9.6

Illustration of the conditions of validity for the diffusion equation. (a) The characteristic time of variation of the energy density must be much longer than $1/(c_n\mu_s')$, the characteristic time between effectively isotropic scattering events. (b) The characteristic frequencies of energy density variation must be much less than the characteristic frequency of effectively isotropic scattering events. (c) The absorption coefficient must be much smaller than the reduced scattering coefficient ($\mu_a \ll \mu_s'$), as compared with the less restrictive condition $\mu_a \ll \mu_s$ for the validity of the P_1 approximation (we have assumed a g value of 0.9, which is representative of tissue, so that $\mu_s = 10\mu_s'$). Here, we have used "$>10\times$" for "$\gg$" and "$<0.1\times$" for "$\ll$."

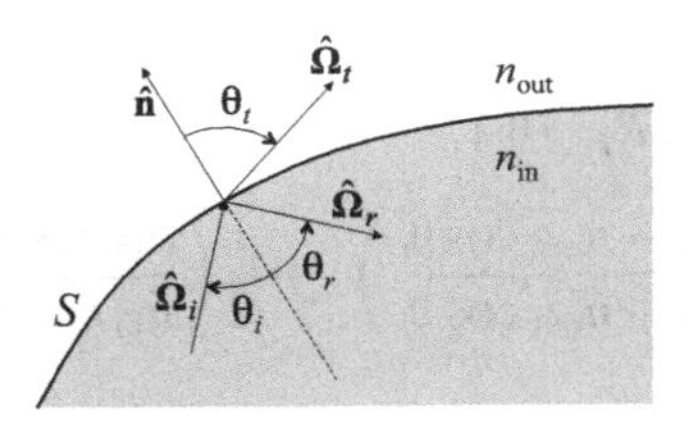

Figure 9.7

Illustration of the angles of incidence (θ_i), reflection (θ_r), and transmission (θ_t) for light incident from direction $\hat{\Omega}_i$ on a surface boundary S between a scattering medium with refractive index n_{in} and a non-scattering medium with refractive index n_{out}. The reflected and transmitted light travel along directions $\hat{\Omega}_r$ and $\hat{\Omega}_t$, respectively, and $\hat{\mathbf{n}}$ is the outer normal to the surface S.

9.6.2 Boundary between a scattering medium and a non-scattering medium: mismatched refractive index conditions

The case of a boundary between a scattering medium (tissue) and a non-scattering medium (vacuum, air, non-scattering material, etc.) with different indices of refraction, say n_{in} for the scattering medium of interest and n_{out} for the outside non-scattering medium, requires taking into account the reflection and refraction of photons at the boundary. By introducing the outer normal, $\hat{\mathbf{n}}$, to the surface S, the incident angle θ_i for a photon traveling along direction $\hat{\Omega}$ is defined by $\cos\theta_i = \hat{\mathbf{n}} \cdot \hat{\Omega}$. The reflection angle, θ_r, for reflected photons, and the refraction angle, θ_t, for transmitted photons are given by the law of reflection ($\theta_i = \theta_r$) and by Snell's law (Eq. (4.5)). As discussed in Chapter 4, total internal reflection occurs at incident angles greater than the critical angle θ_c defined by $n_{in} \sin\theta_c = n_{out}$. The angles are illustrated in Figure 9.7 in conjunction with the outer normal $\hat{\mathbf{n}}$ and the incident, reflected, and transmitted directions of propagation, $\hat{\Omega}_i$, $\hat{\Omega}_r$, and $\hat{\Omega}_t$, respectively.

The fraction of optical energy reflected at the interface is given by the Fresnel reflection coefficient $R_F(\theta_i)$ (named after the French physicist Augustin-Jean Fresnel [1788–1827]), which depends on the polarization of the incident light. Because light tends to become unpolarized as a result of multiple scattering in tissues, one can consider the Fresnel reflection coefficient for unpolarized light, which is given by the average of the reflection coefficients for light polarized parallel ($R_{F\parallel}(\theta_i)$) and perpendicular ($R_{F\perp}(\theta_i)$) to the plane formed by the incident and reflected or refracted rays (these polarization components are referred to as p-polarization and s-polarization, respectively, in Chapter 7,

Section 7.2):

$$R_F(0 \le \theta_i < \theta_c) = \tfrac{1}{2}R_{F\|}(\theta_i) + \tfrac{1}{2}R_{F\perp}(\theta_i)$$
$$= \frac{1}{2}\left(\frac{n_{\text{in}}\cos\theta_t - n_{\text{out}}\cos\theta_i}{n_{\text{in}}\cos\theta_t + n_{\text{out}}\cos\theta_i}\right)^2 + \frac{1}{2}\left(\frac{n_{\text{in}}\cos\theta_i - n_{\text{out}}\cos\theta_t}{n_{\text{in}}\cos\theta_i + n_{\text{out}}\cos\theta_t}\right)^2 \tag{9.40}$$

$$R_F(\theta_c \le \theta_{\text{in}} \le \pi/2) = 1 \tag{9.41}$$

The boundary condition at the surface S for the Boltzmann transport equation can be written in terms of the angular energy density $u(\mathbf{r}, \hat{\mathbf{\Omega}}, t)$ as follows:

$$[u(\mathbf{r}_S, \hat{\mathbf{\Omega}}, t) - R_F(-\hat{\mathbf{\Omega}}_{\supset\pi\hat{\mathbf{n}}} \cdot \hat{\mathbf{n}})u(\mathbf{r}_S, -\hat{\mathbf{\Omega}}_{\supset\pi\hat{\mathbf{n}}}, t)]_{\hat{\mathbf{\Omega}}\cdot\hat{\mathbf{n}}<0} = 0, \tag{9.42}$$

where $\mathbf{r}_S$ is the position vector for a point on the boundary S, and $\hat{\mathbf{\Omega}}_{\supset\pi\hat{\mathbf{n}}}$ indicates the unit vector obtained by rotating $\hat{\mathbf{\Omega}}$ by π radians around the axis defined by $\hat{\mathbf{n}}$ (so that, using the notation of Figure 9.7, if $\hat{\mathbf{\Omega}} = \hat{\mathbf{\Omega}}_r$ then $\hat{\mathbf{\Omega}}_{\supset\pi\hat{\mathbf{n}}} = -\hat{\mathbf{\Omega}}_i$).

In the P_1 approximation, and its further development into the diffusion equation, the angular energy density is integrated over the angular coordinates. In fact, we have seen (Eqs. (9.20) and (9.21)) that the $l = 0$ coefficient $u_{0,0}(\mathbf{r}, t)$ is proportional to the energy density $U(\mathbf{r}, t)$ (or the fluence rate $\phi = c_n U(\mathbf{r}, t)$), and that the $l = 1$ coefficients, specifically $\text{Re}[u_{1,1}(\mathbf{r}, t)]$, $\text{Im}[u_{1,1}(\mathbf{r}, t)]$, and $u_{1,0}(\mathbf{r}, t)$, are proportional to the x, y, and z components, respectively, of the net flux vector $\mathbf{F}(\mathbf{r}, t)$. Consequently, Eq. (9.42), which sets conditions for each direction $\hat{\mathbf{\Omega}}$ pointing inside the medium ($\hat{\mathbf{\Omega}} \cdot \hat{\mathbf{n}} < 0$), cannot be fulfilled by the fixed angular dependence, through $\mathbf{F} \cdot \hat{\mathbf{\Omega}}$, of the P_1 radiance (Eq. (9.24)). Instead, the boundary conditions for the diffusion equation consider the total inward intensity, requiring that it be equal to the integral over the half solid angle of the normal component of the vector radiance reaching the boundary from inside the medium along direction $\hat{\mathbf{\Omega}}$ (i.e., $L(\mathbf{r}, \hat{\mathbf{\Omega}}, t)\hat{\mathbf{\Omega}} \cdot \hat{\mathbf{n}}$ with $\hat{\mathbf{\Omega}} \cdot \hat{\mathbf{n}} > 0$) multiplied by the Fresnel reflection coefficient $R_F(\hat{\mathbf{\Omega}} \cdot \hat{\mathbf{n}})$ (Haskell et al., 1994):

$$\int_{\hat{\mathbf{\Omega}}\cdot\hat{\mathbf{n}}<0} L(\mathbf{r}_S, \hat{\mathbf{\Omega}}, t)\hat{\mathbf{\Omega}} \cdot (-\hat{\mathbf{n}})d\Omega = \int_{\hat{\mathbf{\Omega}}\cdot\hat{\mathbf{n}}>0} R_F(\hat{\mathbf{\Omega}} \cdot \hat{\mathbf{n}})L(\mathbf{r}_S, \hat{\mathbf{\Omega}}, t)\hat{\mathbf{\Omega}} \cdot \hat{\mathbf{n}}d\Omega. \tag{9.43}$$

The direction-dependent reflection coefficient $R_F(\hat{\mathbf{\Omega}} \cdot \hat{\mathbf{n}})$ may be taken out of the integral by replacing it with an effective reflection coefficient, $R_{2\pi}$, for the half-solid-angle integrated radiance:

$$\int_{\hat{\mathbf{\Omega}}\cdot\hat{\mathbf{n}}<0} L(\mathbf{r}_S, \hat{\mathbf{\Omega}}, t)\hat{\mathbf{\Omega}} \cdot (-\hat{\mathbf{n}})d\Omega = R_{2\pi} \int_{\hat{\mathbf{\Omega}}\cdot\hat{\mathbf{n}}>0} L(\mathbf{r}_S, \hat{\mathbf{\Omega}}, t)\hat{\mathbf{\Omega}} \cdot \hat{\mathbf{n}}d\Omega. \tag{9.44}$$

By using the P_1 radiance $L_{P_1}(\mathbf{r}, \hat{\mathbf{\Omega}}, t)$ (Eq. (9.24)) and the diffusion-limit Fick's-law relationship between the net vector flux $\mathbf{F}$ and the energy density U (Eq. (9.37)), the integral for $\hat{\mathbf{\Omega}} \cdot \hat{\mathbf{n}} < 0$ in Eq. (9.44) becomes:

$$\int_{\hat{\mathbf{\Omega}}\cdot\hat{\mathbf{n}}<0} L_{P_1}(\mathbf{r}_S, \hat{\mathbf{\Omega}}, t)\hat{\mathbf{\Omega}} \cdot (-\hat{\mathbf{n}})d\Omega$$

$$= -\frac{c_n U(\mathbf{r}_S, t)}{4\pi} \int_{\hat{\mathbf{\Omega}}\cdot\hat{\mathbf{n}}<0} \hat{\mathbf{\Omega}} \cdot \hat{\mathbf{n}} d\Omega - \frac{3}{4\pi}\mathbf{F}(\mathbf{r}_S, t) \cdot \int_{\hat{\mathbf{\Omega}}\cdot\hat{\mathbf{n}}<0} \hat{\mathbf{\Omega}}(\hat{\mathbf{\Omega}} \cdot \hat{\mathbf{n}})d\Omega$$

$$= -\frac{c_n U(\mathbf{r}_S, t)}{4\pi} \int_0^{2\pi}\int_{\pi/2}^{\pi} \cos\theta \sin\theta d\theta d\varphi$$

$$- \frac{3}{4\pi}\left[F_n(\mathbf{r}_S, t) \int_{\hat{\mathbf{\Omega}}\cdot\hat{\mathbf{n}}<0} (\hat{\mathbf{n}} \cdot \hat{\mathbf{\Omega}})(\hat{\mathbf{\Omega}} \cdot \hat{\mathbf{n}})d\Omega + F_t(\mathbf{r}_S, t) \int_{\hat{\mathbf{\Omega}}\cdot\hat{\mathbf{n}}<0} (\hat{\mathbf{t}} \cdot \hat{\mathbf{\Omega}})(\hat{\mathbf{\Omega}} \cdot \hat{\mathbf{n}})d\Omega\right]$$

$$= \frac{c_n U(\mathbf{r}_S, t)}{4} - \frac{3}{4\pi}\left[F_n(\mathbf{r}_S, t)\int_0^{2\pi}\int_{\pi/2}^{\pi} \cos^2\theta \sin\theta d\theta d\varphi\right.$$

$$\left. + F_t(\mathbf{r}_S, t)\int_0^{2\pi}\int_{\pi/2}^{\pi} (\sin\theta\cos\varphi)\cos\theta \sin\theta d\theta d\varphi\right]$$

$$= \frac{c_n U(\mathbf{r}_S, t)}{4} + \frac{D\hat{\mathbf{n}} \cdot \nabla U(\mathbf{r}_S, t)}{2} \tag{9.45}$$

where F_n and F_t are the components of $\mathbf{F}$ normal (along $\hat{\mathbf{n}}$) and tangential (along $\hat{\mathbf{t}}$), respectively, to the surface S. Also, we have set the z axis and x axis of the spherical coordinate system (defined in Figure 9.1) along the $\hat{\mathbf{n}}$ and $\hat{\mathbf{t}}$ directions, respectively, and we recall that $d\Omega = \sin\theta d\theta d\varphi$. The only difference between the integral for $\hat{\mathbf{\Omega}} \cdot \hat{\mathbf{n}} > 0$ and the integral for $\hat{\mathbf{\Omega}} \cdot \hat{\mathbf{n}} < 0$ is the integration range for θ, which extends from 0 to $\pi/2$ in the first case and from $\pi/2$ to π in the second case. This results in a change of the second sign on the right-hand side of Eq. (9.45), so that the integral for $\hat{\mathbf{\Omega}} \cdot \hat{\mathbf{n}} > 0$ in Eq. (9.44) is:

$$\int_{\hat{\mathbf{\Omega}}\cdot\hat{\mathbf{n}}>0} L_{P_1}(\mathbf{r}_S, \hat{\mathbf{\Omega}}, t)\hat{\mathbf{\Omega}} \cdot \hat{\mathbf{n}} d\Omega = \frac{c_n U(\mathbf{r}_S, t)}{4} - \frac{D\hat{\mathbf{n}} \cdot \nabla U(\mathbf{r}_S, t)}{2}. \tag{9.46}$$

By taking into account Eqs. (9.45) and (9.46) and multiplying Eq. (9.44) by 4, one can write the boundary condition for a refractive-index-mismatch case in the diffusion approximation as follows:

$$c_n U(\mathbf{r}_S, t) + 2D\hat{\mathbf{n}} \cdot \nabla U(\mathbf{r}_S, t) = R_{2\pi,\text{Diff}}[c_n U(\mathbf{r}_S, t) - 2D\hat{\mathbf{n}} \cdot \nabla U(\mathbf{r}_S, t)], \tag{9.47}$$

where the additional subscript "Diff" in $R_{2\pi,\text{Diff}}$ indicates the diffusion approximation case. Equation (9.47) can be further compacted by introducing the *reflection parameter* $A = (1 + R_{2\pi,\text{Diff}})/(1 - R_{2\pi,\text{Diff}})$, and writing the directional derivative of U along direction $\hat{\mathbf{n}}$ (i.e., $\hat{\mathbf{n}} \cdot \nabla U$) as $\partial U/\partial n$:

$$c_n U(\mathbf{r}_S, t) + 2AD\frac{\partial U(\mathbf{r}_S, t)}{\partial n} = 0. \tag{9.48}$$

It is worth emphasizing that the half-space radiance condition of the P_1 and diffusion approximations (Eq. (9.44)) results in a boundary condition (Eq. (9.48)) that involves only the net flux component normal to the boundary. (In fact, the term containing F_t in Eq. (9.45) vanishes.) The reflection parameter A takes values ranging from 1 (totally transmitting or totally absorbing boundary) to infinity (perfectly reflecting boundary). Its value, however, does not depend only on the reflection properties of the boundary (through the Fresnel reflection coefficient $R_F(\hat{\mathbf{\Omega}} \cdot \hat{\mathbf{n}})$): it also depends on the angular distribution of the radiance at the boundary. For the case of the isotropic component of the radiance, which is proportional to the fluence rate $\phi(\mathbf{r}, t)$, the effective reflection coefficient is:

$$R_{2\pi,\phi} = \frac{\int_{\hat{\mathbf{\Omega}}\cdot\hat{\mathbf{n}}>0} R_F(\hat{\mathbf{\Omega}} \cdot \hat{\mathbf{n}})\frac{\phi}{4\pi}\hat{\mathbf{\Omega}} \cdot \hat{\mathbf{n}}d\Omega}{\int_{\hat{\mathbf{\Omega}}\cdot\hat{\mathbf{n}}>0} \frac{\phi}{4\pi}\hat{\mathbf{\Omega}} \cdot \hat{\mathbf{n}}d\Omega} = 2\int_0^{\pi/2} R_F(\cos\theta)\cos\theta\sin\theta d\theta. \tag{9.49}$$

For the $l = 1$ anisotropic component of the radiance, which is proportional to $\mathbf{F} \cdot \hat{\mathbf{\Omega}}$, the effective reflection coefficient is:

$$R_{2\pi,\mathbf{F}\cdot\hat{\mathbf{\Omega}}} = \frac{\int_{\hat{\mathbf{\Omega}}\cdot\hat{\mathbf{n}}>0} R_F(\hat{\mathbf{\Omega}} \cdot \hat{\mathbf{n}})\frac{3}{4\pi}\mathbf{F} \cdot \hat{\mathbf{\Omega}}d\Omega}{\int_{\hat{\mathbf{\Omega}}\cdot\hat{\mathbf{n}}>0} \frac{3}{4\pi}\mathbf{F} \cdot \hat{\mathbf{\Omega}}d\Omega} = 3\int_0^{\pi/2} R_F(\cos\theta)\cos^2\theta\sin\theta d\theta. \tag{9.50}$$

By recalling that the fluence rate, ϕ, is equal to $c_n U$, and that the component of the net flux vector normal to the surface, F_n, is equal to $-D\hat{\mathbf{n}} \cdot \nabla U$, one can rewrite Eq. (9.47) as follows (Haskell et al., 1994):

$$\phi(\mathbf{r}_S, t) - 2F_n(\mathbf{r}_S, t) = R_{2\pi,\phi}\phi(\mathbf{r}_S, t) + 2R_{2\pi,\mathbf{F}\cdot\hat{\mathbf{\Omega}}}F_n(\mathbf{r}_S, t), \tag{9.51}$$

or, by rearranging the terms:

$$\phi(\mathbf{r}_S, t) - 2\frac{1 + R_{2\pi,\mathbf{F}\cdot\hat{\mathbf{\Omega}}}}{1 - R_{2\pi,\phi}}F_n(\mathbf{r}_S, t) = 0, \tag{9.52}$$

which is an alternative way of writing the boundary condition expressed by Eq. (9.48). By comparing Eq. (9.48) and Eq. (9.52), one can deduce that $A = (1 + R_{2\pi,\mathbf{F}\cdot\hat{\mathbf{\Omega}}})/(1 - R_{2\pi,\phi})$, which, when solved for $R_{2\pi,\text{Diff}}$, yields:

$$R_{2\pi,\text{Diff}} = \frac{R_{2\pi,\phi} + R_{2\pi,\mathbf{F}\cdot\hat{\mathbf{\Omega}}}}{2 - R_{2\pi,\phi} + R_{2\pi,\mathbf{F}\cdot\hat{\mathbf{\Omega}}}}. \tag{9.53}$$

Equations (9.48) and (9.52) show that the boundary condition in the diffusion approximation is a condition on the ratio of the amplitudes of the anisotropic to the isotropic components of the P_1 radiance L_{P_1} (as given in Eq. (9.24)). In fact, Eqs. (9.48) and (9.52) show that, at the boundary, $3F_n/\phi = 3/(2A)$. This ratio, which should be small for validity of the P_1 approximation, is typically 10–15% inside strongly scattering tissues. In the case of a typical tissue-air interface, this ratio assumes a much higher value of 50–60% (Haskell et al., 1994), showing that the P_1 approximation is inaccurate near tissue-air boundaries. The case of a 55% anisotropy in the radiance is illustrated in Figure 9.5.

We observe that the reflection parameter A has also been written as $A = (1 + r_d)/(1 - r_d)$, where r_d is the *internal diffuse reflection coefficient* expressed in terms of the relative refractive index ($n_{\text{rel}} = n_{\text{in}}/n_{\text{out}}$) by the empirical relationship $r_d = -1.440n_{\text{rel}}^{-2} + 0.710n_{\text{rel}}^{-1} + 0.668 + 0.0636n_{\text{rel}}$ (Groenhuis et al., 1983).

Extrapolated boundary condition. A practical, albeit approximated, approach to treating boundary conditions for the diffusion equation is that of the *extrapolated boundary condition.* It consists of setting the fluence rate (or the energy density) to zero at an extrapolated boundary located at a distance z_b from the actual boundary. The distance z_b results from the linear extrapolation of the energy density from the inside to the outside of the scattering medium, by using the local normal derivative of U at the boundary, i.e., $\partial U(\mathbf{r}_S, t)/\partial n$, to extend the rate of decrease of U outside the scattering medium. This method leads to an extrapolated energy density that vanishes at a distance z_b given by:

$$z_b = \frac{U(\mathbf{r}_S, t)}{\left|\frac{\partial U(\mathbf{r}_S, t)}{\partial n}\right|}, \tag{9.54}$$

which, by virtue of Eq. (9.48), can be written in terms of the reflection parameter and the optical coefficients of the scattering medium:

$$z_b = \frac{2AD}{c_n} = \frac{2A}{3(\mu_s' + \mu_a)} \cong \frac{2A}{3\mu_s'}. \tag{9.55}$$

One way to satisfy this extrapolated boundary condition is the method of images (Haskell et al., 1994), commonly used in electrostatics, which conceptually extends the scattering medium to the entire space and introduces virtual sources (outside the volume occupied by the scattering medium) to generate a null energy density at the extrapolated boundary. (In the case of light, this artificial but mathematically convenient construct invokes light sources of "negative photons," or light sinks.) A simple example is the case of a half-space scattering medium. A set of negative sources that are mirror images of the actual sources, with respect to the extrapolated boundary plane, combine to generate a zero energy density at the extrapolated boundary. This is illustrated in Figure 9.8, which also shows the linear extrapolation of the energy density outside the scattering medium.

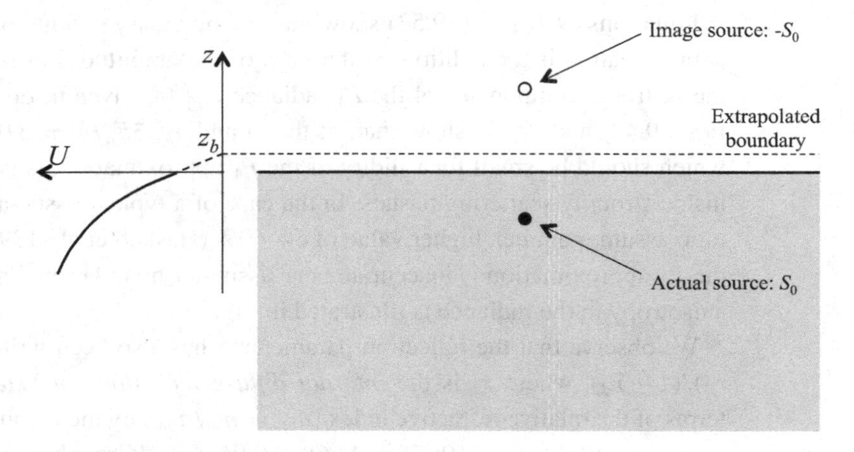

Figure 9.8

Illustration of the extrapolated boundary condition for a semi-infinite scattering medium. A negative source ($-S_0$) that is a mirror image of an actual source (S_0) with respect to the extrapolated boundary realizes the condition of zero energy density (U) at the extrapolated boundary ($z = z_b$).

A further approximation, referred to as *zero boundary condition*, consists of setting $z_b = 0$, i.e., setting the fluence rate to zero at the physical boundary of the scattering medium. We observe that the zero boundary condition is not equivalent to assuming matched refractive index conditions at the boundary, for which the extrapolated boundary is still external to the scattering medium as pointed out in the next section.

9.6.3 Boundary between a scattering medium and a non-scattering medium: matched refractive index conditions

The boundary conditions for the matched refractive index case are directly derived from the mismatched index case by setting the Fresnel reflection coefficient $R_F(\hat{\mathbf{\Omega}} \cdot \hat{\mathbf{n}}) = 0$ and the reflection parameter $A = 1$. The boundary condition for the transport equation (from Eq. (9.42)) becomes:

$$[u(\mathbf{r}_S, \hat{\mathbf{\Omega}}, t)]_{\hat{\mathbf{\Omega}}\cdot\hat{\mathbf{n}}<0} = 0, \tag{9.56}$$

and the boundary condition for the diffusion equation (from Eq. (9.48)) becomes:

$$c_n U(\mathbf{r}_S, t) + 2D\frac{\partial U(\mathbf{r}_S, t)}{\partial n} = 0. \tag{9.57}$$

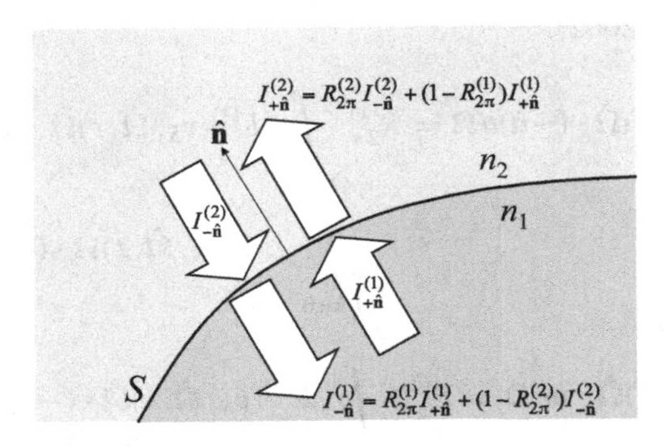

Figure 9.9.

Boundary conditions for an interface (S) across two scattering media with indices of refraction n_1 and n_2. The normal $\hat{\mathbf{n}}$ to the surface points from medium 1 to medium 2. In the P_1 and diffusion approximations, the boundary conditions require that the total intensity entering medium 1 along all directions $\hat{\Omega}$ such that $\hat{\Omega} \cdot \hat{\mathbf{n}} < 0$ ($I_{-\hat{n}}^{(1)}$) be given by the sum of the reflected intensity from medium 1 ($R_{2\pi}^{(1)} I_{+\hat{n}}^{(1)}$) plus the transmitted intensity from medium 2 ($(1 - R_{2\pi}^{(2)}) I_{-\hat{n}}^{(2)}$). Similar conditions yield the total intensity entering medium 2 along all directions $\hat{\Omega}$ such that $\hat{\Omega} \cdot \hat{\mathbf{n}} > 0$ ($I_{+\hat{n}}^{(2)}$).

In the matched conditions ($A = 1$), the ratio of the amplitudes of the anisotropic to the isotropic components of the P_1 radiance, L_{P_1}, at the boundary becomes $3F_n/\phi = 150\%$, clearly outside the range of validity of the P_1 approximation. Such a case, of a 150% anisotropy of the radiance, is illustrated in Figure 9.5.

For extrapolated-boundary conditions in the matched-refractive-index case, the extrapolated boundary (from Eq. (9.55)) is at a distance $z_b = 2/[3(\mu_s' + \mu_a)] \cong 2/(3\mu_s')$ from the actual boundary.

9.6.4 Boundary between two scattering media

The case of an interface between two scattering media characterized by indices of refraction n_1 and n_2, respectively, is illustrated in Figure 9.9. In this case, the entering radiance from the boundary of a medium is determined not just by Fresnel reflection at the boundary (as is the case for an interface between a scattering and a non-scattering medium), but also by transmission of the radiance from the other scattering medium. Therefore, Eq. (9.44) must be expanded by including the radiance transmitted through the boundary from the other medium (using the effective Fresnel transmission coefficient $1 - R_{2\pi}$) and must be written for both media. By using a superscript (1) or (2) to identify the medium, the boundary conditions at the interface of two scattering media are written as follows (Haskell et al., 1994; Faris, 2002):

$$\int_{\hat{\Omega}\cdot\hat{\mathbf{n}}<0} L^{(1)}(\mathbf{r}_S, \hat{\Omega}, t)\hat{\Omega}\cdot(-\hat{\mathbf{n}})d\Omega = R_{2\pi}^{(1)} \int_{\hat{\Omega}\cdot\hat{\mathbf{n}}>0} L^{(1)}(\mathbf{r}_S, \hat{\Omega}, t)\hat{\Omega}\cdot\hat{\mathbf{n}}d\Omega + \left(1 - R_{2\pi}^{(2)}\right) \times \int_{\hat{\Omega}\cdot\hat{\mathbf{n}}<0} L^{(2)}(\mathbf{r}_S, \hat{\Omega}, t)\hat{\Omega}\cdot(-\hat{\mathbf{n}})d\Omega \tag{9.58}$$

$$\int_{\hat{\Omega}\cdot\hat{\mathbf{n}}>0} L^{(2)}(\mathbf{r}_S, \hat{\Omega}, t)\hat{\Omega}\cdot\hat{\mathbf{n}}d\Omega = R_{2\pi}^{(2)} \int_{\hat{\Omega}\cdot\hat{\mathbf{n}}<0} L^{(2)}(\mathbf{r}_S, \hat{\Omega}, t)\hat{\Omega}\cdot(-\hat{\mathbf{n}})d\Omega + \left(1 - R_{2\pi}^{(1)}\right) \times \int_{\hat{\Omega}\cdot\hat{\mathbf{n}}>0} L^{(1)}(\mathbf{r}_S, \hat{\Omega}, t)\hat{\Omega}\cdot\hat{\mathbf{n}}d\Omega \tag{9.59}$$

where $\hat{\mathbf{n}}$ denotes the normal to the boundary pointing from medium 1 to medium 2 (see Figure 9.9). Equations (9.58) and (9.59) can be further developed by:

- considering the expression for the P_1 radiance (Eq. (9.24));
- using the diffusion relationship between the net flux vector and the gradient of the energy density (Eq. (9.37));
- rearranging the terms resulting from the integrals as given by Eqs. (9.45) and (9.46);
- introducing the half-space reflection coefficients for the $l = 0$ (ϕ) and $l = 1$ ($\mathbf{F}\cdot\hat{\Omega}$) terms (i.e., $R_{2\pi,\phi}$ and $R_{2\pi,\mathbf{F}\cdot\hat{\Omega}}$ as defined in Eqs. (9.49) and (9.50)).

As a result, Eqs. (9.58) and (9.59) become:

$$\frac{c_{n1}U^{(1)}(\mathbf{r}_S, t)}{4}\left(1 - R_{2\pi,\phi}^{(1)}\right) + \frac{D^{(1)}\hat{\mathbf{n}}\cdot\nabla U^{(1)}(\mathbf{r}_S, t)}{2}\left(1 + R_{2\pi,\mathbf{F}\cdot\hat{\Omega}}^{(1)}\right) = \frac{c_{n2}U^{(2)}(\mathbf{r}_S, t)}{4}\left(1 - R_{2\pi,\phi}^{(2)}\right) + \frac{D^{(2)}\hat{\mathbf{n}}\cdot\nabla U^{(2)}(\mathbf{r}_S, t)}{2}\left(1 - R_{2\pi,\mathbf{F}\cdot\hat{\Omega}}^{(2)}\right) \tag{9.60}$$

$$\frac{c_{n2}U^{(2)}(\mathbf{r}_S, t)}{4}\left(1 - R_{2\pi,\phi}^{(2)}\right) - \frac{D^{(2)}\hat{\mathbf{n}}\cdot\nabla U^{(2)}(\mathbf{r}_S, t)}{2}\left(1 + R_{2\pi,\mathbf{F}\cdot\hat{\Omega}}^{(2)}\right) = \frac{c_{n1}U^{(1)}(\mathbf{r}_S, t)}{4}\left(1 - R_{2\pi,\phi}^{(1)}\right) - \frac{D^{(1)}\hat{\mathbf{n}}\cdot\nabla U^{(1)}(\mathbf{r}_S, t)}{2}\left(1 - R_{2\pi,\mathbf{F}\cdot\hat{\Omega}}^{(1)}\right) \tag{9.61}$$

The sum of Eqs. (9.60) and (9.61) yields:

$$D^{(1)}\hat{\mathbf{n}}\cdot\nabla U^{(1)}(\mathbf{r}_S, t) = D^{(2)}\hat{\mathbf{n}}\cdot\nabla U^{(2)}(\mathbf{r}_S, t), \tag{9.62}$$

whereas the difference of Eqs. (9.60) and (9.61), after taking into account Eq. (9.62), yields:

$$c_{n1}U^{(1)}(\mathbf{r}_S,t)\left(1-R^{(1)}_{2\pi,\phi}\right)+2D^{(1)}\hat{\mathbf{n}}\cdot\nabla U^{(1)}(\mathbf{r}_S,t)\left(R^{(1)}_{2\pi,\mathbf{F}\cdot\hat{\Omega}}+R^{(2)}_{2\pi,\mathbf{F}\cdot\hat{\Omega}}\right)$$
$$=c_{n2}U^{(2)}(\mathbf{r}_S,t)\left(1-R^{(2)}_{2\pi,\phi}\right) \tag{9.63}$$

By considering that $n_1^2(1-R^{(1)}_{2\pi,\phi})=n_2^2(1-R^{(2)}_{2\pi,\phi})$ (from Snell's law), the diffusion boundary conditions of Eqs. (9.62) and (9.63) for the interface between two scattering media, for the normal component of the flux (F_n) and for the fluence rate ($\phi=c_nU$), can be written as follows (Aronson, 1995):

$$F_n^{(1)}(\mathbf{r}_S,t)=F_n^{(2)}(\mathbf{r}_S,t)=F_n(\mathbf{r}_S,t), \tag{9.64}$$

$$\phi^{(1)}(\mathbf{r}_S,t)-\left(\frac{n_1}{n_2}\right)^2\phi^{(2)}(\mathbf{r}_S,t)=2F_n(\mathbf{r}_S,t)\left(\frac{R^{(1)}_{2\pi,\mathbf{F}\cdot\hat{\Omega}}+R^{(2)}_{2\pi,\mathbf{F}\cdot\hat{\Omega}}}{1-R^{(1)}_{2\pi,\phi}}\right). \tag{9.65}$$

Equation (9.64) states that the normal component of the net flux is continuous across a boundary between two scattering media. Equation (9.65) states that the fluence rate is not continuous and it specifies its discontinuity across the boundary. However, in the case of refractive index matching, $n_1=n_2$, there is no Fresnel reflection at the boundary, so that Eq. (9.65) reduces to a continuity equation for the fluence rate (and the energy density) across an interface between scattering media.

9.7 The microscopic Beer-Lambert law

Consider a non-absorbing medium ($\mu_a=0$) and a pulsed source that emits light for an infinitesimally short time, as represented by the source term $q(\mathbf{r},\hat{\Omega},t)=u_S(\mathbf{r},\hat{\Omega})\delta(t)$, where $u_S(\mathbf{r},\hat{\Omega})$ is the angular energy density generated by the source and δ is the Dirac delta. If we denote the angular energy density in the non-absorbing medium as $u_0(\mathbf{r},\hat{\Omega},t)$, it satisfies the following Boltzmann transport equation:

$$\frac{\partial u_0(\mathbf{r},\hat{\Omega},t)}{\partial t}=-c_n\hat{\Omega}\cdot\nabla u_0(\mathbf{r},\hat{\Omega},t)-c_n\mu_s u_0(\mathbf{r},\hat{\Omega},t)$$
$$+c_n\mu_s\int_{4\pi}u_0(\mathbf{r},\hat{\Omega}',t)p(\hat{\Omega}',\hat{\Omega})d\Omega'+u_S(\mathbf{r},\hat{\Omega})\delta(t) \tag{9.66}$$

Let's now consider a non-zero, spatially uniform absorption coefficient, so that $\nabla\mu_a=0$. Given that the exponential factor $e^{-\mu_a c_n t}$ can be introduced in the source term $u_S(\mathbf{r}, \hat{\Omega})\delta(t)$ without affecting it (since the Dirac delta is always 0 at $t \neq 0$, and the exponential factor is 1 at $t = 0$), by substitution into Eq. (9.2) we can see that the angular photon density,

$$u\left(\mathbf{r}, \hat{\Omega}, t, \mu_a\right) = u_0\left(\mathbf{r}, \hat{\Omega}, t\right) e^{-\mu_a c_n t}, \tag{9.67}$$

is a solution to the Boltzmann transport equation for this pulsed source term and for homogeneous absorption. Multiplication of Eq. (9.67) by c_n and integration over the entire solid angle yields:

$$\int_{4\pi} c_n u(\mathbf{r}, \hat{\Omega}, t, \mu_a) d\Omega = \int_{4\pi} c_n u_0(\mathbf{r}, \hat{\Omega}, t) d\Omega e^{-\mu_a c_n t}, \tag{9.68}$$

which reduces to an equation for the fluence rate (ϕ), after recalling Eqs. (1.15) and (1.17):

$$\phi\left(\mathbf{r}, t, \mu_a\right) = \phi_0\left(\mathbf{r}, t\right) e^{-\mu_a c_n t}. \tag{9.69}$$

Because the product $c_n t$ is equal to the pathlength $L(t)$ traveled by photons over time t, Eqs. (9.67) and (9.69) establish relationships for the angular energy density and the fluence rate, respectively, that are formally identical to the relationship set by the Beer-Lambert law for the intensity (see Eq. (2.8)). However, there is an important conceptual difference. The Beer-Lambert law (Eq. (2.8)) relates the transmitted intensity (I) to the incident intensity (I_0) for a case in which the pathlength of all detected photons is the same and coincides with the medium thickness (L). Instead, Eqs. (9.67) and (9.69) relate the angular energy density and the fluence rate, u and ϕ respectively, at position $\mathbf{r}$ in a scattering medium in the presence of absorption to the corresponding quantities in the case of no absorption (u_0, ϕ_0) at time t following pulsed illumination. In this case, the pathlength of detected photons depends on the time considered and is given by $L(t) = c_n t$. The relationship expressed by Eqs. (9.67) and (9.69) is referred to as the *time-resolved Beer-Lambert law*, or, more commonly, the *microscopic Beer-Lambert law*. The implication of the microscopic Beer-Lambert law is that knowledge of the solution to the Boltzmann transport equation for a non-absorbing case can be generalized to the case of homogeneous absorption by introducing the multiplicative factor $e^{-\mu_a c_n t}$ to the non-absorbing solutions for the angular energy density ($u_0(\mathbf{r}, \hat{\Omega}, t)$) and for the fluence rate ($\phi_0(\mathbf{r}, t)$).

9.8 The fluorescence diffusion equation

The diffusion equation can be generalized to the case in which a fluorophore is present in the scattering medium. We introduce a subscript x to indicate the wavelength of fluorophore excitation (λ_x) and a subscript m to indicate the wavelength of fluorescence emission (λ_m). The diffusion equation for the energy density at the excitation wavelength (U_x) is just Eq. (9.36) with subscripts x for the diffusion coefficient (D), the absorption coefficient (μ_a), and the source term (S_0) to indicate that they refer to λ_x:

$$\frac{\partial U_x(\mathbf{r},t)}{\partial t} = \nabla \cdot [D_x \nabla U_x(\mathbf{r},t)] - c_n \mu_{ax}^{(t)} U_x(\mathbf{r},t) + S_{0x}(\mathbf{r},t). \qquad (9.70)$$

We have not introduced a subscript x in c_n because we approximate the speed of light to be the same at the excitation and emission wavelengths. The superscript (t) in $\mu_{ax}^{(t)}$ specifies a total absorption coefficient due to the fluorophore and to any other chromophores in the medium.

At the emission wavelength, we need to consider a source term that is spatially dependent (according to the spatial distribution of the fluorophore and the spatial dependence of the excitation energy density) and time-dependent (according to the temporal properties of U_x and the decaying fluorescence emission). Furthermore, the amplitude of the source term depends on the fluorophore absorption coefficient at the excitation wavelength ($\mu_{ax}^{(f)}$) and its fluorescence quantum yield (Φ_F). For simplicity let us consider a fluorophore characterized by a single-exponential fluorescence decay, with lifetime τ (so that the temporal decay of the excited fluorophore concentration following excitation at time $t = 0$ by a short pulse is described by $e^{-t/\tau}$). The fluorescence source term at the emission wavelength ($S_{0m}(\mathbf{r},t)$), which represents the energy density of fluorescence emission per unit time, results from the product of four terms:

i. the probability per unit time of absorption of excitation photons by the fluorophore, which, by definition of the absorption coefficient, is given by $c_n \mu_{ax}^{(f)}$;
ii. the fluorescence quantum yield (Φ_F), which is the ratio of the number of emitted fluorescence photons (over the entire emission spectrum) to the number of absorbed photons;
iii. the spectral emission probability within $d\lambda_m$ about λ_m, expressed as $\varphi_m d\lambda_m$ where φ_m is the emission probability per unit wavelength, which takes into account the spectral dependence of fluorescence emission;
iv. the temporal convolution of the excitation energy density with the single-exponential fluorescence decay, which accounts for the fact that the fluorescence source term $S_{0m}(\mathbf{r},t)$ receives contributions from fluorophore excitations at all times $t - t'$ that precede time t. The weight for these contributions originating within the infinitesimal time interval dt' about a time that precedes t by time t' is $\frac{e^{-t'/\tau}}{\tau} dt'$ (because the fluorescent intensity is proportional to the absolute value of the time derivative of the concentration of excited fluorophores [as shown in Section 5.2], resulting in the factor $e^{-t'/\tau}/\tau$).

By consolidating the above factors into the fluorescence emission source term at λ_m, the diffusion equation at the emission wavelength can be written as follows (Cerussi et al., 1997):

$$\frac{\partial U_m(\mathbf{r},t)}{\partial t} = \nabla \cdot [D_m \nabla U_m(\mathbf{r},t)] - c_n \mu_{am}^{(t)} U_m(\mathbf{r},t)$$
$$+ c_n \mu_{ax}^{(f)} \Phi_F \varphi_m d\lambda_m \int_0^\infty U_x(\mathbf{r}, t-t') \frac{e^{-t'/\tau}}{\tau} dt'. \qquad (9.71)$$

The set of Eqs. (9.70) and (9.71) describes the energy density, U (and the fluence rate $\phi = c_n U$), at the wavelengths of fluorophore excitation (subscript x) and fluorescence emission (subscript m) in a highly scattering medium in which the diffusion approximation conditions are satisfied.

9.9 The correlation diffusion equation

In Section 8.9.1, we have seen that the motion of scattering particles determines temporal fluctuations in the intensity of multiply scattered light. These temporal fluctuations carry information about the size and diffusion properties of the scattering particles in the medium. By considering the scalar electric field, $E(\mathbf{r}, \hat{\Omega}, t)$, at position $\mathbf{r}$ and time t that is associated with light propagating along direction $\hat{\Omega}$, one can define the angle-dependent temporal field autocorrelation function as:

$$G_1^T(\mathbf{r}, \hat{\Omega}, \tau) = \langle E(\mathbf{r}, \hat{\Omega}, t) E^*(\mathbf{r}, \hat{\Omega}, t+\tau) \rangle, \qquad (9.72)$$

where * indicates the complex conjugate, and the brackets $\langle\ \rangle$ indicate a time or ensemble average. The physical dimensions of G_1^T are those of the square of the electric field per steradian, which in the Gaussian system are those of an energy density per steradian. $G_1^T(\mathbf{r}, \hat{\Omega}, \tau)$ obeys a *correlation transfer equation* that is formally analogous to the Boltzmann transport equation (Dougherty et al., 1994). As developed in Sections 9.2–9.5 for the Boltzmann transport equation, one can expand the correlation transfer equation in spherical harmonics, retaining only terms of order $l = 0$ or 1, to yield the P_1 approximation. Further, one can consider a nearly isotropic field correlation function (scattering dominating over absorption) and short correlation times (with respect to the time needed for particles to move over a distance equal to the optical wavelength) to yield the diffusion approximation. The angle-independent field autocorrelation function is obtained by integration of the angle-dependent field correlation function over the entire solid angle:

$$G_1(\mathbf{r}, \tau) = \int_{4\pi} G_1^T(\mathbf{r}, \hat{\Omega}, \tau) d\Omega, \qquad (9.73)$$

and it satisfies the *correlation diffusion equation* (Boas and Yodh, 1997):

$$\nabla \cdot [D \nabla G_1(\mathbf{r}, \tau)] - \left[c_n \mu_a + \frac{1}{3} c_n \mu_s' k_0^2 \langle \Delta r^2(\tau) \rangle \right] G_1(\mathbf{r}, \tau) = -S_0(\mathbf{r}), \quad (9.74)$$

where k_0 is the optical wavenumber $(2\pi/\lambda)$, $\langle \Delta r^2(\tau) \rangle$ is the mean square displacement of the particles over time τ, and S_0 is the source term (energy per unit volume per unit time). In the limiting case of $\tau = 0$, Eq. (9.74) reduces to the time-independent diffusion equation for the energy density given by Eq. (9.36), after setting the time-derivative term to zero. Equation (9.74) is the analytical tool for measurements of microvascular blood flow by diffuse correlation spectroscopy (DCS) on the basis of dynamic properties of red blood cells that are typically modeled as Brownian motion expressed by:

$$\langle \Delta r^2(\tau) \rangle = 6 D_B \tau, \quad (9.75)$$

where D_B is the *Brownian diffusion coefficient* (after Scottish botanist Robert Brown [1773–1858]). D_B is proportional to the temporal slope of the mean-squared displacement of the moving particles, and therefore it provides a measure of their collective motion. Even though the motion of red blood cells in the vasculature is not random, as required for a rigorous description by the Brownian model, it is nevertheless well described by Eq. (9.74) and Eq. (9.75), where D_B should be interpreted as an "effective" Brownian diffusion coefficient. It is this effective Brownian diffusion coefficient of red blood cells that is measured by DCS to provide a blood flow index for the study of microvascular circulation.

Appendix 9.A: Low-order spherical harmonics

The spherical harmonics for $l = 0$, $l = 1$, and $l = 2$ are:

$$Y_0^0(\theta, \varphi) = \frac{1}{\sqrt{4\pi}} \quad (9.A.1)$$

$$Y_1^0(\theta, \varphi) = \sqrt{\frac{3}{4\pi}} \cos\theta \quad (9.A.2)$$

$$Y_1^{\pm 1}(\theta, \varphi) = \mp \sqrt{\frac{3}{8\pi}} \sin\theta e^{\pm i\varphi} \quad (9.A.3)$$

$$Y_2^0(\theta, \varphi) = \sqrt{\frac{5}{16\pi}} \left(3\cos^2\theta - 1\right) \quad (9.A.4)$$

$$Y_2^{\pm 1}(\theta, \varphi) = \mp \sqrt{\frac{15}{8\pi}} \sin\theta \cos\theta e^{\pm i\varphi} \quad (9.A.5)$$

$$Y_2^{\pm 2}(\theta, \varphi) = \sqrt{\frac{15}{32\pi}} \sin^2 \theta e^{\pm 2i\varphi} \tag{9.A.6}$$

The angular dependence of their real part squared is represented graphically in Figure 9.4.

Appendix 9.B: The dependence of the diffusion coefficient on absorption

The diffusion coefficient (D) was introduced in Eq. (9.30) and defined as:

$$D = \frac{c_n}{3(\mu_s' + \mu_a)}. \tag{9.B.1}$$

Defined this way, the diffusion coefficient is dependent on both the absorption coefficient, μ_a, and the reduced scattering coefficient, μ_s'. However, there is a theoretical argument against the dependence of D on μ_a. The argument is that the presence of μ_a in the definition of D results in a violation of the microscopic Beer-Lambert law (Eq. (9.69)) in the diffusion approximation. In fact, the diffusion equation for a homogeneous medium (Eq. (9.38)), written for the fluence rate ($\phi = c_n U$) and for a temporal-delta source term, is:

$$\frac{\partial \phi(\mathbf{r}, t)}{\partial t} = D\nabla^2 \phi(\mathbf{r}, t) - c_n \mu_a \phi(\mathbf{r}, t) + c_n U_S(\mathbf{r})\delta(t), \tag{9.B.2}$$

where $U_S(\mathbf{r})$ is the energy density generated by the source distribution. Dependence of D on μ_a would mean that the product of the fluence rate for zero absorption and the exponentially decaying absorption term ($\phi_0(\mathbf{r}, t)e^{-\mu_a c_n t}$) cannot be a solution to Eq. (9.B.2). That consequence is contrary to the case of transport theory that was considered in Section 9.7. In fact, an alternative derivation of the diffusion equation results in the definition of an absorption-independent diffusion coefficient (Furutsu and Yamada, 1994):

$$D_0 = \frac{c_n}{3\mu_s'}. \tag{9.B.3}$$

Theoretical questions have been raised, however, about whether the microscopic Beer-Lambert law must be satisfied in the diffusion regime (Aronson and Corngold, 1999), thus raising doubts on the earlier theoretical argument.

From a practical point of view, this discussion may appear to be somewhat irrelevant since, after all, the diffusion equation only holds if $\mu_s' \gg \mu_a$, so that $D \approx D_0$. However, the discussion does become practically relevant if one is to evaluate the possibility of extending diffusion theory beyond its low-absorption limit of validity. Would the specific dependence of D on μ_a make a difference on the level of accuracy of diffusion theory when applied to cases in which scattering does not dominate over absorption? It has been proposed that diffusion theory can indeed be applied to media with high absorption coefficients (up to $\mu_a \approx \mu_s'$) with the introduction of a diffusion coefficient featuring dependence on absorption that is somewhat intermediate between those of Eqs. (9.B.1) and (9.B.3) (Aronson and Corngold, 1999):

$$D_\alpha = \frac{c_n}{3(\mu_s' + \alpha\mu_a)}, \tag{9.B.4}$$

where α depends on μ_a, μ_s' and the anisotropy factor g as follows (Ripoll et al., 2005):

$$\alpha = 1 - \frac{4}{5}\frac{\mu_s' + \mu_a}{[\mu_s'(1+g) + \mu_a]}. \tag{9.B.5}$$

The factor α, as defined in Eq. (9.B.5), is equal to 0.2 in the case of isotropic scattering ($g = 0$), and it falls within the range 0.55–0.59 when considering values of $\mu_a = 0.1$ cm^{-1}, $\mu_s' = 10$ cm^{-1}, and $0.8 \le g \le 0.95$ that are representative of tissue optical properties at near-infrared wavelengths.

In most of the current literature, the expression for D as defined in Eq. (9.B.1) continues in common use.

Problems – answers to problems with * are on p. 646

9.1 By using Gauss's theorem, which relates the surface integral of a vector **A** to the volume integral of its divergence (namely, $\int_S \mathbf{A}\cdot d\mathbf{S} = \int_V \nabla\cdot\mathbf{A}\,d\mathbf{r}$), show that the diffusion contribution to the rate of change of the angular energy density is $-c_n\hat{\boldsymbol{\Omega}}\cdot\nabla u(r, \hat{\boldsymbol{\Omega}}, t)$, and comment on the reason for the negative sign.

9.2* For the following scattering phase functions $p(\hat{\boldsymbol{\Omega}}'\cdot\hat{\boldsymbol{\Omega}})$ find the normalization factor C required to realize the normalization condition $\int_{4\pi} p(\hat{\boldsymbol{\Omega}}'\cdot\hat{\boldsymbol{\Omega}})d\Omega' = 1$, and the average cosine of the scattering angle Θ (defined as $\langle\cos\Theta\rangle = \int_{4\pi}(\hat{\boldsymbol{\Omega}}'\cdot\hat{\boldsymbol{\Omega}})p(\hat{\boldsymbol{\Omega}}'\cdot\hat{\boldsymbol{\Omega}})d\Omega'$):
(a) isotropic phase function: $p(\hat{\boldsymbol{\Omega}}'\cdot\hat{\boldsymbol{\Omega}}) = C$;
(b) Henyey-Greenstein phase function: $p(\hat{\boldsymbol{\Omega}}'\cdot\hat{\boldsymbol{\Omega}}) = C\frac{1-g^2}{(1+g^2-2g\hat{\boldsymbol{\Omega}}'\cdot\hat{\boldsymbol{\Omega}})^{3/2}}$;
(c) Eddington phase function: $p(\hat{\boldsymbol{\Omega}}'\cdot\hat{\boldsymbol{\Omega}}) = C(1+3g\hat{\boldsymbol{\Omega}}'\cdot\hat{\boldsymbol{\Omega}})$;
(d) Rayleigh phase function: $p(\hat{\boldsymbol{\Omega}}'\cdot\hat{\boldsymbol{\Omega}}) = C[1+(\hat{\boldsymbol{\Omega}}'\cdot\hat{\boldsymbol{\Omega}})^2]$.

9.3 Consider the following source term, expressed by a delta in space, time, and direction, for the Boltzmann transport equation: $q(\mathbf{r}, \hat{\boldsymbol{\Omega}}, t) = Q_s\delta(\mathbf{r})\delta(\hat{\boldsymbol{\Omega}})\delta(t)$, where Q_s is the radiant energy emitted by the source. For this source term, show that if $u(\mathbf{r}, \hat{\boldsymbol{\Omega}}, t)$ is a solution to the Boltzmann transport equation for a total attenuation coefficient μ_t, single-scattering albedo

a, and scattering phase function $p(\hat{\Omega}, \hat{\Omega}')$, then the angular energy density $(\mu_t^*/\mu_t)^3 u(\mathbf{r}, \hat{\Omega}, t)$ is a solution to the Boltzmann transport equation for a different total attenuation coefficient μ_t^*, assuming that the single-scattering albedo and the scattering phase function are unchanged.

9.4* The coefficients u_{lm} of the expansion of the angular energy density into spherical harmonics are given by $u_{lm} = \int_{4\pi} u(\hat{\Omega}) Y_l^{m*}(\hat{\Omega}) d\Omega$. Find the expressions, in terms of the constant A, for the non-zero coefficients u_{lm} of the angular energy density $u(\hat{\Omega}) = A(3 + \sin\theta\cos\varphi)$.

9.5 Consider a new set of axes (x', y', z') resulting from the rotation of the original set of axes (x, y, z) by an angle Φ counterclockwise around the z axis. Find the relationship between the spherical harmonics $Y_l^m(\hat{\Omega})$ and $Y_l^m(\hat{\Omega}')$ in the two coordinate systems.

9.6 Show that $\mathbf{F} \cdot \hat{\Omega} = \frac{4\pi}{3} c_n \sum_{m=-1}^{1} u_{1m} Y_1^m(\hat{\Omega})$.

9.7 An isotropic light source is turned on at time $t = 0$ deeply inside an infinite and strongly scattering medium ($\mu_s' \gg \mu_a$). The time dependence of its emission is described by the temporal function $[1 - \sin(\frac{c_n \mu_s' t_0^2}{t})]$, where c_n is the speed of light in the medium. Does the optical energy distribution in the scattering medium satisfy the requirements for diffusion theory at time $t \sim t_0$?

9.8* Consider a radiant energy density that varies sinusoidally at angular frequency ω in a medium with $n = 1.35$, $\mu_a = 0.055\,\text{cm}^{-1}$, and $\mu_s' = 5.6\,\text{cm}^{-1}$. What is the maximum frequency ($f = \omega/(2\pi)$) for which you would use diffusion theory to model such radiant energy density?

9.9* Find the Fresnel reflection coefficient for normal incidence ($\theta_i = 0$) at a tissue-air interface with $n_{\text{tissue}} = 1.34$ and $n_{\text{air}} = 1$.

9.10* Consider an ideal case in which the Fresnel reflection coefficient for unpolarized light at an interface is given by $R_F(\cos\theta) = k(1 - \cos\theta)$, with k constant. The relative index of refraction of the media at the two sides of the interface is $n_{\text{rel}} = 1.15$.

(a) Find the value of the constant k so that the reflection parameter computed using the expression $A = (1 + R_{2\pi,\mathbf{F}\cdot\hat{\Omega}})/(1 - R_{2\pi,\phi})$ and the empirical formula $A = (1 + r_d)/(1 - r_d)$, with $r_d = -1.440 n_{\text{rel}}^{-2} + 0.710 n_{\text{rel}}^{-1} + 0.668 + 0.0636 n_{\text{rel}}$ yield the same value.

(b) What is the value of A for the constant k found in part (a)?

(c) What is the value of $R_{2\pi,\phi}$ for the constant k found in part (a)?

(d) What is the value of $R_{2\pi,\mathbf{F}\cdot\hat{\Omega}}$ for the constant k found in part (a)?

9.11 The zero boundary conditions require the fluence rate to be zero at the medium boundary. How is a null fluence rate at the boundary consistent with a measurable, non-zero optical signal at the tissue boundary?

9.12 Use the microscopic Beer-Lambert law to find the ratio between the fluence rate measured at a given position within a strongly scattering medium (of refractive index 1.35) in the presence and in the absence of a chromophore (molar extinction coefficient: 3.1×10^5 cm^{-1}M^{-1}; concentration: 0.27 μM) at times 0.58 ns and 1.6 ns after the emission of a short light pulse. Why does the attenuation factor associated with the presence of the chromophore depend on the time of observation?

9.13* Write the source term of the fluorescence diffusion equation (Eq. (9.71)) for an ideal fluorophore of unit quantum yield that emits fluorescence light at a single wavelength λ_m and at a single delay time τ (so that it acts as a source of photons that re-emits all of the photons that it absorbs after a given time τ and at a given wavelength λ_m). Verify that the source term has dimensions of energy per unit time per unit volume.

9.14 The root mean square displacement associated with Brownian motion for a spherical particle of radius R in a medium of viscosity η is $\langle \Delta r^2(\tau) \rangle = 6\frac{k_B T}{6\pi\eta R}\tau$, where k_B is the Boltzmann's constant and T is temperature.

(a) Using this expression for $\langle \Delta r^2(\tau) \rangle$, calculate the Brownian diffusion coefficient D_B defined in Eq. (9.75) considering typical values for the body temperature ($T = 310$ K), the viscosity of plasma ($\eta = 1.3$ mPa·s), and the red blood cell radius ($R = 3$ μm, even though a disk would be a better approximation than a sphere for red blood cells).

(b) Compare the result you found in part (a) with typical values of $D_B \sim 10^{-7}$ cm^2/s measured by diffuse correlation spectroscopy in living tissue and discuss possible reasons that may account for the difference.

9.15* Find the values of the optical diffusion coefficient according to the definitions of Eq. (9.B.1) (D), Eq. (9.B.3) (D_0), and Eq. (9.B.4) (D_α) for a medium with refractive index $n = 1.33$, absorption coefficient $\mu_a = 0.061$ cm^{-1}, scattering coefficient $\mu_s = 150$ cm^{-1}, and anisotropy factor $g = 0.92$.

References

Arfken, G. B., and Weber, H. J. (2001). *Mathematical Methods for Physicists*, 5th edn. San Diego, CA: Academic Press.

Aronson, R. (1995). Boundary conditions for diffusion of light. *Journal of the Optical Society of America A*, 12, 2532–2539.

Aronson, R., and Corngold, N. (1999). Photon diffusion coefficient in an absorbing medium. *Journal of the Optical Society of America A*, 16, 1066–1071.

Boas, D. A., and Yodh, A. G. (1997). Spatially varying dynamical properties of turbid media probed with diffusing temporal light correlation. *Journal of the Optical Society of America A*, 14, 192–215.

Cai, W., Lax, M., and Alfano, R. R. (2000). Analytical solution of the elastic Boltzmann transport equation in an infinite uniform medium using cumulant expansion. *Journal of Physical Chemistry B*, 104, 3996–4000.

Cerussi, A. E., Maier, J. S., Fantini, S., et al. (1997). Experimental verification of a theory for the time-resolved fluorescence spectroscopy of thick tissues. *Applied Optics*, 36, 116–124.

Davison, B. (1958). *Neutron Transport Theory*. Oxford: Clarendon Press.

Dougherty, R. L., Ackerson, B. J., Reguigui, N. M., Dorri-Nowkoorani, F., and Nobbmann, U. (1994). Correlation transfer: development and application. *Journal of Quantitative Spectroscopy and Radiative Transfer*, 52, 713–727.

Faris, G. W. (2002). Diffusion equation boundary conditions for the interface between turbid media: a comment. *Journal of the Optical Society of America A*, 19, 519–520.

Furutsu, K., and Yamada, Y. (1994). Diffusion approximation for a dissipative random medium and the applications. *Physics Review E*, 50, 3634–3640.

Gelbard, E. M. (1968). Spherical harmonics methods: PL and double-PL approximations. In *Computing Methods in Reactor Physics*, ed. Greenspan, H., Kelber, C. N., and Okrent, D. New York: Gordon and Breach.

Groenhuis, R. A. J., Ferwerda, H. A., and Ten Bosch, J. J. (1983). Scattering and absorption of turbid materials determined from reflection measurements. 1: Theory. *Applied Optics*, 22, 2456–2462.

Haskell, R. C., Svaasand, L. O., Tsay, T. T., et al. (1994). Boundary conditions for the diffusion equation in radiative transfer. *Journal of the Optical Society of America A*, 11, 2727–2741.

Liemert, A., and Kienle, A. (2013). Exact and efficient solution of the radiative transport equation for the semi-infinite medium. *Scientific Reports*, 3, 2018 (7pp).

Machida, M., Panasyuk, G. Y., Schotland, J. C., and Markel, V. A. (2010). The Green's function for the radiative transport equation in the slab geometry. *Journal of Physics A: Mathematical and Theoretical*, 43, 065402 (17pp).

Paasschens, J. C. J. (1997). Solution of the time-dependent Boltzmann equation. *Physics Review E*, 56, 1135–1141.

Ripoll, J., Yessayan, D., Zacharakis, G., and Ntziachristos, V. (2005). Experimental determination of photon propagation in highly absorbing and scattering media. *Journal of the Optical Society of America A*, 22, 546–551.

Wyld, H. W. (1994). *Mathematical Methods for Physics*. Reading, MA: Addison-Wesley.

Further reading

Transport theory

Aronson, R. (1997). Radiative transfer implies a modified reciprocity relation. *Journal of the Optical Society of America A*, 14, 486–490.

Case, K. M., and Zweifel, P. F. (1967). *Linear Transport Theory*. Reading, MA: Addison-Wesley.

Chandrasekhar, S. (1960). *Radiative Transfer*. New York: Dover.

Duderstadt, J. J., and Hamilton, L. J. (1976). *Nuclear Reactor Analysis*. New York: Wiley.

Duderstadt, J. J., and Martin, W. R. (1979). *Transport Theory*. New York: Wiley.

Ishimaru, A. (1978). *Wave Propagation and Scattering in Random Media: Single Scattering and Transport Theory (Volume 1)*. New York, NY: Academic Press.

Martelli, F., Del Bianco, S., Ismaelli, A., and Zaccanti, G. (2010). *Light Propagation Through Biological Tissue and Other Diffusive Media: Theory, Solutions, and Software*. Bellingham, WA: SPIE Press.

Wyman, D. R., Patterson, M. S., and Wilson, B. C. (1989). Similarity relations for the interaction parameters in radiation transport. *Applied Optics*, 28, 5243–5249.

Spherical harmonics expansion of the Boltzmann transport equation

Arridge, S. R. (1999). Optical tomography in medical imaging. *Inverse Problems*, 15, R41–R93.

Faris, G. (2005). PN approximation for frequency-domain measurements in scattering media. *Applied Optics*, 44, 2058–2071.

Kaltenbach, J. M., and Kaschke, M. (1993). Frequency- and time-domain modelling of light transport in random media. In *Medical Optical Tomography: Functional Imaging and Monitoring*, ed. Müller, G. J. Bellingham, WA: SPIE Optical Engineering Press, pp. 65–86.

Klose, A. D., and Larsen, E. W. (2006). Light transport in biological tissue based on the simplified spherical harmonics equations. *Journal of Computational Physics*, 220, 441–470.

Liemert, A., and Kienle, A. (2014). Explicit solutions of the radiative transport equation in the P3 approximation. *Medical Physics*, 41, 111916.

Diffusion and P_1 approximations

Fishkin, J. B., Fantini, S., vandeVen, M. J., and Gratton, E. (1996). Gigahertz photon density waves in a turbid medium: theory and experiments. *Physics Review E*, 53, 2307–2319.

Martelli, F., Bassani, M., Alianelli, L., Zangheri, L., and Zaccanti, G. (2000). Accuracy of the diffusion equation to describe photon migration through an infinite medium: numerical and experimental investigation. *Physics in Medicine and Biology*, 45, 1359–1373.

You, J. S., Hayakawa, C. K., and Venugopalan, V. (2005). Frequency domain photon migration in the δ-P1 approximation: analysis of ballistic, transport, and diffuse regimes. *Physics Review E*, 72, 021903.

Microscopic Beer-Lambert law

Montcel, B., Chabrier, R., and Poulet, P. (2006). Time-resolved absorption and hemoglobin concentration difference maps: a method to retrieve depth-related information on cerebral hemodynamics. *Optics Express*, 14, 12271–12287.

Tsuchiya, Y. (2001). Photon path distribution and optical responses of turbid media: theoretical analysis based on the microscopic Beer-Lambert law. *Physics in Medicine and Biology*, 46, 2067–2084.

Fluorescence diffusion theory

Fantini, S., and Gratton, E. (2000). Fluorescence photon-density waves in optically diffusive media. *Optics Communications*, 173, 73–79.

Joshi, A., Bangerth, W., and Sevick-Muraca, E. M. (2004). Adaptive finite element based tomography for fluorescence optical imaging in tissue. *Biomedical Optics Express*, 12, 5402–5417.

Li, X. D., O'Leary, M. A., Boas, D. A., Chance, B., and Yodh, A. G. (1996). Fluorescent diffuse photon density waves in homogeneous and heterogeneous turbid media: analytic solutions and applications. *Applied Optics*, 35, 3746–3758.

Patterson, M. S., and Pogue, B. W. (1994). Mathematical model for time-resolved and frequency-domain fluorescence spectroscopy in biological tissues. *Applied Optics*, 33, 1963–1974.

Reynolds, J. S., Thompson, C. A., Webb, K. J., LaPlant, F. P., and Ben-Amotz, D. (1997). Frequency domain modeling of reradiation in highly scattering media. *Applied Optics*, 36, 2252–2259.

Wu, J., Feld, M. S., and Rava, R. P. (1993). Analytical model for extracting intrinsic fluorescence in turbid media. *Applied Optics*, 32, 3586–3595.

Diffuse correlation spectroscopy

Durduran, T., and Yodh, A. G. (2014). Diffuse correlation spectroscopy for non-invasive, micro-vascular cerebral blood flow measurement. *NeuroImage*, 85, 51–63.

Mesquita, R. C., Durduran, T., Yu, G., et al. (2011). Direct measurement of tissue blood flow and metabolism with diffuse optics. *Philosophical Transactions of the Royal Society A*, 369, 4390–4406.

Absorption dependence of the optical diffusion coefficient

Bassani, M., Martelli, F., Zaccanti, G., and Contini, D. (1997). Independence of the diffusion coefficient from absorption: experimental and numerical evidence. *Optics Letters*, 22, 853–855.

Durduran, T., Yodh, A. G., Chance, B., and Boas, D. A. (1997). Does the photon-diffusion coefficient depend on absorption? *Journal of the Optical Society of America A*, 14, 3358–3365.

Durian, D. J. (1998). The diffusion coefficient depends on absorption. *Optics Letters*, 23, 1502–1504.

Elaloufi, R., Carminati, R., and Greffet, J.-J. (2003). Definition of the diffusion coefficient in scattering and absorbing media. *Journal of the Optical Society of America A*, 20, 678–685.

10 Continuous-wave methods for tissue spectroscopy

In Chapter 9 (Section 9.5), we derived the diffusion equation for spatially uniform turbid media (Eq. (9.38)). If one is interested in measuring an average tissue property (say, the average hemoglobin concentration in muscle tissue, or the average hemoglobin saturation over a certain cerebral cortical volume), then Eq. (9.38) may provide a suitable analytical approach, within the limits of the assumption that the investigated tissue is spatially homogeneous. Of course, strictly speaking, this assumption is not correct, since biological tissues are not spatially homogeneous. At a microscopic level, it is the very presence of molecular species, intracellular organelles, and cellular structures that ultimately accounts for the absorption and scattering of light in tissues. However, these tissue inhomogeneities occur over spatial scales of microns (cellular structures), sub-microns (cellular organelles), and nanometers (structural proteins, organelle ultrastructure, biological macromolecules, etc.) that are not directly relevant to the macroscopic treatment of the absorption coefficient, diffusion coefficient, and optical energy density distribution that are considered by diffusion theory. By contrast, the presence of a macroscopic spatial heterogeneity in tissues, associated, say, with the skin and subcutaneous layers, bones, larger blood vessels, tendons, and multiple tissue types within the optically probed volume, may raise questions about the applicability of a model that assumes the spatial homogeneity of tissue.

In this chapter, we identify the length scale that is relevant in diffusion theory; in other words, the typical diffusion length that one should consider in comparison with the length scale over which the tissue optical properties vary. This chapter also derives expressions for the continuous-wave diffuse reflectance under a variety of conditions, and examines its dependence on the optical properties of the medium. *Diffuse optical spectroscopy* (principles described in Chapters 10–12 and experimental methods in Chapter 13) places an emphasis on the overall optical properties of the examined biological tissue, from which some relevant physiological or diagnostic parameters can be extracted. *Diffuse optical imaging* (described in Chapter 14), instead, places an emphasis on the spatial distribution of the optical properties of tissue at a macroscopic scale and is relevant in the detection of localized abnormalities (tumors, hemorrhagic or ischemic lesions, fluid-filled cysts, etc.) or focal events (targeted brain activation, localized hemodynamic responses, etc.). This chapter is devoted to continuous-wave methods for tissue spectroscopy, where the emission of the light source is constant; i.e., it is not time-varying. Time-resolved methods, which involve time-varying source emission and time-sensitive detection, are covered in Chapter 11 (for the time domain) and Chapter 12 (for the frequency domain).

10.1 The objective of tissue spectroscopy

Tissue spectroscopy (which is referred to as *diffuse optical spectroscopy*, or DOS, in the diffusive regime) can be used to determine the tissue optical properties, and their wavelength dependence and, in turn, to derive parameter values of biological, physiological, functional, metabolic, or diagnostic relevance. The information content that is associated with spectrally resolved measurements of the scattering properties has been presented in Chapter 7 (Section 7.2). Absorption measurements at multiple discrete wavelengths (λ_j, $j = 1, 2, \ldots, N_\lambda$) can be used to quantify the concentration of N_C chromophores in tissue (with $N_C \leq N_\lambda$), as long as the chromophores feature light absorption in the wavelength region considered. The basic idea is that the contribution to the tissue absorption coefficient (μ_a) from the i-th chromophore can be written as the product of the extinction coefficient, ε_i, (i.e., its absorption per unit concentration) and the concentration, C_i, of that chromophore in tissue (Beer's law). As a result, in the presence of N_C chromophores, the absorption coefficient μ_a at wavelength λ_j is given by:

$$\mu_a(\lambda_j) = \sum_{i=1}^{N_C} \varepsilon_i(\lambda_j) C_i, \tag{10.1}$$

as previously described in Chapter 2 (see Eq. (2.2)). If the extinction spectra $\varepsilon_i(\lambda_j)$ of all N_C chromophores are known and each is different from all others, the concentrations C_i can be determined by measuring μ_a at N_C or more different wavelengths, so that the system of N_λ equations of Eq. (10.1) is fully determined. In practice, accurate measurements of the chromophore concentration C_i require a careful selection of the N_λ wavelengths to guarantee the lack of correlation of the optical data at the various wavelengths and their maximal sensitivity to the chromophore concentrations. Relevant near-infrared chromophores in biological tissues include oxyhemoglobin, deoxyhemoglobin, cytochrome *c* oxidase, melanin, water, and lipids, which often carry physiologically relevant information.

This approach assumes that μ_a can be measured independently of μ_s', or that scattering contributions to the measured optical signals can be either ignored or accounted for. The approach of Eq. (10.1), which is based on the absolute determination of μ_a, realizes absolute measurements of chromophore concentrations, thus providing a baseline characterization of biological tissues. For example, if the concentrations C_i include oxy- and deoxyhemoglobin concentrations, one can measure the absolute blood volume fraction or the absolute oxygenation level of blood in tissue, and use these absolute measurements to compare different patient populations (young vs. old, healthy vs. diseased, etc.). Absolute measurements also lend themselves to monitoring applications over an arbitrary period of time (hours, days, months, or years) to study the effect of special diets, exercise regimens, therapeutic treatments, etc.

In cases where scattering contributions to the optical measurements cannot be canceled out or measured, but can be considered constant, changes in absorption coefficients ($\Delta\mu_a(\lambda_j)$) can still be measured and translated into changes of the chromophore concentrations $[\Delta C_i]$ relative to a reference concentration value, according to the following equation:

$$\Delta\mu_a(\lambda_j) = \sum_{i=1}^{N_C} \varepsilon_i(\lambda_j)\Delta C_i. \tag{10.2}$$

In this case, rather than *absolute* measurements, one achieves *relative* measurements that reflect changes recorded in the course of one measurement session. For example, one can measure changes in cerebral hemoglobin concentration associated with brain activation, changes in tissue oxygenation due to perturbations in the blood supply to the tissue, effects of manipulations in the amount of oxygen or carbon dioxide inspired by the subject, hemoglobin concentration changes in skeletal muscle induced by exercise, etc. It should be noted, nonetheless, that Eq. (10.2) becomes a measurement of absolute concentrations in the case where the concentration is initially known to be zero, such as for the case of measurement of an administered chromophoric drug, as discussed in Chapter 8 (Section 8.8.2).

10.2 CW tissue spectroscopy with diffusion theory

10.2.1 CW solution for an infinite, homogeneous medium

Continuous-wave (CW) optical studies employ a light source that emits a constant optical power, which is described by a time-independent source term $S_0(\mathbf{r})$ in the diffusion theory notation of Eq. (9.38). Under these CW conditions, the time-derivative term on the left-hand side of Eq. (9.38) vanishes. Furthermore, if we consider a point source at $\mathbf{r} = 0$, the source term can be written as $S_0(\mathbf{r}) = P_{CW}\,\delta(\mathbf{r})$ where P_{CW} is the source power and δ is the Dirac delta. (The Dirac delta is defined in Appendix 10.A.) Note that the units of the Dirac delta are the inverse of the units of its argument, so that the units of $S_0(\mathbf{r})$ are energy per unit time per unit volume.

In diffuse optics, the measured optical signal in an infinite medium is proportional to the fluence rate ϕ (Liu et al., 1993), which is the radiance integrated over the entire solid angle and represents the optical power traveling per unit cross-sectional area impinging from all directions on a point at a given position. We recall that the fluence rate is given by the product of the speed of light in the medium times the energy density ($\phi = c_n U$). The solution to the CW diffusion equation (i.e., Eq. (9.38) with no time-derivative term) for the fluence rate in the case of a point source at the origin in a homogeneous, infinite medium is given by:

$$\phi_{\mathrm{CW}}(r) = P_{\mathrm{CW}} \frac{3(\mu_s' + \mu_a)}{4\pi} \frac{e^{-r\sqrt{3\mu_a(\mu_s'+\mu_a)}}}{r}, \tag{10.3}$$

which quantifies the attenuation of the optical fluence rate, ϕ_{CW}, as a function of the distance, r, from the point source. In Eq. (10.3) we use the rigorous definition for the diffusion coefficient, $D = c_n/[3(\mu_s' + \mu_a)]$, which is well approximated by $D \approx c_n/(3\mu_s')$ in the diffusion regime as discussed in Chapter 9 (Section 9.5) (see also Appendix 9.B for a discussion of the dependence of D on μ_a). The total reduced attenuation coefficient (μ_t') is introduced in diffusion theory as the sum of the absorption and reduced scattering coefficients ($\mu_t' = \mu_s' + \mu_a$), whereas the effective attenuation coefficient (μ_{eff}) is defined as $\mu_{\mathrm{eff}} = \sqrt{3\mu_a(\mu_s' + \mu_a)}$. These definitions can be used to write the expression in Eq. (10.3) in a more compact form, with the exponential term simply given by $\exp(-r\mu_{\mathrm{eff}})$.

We recall that diffusion theory holds at distances from the source that are greater than the inverse of the reduced scattering coefficient. The approximately isotropic distribution of the optical energy density is a result of the randomization of light propagation directions at distances far enough from the light source – we recall that the source term S_0 in the diffusion equation is assumed to be spherically symmetric – and the assumptions of homogeneity of the medium and infinite boundary conditions. The infinite fluence rate at $r = 0$ is of no concern because Eq. (10.3) is only applicable where the diffusion regime of light propagation is established, which only happens at distances $>1/\mu_s'$ from light sources and tissue boundaries. Equation (10.3) provides a measure of the length scale over which the fluence rate is attenuated in the scattering medium. This typical attenuation length is called the *diffusion length* (L_D) and is given by:

$$L_D = \frac{1}{\sqrt{3\mu_a(\mu_s' + \mu_a)}} = \frac{1}{\mu_{\mathrm{eff}}} \approx \frac{1}{\sqrt{3\mu_a\mu_s'}}, \tag{10.4}$$

and can be expressed in words by saying that the optical energy density is attenuated by a factor of $\sim 1/e$ over a distance L_D. The $\sim$ sign is required because the presence of r in the denominator of Eq. (10.3) accounts for a variable attenuation factor that is, for example, $0.5/e$ at $r = L_D$ and $0.67/e$ at $r = 2L_D$. The point is that L_D provides a measure of the length scale over which the optical fluence rate attenuates in a highly scattering medium characterized by an absorption coefficient μ_a and a reduced scattering coefficient μ_s'. Such a length scale depends on both μ_a and μ_s', with the same functional dependence, given by the inverse square root on both optical coefficients, and it is intuitive that L_D decreases as μ_a and μ_s' increase.

It is important, however, to observe that the mechanisms of light attenuation resulting from absorption and scattering processes are fundamentally different: absorption annihilates the photon and transfers its energy to the medium, while scattering deflects the photon, either driving it away from the collection point or lengthening its path before it can reach the collection point, thereby increasing its absorption probability (this latter phenomenon is referred to as *scattering-induced absorption*). Remembering the condition that $\mu_a \ll \mu_s'$, the different effects of absorption and scattering phenomena on light propagation are reflected in the different dependence of ϕ_{CW} on μ_a (which has a significant effect only in the exponential term of Eq. (10.3)) and μ_s' (which affects both the exponential term and the multiplicative pre-exponential factor of Eq. (10.3)). As a result, ϕ_{CW} can only decrease as a function of absorption, while ϕ_{CW} may also increase as a function of scattering (at relatively low reduced scattering coefficients and/or short distances from the light source) before the exponential term dominates at larger μ_s' and/or r, in which case ϕ_{CW} decreases also as a function of μ_s'. (This effect stems from the same principles that result in the spatial dependence of fluence rate shown in Figure 2.12.) These results derived from diffusion theory can be qualitatively instructive even when they involve scenarios, such as short source-detector separations and low reduced scattering coefficients, that may fall outside the limits of applicability of diffusion theory.

For typical optical properties of blood-perfused biological tissue in the near-infrared spectral range, say $\mu_a \approx 0.1$ cm^{-1} and $\mu_s' \approx 10$ cm^{-1}, the diffusion length $L_D \approx 0.5$ cm results in an attenuation of the optical fluence rate ϕ_{CW} by a factor of about $0.1/\text{cm}$ (in fact, the attenuation factor over a distance of 1 cm is $\sim\exp(-1\ \text{cm}/L_D) = \sim 0.13$). While, of course, the diffusion length depends on the specific tissue under consideration and the wavelength of light, it is useful to keep in mind this order of magnitude of optical attenuation in tissue for near-infrared light: a factor of ~ 0.1 per centimeter. Ultimately, this attenuation factor determines the useful penetration depth of light into tissue, the sensitivity requirements for optical detection at given distances from the illumination point, and the dynamic range requirements for optical detection over a given range of distances from the illumination point.

It is interesting to consider the total optical power flowing out of a sphere of radius r centered at the photon source. The net optical power flowing along the radial coordinate (in the outward direction) per unit area orthogonal to the radial coordinate is given by the component of the net flux along r, which is $F_{r|CW}(r) = -(D/c_n)\nabla\phi_{CW}\cdot\mathbf{r}$ (see also Eq. (9.37)):

$$F_{r|CW}(r) = -\frac{1}{3(\mu_s' + \mu_a)}\frac{d\phi_{CW}(r)}{dr} = \frac{P_{CW}}{4\pi r^2}e^{-r\mu_{eff}}(1 + r\mu_{eff}), \quad (10.5)$$

where $\mu_{eff} = \sqrt{3\mu_a(\mu_s' + \mu_a)}$ is the effective attenuation coefficient. Intuitively, one can think of μ_{eff} , in the exponential term of Eq. (10.3) (which reminds us of the Beer-Lambert law), as accounting for the fact that the absorption is enhanced by the lengthening of the optical pathlength caused by scattering. Integration of $F_{r|CW}(r)$ over the entire sphere surface ($4\pi r^2$) – since there is no dependence on the angular coordinates, the integration over the sphere surface reduces to a product by $4\pi r^2$ – yields the total optical power flowing out of the sphere of radius r:

$$P_{Tot}(r) = P_{CW}e^{-r\mu_{eff}}(1 + r\mu_{eff}). \quad (10.6)$$

As expected, in the non-absorbing case (i.e., $\mu_{eff} = 0$) one finds $P_{Tot}(r) = P_{CW}$, which expresses the conservation of total radiant power in the absence of absorption. Figure 10.1(a) illustrates a sphere of radius r centered on the photon source, indicating the radial component of the net energy flux ($F_{r|CW}(r)$) and the total power ($P_{Tot}(r)$) flowing out of the sphere. Figure 10.1(b) reports the r dependence of $F_{r|CW}(r)$, $P_{Tot}(r)$, and $\phi_{CW}(r)$ for a photon source of power $P_{CW} =$ 1 mW and a medium with $\mu_a = 0.1$ cm^{-1} and $\mu_s' = 10$ cm^{-1}.

10.2.2 Determination of the tissue optical properties with CW diffusion theory

In principle, Eq. (10.3) appears to indicate that CW spectroscopy in the diffusion regime allows for the measurement of both μ_a and μ_s', thanks to their appearance in the exponential term (combined into the product $\mu_a(\mu_s' + \mu_a)$) and in the multiplicative pre-factor (which essentially depends only on μ_s' since $\mu_s' + \mu_a \approx \mu_s'$). In reality, an accurate measurement of the multiplicative pre-exponential factor is challenging, if not impossible, because, in addition to the difficulty of an absolute measurement of P_{CW}, the pre-exponential factor also contains unknown terms associated with the optical coupling between the light source and the medium, the detector sensitivity, the transmission properties of any optical fibers used, etc. These experimental factors are not explicitly indicated in Eq. (10.3) but must be taken into account in any practical measurement. One way to address this issue is

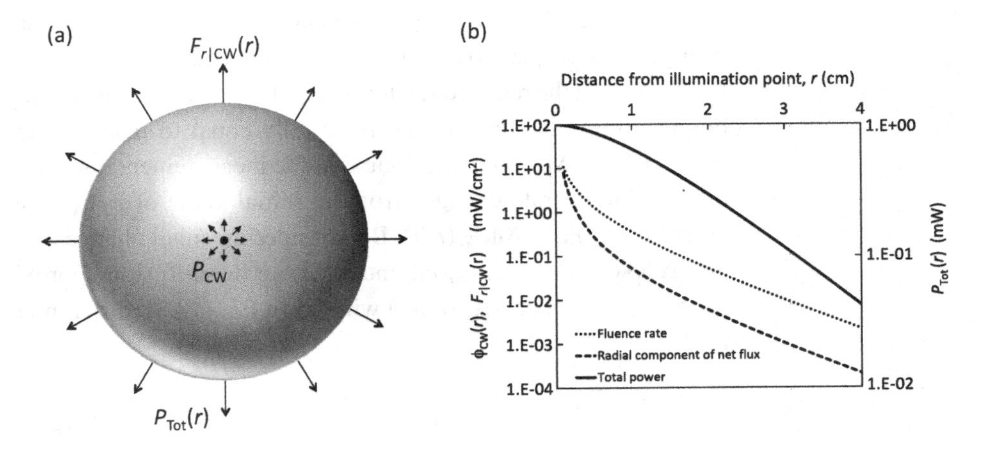

Figure 10.1

(a) A sphere of radius r centered at the isotropic photon source (emitting a power P_{CW}) in a scattering medium. The radial component of the net energy flux ($F_{r|CW}(r)$) at the surface of the sphere is indicated by the arrows. The total power ($P_{Tot}(r)$) flowing out of the sphere is given by the sphere surface area ($4\pi r^2$) multiplied by $F_{r|CW}(r)$. (b) r dependence of the radial component of the net flux ($F_{r|CW}(r)$), the total power flowing out of the sphere ($P_{Tot}(r)$), and the fluence rate ($\phi_{CW}(r)$) for a photon source power $P_{CW} = 1$ mW and a medium with $\mu_a = 0.1$ cm^{-1} and $\mu_s' = 10$ cm^{-1}.

to cancel out the pre-exponential factor by considering the spatial dependence of ϕ_{CW}, which can be robustly measured from the slope of the linear dependence of $\ln[r\,\phi_{CW}(r)]$ on r. This slope is equal to $-\sqrt{3\mu_a(\mu_s' + \mu_a)}$ so that, following this approach, CW measurements only afford a combined measurement of absorption and reduced scattering coefficients. Either an assumption about the value of μ_s' or an independent measurement of μ_s' is needed to achieve absolute μ_a measurements with CW spectroscopy in the diffusion regime.

Nonetheless, there are cases in which specific boundary conditions (which may introduce a nonlinear dependence of $\ln[r\,\phi_{CW}(r)]$ on r) or prior knowledge of the sources of absorption in the medium (which would allow for a decomposition of measured optical spectra into a set of extinction and scattering spectral contributions) may allow for absolute measurements of both μ_a and μ_s' by spatially resolved and/or spectrally resolved CW measurements in highly scattering media. It is also possible to use Eq. (10.3) to perform absolute absorption spectroscopy of tissue at a single distance from the illumination point, provided that the reduced scattering coefficient is known at all wavelengths and that the absorption coefficient is known at a given wavelength. In fact, the known values of μ_a and μ_s' at this given wavelength can be used to determine the pre-exponential factor (including unknown instrumental and optical coupling terms), so that μ_a is the only unknown in Eq. (10.3) at all other wavelengths considered.

In the case of relative measurements of μ_a, i.e., measurements of absorption changes with respect to an initial or baseline condition, it is sometimes reasonable to assume that the reduced scattering coefficient stays constant ($\mu_s' = \mu_{s0}'$), whereas the absorption coefficient that is initially equal to μ_{a0} assumes time-dependent values $\mu_{a0} + \Delta\mu_a(t)$ in the course of the measurement; and, correspondingly, the optical fluence rate changes from an initial value of $\phi_{CW0}(r)$ to time-dependent values $\phi_{CW0}(r) + \Delta\phi_{CW}(r, t)$. By considering small changes (i.e., $\Delta\mu_a(t) \ll \mu_{a0}$ and $\Delta\phi_{CW}(r, t) \ll \phi_{CW0}(r)$) and by using the diffusion approximation condition $\mu_s' \gg \mu_a$ to replace $(\mu_s' + \mu_a)$ with μ_s' in Eq. (10.3), differentiation of Eq. (10.3) leads to:

$$\Delta\phi_{CW}(r, t) \approx -P_{CW} \frac{[3\mu_{s0}']^{3/2}}{8\pi\sqrt{\mu_{a0}}} e^{-r\sqrt{3\mu_{a0}\mu_{s0}'}} \Delta\mu_a(t). \tag{10.7}$$

Then, taking the ratio between the energy density perturbation described by Eq. (10.7) and the initial energy density of Eq. (10.3) (with $\mu_s' + \mu_a \approx \mu_{s0}'$) yields:

$$\frac{\Delta\phi_{CW}}{\phi_{CW0}}(r, t) \approx -r\frac{\sqrt{3\mu_{s0}'}}{2\sqrt{\mu_{a0}}} \Delta\mu_a(t), \tag{10.8}$$

which expresses the relationship between the relative change in fluence rate and the change in the absorption coefficient in the diffusion regime. The minus sign in Eq. (10.8) tells us that an *increase* in absorption (positive $\Delta\mu_a$) corresponds to a *decrease* in the fluence rate (negative $\Delta\phi_{CW}$). From Eq. (10.8), it follows that the ratio between the percent changes in the fluence rate ($\Delta\phi_{CW}/\phi_{CW0}$) and in the absorption coefficient ($\Delta\mu_a/\mu_{a0}$) is given by minus one-half the ratio of the distance from the illumination point (r) to the diffusion length (L_D). It is intuitive that for a given change in the absorption coefficient, the fluence rate will experience larger changes (in absolute value) at larger distances from the source, because of the longer photon pathlengths. The proportionality between $\Delta\phi_{CW}/\phi_{CW0}$ and r, as indicated by Eq. (10.8), is a consequence of the assumption of spatial uniformity of the absorption perturbation $\Delta\mu_a$ in the derivation of Eq. (10.8).

10.3 The modified Beer-Lambert law

The basic absorption law that describes the exponential attenuation of the optical intensity (I) as light propagates through a non-scattering, dilute medium is referred to as the Beer-Lambert law, which was introduced in Section 2.5 (see Eq. (2.8)). The Beer-Lambert law holds for non-scattering media (so that there are no scattering contributions to the attenuation of the intensity) and dilute solutions (so that the

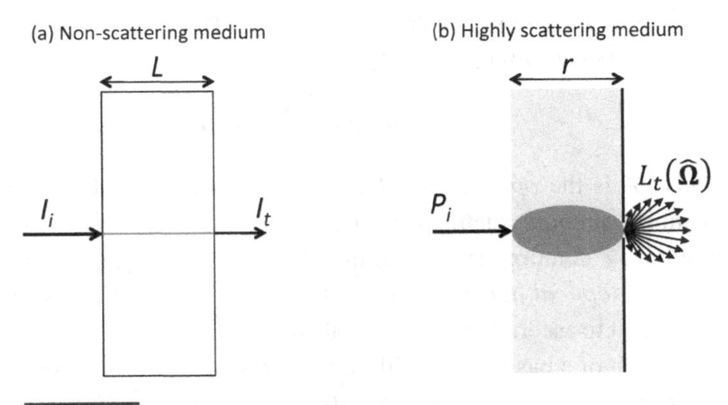

Figure 10.2

(a) The optical transmission through a non-scattering medium of thickness L is described in terms of the transmitted intensity I_t in relation to the incident intensity I_i. (b) The optical transmission measured at a given location through a distance r in a highly scattering medium is described in terms of the incident optical power P_i and the transmitted radiance $L_t(\hat{\Omega})$, which may be integrated over the solid angle either as a scalar (to yield the fluence rate $\phi(r)$) or as a vector (to yield the flux $\mathbf{F}(r)$). The dark-shaded area in (b) qualitatively indicates the volume sampled by the transmitted diffuse radiance, and the black line on the right side of the sample indicates a light block with an opening through which the transmitted radiance is measured.

absorption contributions from the individual absorbing units [molecules, particles, etc.] are independent of each other).

The Beer-Lambert law follows from the fact that the differential change in intensity (dI, with units of W/m^2) is proportional to the absorption coefficient (μ_a), the intensity itself, and the differential optical pathlength (dx) in the medium. This can be written as:

$$dI = -\mu_a I dx, \tag{10.9}$$

where the minus sign indicates the decrease in the optical intensity as the light propagates through the medium. If we consider the propagation of light through a total medium length L, by integrating Eq. (10.9) between 0 and L in dx, and, correspondingly, between the incident intensity I_i and the transmitted intensity I_t, one obtains Eq. (2.8), which we restate here:

$$I_t = I_i e^{-\mu_a L}. \tag{10.10}$$

This non-scattering case, in which the incident and transmitted intensities are the relevant optical quantities, is illustrated in Figure 10.2(a). Equation (10.10) can be written in the form of the Beer-Lambert law by considering Eq. (2.1) ($\mu_a = \varepsilon C$), where ε and C are the molar extinction coefficient and the concentration of the chromophore, respectively (as given by Eq. (10.1) for the case of a single chromophore), and by rearranging it as follows:

$$A = \ln\left(\frac{I_i}{I_t}\right) = \mu_a L = \varepsilon CL, \tag{10.11}$$

where A is the *optical absorbance*. The reader should note that the absorbance is most commonly defined in terms of the $\log_{10}$, but in this book we consistently use base-e notation. (When defined in terms of $\log_{10}$, absorbance is also referred to as the *optical density*, or OD.) Given the potential for confusion, it is always important to ascertain whether an absorbance value is reported in terms of a natural logarithm or a base-10 logarithm. (This distinction was addressed for the ordinate scale of Figure 2.4.) From Eq. (10.11) it follows immediately that, according to the Beer-Lambert law, an infinitesimal change in absorption ($d\mu_a$) is given by:

$$d\mu_a = \frac{dA}{L} = -\frac{1}{L}\frac{dI_t}{I_t}. \tag{10.12}$$

The CW solution to the diffusion equation in an infinite geometry (Eq. (10.3)) yields a formally similar result, under the assumption that the reduced scattering coefficient is constant and much greater than the absorption coefficient. In fact, by considering the variation of ϕ_{CW} associated with a variation of μ_a, differentiation of Eq. (10.3) (with the replacement of μ_s' for $\mu_s' + \mu_a$) yields the following result under the diffusion approximation conditions, as derived in Eq. (10.8):

$$d\mu_a = -\frac{2\sqrt{\mu_a}}{r\sqrt{3\mu_s'}}\frac{d\phi_{CW}}{\phi_{CW}}. \tag{10.13}$$

A comparison between Eq. (10.12) and Eq. (10.13) shows that the Beer-Lambert relationship of Eq. (10.12) can be extended to the highly scattering case described by diffusion theory, provided that a new parameter, the *differential pathlength factor* (DPF) (Delpy et al., 1988), is introduced to scale the direct distance r to the mean optical pathlength $\langle L \rangle$. By introducing the DPF, Eq. (10.13) becomes the *modified Beer-Lambert law*:

$$d\mu_a = -\frac{1}{r\,\mathrm{DPF}}\frac{d\phi_{CW}}{\phi_{CW}} = -\frac{1}{\langle L \rangle}\frac{d\phi_{CW}}{\phi_{CW}}, \tag{10.14}$$

where:

$$\mathrm{DPF} = \frac{\sqrt{3\mu_s'}}{2\sqrt{\mu_a}}. \tag{10.15}$$

The differential pathlength factor is a dimensionless quantity that represents the ratio between the mean optical pathlength and the direct distance r between a point light source and the observation point. It depends on the optical properties, the sample geometry, and the boundary conditions, and, in general, it is defined as

DPF $= \frac{1}{r}\frac{\partial A}{\partial \mu_a}$. The expression in Eq. (10.15) holds only for the diffusion approximation.

The case of a slab geometry, which was considered in the derivation of the Beer-Lambert law for non-scattering media, is depicted in Figure 10.2(b), which illustrates that the incident power and the transmitted radiance are the relevant optical quantities in the case of highly scattering media. In the case of a homogeneous infinite medium, Eq. (10.15) expresses the DPF in terms of the medium optical properties. Since the DPF represents the ratio between the actual length traveled by photons (on average) and the geometrical distance between the illumination and detection points, the DPF is always greater than 1.

In an infinite geometry, and for the conditions of the diffusion approximation, the DPF always increases with μ_s' and decreases with μ_a. For typical optical properties of tissue, say $\mu_a = 0.1$ cm^{-1} and $\mu_s' = 10$ cm^{-1}, Eq. (10.15) yields an infinite medium DPF of 8.7, which states that a photon, on average, travels 8.7 times the geometrical distance between the illumination and the detection points. The DPF may vary significantly for different wavelengths, biological tissues and boundary conditions; typical values are in the broad range 4–9 for diffuse reflectance measurements in living tissues (Duncan et al., 1995; Zhao et al., 2002). An analytical expression for the DPF in the case of the single-distance diffuse reflectance from a semi-infinite medium can be readily derived from Eq. (11.23), and the associated discussion in Chapter 11 comments on the intriguing result that the DPF is smaller in a semi-infinite geometry than in an infinite geometry.

The modified Beer-Lambert law of Eq. (10.14) is a generalized form of the Beer-Lambert law expression of Eq. (10.12). It can be used to translate measured changes in diffuse optical signals into changes in the tissue absorption coefficient, which can in turn be converted into concentration changes of tissue chromophores according to Eq. (10.2).

10.4 Continuous-wave diffuse reflectance

10.4.1 Total diffuse reflectance vs. single-distance diffuse reflectance

Diffuse optical measurements on human subjects are most often performed in a reflectance geometry, with the illumination and optical collection occurring on the same side of the surface of an investigated tissue. The most common exceptions are measurements on the human breast, fingers, ear lobe, and the neonate's head, where a measurable optical signal can be detected in transmission. For reflectance geometry, a first approximation for diffuse reflectance is to treat the investigated tissue as a semi-infinite medium, with the illumination and optical detection occurring at the interface between the tissue and the outside medium, which is typically air. One can consider several different scenarios.

In a first scenario, the sample is illuminated at a specific location and the diffuse reflectance is collected at a certain distance along the surface from the illumination point. If ρ denotes the distance, measured on the sample surface, between the illumination and the collection points, the quantity of interest is the diffuse reflectance as a function of distance, $R(\rho)$.

In a second scenario, the sample is also illuminated at a specific location, but the total diffuse reflectance ($\dot{R}_d$) is measured over the entire sample surface (here, the dot over the symbol for the total diffuse reflectance R_d indicates point illumination). By definition, $\dot{R}_d = \int_0^\infty R(\rho) 2\pi\rho \, d\rho$.

A third scenario involves broad-area illumination of the entire sample surface with a plane wave incident normally onto the sample surface, and the collection of the total diffuse reflectance $\bar{R}_d$. (Here, the bar over the symbol for the total diffuse reflectance R_d indicates broad, plane-wave illumination.) These three scenarios are illustrated in Figure 10.3.

A fourth scenario, for which we use the symbol $\ddot{R}_d$, involves the collection of total diffuse reflectance under conditions of diffuse, broad-beam illumination, i.e., for incoming optical radiation at all possible angles of incidence (the multiple dots over R_d in this case indicate the broad spatial distribution of isotropic illumination points).

A fifth scenario involves spatially modulated illumination, and we use the symbol $\tilde{R}_d$ to indicate the total diffuse reflectance resulting from such structured illumination. (The wave over R_d indicates here the spatially modulated conditions for the illumination beam.)

In the case of diffuse reflectance measured at the surface of the scattering medium, the measured optical signal is described by the normal component of the net flux, $\hat{\mathbf{n}} \cdot \mathbf{F}(\rho)$, (with $\hat{\mathbf{n}}$, the outward-pointing normal unit vector), which has units of W/m^2 like the fluence rate, but describes a net energy flow in the outward direction from the scattering medium. In the case of point illumination, the relevant optical input is the illumination power (P, energy per unit time); in fact, what matters is the amount of energy delivered to the illumination spot per unit time, as opposed to the intensity (energy per unit time per unit area). This applies to the single-distance diffuse reflectance, $R(\rho)$, and to the total diffuse reflectance under point illumination conditions, $\dot{R}_d$. The two measures, however, differ in the relevant quantity to describe the collected diffuse reflectance. In the case of $R(\rho)$, the detected optical signal is the normal component of the net flux, so that the single-distance diffuse reflectance, $R(\rho)$, is defined as the ratio $\hat{\mathbf{n}} \cdot \mathbf{F}(\rho)/P$

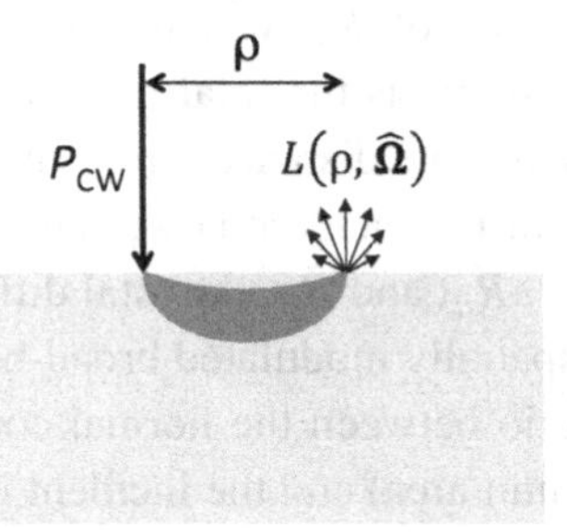

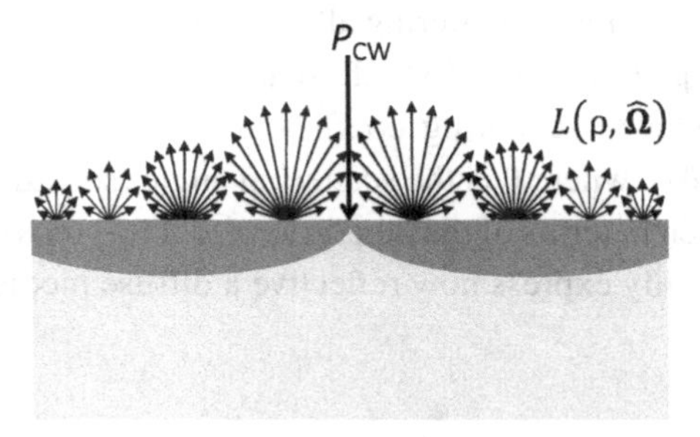

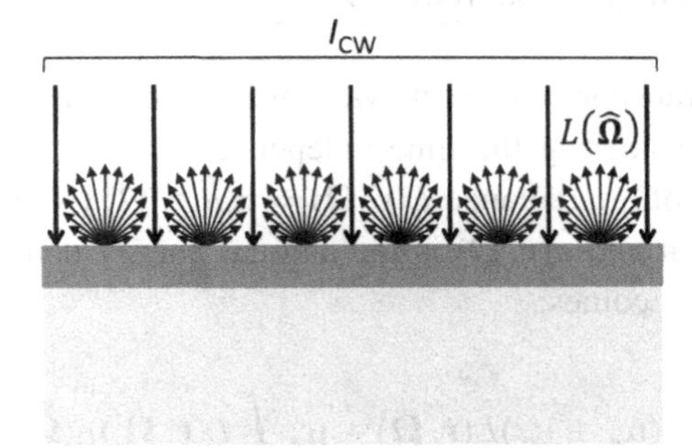

Figure 10.3

Schematic representation of the configurations for (a) single-distance diffuse reflectance ($R(\rho)$) under point illumination; (b) total diffuse reflectance ($\dot{R}_d$) under point illumination; (c) total diffuse reflectance ($\bar{R}_d$) under broad-beam collimated illumination. The distance from the illumination point, as measured on the sample surface, is indicated with ρ. The diffusely reflected radiance L (units of W-m^{-2}-sr^{-1}) depends on both ρ and $\hat{\Omega}$ in panels (a) and (b), and only on $\hat{\Omega}$ in panel (c). The relevant quantity to describe the optical input is the power (P_{CW}) in the case of point illumination (panels (a) and (b)), and the intensity (I_{CW}) in the case of broad-beam collimated illumination (panel (c)). As discussed in the text, the units of $R(\rho)$ are m^{-2} whereas $\dot{R}_d$ and $\bar{R}_d$ are dimensionless. The dark-shaded areas qualitatively indicate the sample regions probed by the diffuse reflectance in the various conditions.

and has units of m^{-2}. Therefore, $R(\rho)$ represents the detected intensity per unit input power. In the case of $\dot{R}_d$, which is given by the surface integral of $R(\rho)$, the detected optical signal is the total reflected power, which is a finite quantity since $R(\rho)$ vanishes at large distances from the illumination point. Therefore, $\dot{R}_d$ is defined as a ratio of the incident power and the total reflected power, so that it is dimensionless. $\bar{R}_d$, $\ddot{R}_d$, and $\tilde{R}_d$, the total diffuse reflectance values under plane wave, diffuse, and spatially modulated broad-beam illumination, respectively, are all defined as the ratio between the normal component of the diffusely reflected net flux (power per unit area) and the incident intensity or fluence rate (power per unit area). Consequently, $\bar{R}_d$, $\ddot{R}_d$, and $\tilde{R}_d$ are all dimensionless.

An important parameter used in conjunction with the diffuse reflectance of scattering media is the single-scattering albedo (a), which was defined in Chapter 1 as $a = \mu_s/(\mu_s + \mu_a)$; a reduced single-scattering albedo (a') was also defined in terms of the reduced scattering coefficient as $a' = \mu_s'/(\mu_s' + \mu_a)$. Because $\mu_s' = \mu_s(1 - \langle\cos\theta\rangle)$, where θ is the scattering angle (see Eq. (1.30)), the reduced albedo can be written in terms of the albedo as $a' = (1 - \langle\cos\theta\rangle)a/(1 - a\langle\cos\theta\rangle)$. These terms essentially express how reflective a diffuse medium is, in terms of its optical parameters.

10.4.2 Diffuse reflectance with transport theory

The Boltzmann transport equation was introduced and derived in Chapter 9 (Eq. (9.2)). By considering the time-independent case of interest here, and by writing it in terms of the radiance $L(\mathbf{r}, \hat{\Omega}) = c_n u(\mathbf{r}, \hat{\Omega})$ (where c_n is the speed of light in the medium and $u(\mathbf{r}, \hat{\Omega})$ is the angular energy density), the Boltzmann transport equation becomes:

$$\hat{\Omega} \cdot \nabla L(\mathbf{r}, \hat{\Omega}) = -(\mu_a + \mu_s)L(\mathbf{r}, \hat{\Omega}) + \mu_s \int_{4\pi} L(\mathbf{r}, \hat{\Omega}')p(\hat{\Omega}', \hat{\Omega})d\Omega' + q(\mathbf{r}, \hat{\Omega}), \tag{10.16}$$

where $p(\hat{\Omega}', \hat{\Omega})$ is the familiar scattering phase function (i.e., the probability per unit solid angle of scattering from direction $\hat{\Omega}'$ into direction $\hat{\Omega}$), and $q(\mathbf{r}, \hat{\Omega})$ is the source term describing the spatial and angular distribution of the source power per unit volume per unit solid angle. The transport equation (Eq. (10.16)) provides a general treatment of light propagation in scattering and absorbing media when fluorescence, polarization effects, interference and diffraction phenomena can all be ignored. Equation (10.16), with the proper boundary conditions, can be solved numerically to quantitatively describe the total diffuse reflectance from a semi-infinite scattering medium under broad-beam illumination conditions and with a phase function $p(\hat{\Omega}', \hat{\Omega}) = (1 + x\, \hat{\Omega}' \cdot \hat{\Omega})/(4\pi)$ with x ranging from 0 (isotropic scattering) to 1 (Chandrasekar, 1960; Giovanelli, 1955). This phase function is referred to as the Eddington phase function, after the English astrophysicist Arthur Eddington [1882–1944], and the parameter x is equal to three times the average cosine of the scattering angle ($x = 3g$).

Transport theory does not yield analytical solutions for the total diffuse reflectance $\bar{R}_d$, which is typically tabulated or solved numerically, so the practical utility of the theory is limited, but its numerical solution generates important results that have general validity in terms of the dependence of $\bar{R}_d$ on the medium optical properties. One result is that *the total diffuse reflectance only depends on the absorption and scattering coefficients through the single-scattering albedo* $a = \mu_s/(\mu_a + \mu_s)$. By dividing numerator and denominator by μ_s, the albedo can be written as $a = 1/(1 + \mu_a/\mu_s)$, thus showing that it is only the ratio μ_a/μ_s that matters for the total diffuse reflectance $\bar{R}_d$. This general result establishes a scaling relationship that states that multiplying the absorption and scattering coefficients by the same factor has no effect on $\bar{R}_d$. Another important result is that *the index of refraction mismatch at the interface between a scattering medium and a non-scattering medium does affect* $\bar{R}_d$, because of the role of the index in determining the boundary conditions.

A method to solve the time-independent Boltzmann transport equation (Eq. (10.16)) to yield the diffuse reflection and transmission for the one-dimensional case of a plane-parallel medium and uniform plane illumination is the *adding-doubling method* (Prahl, 1995). Despite the limiting assumption of a one-dimensional case, which means that the medium consists of a combination of plane-parallel layers of uniform optical properties, this method does have the advantages of (1) being valid for all ratios of scattering to absorption coefficients, (2) accommodating arbitrary scattering phase functions, and (3) accounting for internal reflections at boundaries. The basic approach of the adding-doubling method is to compute the diffuse reflection and transmission of a combination of two layers with different ("adding") or identical ("doubling") optical properties from the known reflection and transmission of the two individual layers. By repeating these adding or doubling steps as many times as needed, it is possible to find the diffuse reflectance and transmittance for an arbitrary combination of layers.

The mathematical scheme of the adding-doubling method is that of matrix operator theory, where the diffuse radiance (units of power per unit area, per unit solid angle) is expressed as an $M \times 1$ column vector (**L**), whose elements are the radiance values along a finite number M of directions, and the reflection and transmission operators are $M \times M$ matrices **R** and **T** that transform the incident radiance into diffusely reflected and transmitted radiances, respectively (Plass et al., 1973). The starting point of the method is the determination of the reflection and transmission operators for a thin slab (by solving the time-independent, one-dimensional Boltzmann transport equation) which are then used to find **R** and **T** for a layer of arbitrary thickness by proper adding-doubling combinations. This method to solve the transport equation does not lead to an analytical solution, so results have been tabulated for a set of values of the single-scattering albedo (a), slab thickness, isotropic and anisotropic (Henyey-Greenstein) phase function, and either matched or mismatched boundary conditions (Prahl, 1995).

Because of the complexity of the transport equation, a number of approximations or simplifications have been considered. One of them is the so-called *three-flux theory*, which is based on approximating the phase function in Eq. (10.16) with the sum of an isotropic term (describing diffuse radiation) and a δ peak in the forward direction (describing forward scattered radiation) (Burger et al., 1997). This approach results in a set of differential equations for three intensities that represent three optical fluxes: the incident illumination (which may be either collimated or diffuse), the diffuse flux propagating forward (transmitted flux), and the diffuse flux propagating backward (reflected flux). The three-flux theory yields the following relationship between the total diffuse reflectance ($\bar{R}_d$) and the absorption-to-scattering ratio ($\mu_a/\mu_s = (1-a)/a$) for the case of collimated, directional illumination (Burger et al., 1997):

$$\frac{\mu_a}{\mu_s} = \frac{6}{5(\bar{R}_d + 4)} \frac{(1 - \bar{R}_d)^2}{2\bar{R}_d}. \tag{10.17}$$

In the case of diffuse illumination, the result of three-flux theory is (Burger et al., 1997):

$$\frac{\mu_a}{\mu_s} = \frac{3}{8} \frac{(1 - \ddot{R}_d)^2}{2\ddot{R}_d}. \tag{10.18}$$

Other simplified approaches include the two-flux Kubelka-Munk theory and diffusion theory, which are described next.

10.4.3 Two-flux Kubelka-Munk theory

The Czech chemists Paul Kubelka [1900–1954] and Franz Munk [1900–1964] published a seminal work in 1931, in which they described a method, further refined in 1948 (Kubelka, 1948), that has been extensively used to assess the ratio of absorption to scattering of optically turbid samples from diffuse reflectance measurements. Under the assumption of isotropic scattering, matched boundaries (no specular reflection at the sample interface), and diffuse illumination, the basic idea of the Kubelka-Munk theory is to treat light propagation in an infinite slab in terms of two optical fluxes, one (indicated with i) propagating toward the unilluminated surface, and the other (indicated with j) propagating toward the illuminated surface. This is a one-dimensional problem in the coordinate z normal to the infinite slab, where i is assumed to propagate in the negative z direction and j in the positive z direction. Two specially defined coefficients are introduced to describe the absorption (K) and scattering (S) of these two fluxes, so that over an infinitesimal depth, dz, flux j (which propagates along $+z$) decreases by $Kjdz$ due to absorption and by $Sjdz$ due to scattering into the opposite direction ($-z$), whereas it increases by $Sidz$ due to scattering of i from $-z$ into $+z$. Similarly, over an infinitesimal depth, dz, flux i (which propagates along $-z$) increases by $Kidz$ due to absorption and by $Sidz$ due to scattering into the opposite direction ($+z$), whereas it decreases by $Sjdz$ due to scattering of j from $+z$ into $-z$ (for flux i, a decrease along its direction of propagation, i.e., along $-dz$, corresponds to an increase along $+dz$, and vice versa). This approach leads to the following system of differential equations for the two fluxes i and j:

$$\begin{cases} \dfrac{di}{dz} = (S + K)i - Sj \\ \dfrac{dj}{dz} = -(S + K)j + Si \end{cases}. \tag{10.19}$$

In the case of a semi-infinite scattering medium, the Kubelka-Munk theory yields the following relationship between the total diffuse reflectance $\ddot{R}_d$ and the ratio of the Kubelka-Munk absorption and scattering coefficients (K/S):

$$\frac{K}{S} = \frac{(1 - \ddot{R}_d)^2}{2\ddot{R}_d}. \tag{10.20}$$

The Kubelka-Munk absorption and scattering coefficients K and S do not coincide with the absorption and scattering coefficients μ_a and μ_s defined previously. A comparison of Eqs. (10.18) and (10.20) shows that these two sets of absorption and scattering coefficients are related as follows:

$$\frac{\mu_a}{\mu_s} = \frac{3}{8}\frac{K}{S}. \tag{10.21}$$

The restrictive assumptions of the Kubelka-Munk theory (isotropic scattering, perfectly diffuse illumination, matched boundary conditions) have limited its practical applicability in the optical study of biological tissues.

10.4.4 Diffusion theory for single-distance reflectance and total diffuse reflectance

In diffusion theory, the scattering coefficient μ_s is replaced by the reduced scattering coefficient μ_s', which does not depend on the specific angular dependence of the phase function but only on the mean value of the cosine of the scattering angle ($g = \langle \cos\theta \rangle$). Correspondingly, the single-scattering albedo a is replaced by the reduced single-scattering albedo a'. The diffuse reflectance for point illumination can be accurately calculated with diffusion theory at distances ρ from the illumination point that are greater than several times the reduced scattering length ($1/\mu_s'$). Finding $R(\rho)$ with diffusion theory requires one to solve the CW diffusion equation (Eq. (9.38) with no time-derivative term) with the proper boundary conditions.

A common approach for a semi-infinite medium is to use extrapolated boundary conditions with the method of image sources as described in Section 9.6.2 and illustrated in Figure 9.8. By using this method, the solution to the diffusion equation for a semi-infinite medium and a single point source is simply given by the superposition of two infinite medium solutions (given by Eq. (10.3)), one for the real (positive) light source and one for the image (negative) light source. In the case of point illumination with power P_{CW} at the medium surface, the real source is typically taken to have a power $a'P_{CW}$ and to be located at a distance $z_0 = 1/\mu_s'$ from the surface in the medium. (Recall that the source term in the diffusion equation, S_0, is isotropic, and $1/\mu_s'$ is a measure of the distance over which the photons lose memory of their original direction of propagation.) The image source is in a symmetrical position with respect to an extrapolated boundary at a distance $z_b = 2A/(3\mu_t') \approx 2A/(3\mu_s')$, where A is the reflection parameter (not to be confused with the absorbance, as defined by Eq. (10.11)) defined in Section 9.6.2 ($A = 1$ for a perfectly transmitting boundary and $A \to \infty$ for a perfectly reflecting boundary). For typical optical reduced scattering coefficients in tissue and for a tissue-air refractive index mismatch, one finds values $z_0 \sim$ 1 mm, and $z_b \sim 1.8$ mm. The configuration for the application of the extrapolated boundary condition is illustrated in Figure 10.4.

Given the cylindrical symmetry of the problem, it is convenient to consider a set of cylindrical coordinates (ρ, z) with the longitudinal axis at the illumination point and perpendicular to the semi-infinite medium boundary. By setting the origin of the coordinate system at the illumination point on the medium surface, the real source coordinates are $(0, -z_0)$ and the image source coordinates are $(0, 2z_b + z_0)$, so that a generic point (ρ, z) is at a distance $r_1 = \sqrt{(z + z_0)^2 + \rho^2}$ from the real source and $r_2 = \sqrt{(z - 2z_b - z_0)^2 + \rho^2}$ from the imaginary source. The fluence rate in the semi-infinite scattering medium is given by the sum of the fluence rates associated with the real and imaginary sources in an infinite medium (so

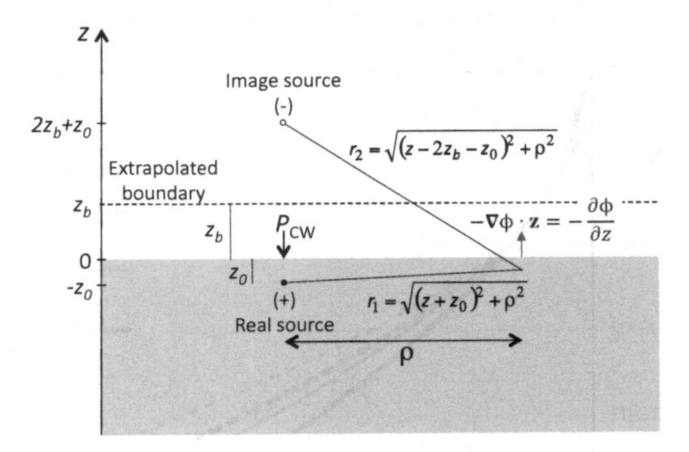

Figure 10.4

Illustration of the extrapolated boundary conditions to solve the diffusion equation for a semi-infinite medium. An isotropic photon source at a depth z_0 inside the medium is associated with the optical power P_{CW} that is delivered by a collimated laser beam at a specific surface location. An equal-strength but negative photon source is located symmetrically with respect to an extrapolated boundary that is separated by z_b from the actual sample surface. The fluence rate (ϕ) at a distance ρ (as measured on the sample surface) from the illumination point results from the superposition of the contributions from the real and image sources (which are at distance r_1 and r_2, respectively, from points at coordinates (ρ, z)) computed for an infinite medium. Finally, the measured signal at the surface is proportional to the normal component of the negative gradient of ϕ, i.e., to its negative derivative with respect to z.

that the solution of Eq. (10.3) applies, with the distances r_1 and r_2 for the two sources):

$$\phi_{CW}(\rho, z \leq 0) = a' P_{CW} \frac{3(\mu_s' + \mu_a)}{4\pi} \left(\frac{e^{-r_1 \mu_{eff}}}{r_1} - \frac{e^{-r_2 \mu_{eff}}}{r_2} \right). \tag{10.22}$$

The single-distance diffuse reflectance at the surface, $R(\rho)$, (see Section 10.4.1) is defined as the ratio $\hat{\mathbf{n}} \cdot \mathbf{F}_{CW}(\rho, z = 0)/P_{CW}$, where $\hat{\mathbf{n}}$ is the outward-pointing normal to the medium boundary, which coincides with $\hat{\mathbf{z}}$ in the case of Figure 10.4, and in the diffusion approximation $\mathbf{F}_{CW}(\rho, z = 0) = -\nabla\phi_{CW}(\rho, z)|_{z=0}/[3(\mu_s' + \mu_a)]$ (see Eq. (9.37)). Consequently, diffusion theory yields the following result for the single-distance, CW diffuse reflectance for a semi-infinite medium (Farrel et al., 1992):

$$\begin{aligned} R_{CW}(\rho) &= -\frac{1}{3(\mu_s' + \mu_a) P_{CW}} \left. \frac{\partial \phi_{CW}(\rho, z)}{\partial z} \right|_{z=0} \\ &= \frac{a'}{4\pi} \left[z_0 \left(\mu_{eff} + \frac{1}{r_1} \right) \frac{e^{-r_1 \mu_{eff}}}{r_1^2} + (2z_b + z_0) \left(\mu_{eff} + \frac{1}{r_2} \right) \frac{e^{-r_2 \mu_{eff}}}{r_2^2} \right]. \end{aligned} \tag{10.23}$$

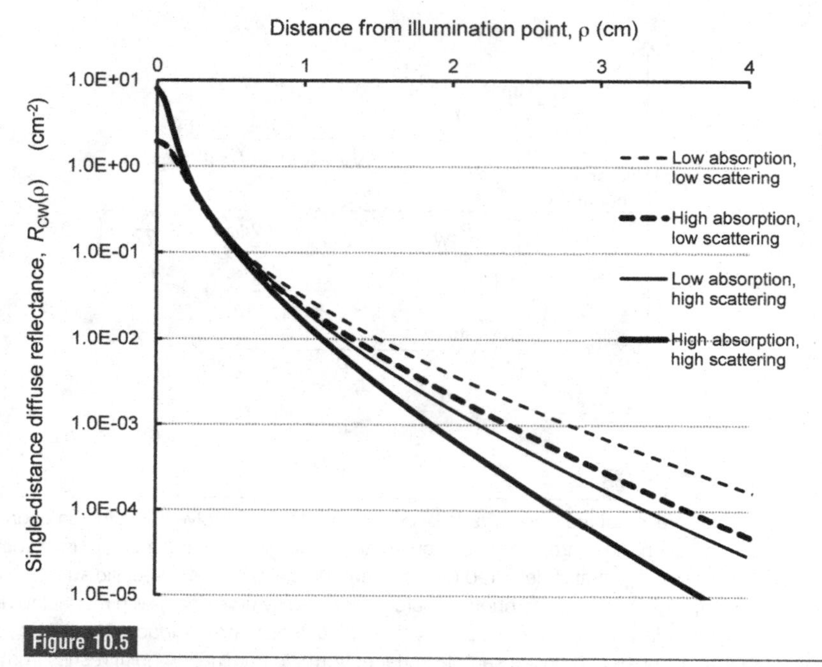

Figure 10.5

Single-distance diffuse reflectance, $R_{CW}(\rho)$, obtained with diffusion theory for four combinations of low absorption ($\mu_a = 0.05$ cm^{-1}), high absorption ($\mu_a = 0.10$ cm^{-1}), low scattering ($\mu_s' = 5$ cm^{-1}), and high scattering ($\mu_s' = 10$ cm^{-1}) of a semi-infinite medium. Mismatched boundary conditions ($n_{in} = 1.33$, $n_{out} = 1$) have been treated with the method of image sources and extrapolated boundary as described in the text. ρ indicates the distance from the illumination point, measured on the surface of the scattering medium. The units of cm^{-2} for $R(\rho)$ result from its definition as a ratio between the detected intensity (or radiant energy flux) and the incident optical power.

The single-distance diffuse reflectance expressed by Eq. (10.23) is plotted as a function of ρ in Figure 10.5 for four different combinations of μ_a (0.05 or 0.10 cm^{-1}) and μ_s' (5 or 10 cm^{-1}). The figure illustrates some of the features discussed above, namely the approximate attenuation of $R(\rho)$ by one order of magnitude per centimeter, and the fact that an increase in scattering determines an increase in the diffuse reflectance at short distances (ρ less than ~0.3 cm in Figure 10.5), whereas an increase in either absorption or scattering results in a decrease of the diffuse reflectance at large distances. The relative insensitivity of the single-distance diffuse reflectance to the reduced scattering coefficient at a distance $\rho \approx 0.3$ cm is in agreement with the discussion in Section 8.6.2. It is also worth noticing that diffusion theory predicts that the diffuse reflectance at short distances is largely insensitive to absorption (the two dashed lines, which correspond to different absorption values at low scattering, overlap at short ρ, and the same is true for the two continuous lines, which correspond to different absorption values at high scattering).

By integrating the single-distance diffuse reflectance of Eq. (10.23) over the entire sample surface, one obtains the total diffuse reflectance for point-illumination on the surface of a semi-infinite medium:

$$\dot{R}_{d|\mathrm{CW}} = \int_0^\infty R_{\mathrm{CW}}(\rho) 2\pi\rho \, d\rho = \frac{a'}{2}\left(e^{-z_0\mu_{\mathrm{eff}}} + e^{-(2z_b+z_0)\mu_{\mathrm{eff}}}\right), \qquad (10.24)$$

which can be written as follows by recalling that $a' = \mu_s'/(\mu_a + \mu_s')$, $z_0 = 1/\mu_s'$ and $z_b = 2A/[3(\mu_a + \mu_s')]$:

$$\dot{R}_{d|\mathrm{CW}} = \frac{a'}{2} e^{-\sqrt{3(1-a')}}\left(1 + e^{-\frac{4A}{3}\sqrt{3(1-a')}}\right). \qquad (10.25)$$

This shows, once more, that the total diffuse reflectance depends on the optical properties only through their ratio μ_a/μ_s' (since $a' = 1/(1 + \mu_a/\mu_s')$).

The description of the total diffuse reflectance under broad-beam illumination may fall outside the limits of applicability of diffusion theory. In fact, a significant portion of $\bar{R}_d$ results from photons that have traveled only within a superficial depth of the scattering medium, for which the conditions of diffusive light propagation may not apply. Nevertheless, the formalism of diffusion theory can be applied, and the solution of the diffusion equation for an infinitely broad-beam illumination impinging on the surface of a semi-infinite medium with mismatched boundary conditions (as given by Eq. (9.52)) leads to the following result for the total diffuse reflectance (Flock et al., 1989):

$$\bar{R}_{d|\mathrm{CW}} = \frac{a'}{1 + 2A(1-a') + \left(1 + \frac{2A}{3}\right)\sqrt{3(1-a')}}. \qquad (10.26)$$

It is interesting to note that the total diffuse reflectance for point illumination ($\dot{R}_{d|\mathrm{CW}}$) and for broad-beam illumination ($\bar{R}_{d|\mathrm{CW}}$) obtained with diffusion theory are coincident under the diffusion condition $\mu_a \ll \mu_s'$. In fact, by considering the lowest orders in a Taylor expansion of the right-hand side of Eq. (10.25) in terms of $\mu_a/\mu_s' \ll 1$ (in which case $a' \cong 1 - \mu_a/\mu_s'$), it can be easily shown that the term reduces to the right-hand side of Eq. (10.26), so that $\dot{R}_{d|\mathrm{CW}} = \bar{R}_{d|\mathrm{CW}}$. The Taylor series is named after the English mathematician Brook Taylor [1685–1731].

Figure 10.6 shows a comparison of the CW total diffuse reflectance obtained with diffusion theory, adding-doubling method, three-flux theory, and diffusion theory from a semi-infinite medium under broad-beam, plane-wave illumination, matched boundary conditions, and isotropic scattering (so that $\langle\cos\theta\rangle = 0$ and $a' = a$). Two cases of the diffusion theory solution are reported in Figure 10.6, one considering

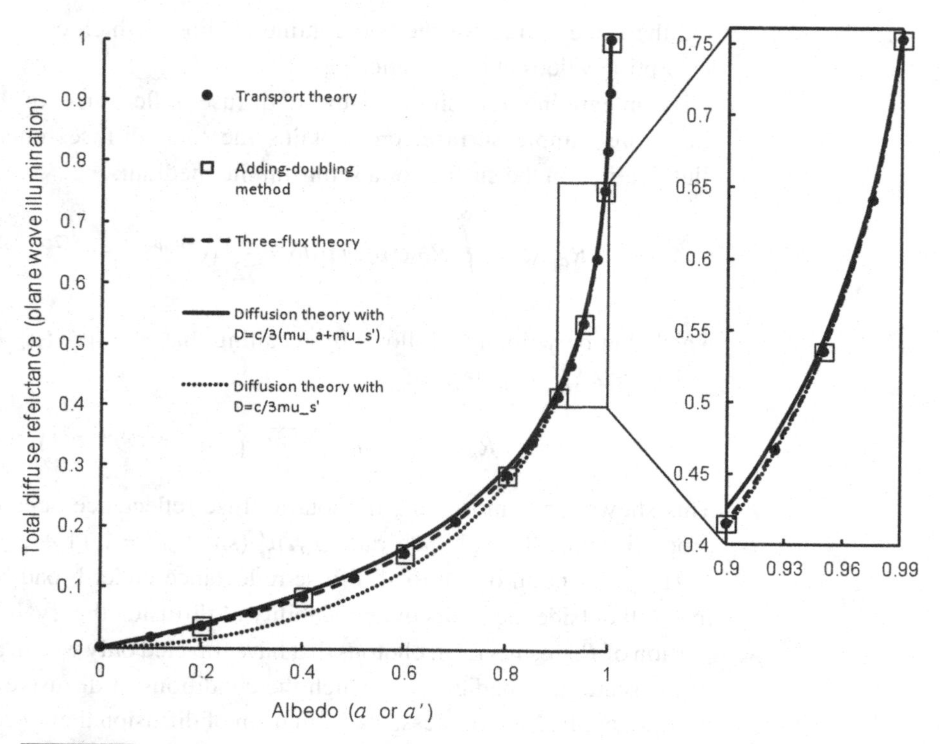

Figure 10.6

Continuous-wave, total diffuse reflectance under conditions of broad-beam, plane-wave illumination ($\bar{R}_{d|CW}$) for a semi-infinite medium featuring isotropic scattering (so that $\mu_s' = \mu_s$ and the reduced single-scattering albedo a' is equal to the single-scattering albedo a), as obtained from transport theory (Giovanelli, 1955), adding-doubling method (Prahl, 1995), three-flux theory (Burger et al., 1997), and diffusion theory (Flock et al., 1989) in the two cases $D = c_n/[3(\mu_a + \mu_s')]$ and $D = c_n/(3\mu_s')$. All cases refer to conditions of matched boundary conditions (i.e., $n_{in} = n_{out}$ and $A = 1$). The inset shows a magnified view of the high albedo limit, which corresponds to the diffusion regime, in which the absence of μ_a in the definition of D leads to a closer agreement between diffusion theory and transport theory. By contrast, at low albedo values, the inclusion of μ_a in the definition of D leads to a more accurate extrapolation of diffusion theory to high absorption or low scattering conditions.

the strict definition of the diffusion coefficient $D = c_n/[3(\mu_a + \mu_s')]$ and the other considering the diffusion approximation $D \cong c_n/(3\mu_s')$. Figure 10.6 shows that the presence of μ_a in the strict definition of D accounts for a better extrapolation of the diffuse reflectance expression obtained with diffusion theory beyond the diffusion conditions (i.e., for low values of the single-scattering albedo). The issue of the presence of the absorption coefficient in the definition of the diffusion coefficient is discussed in Appendix 9.B.

10.4.5 Spatially modulated spectroscopy

To increase the information content of CW data, it is possible to use spatially structured illumination (for example, modulated light intensity at a set of spatial frequencies) and to measure the spatial modulation of the CW diffuse reflectance $\tilde{R}_{d|\mathrm{CW}}$ as a function of spatial frequency k. Such a measurement, when spectrally resolved over a range of illumination wavelengths and calibrated with an optically characterized reference medium, achieves absolute measurements of absorption and reduced scattering coefficients. The method is referred to as *spatially modulated quantitative spectroscopy* (Saager et al., 2010) and can be extended into an imaging approach called *modulated imaging* or *spatial frequency-domain imaging* (SFDI) (Cuccia et al., 2005), whose basic ideas will be presented in Section 14.4. To a first approximation, this method can be treated with diffusion theory. We consider a semi-infinite geometry as in Figure 10.4 (with a Cartesian coordinate system, where the medium surface is on the xy plane), and the following source term that is sinusoidally modulated along the x and y axes, has an intensity amplitude of $a'I_{\mathrm{CW}}$, and is localized at a distance $z_0 = 1/\mu_s'$ into the medium. (The results do not change by considering a source term that is exponentially attenuating into the medium.)

$$S_0(\mathbf{r}) = a'I_{\mathrm{CW}}\delta\left(z - \frac{1}{\mu_s'}\right)\cos(k_x x + \alpha)\cos(k_y y + \beta), \tag{10.27}$$

where δ is the Dirac delta, k_x and k_y are the spatial modulation frequencies along x and y, and α and β are constant phase terms. By solving the CW diffusion equation with the spatially modulated source term of Eq. (10.27), and with mismatched boundary conditions (expressed by Eq. (9.48)), one finds the following result for the total diffuse reflectance under spatially modulated illumination conditions (Cuccia et al., 2009):

$$\tilde{R}_{d|\mathrm{CW}}(k_x, k_y) = \frac{a'}{1 + 2A(1-a')\left[1 + \left(\frac{k_x}{\mu_{\mathrm{eff}}}\right)^2 + \left(\frac{k_y}{\mu_{\mathrm{eff}}}\right)^2\right] + \left(1 + \frac{2A}{3}\right)\sqrt{3(1-a')}\sqrt{1 + \left(\frac{k_x}{\mu_{\mathrm{eff}}}\right)^2 + \left(\frac{k_y}{\mu_{\mathrm{eff}}}\right)^2}}. \tag{10.28}$$

This reduces to the expression for $\tilde{R}_{d|\mathrm{CW}}$ of Eq. (10.26) in the case of unmodulated broad-beam illumination: i.e., $k_x = k_y = 0$. Equation (10.28) shows how the dependence of $\tilde{R}_{d|\mathrm{CW}}(k_x, k_y)$ on k_x and k_y breaks the scaling relationship between μ_a and μ_s'. That is, the spatially modulated diffuse refelectance no longer depends only on the ratio μ_a/μ_s', because, unlike a', $\mu_{\mathrm{eff}} = \sqrt{3\mu_a(\mu_s' + \mu_a)}$ does not depend only on the ratio of the optical coefficients. Consequently, spatially modulated spectroscopy can achieve independent measurements of the absorption and reduced scattering coefficients.

10.4.6 Monte Carlo simulations: limits of validity of continuous-wave diffusion theory

Monte Carlo numerical simulations have been extensively used to derive the diffuse reflectance of turbid media because they offer a number of advantages over transport theory and its various approximations (see Section 8.1.1). Like transport theory, Monte Carlo methods provide a general description of light propagation in scattering and absorbing media. Furthermore, Monte Carlo simulations can model arbitrary scattering wave functions, irregular medium boundaries, macroscopic optical inhomogeneities in the medium, a variety of boundary conditions, and a range of illumination and collection geometries. The major drawback of Monte Carlo simulations is their computational inefficiency – a significant amount of computation time may be devoted to tracking photons that do not contribute to the signal of interest – especially under conditions of low absorption and large distances from the light sources. The fact that Monte Carlo simulations are applicable to situations that fall outside the limits of validity of diffusion theory (short distances from light sources, weakly scattering media, strong absorption, rapid spatial variations of the energy density, etc.) has led to a number of studies devoted to validating and extending diffusion theory predictions on the basis of Monte Carlo simulations. Some representative results are reported in Figure 10.7.

As expected, the single-distance diffuse reflectance ($R_{\mathrm{CW}}(\rho)$) is accurately described by diffusion theory only at distances from the source that are greater than several times $1/\mu_s'$. This result is shown in Figure 10.7(a), which reports the single-distance diffuse reflectance for a slab (thickness: 4 cm) with $\mu_a = 0.1\ \mathrm{cm}^{-1}$, $\mu_s' = 5\ \mathrm{cm}^{-1}$ and mismatched boundary conditions ($n_{\mathrm{in}} = 1.4$, $n_{\mathrm{out}} = 1.0$) (Martelli et al., 1997). The single-distance diffuse reflectance obtained with diffusion theory agrees well with Monte Carlo simulations at distances $\rho \geq 0.6$ cm. Considering that $\mu_s' = 5\ \mathrm{cm}^{-1}$, diffusion theory is accurate for $\rho \geq 3/\mu_s'$ in this case. The detailed derivation of the diffusion equation from the Boltzmann transport equation (see Chapter 9) has invoked the requirement of large distances ($\gg 1/\mu_s$) from the source only when the $l > 1$ orders in the spherical expansion of the transport equation have been dropped in the P_1 approximation (see Section 9.4). In terms of the reduced scattering coefficient ($\mu_s' = \mu_s(1 - \langle\cos\theta\rangle)$) this condition translates into a requirement of distances that are $\gg (1 - \langle\cos\theta\rangle)/\mu_s'$ for the validity of diffusion theory, which shows how this condition depends on the specifics of the scattering phase function through the average cosine of the scattering angle. Because light scattering in biological tissue is typically forward directed, with

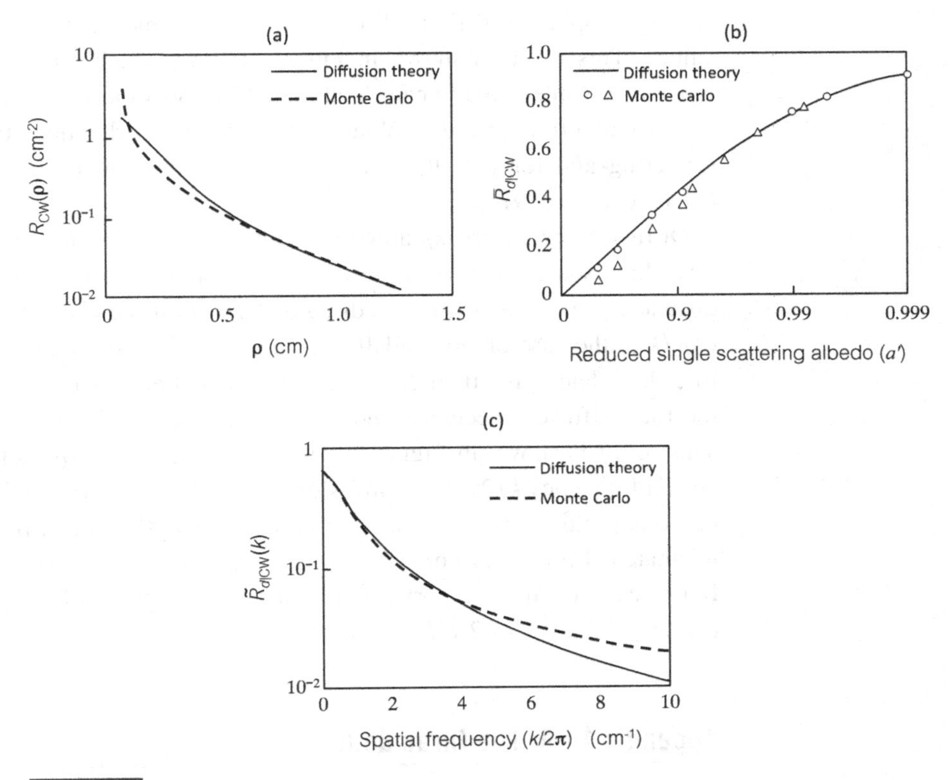

Figure 10.7

Comparison between diffusion theory results (solid lines) and Monte Carlo simulation (dashed lines or symbols) for (a) single-distance diffuse reflectance $R_{CW}(\rho)$ (figure generated from data reported in Martelli et al. (1997)), (b) total diffuse reflectance for broad-beam, collimated illumination $\bar{R}_{d|CW}$ (figure generated from data reported in Flock et al. (1989)), and (c) total diffuse reflectance for spatially modulated illumination $\tilde{R}_{d|CW}(k)$ (figure generated from data reported in Cuccia et al. (2009)). The optical properties and boundary conditions are specified in the text. The circles and triangles in panel (b) refer to isotropic and forward scattering ($\cos\theta = 0.875$–0.99), respectively.

typical values of $\langle \cos\theta \rangle \sim 0.8$–$0.9$ (Cheong et al., 1990), the condition $\rho \geq \alpha/\mu_s'$ can be adopted with values of α that may range from 1 to 10 depending on the details of the scattering phase function.

The total diffuse reflectance under broad-beam, collimated illumination conditions ($\bar{R}_{d|CW}$) includes significant contributions from light that has only visited the most superficial depth of the scattering medium, thereby not having necessarily achieved the conditions for diffusion theory. Nevertheless, Monte Carlo simulations have shown that the expression of $\bar{R}_{d|CW}$ obtained with diffusion theory (Eq. (10.26)) is accurate for all values of the reduced single-scattering albedo, as long as an isotropic scattering phase function is considered (so that $\mu_s' = \mu_s$). For anisotropic phase functions, diffusion theory becomes less accurate at low albedo values. This result is shown in Figure 10.7(b), which reports the total diffuse reflectance under broad-beam, collimated illumination as predicted by diffusion theory (Eq. (10.26)) and as obtained with Monte Carlo simulations for isotropic scattering and for $\langle \cos\theta \rangle = 0.875$–$0.99$, under matched boundary conditions (Flock et al., 1989).

Diffusion theory breaks down in the case of fast spatial variations of a structured light source. This is because the diffusion conditions are established only at distances greater than $1/\mu_s'$, as discussed above. If the characteristic length scale ($2\pi/k$ in the case of sinusoidally varying structured illumination) is comparable to or less than $1/\mu_s'$, then diffusion theory is not expected to accurately describe the total diffuse reflectance under spatially modulated illumination ($\tilde{R}_{d|\mathrm{CW}}(k)$). This result is shown in Figure 10.7(c), which reports $\tilde{R}_{d|\mathrm{CW}}(k)$ as a function of spatial frequency $k/2\pi$ from diffusion theory (Eq. (10.28)) and Monte Carlo simulations (with $\cos\theta = 0.71$), for a case of $\mu_a = 0.1$ cm^{-1}, $\mu_s' = 10$ cm^{-1} and mismatched boundary conditions ($n_{\mathrm{in}} = 1.33$, $n_{\mathrm{out}} = 1.0$) (Cuccia et al., 2009). In this case, diffusion theory provides accurate results for low spatial frequencies ($k/(2\pi) < 0.5\mu_s'$) or $2\pi/k > 2/\mu_s'$.

Appendix 10.A: The Dirac delta

Strictly speaking, the Dirac delta is not a function, and it can be more accurately described as an improper function, a virtual function, a generalized function, a distribution, or a measure. It is named after the English physicist Paul Adrien Maurice Dirac [1902–1984], who received the Nobel Prize for physics in 1933. The one-dimensional Dirac delta is defined by the following properties:

$$\begin{cases} \delta(x) = 0 & \text{for } x \neq 0 \\ \delta(x) = \infty & \text{for } x = 0 \end{cases}, \tag{10.A.1}$$

and:

$$\int_{-\infty}^{+\infty} \delta(x)dx = 1, \tag{10.A.2}$$

which yield the following important relationship:

$$\int_{-\infty}^{+\infty} f(x)\delta(x - x_0)dx = f(x_0), \tag{10.A.3}$$

where $f(x)$ is any continuous function and x_0 is a constant. The extension of Eq. (10.A.3) to three dimensions is as follows:

$$\iiint_{-\infty}^{+\infty} f(\mathbf{r})\delta(\mathbf{r}-\mathbf{r}_0)d\mathbf{r} = f(\mathbf{r}_0), \tag{10.A.4}$$

and the 3D extension of the normalization condition of Eq. (10.A.2) leads to the following expressions for the three-dimensional Dirac delta in Cartesian coordinates:

$$\delta(\mathbf{r}-\mathbf{r}_0) = \delta(x-x_0)\delta(y-y_0)\delta(z-z_0), \tag{10.A.5}$$

and in spherical coordinates:

$$\delta(\mathbf{r}-\mathbf{r}_0) = \frac{1}{r^2 \sin\theta}\delta(r-r_0)\delta(\theta-\theta_0)\delta(\varphi-\varphi_0). \tag{10.A.6}$$

An examination of Eqs. (10.A.3) and (10.A.4) shows that the Dirac delta is associated with physical dimensions that are the inverse of the dimensions of its argument. For example, if t is time in s, then the dimensions of $\delta(t)$ are inverse time in s^{-1}, and the dimensions of $\delta(\mathbf{r})$ are inverse volume (m^{-3}).

Problems – answers to problems with * are on p. 647

10.1* Consider a mixture of two chromophores A and B that have extinction coefficients $\varepsilon_A(\lambda_1) = 2.5 \times 10^4\ M^{-1}cm^{-1}$, $\varepsilon_A(\lambda_2) = 8.2 \times 10^4\ M^{-1}cm^{-1}$, $\varepsilon_B(\lambda_1) = 6.3 \times 10^5\ M^{-1}cm^{-1}$, $\varepsilon_B(\lambda_2) = 1.4 \times 10^5\ M^{-1}cm^{-1}$ at two wavelengths λ_1 and λ_2. The absorption coefficients of the mixture at the two wavelengths are $\mu_a(\lambda_1) = 1.8\ cm^{-1}$ and $\mu_a(\lambda_2) = 3.1\ cm^{-1}$.

(a) What are the concentrations of chromophores A and B in the mixture?

(b) What are the changes in the absorption coefficient of the mixture at the two wavelengths if the concentration of chromophore A is doubled and that of chromophore B is tripled?

10.2 Consider an isotropic point source with a power of 12 mW embedded in an infinite scattering medium. Find the total power flowing out of a sphere of radius $r_0 = 3$ cm centered at the photon source, for the following cases:

(a) $\mu_a = 0$, $\mu_s' = 5\ cm^{-1}$;

(b) $\mu_a = 0$, $\mu_s' = 10\ cm^{-1}$;

(c) $\mu_a = 0.1\ cm^{-1}$, $\mu_s' = 5\ cm^{-1}$;

(d) $\mu_a = 0.1\ cm^{-1}$, $\mu_s' = 10\ cm^{-1}$;

(e) What is the physical mechanism by which the total power changes between cases (c) and (d), considering that the absorption coefficient of the medium is the same in the two cases, and that all scattered light is collected since the power is integrated over the entire sphere surface?

10.3* What is the reduced scattering coefficient μ_s' of a highly scattering, infinite medium that has an absorption coefficient $\mu_a = 0.12\ \text{cm}^{-1}$ and in which $[r_1\phi_{\text{CW}}(r_1)]/[r_2\phi_{\text{CW}}(r_2)] = 18$ for two distances from a point source $r_1 = 2.5$ cm and $r_2 = 4.0$ cm?

10.4 A slab of thickness $L = 1.5$ cm contains a non-scattering solution of a 3.7 μM concentration of a chromophore that has an extinction coefficient $\varepsilon = 3.2 \times 10^5\ \text{M}^{-1}\text{cm}^{-1}$ (ignore the absorption of the solvent). If one side of the slab is illuminated with an intensity of 48 mW/cm^2, what is the transmitted intensity on the other side of the slab?

10.5* According to the Beer-Lambert law, what is the percent change in the transmitted intensity through a non-scattering medium of thickness 2.1 cm if its absorption coefficient changes from 0.0240 cm^{-1} to 0.0235 cm^{-1}?

10.6 According to the modified Beer-Lambert law, what is the percent change in the fluence rate at a distance of 2.1 cm from a point source in an infinite medium with $\mu_s' = 8.2\ \text{cm}^{-1}$ if the absorption coefficient changes from 0.0240 cm^{-1} to 0.0235 cm^{-1}? Discuss the physical reason for the difference between this result and the one found in Problem 10.5.

10.7 What is the average distance traveled by photons from a point source at the origin of a Cartesian coordinate system (0,0,0) to point (2 cm, 2 cm, 2 cm) in an infinite medium with $\mu_a = 0.11\ \text{cm}^{-1}$ and $\mu_s' = 6.4\ \text{cm}^{-1}$? How much longer is this distance with respect to the diffusion length in this medium?

10.8 From Eq. (10.17), find the expression of the three-flux total diffuse reflectance for broad-beam, collimated illumination as a function of the single-scattering albedo ($\bar{R}_d(a)$), and find its values at $a = 0$, 0.5, and 1.0.

10.9* From Eq. (10.18), find the expression of the three-flux total diffuse reflectance for diffuse illumination as a function of the single-scattering albedo $\ddot{R}_d(a)$, and find its values at $a = 0$, 0.5, and 1.0.

10.10 Verify Eq. (10.24) by calculating the integral $\int_0^\infty R_{\text{CW}}(\rho)2\pi\rho\, d\rho$ with $R_{\text{CW}}(\rho)$ given by Eq. (10.23).

10.11* For a scattering phase function with $\langle\cos(\theta)\rangle = 1/3$, find the total diffuse reflectance for point illumination predicted by diffusion theory ($\dot{R}_{d|\text{CW}}$ of Eq. (10.25)) for a single-scattering albedo $a = 0.5$ under matched boundary conditions.

10.12 Show that, considering $\mu_a/\mu_s' \ll 1$, the lowest orders of a Taylor expansion of $\dot{R}_{d|\text{CW}}$ in Eq. (10.25) lead to the expression of $\bar{R}_{d|\text{CW}}$ in Eq. (10.26).

References

Burger, T., Kuhn, J., Caps, R., and Fricke, J. (1997). Quantitative determination of the scattering and absorption coefficients from diffuse reflectance and transmittance measurements: Application to pharmaceutical powders. *Applied Spectroscopy*, 51, 309–317.

Chandrasekhar, S. (1960). *Radiative Transfer*. New York: Dover.

Cheong, W. F., Prahl, S. A., and Welch, A. J. (1990). A review of the optical properties of biological tissues. *IEEE Journal of Quantum Electronics*, 26, 2166–2185.

Cuccia, D. J., Bevilacqua, F., Durkin, A. J., and Tromberg, B. J. (2005). Modulated imaging: quantitative analysis and tomography of turbid media in the spatial-frequency domain. *Optics Letters*, 30, 1354–1356.

Cuccia, D. J., Bevilacqua, F., Durkin, A. J., Ayers, F. R., and Tromberg, B. J. (2009). Quantitation and mapping of tissue optical properties using modulated imaging. *Journal of Biomedical Optics*, 14, 024012.

Delpy, D. T., Cope, M., van der Zee, P., Arridge, S., Wray, S., and Wyatt, J. (1988). Estimation of optical pathlength through tissue from direct time of flight measurement. *Physics in Medicine and Biology*, 33, 1433–1442.

Duncan, A., Meek, J. H., Clemence, M., et al. (1995). Optical pathlength measurements on adult head, calf and forearm and the head of the newborn infant using phase resolved optical spectroscopy. *Physics in Medicine and Biology*, 40, 295–304.

Farrel, T. J., Patterson, M. S., and Wilson, B. (1992). A diffusion theory model of spatially resolved, steady-state diffuse reflectance for the noninvasive determination of tissue optical properties in vivo. *Medical Physics*, 19, 879–888.

Flock, S. T., Patterson, M. S., Wilson, B. C., and Wyman, D. R. (1989). Monte Carlo modeling of light propagation in highly scattering tissues – I: Model predictions and comparison with diffusion theory. *IEEE Transactions on Biomedical Engineering*, 36, 1162–1168.

Giovanelli, R. G. (1955). Reflection by semi-infinite diffusers. *Optica Acta*, 2, 153–162.

Kubelka, P. (1948). New contributions to the optics of intensely light-scattering materials. Part I. *Journal of the Optical Society of America*, 38, 448–457.

Liu, F., Yoo, K. M., and Alfano, R. R. (1993). Should the photon flux or the photon density be used to describe the temporal profiles of scattered ultrashort laser pulses in random media? *Optics Letters*, 18, 432–434.

Martelli, F., Contini, D., Taddeucci, A., and Zaccanti, G. (1997). Photon migration through a turbid slab described by a model based on diffusion approximation. II. Comparison with Monte Carlo results. *Applied Optics*, 36, 4600–4612.

Plass, G. N., Kattawar, G. W., and Catchings, F. E. (1973). Matrix operator theory of radiative transfer. 1: Rayleigh scattering. *Applied Optics*, 12, 314–329.

Prahl, S. A. (1995). The adding-doubling method. In *Optical-Thermal Response of Laser-Irradiated Tissue*, ed. A. J. Welch and M. J. C. van Gemert. New York: Plenum Press.

Saager, R. B., Cuccia, D. J., and Durkin, A. J. (2010). Determination of optical properties of turbid media spanning visible and near-infrared regimes via spatially modulated quantitative spectroscopy. *Journal of Biomedical Optics*, 15, 01712.

Zhao, H., Tanikawa, Y., Gao, F., et al. (2002). Maps of optical differential pathlength factor of human adult forehead, somatosensory motor and occipital regions at multi-wavelengths in NIR. *Physics in Medicine and Biology*, 47, 2075–2093.

Further reading

Differential pathlength factor

Kohl, M., Nolte, C., Heekeren, H. R., et al. (1998). Determination of the wavelength dependence of the differential pathlength factor from near-infrared pulse signals. *Physics in Medicine and Biology*, 43, 1771–1782.

Sassaroli, A., and Fantini, S. (2004). Comment on the modified Beer-Lambert law for scattering media. *Physics in Medicine and Biology*, 49, N255-N257.

Adding-doubling method

van de Hulst, H. C. (1980). *Multiple Light Scattering: Tables, Formulas, and Applications*, Vol. 1. New York: Academic Press.

Kattawar, G. W., Plass, G. N., and Catchings, F. E. (1973). Matrix operator theory of radiative transfer. 2: Scattering from maritime haze. *Applied Optics*, 12, 1071–1084.

Spatial frequency-domain reflectometry

Dögnitz, N., and Wagnières, G. (1998). Determination of tissue optical properties by steady-state spatial frequency-domain reflectometry. *Lasers in Medical Science*, 13, 55–65.

Diffuse reflectance with Monte Carlo simulations

Bevilacqua, F., and Depeursinge, C. (1999). MonteCarlo study of diffuse reflectance at source-detector separations close to one transport mean free path. *Journal of the Optical Society of America A*, 16, 2935–2945.

Wang, L. V., and Jacques, S. L. (2000). Source of error in calculation of optical diffuse reflectance from turbid media using diffusion theory. *Computer Methods and Programs in Biomedicine*, 61, 163–170.

Limits of validity of continuous-wave diffusion theory

Martelli, F., Bassani, M., Alianelli, L., Zangheri, L., and Zaccanti, G. (2000). Accuracy of the diffusion equation to describe photon migration through an infinite medium: numerical and experimental investigation. *Physics in Medicine and Biology*, 45, 1359–1373.

Vitkin, E., Turzhitsky, V., Qiu, L., et al. (2011). Photon diffusion near the point-of-entry in anisotropically scattering turbid media. *Nature Communications*, 2, Art. No. 587.

Yoo, K. M., Liu, F., and Alfano, R. R. (1990). When does the diffusion approximation fail to describe photon transport in random media? *Physical Review Letters*, 64, 2647–2649.

11 Time-domain methods for tissue spectroscopy in the diffusion regime

The information content of the optical signals collected with continuous-wave (CW) methods is restricted by the constant intensity. In non-scattering media, the optical properties are fully characterized by one parameter, the absorption coefficient, which can be determined by application of the Beer-Lambert law (Eqs. (2.8) and (10.11)). In scattering media, which are described by two optical coefficients (the absorption and scattering coefficients) and by the scattering phase function, the information content of CW methods (measured at a single location and a single wavelength) is insufficient for a full optical characterization of the material. Even in the diffusion-approximation regime, where the details of the scattering phase function are not relevant, since it is only the average cosine of the scattering angle ($g = \langle\cos\theta\rangle$) that matters, and where two parameters (the absorption and the reduced scattering coefficient) fully characterize the medium, a single measured quantity, the constant intensity, provides limited information content. Such information content may be enriched by measuring the continuous-wave intensity under different conditions (e.g., multiple angles in the non-diffusive regime, or multiple distances from a point source, or multiple wavelengths of light, or multiple spatial frequencies of structured illumination, etc.), and this may provide valuable additional information for a more complete optical characterization of the medium.

Time-resolved methods, however, in which the detected optical signal is not constant in time, provide additional information related to the distribution of the optical pathlengths in scattering media, which is not accessible to CW methods. Time-resolved methods are typically implemented either in the time domain, with a pulsed light source and time-resolved detection, or in the frequency domain, with an intensity-modulated light source and phase-sensitive detection. Theoretical and general concepts for time-domain methods are described in this chapter, whereas frequency-domain methods are described in Chapter 12. Experimental and instrumentation aspects of time-domain spectroscopy are presented in Chapter 13 (Section 13.4.1). The reader will note that the index of refraction (n) has not been mentioned above in the description of the optical quantities that characterize a scattering medium. The reason is that for light propagation in scattering media, the index of refraction only plays the role of scaling the speed of light in the medium, since $c_n = c/n$. Nonetheless, the index of refraction mismatch at boundaries does play a key role in determining boundary conditions, as described in Section 9.6. Also, while gradients in the refractive index are key to the microscopic description of scattering, as discussed in Chapter 7, such detail is not necessary for describing bulk scattering properties of the medium.

11.1 Diffusion theory: time-domain solution for an infinite, homogeneous medium

The time-domain method involves illumination by short light pulses, at a certain repetition rate, and time-sensitive optical detection. On the basis of the material presented in Chapter 10, we are able to quantify the words "short" and "certain" in the previous sentence. Scattering events, which happen much more frequently than absorption events in the diffusion-approximation regime of light propagation, determine a distribution of photon paths in a scattering medium. This is in contrast to the case of non-scattering media, in which photons travel as rays along a given direction until they are absorbed. The differential pathlength factor (DPF), which was introduced in Section 10.3, provides a measure of the average pathlength for a multitude of photons that have traveled between two points in a turbid medium separated by a distance r. This average pathlength (which we have denoted as $\langle L \rangle$ in earlier chapters) is determined by the product $r \times$ DPF. Since the DPF in biological tissues is, typically, about 8, if we consider an experimentally common source-detector separation for diffuse tissue spectroscopy of about 3 cm, one can estimate a typical average distance traveled by photons in diffuse optical spectroscopy measurements to be $\sim$24 cm. Dividing this distance by the speed of light in the medium ($\sim 2.2 \times 10^{10}$ cm/s, assuming $n = 1.35$) yields the average time-of-flight of collected photons of $\sim$1.1 ns.

So, the photon time-of-flight distribution in a typical tissue spectroscopy measurement is on a time scale of nanoseconds. To perform accurate measurements of such a time distribution, the light source pulse duration should be much shorter: say, picoseconds. Repeated measurements should not happen before all photons from a pulse have passed by the point of detection (to avoid cross-talk between measurements from different pulses), which means that the time separation between successive light pulses should be much greater than 1 ns, say 10–100 ns, corresponding to a pulse repetition rate of 10–100 MHz. The time-frame of the time-domain approach is illustrated in Figure 11.1, which plots the pulsed illumination, the detected time-dependent optical fluence rate, and the sequence of repeated pulsed illumination and time-resolved detection.

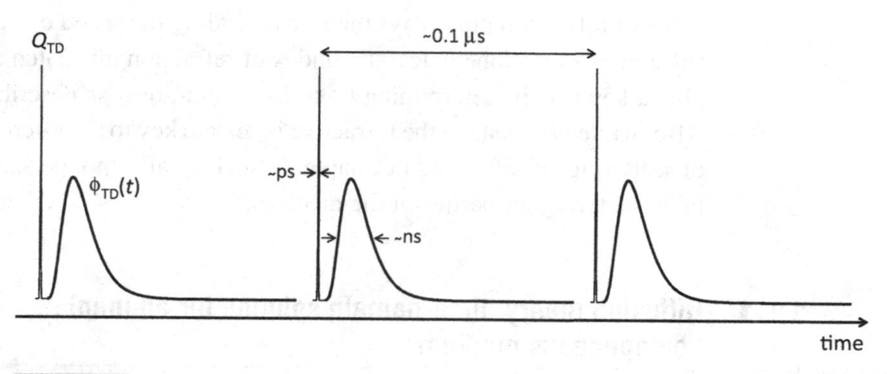

Figure 11.1
Schematic representation of the time-domain approach showing the pulsed optical energy (Q_{TD}) (typical time scale: ~ps), the time-domain fluence rate (ϕ_{TD}) (typical time scale: ~ns), and the pulse repetition over a typical time scale of ~0.1 μs.

Because the light source is emitting an energy pulse over a time (ps) that is much shorter than the time scale of the fluence rate dynamics (ns), it is appropriate to express the source term with a temporal delta function, $\delta(t)$ (taking $t = 0$ as the time of the light pulse). A spatial delta function, $\delta(\mathbf{r})$, is also introduced to describe a point source at the origin, $\mathbf{r} = 0$. (The Dirac delta is defined in Appendix 10.A.) As a result, the source term for the diffusion equation (Eq. (9.38)) to describe a pulsed, point source that emits an energy Q_{TD} at the location $\mathbf{r} = 0$ can be written as $S_0(\mathbf{r}, t) = Q_{TD}\delta(\mathbf{r})\delta(t)$. Using this source term, the solution of the diffusion equation for the time-domain fluence rate in a homogeneous, infinite medium is:

$$\phi_{TD}(r, t) = Q_{TD}c_n\left[\frac{3(\mu_s' + \mu_a)}{4\pi c_n t}\right]^{3/2} e^{-\frac{3(\mu_s'+\mu_a)r^2}{4c_n t} - \mu_a c_n t}, \qquad (11.1)$$

where c_n is the speed of light in the medium and t is the time after the pulse emission (at $t = 0$). Similar to the continuous-wave fluence rate (ϕ_{CW}) of Eq. (10.3), the time-domain fluence rate (ϕ_{TD}) has a pre-exponential factor that mostly depends on μ_s' (since $\mu_s' \gg \mu_a$ in the diffusion-approximation regime), and an exponential factor that depends on both μ_a and μ_s'. Both the pre-exponential and the exponential factors in ϕ_{TD} depend on time.

The time dependence of ϕ_{TD} provides a measure of the time-of-flight distribution at any distance r from the point source, which can be directly translated into a photon pathlength distribution, since each photon pathlength is given by its time-of-flight multiplied by the speed of light in the medium (c_n). The ability of time-domain methods to measure the distribution of photon pathlengths in the

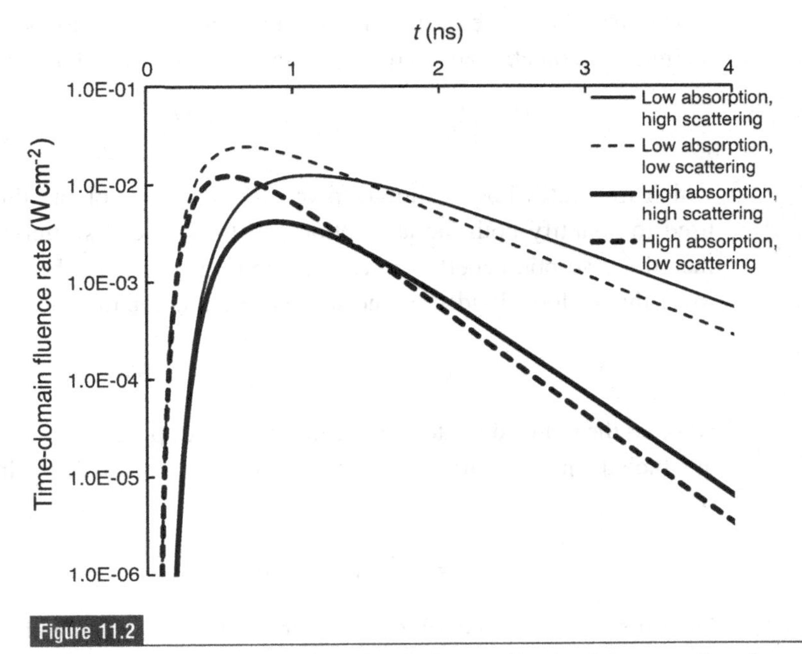

Figure 11.2

Diffusion theory results for the time-domain fluence rate at a distance $r = 3$ cm from a point source emitting a 1 nJ pulse in an infinite medium for four combinations of low absorption ($\mu_a = 0.05$ cm^{-1}), high absorption ($\mu_a = 0.10$ cm^{-1}), low scattering ($\mu_s' = 5$ cm^{-1}), and high scattering ($\mu_s' = 10$ cm^{-1}). Notice how the time dependence of the fluence rate is mostly determined by scattering properties at short times (rising portion of the fluence rate) and by absorption properties at long times (slope of decaying portion of the fluence rate).

medium renders them a powerful approach for the optical study of scattering media and biological tissue. Equation (11.1) shows that the fluence rate at short times is mostly determined by the scattering properties, because the first term of the exponent (the one dependent on μ_s') dominates for small values of t, since t is in the denominator; conversely, at longer times the fluence rate is mostly determined by the absorption properties, because at long times the second term of the exponent (the one dependent on μ_a) dominates. This is illustrated in Figure 11.2, which reports the time-domain fluence rate of Eq. (11.1) at a distance of 3 cm from a point source emitting a delta-pulse of energy of 1 nJ (corresponding to an average power of 10 mW at a repetition rate of 10 MHz), and for different combinations of absorption and reduced scattering coefficients. From Figure 11.2, it is apparent how the rising portion of the fluence rate at $t < 0.5$ ns mostly depends on scattering (low scattering: dashed lines; high scattering: solid lines), whereas the logarithmic slope of the fluence rate at $t \geq 3$ ns mostly depends on absorption (low absorption: thin lines; high absorption: thick lines).

The time derivative of the natural logarithm of the time-domain fluence rate in an infinite geometry can be readily obtained from Eq. (11.1):

$$\frac{d}{dt}\ln\left[\phi_{\mathrm{TD}}(r,t)\right]=\frac{3\left(\mu_s'+\mu_a\right)r^2}{4c_n t^2}-\frac{3}{2t}-\mu_a c_n, \tag{11.2}$$

which illustrates how a time-resolved measurement of the fluence rate can be used to quantify both the absorption and the reduced scattering coefficients. In fact, the absorption coefficient can be approximated from the negative asymptotic logarithmic slope divided by the speed of light in the medium:

$$\mu_a=-\frac{1}{c_n}\lim_{t\to\infty}\frac{d}{dt}\ln\left[\phi_{\mathrm{TD}}(r,t)\right], \tag{11.3}$$

whereas the reduced scattering coefficient can be approximated from the time, $t_{\max}$, at which the photon fluence rate reaches a maximum (so that $\frac{d}{dt}\ln[\phi_{\mathrm{TD}}(r,t)]|_{t_{\max}}=0$):

$$\mu_s'=\frac{2}{3r^2}\left(2c_n^2 t_{\max}^2\mu_a+3c_n t_{\max}\right)-\mu_a. \tag{11.4}$$

Of course, it is also possible to determine the absorption and reduced scattering coefficients by fitting the measured data of ϕ_{TD} with the diffusion theory expression of Eq. (11.1). We emphasize that, for accurate measurements of the optical properties in the time domain, it is important to have precise and accurate measures of times t relative to time zero (sometimes indicated as t_0), the time at which the source emits its pulse.

In common time-domain methods, the pulse energy Q_{TD} is emitted at time intervals separated by the repetition time t_r , and the time-averaged source power is Q_{TD}/t_r. The total fluence (J/m^2) resulting from a single source pulse, measured at the location **r**, can be obtained by integrating the time-domain fluence rate from 0 to t_r, or 0 to ∞, given that t_r is chosen such that the fluence rate is negligible after time t_r. It is instructive to note that the continuous-wave fluence rate of Eq. (10.3) is thereby reproduced as the average time-domain fluence rate as follows:

$$\begin{aligned}\phi_{\mathrm{CW}}(r)=\langle\phi_{\mathrm{TD}}(r,t)\rangle&=\frac{\int_0^\infty \phi_{\mathrm{TD}}(r,t)\,dt}{t_r}\\&=\frac{Q_{\mathrm{TD}}}{t_r}\frac{3\left(\mu_s'+\mu_a\right)}{4\pi}\frac{e^{-r\sqrt{3\mu_a(\mu_s'+\mu_a)}}}{r},\end{aligned} \tag{11.5}$$

which coincides with Eq. (10.3), recognizing that the average time-domain source power is $\langle P_{\mathrm{TD}}\rangle=Q_{\mathrm{TD}}/t_r$. The approach to solving the integral in Eq. (11.5) is described in Appendix 12.A. It is important to observe that the meaningful source emission quantity is power (P_{CW}) for continuous-wave methods, while it is single-pulse energy (Q_{TD}) for time-domain methods.

It is interesting to consider the total optical energy flowing out of a sphere of radius r centered on the photon source. The net optical power flowing along the radial coordinate (in the outward direction), per unit area orthogonal to the radial coordinate, is given by the component of the net time-domain flux along r ($F_{r|\mathrm{TD}}(r,t)$) (see also Eq. (9.37)):

$$F_{r|\mathrm{TD}}(r,t) = -\frac{1}{3(\mu_s' + \mu_a)} \frac{d\phi_{\mathrm{TD}}(r,t)}{dr} = Q_{\mathrm{TD}} \frac{r}{2} \left[\frac{3(\mu_s' + \mu_a)}{4\pi c_n} \right]^{3/2} t^{-5/2} e^{-\frac{3(\mu_s'+\mu_a)r^2}{4c_n t} - \mu_a c_n t}. \tag{11.6}$$

Noting that the repetition time, t_r, is effectively infinity, integration of $F_{r|\mathrm{TD}}(r,t)$ over time (from 0 to ∞) and over the entire sphere surface ($4\pi r^2$) yields the total optical energy flowing out of the sphere of radius r:

$$Q_{\mathrm{Tot}}(r) = 4\pi r^2 \int_0^\infty F_{r|\mathrm{TD}}(r,t)dt = Q_{\mathrm{TD}} e^{-r\mu_{\mathrm{eff}}} (1 + r\mu_{\mathrm{eff}}), \tag{11.7}$$

where the integral has been solved as explained in Appendix 12.A, and $\mu_{\mathrm{eff}} = \sqrt{3\mu_a(\mu_s' + \mu_a)}$ is the effective attenuation coefficient, introduced in Section 10.2.1. As expected, the total energy flowing out of the sphere in the time domain, after normalization by the source energy, coincides with the total power flowing out of the sphere in the continuous-wave case after normalizing by the source power (see Eq. (10.6)). Furthermore, in the non-absorbing case (i.e., $\mu_{\mathrm{eff}} = 0$), one finds, as expected, $Q_{\mathrm{Tot}}(r) = Q_{\mathrm{TD}}$, which expresses the conservation of total radiant energy in the absence of absorption.

Figure 11.3(a) illustrates a sphere of radius r centered at the photon source, the radial component of the net energy flux ($F_{r|\mathrm{TD}}(r,t)$), and the total energy ($Q_{\mathrm{Tot}}(r)$) flowing out of the sphere. (The reader will note that Figure 11.3(a) is analogous to Figure 10.1(a), with Q_{TD} and Q_{Tot} replacing P_{CW} and P_{Tot}, respectively.) Figure 11.3(b) reports the time dependence of $F_{r|\mathrm{TD}}(r_0,t)$ and $\phi_{\mathrm{TD}}(r_0,t)$ (for $r_0 = 3$ cm), and the radial dependence of $Q_{\mathrm{Tot}}(r)$ for an isotropic photon source of energy $Q_{\mathrm{TD}} = 1$ nJ, in a medium with $\mu_a = 0.1\ \mathrm{cm}^{-1}$ and $\mu_s' = 10\ \mathrm{cm}^{-1}$. As in the continuous-wave case, the radial component of the net flux is much smaller than the fluence rate at any specific location, because photons traveling in all directions contribute to the fluence rate, whereas only the net components along the radial coordinate of the photon velocities, which may be positive or negative, contribute to the net flux along r. The time-domain case further illustrates that the net flux peaks at an earlier time than the fluence rate (see Figure 11.3(b)), *an indication of the different temporal dynamics of the directional radial component of the net flux versus the omnidirectional fluence rate*.

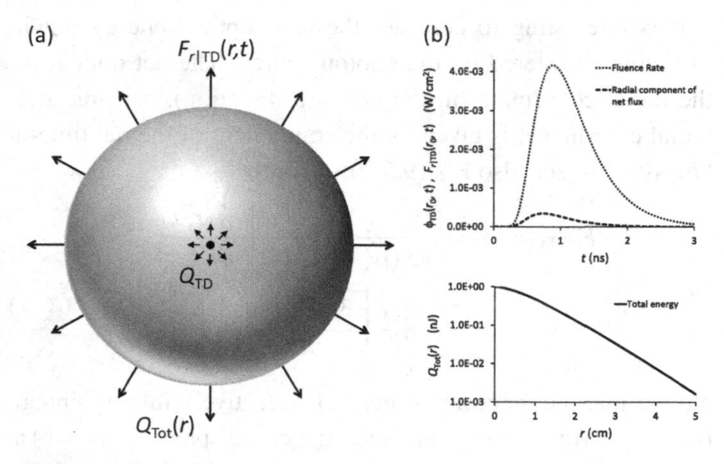

Figure 11.3

(a) A sphere of radius r centered at the isotropic photon source (emitting an energy Q_{TD}) in a scattering medium. The radial component of the net time-domain energy flux ($F_{r|TD}(r,t)$) at the surface of the sphere is indicated by the arrows. The total energy ($Q_{Tot}(r)$) flowing out of the sphere is given by the sphere surface area ($4\pi r^2$) multiplied by the time integral (from 0 to ∞) of $F_{r|TD}(r,t)$. (b) Time dependence of the radial component of the energy flux ($F_{r|TD}(r_0,t)$) and the fluence rate ($\phi_{CW}(r_0,t)$) (with $r_0 = 3$ cm) (upper panel) and of the total energy flowing out of the sphere ($Q_{Tot}(r)$) (lower panel), for a photon source energy $Q_{TD} = 1$ nJ and a medium with $\mu_a = 0.1$ cm^{-1} and $\mu_s' = 10$ cm^{-1}.

11.2 Moments of the photon time-of-flight distribution

The photon time-of-flight distribution, $\mathcal{D}_{TOF}(r,t)$, (which depends on t, r, μ_a, and μ_s') represents the probability density (or probability per unit time) that photons take a time t to travel a distance r (from a point source) in a medium of optical coefficients μ_a and μ_s'. In an infinite medium, the photon time-of-flight (TOF) distribution is defined as:

$$\mathcal{D}_{TOF}(r,t) = \frac{\phi_{TD}(r,t)}{\int_0^\infty \phi_{TD}(r,t)\,dt}\,. \tag{11.8}$$

The time-of-flight distribution can be described in terms of its moments. The first two moments (orders 0 and 1) are algebraic moments about zero, and they are defined as:

$$m_i(r) = \int_0^\infty t^i \mathcal{D}_{TOF}(r,t)dt \quad i = 0, 1 \tag{11.9}$$

By definition, the zeroth moment, which is dimensionless, is equal to 1 as a result of the normalization factor in Eq. (11.8): $m_0 = 1$. The first moment, which has units of time, is the *mean time-of-flight* for detected photons, which depends on the distance from the source as well as the optical properties of the medium: $m_1(r) = \langle t \rangle$. The higher-order moments (orders ≥ 2) are central moments about the first moment ($\langle t \rangle$), and they are defined as:

$$m_j(r) = \int_0^\infty (t - \langle t \rangle)^j \mathcal{D}_{\text{TOF}}(r, t) dt \quad j = 2, 3, 4 \ldots \tag{11.10}$$

The second moment, which has units of time squared, is the *variance* of the time-of-flight distribution: $m_2(r) = \sigma^2(t)$. Its square root, or *standard deviation* ($\sigma(t)$), quantifies the variability of the photon time-of-flights. The third moment, which has units of time to the third power, is related to the *skewness*, γ, of the $\mathcal{D}_{\text{TOF}}$ distribution by writing it as: $m_3(r) = \gamma\sigma^3(t)$. The skewness γ is a dimensionless parameter that is negative for distributions that are skewed toward values lower than $\langle t \rangle$, positive for distributions that are skewed toward values greater than $\langle t \rangle$, and zero for symmetrical distributions. (This parameter should not be confused with the similarity parameter γ defined in Chapter 8, Eq. (8.9), in terms of the first and second moments of the scattering phase function.) As a result of its asymmetry weighted to the right (i.e., toward longer times), the skew of the photon time-of-flight distribution in scattering media is positive, and it decreases with r.

The fourth moment, $m_4(r)$, provides a measure of the *kurtosis*, or peakedness, of the distribution by writing it as: $m_4(r) = (\kappa + 3)\sigma^4(t)$, where κ is defined as the excess kurtosis. The excess kurtosis of a Gaussian distribution is zero. Distributions with zero excess kurtosis are referred to as *mesokurtic*, since meso- means intermediate. The Student t-distribution and the Poisson distribution are examples of positive excess kurtosis distributions, which are referred to as super-Gaussian or *leptokurtic*, since lepto- means slender or narrow. The discrete and continuous uniform distributions are examples of negative excess kurtosis distribution, which are referred to as sub-Gaussian or *platykurtic*, since platy- means broad or flat. As a result of its higher peakedness than a Gaussian and a long tail at long times, the excess kurtosis of the photon time-of-flight distribution in scattering media is positive (leptokurtic distribution) and it decreases with r.

Remembering that, in the diffusion regime, scattering media are fully characterized by two optical coefficients, μ_a and μ_s', one can perform absolute spectroscopy of tissue by relying on two moments of the time-of-flight distribution, whose dependence on the optical coefficients is fully specified by the analytical expressions derived with diffusion theory. By combining the definitions of the first- and second-order moments (Eq. (11.9) for $i = 1$, and Eq. (11.10) for $j = 2$), the definition of the photon time-of-flight distribution (Eq. (11.8)), and the expression for the time-domain fluence rate for a point source in an infinite medium (Eq. (11.1)), the first moment ($m_1(r) = \langle t \rangle$) and the second moment ($m_2(r) = \sigma^2(t)$) can be written in terms of the optical coefficients of a macroscopically homogeneous, infinite medium (Arridge et al., 1992; Liebert et al., 2003):

$$\langle t \rangle = r \frac{\sqrt{3\left(\mu_s' + \mu_a\right)}}{2c_n\sqrt{\mu_a}} = \frac{r\mu_{\text{eff}}}{2c_n\mu_a}, \tag{11.11}$$

$$\sigma^2(t) = r \frac{\sqrt{3\left(\mu_s' + \mu_a\right)}}{4c_n^2\sqrt{\mu_a^3}} = \frac{r\mu_{\text{eff}}}{(2c_n\mu_a)^2}. \tag{11.12}$$

Since $\mu_a \ll \mu_s'$ in diffusion theory, Eqs. (11.11) and (11.12) show that an increase in the absorption coefficient decreases both $\langle t \rangle$ and $\sigma^2(t)$, as a result of the suppression of longer photon paths, whereas an increase in scattering increases both $\langle t \rangle$ and $\sigma^2(t)$, as a result of more tortuous photon routes between source and detector. It is interesting to observe, however, that $\langle t \rangle$ and $\sigma^2(t)$ feature a different dependence on the absorption coefficient ($\sim\mu_a^{-1/2}$ and $\sim\mu_a^{-3/2}$, respectively), whereas they feature the same dependence on the reduced scattering coefficient ($\sim\mu_s'^{1/2}$). Furthermore, the relative standard deviation (i.e., the standard deviation divided by the mean) of the photon time-of-flight distribution decreases with greater absorption, scattering, or distance from the source:

$$\frac{\sigma(t)}{\langle t \rangle} = \frac{1}{\sqrt{r\sqrt{3\mu_a\left(\mu_s' + \mu_a\right)}}} = \frac{1}{\sqrt{r\mu_{\text{eff}}}} \tag{11.13}$$

The differential pathlength factor (DPF), introduced in Section 10.3, may be loosely defined as the ratio of the mean photon pathlength to the geometric distance from the source: i.e., DPF $= c_n\langle t \rangle / r$. Equation (11.11) shows that, in an infinite medium, DPF $= \mu_{\text{eff}}/(2\mu_a)$, in agreement with Eq. (10.15) under the approximation $\mu_{\text{eff}} \approx \sqrt{3\mu_a\mu_s'}$. From Eqs. (11.11) and (11.12), it is possible to express the absorption and reduced scattering coefficients in terms of the two moments of the time-of-flight distribution:

$$\mu_a = \frac{\langle t \rangle}{2c_n\sigma^2(t)}, \tag{11.14}$$

$$\mu_s' = \frac{2c_n\langle t \rangle^3}{3r^2\sigma^2(t)}, \tag{11.15}$$

where Eq. (11.15) has been derived under the diffusion approximation $\mu_a \ll \mu_s'$. Equations (11.14) and (11.15) show that time-domain measurements of the photon time-of-flight distribution and its moments in the diffusion regime can yield absolute measurements of the absorption and reduced scattering coefficients of tissue.

The first moment of the time-of-flight distribution, the mean time of flight, and its relationship to the medium optical properties (Eq. (11.11)) allow us to gain better insight into the determination that, at long times, the logarithm of the time-domain fluence rate decays linearly, with a slope given by $-\mu_a c_n$ (see also Figure 11.2). The mathematical limit $t \to \infty$ in Eq. (11.3) is another formal way of expressing this fact. In practice, the fluence rate tends to 0 as $t \to \infty$, and under typical tissue spectroscopy conditions it is vanishingly small (not measurable) beyond a time scale of nanoseconds, as discussed above. The practical meaning of long times ($t \to \infty$) in Eq. (11.3) must therefore be specified by quantifying the reference time that t must significantly exceed for Eq. (11.3) to be applicable. In fact, there are two requirements that guarantee that the term $\mu_a c_n$ is the dominant term in Eq. (11.2): the first requirement is that $\frac{3(\mu_s'+\mu_a)r^2}{4c_n t^2} \ll \mu_a c_n$; and the second requirement is that $\frac{3}{2t} \ll \mu_a c_n$. By considering the expression for the mean time of flight in terms of the optical coefficients (Eq. (11.11), these two conditions can be written as $t \gg \langle t \rangle$ and $t \gg 1/(\mu_a c_n)$, respectively. They can be consolidated into the single condition that enables the logarithmic slope of $\phi_{\mathrm{TD}}(r, t)$ to be well approximated by $-\mu_a c_n$:

$$t \gg \max\left(\langle t \rangle, \frac{1}{\mu_a c_n}\right). \tag{11.16}$$

The first condition, $t \gg \langle t \rangle$, is a way of identifying the long-time tail of the photon time-of-flight distribution. (This condition is the same as the requirement for the minimum repetition time, as discussed above.) The second condition indicates the requirement that the photons have traveled a distance much longer than the mean free path for absorption, to guarantee that their temporal decay be dominated by absorption. Under low absorption conditions, the second requirement ($t \gg 1/(\mu_a c_n)$) becomes the relevant one since $\langle t \rangle$ increases more slowly (with the square root of μ_a) than $1/(\mu_a c_n)$ as the absorption decreases. For a typical reduced scattering coefficient of tissue (10 cm^{-1}) and a distance $r = 3$ cm, the value of μ_a for which $\langle t \rangle = 1/(\mu_a c_n)$ is ~0.015 cm^{-1}. This value is typically lower than the absorption coefficient of most tissues in the near-infrared, so that in most cases the condition $t \gg \langle t \rangle$ is sufficient to translate into practical terms the mathematical limit ($t \to \infty$) in Eq. (11.3).

11.3 Time-domain diffuse reflectance

Time-domain measurements in biological tissues are typically performed in a reflectance geometry: i.e., with the light source and the optical detector on the same side of a tissue surface (with the same exceptions noted in Section 10.4.1).

The single-distance, time-domain diffuse reflectance, $R_{TD}(\rho, t)$, with ρ indicating the surface distance between the illumination and detection locations, is typically of interest in animal and human studies, simply because one typically measures the optical signal from the tissue surface to minimize invasiveness. As described in Section 10.4.1, the diffuse reflectance is defined as the ratio between a measure of the diffusely reflected optical signal, and the incident optical signal. In the case of both continuous-wave and time-domain diffuse reflectance, the net optical flux (power per unit area) exiting the tissue boundary is the measured optical quantity. By contrast, as already observed, the significant input quantity is the optical power in continuous-wave methods, whereas it is pulse energy in time-domain measurements. Consequently, the units of the single-distance diffuse reflectance are different in continuous wave (m^{-2}) and in the time domain ($s^{-1}m^{-2}$).

The time-domain diffuse reflectance may be readily obtained with the method of image sources in the case of a semi-infinite medium, following the same procedure described in Section 10.4.4 for the continuous-wave case. The method of image sources allows expression of the fluence rate in a semi-infinite medium as the superposition of two infinite-medium fluence rates. By considering zero boundary conditions (see Section 9.6.2) – it turns out that the shape of the time-of-flight distribution is relatively insensitive to the details of the boundary conditions – the first fluence rate results from a positive source located one reduced scattering mean free path inside the medium (at $-z_0 = -1/\mu_s'$), and the second fluence rate results from an image (negative) photon source outside the medium (at $+z_0$):

$$\phi_{TD}(\rho, t) = Q_{TD} c_n \left[\frac{3(\mu_s' + \mu_a)}{4\pi c_n t} \right]^{3/2} \times \left[e^{-\frac{3(\mu_s'+\mu_a)\left[(z+z_0)^2+\rho^2\right]}{4c_n t}} - e^{-\frac{3(\mu_s'+\mu_a)\left[(z-z_0)^2+\rho^2\right]}{4c_n t}} \right] e^{-\mu_a c_n t}, \tag{11.17}$$

where the cylindrical coordinate system is the same as in Figure 10.4. The time-domain single-distance diffuse reflectance $R_{TD}(\rho, t)$ is defined as the ratio $\hat{\mathbf{n}} \cdot \mathbf{F}_{TD}(\rho, t, z = 0)/Q_{TD}$, where $\hat{\mathbf{n}}$ is the outward-pointing normal to the medium boundary, which coincides with $\hat{\mathbf{z}}$ in this case (see Figure 10.4); and in the diffusion approximation $\mathbf{F}_{TD}(\rho, t, z = 0) = -\nabla\phi_{TD}(\rho, t, z)|_{z=0}/[3(\mu_s' + \mu_a)]$ (see Eq. (9.37)). Consequently, diffusion theory yields the following result for the single-distance time-domain diffuse reflectance for a semi-infinite medium (Patterson et al., 1989):

$$R_{TD}(\rho, t) = \left[\frac{3\left(\mu_s' + \mu_a\right)}{4\pi c_n} \right]^{3/2} \frac{z_0}{t^{5/2}} e^{-\frac{3(\mu_s'+\mu_a)(z_0{}^2+\rho^2)}{4c_n t} - \mu_a c_n t}. \tag{11.18}$$

At large distances from the source ($\rho^2 \gg z_0^2$), the time derivative of the natural logarithm of the time-domain diffuse reflectance is:

$$\frac{d}{dt}\ln\left[R_{\mathrm{TD}}(\rho,t)\right] = \frac{3\left(\mu_s' + \mu_a\right) r^2}{4c_n t^2} - \frac{5}{2t} - \mu_a c_n, \tag{11.19}$$

which differs from the corresponding expression in the infinite medium (Eq. (11.2)) only for the $5/(2t)$ term in place of the $3/(2t)$ term. Correspondingly, the expressions for μ_a and μ_s' in terms of the semi-infinite, time-domain diffuse reflectance are as follows (Patterson et al., 1989):

$$\mu_a = -\frac{1}{c_n}\lim_{t\to\infty}\frac{d}{dt}\ln\left[R_{\mathrm{TD}}(\rho,t)\right], \tag{11.20}$$

and:

$$\mu_s' = \frac{2}{3r^2}\left(2c_n^2 t_{\max}^2 \mu_a + 5c_n t_{\max}\right) - \mu_a. \tag{11.21}$$

For a diffuse reflectance measurement from a semi-infinite medium, the photon time-of-flight distribution is defined in a manner analogous to that for an infinite medium (Eq. (11.8)) in terms of the diffuse reflectance (instead of the fluence rate):

$$\mathcal{D}_{\mathrm{TOF}}(\rho,t) = \frac{R_{\mathrm{TD}}(\rho,t)}{\int_0^\infty R_{\mathrm{TD}}(\rho,t)dt}, \tag{11.22}$$

and the resulting first- and second-order moments, for $\rho^2 \gg z_0^2$, are (Liebert et al., 2003):

$$\langle t\rangle = \frac{3\left(\mu_s'+\mu_a\right)\rho^2}{2c_n\left(1+\rho\sqrt{3\mu_a\left(\mu_s'+\mu_a\right)}\right)} = \frac{\rho^2\mu_{\mathrm{eff}}^2}{2c_n\mu_a\left(1+\rho\mu_{\mathrm{eff}}\right)}, \tag{11.23}$$

$$\sigma^2(t) = \frac{\left[3\left(\mu_s'+\mu_a\right)\right]^{3/2}\rho^3}{4c_n^2\sqrt{\mu_a}\left(1+\rho\sqrt{3\mu_a\left(\mu_s'+\mu_a\right)}\right)^2} = \frac{\rho^3\mu_{\mathrm{eff}}^3}{\left(2c_n\mu_a\right)^2\left(1+\rho\mu_{\mathrm{eff}}\right)}. \tag{11.24}$$

These expressions for the first- and second-moments of the photon time-of-flight distribution for the diffuse reflectance allow for a quantitative comparison with the case of an infinite medium geometry. A comparison of Eqs. (11.11) and (11.23) shows that for a given distance from the source (r in an infinite medium, ρ on the surface of a semi-infinite medium) the mean photon time-of-flight is longer in an infinite medium than in a semi-infinite medium. Consequently, the differential pathlength factor (DPF) defined in Section 10.3 is smaller in a semi-infinite geometry than in the infinite geometry. This result may be counter-intuitive, given that in an infinite medium photons have access to paths that are closer to the geometrical line between source and detector, whereas photons traveling close

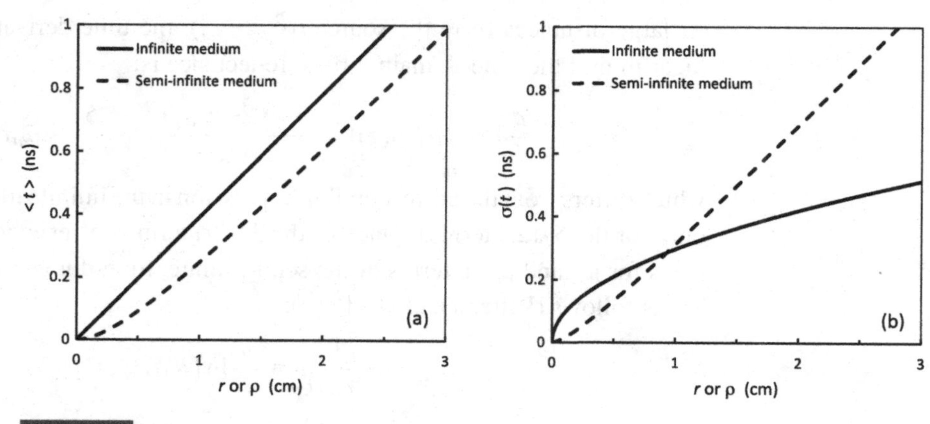

Figure 11.4

(a) Mean time-of-flight, $\langle t \rangle$, and (b) standard deviation, $\sigma(t)$, of the photon time-of-flight distribution, $\mathcal{D}_{TOF}$, for an infinite medium (solid lines) and a semi-infinite medium (dashed lines). The *x* axis reports the distance from the point source, which is indicated with *r* for the infinite medium and ρ (measured on the medium surface) for the semi-infinite medium. The optical properties of the medium are $\mu_a = 0.1$ cm^{-1} and $\mu_s' = 10$ cm^{-1}.

to the surface of a semi-infinite medium have a significant probability of exiting the medium and becoming "lost." In the infinite geometry, however, the following must be considered: the fact that no photons can escape the medium, together with the fact that the photon paths are strongly tortuous (because the mean free path for scattering is much less than r or ρ), results in the possibility of longer paths from the illumination point to any given point, and those long paths contribute to the mean photon pathlength. Equation (11.23) shows that the mean photon pathlength $c_n\langle t \rangle$, which is proportional to r in the infinite geometry, is not simply proportional to ρ in the semi-infinite geometry. Nonetheless, the DPF in diffuse reflectance (DPF_R) may still be defined as the ratio of the mean photon pathlength to the geometrical distance from the source, yielding $\mathrm{DPF}_R = \rho\mu_{\mathrm{eff}}^2/[2\mu_a(1+\rho\mu_{\mathrm{eff}})]$, whereas, in an infinite medium, $\mathrm{DPF} = \mu_{\mathrm{eff}}/(2\mu_a)$ (see Eq. (10.15)). The ratio of the DPFs for the cases of diffuse reflectance and infinite media ($\mathrm{DPF}_R/\mathrm{DPF}$) is therefore always less than 1, being given by $\rho\mu_{\mathrm{eff}}/(1+\rho\mu_{\mathrm{eff}})$, and approaches 1 when $\rho\mu_{\mathrm{eff}} \gg 1$.

The first moment (mean time of flight) and second moment (variance of the time-of-flight distribution) for infinite and semi-infinite media (with $\mu_a = 0.1$ cm^{-1} and $\mu_s' = 10$ cm^{-1}) are illustrated in Figure 11.4 as a function of the distance from the point source. Similar to the infinite medium case, the relative standard deviation (i.e., the standard deviation divided by the mean) of the photon time-of-flight distribution for the diffuse reflectance decreases at greater absorption, scattering, or distances from the source. Its expression, directly obtained from Eqs. (11.23) and (11.24), is as follows:

$$\frac{\sigma(t)}{\langle t\rangle} = \sqrt{1+\frac{1}{\rho\sqrt{3\mu_a\left(\mu_s'+\mu_a\right)}}} = \sqrt{1+\frac{1}{\rho\mu_{\mathrm{eff}}}}. \tag{11.25}$$

From Eqs. (11.23) and (11.24), the absorption and reduced scattering coefficients of a semi-infinite medium are given by the following expressions in terms of the first two moments of the time-of-flight distribution for the time-domain diffuse reflectance:

$$\mu_a = \frac{\langle t\rangle^3}{2c_n\sigma^2(t)(\langle t\rangle^2+\sigma^2(t))}, \tag{11.26}$$

$$\mu_s' = \frac{2c_n\langle t\rangle(\langle t\rangle^2+\sigma^2(t))}{3\rho^2\sigma^2(t)}, \tag{11.27}$$

where Eq. (11.27) has been derived under the diffusion approximation $\mu_a \ll \mu_s'$.

The total time-domain diffuse reflectance for point illumination is given by (Patterson et al., 1989):

$$\dot{R}_{d|\mathrm{TD}}(t) = \int_0^\infty R_{\mathrm{TD}}(\rho,t)\,2\pi\rho\,d\rho = \left[\frac{3\left(\mu_s'+\mu_a\right)}{4\pi c_n}\right]^{1/2}\frac{z_0}{t^{3/2}}e^{-\frac{3\left(\mu_s'+\mu_a\right)z_0{}^2}{4c_nt}-\mu_ac_nt}. \tag{11.28}$$

In the case of zero absorption, the total diffuse reflectance decreases with time as $t^{-3/2}$ (once sufficiently long times have been reached so that the $-1/t$ dependence of the exponent sets the exponential function to $\sim$1). Note that $\dot{R}_{d|\mathrm{TD}}(t)$ has units of s^{-1} and represents the diffusely reflected power per unit energy of the pulsed source.

11.4 Limits of validity of time-domain diffusion theory

The application of diffusion theory requires that light propagation has reached conditions where the diffusion approximation holds: i.e., a nearly isotropic radiance, which happens at a distance from the source that exceeds several times the inverse of the reduced scattering coefficient (i.e., $r > \alpha/\mu_s'$ with $\alpha > 1$). This requirement has been discussed in Section 10.4.6 in relation to continuous-wave diffuse reflectance. In the time domain, there is the additional requirement that the photon time-of-flight be much greater than the time-of-flight of unscattered photons (which is given by the distance from the source, r or ρ, divided by the speed of light in the medium, c_n). This additional requirement has been investigated with Monte Carlo simulations, which showed that the diffusion theory solution for the fluence rate (Eq. (11.1)) in an infinite medium is accurate at times $t > 4r/c_n$ (Martelli et al., 2000) for optical properties relevant to tissue. By combining this condition with the previous condition $r > \alpha/\mu_s'$, the times at which time-domain diffusion theory is valid become $t \gg 1/(c_n\mu_s')$. This condition can be generalized to saying that diffusion theory applies on a time scale that is longer than the mean time between effectively isotropic scattering events, which reiterates one of the conditions for diffusion theory, discussed in Section 9.5, that the typical time scale of the fluence rate variations must be much longer than $1/(c_n\mu_s')$ (see Eqs. (9.34) and (9.35)). For typical biological tissues, $1/(c_n\mu_s') \approx 5$ ps, so that the representative fluence rate dynamics of Figure 11.2 for time-domain tissue spectroscopy, which occurs on a time scale of a nanosecond, meets the conditions of diffusion theory, with the possible exception of the very early rising edge.

Problems – answers to problems with * are on p. 647

11.1* Consider a photon source that emits short pulses at a repetition rate of 37 MHz, resulting in an average power of 23.2 mW. What is the energy emitted by each pulse?

11.2 Derive Eq. (11.2) from Eq. (11.1).

11.3 A measurement of the time-domain fluence rate in a uniform, infinite medium at a distance of 2.4 cm from a pulsed, point source yields the following data set:

Time (ns)	Fluence rate (mW/cm^2)
0.3	4.2
0.4	7.7
0.5	8.9 (max value)
0.6	8.1
0.7	6.6
0.8	5.0
0.9	3.6
1.0	2.5
3.0	7.3×10^{-4}
3.1	4.9×10^{-4}

Assuming that the index of refraction of the medium is 1.35, find the absorption coefficient and the reduced scattering coefficient of the medium.

11.4* Find the total energy flowing out of a sphere of radius 3.4 cm centered at an isotropic point source that has emitted one pulse of energy 9.1 nJ in an infinite medium with $\mu_a = 0.12\ \text{cm}^{-1}$ and $\mu_s' = 15\ \text{cm}^{-1}$.

11.5 By using $\dot{R}_d(t)$ as given by Eq. (11.28), find the total energy Q_d diffusely reflected from the surface of a semi-infinite medium (optical properties: $\mu_a = 0.065\ \text{cm}^{-1}$, $\mu_s' = 11.3\ \text{cm}^{-1}$) following the illumination with a short pulse of energy 17 nJ. (Hint: See Appendix 12.A.)

11.6* Consider a non-absorbing ($\mu_a = 0$), scattering semi-infinite medium and a pulsed source of energy Q_{TD} at its boundary. What do you expect the total diffusely reflected energy to be? Verify your answer by integrating the diffusely reflected power $Q_{\text{TD}}\dot{R}_d(t)$ over the time interval from 0 to ∞. (Hint: See Appendix 12.A.)

11.7 Consider a homogeneous, non-absorbing infinite medium and an embedded point source that emits a short pulse of light. Now consider the photons whose time-of-flight to a point at a given distance r from the source has maximal probability density. How does the number of scattering events experienced by such photons depend on (a) μ_s' and (b) r?

References

Arridge, S. R., Cope, M., and Delpy, D. T. (1992). The theoretical basis for the determination of optical pathlengths in tissue: temporal and frequency analysis. *Physics in Medicine and Biology*, 37, 1531–1560.

Liebert, A., Weibnitz, H., Grosenick, D., et al. (2003). Evaluation of optical properties of highly scattering media by moments of distributions of times of flight of photons. *Applied Optics*, 42, 5785–5792.

Martelli, F., Bassani, M., Alianelli, L., Zangheri, L., and Zaccanti, G. (2000). Accuracy of the diffusion equation to describe photon migration through an infinite medium: numerical and experimental investigation. *Physics in Medicine and Biology*, 45, 1359–1373.

Patterson, M. S., Chance, B., and Wilson, B. C. (1989). Time resolved reflectance and transmittance for the non-invasive measurement of optical properties. *Applied Optics*, 28, 2331–2336.

Further reading

Photon pathlength distribution in scattering media

Patterson, M. S., Andersson-Engels, S., Wilson, B. C., and Osei, E. K. (1995). Absorption spectroscopy in tissue-simulating materials: a theoretical and experimental study of photon paths. *Applied Optics*, 34, 22–30.

Time-domain diffuse reflectance

Kienle, A., and Patterson, M. S. (1997). Improved solutions of the steady-state and the time-resolved diffusion equations for reflectance from a semi-infinite turbid medium. *Journal of the Optical Society of America A*, 14, 246–254.

Madsen, S. J., Wilson, B. C., Patterson, M. S., et al. (1992). Experimental tests of a simple diffusion model for the estimation of scattering and absorption coefficient of turbid media from time-resolved diffuse reflectance measurements. *Applied Optics*, 31, 3509–3517.

Moments of the photon time-of-flight distribution

Arridge, S. R., and Schweiger, M. (1995). Direct calculation of the moments of the distribution of photon time of flight in tissue with a finite-element method. *Applied Optics*, 34, 2683–2687.

12 Frequency-domain methods for tissue spectroscopy in the diffusion regime

In Chapter 11, we considered time-domain methods as one form of time-resolved tissue spectroscopy that is based on pulsed illumination. In this chapter, we consider an alternative type of time-resolved spectroscopy that operates in the frequency space by employing temporally modulated illumination. While these two variations of time-resolved spectroscopy are mathematically equivalent, being related by Fourier transformation, they invoke different concepts, instrumentation, and methods of data collection. The conceptual and theoretical frameworks of *frequency-domain spectroscopy* are described and developed in this chapter. Experimental and instrumentation aspects of frequency-domain spectroscopy are presented in Chapter 13 (Section 13.4.2).

12.1 Basic concepts of frequency-domain tissue spectroscopy

In frequency-domain diffuse spectroscopy of tissue, the optical power incident from the light source is temporally modulated at a sinusoidal angular frequency ω, and phase-resolved detection yields the amplitude and phase of the resulting fluence rate or flux measured in the scattering medium. The source power, and the fluence rate at any point, are both modulated about an average value. By applying the nomenclature of electrical direct current (DC) and alternating current (AC) to indicate the average value and the amplitude of the modulated component of the optical power, respectively, the *modulation depth* (or, simply, *modulation*) (m) is defined as the AC/DC ratio. The modulation is a measure of the amplitude of the oscillations normalized to their average value, and as such is a dimensionless quantity that represents the relative depth of optical oscillations.

This approach can be thought of as analogous to the modulation and carrier frequencies of an amplitude modulation (AM) radio signal, with a difference that the carrier signal is at zero frequency in the case of frequency-domain optical spectroscopy. Given that the optical power and the fluence rate are always positive, the modulation is bounded between 0 and 1 ($0 \leq m \leq 1$). There is, therefore, a fundamental difference between alternating electrical current, which is bipolar, thereby oscillating between positive and negative values, and modulated optical power and fluence rate, which oscillates about an average value, remaining always positive. The average value, or DC component, corresponds to the time-independent optical power and fluence rate considered in Chapter 10 for continuous-wave (CW) spectroscopy. The oscillating optical signal, or AC component, at angular frequency ω is related to the time-domain optical signal of Chapter 11 through a temporal Fourier transform.

What values of angular frequency ω are appropriate for frequency-domain tissue spectroscopy? The requirement to measure the phase delay of the detected optical signal (fluence rate or flux) with respect to the input optical power means that the time delay between the two optical signals, $\langle t \rangle$, should be such that $\omega \langle t \rangle \approx 1$. The reason for this condition is that phase delays of the order of 1 rad (or $\sim 57°$) are large enough to be measured yet small enough to avoid phase-wrapping ambiguity. Considering typical values for the tissue optical properties in the near-infrared spectral region (say, $\mu_a = 0.1\ \text{cm}^{-1}$ and $\mu_s' = 10\ \text{cm}^{-1}$) and for the source-detector separations used in diffuse optical spectroscopy (say, $r = 3$ cm), the expression for the mean photon time-of-flight derived in Chapter 11 (Eq. (11.11) for an infinite medium) results in $\langle t \rangle \approx 1.2$ ns. Consequently, the condition $\omega \langle t \rangle \approx 1$ translates into typical modulation frequencies $\omega/(2\pi) \approx 1/(2\pi \langle t \rangle) \approx 130$ MHz. In fact, frequency-domain tissue spectroscopy typically employs modulation frequencies of the order of 100 MHz. Much lower frequencies (say, 10 MHz or less) would result in small phase shifts, which are difficult to measure accurately, whereas much higher frequencies (say, 1 GHz or more) would result in small modulation depths, as we will see in the next section, as well as possible confounding phase-wrapping.

It is important not to confuse the frequency-domain modulation frequency, which is of the order of 100 MHz, with the optical frequency of the light, which is of the order of 10^{14} Hz in the near-infrared. To avoid any potential confusion, in frequency-domain spectroscopy it is convenient to use "frequency" to indicate the modulation frequency, and "wavelength" to indicate the color of the light, determined by its optical frequency. As a final note, we point out that frequency-domain methods are also referred to as *amplitude modulation methods*, to indicate that they are based on modulating the amplitude of the source power, and *phase-modulation methods* to indicate that they are based on measurements of phase and modulation. Sometimes the expression "phase modulation" (without the hyphen) is also used, but this is inappropriate because frequency-domain tissue spectroscopy is not based on modulating the phase of any signal. (Such a technique is the basis of FM radio signals, but does not apply here.)

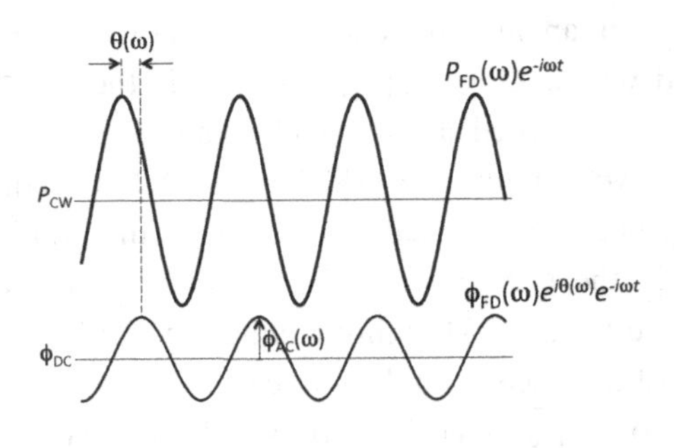

Figure 12.1

Illustration of the frequency-domain approach to tissue spectroscopy. The optical power of the light source ($P_{FD}(\omega)e^{-i\omega t}$) is sinusoidally modulated at angular frequency ω about an average value ($P_{CW} = P_{FD}(\omega = 0)$). The fluence rate ($\phi_{FD}(\omega)e^{-i[\omega t - \theta(\omega)]}$) is also modulated at angular frequency ω, and it is phase shifted by $\theta(\omega)$ with respect to the modulated source power. The average value and the amplitude of the fluence rate oscillations are indicated with ϕ_{DC} and $\phi_{AC}(\omega)$, respectively.

The basic approach of frequency-domain spectroscopy is illustrated in Figure 12.1.

12.2 Diffusion theory: frequency-domain solution for an infinite, homogeneous medium

Consider a point light source at $\mathbf{r} = 0$ that emits a sinusoidally modulated power at angular frequency ω. This source is expressed as $S_0(\mathbf{r}, \omega, t) = P_{FD}(\omega)\delta(\mathbf{r})e^{-i\omega t}$, where $P_{FD}(\omega)$ is the frequency-dependent amplitude of the source power and $e^{-i\omega t}$ is the conventional complex representation of sinusoidal oscillations. Note that, by this expression, the depth of modulation of the source term is given by $P_{FD}(\omega)/P_{FD}(0)$. Because the time dependence is solely described by the sinusoidal term $e^{-i\omega t}$, the source term can be written as $S_0(\mathbf{r}, \omega, t) = S_0(\mathbf{r}, \omega)e^{-i\omega t}$, with $S_0(\mathbf{r}, \omega) = P_{FD}(\omega)\delta(\mathbf{r})$, the frequency-domain energy density can be written as $U_{FD}(\mathbf{r}, \omega, t) = U_{FD}(\mathbf{r}, \omega)e^{-i\omega t}$, and time derivatives $(\partial/\partial t)$ become multiplications by a factor $-i\omega$. As a result, the diffusion equation for a homogeneous medium (Eq. (9.38)) is written as follows in the frequency domain:

$$[\nabla^2 + k^2(\omega)]U_{FD}(\mathbf{r}, \omega) = -\frac{P_{FD}(\omega)\delta(\mathbf{r})}{D}, \tag{12.1}$$

which is in the form of an inhomogeneous Helmholtz equation with $k^2(\omega) = (i\omega - c_n\mu_a)/D$, and where $D = c_n/[3(\mu_s' + \mu_a)]$ is the optical diffusion coefficient. (The equation is named after the German physiologist, physicist, and philosopher Hermann von Helmholtz [1821–1894].) The frequency-domain diffusion equation (Eq. (12.1)) is formally identical to the continuous-wave (CW) diffusion equation. In the CW case, k^2 is negative, being given by $-\mu_{\text{eff}}^2$ (or $-c_n\mu_a/D$). A negative k^2 in the Helmholtz equation results in a fluence rate that decays as a function of distance from the source, with a typical attenuation length, the diffusion length, of $1/\mu_{\text{eff}}$ (see Eq. (10.4)). A positive k^2 in the Helmholtz equation results in the wave equation, for which k takes on the meaning of the wavenumber, $2\pi/\lambda$, already introduced in Eq. (4.4). In the case of the frequency-domain diffusion equation, k^2 is complex, since it contains the imaginary term $i\omega$, which comes from the first derivative of the frequency-domain energy density. By considering the fluence rate $\phi_{\text{FD}}(\mathbf{r}, \omega) = c_n U_{\text{FD}}(\mathbf{r}, \omega)$, the solution to Eq. (12.1) is:

$$\phi_{\text{FD}}(r, \omega) = P_{\text{FD}}(\omega)\frac{3\left(\mu_s' + \mu_a\right)}{4\pi}\frac{e^{-r\sqrt{3(\mu_s'+\mu_a)\left(\mu_a - \frac{i\omega}{c_n}\right)}}}{r}, \tag{12.2}$$

which gives the frequency-domain fluence rate in a homogeneous infinite medium with a point source at $\mathbf{r} = 0$ that emits an optical power that is sinusoidally modulated at angular frequency ω and has an amplitude $P_{\text{FD}}(\omega)$. Of course, a comparison of Eq. (12.2) and Eq. (10.3) shows that $\phi_{\text{FD}}(r, \omega = 0) = \phi_{\text{CW}}(r)$, after replacing $P_{\text{FD}}(\omega = 0)$ with P_{CW}. We observe that the spatial dependence of the frequency-domain fluence rate in the exponential function of Eq. (12.2) is expressed as $\exp(ikr)$, which shows that a purely imaginary k, as in the CW case, results in a purely attenuating fluence rate, whereas a real component of k, which appears in the FD case, also contributes a phase term in the complex fluence rate. This latter point will be explored in detail in Section 12.4.

Given the presence of the imaginary unit i, Eq. (12.2) provides a complex representation of $\phi_{\text{FD}}(r, \omega)$, which reduces to a real representation in the limiting continuous-wave case $\omega = 0$. A powerful approach is to interpret the expression of Eq. (12.2) as a *phasor*, i.e., a two-dimensional vector in the complex plane whose amplitude and phase represent the amplitude and phase of the associated oscillations at angular frequency ω. The DC component ($\omega = 0$) of the fluence rate is real, and it coincides with ϕ_{CW}; the AC component is the modulation amplitude of the fluence rate, and the phase (relative to the source) is the argument (indicated by the function "Arg", i.e., the arctan of the ratio of the imaginary part to the real part) of the complex fluence rate. The expressions for the DC, AC, and phase of the fluence rate in an infinite medium, which we indicate with ϕ_{DC}, ϕ_{AC}, and θ, are as follows:

$$\phi_{\mathrm{DC}}(r) = P_{\mathrm{CW}} \frac{3\left(\mu_s' + \mu_a\right)}{4\pi} \frac{e^{-r\sqrt{3\mu_a(\mu_s'+\mu_a)}}}{r}, \tag{12.3}$$

$$\phi_{\mathrm{AC}}(r,\omega) = |\phi_{\mathrm{FD}}(r,\omega)| = P_{\mathrm{FD}}(\omega) \frac{3\left(\mu_s' + \mu_a\right)}{4\pi} \frac{e^{-r\sqrt{\frac{3\mu_a(\mu_s'+\mu_a)}{2}}\sqrt{\sqrt{1+\left(\frac{\omega}{c_n\mu_a}\right)^2}+1}}}{r}, \tag{12.4}$$

$$\theta(r,\omega) = \mathrm{Arg}\left[\phi_{\mathrm{FD}}(r,\omega)\right] = r\sqrt{\frac{3\mu_a\left(\mu_s'+\mu_a\right)}{2}}\sqrt{\sqrt{1+\left(\frac{\omega}{c_n\mu_a}\right)^2}-1}, \tag{12.5}$$

where the phase in Eq. (12.5) is in radians. As expected, $\phi_{\mathrm{AC}}(r, \omega = 0) = \phi_{\mathrm{DC}}(r)$ and $\theta(r, \omega = 0) = 0$. The ratio of Eq. (12.4) to Eq. (12.3), assuming a uniform source power over frequency ($P_{\mathrm{FD}}(\omega) = P_{\mathrm{CW}} = P_0$), results in the following expression for the modulation depth $m(r, \omega)$:

$$m(r,\omega) = \frac{\phi_{\mathrm{AC}}(r,\omega)}{\phi_{\mathrm{DC}}(r)} = \frac{e^{-r\sqrt{\frac{3\mu_a(\mu_s'+\mu_a)}{2}}\sqrt{\sqrt{1+\left(\frac{\omega}{c_n\mu_a}\right)^2}+1}}}{e^{-r\sqrt{3\mu_a(\mu_s'+\mu_a)}}} = \frac{e^{-r\frac{\mu_{\mathrm{eff}}}{\sqrt{2}}\sqrt{\sqrt{1+\left(\frac{\omega}{c_n\mu_a}\right)^2}+1}}}{e^{-r\mu_{\mathrm{eff}}}}. \tag{12.6}$$

Equations (12.4), (12.5), and (12.6) show that the AC component, the phase, and the modulation depth of the fluence rate depend on the distance from the point source (r), the modulation frequency of the source power (ω), and the optical properties of the medium (μ_a, μ_s'). Figure 12.2 shows the phase (panel (a)) and modulation (panel (b)) as a function of frequency for a fixed distance ($r = 3$ cm) from the source and four sets of optical properties. Figure 12.2(a) shows that, in general, the phase features a nonlinear increase with frequency. In the low-frequency limit, when $\omega \ll c_n\mu_a$, Eq. (12.5) can be linearized to yield:

$$\theta(r, \omega \ll c_n\mu_a) \cong \omega r \frac{\sqrt{3\left(\mu_s'+\mu_a\right)}}{2c_n\sqrt{\mu_a}} = \omega\langle t\rangle, \tag{12.7}$$

where $\langle t\rangle$ is the mean photon time-of-flight given by Eq. (11.11). The phase of the frequency-domain fluence rate is therefore associated with the mean photon time-of-flight, but only at low frequencies is the phase truly proportional to the time-of-flight. The deviation from linearity in the dependence of phase versus frequency is a result of the positive skewness of the photon time-of-flight distribution. The low-frequency condition for linearity ($\omega \ll c_n\mu_a$) requires that many absorption

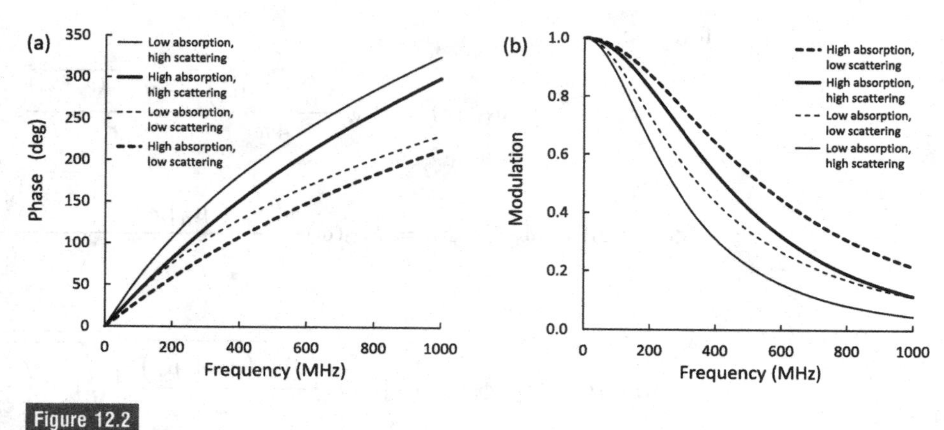

Figure 12.2

(a) Phase and (b) modulation of the frequency-domain fluence rate calculated with diffusion theory at a distance $r = 3$ cm from a point source in an infinite medium for four combinations of low absorption ($\mu_a = 0.05$ cm^{-1}), high absorption ($\mu_a = 0.10$ cm^{-1}), low scattering ($\mu_s' = 5$ cm^{-1}), and high scattering ($\mu_s' = 10$ cm^{-1}).

events take place within a modulation period. For near-infrared tissue absorption coefficients within the range 0.05–0.15 cm^{-1}, this condition requires that the modulation frequency ($\omega/(2\pi)$) be much less than ~180–500 MHz, so that typical frequencies of ~100 MHz are usually within the linear portion of the phase-vs-frequency curve (as confirmed by Figure 12.2(a)). Figure 12.2(b) shows the demodulation of the fluence rate at higher modulation frequencies, indicating the significantly reduced frequency-domain signal at frequencies of ~1 GHz or greater. In general, a greater scattering coefficient results in an increase in the phase and a decrease in the modulation of ϕ_{FD}, whereas a greater absorption has the opposite effects of a decrease in the phase and an increase in the modulation of ϕ_{FD}.

Figure 12.3 reports the linear functions (in a homogeneous, infinite medium) of $\ln[r\phi_{AC}(r, \omega)]$ and $\theta(r, \omega)$ versus r for a fixed modulation frequency (100 MHz) and for four sets of optical properties, thus showing the sensitivity of the slopes of these straight lines to the optical properties of the medium. As intuition dictates, the slope of $\ln[r\phi_{AC}]$ becomes more negative for higher absorption and higher scattering values, reflecting the stronger optical attenuation associated with an increase in either μ_a or μ_s'. In fact, both coefficients appear in the effective attenuation coefficient (μ_{eff}).

It is also interesting, yet intuitive, to observe that in the short-distance limit (where, however, the diffusion approximation may be inaccurate) the AC component of the frequency-domain diffuse reflectance is dominated by scattering

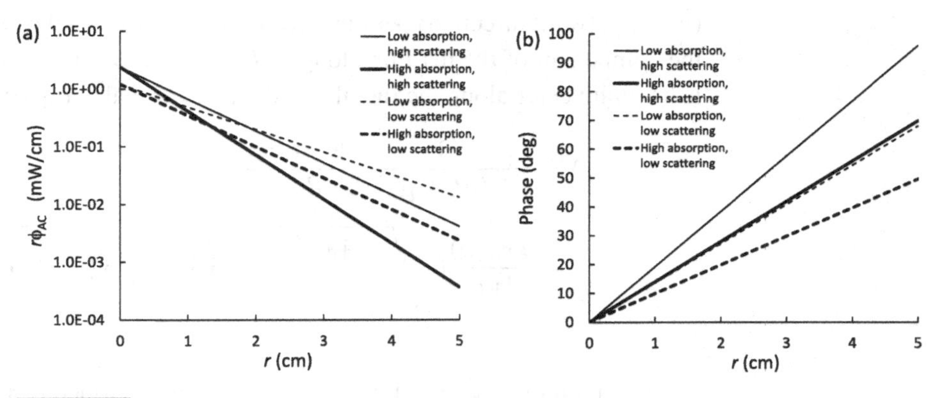

Figure 12.3

Dependence on the distance (r) from a point source (emitting a modulated optical power of amplitude 1 mW) in an infinite medium of (a) the product of r times the AC component of the frequency-domain fluence rate (ϕ_{AC}) (on a logarithmic scale), and (b) the phase of the frequency-domain fluence rate (θ). The modulation frequency ($\omega/(2\pi)$) is 100 MHz. The slopes of the linear functions of ln [$r\,\phi_{AC}$] and θ vs. r are determined by the optical properties of the medium, as shown by the four combinations of low absorption ($\mu_a = 0.05$ cm^{-1}), high absorption ($\mu_a = 0.10$ cm^{-1}), low scattering ($\mu_s' = 5$ cm^{-1}), and high scattering ($\mu_s' = 10$ cm^{-1}). Notice how, similar to the short-time limit in the time-domain case, the short-distance limit for $r\,\phi_{AC}$ in the frequency domain is dominated by scattering properties.

rather than absorption properties (see Figure 12.3(a)). This result mirrors the similar result observed for the short-time limit of the time-domain fluence rate (see Figure 11.2), and it follows from the fact that the shorter optical pathlengths associated with short times (in the time-domain) or short distances from the source are more affected by scattering than by absorption phenomena, especially under the diffusion condition $\mu_a \ll \mu_s'$. The effects of absorption and scattering on the phase slope, instead, are opposite: an increase in scattering increases the phase slope, whereas an increase in absorption decreases the phase slope. This result is in line with the fact that the phase slope is closely related to the mean photon time-of-flight, $\langle t \rangle$: it is simply proportional to it in the low-frequency limit $\omega \ll c_n\mu_a$ as shown by Eq. (12.7), and we have already found $\langle t \rangle$ to increase with μ_s' and decrease with μ_a (see Eq. (11.11)). We also point out that Figure 12.3(b) shows that the phase of the frequency-domain fluence rate typically increases by 10–20 degrees/cm in media with optical properties similar to those of biological tissues in the near-infrared.

As in the continuous-wave and time-domain cases, we now consider the frequency-domain total optical power flowing out of a sphere of radius r centered at the photon source. The net optical power flowing along the radial coordinate (in the outward direction) per unit area normal to the radial coordinate is given by the component of the net flux along r ($F_{r|\mathrm{FD}}(r,\omega) = -(D/c_n)\,\nabla\phi_{\mathrm{FD}}\cdot\hat{\mathbf{r}}$, where $\hat{\mathbf{r}}$ is the unit vector along the radial coordinate r) (see also Eq. (9.37)):

$$F_{r|\mathrm{FD}}(r,\omega) = -\frac{1}{3\left(\mu_s' + \mu_a\right)}\frac{d\phi_{\mathrm{FD}}(r,\omega)}{dr}$$
$$= \frac{P_{\mathrm{FD}}(\omega)}{4\pi r^2} e^{-r\sqrt{3(\mu_s'+\mu_a)\left(\mu_a - \frac{i\omega}{c_n}\right)}}\left(1 + r\sqrt{3\left(\mu_s' + \mu_a\right)\left(\mu_a - \frac{i\omega}{c_n}\right)}\right). \tag{12.8}$$

Integration of $F_{r|\mathrm{FD}}(r,\omega)$ over the entire sphere surface ($4\pi r^2$) yields the total optical power flowing out of the sphere of radius r. In the frequency domain, the total optical power oscillates at angular frequency ω, with an amplitude ($P_{\mathrm{Tot|AC}}(\omega)$) and a phase ($\theta_{P_{\mathrm{Tot}}}(\omega)$) that are given by the magnitude and the argument, respectively, of Eq. (12.8) multiplied by $4\pi r^2$. In the non-absorbing case ($\mu_a = 0$), the amplitude and phase of the total optical power flowing out of the sphere are:

$$P_{\mathrm{Tot|AC}}(\omega)\Big|_{\mu_a=0} = r\,P_{\mathrm{FD}}(\omega)\sqrt{\frac{3\mu_s'\omega}{c_n} + \frac{2}{r}\sqrt{\frac{3\mu_s'\omega}{2c_n}} + \frac{1}{r^2}}\;e^{-r\sqrt{\frac{3\mu_s'\omega}{2c_n}}}, \tag{12.9}$$

$$\theta_{P_{\mathrm{Tot}}}(\omega)\Big|_{\mu_a=0} = r\sqrt{\frac{3\mu_s'\omega}{2c_n}} - \tan^{-1}\left(\frac{r\sqrt{\frac{3\mu_s'\omega}{2c_n}}}{1 + r\sqrt{\frac{3\mu_s'\omega}{2c_n}}}\right). \tag{12.10}$$

The case $\omega = 0$ coincides with the CW solutions of Eqs. (10.5) and (10.6) where, in non-absorbing conditions, Eq. (10.6) expresses the conservation of the total non-modulated optical power flowing out of a sphere containing the source.

The case $\omega \neq 0$, on the other hand, shows that even in the absence of absorption, the amplitude of the total modulated power flowing out of the sphere is attenuated as r increases. This raises the question: where does this modulated power go? It cannot be transferred to the medium, since we are assuming elastic scattering and no absorption, nor can it escape the medium, which extends indefinitely. The fact is that (in a limit to the analogy with electrical current) there is no real optical power associated with the AC component of the photon flux, and the power that must be conserved is the one associated with the *average* number of photons exiting the sphere per unit time, which is the total DC power (which is indeed conserved). Instead, the amplitude of the oscillation of the photon density due to the source modulation can be attenuated even in the absence of dissipation (zero absorption)

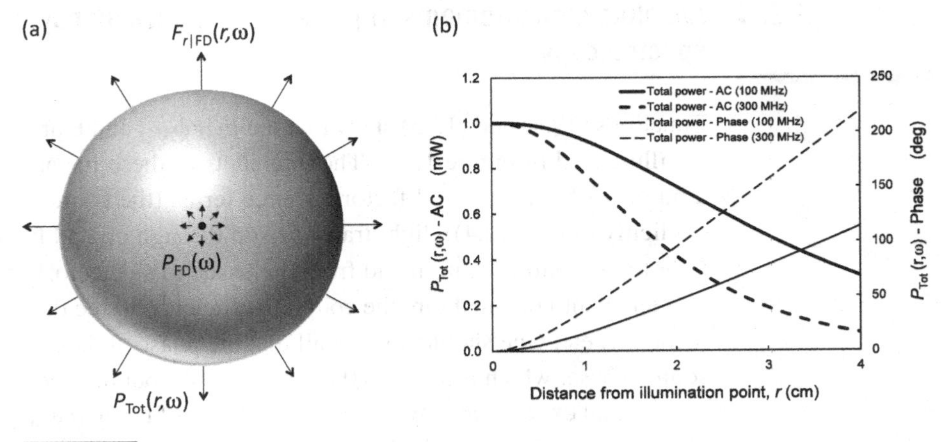

Figure 12.4

(a) A sphere of radius r centered at the isotropic, modulated photon source (emitting a power $P_{FD}(\omega)$) in a scattering medium. The radial component of the net frequency-domain energy flux ($F_{r|FD}(r, \omega)$) at the surface of the sphere is indicated by the arrows. The total frequency-domain power ($P_{Tot}(r, \omega)$) flowing out of the sphere is given by the sphere surface area ($4\pi r^2$) multiplied by $F_{r|FD}(r, \omega)$. (b) r dependence of the amplitude (AC) and phase of the modulated total power flowing out of the sphere ($P_{Tot}(r, \omega)$) for a modulated source of power $P_{FD}(\omega) = 1$ mW, a non-absorbing, scattering medium with $\mu_a = 0$ and $\mu_s' = 10$ cm^{-1}, and two modulation frequencies of 100 and 300 MHz.

as a result of the scattering effects that progressively randomize the photon paths and homogenize the angular distribution of the optical radiance. In fact, in the limit of large distances from the source ($r \to \infty$), the amplitude of the modulation (of the total power) approaches zero for every $\omega \neq 0$, even in the absence of absorption, whereas the DC total power ($\omega = 0$) stays constant and equal to the DC (or CW) source power. Another way to think of this is by remembering that, due to scattering and at large distances from the source, arriving photons have traveled a large range of distances, so many different modulation phases have been randomly blended, effectively eliminating the modulation.

Analogous with Figures 10.1(a) and 11.3(a), Figure 12.4(a) illustrates a sphere of radius r centered at the modulated photon source, the radial component of the net frequency-domain energy flux ($F_r(r, \omega)$), and the total frequency-domain power ($P_{Tot}(r, \omega)$) flowing out of the sphere. Figure 12.4(b) reports the r dependence of the amplitude and phase of $P_{Tot}(r, \omega)$ for a non-absorbing, scattering medium ($\mu_a = 0$, $\mu_s' = 10$ cm^{-1}), a photon source of power $P_{FD}(\omega) = 1$ mW and two modulation frequencies of 100 and 300 MHz. Notice in Figure 12.4(b) the stronger attenuation and the greater phase shift of the total power at higher modulation frequencies.

12.3 Absolute measurements of μ_a and μ_s' with frequency-domain spectroscopy

Equations (12.4) and (12.5) show that the ln $[r\phi_{AC}]$ and θ are linear functions of r (as illustrated in Figure 12.3). The intercepts of these linear functions depend on a number of experimental factors: source terms (the source power, P_{FD}, appears explicitly in Eq. (12.4)), light transmission through optical fibers (used to convey light to the source location and from the tissue to a detector), the optical coupling efficiency into the medium, the coupling efficiency to the detector at the measurement site, etc. One should also recall that the phase, θ, is measured relative to the source phase, which is taken as 0 in this analysis, but may be unknown or variable in an actual experiment. By contrast, the slopes of the ln $[r\phi_{AC}]$ and θ with respect to r (S_{AC} and S_θ, respectively) are solely determined by the optical properties of the medium. Specifically:

$$S_{AC} = \frac{d}{dr}\ln[r\phi_{AC}(r,\omega)] = -\sqrt{\frac{3\mu_a(\mu_s'+\mu_a)}{2}}\sqrt{\sqrt{1+\left(\frac{\omega}{c_n\mu_a}\right)^2}+1}, \tag{12.11}$$

$$S_\theta = \frac{d}{dr}\theta(r,\omega) = \sqrt{\frac{3\mu_a\left(\mu_s'+\mu_a\right)}{2}}\sqrt{\sqrt{1+\left(\frac{\omega}{c_n\mu_a}\right)^2}-1}. \tag{12.12}$$

The system of two Eqs. (12.11) and (12.12) can be inverted to yield expressions for the absorption and reduced scattering coefficients in terms of S_{AC} and S_θ (Fantini et al., 1994):

$$\mu_a = \frac{\omega}{2c_n}\left(\frac{S_\theta}{S_{AC}} - \frac{S_{AC}}{S_\theta}\right), \tag{12.13}$$

$$\mu_s' = -\frac{2c_n}{3\omega}S_{AC}S_\theta - \frac{\omega}{2c_n}\left(\frac{S_\theta}{S_{AC}} - \frac{S_{AC}}{S_\theta}\right), \tag{12.14}$$

which show how frequency-domain methods can determine the absolute values of μ_a and μ_s' on the basis of optical measurements at multiple distances from a point source. In fact, the determination of the slopes S_{AC} and S_θ requires measurements at a minimum of two values of r. As a practical note, the units of S_{AC} are typically cm^{-1}, and they are not affected by the units of ϕ_{AC} because $\frac{d}{dr}\ln[r\phi_{AC}] = \frac{1}{r} + \frac{1}{\phi_{AC}}\frac{d\phi_{AC}}{dr}$, so that the units of ϕ_{AC} cancel out. By contrast, to apply Eqs. (12.13) and (12.14) the units of θ must be radians, so that the units of S_θ are rad/cm.

It is also possible to perform multi-frequency measurements of the frequency-domain fluence rate, and to obtain the optical coefficients with a nonlinear fit of the frequency dependence of the phase and modulation (as shown in Figure 12.2), or the phase and AC amplitude (Pham et al., 2000). The advantage of this approach is that it can be applied to data collected at a single distance from the illumination point – a single collection point would not be enough with CW methods – thereby being less sensitive to macroscopic medium inhomogeneities, which may instead have a significant confounding effect on a multi-distance approach based on Eqs. (12.13) and (12.14). Nonetheless, calibration measurements on phantoms of known optical properties are required to correct for frequency-dependent instrumental factors.

12.4 Photon-density waves

On the basis of the above description, for an infinite homogeneous medium it is possible to express the frequency-domain optical energy density at a specific time point, t, ($U_{\mathrm{FD}}(r, t, \omega)$) as the sum of a constant DC component and a sinusoidally varying AC component as follows:

$$U_{\mathrm{FD}}(r,t,\omega) = \frac{3\left(\mu_s' + \mu_a\right)}{4\pi c_n} \left\{ \underbrace{P_{\mathrm{DC}} \frac{e^{ik(0)r}}{r}}_{\mathrm{CW}} + \underbrace{P_{\mathrm{FD}}(\omega) \frac{e^{-k_{\mathrm{Im}}(\omega) r}}{r} e^{i[k_{\mathrm{Re}}(\omega) r - \omega t]}}_{\mathrm{PDW}} \right\}, \tag{12.15}$$

where $k(0) = i\mu_{\mathrm{eff}}$, and $k_{\mathrm{Re}}(\omega)$ and $k_{\mathrm{Im}}(\omega)$ are the real and imaginary parts, respectively, of $k(\omega)$, which are given by:

$$k_{\mathrm{Re}}(\omega) = \frac{\mu_{\mathrm{eff}}}{\sqrt{2}} \sqrt{\sqrt{1 + \left(\frac{\omega}{c_n \mu_a}\right)^2} - 1}, \tag{12.16}$$

$$k_{\mathrm{Im}}(\omega) = \frac{\mu_{\mathrm{eff}}}{\sqrt{2}} \sqrt{\sqrt{1 + \left(\frac{\omega}{c_n \mu_a}\right)^2} + 1}. \tag{12.17}$$

(We recall that $\mu_{\mathrm{eff}} = \sqrt{3\mu_a(\mu_s' + \mu_a)}$.) The time-independent term in the brackets of Eq. (12.15) is the continuous-wave (CW) term that describes the attenuation of the DC component of the energy density as a function of r. The time-dependent term in the brackets of Eq. (12.15) has the mathematical form of a damped spherical wave propagating away from the light source. This spherical wave describes the time and space dependence of the energy density in the scattering medium. Since the energy density is directly related to the photon density, it is referred to as *photon-density wave* (PDW) (Fishkin and Gratton, 1993). The complex wavenumber $k(\omega)$ of the PDW yields the definitions of the

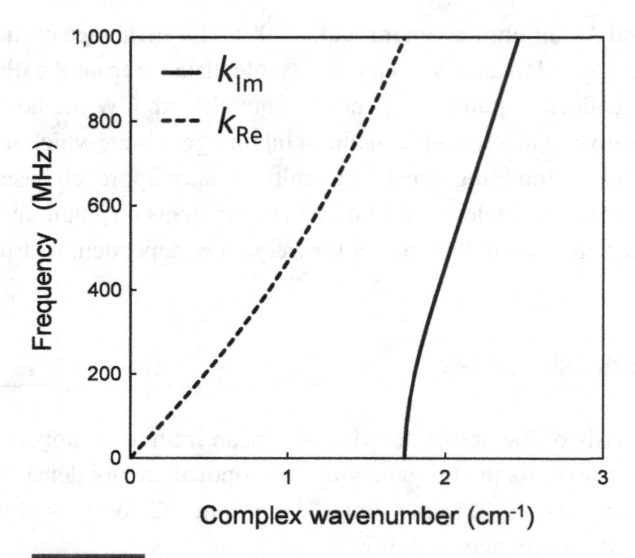

Figure 12.5

Relationships between modulation frequency ($\omega/(2\pi)$) and the real part (k_{Re}) and imaginary part (k_{Im}) of the complex wavenumber of the photon-density wave (PDW). Here, $\mu_a = 0.10$ cm^{-1} and $\mu_s' = 10$ cm^{-1}.

wavelength (λ_{PDW}), phase velocity (v_{PDW}), and attenuation length (L_{PDW}) of the photon-density waves:

$$\lambda_{PDW} = \frac{2\pi}{k_{Re}(\omega)} = \frac{2\pi}{\frac{\mu_{eff}}{\sqrt{2}}\sqrt{\sqrt{1+\left(\frac{\omega}{c_n\mu_a}\right)^2}-1}}, \tag{12.18}$$

$$v_{PDW} = \frac{\omega}{k_{Re}(\omega)} = \frac{\omega}{\frac{\mu_{eff}}{\sqrt{2}}\sqrt{\sqrt{1+\left(\frac{\omega}{c_n\mu_a}\right)^2}-1}}, \tag{12.19}$$

$$L_{PDW} = \frac{1}{k_{Im}(\omega)} = \frac{1}{\frac{\mu_{eff}}{\sqrt{2}}\sqrt{\sqrt{1+\left(\frac{\omega}{c_n\mu_a}\right)^2}+1}}. \tag{12.20}$$

(It is important not to confuse λ_{PDW} and v_{PDW} with the optical wavelength and phase velocity of the light itself.) The dispersion relation between frequency and wavenumber (whose real part [$k_{Re}(\omega)$] and imaginary part [$k_{Im}(\omega)$] are associated with propagation [i.e., the phase of photon-density waves], and attenuation [i.e., the reduction in amplitude of photon-density waves], respectively) is illustrated in Figure 12.5.

Photon-density waves are described by the same mathematical form as general damped waves, and were shown to undergo typical wave phenomena such as reflection (Fishkin and Gratton, 1993), refraction (O'Leary et al., 1992), and diffraction (Fishkin and Gratton, 1993). However, they feature two properties that distinguish photon density waves from more common optical or ultrasound waves:

1. They are strongly damped over distances short compared to their wavelength. Photon-density waves are attenuated by many orders of magnitude over one wavelength, so that they are observable only over a fraction of the wavelength, rendering them analogous to near-field waves.
2. They are energy waves. Their amplitude relates linearly to an energy, as opposed to more common cases in which the energy associated with the wave is related to the square of the amplitude of oscillating quantities (for example, the electric field for electromagnetic waves or the pressure for ultrasound waves).

The wave properties (wavelength and phase velocity) and the characteristic attenuation length of the PDW depend on the optical properties of the medium and on the modulation frequency of the source power. By considering typical values for frequency-domain spectroscopy of tissue, say $\mu_a = 0.1$ cm^{-1}, $\mu_s' = 10$ cm^{-1}, $n = 1.35$, and $f = \omega/(2\pi) = 100$ MHz, one obtains representative values of $\lambda_{\text{PDW}} \sim 25$ cm, $v_{\text{PDW}} \sim c_n/10$, and $L_{\text{PDW}} \sim 0.6$ cm. We now consider the three PDW properties (λ_{PDW}, v_{PDW}, and L_{PDW}) in more detail.

12.4.1 The wavelength of photon-density waves

The PDW wavelength, which, again, should not be confused with the wavelength of the light, depends not only on the modulation frequency, but also on the absorption and scattering properties of the medium. For a typical modulation frequency of 100 MHz, over a broad range of absorption coefficients (0.02–0.2 cm^{-1}) and reduced scattering coefficients (2–20 cm^{-1}), λ_{PDW} ranges from 10 to 80 cm. The attenuation factor of PDWs over one wavelength, i.e., the ratio between the PDW amplitudes at $r + \lambda_{\text{PDW}}$ and r, is given by:

$$\frac{r}{r + \lambda_{\text{PDW}}} e^{-k_{\text{Im}}(\omega)\lambda_{\text{PDW}}} \approx \frac{r k_{\text{Re}}(\omega)}{2\pi} e^{-2\pi \frac{k_{\text{Im}}(\omega)}{k_{\text{Re}}(\omega)}}. \tag{12.21}$$

The real and imaginary parts of the complex wavenumber $k(\omega)$ are both positive, and $k_{\text{Im}}(\omega) > k_{\text{Re}}(\omega)$ at all frequencies (which means that the rate of attenuation of PDWs with r, as given by k_{Im}, is always greater than the rate of increase of their phase with r, as given by k_{Re}). This translates into a large attenuation factor over one wavelength. In fact, this factor (given by Eq. (12.21)) ranges from $\sim 10^{-6}$ to $\sim 10^{-40}$ (!) for low absorption (0.02 cm^{-1}) and high absorption (0.2 cm^{-1}) cases, respectively, at a modulation frequency of 100 MHz. The absorption coefficient is the dominant parameter here since r and μ_s' do not appear in the exponential decay factor of Eq. (12.21), which is the one that dominates in determining the PDW attenuation. The fact that PDWs are strongly damped over one wavelength limits their measurement to within a fraction of their wavelength, so that they are near-field waves. This means that, in diffuse-field imaging applications, the spatial resolution achievable with PDWs is not diffraction-limited, but is rather determined by the properties of diffusive light propagation and signal-to-noise levels.

In the limit $\frac{\omega}{c_n\mu_a} \ll 1$, which may indicate either low frequencies ($\omega \ll c_n\mu_a$) or high absorption ($\mu_a \gg \omega/c_n$), the PDW wavelength becomes:

$$\lim_{\frac{\omega}{c_n\mu_a}\ll 1} \lambda_{\mathrm{PDW}} = 2\pi\sqrt{\frac{4\mu_a}{3\left(\mu_s' + \mu_a\right)}\frac{c_n}{\omega}}, \tag{12.22}$$

showing that the wavelength increases with absorption, decreases with scattering, and decreases with ω as $1/\omega$. In the opposite limiting case $\frac{\omega}{c_n\mu_a} \gg 1$, which may indicate either high frequencies ($\omega \gg c_n\mu_a$) or low absorption ($\mu_a \ll \omega/c_n$), the PDW wavelength becomes:

$$\lim_{\frac{\omega}{c_n\mu_a}\gg 1} \lambda_{\mathrm{PDW}} = 2\pi\sqrt{\frac{2c_n}{3\omega\left(\mu_s' + \mu_a\right)}}, \tag{12.23}$$

indicating that (in this limit) λ_{PDW} is approximately independent of absorption (since $\mu_a \ll \mu_s'$) and has an inverse square-root dependence on both scattering and frequency. The non-absorbing case is given by Eq. (12.23) with $\mu_a = 0$.

12.4.2 The phase velocity of photon-density waves

The PDW phase velocity is given by the product of the modulation frequency times the PDW wavelength. For a typical modulation frequency of 100 MHz, the range of values of λ_{PDW} discussed above (10–80 cm) corresponds to a range of phase velocities of $(0.1–0.8) \times 10^{10}$ cm/s. This is about one order of magnitude less than the speed of light in the medium ($\sim 2.2 \times 10^{10}$ cm/s), indicating that photon density waves, which describe the collective migration of photons in scattering media, propagate much slower than the local velocity of their individual photon components. The two limiting cases considered in Eqs. (12.22) and (12.23) result in the following limiting expressions for the PDW phase velocity:

$$\lim_{\frac{\omega}{c_n\mu_a}\ll 1} v_{\mathrm{PDW}} = \sqrt{\frac{4\mu_a}{3\left(\mu_s' + \mu_a\right)}}c_n, \tag{12.24}$$

$$\lim_{\frac{\omega}{c_n \mu_a} \gg 1} v_{PDW} = \sqrt{\frac{2 c_n \omega}{3\left(\mu_s' + \mu_a\right)}}. \tag{12.25}$$

Equations (12.24) and (12.25) show that v_{PDW} increases for greater c_n and smaller μ_s', both intuitive results associated with a faster photon speed and less tortuous photon paths, respectively. An increase in absorption results in a greater v_{PDW} in the low-frequency or high-absorption limit (by suppressing longer photon paths), whereas it has a negligible effect in the high-frequency or low-absorption limit. We observe that, in the low-frequency limit of Eq. (12.24), the ratio c_n/v_{PDW} is equal to the ratio of the geometrical distance from the source (r) to the mean photon pathlength to reach that distance ($\langle L \rangle$). Recalling the definition of the differential pathlength factor (DPF) in Eq. (10.14), this means that $\lim_{\frac{\omega}{c_n \mu_a} \ll 1} v_{PDW} = c_n/\text{DPF}$, as one can verify by comparing Eq. (12.24) (with $\mu_s' \gg \mu_a$) with Eq. (10.15), which reports the DPF for infinite media.

It is important to recall that the diffusion approximation requires that the time scale of variations of the fluence rate be much greater than the mean time between photon interactions (effectively isotropic scattering or absorption events) (see Section 9.5). In the frequency domain, this requirement translates into the condition $\omega \ll c_n(\mu_s' + \mu_a)$. Consequently, the expressions of Eq. (12.25) for v_{PDW} (and Eq. (12.23) for λ_{PDW}) are not valid at arbitrarily high frequencies. The P_1 approximation (presented in Section 9.4) is more general than the diffusion approximation and simply requires that photons undergo multiple scattering events, so that the angular energy density can be approximated by the first two terms in a spherical harmonics expansion (these two terms correspond to $L = 0$ and $L = 1$ in the formalism of Section 9.3). Because the P_1 approximation does not impose constraints on the modulation frequency, in the P_1 treatment is possible to extend the limit of Eq. (12.25) to the mathematical limit $\omega \to \infty$, leading to the following result (Fishkin et al., 1996):

$$\lim_{\omega \to \infty} v_{PDW} = \frac{c_n}{\sqrt{3}}. \quad (P_1 \text{ approximation}) \tag{12.26}$$

The result of Eq. (12.26) implies that, in the P_1 approximation, the leading edge of a light pulse travels a minimum distance $\sqrt{3}r$ to reach a point at a distance r from the source (Fishkin et al., 1996).

12.4.3 The attenuation length of photon-density waves

The attenuation length of the PDW (the distance over which the modulation amplitude of the PDW has decreased by a factor of $\sim 1/e$) depends on the modulation frequency, and on both absorption and scattering properties of the medium. For a typical modulation frequency of 100 MHz, over the broad range of absorption coefficients (0.02–0.2 cm^{-1}) and reduced scattering coefficients (2–20 cm^{-1}), L_{PDW} ranges from 0.3 to 2.5 cm. As discussed above, the PDW attenuation length is much shorter than the PDW wavelength. Of course, at $\omega = 0$ it reduces to the CW attenuation length, or diffusion length $L_D = 1/\mu_{eff}$ (see Eq. (12.20)). The two limiting cases considered above result in the following limiting expressions for the PDW attenuation length:

$$\lim_{\frac{\omega}{c_n \mu_a} \ll 1} L_{PDW} = L_D \left[1 - \frac{1}{8} \left(\frac{\omega}{c_n \mu_a} \right)^2 \right], \tag{12.27}$$

$$\lim_{\frac{\omega}{c_n \mu_a} \gg 1} L_{PDW} = L_D \sqrt{\frac{2 c_n \mu_a}{\omega}}. \tag{12.28}$$

Equations (12.27) and (12.28) show that L_{PDW} always decreases with increasing ω (as a result of the stronger attenuation of the modulation of the fluence rate at higher frequencies) and with increasing μ_s'. Conversely, L_{PDW} decreases with increasing μ_a only in the low-frequency or high-absorption limit, whereas it is largely insensitive to μ_a in the high-frequency or low-absorption limit. (Recall that L_D contains $\sqrt{\mu_a}$ in the denominator, thus canceling out the $\sqrt{\mu_a}$ on the right-hand side of Eq. (12.28).)

12.5 The frequency domain as the Fourier transform of the time domain

The frequency-domain and the time-domain methods are related by a temporal Fourier transformation. Conceptually, this is intuitive because the sinusoidal oscillations considered in the frequency domain can be seen as the Fourier components of the general time-varying signals observed in the time domain in response to a Dirac delta pulse of light (which corresponds to a flat frequency spectrum extending infinitely). One might think of the time-domain signal as a collection of frequency-domain signals at all possible frequencies from $-\infty$ to $+\infty$. In fact, the Green's functions (named after the British mathematician George Green [1793–1841]) for the time-domain fluence rate ($\phi_{TD}(r,t)/Q_{TD}$) and for the frequency-domain fluence rate ($\phi_{FD}(r,\omega)/P_{FD}(\omega)$) are related by a Fourier transformation. The fluence rate Green's function is a solution to a unitary, point-source term. In the time-domain, the Green's function is the time-domain fluence rate (Eq. (11.1)) divided by the source energy Q_{TD} (in which case, a "unitary" source is one that emits a unitary energy). In the frequency-domain, the Green's function is given by frequency-domain fluence rate (Eq. (12.2)) divided by the source power $P_{FD}(\omega)$ (in which case, a "unitary" source is one that emits a unitary power). Notice that the Green's functions for the time-domain and frequency-domain fluence rates have different units of $m^{-2}s^{-1}$ and m^{-2}, respectively. The Fourier relationships (in an infinite medium) between them are as follows:

$$\frac{\phi_{FD}(r,\omega)}{P_{FD}(\omega)} = \int_{-\infty}^{+\infty} \frac{\phi_{TD}(r,t)}{Q_{TD}} e^{i\omega t} dt, \tag{12.29}$$

$$\frac{\phi_{TD}(r,t)}{Q_{TD}} = \frac{1}{2\pi} \int_{-\infty}^{+\infty} \frac{\phi_{FD}(r,\omega)}{P_{FD}(\omega)} e^{-i\omega t} d\omega. \tag{12.30}$$

The corresponding Fourier relationships for the single-distance diffuse reflectance, which also has different units in the time domain ($m^{-2}s^{-1}$) and in the frequency domain (m^{-2}), are as follows:

$$R_{FD}(\rho,\omega) = \int_{-\infty}^{+\infty} R_{TD}(\rho,t) e^{i\omega t} dt, \tag{12.31}$$

$$R_{TD}(\rho,t) = \frac{1}{2\pi} \int_{-\infty}^{+\infty} R_{FD}(\rho,\omega) e^{-i\omega t} d\omega, \tag{12.32}$$

where the spatial variable (ρ) indicates a distance measured on the surface of the semi-infinite medium.

In Appendix 12.A, we explicitly calculate the integrals in Eqs. (12.29) and (12.31) to show how the frequency-domain fluence rate (in an infinite medium) and diffuse reflectance (in a semi-infinite medium) can be derived from the corresponding time-domain expressions. Because the Fourier relationships of Eqs. (12.29)–(12.32) are true in general, given the fluence rate in one domain for any boundary conditions and medium geometry, it is possible to determine the fluence rate in the other domain by Fourier or inverse Fourier transformation.

12.6 Frequency-domain diffuse reflectance

The single-distance, frequency-domain diffuse reflectance from the surface of a semi-infinite scattering medium ($R(\rho,\omega)$) is derived, in Appendix 12.A, as the Fourier transform of the corresponding time-domain diffuse reflectance (see Eq. (12.A.7)). In the DC (or CW) case corresponding to $\omega = 0$, Eq. (12.A.7) reduces to:

$$R_{CW}(\rho) = \frac{z_0}{2\pi}\left(\mu_{eff} + \frac{1}{\sqrt{z_0^2+\rho^2}}\right)\frac{e^{-\sqrt{z_0^2+\rho^2}\,\mu_{eff}}}{z_0^2+\rho^2}. \tag{12.33}$$

Considering zero boundary conditions (no extrapolation of the boundary: $z_b = 0$), the distances r_1 and r_2 introduced for the derivation of the CW single-distance reflectance in Section 10.4.4 are the same for points on the tissue boundary ($z = 0$, $r_1 = r_2 = \sqrt{z_0^2 + \rho^2}$). Therefore, the single-distance CW diffuse reflectance of Eq. (12.33) coincides with the expression of Eq. (10.23) after setting the source weight $a' = 1$, since albedo effects are not considered here (corresponding to negligible absorption).

Let's now consider distances from the source (ρ) along the surface of the medium that are much greater than the depth of the effective light source inside the medium ($z_0 = 1/\mu_s'$). (Remembering the representation in which a light source that is incident on the surface of a semi-infinite scattering medium can be replaced by an effective source location at a small distance below the surface, as determined by the scattering properties of the medium, hence at a depth of $\sim 1/\mu_s'$.) In fact, diffuse tissue spectroscopy is typically conducted at distances ρ of centimeters, whereas $1/\mu_s'$ is ~ 1 mm. By considering $\rho \gg z_0$, the frequency-domain diffuse reflectance of Eq. (12.A.7) becomes:

$$R_{\mathrm{FD}}(\rho, \omega) = \frac{z_0}{2\pi}\sqrt{3\left(\mu_s' + \mu_a\right)}\sqrt{\left(\mu_a - \frac{i\omega}{c_n}\right)} \times \left(1 + \frac{1}{\rho\sqrt{3\left(\mu_s' + \mu_a\right)}\sqrt{\left(\mu_a - \frac{i\omega}{c_n}\right)}}\right)\frac{e^{-\rho\sqrt{3(\mu_s'+\mu_a)}\sqrt{\left(\mu_a - \frac{i\omega}{c_n}\right)}}}{\rho^2}. \tag{12.34}$$

After some lengthy complex algebra, the amplitude and phase of $R_{\mathrm{FD}}(\rho, \omega)$ emerge as follows:

$$R_{\mathrm{FD|AC}}(\rho, \omega) = \frac{z_0}{2\pi}\sqrt{\left[\mu_{\mathrm{eff}}^2\sqrt{1 + \left(\frac{\omega}{c_n\mu_a}\right)^2} + \frac{\sqrt{2}\mu_{\mathrm{eff}}\sqrt{\sqrt{1 + \left(\frac{\omega}{c_n\mu_a}\right)^2} + 1}}{\rho} + \frac{1}{\rho^2}\right]} \times \left(\frac{e^{-\rho\frac{\mu_{\mathrm{eff}}}{\sqrt{2}}\sqrt{\sqrt{1+\left(\frac{\omega}{c_n\mu_a}\right)^2}+1}}}{\rho^2}\right), \tag{12.35}$$

$$\theta_{R_{\mathrm{FD}}}(\rho, \omega) = \rho\frac{\mu_{\mathrm{eff}}}{\sqrt{2}}\sqrt{\sqrt{1 + \left(\frac{\omega}{c_n\mu_a}\right)^2} - 1} - \tan^{-1}\frac{\rho\frac{\mu_{\mathrm{eff}}}{\sqrt{2}}\sqrt{\sqrt{1 + \left(\frac{\omega}{c_n\mu_a}\right)^2} - 1}}{1 + \rho\frac{\mu_{\mathrm{eff}}}{\sqrt{2}}\sqrt{\sqrt{1 + \left(\frac{\omega}{c_n\mu_a}\right)^2} - 1}}. \tag{12.36}$$

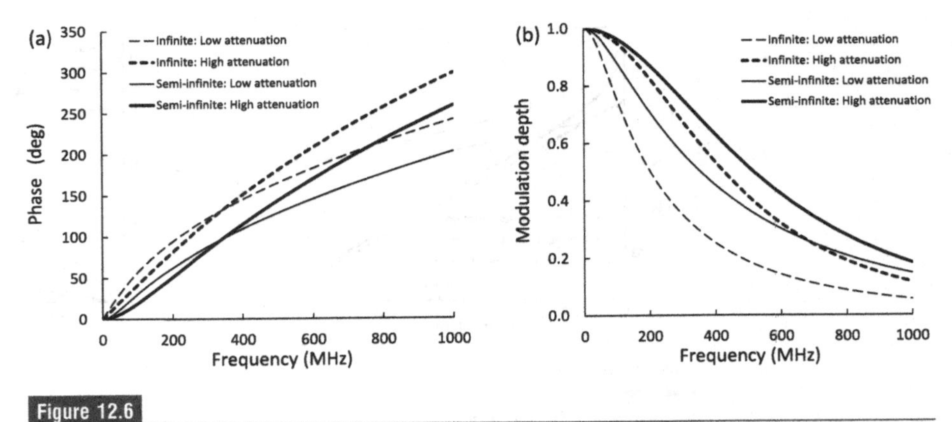

Figure 12.6

(a) Phase and (b) modulation depth of the frequency-domain fluence rate (infinite medium) and diffuse reflectance (semi-infinite medium) calculated with diffusion theory at a distance (r or ρ) of 3 cm from a point source. Two cases of optical properties are shown: low attenuation ($\mu_a = 0.02$ cm^{-1}, $\mu_s' = 5$ cm^{-1}) and high attenuation ($\mu_a = 0.10$ cm^{-1}, $\mu_s' = 10$ cm^{-1}).

One can verify that the expression for $R_{FD|AC}(\rho, \omega)$ in Eq. (12.35) reduces to $R_{CW}(\rho)$ for $\omega = 0$. The phase of the diffuse reflectance in Eq. (12.36) results from the sum of two terms: a first term that coincides with the phase of the fluence rate in an infinite medium (Eq. (12.5)), and a negative second term that ranges from 0 (at $\rho = 0$) to $-\pi/4$ (or $-45°$) (at large distances or strong attenuation). The phase ($\theta_{R_{FD}}(\rho, \omega)$) and modulation depth ($R_{FD|AC}(\rho, \omega)/R_{DC}(\rho)$) of the single-distance frequency-domain diffuse reflectance are plotted as a function of frequency ($\omega/(2\pi)$) for $\rho = 3$ cm in Figure 12.6. For comparison, Figure 12.6 also plots the corresponding fluence rate in an infinite medium (dashed lines). Figure 12.7 shows the $\ln[\rho^2 R_{FD|AC}(\rho, \omega)]$ and $\theta_{R_{FD}}(\rho, \omega)$ as a function of distance (ρ) for a modulation frequency of 100 MHz. For comparison, Figure 12.7 also shows the infinite-medium case (dashed lines), where $\ln[r\phi_{AC}]$ and the phase of ϕ_{AC} are linear functions of the distance (r). Two cases of low attenuation ($\mu_a = 0.02$ cm^{-1}, $\mu_s' = 5$ cm^{-1}) and high attenuation ($\mu_a = 0.1$ cm^{-1}, $\mu_s' = 10$ cm^{-1}) are depicted.

At distances ρ from the light source that are large enough to suppress the terms with factors $1/\rho$ and $1/\rho^2$ inside the square root in Eq. (12.35), the $\ln[\rho^2 R_{FD|AC}(\rho, \omega)]$ reduces to a linear function of ρ, with a slope that leads to the same value of the AC slope (S_{AC}) defined in Eq. (12.11) for the infinite geometry. Similarly, at distances that are large enough for the arctangent term in Eq. (12.36) to become a constant ($-45°$), the phase of the diffuse reflectance $\theta_{R_{FD}}(\rho, \omega)$ becomes a linear function of ρ with a slope that approaches the same value of the phase slope (S_θ) defined in Eq. (12.12) for the infinite geometry. Therefore, at sufficiently large distances from the source, one can determine the medium optical properties by

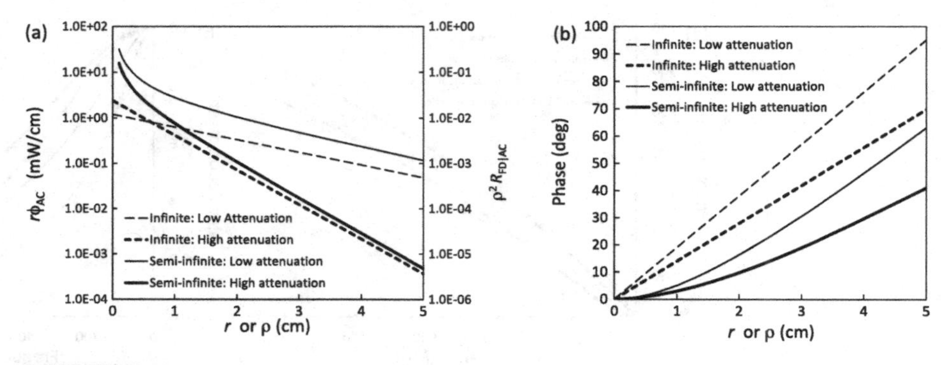

Figure 12.7

Dependence of (a) the amplitude and (b) the phase of the frequency-domain diffuse reflectance from a semi-infinite medium ($R_{\mathrm{FD|AC}}$) and fluence rate in an infinite medium (ϕ_{AC}) as a function of the distance from a point source modulated at 100 MHz (1 mW power amplitude). The distance is indicated with r in the infinite medium and with ρ on the surface of the semi-infinite medium. Two cases of optical properties are shown: low attenuation ($\mu_a = 0.02$ cm^{-1}, $\mu_s' = 5$ cm^{-1}) and high attenuation ($\mu_a = 0.10$ cm^{-1}, $\mu_s' = 10$ cm^{-1}). The amplitude is plotted in a log scale in panel (a) to highlight the linearity of $r\phi_{\mathrm{AC}}$ (at all distances) and $\rho^2 R_{\mathrm{FD|AC}}$ at long distances, with slopes that tend to the same value at long distances. The phase of the frequency-domain diffuse reflectance also becomes linear with ρ at large distances, with a slope that tends to the same value as in the infinite medium.

using the very same equations introduced for the infinite medium (Eqs. (12.13) and (12.14)), with the only difference being that the AC slope is defined in terms of the $\ln[\rho^2 R_{\mathrm{FD|AC}}(\rho, \omega)]$ instead of the $\ln[r\phi_{\mathrm{AC}}(r, \omega)]$ of the infinite-medium case. The condition for linearity of $\ln[\rho^2 R_{\mathrm{FD|AC}}(\rho, \omega)]$ is essentially that $\rho \gg 1/\mu_{\mathrm{eff}}$. Since $1/\mu_{\mathrm{eff}}$ for tissue is of the order of 1 cm in the near-infrared wavelength range, this condition is fulfilled at a distance of several centimeters from the source. The condition for linearity of $\theta_{R_{\mathrm{FD}}}(\rho, \omega)$, namely $\rho \gg \sqrt{2}/(\mu_{\mathrm{eff}}\sqrt{\sqrt{1 + (\frac{\omega}{c_n \mu_a})^2} - 1})$ is more demanding, since the right-hand side of this inequality is typically 3–4 cm for typical conditions of frequency-domain tissue spectroscopy. Figure 12.7 illustrates the fact that the linearity of the phase of the diffuse reflectance is achieved at longer distances from the source than required to achieve linearity of the log of ρ^2 times the amplitude. Beyond a distance of ~3 cm, however, both $\ln[\rho^2 R_{\mathrm{FD|AC}}(\rho, \omega)]$ and $\theta_{R_{\mathrm{FD}}}(\rho, \omega)$ show an approximately linear behavior in Figure 12.7, with a slope closely matching the infinite medium slopes S_{AC} and S_θ. Of course, in the absence of linearity, rather than using Eqs. (12.13) and (12.14) to retrieve the optical properties of a semi-infinite medium from frequency-domain diffuse reflectance measurements, one should consider the full, nonlinear expressions of Eqs. (12.35) and (12.36).

12.7 Limits of validity of frequency-domain diffusion theory

In addition to the limitations of diffusion theory already considered in the CW case (Section 10.4.6) in relation to short distances from the source, the case of high modulation frequencies in the frequency domain also falls outside the limits of applicability of diffusion theory. The diffusion theory requirement is that the modulated signal should not change significantly over the mean time between effectively isotropic scattering or absorption events (see Section 9.5). This requirement imposes a limitation on the modulation angular frequency (namely $\omega \ll c_n \mu_s'$) to guarantee that the modulation period be much longer than the mean time between scattering events. (Absorption events can be ignored here, since they occur much more rarely than scattering events within the applicability of the diffusion approximation.) As expected, greater reduced scattering coefficients result in less stringent requirements on the modulation frequency for the validity of diffusion theory. For μ_s' of the order of 1 cm^{-1}, the requirement on the modulation frequency ($\omega/(2\pi)$) is that it be much less than a few GHz, and in fact it was found that diffusion theory is accurate up to frequencies of several hundred MHz under these conditions (Fishkin et al., 1996). Even higher modulation frequencies can be accurately described by diffusion theory for reduced scattering coefficients of 5–10 cm^{-1} or more.

Appendix 12.A: Fourier integrals of time-domain Green's functions

The relevant integral required to solve the Fourier integrals for transforming time-domain fluence rates into frequency-domain fluence rates is (Gradshteyn and Ryzhik, 2007):

$$\int_0^\infty t^{\nu-1} e^{-\frac{\beta}{t}-\gamma t} dt = 2\left(\frac{\beta}{\gamma}\right)^{\nu/2} K_\nu\left(2\sqrt{\beta\gamma}\right); \quad [\mathrm{Re}(\beta) > 0, \mathrm{Re}(\gamma) > 0] \tag{12.A.1}$$

where ν, β, and γ are arbitrary complex numbers, and $K_\nu(z)$ is the modified Bessel function of imaginary argument. In the cases considered here, ν takes values of $-1/2$ and $-3/2$, β is real and positive, and γ is complex with a positive real part and a negative imaginary part. Under these conditions, the modified Bessel functions of interest are:

$$K_{-1/2}(z) = \sqrt{\frac{\pi}{2z}} e^{-z}, \tag{12.A.2}$$

$$K_{-3/2}(z) = \sqrt{\frac{\pi}{2z}} e^{-z}\left(1 + \frac{1}{z}\right), \tag{12.A.3}$$

12.A.1 Frequency-domain fluence rate in an infinite medium

In the case of an infinite medium, the Fourier integral that yields the frequency-domain fluence rate, according to Eq. (12.29) and the time-domain fluence rate of Eq. (11.1), is:

$$\phi_{FD}(r,\omega) = P_{FD}(\omega)c_n\left[\frac{3\left(\mu_s'+\mu_a\right)}{4\pi c_n}\right]^{3/2}\int_0^{\infty} t^{-3/2}e^{-\frac{3(\mu_s'+\mu_a)r^2}{4c_n t}-\mu_a c_n t}e^{i\omega t}\,dt, \tag{12.A.4}$$

where the lower integration limit has been set to 0 (the time of the source pulse emission) because the time-domain fluence rate is 0 at $t < 0$. In this case, $\nu = -1/2$, $\beta = 3(\mu_s' + \mu_a)r^2/(4c_n)$, and $\gamma = \mu_a c_n - i\omega$. The solution of Eq. (12.A.4), in light of Eq. (12.A.1) is therefore:

$$\begin{aligned}\phi_{FD}(r,\omega) &= P_{FD}(\omega)c_n\left[\frac{3\left(\mu_s'+\mu_a\right)}{4\pi c_n}\right]^{3/2} 2\left[\frac{3\left(\mu_s'+\mu_a\right)r^2}{4c_n\left(\mu_a c_n - i\omega\right)}\right]^{-1/4} \\ &\quad\times K_{-1/2}\left(2\sqrt{\frac{3\left(\mu_s'+\mu_a\right)r^2}{4c_n}\left(\mu_a c_n - i\omega\right)}\right) \\ &= P_{FD}(\omega)\frac{\left[3\left(\mu_s'+\mu_a\right)\right]^{5/4}}{(2\pi)^{3/2}r^{1/2}}\left(\frac{\mu_a c_n - i\omega}{c_n}\right)^{1/4} \\ &\quad\times\sqrt{\frac{\pi}{4\sqrt{\frac{3(\mu_s'+\mu_a)r^2}{4c_n}\left(\mu_a c_n - i\omega\right)}}}\,e^{-2\sqrt{\frac{3(\mu_s'+\mu_a)r^2}{4c_n}(\mu_a c_n - i\omega)}} \\ &= P_{FD}(\omega)\frac{3\left(\mu_s'+\mu_a\right)}{4\pi}\frac{e^{-r\sqrt{3(\mu_s'+\mu_a)\left(\mu_a-\frac{i\omega}{c_n}\right)}}}{r},\end{aligned} \tag{12.A.5}$$

which coincides with the expression in Eq. (12.2) obtained by solving the frequency-domain diffusion equation.

12.A.2 Frequency-domain diffuse reflectance from a semi-infinite medium

In the case of a semi-infinite medium with zero boundary conditions ($z_b = 0$), the Fourier integral that yields the frequency-domain reflectance, according to Eq. (12.31) and the time-domain diffuse reflectance of Eq. (11.18), is:

$$R_{FD}(\rho,\omega) = z_0\left[\frac{3\left(\mu_s'+\mu_a\right)}{4\pi c_n}\right]^{3/2}\int_0^{\infty} t^{-5/2}e^{-\frac{3(\mu_s'+\mu_a)\left(z_0^2+\rho^2\right)}{4c_n t}-\mu_a c_n t}e^{i\omega t}\,dt, \tag{12.A.6}$$

where, again, the lower integration limit has been set to 0 because the time-domain reflectance is 0 at $t < 0$ ($t = 0$ is the time of the source pulse emission). In this case, $\nu = -3/2$, $\beta = 3(\mu_s' + \mu_a)\left(z_0^2 + \rho^2\right)/(4c_n)$, and $\gamma = \mu_a c_n - i\omega$. The solution of Eq. (12.A.6), in light of Eq. (12.A.1) is therefore:

$$
\begin{aligned}
&R_{\mathrm{FD}}(\rho, \omega) \\
&= z_0\left[\frac{3\left(\mu_s' + \mu_a\right)}{4\pi c_n}\right]^{3/2} 2\left[\frac{3\left(\mu_s' + \mu_a\right)\left(z_0^2 + \rho^2\right)}{4c_n\left(\mu_a c_n - i\omega\right)}\right]^{-3/4} \\
&\quad \times K_{-3/2}\left(2\sqrt{\frac{3\left(\mu_s' + \mu_a\right)\left(z_0^2 + \rho^2\right)}{4c_n}\left(\mu_a c_n - i\omega\right)}\right) \\
&= z_0 \frac{\left[3\left(\mu_s' + \mu_a\right)\right]^{\frac{3}{4}}}{2^{\frac{1}{2}}\pi^{\frac{3}{2}}\left(z_0^2 + \rho^2\right)^{\frac{3}{4}}}\left(\frac{\mu_a c_n - i\omega}{c_n}\right)^{\frac{3}{4}} \sqrt{\frac{\pi}{4\sqrt{\frac{3(\mu_s'+\mu_a)(z_0^2+\rho^2)}{4c_n}}\left(\mu_a c_n - i\omega\right)}} \\
&\quad \times e^{-2\sqrt{\frac{3(\mu_s'+\mu_a)(z_0^2+\rho^2)}{4c_n}\left(\mu_a c_n - i\omega\right)}}\left(1 + \frac{1}{2\sqrt{\frac{3(\mu_s'+\mu_a)(z_0^2+\rho^2)}{4c_n}}\left(\mu_a c_n - i\omega\right)}\right) \\
&= \frac{z_0\sqrt{3\left(\mu_s' + \mu_a\right)}}{2\pi} \frac{\sqrt{\left(\mu_a - \frac{i\omega}{c_n}\right)}}{\left(z_0^2 + \rho^2\right)}\left(1 + \frac{1}{\sqrt{z_0^2 + \rho^2}\sqrt{3\left(\mu_s' + \mu_a\right)}\sqrt{\left(\mu_a - \frac{i\omega}{c_n}\right)}}\right) \\
&\quad \times e^{-\sqrt{z_0^2+\rho^2}\sqrt{3(\mu_s'+\mu_a)}\sqrt{\left(\mu_a - \frac{i\omega}{c_n}\right)}}. \qquad (12.A.7)
\end{aligned}
$$

Equation (12.A.7) gives the frequency-domain single-distance diffuse reflection from a semi-infinite medium with zero boundary conditions (i.e., $z_b = 0$, meaning that the extrapolated boundary introduced in Section 9.6.2 coincides with the physical tissue boundary).

Problems – answers to problems with * are on p. 647

12.1 Consider two modulated light sources (at the same modulation frequency) and the associated fluence rates having amplitudes $\phi_{\mathrm{AC}}^{(1)}(\mathbf{r}, \omega)$ (for source 1), $\phi_{\mathrm{AC}}^{(2)}(\mathbf{r}, \omega)$ (for source 2) and phases $\theta^{(1)}(\mathbf{r}, \omega)$ (for source 1), $\theta^{(2)}(\mathbf{r}, \omega)$ (for source 2) at a given position $\mathbf{r}$ in a scattering medium. What are the amplitude and phase of the total fluence rate resulting from the superposition of the fluence rates associated with the two individual sources?

12.2 Find the expressions for the DC, AC, and phase of the fluence rate in a non-absorbing, uniform, infinite turbid medium (i.e., for $\mu_a = 0$ and

$\mu_s' \neq 0$) for a point source emitting a power that is sinusoidally modulated at angular frequency ω.

12.3* Using the expressions derived in Problem 12.2, and considering a non-absorbing medium with $n = 1.38$, a source modulation frequency ($\omega/(2\pi)$) of 115 MHz, and a distance from the point source of 2.8 cm, find:

(a) the modulation depths for two different media with $\mu_s' = 5\ \text{cm}^{-1}$ and $\mu_s' = 15\ \text{cm}^{-1}$, respectively (assume $P_{CW} = P_{FD}(\omega)$);

(b) the phase difference between the fluence rates measured in the media with $\mu_s' = 5\ \text{cm}^{-1}$ and $\mu_s' = 15\ \text{cm}^{-1}$.

12.4 Equations (12.13) and (12.14) express the absorption and the reduced scattering coefficients of a uniform, infinite medium as a function of the AC and phase slopes, as defined in the text. Find the corresponding expressions as a function of (a) the DC and phase slopes, and (b) the DC and AC slopes.

12.5* In a frequency-domain measurement (source modulation frequency: 80 MHz) conducted in a macroscopically homogeneous, infinite turbid medium with $n = 1.33$, an investigator measures AC and phase linear trends as a function of the distance from a point source. The slopes of these linear trends are $-2.436\ \text{cm}^{-1}$ for the $\ln[r\phi_{AC}(r, \omega)]$, and 10.88°/cm for the phase. Find the absorption and the reduced scattering coefficients of the medium.

12.6* The law of propagation of error for a quantity z that has a functional dependence $z(x, y)$ on uncorrelated measured quantities x and y states that $\Delta z = [\frac{\partial z}{\partial x}|^2_{x_0,y_0}(\Delta x)^2 + \frac{\partial z}{\partial y}|^2_{x_0,y_0}(\Delta y)^2]^{1/2}$, where x_0 and y_0 are the measured values of x and y, respectively, and Δx, Δy, Δz are the errors in x, y, and z, respectively. If the errors on the measured AC and phase slopes (assumed to be uncorrelated) of Problem 12.5 are $0.03\ \text{cm}^{-1}$ and 0.2°/cm, respectively, what are the errors on the absorption and reduced scattering coefficients found in Problem 12.5?

12.7* Find the wavelength, phase velocity, and attenuation length for a 120 MHz photon-density wave that is propagating away from a point source in an infinite medium with $n = 1.35$, $\mu_a = 0.082\ \text{cm}^{-1}$, and $\mu_s' = 11.3\ \text{cm}^{-1}$.

12.8 The P_1 approximation yields the result that the leading edge of an optical pulse emitted by a point source travels a minimum distance of $\sqrt{3}r$ to reach a point at a distance r from the source. Considering a cubic lattice of step $L = r/n$ (with n integer), what is a geometrical interpretation of this result?

12.9* Two point sources are modulated at the same frequency and are located at a distance of 4 cm from each other in a macroscopically homogeneous infinite medium with $\mu_a = 0.054\ \text{cm}^{-1}$, and $\mu_s' = 9.3\ \text{cm}^{-1}$. Their modulated power emissions have the same amplitude ($P_{FD} = 0.82$ mW) and modulation depth ($m = 0.6$) but are in opposition of phase (i.e., they have a phase difference of π). Find the fluence rate resulting from the superposition of the two photon-density waves at the mid-point of the source-to-source line (i.e., at the point that is half-way between the two sources at a distance of 2 cm from each source).

12.10 Show that, for $\omega = 0$, $R_{FD|AC}(\rho, \omega = 0)$ of Eq. (12.35) reduces to $R_{CW}(\rho)$ of Eq. (12.33).

References

Fantini, S., Franceschini, M. A., Fishkin, J. B., Barbieri, B., and Gratton, E. (1994). Quantitative determination of the absorption spectra of chromophores in strongly scattering media: a light-emitting-diode based technique. *Applied Optics*, 33, 5204–5213.

Fishkin, J. B., and Gratton, E. (1993). Propagation of photon-density waves in strongly scattering media containing an absorbing semi-infinite plane bounded by a straight edge. *Journal of the Optical Society of America A*, 10, 127–140.

Fishkin, J. B., Fantini, S., vandeVen, M. J., and Gratton, E. (1996). Gigahertz photon density waves in a turbid medium: theory and experiments. *Physics Review E*, 53, 2307–2319.

Gradshteyn, I. S., and Ryzhik, I. M. (2007). *Table of Integrals, Series, and Products*, 7th edn. Burlington, MA: Academic Press, Ch. 3: 3.472 #9.

O'Leary, M. A., Boas, D. A., Chance, B., and Yodh, A. G. (1992). Refraction of diffuse photon density waves. *Physical Review Letters*, 69, 2658–2661.

Pham, T. H., Coquoz, O., Fishkin, J. B., Anderson, E., and Tromberg, B. J. (2000). Broad bandwidth frequency domain instrument for quantitative tissue optical spectroscopy. *Review of Scientific Instruments*, 71, 2500–2513.

Further reading

General description of frequency-domain spectroscopy

Cerussi, A. E., and Tromberg, B. J. (2003). Photon migration spectroscopy frequency-domain techniques. In *Biomedical Photonics Handbook*, ed. T. Vo-Dinh. Boca Raton, FL: CRC Press, Ch. 22.

Absolute measurements of optical properties with frequency-domain methods

Fantini, S., Franceschini, M. A., and Gratton, E. (1994). Semi-infinite-geometry boundary problem for light migration in highly scattering media: a frequency-domain study in the diffusion approximation. *Journal of the Optical Society of America B*, 11, 2128–2138.

Pham, T. H., Coquoz, O., Fishkin, J. B., Anderson, E., and Tromberg, B. J. (2000). Broad bandwidth frequency-domain instrument for quantitative tissue optical spectroscopy. *Review of Scientific Instruments*, 71, 2500–2513.

Sultan, E., Najafizadeh, L., Gandjbakhche, A. H., Pourrezaei, K., and Daryoush, A. (2013). Accurate optical parameter extraction procedure for broadband near-infrared spectroscopy of brain matter. *Journal of Biomedical Optics*, 18, 017008.

Solutions of the frequency-domain diffusion equation in various geometries

Arridge, S. R., Cope, M., and Delpy, D. T. (1992). The theoretical basis for the determination of optical pathlengths in tissue: temporal and frequency analysis. *Physics in Medicine and Biology*, 37, 1531–1560.

Photon-density waves

Boas, D. A., O'Leary, M. A., Chance, B., and Yodh, A. G. (1994). Scattering of diffuse photon density waves by spherical inhomogeneities within turbid media: analytic solution and applications. *Proceedings of the National Academy of Sciences USA*, 91, 4887–4891.

Svaasand, L. O., Tromberg, B. J., Haskell, R. C., Tsay, T. T., and Berns, M. W. (1993). Tissue characterization and imaging using photon density waves. *Optical Engineering*, 32, 258–266.

Tromberg, B. J., Svaasand, L. O., Tsay, T. T., and Haskell, R. C. (1993). Properties of photon density waves in multiple-scattering media. *Applied Optics*, 32, 607–616.

13 Instrumentation and experimental methods for diffuse tissue spectroscopy

Optical measurements on biological tissues require four major components: a light source, a means to deliver light to and from the investigated tissue, an optical detector, and a method to process the signals generated by the optical detectors. In this chapter, we build upon the general description of spectroscopy instrumentation in Section 4.7, and we add detail about all four components as applicable to systems devoted to diffuse optical measurements on biological tissue. We also depict typical configurations of time-resolved instrumentation in the time domain and frequency domain. Additionally, we address the safe limits of skin exposure to light, which ultimately set the maximum levels of radiant exposure or optical intensity that can be delivered to tissue and still be considered noninvasive for diagnostic applications. (In Chapter 19, we will consider higher exposure levels, which may be encountered with therapeutic applications.)

13.1 Light sources

13.1.1 Relevant properties of light sources for diffuse optical spectroscopy

Optical spectroscopy in the diffusion regime does not involve specific requirements for the illumination source in terms of its coherence (except for diffuse correlation spectroscopy, as discussed in Section 8.9.1), polarization, directionality, and bandwidth. In fact, polarization and directionality of the incident light are quickly lost, as multiple scattering events result in a distribution of photon paths, and randomize the photon polarization and direction of propagation. Coherence (to be discussed extensively in Chapter 18) is reduced by virtue of the range of optical pathlengths between the source and a detection point. Additionally, the broad spectral features typically associated with absorption and scattering of biological tissue do not translate into stringent spectral resolution requirements, so that highly monochromatic illumination is usually not required. The relevant properties of light illumination for optical tissue spectroscopy in the diffusion regime are mostly:

- the spectral distribution, i.e., the set of wavelengths or wavelength bands of illumination;
- the temporal profile of the source emission, which is constant in the continuous-wave (CW) domain, pulsed in the time domain (TD), or intensity-modulated in the frequency domain (FD);
- the radiant exposure (H) or intensity (I) delivered to the tissue surface.

13.1.2 Spectral distribution of illumination

Quantifying the concentrations of multiple tissue chromophores, or their changes due to physiological processes, requires measurements at multiple wavelengths. At the very least, the number of wavelengths must equal the number of chromophores, as discussed in Chapter 10 in relation to Eq. (10.1). In the near-infrared spectral range, the optical absorption of most biological tissues results from a small number of chromophores: heme proteins such as hemoglobin, myoglobin, and cytochrome *c* oxidase, as well as water, melanin, and lipids. Due to the broad extinction spectra of these compounds, the individual nominal wavelengths of illumination (λ_j) do not usually need to be associated with truly monochromatic light. Helium-neon (He-Ne) lasers, emitting in the red spectral region (632.8 nm) with a wavelength bandwidth of <0.1 nm can be used, but laser diodes (LDs), with center wavelengths in the red/near-infrared spectral region and larger bandwidths of a few nanometers, are equally suitable light sources for diffuse optical spectroscopy. Even light-emitting diodes (LEDs) with significantly broader bandwidths of a few tens of nanometers can still be used, although such large bandwidths may introduce some errors if the measured optical signals are assigned to the specific center wavelength of the LED's emission band in quantitative analyses. However, LEDs can also be used as light sources for spectrally resolved measurements, thus truly taking advantage of their relatively broad bandwidth of emission (Fantini et al., 1994).

For even broader wavelength bandwidths, xenon arc lamps and quartz tungsten halogen lamps cover the entire visible through near-infrared spectral range continuously, and can also be used for spectroscopic measurements by CW diffuse optical methods over the entire wavelength range (Mourant et al., 1997). Tunable lasers also allow for spectral measurements over broad wavelength ranges. For example, titanium-doped sapphire (Ti:sapphire) lasers can be tuned in the wavelength range 650–1100 nm, whereas dye lasers can be tuned anywhere in the visible and near-infrared through appropriate dye selection. As introduced in Section 4.8.6, a more recent development for broadband tissue spectroscopy is supercontinuum generation by nonlinear propagation of ultrashort laser pulses in photonic crystal fibers, which results in illumination over the wavelength band 550–1000 nm (Bassi et al., 2004) or 500–1200 nm (Abrahamsson et al., 2004).

13.1.3 Pulsed and modulated sources

Time-domain methods are based on short-pulse light sources (pulse durations of picoseconds) with pulse repetition rates of ~10–80 MHz. Frequency-domain methods require intensity-modulated sources with modulation frequencies of ~100 MHz. Some light sources are intrinsically pulsed (for example, Ti:sapphire lasers), and power modulation at angular frequency ω can be achieved by pulse trains with repetition times (t_{rep}) of $2\pi/\omega$ (where, here, $1/t_{rep}$ represents the desired fundamental modulation frequency). In the case of Ti:sapphire lasers, the individual light pulses can have durations (t_{pulse}) as short as ~0.01–1 ps, and the repetition rate is typically ~80 MHz. Therefore, the individual pulses lend themselves to time-domain diffuse optical measurements, since the time between pulses (12.5 ns at a pulse repetition rate of 80 MHz) is usually long enough to guarantee that no photons generated by a given pulse are left to be detected after the next pulse is emitted.

The typical repetition rate of 80 MHz for Ti:sapphire pulses is also ideal for frequency-domain diffuse optical measurements, for which, as mentioned above, the repetition rate represents the fundamental modulation frequency (~80 MHz), whereas the individual pulse duration is inversely related to the modulation-frequency bandwidth of the pulse train emission (GHz bandwidth for a ps pulse duration). This result follows from the fact that the Fourier transform of a train of pulses with individual duration t_{pulse} and repetition time t_{rep} is given by a train of frequency spikes (a comb function) at frequencies n/t_r ($n = 0,1,2\ldots$), weighted by an envelope function given by the Fourier transform of the individual pulse shape. For the case of a Gaussian pulse shape with full-width-half-maximum of t_{pulse}, its Fourier transform is also a Gaussian with a frequency standard deviation of $1/(2\pi t_{pulse})$. As a result, typical values of $t_{pulse} = 1$ ps and $t_{rep} = 12.5$ ns for a Ti:sapphire laser result in emission frequencies of $n \times 80$ MHz over a bandwidth of ~150 GHz. The weighted comb function representing the frequency-domain emission associated with a temporal pulse train, and the inverse relationship between the individual pulse duration and the corresponding frequency bandwidth are illustrated in Figure 13.1.

Laser diodes and light-emitting diodes (LEDs) can be pulsed and intensity-modulated by driving them with a pulsed or modulated current; however, their time response poses a limit to the shortest pulse or highest modulation frequency that can be achieved. Laser diodes feature faster responses (rise time: ~0.1 ns) than

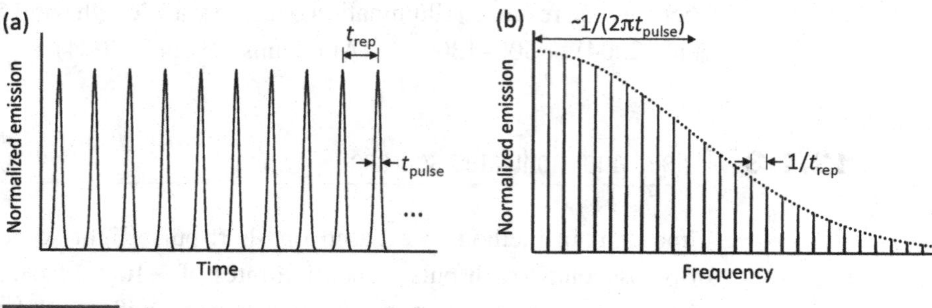

Figure 13.1

(a) Time-domain representation of a train of light pulses of duration t_{pulse} and repetition time t_{rep}, and (b) corresponding frequency-domain representation consisting of a collection of delta-like spikes at frequencies n/t_{rep} ($n = 0,1,2\ldots$) over a bandwidth $\sim 1/(2\pi t_{pulse})$.

light-emitting diodes (rise time: ~1 ns), and are therefore more appropriate for time-resolved applications, while both LEDs and laser diodes are suitable sources for frequency-domain methods. Laser diodes can emit ~100 ps pulses and can be modulated at frequencies up to several GHz (the −3 dB bandwidth is given by 0.35 divided by the rise time [see Section 16.1.1 for the definition of decibel, dB]), whereas LEDs can be modulated at frequencies up to ~50–100 MHz.

13.1.4 Maximum permissible levels of optical exposure of the skin

Safety considerations pose a limit to the amount of optical energy that can be delivered to tissues in a given time period. For diffuse tissue spectroscopy studies, one must consider the maximum permissible exposure (MPE) of the skin to optical irradiation in the red and near-infrared spectral regions. The skin MPE depends on wavelength (it is higher in the near-infrared than in the visible) and on the exposure duration (it is higher at a shorter exposure time, for which the energy per unit area, i.e., the radiant exposure, becomes the relevant parameter in place of the power per unit area, i.e., the intensity).

Table 13.1 lists the skin MPE for a laser beam, as reported by the American National Standards Institute (ANSI) (ANSI Z136.1, 2007), whereas Figure 13.2 reports the skin maximum permissible intensity as a function of exposure time to laser radiation for two different wavelength bands (0.4–0.7 μm and 1.05–1.40 μm), and for a wavelength of 0.85 μm. There are also safety guidelines for maximum permissible exposure of the skin to incoherent optical radiation emitted by LEDs, arc lamps, halogen tungsten lamps, etc., compiled by the International Electrotechnical Commission. These guidelines are based on preventing thermal

Table 13.1 Maximum permissible skin exposure to a laser beam with <3.5 mm aperture (adapted from Tables 6 and 7 of the American National Standard for Safe Use of Lasers [ANSI Z136.1, 2007]). The wavelength (λ) is expressed in μm, and the exposure time t is expressed in s.

		Maximum permissible exposure	
Wavelength λ (μm)	**Exposure duration t(s)**	**Radiant exposure H(J/cm^2)**	**Intensity I(W/cm^2)**
0.400–0.700	10^{-9}–10^{-7}	2×10^{-2}	–
	10^{-7}–10	$1.1 \times t^{0.25}$	–
	$10 - 3 \times 10^4$	–	0.2
0.700–1.050	10^{-9}–10^{-7}	$2 \times 10^{2(\lambda-1.7)}$	–
	10^{-7}–10	$1.1 \times 10^{2(\lambda-0.7)} \times t^{0.25}$	–
	$10 - 3 \times 10^4$	–	$0.2 \times 10^{2(\lambda-0.7)}$
1.050–1.400	10^{-9}–10^{-7}	10^{-1}	–
	10^{-7}–10	$5.5 \times t^{0.25}$	–
	$10 - 3 \times 10^4$	–	1

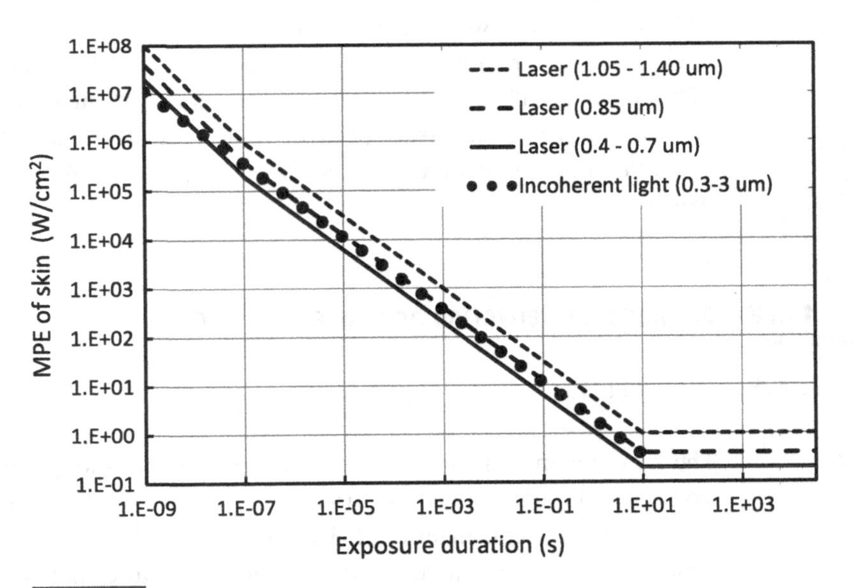

Figure 13.2

Maximum permissible exposure (MPE) of the skin to laser and incoherent optical radiation as a function of exposure duration for different wavelengths (laser: 0.4–0.7 μm, 0.85 μm, 1.05–1.40 μm; incoherent light: 0.3–3 μm). Based on data from Tables 6 and 7 of the American National Standard for Safe Use of Lasers (ANSI Z136.1, 2007) and Section 4.9.2 of the International Electrotechnical Commision technical report on maximum permissible exposure to incoherent optical radiation (IEC TR 60825-9, 1999).

damage to the skin. In the visible to near-infrared spectral range of 380–3000 nm, the maximum permissible radiant exposure (H) of the skin for exposure duration $t < 10$ s is given by $H = 2 \times t^{0.25}$ J/cm^2 (with t in seconds and H in J/cm^2) (IEC TR 60825–9, 1999). The MPE of the skin for incoherent optical radiation coincides with that for laser radiation in the near-infrared spectral band (~0.85 μm) as shown in Figure 13.2, where the MPE for incoherent radiation is shown as a dotted line.

It is instructive to consider that the intensity of the solar emission that reaches the earth's surface is of the order of 0.1 W/cm^2. Less than 5% of this energy is in the ultraviolet, about 45% in the visible, and the remainder in the infrared spectral regions. The MPE of the skin for long laser irradiation times (minutes or hours) in the visible portion of the spectrum (0.2 W/cm^2) is only about twice the intensity from the sun that reaches the earth's surface. High-altitude hiking can significantly increase the exposure risk, given that the UVB intensity (around 310 nm) at 3000 m altitude (~10,000 ft) is ~50% higher than at sea level (Askew, 2002).

The intensity limit of 0.2 W/cm^2 (applicable to visible light illumination for exposure of 10 s or longer) is a good reference point when considering the optical radiation levels on the skin for diffuse tissue spectroscopy. The maximum permissible illumination power depends on the surface area of illumination. An intensity of 0.2 W/cm^2 over a 1 mm^2 illumination spot corresponds to an illumination power of 2 mW. In fact, the average illumination power in continuous-wave, time-domain, and frequency-domain tissue spectroscopy is typically of the order of milliwatts, given that illumination areas are governed by the sizes of optical fibers used for delivery.

13.2 Methods for delivering and collecting light

13.2.1 Optical fibers

The most common method of light delivery and collection in diffuse tissue spectroscopy is through optical fibers (which are described in Section 4.8). There are no specific requirements in terms of size (except for diffuse correlation spectroscopy, which requires single-mode fibers), numerical aperture, and material, so that they can be selected to optimize each specific application on the basis of the efficiency of optical coupling with the source or detector, intensity requirements on the tissue, levels of required optical collection from the tissue, transmission properties at the wavelengths used, and mechanical properties such as stiffness, etc. For maximizing delivery or collection of NIR light in applications of diffuse optical spectroscopy, large multimode glass optical fibers or fiber bundles (0.4–4 mm in total active diameter) with high numerical apertures of 0.5–0.6 (corresponding to an acceptance angle of 30–37°) are common. In particular, a high numerical aperture is desirable for the collection of diffuse light from tissue, where the radiance is distributed over the entire solid angle. However, for light collection, a fiber with a high numerical aperture may not satisfy the requirement of *f*-number matching between the optical fiber and a spectrometer (see Section 4.7.2), so that best compromise solutions may have to be identified based on specifications of instrumentation.

13.2.2 Direct illumination and light collection

The collimated or focused emission of laser sources lends itself to direct tissue illumination in a non-contact fashion, and even the broader angular emissions of laser diodes (~10–40°) and LEDs (~50°) can be directly shone on tissues, provided that the light sources are either in contact or at a short distance from the tissue. In particular, methods that are based on structured illumination of the tissue surface (see Section 10.4.5) invoke non-contact illumination of the tissue from a "light engine" that projects the spatially modulated pattern of light onto the tissue surface. The light sources used for structured illumination are typically CW, and can be LEDs or lasers, and the spatial modulation is often achieved with the types of optical components that are used in digital projectors to generate the image that is projected.

Some optical detectors, for example photodiodes and avalanche photodiodes, are suitable for direct placement on the examined tissue for collection of diffuse light emerging from the tissue, without a need for optical fibers. Non-contact optical collection can also be performed by means of objective or camera lenses that image light from a specific tissue location onto the sensitive element of an optical detector or camera.

13.3 Optical detectors

13.3.1 General description of optical detectors

The basic approach of light detectors in tissue spectroscopy is to transform incoming photons (which make up the detected optical signal) into electrons (which result in an electrical signal). In this fashion, the detected optical energy is converted into an electric charge, and the detected optical power is converted into an electric current. These electric charges and currents give rise to the output signals of the optical detectors. The "conversion" of photons into electrons is based on the *photoelectric effect*. In the case of photomultiplier tubes (PMTs), intensified charge-coupled device cameras (ICCD), or streak cameras, the conversion results from the photon-induced emission of electrons from a photocathode. In the case of solid-state photodiodes (PD), p-i-n photodiodes (PIN PD), avalanche photodiodes (APD), and charge-coupled device (CCD) cameras, electrons are injected into the conduction band of a semiconductor material during the photon-induced creation of electron-hole pairs in the depletion or photo-sensitive regions of p-n junctions. The emission of electrons from a photocathode is referred to as *external photoelectric effect*, whereas the excitation of electrons into the conduction band of a semiconductor is referred to as *internal photoelectric effect*.

A p-n junction is an interface between two layers of a semiconductor that are doped to generate either a deficiency of valence electrons (associated with positively charged holes, in a p-type material) or an excess of valence electrons (negative particles, in an n-type material). In p-i-n photodiodes, an undoped layer of intrinsic material, i-type, is inserted between the p and n layers to increase the photon-sensitive area (for higher sensitivity) and reduce the junction capacitance (for faster response). A p-n junction is forward biased when the p-side and the n-side are connected to the positive and negative terminals, respectively, of a voltage source. In this forward-biased mode, the device conducts current regardless of illumination. In the case of zero bias (no applied voltage), the p-n junction operates in the *photovoltaic mode*, which is the basis for the operation of solar cells.

A p-n junction is reverse biased when the p-side and the n-side are connected to the negative and positive terminals of a voltage source, respectively, resulting in a small back current, which is referred to as *dark current* in the absence of illumination. This is the so-called *photoconductive mode*, in which the device operates as a photodetector, and its current is proportional to the incident optical power. PDs, PIN PDs, APDs, and CCDs all essentially use reverse-biased p-n junctions in different flavors and configurations. The reverse bias is kept well below the *breakdown voltage*, which is the maximum reverse voltage that the device can withstand before acting as a conductor. The exception is the APD, where the strong reverse bias approaches the breakdown voltage and accelerates photoelectrons to the point that they have enough energy to generate additional electron-hole pairs (by impact ionization), resulting in an avalanche multiplication of charge carriers (in a manner not unlike a nuclear chain-reaction). Schematic diagrams of the p-n and p-i-n junctions in PDs, PIN PDs, and APDs are illustrated in Figure 13.3, which also shows the device polarization and the conversion of photons into electron (filled circles) and holes (empty circles).

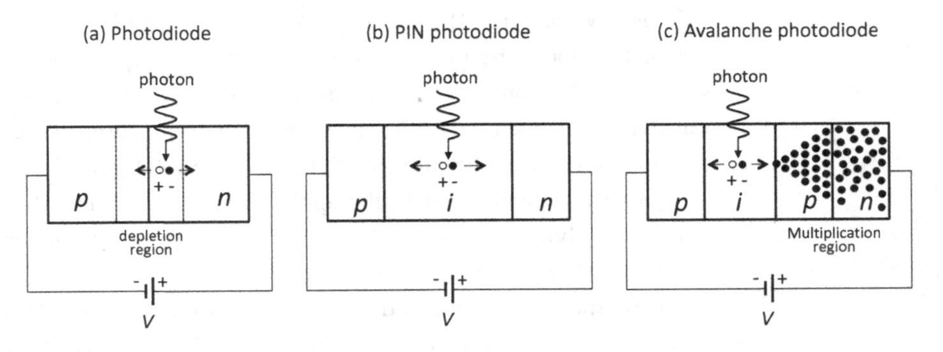

Figure 13.3

Schematic diagrams of semiconductor-based photodetectors: (a) photodiode (PD), (b) p-i-n photodiode (PIN PD), (c) avalanche photodiode (APD). A detected photon generates an electron-hole pair (electron (−): solid circle; hole (+): empty circle) that results in the detector photocurrent. The photon detection occurs in the depletion region (the region from where mobile charge carriers have diffused away) of the p-n junction in PDs (a), and in an intrinsic (undoped) semiconductor region in PIN PDs (b) and APDs (c). The avalanche multiplication of charge carriers in APDs occurs in a multiplication region consisting of a p-layer and a heavily doped n-layer. The operating voltage (V) is of the order of ~10 V in PDs and PIN PDs, and ~100 V in APDs.

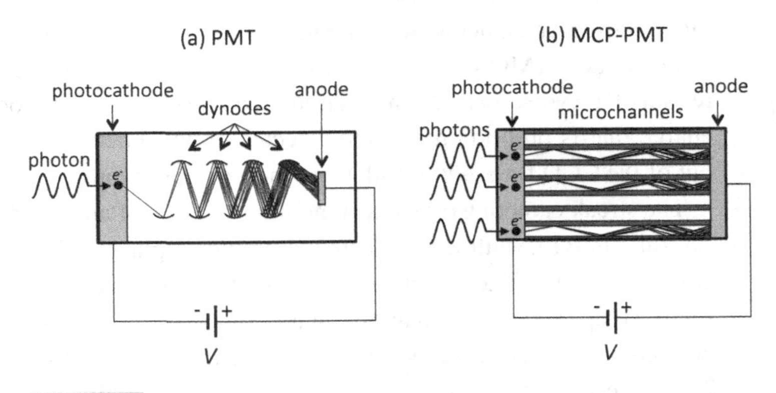

Figure 13.4

Schematic diagrams of (a) a photomultiplier tube (PMT) and (b) a microchannel plate photomultiplier tube (MCP-PMT). In both cases, photoelectrons generated in the photocathode are multiplied to generate relatively large photocurrents. The electron multiplication takes place over multiple dynodes (typically ~10) in PMTs, and within glass microchannels coated with an electron-emissive material in MCP-PMTs. The operating voltage (V) is of the order of kilovolts.

Photomultiplier tubes, schematically illustrated in Figure 13.4(a), feature high electron gain, and are therefore predominantly used for detection of low light levels, down to single-photon detection. The photoelectron generated at the photocathode of a PMT is accelerated by an applied high voltage, which is divided (with a voltage-divider circuit) over a set of electrodes called dynodes. Upon striking the first dynode, the photoelectron generates multiple secondary electrons. The secondary electrons in turn generate additional secondary electrons at the following dynodes, resulting in the mentioned chain-reaction-like cascade of electron multiplication over the multiple dynodes. The total gain depends on the number of dynodes (typically 6–10), the applied voltage, and how the voltage is divided among the dynode string. With a typical gain-per-stage of the order of ~ 4, the resulting overall gain can be a factor of 10^3–10^6.

An alternative design to a dynode cascade of a PMT is a set of glass capillaries, which are internally coated with a photocathode material. A large number of such capillaries are assembled into a two-dimensional array called a *microchannel plate* (MCP), as illustrated in Figure 13.4(b). MCPs are used in both intensified CCD cameras (ICCD) and streak cameras to multiply photoelectrons that are then converted back to photons by a phosphor screen, so that they can be detected by a regular CCD. When an MCP is used with a CCD, it can provide gain for as many pixels in an image as there are capillaries in the MCP, thus amplifying an entire image (megapixel MCPs are not uncommon). The MCP can also serve as a fast gate switch for the camera, if the applied voltage is rapidly switched.

CCDs are the most familiar two-dimensional detectors, as they are found in most consumer cameras and mobile phones. CCDs are arrays of metal oxide semiconductor (MOS) capacitors, where light detection also occurs in the depletion region of a reverse-biased p-n junction. Each pixel of a CCD produces an electrical charge that is proportional to the detected optical energy, and that is transferred out of the CCD for readout following the exposure time.

The streak camera works like a modified version of an "old-fashioned" cathode ray tube (CRT), with a photocathode emitter replacing the thermionic electron emitter of a CRT. The "front end" of a streak camera features an entrance slit and a set of sweep electrodes that apply a high-voltage ramp to deflect the photoelectrons emitted by the photocathode along the direction perpendicular to the entrance slit. The net result is that photons that are detected at earlier times are deflected at different angles than photons detected at later times, realizing a two-dimensional image, in which one dimension represents the distribution of light along the entrance slit, while the other dimension represents time of detection. The ICCD and the streak camera are schematically illustrated in Figure 13.5.

We now discuss key performance specifications of optical detectors for diffuse tissue spectroscopy:

- spectral sensitivity;
- linearity and dynamic range;
- temporal response;
- responsivity and sensitivity.

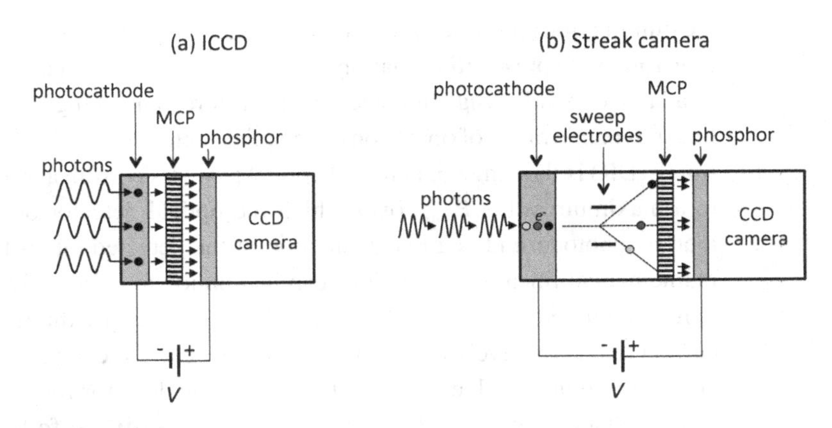

Figure 13.5

(a) Intensified charge-coupled device (ICCD) camera and (b) streak camera. Both detector systems are based on photoelectron generation in a photocathode, electron multiplication at a microchannel plate (MCP), and a phosphor screen that converts the electrons emitted by the MCP back into photons that are detected by a charge-coupled device (CCD) camera. The sweep electrodes in the streak camera (b) apply a high-voltage ramp that serves the purpose of deflecting photoelectrons by a variable amount over a time sweep, to separate in space photoelectrons that are arriving at different times.

13.3.2 Spectral sensitivity

The spectral response of optical detectors in the visible and near-infrared spectral regions is mostly determined by the materials used for the photocathode or semiconductor p-n junctions. PMTs often utilize photocathodes based on multi-alkali (Sb-Na-K-Cs) and cesium-activated gallium arsenide (GaAs:Cs), which allow for a spectral response in the wavelength range ~200–900 nm. Semiconductor p-n junctions feature sensitivity from 300 to 1100 nm (silicon, Si), and 800 to 1700 nm (germanium, Ge; indium gallium arsenide, InGaAs). Consequently photodiodes (PD) and avalanche photodiodes (APDs) feature sensitivity deeper in the near-infrared than PMTs. CCD cameras are based on silicon substrates, which pose an upper limit of ~1000 nm to their measurable wavelength range. Photocathodes used in MCPs (hence intensified CCD cameras and streak cameras) allow for measurements in the wavelength range 300–900 nm. Photocathodes sensitive at longer wavelengths, up to ~1700 nm, are available but at significantly lower responsivity levels.

13.3.3 Linearity and dynamic range

Quantitative measurements rely on the linearity of the optical detector. The strong attenuation of the optical fluence rate in tissue as a function of the distance between the illumination and collection points (by a factor of ~10/cm), or as a function of the time following pulsed illumination (by a factor of ~10/ns), leads to the requirement of a large dynamic range of detection. Therefore, a meaningful specification is the linear dynamic range of operation of an optical detector. In this context, the *dynamic range* (DR) is the range of detected optical power, say from a minimum value P_{min} to a maximum value P_{max}, over which the optical detector operates linearly, so that the photocurrent is a linear function of the detected intensity. In some cases, mathematical linearity corrections can be applied to effectively extend the linear dynamic range of detection. For example, if a detector exhibits a degree of dark current that is relatively constant, this baseline current can be subtracted from the detector output signal, effectively linearizing the detector for small signal levels.

The dynamic range can be expressed in decibels (dB) as follows:

$$\mathrm{DR_{dB}} = 10\mathrm{Log}_{10}\left(\frac{P_{max}}{P_{min}}\right). \tag{13.1}$$

The dynamic range $\mathrm{DR_{dB}}$ represents 10 times the number of factors of ten over which a detected optical power results in a photocurrent that is linearly related to it. Usually, the dynamic range is inversely related to the responsivity of a detector (which is defined below, in Section 13.3.5). The dynamic range of a CCD camera, which is associated with the ratio of the full well capacity (the number of electrons that can be accommodated in each potential well) to the readout noise (the number of electrons associated with electronic noise), ranges from 20 to 40 dB. The linear dynamic range is ~60–70 dB for PDs and PIN PDs, ~40 dB for APDs, 30–40 dB for streak cameras, and ~30 dB for PMTs. The dynamic range of an ICCD camera coincides with the dynamic range of the CCD detector at low gain of the image intensifier (consisting of a photocathode, an MCP, and a phosphor screen as shown in Figure 13.5(a)) and decreases at higher gains of the image intensifier.

13.3.4 Temporal response

The response time of an optical detector can be defined in terms of the *rise time* (t_r), defined as the time from 10% to 90% of the final signal level, in response to a step input. The detector bandwidth (BW) and the rise time are related as follows:

$$\mathrm{BW} \cong \frac{0.35}{t_r}, \tag{13.2}$$

where the "≅" sign in Eq. (13.2) indicates that the factor 0.35, which is accurate for the case of a one-stage RC (resistor-capacitor) low-pass network, may vary for different systems. (For example, for a Gaussian response system it is 0.34.) This concept of the detector bandwidth (which suggests a frequency response) assumes that the detector's fall time is not slower than its rise time, and that there is no limitation in the length of a measurement or repetition rate of a measurement that might be associated with a charge replenishment time of depleted photoemitters. Such a limitation is not uncommon with PMTs, which, if they are optimized for very fast detection of single photons, cannot provide sustained current for longer-duration optical signals.

The temporal response of a photodiode is mostly determined by the photodiode capacitance and by the diffusion time of charge carriers (electrons and holes) outside the depletion layer (which determines the replenishment rate or carriers). PDs and PIN PDs have typical rise times of ~10–100 ns (with typically faster rise times in PIN PDs because of their smaller photodiode capacitance), corresponding to a detector bandwidth of ~1–30 MHz (up to ~60 MHz in PIN PDs). APDs feature even faster rise times (~0.1–1 ns) and, consequently, a broader detector bandwidth (~1 GHz). There is typically a trade-off between the speed of response and the size of the active area (hence, capacitance) of PIN PDs and APDs, with a faster response for a smaller active area. The temporal response of PMTs is limited to ~1 ns (bandwidth: ~300 MHz) by the spread in photoelectron transit times from the photocathode to the anode. The coated capillary microchannels in an MCP act as continuous photoelectron multipliers, decreasing the spread in photoelectron transit times and achieving a rise time as short as ~0.2 ns (bandwidth: ~1.8 GHz). Streak cameras, which commonly employ a microchannel plate for photoelectron multiplication, have comparable rise times of ~0.2–0.8 ns, resulting in a temporal resolution of ~10 ps, limited by shot-to-shot jitter.

13.3.5 Responsivity and sensitivity

The *responsivity* (R) of an optical detector is a measure of the photocurrent generated per unit detected optical power, and is expressed as:

$$R = G\frac{\eta e}{\hbar\omega}, \tag{13.3}$$

where G is the internal gain of the detector, η is the *quantum efficiency* (defined as the ratio of the number of generated photoelectrons to the number of detected photons), e is the electron charge, $\hbar$ is Planck's constant divided by 2π, and ω is the angular frequency of the detected photon. The quantum efficiency of photodiodes (PD, PIN PD, APD) is typically >80% (except for germanium at <50%), whereas the photocathode quantum efficiencies are lower in ICCD cameras (10–40%), PMTs (~20%), and streak cameras (~5%). The expression of R in Eq. (13.3) represents the electrical charge (ηe) resulting from the photoelectric conversion of one photon divided by the photon energy ($\hbar\omega$), amplified by the detector gain (G). The resulting units are those of an electric charge divided by energy. In practice, however, both the numerator and the denominator of the expression on the right are divided by time (representing the integration time of the measurement), such that the typically quoted units are current divided by power (A/W). The lack of a gain mechanism ($G = 1$) capable of generating secondary photoelectrons in photodiodes and PIN photodiodes results in their smaller responsivity of ~0.3–0.5 A/W, when compared with APDs (~50 A/W) and PMTs (10^5 A/W), which invoke photocurrent amplification. (It may be noted that, due to this intrinsic limitation of PDs, semiconductor photodiode devices are commonly fabricated and sold with integrated current amplifier circuits.) The internal current amplification in APDs is of the order of $\sim 10^2$, in microchannel plates (MCPs) is $\sim 10^3$–10^5, and in PMTs is of the order of 10^6. The responsivity of MCP-PMTs and streak cameras is significantly reduced to values of 0.01–0.05 A/W by the lower photocathode quantum yield and further loss mechanisms.

The sensitivity of an optical detector, a gain-independent measure, can be specified by the *noise equivalent power* (NEP), which is defined as the optical power that results in a photocurrent that is equal to the root-mean-square (RMS) noise current of the detector: a high-sensitivity detector has a low NEP. By this definition, the sensitivity (or NEP) is specified at the wavelength that corresponds to the peak of the spectral responsivity, and the bandwidth is taken to be 1 Hz. (In practical terms, this is equivalent to specifying that the integration time be 0.5 s, since the detection bandwidth associated with integrated detection over time T is given by $\frac{1}{2T}$.) The NEP is usually expressed in units of W/$\sqrt{\text{Hz}}$ (or fW/$\sqrt{\text{Hz}}$, where "f" stands for femto-, or 10^{-15}) to indicate that it scales with the square root of the detection bandwidth or, in other words, with the inverse of the square root of the integration time. The NEP provides a measure of the minimum detectable power, and its inverse is defined as the *detectivity* (D = 1/NEP), a measure of the ability to detect small optical signals. The NEP decreases, and therefore detectivity increases, going from photodiodes, PDs, and PIN PDs (NEP ~ 10–100 fW/$\sqrt{\text{Hz}}$), to avalanche photodiodes (NEP ~ 1–10 fW/$\sqrt{\text{Hz}}$), to photomultiplier tubes (NEP ~ 0.1 fW/$\sqrt{\text{Hz}}$). CCD detectors generally exhibit higher NEP than any of these. It should be noted that in some detectors (especially PMTs and CCDs) the NEP is dominated by thermal "dark" emission, which can be lowered by cooling the photocathode or semiconductor junction of the detector. Cooled detectors can be enabling when measuring weak optical signals over long integration times.

As mentioned above, the sensitivity can be pushed to the single-photon limit in PMTs thanks to their large internal gain ($\sim 10^6$) for photoelectrons. In fact, PMTs have the ability to detect individual photons, enabling what is known as photon counting. The capability of single-photon detection is marginally within the reach of APDs, which are reverse biased just below the breakdown voltage, resulting in gain factors of the order of 10^2. Single-photon avalanche diodes (SPAD), also called Geiger-mode APDs (after the German physicist Hans Geiger [1882–1945]), are avalanche photodiodes that are reverse biased just above the breakdown voltage, resulting in a much larger gain of 10^5–10^6, rendering them suitable for single-photon detection. A recent and ongoing development in optical detection technology is the silicon photomultiplier tube (SiPMT), which is a densely packed array of SPADs (100–1000/mm^2), whose output is the analog sum of the outputs from all of its individual SPAD elements. The SiPMT realizes the solid-state equivalent of a vacuum photomultiplier tube. In fact, the gain factor (10^5–10^6), the dynamic range (30 dB), the frequency bandwidth (~350 MHz), and the responsivity (10^4–10^5 A/W) of the SiPMT are all comparable to those of the vacuum PMT. Attractive features of the SiPMT with respect to the PMT are the much lower operating voltage (~50 V vs. ~1 kV) and the insensitivity to magnetic fields (by contrast, PMTs are strongly affected by magnetic fields, although the MCP version of PMTs can be operated in the presence of magnetic fields).

Table 13.2 reports a summary of the relevant properties discussed above for optical detectors used for diffuse optical spectroscopy of tissue.

13.3.6 Shot noise

The fact that optical energy is exchanged by small packets, the photons, results in a fundamental source of noise in optical measurements. Light detection may be seen as more similar to collecting water from a surface hit by rain, with randomly arriving water droplets, than from a pipe through which water flows continuously. The stochastic nature of the arrival times of photons contributes intrinsic fluctuations in the detected optical signal. Such fluctuations result in the so-called *shot noise*. Although shot noise is a fundamental property of the measured signal and not the detector itself, the shot-noise limit is a useful measure to relate to the detector specifications. If one considers a constant average optical power (i.e., a constant average rate of photon arrivals) and uncorrelated events of photon arrivals, the probability of detecting one photon within an infinitesimally short amount of time, dt, can be written as:

$$p(1, dt) = \alpha dt, \tag{13.4}$$

where α is the rate constant of photon arrivals. It is intuitive that the average number of photons detected within a finite time interval Δt is given by $\bar{N} = \alpha\Delta t$. However, the actual number of photons, N, detected over individual time intervals, all having the same duration Δt, varies randomly as a result of shot noise. Therefore,

Table 13.2 Summary of relevant properties of optical detectors used in near-infrared tissue spectroscopy and imaging. (PD: photodiode; PIN PD: p-i-n photodiode; APD: avalanche photodiode; SPAD: single-photon avalanche diode; SiPMT: silicon photomultipier tube; MCP-PMT: microchannel plate photomultiplier tube; CCD: charge-coupled device; ICCD: intensified charge-coupled device.)

	Optical detectors									
Property	PD	PIN PD	APD	SPAD	SiPMT	PMT	MCP-PMT	CCD	ICCD	Streak camera
Spectral range (μm)	0.3–1.7	0.3–1.7	0.3–1.7	0.3–1.7	0.3–1.1	0.2–0.9	0.2–0.9	0.3–1.0	0.3–0.9	0.3–0.9
Dynamic range (dB)	60–70	60–70	40	–	30	30	30	30–60	20–60	30–40
Rise time (ns)	10–100	10	0.1–1	0.4	1	1	0.2	–	–	0.2–0.8
Frequency bandwidth (MHz)	1–30	30–60	1000	1000	350	350	1800	–	–	1000
Internal gain	1	1	10^2	10^5–10^6	10^5–10^6	10^6	10^3–10^5	–	10^3	10^3–10^5
Responsivity (A/W)	0.3–0.5	0.3–0.5	50	–	10^4–10^5	10^5	10^2–10^4	–	–	–
Noise equivalent power ($fW/\sqrt{Hz}$)	10–100	10–100	1–10	–	1	0.1	–	–	–	–
Operating voltage (V)	5–60	5–60	100–400	400–500	30–100	1k-3k	1k-3k	–	6k	15k
Single photon detection	N	N	Possibly	Y	Y	Y	Y	N	Y	Y

even in the case of a constant radiant power reaching the detector, the number of photons, N, detected in a time interval of given width Δt behaves as a random variable.

The probability distribution of this random variable is given by:

$$p(N, \Delta t) = \frac{(\bar{N})^N e^{-\bar{N}}}{N!} \tag{13.5}$$

which is called the *Poisson probability distribution*. Its normalization condition is such that $\sum_{N=0}^{\infty} p(N) = 1$. It is worth pointing out that the Poisson distribution depends on only one parameter, the rate constant α that appears in $\bar{N} = \alpha \Delta t$. The mean of this distribution is $\bar{N}$, and its variance is also $\bar{N}$, so that the standard deviation of the number of detected photons in interval Δt is given by $\sqrt{\bar{N}}$. Therefore, in a shot-noise limited case (where shot noise is the only source of noise) the *signal-to-noise ratio (SNR)*, defined here as the ratio of the mean value to the standard deviation of the optical signal, is given by $\frac{\bar{N}}{\sqrt{\bar{N}}} = \sqrt{\bar{N}}$. The SNR in a shot-noise limited case, also called the signal-to-shot-noise ratio, increases with the square root of the signal.

We note that in practice, and as a practical matter when choosing a detector, the quantum efficiency comes into play because a detector actually measures the photoelectrons initially created at the photocathode (of a PMT, intensified CCD, or streak camera) or semiconductor junction (of a PD or other solid-state device), and it is the number of photoelectrons that determines the statistics. An interesting case worth examining is that of a CCD detector, in which the photoelectrons are accumulated in the well of each pixel until readout. A common value for the pixel well-depth (the maximum storable charge per pixel) in a CCD detector might be, say, 40,000 electrons. If 75% of the dynamic range of the detector is used, this means that an individual measurement may read ~30,000 electrons from a pixel. Since the square root of that number is ~173, this means that the shot noise for the measurement is ~0.6% of the signal. (Of course, readout noise, dark current, and other sources of noise add to this theoretically minimum noise level.)

13.4 Experimental approaches for time-resolved spectroscopy

13.4.1 Time domain

The goal of time-domain measurements is to measure the time-of-flight distribution ($\mathcal{D}_{\mathrm{TOF}}(t)$) of the photons that propagate from the illumination point to the collection point in the scattering medium.

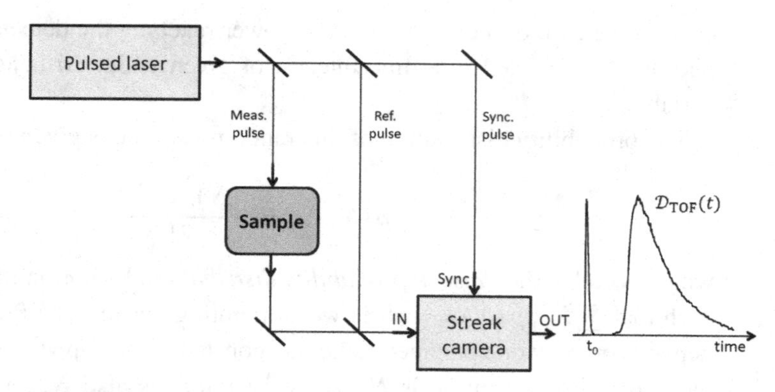

Figure 13.6

Time domain approach based on a streak camera detector. A pulsed laser (Ti:sapphire, Nd:YAG, or dye laser) emits short pulses (~1 ps) that are split into a measurement pulse (Meas. pulse: directed to the scattering sample), a reference pulse (Ref. pulse: used to identify the time of the pulse emission, t_0), and a synchronization pulse (Sync. pulse: to synchronize the pulse repetition frequency with the streak camera scanning frequency). The measured output is the average of the measured pulses over the integration time T (for a typical pulse repetition frequency of 80 MHz, the output results from 80 million pulses per second of integration time).

13.4.1.1 *Streak camera*

In principle, a streak camera is the ideal instrument for this task, since it is able to perform measurements with a temporal resolution of 1–10 ps (which is excellent, given the nanosecond time scale of $\mathcal{D}_{TOF}$ in diffuse tissue spectroscopy). Furthermore, a streak camera affords time-resolved measurements of the spatial intensity distribution over the linear dimension of the entrance slit, and this capability can be used to map an optical signal over a spatial (for imaging) or spectral (for spectroscopy) coordinate. Figure 13.6 is a block diagram of a time-domain measurement system based on a streak camera, illustrating the decomposition of the laser pulse into a synchronization pulse (to synchronize the laser repetition rate with the scanning rate of the streak camera), a reference pulse (to provide a fiducial marker to identify t_0, the time of the laser pulse emission), and a measurement pulse that propagates through the scattering sample of interest.

The output signals of the streak camera for a series of pulses (single shots) can be accumulated to improve the signal-to-noise ratio. The resulting multi-shot time-of-flight distribution represents an average over a large number of pulses, that number given by the product of the laser repetition rate times the accumulation time. A typical repetition rate of 80 MHz corresponds to 80 million pulses per second of accumulation time. Streak cameras have been used for time-domain optical studies of highly scattering media and tissue (Hebden et al., 1991; Mitic et al., 1994; Abrahamsson et al., 2004). However, the major strength of streak cameras, their high temporal resolution, is outweighed by technical and practical limitations (low sensitivity, limited dynamic range, bulkiness, expense), which account for their limited use for time-domain tissue spectroscopy.

13.4.1.2 *Time-correlated single-photon counting*

The most common detection approach in time-domain diffuse tissue spectroscopy is time-correlated single-photon counting (TCSPC). The idea is to measure a single photon arriving at the detector from the collection of diffuse photons associated with each illumination light pulse, and to build the time-of-flight distribution ($\mathcal{D}_{TOF}(t)$) as the histogram of the single-photon arrival times with respect to the time of the associated illumination pulse (t_0). In TCSPC, it is extremely important that no more than one photon reaches the detector for each illumination pulse. This exigency is due to the fact that only the arrival time of the first photon would be recorded, as a result of the dead recording time after each pulse detection, which would lead to a bias of $\mathcal{D}_{TOF}$ toward shorter times if more than one photon were to reach the detector after a given illumination pulse. Additionally, it is essential that no photons from one illumination pulse are detected after the next illumination pulse is emitted (because, in this case, the arrival time of the photon would be underestimated by the time separation between the two successive illumination pulses).

The first condition is met by ensuring that only 1–10% of the illumination pulses result in the detection of a photon. In practice, this is accomplished by controlling the energy, i.e., the number of photons, of the illumination pulses that illuminate the sample. The second condition is met by using pulse repetition rates (f_{rep}) such that $1/f_{rep}$ is much longer than the time scale of $\mathcal{D}_{TOF}(t)$, a requirement that was discussed in Section 11.2. The typical time scale of ~1 ns for $\mathcal{D}_{TOF}(t)$ in diffuse tissue spectroscopy allows for pulse repetition frequencies up to ~100 MHz. It is important to note that the way in which the time-of-flight distribution is measured by TCSPC is conceptually different from the method using a streak camera. In TCSPC, the $\mathcal{D}_{TOF}(t)$ is built one photon at a time, by detecting one photon (at most) per illumination pulse, and by building a histogram of arrival times associated with the pulses collected during the overall measurement time. A streak camera, in contrast, measures the entire $\mathcal{D}_{TOF}(t)$ for each illumination pulse (single shot), and one accumulates the multiple individual $\mathcal{D}_{TOF}(t)$ measurements for the pulses measured during the desired accumulation time (multiple shots). Optical detectors for TCSPC are typically photomultiplier tubes (PMTs), microchannel plate photomultiplier tubes (MCP-PMTs), or single-photon avalanche diodes (SPADs). The general approach of TCSPC is illustrated in Figure 13.7. The $\mathcal{D}_{TOF}(t)$ histogram represents the number of photons detected over a set of intervals of $t - t_0$, i.e., the difference between the detection time of each photon (t) and the time of the

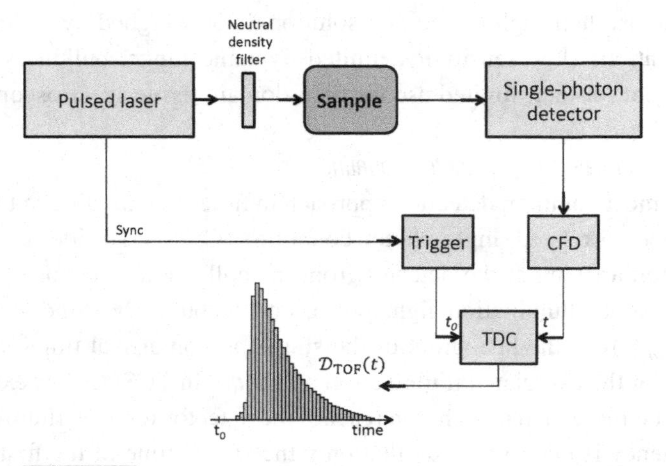

Figure 13.7

Time-correlated single-photon counting (TCSPC) approach for time-domain measurements of the time of flight distribution ($\mathcal{D}_{TOF}(t)$). A pulsed laser (repetition frequency: ~10–80 MHz) emits light pulses that are attenuated (with a neutral density filter) to ensure that no more than 1–10% of the illumination pulses, after propagation through the sample, lead to the detection of one photon by the single-photon detector (which may be, for example, a photomultiplier tube [PMT] or a single-photon avalanche diode [SPAD]). As a result, the detected photon count rate is ~1 MHz or less. A trigger identifies the pulse emission time (t_0) from a synchronization signal from the pulsed laser, whereas a constant fraction discriminator (CFD) identifies the time (t) of the detected photon. A time to digital converter (TDC), which combines the functions of a time to amplitude converter (TAC) and an analog to digital converter (ADC), provides a digital reading of the time of flight ($t - t_0$) of each detected photon, leading to a histogram representation of $\mathcal{D}_{TOF}(t)$.

corresponding illumination pulse (t_0). A constant fraction discriminator (CFD) extracts accurate timing information from single-photon pulses that have variable amplitude, triggered by a synchronization pulse that provides the pulse illumination time. After the time difference $t - t_0$ is converted to a digital reading, the histogram of the $\mathcal{D}_{TOF}(t)$ can be built.

13.4.1.3 *Time gating*

Time-domain measurements provide a wealth of information through the time-of-flight distribution ($\mathcal{D}_{TOF}$) of detected photons. The $\mathcal{D}_{TOF}$ can be separated into its time components at arbitrary intervals of the time-of-flight, so that one can investigate early-arriving photons, late-arriving photons, or photons arriving within any specific time window. The observation of photons within a specific time interval is called *time gating*. Time gating can be performed by post-processing $\mathcal{D}_{TOF}(t)$ traces detected with a streak camera or with TCSPC. However, there are cases where it is more useful to have the ability to perform time gating at the time of collection, limiting the measurement to photons within a specific time-of-flight range. This can be accomplished with ICCDs and with SPADs by switching on and off the image intensifier (for the ICCD) or by switching the supply voltage below and above the breakdown voltage (for the SPAD). In the case of ICCDs, the gating speed is limited by the high resistance of the photocathode material (and the associated RC time-constant), leading to a minimum time gate of ~0.2 ns. In the case of SPADs, the gating speed is limited to a minimum time gate of ~2 ns, but the fast rise-time of ~0.2 ns has led to their use as a fast-switch optical detector with a high efficiency of rejection of early arriving photons (Dalla Mora et al., 2010), thus allowing for optical detection in the vicinity of the illumination point.

13.4.2 Frequency domain

The goal of frequency-domain measurements is to measure the amplitude and phase of the photon-density wave resulting from the illumination of a highly scattering medium with a modulated optical power. As discussed in Section 12.1, the modulation frequency is of the order of ~100 MHz for tissue spectroscopy. This means that the light source emission must be modulated at ~100 MHz and that the optical detector must have a bandwidth extending to hundreds of MHz.

As discussed above (Section 13.1.3), the pulsed emissions of Ti:sapphire lasers and dye lasers at repetition frequencies of ~80 MHz lend themselves to frequency-domain spectroscopy, as the fundamental frequency and first few harmonics of the corresponding frequency spectrum are suitable for frequency-domain measurements (see Figure 13.1). Laser diodes and light-emitting diodes can be directly modulated at frequencies up to several GHz and ~100 MHz, respectively, by driving them with a modulated current.

It is also possible to externally modulate the emission of CW light sources (for example He-Ne lasers) at a few hundred MHz by means of acousto-optic modulators (AOMs) or electro-optic modulators (EOMs). In acousto-optic modulators, a piezoelectric transducer converts a supplied oscillating electric field (voltage potential) into mechanical oscillations that induce acoustic waves into an optically clear material (often glass or quartz). These acoustic waves modulate the index of refraction of the material by the photoelastic effect, thus realizing a time-dependent volumetric diffraction grating that can be used to modulate the deflection of a light beam. Electro-optic modulators are based on Pockels cells (named after Friedrich Pockels [1865–1913], a German physicist who studied electro-optic effects), in which an applied voltage induces birefringence (a difference in refractive indices along different axes in a medium), thus inducing a voltage-controlled rotation of the optical polarization angle of light propagating through the cell. The transmission of a light beam can be modulated by sandwiching a Pockels cell between two crossed linear polarizers, and by driving the Pockels cell with an oscillating voltage. AOMs and EOMs require the light beam to be collimated; hence, they are appropriate for modulating the emission of CW lasers. However, given the convenience of modulation of laser diodes, LEDs, and the availability of mode-locked Ti:sapphire lasers at suitable pulse repetition frequencies, AOMs and EOMs are less commonly used to modulate light sources in frequency-domain optical studies of tissue.

The most common optical detectors for frequency-domain optical studies of tissue are photomultiplier tubes (PMTs) and avalanche photodiodes (APDs). As shown in Table 13.2, their bandwidths extending to several hundred MHz (PMTs) and ~1 GHz (APDs) are appropriate for frequency-domain studies, with microchannel plate photomultipliers (MCP-PMTs) offering the option of extending the bandwidth to several GHz (Fishkin et al., 1996). As will be discussed below for heterodyne detection schemes (Section 13.4.2.2), it may also be helpful to modulate the gain of the detector. It is possible to modulate the gain of the image intensifier of ICCD cameras in the 100–1000 MHz frequency range, to allow for frequency-domain measurements at each CCD pixel (Netz et al., 2008).

There are two broad methods to measure the phase and amplitude of the modulated light signals detected in the frequency domain, based on either homodyne or heterodyne detection (Chance et al., 1998), and these are described in the next two sections.

13.4.2.1 *Homodyne detection*

Homodyne systems perform amplitude and phase measurements at the same angular frequency (ω) of the light source modulation. *Homodyne detection* is based on an *in-phase/quadrature (I/Q) demodulator*, whose inputs are the source modulation signal ($\sim\sin(\omega t)$) and the modulated signal detected from a scattering sample, which is at the same frequency ω but includes an amplitude factor $2A$ (the factor 2 is introduced to cancel out with the factor 2 in the addition trigonometric formulas – see below) and a phase shift θ with respect to the source modulation signal, yielding $2A\sin(\omega t + \theta)$. The I/Q demodulator combines the two inputs, namely the source modulation signal and the detector signal, through two electronic mixers, which are multiplier circuits with two inputs and one output, where the output signal is equivalent to the product of the two input signals. One mixer receives the two signals in phase (0° phase difference), whereas the other mixer receives the two signals in quadrature of phase (90° phase difference). Consequently, the outputs of the I/Q demodulator are in-phase and quadrature signals, $I(t)$ and $Q(t)$ respectively, which consist of a direct current (DC) component (at 0 frequency)

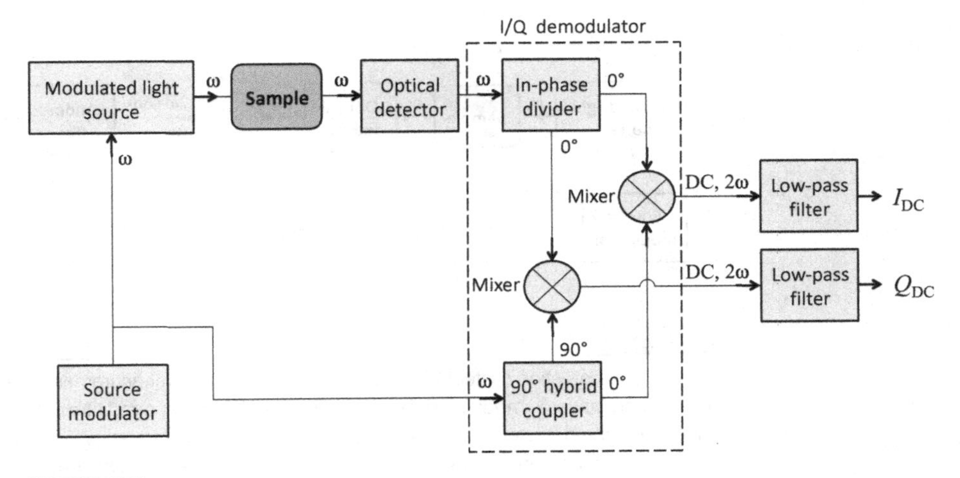

Figure 13.8

Homodyne detection system for frequency-domain optical measurements on a scattering sample. The source power is modulated at a radiofrequency (ω) in the 100 MHz range, and the detected signal, which is also modulated at frequency ω, is split into two in-phase components that are fed to the radiofrequency inputs of two electronic mixers. In one of the two mixers, the local oscillator input is a signal in phase with the source modulation at frequency ω. In the other mixer, the local oscillator input is a signal in quadrature of phase (90° phase difference) with the source modulation at frequency ω. The signals in phase and in quadrature of phase with respect to the source modulation signal are the outputs of a 90° hybrid coupler. The high-frequency components (at 2ω) of the two electronic mixers are filtered out by two low-pass filters, whose direct current (DC) outputs are the in-phase and quadrature DC signals (I_{DC} and Q_{DC}) from which the amplitude and phase of the desired signal can be derived. The components that make up the in-phase/quadrature (I/Q) demodulator are within the dashed rectangle.

and a radiofrequency component at twice the source modulation frequency (2ω) (Yang et al., 1997):

$$I(t) = 2A\sin(\omega t + \theta)\sin(\omega t) = A\cos\theta - A\cos(2\omega t + \theta), \tag{13.6}$$

$$Q(t) = 2A\sin(\omega t + \theta)\cos(\omega t) = A\sin\theta + A\sin(2\omega t + \theta). \tag{13.7}$$

The DC components $I_{DC} = A\cos\theta$ and $Q_{DC} = A\sin\theta$ can be isolated with a low-pass filter, and they yield the amplitude and phase of the detected signal of interest from the following relationships:

$$A = \sqrt{I_{DC}^2 + Q_{DC}^2}, \tag{13.8}$$

$$\theta = \tan^{-1}\left(\frac{Q_{DC}}{I_{DC}}\right). \tag{13.9}$$

The block diagram of a homodyne detection system is shown in Figure 13.8.

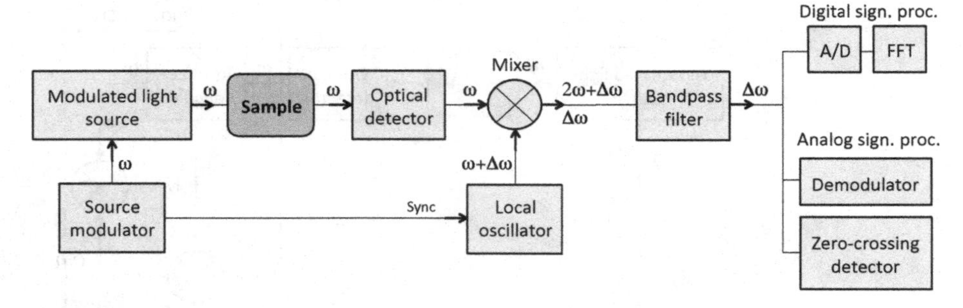

Figure 13.9

Heterodyne detection system for frequency-domain optical measurements on a scattering sample. The source power is modulated (by the source modulator) at a radiofrequency (ω) in the 100 MHz range, and the detected signal, which is also modulated at frequency ω, is electronically mixed (i.e., multiplied) with a signal (generated by the local oscillator) at an offset frequency $\omega + \Delta\omega$, where $\Delta\omega$ is in the kHz range. The low-frequency component ($\Delta\omega$) of the mixer output is isolated by a bandpass filter. Amplitude and phase values can be obtained by digital signal processing via analog-to-digital conversion (A/D) and fast Fourier transformation (FFT), or by analog signal processing via demodulation (for amplitude measurements) and zero-crossing detection (for phase measurements).

13.4.2.2 *Heterodyne detection*

In contrast with homodyne systems, *heterodyne detection* systems perform amplitude and phase measurements at a down-converted frequency that is much lower (~kHz) than the frequency of light source modulation (~100 MHz). This method was invented in 1918 by the American engineer Edwin Armstrong [1890–1954] to down-convert the short-wave radiofrequencies of radio transmissions (>500 kHz) to a lower frequency above the audible range (~50 kHz) that he termed *intermediate frequency* (IF). This is achieved by electronically mixing (i.e., multiplying) the radiofrequency (RF) signal with the signal generated by a local oscillator (LO) at a frequency that is given by the radiofrequency ± the intermediate frequency. Because the IF was in the supersonic range, the technique was called *superheterodyne*, and it is still widely used in amplitude modulation (AM) radio receivers. This terminology is still used for electronic mixers to indicate the two inputs (RF, LO) and the output (IF), where the output signal is equivalent to the mathematical product of the two input signals and thus contains components at the sum and difference of their frequencies.

In a typical frequency-domain heterodyne system (illustrated in Figure 13.9), the source modulator (frequency ω) and the local oscillator (frequency $\omega + \Delta\omega$) are synchronized, which means that the relative phase of their signals is linearly dependent on time as $(\Delta\omega)t$. (This is a generalization of the concept of phase synchronization for signals at different frequencies.) The signal detected from a scattering sample is at the same frequency of the source modulation (ω), and it includes an amplitude factor $2A$ (with the factor 2 introduced for the same reason as for the homodyne case above) and a phase shift θ with respect to the source modulation signal, $2A\sin(\omega t + \theta)$. The detected signal at frequency ω is multiplied (with an electronic mixer) by the local oscillator signal at frequency $\omega + \Delta\omega$, $\sin[(\omega + \Delta\omega)t)]$, to result in components at the sum $(2\omega + \Delta\omega)$ and difference $(\Delta\omega)$ of the frequencies of the two signals. This multiplication can happen externally, after the optical detector via an electronic mixer, as shown in Figure 13.9, or internally by modulating the detector response function at the local oscillator frequency. For example, this can be achieved by modulating the gain of a photomultiplier tube or the image intensifier of an ICCD camera at the local oscillator frequency $\omega + \Delta\omega$. In either case, the signal (S) at the output of the electronic mixer (for mixing external to the detector) or at the output of the optical detector (for mixing internal to the detector) is given by the product of the source modulator signal and the local oscillator signal:

$$\begin{aligned} S(t) &= 2A\sin(\omega t + \theta)\sin[(\omega + \Delta\omega)t] \\ &= A\cos(\Delta\omega t - \theta) - A\cos[(2\omega + \Delta\omega)t + \theta]. \end{aligned} \tag{13.10}$$

A bandpass filter centered at frequency $\Delta\omega$ isolates the low-frequency component. As mentioned above, this low-frequency component was termed "intermediate frequency" in radio communications, but in tissue spectroscopy and fluorescence spectroscopy it is called *cross-correlation frequency* (Spencer and Weber, 1969). The amplitude and phase of the signal component at the cross-correlation frequency (which is typically in the kHz range) coincide with the amplitude and phase of the signal of interest at radio frequency (in the 100 MHz range) and can be processed with either digital or analog methods. Digital signal processing typically involves analog-to-digital (A/D) conversion and fast Fourier transformation to yield amplitude and phase readings. Analog signal processing yields amplitude measurements with a demodulator, which performs envelope detection, consisting of a rectifier (to convert the positive and negative cycles of the signal into positive cycles only) and a low-pass filter. Analog phase measurements are obtained with a zero-crossing detector that identifies the time at which the sinusoidal signal crosses a specific threshold and compares it with a reference time.

Heterodyne detection has developed into the most common method of frequency-domain optical studies of tissues. Alternative implementations, with respect to the general approach illustrated in Figure 13.9, include frequency sweeps (300 kHz – 3 GHz) using a network analyzer (Pham et al., 2000), and digital parallel heterodyning to replace analog mixing with time-gated detection over sampling windows at a frequency given by a specific factor times the cross-correlation frequency (Arnesano et al., 2012). We conclude by observing that fast analog-to-digital converters have been shown to allow for direct digital sampling of radiofrequency signals at frequencies up to 400 MHz, realizing an alternative approach to frequency-domain tissue spectroscopy with respect to homodyne and heterodyne detection (Roblyer et al., 2013).

Problems – answers to problems with * are on p. 647

13.1* A point light source emitting a power of 88 W that is uniformly distributed over the entire solid angle is used to illuminate a sample surface area of 2 mm^2 (orthogonal to the direction of illumination) that is 4 cm away from the light source. What is the optical intensity at the sample?

13.2 A 37 mW CW collimated diode laser beam (wavelength: 830 nm), with a beam diameter of 1.4 mm, illuminates the skin of a human subject for 3 s. Is this level of illumination below or above the maximum permissible exposure of skin?

13.3* A pulsed Ti:sapphire laser irradiates the skin of a human subjects at a wavelength of 780 nm with a train of pulses with a repetition frequency of 80 MHz over a total time of 30 s. Each pulse delivers an energy of 50 pJ.

(a) Find the average power delivered to the skin over the 30 s irradiation time.

(b) If the irradiation area on the skin is 1.1 mm^2, determine whether this configuration is within the limits of maximum permissible exposure of the skin.

13.4 Find the quantum efficiency at a wavelength of 685 nm of a photodiode detector that has a responsivity of 0.43 A/W.

13.5* Find the anode current in a photomultiplier tube (PMT) that detects an optical power of 53 pW at a wavelength of 750 nm. Assume an internal gain of 6×10^6 and a quantum yield of 12%.

13.6 An optical detector has a noise equivalent power of 7 fW for an integration time of 400 ms. Find the noise equivalent power in the case of an integration time of 800 ms.

13.7* What is the maximum optical power that can reach an optical detector for time-domain measurements with time-correlated single photon counting if the illumination wavelength is 692 nm, the pulse repetition rate is 40 MHz, and no more than one photon every 10 pulses is to be detected?

13.8 In frequency-domain heterodyne detection, the signal at the cross-correlation angular frequency ($\Delta\omega$) may be digitally processed with a fast-Fourier transform to obtain its amplitude (A) and phase (θ).

Consider synchronous digital sampling at a rate that is four times the cross-correlation frequency, so that the cross-correlation signal is sampled four times in each period. If we indicate the four samples in each period as I_1, I_2, I_3, and I_4, show that:

$$A = \frac{\sqrt{(I_1 - I_3)^2 + (I_2 - I_4)^2}}{2}$$

$$\theta = -\tan^{-1}\left(\frac{I_2 - I_4}{I_1 - I_3}\right)$$

References

Abrahamsson, A., Svensson, T., Svanberg, S., et al. (2004). Time and wavelength resolved spectroscopy of turbid media using light continuum generate in a crystal fiber. *Optics Express*, 12, 4103–4112.

American National Standards for Safe Use of Lasers. (2007). ANSI Z136.1–2007. Orlando, FL: Laser Institute of America.

Arnesano, C., Santori, Y., and Gratton, E. (2012). Digital parallel frequency-domain spectroscopy for tissue imaging. *Journal of Biomedical Optics*, 17, 096014.

Askew, E.W. (2002). Work at high altitude and oxidative stress: antioxidant nutrients. *Toxicology*, 180, 107–119.

Bassi, A., Swartling, J., D'Andrea, C., et al. (2004). Time-resolved spectrophotometer for turbid media based on supercontinuum generation in a photonic crystal fiber. *Optics Letters*, 29, 2405–2407.

Chance, B., Cope, M., Gratton, E., Ramanujam, N., and Tromberg, B. (1998). Phase measurement of light absorption and scatter in human tissue. *Review of Scientific Instruments*, 69, 3457–3481.

Dalla Mora, A., Tosi, A., Zappa, F., et al. (2010) Fast-gated single-photon avalanche diode for wide dynamic range near-infrared spectroscopy. *IEEE Journal of Selected Topics in Quantum Electronics*, 16, 1023–1030.

Fantini, S., Franceschini, M. A., Fishkin, J. B., Barbieri, B., and Gratton, E. (1994). Quantitative determination of the absorption spectra of chromophores in strongly scattering media: a light-emitting-diode based technique. *Applied Optics*, 33, 5204–5213.

Fishkin, J. B., Fantini, S., vandeVen, M. J., and Gratton, E. (1996). Gigahertz photon density waves in a turbid medium: theory and experiments. *Physics Review E*, 53, 2307–2319.

Hebden, J. C., Kruger, R. A., and Wong, K. S. (1991). Time resolved imaging through a highly scattering medium. *Applied Optics*, 30, 788–794.

International Electrotechnical Commission. (1999). Technical Report on *Safety of laser products-Part 9: Compilation of maximum permissible exposure to incoherent optical radiation*, IEC TR 60825–9. Geneva: International Electrotechnical Commission.

Mitic, G., Kölzer, J., Otto, J., et al. (1994). Time-gated transillumination of biological tissues and tissuelike phantoms. *Applied Optics*, 33, 6699–6710.

Mourant, J. R., Fuselier, T., Boyer, J., Johnson, T. M., and Bigio, I. J. (1997). Predictions and measurements of scattering and absorption over broad wavelength ranges in tissue phantoms. *Applied Optics*, 36, 949–957.

Netz, U. J., Beuthan, J., and Hielscher, A. H. (2008). Multipixel system for gigahertz frequency-domain optical imaging of finger joints. *Review of Scientific Instruments*, 79, 034301.

Pham, T. H., Coquoz, O., Fishkin, J. B., Anderson, E., and Tromberg, B. J. (2000). Broad bandwidth frequency domain instrument for quantitative tissue optical spectroscopy. *Review of Scientific Instruments*, 71, 2500–2513.

Roblyer, D., O'Sullivan, T. D., Warren, R. V., and Tromberg, B. J. (2013). Feasibility of direct digital sampling for diffuse optical frequency domain spectroscopy in tissue. *Measurement Science and Technology*, 24, 045501.

Spencer, R. D., and Weber, G. (1969). Measurement of subnanosecond fluorescence lifetimes with a cross-correlation phase fluorometer. *Annals of the New York Academy Sciences*, 158, 361–376.

Yang, Y., Liu, H., Li, X., and Chance, B. (1997). Low-cost frequency-domain photon migration instrument for tissue spectroscopy, oximetry, and imaging. *Optical Engineering*, 36, 1562–1569.

Further reading

Diode lasers and semiconductor photodetectors

Razeghi, M. (2010). *Technology of Quantum Devices*. New York: Springer.

Photodetectors

Donati, S. (2000) Photodetectors: devices, circuits and applications. Upper Saddle River, NJ: Prentice Hall.

Streak camera applications in time-domain diffuse optical measurements

Heusmann, H., Kölzer, J., and Mitic, G. (1996). Characterization of female breasts in vivo by time resolved and spectroscopic measurements in near infrared spectroscopy. *Journal of Biomedical Optics*, 1, 425–434.

Tualle, J.-M., Gélébart, B., Tiner, E., Avrillier, S., and Ollivier, J. P. (1996). Real time optical coefficients evaluation from time and space resolved reflectance measurements in biological tissues. *Optics Communications*, 124, 216–221.

Zint, C. V., Uhring, W., Torregrossa, M., Cunin, B., and Poulet, P. (2003). Streak camera: a multidetector for diffuse optical tomography. *Applied Optics*, 42, 3313–3320.

Homodyne systems for frequency-domain optical measurements

Culver, J. P., Choe, R., Holboke, M. J., et al. (2003). Three-dimensional diffuse optical tomography in the parallel plane transmission geometry: evaluation of a hybrid frequency domain/continuous wave clinical system for breast imaging. *Medical Physics*, 30, 235–247.

Yu, G., Durduran, T., Furuya, D., Greenberg, J. H., and Yodh, A. G. (2003). Frequency-domain multiplexing system for *in vivo* diffuse light measurements of rapid cerebral hemodynamics. *Applied Optics*, 42, 2931–2939.

Modulated intensified CCD camera for frequency-domain measurements

French, T., Gratton, E., and Maier, J. (1992). Frequency domain imaging of thick tissues using a CCD. *Proceedings of SPIE*, 1640, 254–261.

Godavarty, A., Eppstein, M. J., Zhang, C., et al. (2003). Fluorescence-enhanced optical imaging in large tissue volumes using a gain-modulated ICCD camera. *Physics in Medicine and Biology*, 48, 1701–1720.

Lakowicz, J. R., and Berndt, K. W. (1991). Lifetime-selective fluorescence imaging using an rf phase-sensitive camera. *Review of Scientific Instruments*, 62, 1727–1734.

Thompson, A. B., and Sevick-Muraca, E. M. (2003). Near-infrared fluorescence contrast-enhanced imaging with intensified charge-coupled device homodyne detection: measurement precision and accuracy. *Journal of Biomedical Optics*, 8, 111–120.

Heterodyne systems for frequency-domain optical measurements

Chen, N. G., Huang, M., Xia, H., Piao, D., Cronin, E., and Zhu, Q. (2004). Portable near-infrared diffusive light imager for breast cancer detection. *Journal of Biomedical Optics*, 9, 504–510.

McBride, T. O., Pogue, B. W., Jiang, S., Österberg, U. L., and Paulsen, K. D. (2001). A parallel-detection frequency-domain near-infrared tomography system for hemoglobin imaging of the breast in vivo. *Review of Scientific Instruments*, 72, 1817–1824.

No, K.-S., Kwong, R., Chou, P. H., and Cerussi, A. (2008). Design and testing of a miniature broadband frequency domain photon migration instrument. *Journal of Biomedical Optics*, 050509 (3pp).

Orlova, A. G., Turchin, I. V., Plehanov, V. I., et al. (2008) Frequency-domain diffuse optical tomography with single source-detector pair for breast cancer detection. *Laser Physics Letters*, 5, 321–327.

14 Diffuse optical imaging and tomography

Diffuse optical spectroscopy (DOS) is applied to the local measurement of the optical properties of scattering media, whereas diffuse optical imaging (DOI) and diffuse optical tomography (DOT) are used to characterize or image the spatial distribution of the properties. These methods can be converged into powerful spectral imaging measurements that combine information on the wavelength dependence of the optical properties of the scattering medium (which are relevant for the characterization of its composition and microscopic scattering properties) and on their spatial dependence (which provides localization of regions of interest and indications on large-scale organization). A quantitative description of the volume probed by an interrogating optical signal in a scattering medium is significant for both spectroscopy and imaging measurements. In spectroscopy (localized measurements) it specifies the sampled region, whereas in imaging (spatially resolved measurements) it gives information on the spatial resolution. Such a quantitative description, leading to the definition of a region of sensitivity in the scattering medium, is presented in detail in this chapter.

There are numerous methods for diffuse optical imaging, ranging from straightforward backprojection and optical topography (i.e., 2D projections from diffuse reflectance data) to more complex linear and nonlinear image reconstruction schemes. Some of these methods (for example topography and backprojection) directly translate the measured optical data into an image of the investigated tissue. Other methods (such as linear image reconstruction approaches) translate the imaging problem into an algebraic matrix inversion that can be performed with numerical methods. The most powerful tomographic approaches are nonlinear iterative techniques that make use of forward solvers to calculate the optical signals that correspond to a given spatial distribution of the tissue optical properties. An initial estimate of a spatial distribution is iteratively updated until its associated optical signals match, to within a given tolerance level, the measured optical signals and constitute the desired reconstructed optical image.

These imaging methods have their relative strengths and weaknesses, and are subject to different trade-offs among robustness, simplicity, power, and accuracy. Our goal in this chapter is to describe the basic principles and quantitative methods for diffuse optical imaging techniques that are frequently used in the study of biological tissues.

14.1 Collective photon paths in a scattering medium

14.1.1 Sensitivity function of a given optical signal to a specific optical property

Let $\boldsymbol{\mu}(\mathbf{r})$ denote the spatial distribution of the optical properties (i.e., the spatially dependent vector field of absorption and scattering parameters) and let's consider an optical signal $\mathcal{S}[\mathbf{r}_d, \boldsymbol{\mu}(\mathbf{r})]$ that is measured at a detector position $\mathbf{r}_d$ for an arbitrary distribution of light sources. In diffuse optical tomography, the vector $\boldsymbol{\mu}$ typically has two components, namely the absorption coefficient (μ_a) and either the reduced scattering coefficient or the diffusion coefficient (μ_s' or D, respectively). In general, however, the vector $\boldsymbol{\mu}$ may have multiple components that represent any of the optical properties that characterize biological tissue (index of refraction, absorption coefficient, scattering coefficient, anisotropy factor, etc.). The optical signal $\mathcal{S}$ may represent any type of optical measurement; for example, the steady-state intensity in continuous-wave (CW) methods, the modulation depth or the phase in frequency-domain (FD) methods, the time-gated optical signal or moments of the photon time-of-flight distribution in time-domain (TD) methods, etc.

Let's now consider one specific optical property (for example absorption or scattering) and let's identify it with μ_i, i.e., the i-th component of the optical-property vector $\boldsymbol{\mu}$. One can introduce a general concept of sensitivity of the optical signal $\mathcal{S}[\mathbf{r}_d, \boldsymbol{\mu}(\mathbf{r})]$ to a change $\delta\mu_i(\mathbf{r}')$ in any such specific optical property μ_i at position $\mathbf{r}'$ (or, more accurately, within a small volume about $\mathbf{r}'$), by defining the following *optical sensitivity function*:

$$s_{\mathcal{S},\mu_i}(\mathbf{r}') = \lim_{\delta\mu_i \to 0} \left| \frac{\mathcal{S}[\mathbf{r}_d, \boldsymbol{\mu}(\mathbf{r}) + \mathbf{e}_i \delta\mu_i(\mathbf{r}')] - \mathcal{S}[\mathbf{r}_d, \boldsymbol{\mu}(\mathbf{r})]}{\delta\mu_i(\mathbf{r}')} \right|, \tag{14.1}$$

where $\mathbf{e}_i$ is the unit vector whose components are all zero except for the i-th component (the one that represents the optical property of interest μ_i) that is 1. Equation (14.1) provides a definition of sensitivity that depends on the type of optical signal $\mathcal{S}$ and on the nature of the optical perturbation $\delta\mu_i$ (absorption, scattering, etc.). A way to read Eq. (14.1) is that $s_{\mathcal{S},\mu_i}(\mathbf{r}')$ provides a measure of the impact that a given optical perturbation $\delta\mu_i$, at position $\mathbf{r}'$, has on a measured signal $\mathcal{S}(\mathbf{r}_d)$. The sensitivity function defined by Eq. (14.1) can be significantly different for different types of optical measurements. For example, measurements of time-gated optical signals have different regions of sensitivity for different time-gating windows (Schotland et al., 1993; Sawosz et al., 2012); the amplitude and the phase of modulated optical signals have unique regions of sensitivity (Fantini et al., 1995) that depend on the modulation frequency (Sevick et al., 1994), etc. This dependence allows for some control over the spatial region probed with optical methods, in addition to the obvious effect of the placement and relative distance between the illumination and optical collection points.

14.1.2 The CW region of sensitivity

Let's now consider the specific case of a continuous-wave (CW) point light source of unit power at position $\mathbf{r}_s$ in a macroscopically homogeneous scattering medium. The resulting photon fluence rate at an arbitrary position $\mathbf{r}'$ can be interpreted as a measure of the probability that a photon emitted at $\mathbf{r}_s$ reaches a small volume element about $\mathbf{r}'$. The fluence rate at $\mathbf{r}'$ resulting from a point light source at $\mathbf{r}_s$ is proportional to the Green's function $G_{\phi\mathrm{CW}}(\mathbf{r}_s, \mathbf{r}')$. The Green's function is the solution to an inhomogeneous differential equation for a point source and a given set of boundary conditions. Mathematically, the spatial dependence of a "point source" is described by a Dirac delta ($\delta(\mathbf{r})$). Therefore, the fluence rate Green's function $G_{\phi\mathrm{CW}}(\mathbf{r}_s, \mathbf{r}')$ can be related to the probability that a photon emitted at $\mathbf{r}_s$ reaches a small volume element about $\mathbf{r}'$.

Similarly, the fluence rate Green's function for a source at position $\mathbf{r}'$ evaluated at a detector position $\mathbf{r}_d$ ($G_{\phi\mathrm{CW}}(\mathbf{r}', \mathbf{r}_d)$) can be related to the probability that a photon emitted at position $\mathbf{r}'$ reaches a small volume element about the detector position $\mathbf{r}_d$. The product $G_{\phi\mathrm{CW}}(\mathbf{r}_s, \mathbf{r}')G_{\phi\mathrm{CW}}(\mathbf{r}', \mathbf{r}_d)$ is, in fact, associated with the *photon visiting probability*, which is defined as the probability that a photon that is injected in the scattering medium at position $\mathbf{r}_s$ and detected at position $\mathbf{r}_d$ visited position $\mathbf{r}'$ (or a small volume about $\mathbf{r}'$). Nonetheless, we refrain from using the term probability in the following treatment for two reasons: (1) expressing the photon visiting probability in terms of the product $G_{\phi\mathrm{CW}}(\mathbf{r}_s, \mathbf{r}')G_{\phi\mathrm{CW}}(\mathbf{r}', \mathbf{r}_d)$ requires the definition of a proper normalization factor, which would depend on the specific definition of the "photon visitation" event, which is not addressed here; and (2) the fact that rather than the spatial distribution of the photon visiting probability density, it is more relevant to consider a quantitative measure of the sensitivity of the detected optical signal to the optical properties at an arbitrary position $\mathbf{r}'$ in the scattering medium, as represented by Eq. (14.1) for an arbitrary source distribution and for a given optical signal $\mathcal{S}(\mathbf{r}_d)$. In some cases, depending on the specific definition of the visiting probability density and the particular optical signal measured, the two concepts may yield the same information, but, in general, visitation probability and sensitivity are not equivalent, and Eq. (14.1) is the most general definition of optical sensitivity.

We now turn to a specific definition of the *CW region of sensitivity* ($s_{\phi_{CW},\mu_a}(\mathbf{r})$), which is indeed based on the product of the two Green's functions, $G_{\phi_{CW}}(\mathbf{r}_s, \mathbf{r})$ and $G_{\phi_{CW}}(\mathbf{r}, \mathbf{r}_d)$, considered above. Here, we indicate with $\mathbf{r}$ (instead of $\mathbf{r}'$) the arbitrary position in the scattering medium. As mentioned above, the first Green's function provides a measure of the number of photons originating from the source (at $\mathbf{r}_s$) that reach the general position $\mathbf{r}$ in the medium. The second Green's function provides a measure of how many of these photons that have reached position $\mathbf{r}$ go on to reach the detector position $\mathbf{r}_d$. As we will see in Section 14.6.1.2, the product $G_{\phi_{CW}}(\mathbf{r}_s, \mathbf{r})G_{\phi_{CW}}(\mathbf{r}, \mathbf{r}_d)$ is indeed a measure of the sensitivity function (as defined by Eq. (14.1)) for the case of the CW fluence rate (ϕ_{CW}) and for a perturbation in the absorption coefficient (μ_a) at position $\mathbf{r}$; and this motivates the notation $s_{\phi_{CW},\mu_a}(\mathbf{r})$. This CW sensitivity function to an absorption perturbation has been designated with different terms in the literature, such as *photon hitting density*, *photon weight function*, *photon sampling function*, or *line spread function*:

$$s_{\phi_{CW},\mu_a}(\mathbf{r}) = kG_{\phi_{CW}}(\mathbf{r}_s, \mathbf{r})G_{\phi_{CW}}(\mathbf{r}, \mathbf{r}_d). \tag{14.2}$$

The factor k (not to be confused with the wavenumber) in Eq. (14.2) is a spatially independent normalization factor. In the remainder of this section and in Section 14.6.1 we quantitatively describe a number of critical features of the CW region of sensitivity defined by Eq. (14.2) for the cases of homogeneous infinite and semi-infinite scattering media. In this discussion, we also ignore any dependence on the optical wavelength or bandwidth of the source.

14.1.2.1 *Infinite geometry*

In the case of a macroscopically homogeneous and infinite medium, the CW fluence rate Green's function only depends on the distance from the point source, i.e., $G^{(\mathrm{inf})}_{\phi_{CW}}(\mathbf{r}_s, \mathbf{r}) = G^{(\mathrm{inf})}_{\phi_{CW}}(|\mathbf{r} - \mathbf{r}_s|)$. In the diffusion approximation, the CW fluence rate Green's function is given by the CW fluence rate for a point source (as given by Eq. (10.3)) divided by the source power, the result being a fluence rate per unit source power. By using Eq. (10.3), where r indicates the distance from the point source, the sensitivity function at a position $\mathbf{r}$ can be written as follows in an infinite medium (Feng et al., 1995):

$$s^{(\mathrm{inf})}_{\phi_{CW},\mu_a}(\mathbf{r}) = kG^{(\mathrm{inf})}_{\phi_{CW}}(|\mathbf{r} - \mathbf{r}_s|)G^{(\mathrm{inf})}_{\phi_{CW}}(|\mathbf{r}_d - \mathbf{r}|) = k^{(\mathrm{inf})}\frac{e^{-\mu_{\mathrm{eff}}|\mathbf{r}-\mathbf{r}_s|}}{|\mathbf{r} - \mathbf{r}_s|}\frac{e^{-\mu_{\mathrm{eff}}|\mathbf{r}_d-\mathbf{r}|}}{|\mathbf{r}_d - \mathbf{r}|}, \tag{14.3}$$

where $k^{(\mathrm{inf})}$ includes the spatially independent factors of the fluence rate Green's functions, and we recall that the effective attenuation coefficient is defined in terms of the absorption and reduced scattering coefficients as $\mu_{\text{eff}} = \sqrt{3\mu_a(\mu'_s + \mu_a)}$. By using a cylindrical coordinate system with radial coordinate ρ, azimuth angle φ, and the z axis along the source-detector line (with the origin at the point source), the general position vector is $\mathbf{r} = (\rho, \varphi, z)$, whereas the source and detector position vectors are $\mathbf{r}_s = (0, 0, 0)$ and $\mathbf{r}_d = (0, 0, r_{sd})$, where r_{sd} is the source-detector distance. In this coordinate system, Eq. (14.3) becomes:

$$s^{(\text{inf})}_{\phi\text{CW},\mu_a}(\rho, \varphi, z) = k^{(\text{inf})} \frac{e^{-\mu_{\text{eff}}\left[\sqrt{\rho^2+z^2}+\sqrt{\rho^2+(r_{sd}-z)^2}\right]}}{\sqrt{\rho^2 + z^2}\sqrt{\rho^2 + (r_{sd} - z)^2}}. \tag{14.4}$$

As intuition dictates, for any cross-sectional plane normal to the z axis, the sensitivity of the CW fluence rate to a localized absorption perturbation is maximal at $\rho = 0$, i.e., on the geometrical line between the source and the detector, and is azimuthally symmetric. We may determine $k^{(\text{inf})}$ by normalizing the sensitivity to that maximum value at any plane of constant z. As a result, we determine a CW sensitivity function for a cross-sectional plane perpendicular to the z axis ($s^{(\text{inf})}_{\phi\text{CW},\mu_a}(\rho, z)$) that specifies the relative sensitivity for each point on that plane with respect to the maximum value on the z axis (i.e., at $\rho = 0$):

$$s^{(\text{inf})}_{\phi\text{CW},\mu_a}(\rho, z) = \frac{e^{-\mu_{\text{eff}}\left[\sqrt{\rho^2+z^2}+\sqrt{\rho^2+(r_{sd}-z)^2}-r_{sd}\right]}}{\sqrt{\left[1+\left(\frac{\rho}{z}\right)^2\right]\left[1+\left(\frac{\rho}{r_{sd}-z}\right)^2\right]}}. \tag{14.5}$$

For each value of z, the sensitivity function of Eq. (14.5) is such that $0 \leq s^{(\text{inf})}_{\phi\text{CW},\mu_a}(\rho, z) \leq 1$.

Examination of Eq. (14.5) reveals that the CW region of sensitivity extends significantly away from the geometrical line joining the source and the detector. It is possible to get a graphical representation of the region of optical sensitivity by introducing a threshold value $\alpha < 1$, for the normalized optical sensitivity function $s^{(\text{inf})}_{\phi\text{CW},\mu_a}(\rho, z)$, and plot the surface (ρ_α, z) defined by $s^{(\text{inf})}_{\phi\text{CW},\mu_a}(\rho_\alpha, z) = \alpha$. For each value of z, all points that are at a distance from the axis smaller than ρ_α feature a sensitivity value that is at least α times the axial sensitivity at that z coordinate.

Figure 14.1(a) shows the $\alpha = 0.5$ region of sensitivity for an infinite medium with $\mu_a = 0.10$ cm^{-1} and $\mu'_s = 10$ cm^{-1}, and for three source-detector distances (r_{sd}) of 1, 3, and 5 cm. Figure 14.1(b) shows the $\alpha = 0.5$ region of sensitivity for an infinite medium with three sets of optical coefficients ((μ_a, μ'_s) = (0.05, 10) cm^{-1}, (0.05, 5) cm^{-1}, (0.10, 10) cm^{-1}), all with the same source-detector distance of 3 cm. From Figure 14.1(a), it is evident that the CW region of sensitivity becomes wider at greater source-detector distances, r_{sd}, although the widening of the region of sensitivity does not scale linearly with r_{sd}.

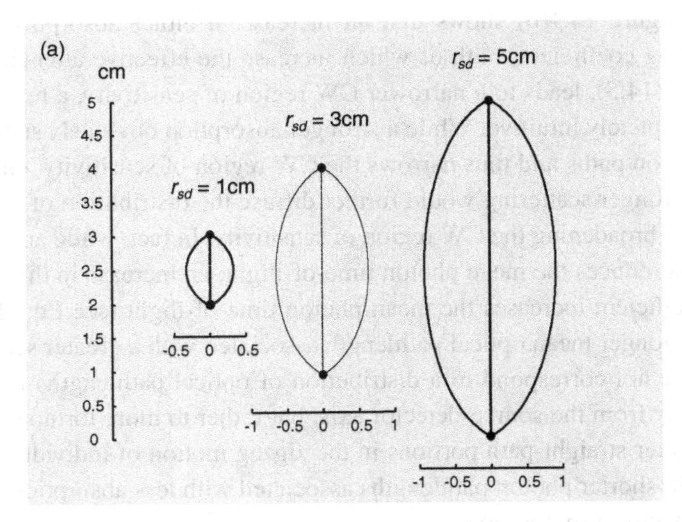

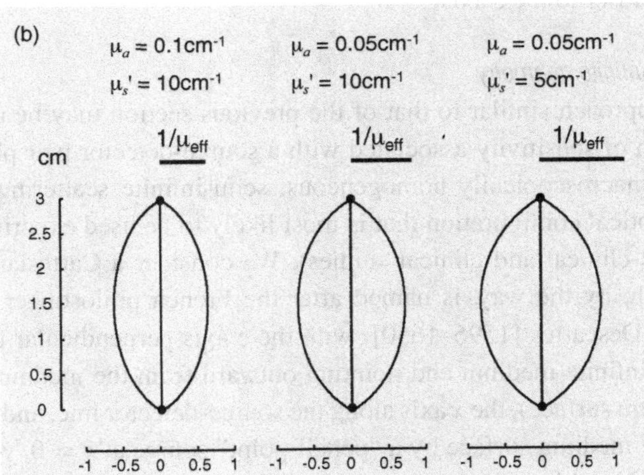

Figure 14.1

CW regions of sensitivity in an infinite scattering medium, defined by the lines where the sensitivity function $s^{(inf)}_{\phi_{CW},\mu_a}$, given by Eq. (14.5), is 0.5 times its maximum value for each cross section orthogonal to the source-detector line. Such maximum values of $s^{(inf)}_{\phi_{CW},\mu_a}$ are on the geometrical line joining the source and the detector for each cross-sectional plane. The source and detector are indicated by filled circles in the figure. (a) CW regions of sensitivity for a medium with $\mu_a = 0.10\ \text{cm}^{-1}$ and $\mu_s' = 10\ \text{cm}^{-1}$, and for three source-detector distances (r_{sd}) of 1, 3, and 5 cm. (b) CW regions of sensitivity for a source-detector distance of 3 cm, and three sets of optical coefficients (μ_a, μ_s') of (0.10, 10), (0.05, 10), and (0.05, 5) cm^{-1}. The horizontal bars in (b) indicate the inverse of the effective attenuation coefficient, which scales with the width of the region of sensitivity. All axes are in cm.

Figure 14.1(b) shows that an increase in either absorption or reduced scattering coefficient, both of which increase the effective attenuation coefficient in Eq. (14.5), leads to a narrower CW region of sensitivity, a result that may not be completely intuitive. While a stronger absorption obviously suppresses the longer photon paths and thus narrows the CW region of sensitivity, one may expect that a stronger scattering would further diffuse the distribution of optical pathlengths, thus broadening the CW region of sensitivity. In fact, while an increase in absorption reduces the mean photon time-of-flight, an increase in the reduced scattering coefficient increases the mean photon time-of-flight (see Eq. (11.11)). However, the longer mean optical pathlength associated with a greater scattering coefficient does not correspond to a distribution of optical pathlengths that extends further away from the source-detector axis, but rather to more tortuous photon paths and shorter straight-path portions in the zigzag motion of individual photons. In general, shorter photon pathlengths associated with less absorption are still those that stay closer to the z axis.

14.1.2.2 *Semi-infinite geometry*

An approach similar to that of the previous section may be used to find the CW region of sensitivity associated with a source-detector pair placed on the surface of a macroscopically homogeneous, semi-infinite scattering medium, which is the optical configuration that is most likely to be used experimentally (especially in pre-clinical and clinical studies). We consider a Cartesian coordinate system (which, by the way, is named after the French philosopher and mathematician René Descartes [1596–1650]) with the z axis perpendicular to the surface of the semi-infinite medium and pointing outward from the medium (with $z = 0$ at the medium surface), the x axis along the source-detector line, and optical illumination on the medium surface by a "pencil-point" source at $x = 0$, $y = 0$. Therefore, the general position vector is $\mathbf{r} = (x, y, z)$, whereas the source and detector position vectors are $\mathbf{r}_s = (0, 0, -z_0)$, recalling that $-z_0 = -1/\mu_s'$ is the location of the effective photon source inside the scattering medium, and $\mathbf{r}_d = (\rho_{sd}, 0, 0)$. As we have done in Chapters 10–12, we identify with ρ the distance between a source and a detector on the surface of a semi-infinite scattering medium.

The fluence rate Green's function for a semi-infinite medium is given by Eq. (10.22) divided by $a'P_{\mathrm{CW}}$ (the product of the reduced single-scattering albedo times the source power), and we consider zero boundary conditions, for which the extrapolated boundary coincides with the physical boundary, i.e. $z_b = 0$ (see Section 9.6.2). Furthermore, we consider distances from the light source that are much greater than z_0 (which is consistent with the limit of applicability of diffusion theory). Under these conditions, we show in Appendix 14.A that the fluence rate Green's function (i.e., the fluence rate per unit source power) for the case of CW illumination at the boundary of a semi-infinite scattering medium becomes:

$$G_{\phi\mathrm{CW}}^{(\mathrm{seminf})}(\mathbf{r}_s, \mathbf{r}) \cong -\frac{3(\mu_s' + \mu_a)}{4\pi} 2z_0 z \left(\mu_{\mathrm{eff}} + \frac{1}{\sqrt{x^2+y^2+z^2}}\right) \frac{e^{-\mu_{\mathrm{eff}}\sqrt{x^2+y^2+z^2}}}{(x^2+y^2+z^2)}, \tag{14.6}$$

where the minus sign takes into account the fact that $z < 0$ for all points inside the scattering medium. As discussed in Section 10.4.1, the optical signal measured at the surface of a scattering medium is proportional to the normal component of the photon flux ($\hat{\mathbf{n}} \cdot \mathbf{F}$), which, in the diffusion approximation, is proportional to the gradient of the fluence rate $\mathbf{F} = -\nabla\phi/[3(\mu_s' + \mu_a)]$ (see Eq. (9.37)). Consequently, at the detector side, it is the normal component of the gradient of the fluence rate Green's function that determines the measured optical intensity, and it is given by Eq. (10.23) (with $a' = 1$).

By considering, as presented above, zero boundary conditions ($z_b = 0$), and a detector location at $z = 0$ (i.e., on the medium surface), the two distances r_1 and r_2 in Eq. (10.23) are equal. In this case, they represent the distances between the detector location (ρ_{sd}, 0, 0) and a general point inside the medium (x, y, z), distance r_1, and its mirror image with respect to the medium surface at (x, y, $-z$) distance r_2. With these notations, the normal component (here, $\hat{\mathbf{n}} = \hat{\mathbf{z}}$) of the gradient of the fluence rate Green's function at the location of the detector is:

$$-\frac{1}{3(\mu_s' + \mu_a)} \left(\frac{\partial}{\partial z_d} G_{\phi\mathrm{CW}}^{(\mathrm{seminf})}(\mathbf{r}, \mathbf{r}_d)\right)_{z_d=0}$$
$$= -\frac{2z}{4\pi} \left(\mu_{\mathrm{eff}} + \frac{1}{\sqrt{(\rho_{sd}-x)^2+y^2+z^2}}\right) \frac{e^{-\mu_{\mathrm{eff}}\sqrt{(\rho_{sd}-x)^2+y^2+z^2}}}{[(\rho_{sd}-x)^2+y^2+z^2]}, \tag{14.7}$$

where z_d is the z coordinate of the detector, and z_0, the depth coordinate of the effective source in Eq. (10.23), is replaced by the z value of the general coordinate, which itself represents a conceptual source location in Eq. (14.7).

The CW sensitivity function in a semi-infinite medium, for illumination and optical detection at the surface, can now be written as follows:

$$s_{\phi\mathrm{CW},\mu_a}^{(\mathrm{seminf})}(\mathbf{r}) = kG_{\phi\mathrm{CW}}^{(\mathrm{seminf})}(\mathbf{r}_s, \mathbf{r}) \left[-\frac{1}{3(\mu_s' + \mu_a)} \left(\frac{\partial}{\partial z_d} G_{\phi\mathrm{CW}}^{(\mathrm{seminf})}(\mathbf{r}, \mathbf{r}_d)\right)_{z_d=0}\right]$$
$$= k^{(\mathrm{seminf})} z^2 \left(\mu_{\mathrm{eff}} + \frac{1}{\sqrt{x^2+y^2+z^2}}\right)\left(\mu_{\mathrm{eff}} + \frac{1}{\sqrt{(\rho_{sd}-x)^2+y^2+z^2}}\right)$$
$$\times \frac{e^{-\mu_{\mathrm{eff}}\left(\sqrt{x^2+y^2+z^2}+\sqrt{(\rho_{sd}-x)^2+y^2+z^2}\right)}}{(x^2+y^2+z^2)[(\rho_{sd}-x)^2+y^2+z^2]} \tag{14.8}$$

where $k^{(\text{seminf})}$ includes spatially independent factors in $G_{\phi\text{CW}}^{(\text{seminf})}$ and $\frac{1}{3(\mu_s'+\mu_a)} \times \frac{\partial G_{\phi\text{CW}}^{(\text{seminf})}}{\partial z_d}$. The normalization factor $k^{(\text{seminf})}$ may be specified by normalizing the photon visiting probability to its maximum value at $(x, 0, z_{\max}(x))$ for each cross-sectional plane of $x = \text{const}$. This leads to the following definition of the normalization factor $k^{(\text{seminf})}(x)$ for each cross section perpendicular to the x axis:

$$k^{(\text{seminf})}(x) = \frac{1}{s_{\phi\text{CW},\mu_a}^{(\text{seminf})}(x, 0, z_{\max}(x))}. \tag{14.9}$$

With the normalization factor expressed by Eq. (14.9), the CW sensitivity function $s_{\phi\text{CW},\mu_a}^{(\text{seminf})}(x, y, z)$ (Eq. (14.8)) specifies the optical sensitivity of each point on a plane of constant x with respect to the maximum value on that plane. By definition, this normalization leads to a sensitivity function $s_{\phi\text{CW},\mu_a}^{(\text{seminf})}(x, y, z)$ that is bound between 0 and 1. Analogous to the infinite-medium case, by introducing a threshold value $\alpha < 1$ for the normalized CW sensitivity function $s_{\phi\text{CW},\mu_a}^{(\text{seminf})}(x, y, z)$, the set of points $(x, 0, z_\alpha)$ designated by $s_{\phi\text{CW},\mu_a}^{(\text{seminf})}(x, 0, z_\alpha) = \alpha$ defines the boundary of the α-region of sensitivity for the semi-infinite geometry. Figure 14.2(a) shows the $\alpha = 0.5$ region of sensitivity for a semi-infinite medium with $\mu_a = 0.10\ \text{cm}^{-1}$ and $\mu_s' = 10\ \text{cm}^{-1}$, and for three source-detector distances (ρ_{sd}) of 1, 3, and 5 cm.

The shape of the CW region of sensitivity resembles a banana, and it is therefore commonly referred to as the *banana region of sensitivity* or, sometimes, simply the *photon banana*. In the biomedical optics field, it is commonly described qualitatively, but we provide here a quantitative description that provides insight into its dependence on the optical properties of the medium and on the source-detector distance. From Figure 14.2(a), as one would intuitively expect, it is evident that the CW region of sensitivity extends to deeper regions in the medium for greater source-detector distances; however, the deepening of the CW region of sensitivity does not scale linearly with source-detector distance. In fact, it was found that the mean photon-visitation depth, $\langle z \rangle_{\rho_{sd}}$, i.e., the average depth of the CW region of sensitivity, is well approximated by the following expression (Patterson et al., 1995):

$$\langle z \rangle_{\rho_{sd}} = \frac{1}{2}\sqrt{\frac{\rho_{sd}}{\mu_{\text{eff}}}}, \tag{14.10}$$

showing that the mean penetration depth scales with the square root of the source-detector separation. Figure 14.2(b) shows the $\alpha = 0.5$ region of sensitivity for a semi-infinite medium with three sets of optical coefficients ($(\mu_a, \mu_s') = (0.05, 10)\ \text{cm}^{-1}$, $(0.05, 5)\ \text{cm}^{-1}$, and $(0.10, 10)\ \text{cm}^{-1}$), and for a source-detector distance of 3 cm. Figure 14.2(b) also exhibits a trend qualitatively similar to that

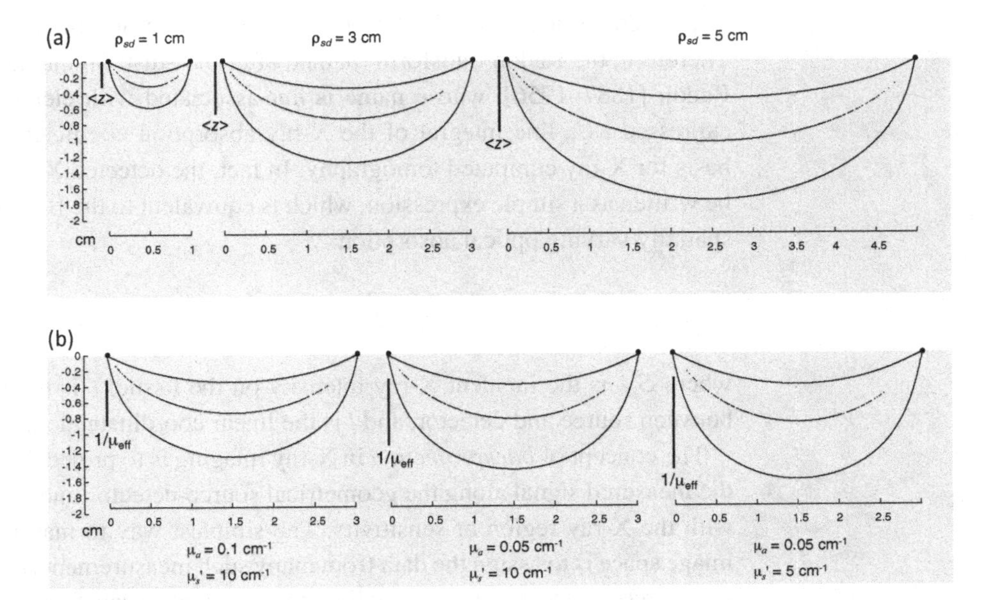

Figure 14.2

CW regions of sensitivity in a semi-infinite scattering medium, defined by the lines where the normalized sensitivity function $s^{(\mathrm{seminf})}_{\Phi\mathrm{CW},\mu_a}$, given by Eq. (14.8) and the normalization condition of Eq. (14.9), is equal to 0.5. The source and detector are indicated by filled circles, and the line where $s^{(\mathrm{seminf})}_{\Phi\mathrm{CW},\mu_a}$ is maximal for each cross-sectional plane orthogonal to the source-detector line is also indicated in the figure. (a) CW regions of sensitivity for a medium with $\mu_a = 0.10\ \mathrm{cm}^{-1}$ and $\mu_s' = 10\ \mathrm{cm}^{-1}$, and for three source-detector distances (ρ_{sd}) of 1, 3, and 5 cm. The vertical bars in (a) indicate the mean photon-visit depth given by Eq. (14.10). (b) CW regions of sensitivity for a source-detector distance of 3 cm, and three sets of optical coefficients (μ_a, μ_s') of (0.10, 10), (0.05, 10), and (0.05, 5) cm^{-1}. The vertical bars in (b) indicate the inverse of the effective attenuation coefficient, which scales with the width of the region of sensitivity. All axes are in cm.

observed in the infinite geometry: an increase in either absorption or reduced scattering coefficient determines a narrower (and shallower) CW region of sensitivity.

14.2 Backprojection methods

The region of sensitivity discussed in Section 14.1.2 is highly significant for the optical study of scattering media because it specifies the optically probed volume in the medium. In optical spectroscopy measurements, the region of sensitivity represents the tissue volume that is interrogated by photons. In optical imaging measurements, the region of sensitivity is associated with the *point spread function* (PSF), which specifies the response to a point object in the scattering medium. In the case of X-ray photons, the region of sensitivity in tissues is essentially limited to the geometrical line-of-sight between the source and the detector of X-rays. Therefore, the Radon transform (named after the Austrian mathematician Johann Radon [1887–1956], whose name is *not* associated with element 86), which is expressed as a line integral of the X-ray absorption coefficient (μ_x), forms the basis for X-ray computed tomography. In fact, the detected X-ray signal ($\mathcal{S}_x$) can be written as a simple expression, which is equivalent to the Beer-Lambert law for spatially varying optical absorption:

$$\mathcal{S}_x = \mathcal{S}_{x0} e^{-\int_{\ell} \mu_x(\mathbf{r}) dl}, \tag{14.11}$$

where $\mathcal{S}_{x0}$ is the incident X-ray intensity on the tissue, ℓ is the geometrical line between source and detector, and l is the linear coordinate along ℓ.

The concept of *backprojection* in X-ray imaging is to project back, or distribute, the measured signal along the geometrical source-detector line ℓ, which coincides with the X-ray region of sensitivity. The simplest way to map the data onto the image space is to assign the data from many such measurements to the corresponding source-detector lines ℓ, and to linearly combine the data from the measurements over multiple, intersecting source-detector lines at varying directions in two or three coordinates, to yield images in two dimensions or three dimensions, respectively. This straightforward approach is often called *simple backprojection*, and it is also referred to as *optical topography* when it generates 2D images by direct superposition of diffuse reflectance data. A more accurate imaging approach is based on inverting the Radon transform of Eq. (14.11) by *filtered backprojection*, in which the measured signal is convolved with a filter kernel before being back-projected. The purpose of the filtering procedure is to introduce appropriate weights to the spatial frequency components of the measured data to minimize blurring in the backprojection image. A ramp filter proportional to the spatial frequency, or some variations to reduce amplification at higher spatial frequencies, is used to invert the Radon transform by filtered backprojection.

In the case of optical signals from scattering media, the region of sensitivity extends over a three-dimensional volume within the medium, as we have seen in the previous section. Backprojection imaging schemes can still be applied following two different methods. In the first method, the optical signal is back-projected over the line between source and detector, comprising the points in each cross-sectional plane for which the sensitivity of the optical signal is maximal. In an infinite medium, this is the geometrical line joining the source and the detector; in a semi-infinite medium, this is the arched line $z = z_{\max}(x)$ at $y = 0$, according to the notation and coordinate system used in the previous section. This first method has been formulated in terms of the most favorable photon path (associated with so-called "Fermat photons" in a generalization of the Fermat principle, according to which photons travel along a path that minimizes their time-of-flight) (Polishchuk and Alfano, 1995), or a classical photon path based on a Lagrangian approach (Perelman et al., 1997). (Fermat photons are named after the French mathematician Pierre de Fermat [1601–1665], whereas the Lagrangian approach, which consists of conceptually keeping track of individual flowing particles, or photons, is named after the Italian mathematician Giuseppe Luigi Lagrangia, or Joseph-Louis Lagrange [1736–1813].) In the second method of backprojection, the optical signal is back-projected over the volume described by the region of sensitivity for the particular geometry of the scattering medium or tissue examined.

We recall that the region of sensitivity defined in Section 14.1.2 refers to a CW optical signal. One can consider a more general backprojection volume that is defined in terms of the sensitivity function of Eq. (14.1) and that depends on the measured optical signal. For example, the backprojection volume can be narrowed by time-gated detection that selects shorter transit times and rejects later-arriving photons in the time domain (Mitic et al., 1994), or by increasing the modulation frequency of amplitude measurements in the frequency-domain (Fishkin and Gratton, 1993).

While simple backprojection methods are relatively straightforward to implement and can generate qualitatively reliable images, their utility for reconstructing quantitative images of optical property distributions is limited. Furthermore, they are mainly relevant to diffuse transmission geometries and less effective for source-detector geometries used in diffuse reflectance measurements.

14.3 Diffuse optical imaging with time-gated approaches

In a transmission geometry, in theory, it should be possible to optimize the resolution of backprojection methods by restricting the measurement to photons that have scattered the least and are closest to X-ray-like straight-through propagation. In the literature, photons that have traversed a turbid medium essentially without scattering (in a straight line) are referred to as *ballistic photons*, and photons that have undergone slight scattering that is near-forward (with only small deviations from the geometrical source-detector line) are often called *snake-like photons* or *quasi-ballistic photons*. Ballistic photons and snake-like photons can be quantitatively defined in terms of their time-of-flight in the medium, the straight-line distance between source and detector divided by c_n for ballistic photons, and slightly longer times for snake-like photons. Ballistic and snake-like photons do not really play a role in diffuse optical imaging because of their paucity, which renders them practically undetectable for tissue thicknesses ≥ 1 cm. In fact, as described in

Section 1.3, the probability that a photon is neither absorbed nor scattered when traveling over a distance of 1 cm in a typical tissue with $\mu_a = 0.1$ cm^{-1} and $\mu_s = 100$ cm^{-1} is $e^{-100.1} \sim 3 \times 10^{-44}$! However, performing selective detection of photons on the basis of their time of flight in tissues has significant implications for diffuse optical imaging. This can be seen by replacing μ_s with μ_s' in the above estimate (with $\mu_s' \cong 10$ cm^{-1}), which leads to a probability of $\sim 4 \times 10^{-5}$ that photons travel over a distance of 1 cm while still retaining some level of directional information (a softer condition for snake-like photons).

Time-domain measurements generate the temporal distribution of collected photons from a pulsed light source, as described in Chapter 11. Time-gating techniques allow for the selective detection of photons whose times of flight are within a specific temporal window, which can be defined in absolute terms (say, 1.3–1.5 ns following the source emission) or relative to the detected optical pulse (say, the time window corresponding to the collection of the first-arriving 10% of the total detected photons). As a result, in the time-domain one can define optical signals given by a time-gated fluence rate that we indicate as $\phi_{\mathrm{TD,abs}}$ and $\phi_{\mathrm{TD,rel}}$, in the case of absolute (Δt) and relative (temporal percentile range, $\Delta\%$, of detected photons) time windows, respectively. These time-gated optical signals are characterized by individual sensitivity functions, according to the definition stated by Eq. (14.1) and, correspondingly, individual spatial regions of sensitivity.

It is intuitive that photons detected in early time windows, having traveled shorter pathlengths, will be characterized by a narrower region of sensitivity than late time windows. This principle can be exploited, especially for a transmission geometry, to enhance the spatial resolution and contrast of diffuse optical imaging by selectively detecting early arriving photons. However, the improvement in spatial resolution, in thick tissues, may be limited to, at most, a factor of ~ 2 with respect to continuous-wave illumination (Mitic et al., 1994), whereas the contrast enhancement may depend on the specific optical features (absorbing versus scattering) of localized optical inhomogeneities (Hall et al., 1997). There is an additional limitation in that, especially for highly scattering media and/or large source-detector distances, the number of early-arriving photons (with significantly shorter pathlengths) may be too small to provide an adequate signal-to-noise ratio, thus limiting the advantage over CW illumination.

In the case of a semi-infinite geometry (source and detector at the tissue surface), however, there is a complementary time-gating approach to that of selectively detecting the early-arriving photons. It consists of selectively suppressing the early-arriving photons to detect only those photons that have traveled over longer pathlengths. In this case, the idea is to use short source-detector separations and discard the early photons that have only probed the most superficial tissue layers, thus retaining sensitivity to deeper tissue regions. This method has been realized by single-photon avalanche diode (SPAD) detectors, which feature a fast gating time of a few hundred picoseconds. The fast gating time enables the detector to reject the large number of early photons, preventing them from causing damage or contributing to background noise (Tosi et al., 2009).

The ability to effectively reject early photons, with sharp turn-on time, has made it possible to collapse the source-detector distance in a so-called null source-detector separation (NSDS) imaging method. In this approach, the two ends of the banana-shaped region of sensitivity characteristic of the $\mathbf{r}_s \neq \mathbf{r}_d$ case (see Figure 14.2) coalesce and realize a region of sensitivity that resembles a pear (to stick with fruit-based descriptions!). This pear-like region of sensitivity of the NSDS ($\mathbf{r}_s = \mathbf{r}_d$) approach is typically smaller than the banana-like region of sensitivity of the $\mathbf{r}_s \neq \mathbf{r}_d$ case, thus enhancing the sensitivity of the NSDS time-gated optical signal with respect to optical signals collected at a non-zero distance from the illumination point. Such enhanced sensitivity over a smaller spatial volume may result in an enhancement of contrast and spatial resolution when the NSDS method is applied to generate optical images by scanning the collocated source and detector (Torricelli et al., 2005).

14.4 Spatial frequency-domain imaging

In Section 10.4.5, we have introduced the concept of spatially modulated spectroscopy, which is based on structured illumination. The relevant measurements of the spatial modulation depth of the diffuse reflectance can be performed at each point of the sample surface by sequential illumination of sinusoidal spatial patterns with different spatial phases. Spatially modulated reflectance data at a minimum of two frequencies, once properly calibrated via measurements on a reference sample, can be fit to the expression of Eq. (10.28) to generate maps of absorption and reduced scattering coefficients (Cuccia et al., 2009) or tissue oxygenation (Gioux et al., 2011). Spatial frequency-domain imaging is a fast reflectance-mode technique (temporal resolution down to $\sim$1 s or less) that can acquire images of tissue optical properties over a large field of view (tens to hundreds of cm^2), but at depths limited to $\lesssim$1 cm (Konecky et al., 2009) rather than volumetric imaging at larger depths.

14.5 Diffuse optical tomography: the forward problem

The *forward problem* for diffuse tissue imaging consists of the estimation of the measured optical signals for a specified arrangement of light sources and for an

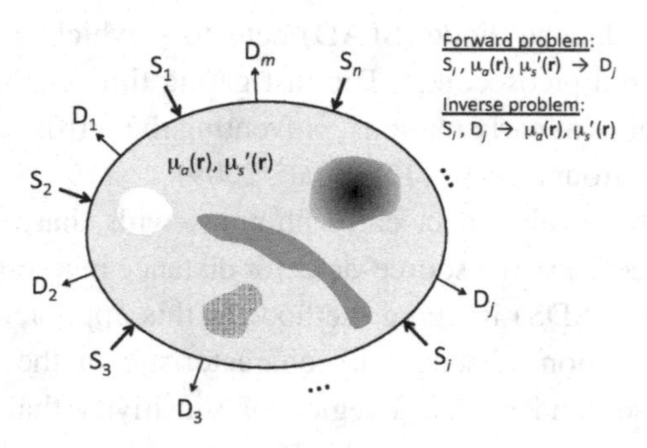

Figure 14.3

Schematic representation of the forward problem and inverse imaging problem. Given a distribution of light sources S_i ($i = 1,2\ldots n$) of known spatial location and emission properties, and a given spatial distribution of the medium absorption and reduced scattering coefficients ($\mu_a(\mathbf{r})$ and $\mu_s'(\mathbf{r})$), the forward problem generates the optical signals measured at a set of optical detectors D_j ($j = 1,2\ldots m$). The inverse problem generates a spatial distribution of $\mu_a(\mathbf{r})$ and $\mu_s'(\mathbf{r})$ from the optical signals generated by a given set of light sources S_i and optical detectors D_j.

a priori (known) spatial distribution of the medium optical properties. This is illustrated in Figure 14.3. The arrangement of light sources and optical detectors on the scattering medium (typically on its surface) is set by the user. For a given spatial distribution of the optical coefficients in the medium and a given set of light sources, the forward problem models the optical signals measured by the detectors.

Solving the forward problem by itself does not generate an image of unknown tissue structures, because it is assumed that the spatial distribution of tissue optical properties is already known. In contrast, solving the *inverse problem* constitutes determining the spatial distribution (image) of tissue properties when the source locations and detector signals are known. This often invokes iterative application of solutions to the forward problem as follows: one starts with an initial estimate (educated guess) of the spatial distribution of tissue optical properties; then a specific forward-problem solution is obtained (given the known light sources) resulting in calculated signals at the detector locations; the differences between the calculated and the actual measured signals at the detectors are then used to iterate a revised estimate of the distribution of tissue properties, and the forward problem is run again. This cycle is iterated until the difference between calculated and measured signals at the detectors is small, hopefully yielding a more accurate estimate of the spatial distribution of tissue properties. Hence, understanding different methods to tackle the forward problem can facilitate developing solutions to the inverse problem.

The approaches to solving the forward problem can be broadly classified into deterministic and stochastic methods.

14.5.1 Deterministic forward models: transport theory and diffusion theory

The Boltzmann transport equation (BTE) (Eq. (9.2)), its P_N approximations (the P_1 approximation is given by Eq. (9.32)), and the diffusion equation (Eq. (9.36)) provide a deterministic framework for the quantitative description of light propagation in scattering media. The Boltzmann transport equation provides the most general treatment, and is applicable to cases in which its diffusion approximation fails: for example, in the vicinity of light sources or boundaries in the study of small tissue volumes, in the presence of tissues featuring low scattering such as the cornea, cerebrospinal fluid, or fluid-filled cysts, or in the case of highly absorbing tissues. The BTE is re-written here in terms of the radiance (L) for the time-domain (L_{TD}), frequency-domain (L_{FD}), and continuous-wave (L_{CW}) cases, respectively:

$$\left[\frac{1}{c_n}\frac{\partial}{\partial t}+\hat{\boldsymbol{\Omega}}\cdot\nabla+\mu_t(\mathbf{r})\right]L_{\mathrm{TD}}(\mathbf{r},\hat{\boldsymbol{\Omega}},t)$$
$$=\mu_s(\mathbf{r})\int_{4\pi}L_{\mathrm{TD}}(\mathbf{r},\hat{\boldsymbol{\Omega}}',t)p(\hat{\boldsymbol{\Omega}}',\hat{\boldsymbol{\Omega}})d\hat{\Omega}'+q(\mathbf{r},\hat{\boldsymbol{\Omega}},t), \tag{14.12}$$

$$\left[-i\frac{\omega}{c_n}+\hat{\boldsymbol{\Omega}}\cdot\nabla+\mu_t(\mathbf{r})\right]L_{\mathrm{FD}}(\mathbf{r},\hat{\boldsymbol{\Omega}},\omega)$$
$$=\mu_s(\mathbf{r})\int_{4\pi}L_{\mathrm{FD}}(\mathbf{r},\hat{\boldsymbol{\Omega}}',\omega)p(\hat{\boldsymbol{\Omega}}',\hat{\boldsymbol{\Omega}})d\Omega'+q(\mathbf{r},\hat{\boldsymbol{\Omega}},\omega), \tag{14.13}$$

$$\left[\hat{\boldsymbol{\Omega}}\cdot\nabla+\mu_t(\mathbf{r})\right]L_{\mathrm{CW}}(\mathbf{r},\hat{\boldsymbol{\Omega}})=\mu_s(\mathbf{r})\int_{4\pi}L_{\mathrm{CW}}(\mathbf{r},\hat{\boldsymbol{\Omega}}')p(\hat{\boldsymbol{\Omega}}',\hat{\boldsymbol{\Omega}})d\hat{\Omega}'+q(\mathbf{r},\hat{\boldsymbol{\Omega}}), \tag{14.14}$$

where we recall that the total attenuation coefficient is defined as $\mu_t=\mu_a+\mu_s$, c_n is the speed of light in the medium, $\hat{\boldsymbol{\Omega}}$ is the directional unit vector, $p(\hat{\boldsymbol{\Omega}}',\hat{\boldsymbol{\Omega}})$ is the scattering phase function, and q is the source term (representing the energy per unit volume per unit time per unit solid angle delivered by the source). The absorption and scattering coefficients ($\mu_a(\mathbf{r})$, $\mu_s(\mathbf{r})$), and therefore the total attenuation coefficient ($\mu_t(\mathbf{r})$), are spatial distributions of tissue properties, hence are functions of the spatial coordinate $\mathbf{r}$.

Diffusion theory is the most commonly employed forward-problem analytical tool for diffuse optical tomography. The diffusion equation for inhomogeneous media, derived in Chapter 9 as Eq. (9.36) for the optical energy density (U), is re-written here for the fluence rate ($\phi = c_n U$) in the time-domain (ϕ_{TD}), frequency-domain (ϕ_{FD}), and continuous-wave (ϕ_{CW}) cases (invoking the identity $\nabla \cdot (D\nabla U) = D\nabla^2 U + \nabla D \cdot \nabla U$):

$$D(\mathbf{r})\nabla^2\phi_{TD}(\mathbf{r},t) + \nabla D(\mathbf{r}) \cdot \nabla\phi_{TD}(\mathbf{r},t) - \left[c_n\mu_a(\mathbf{r}) + \frac{\partial}{\partial t}\right]\phi_{TD}(\mathbf{r},t) = -c_n S_0(\mathbf{r},t), \tag{14.15}$$

$$D(\mathbf{r})\nabla^2\phi_{FD}(\mathbf{r},\omega) + \nabla D(\mathbf{r}) \cdot \nabla\phi_{FD}(\mathbf{r},\omega) - [c_n\mu_a(\mathbf{r}) - i\omega]\phi_{FD}(\mathbf{r},\omega) = -c_n S_0(\mathbf{r},\omega), \tag{14.16}$$

$$D(\mathbf{r})\nabla^2\phi_{CW}(\mathbf{r}) + \nabla D(\mathbf{r}) \cdot \nabla\phi_{CW}(\mathbf{r}) - c_n\mu_a(\mathbf{r})\phi_{CW}(\mathbf{r}) = -c_n S_0(\mathbf{r}), \tag{14.17}$$

where S_0 is the isotropic source term (representing the energy per unit volume per unit time delivered by the source, isotropically). The absorption and reduced scattering coefficients ($\mu_a(\mathbf{r})$, $\mu_s'(\mathbf{r})$), and therefore the optical diffusion coefficient, $D(\mathbf{r}) = c_n/[3(\mu_s'(\mathbf{r}) + \mu_a(\mathbf{r}))]$, are functions of the spatial coordinate $\mathbf{r}$.

14.5.1.1 *Analytical solutions*

The method of Green's functions is a powerful analytical tool to solve the Boltzmann transport equation and the diffusion equation. As discussed in Section 14.1.2, the Green's function is the solution to an inhomogeneous differential equation for a point source described by a spatial delta ($\delta(\mathbf{r})$) and, in the time-domain, a temporal delta $\delta(t)$. The solution for an arbitrary source term (which features a certain spatial distribution and, in the time-domain, a certain temporal dependence) is given by an integral of the Green's function times the source term, which reduces to a convolution product in the case of a homogeneous, infinite medium. Consequently, the Green's function for a differential equation under given boundary conditions allows for the analytical derivation of the solution for any arbitrary source term.

For the BTE, the analytical Green's function is available for homogeneous, infinite media (Paaschens, 1997; Cai et al., 2000), with more recent derivations in semi-infinite media (Liemert and Kienle, 2013) and slab geometries (Machida et al., 2010); nevertheless, use of the BTE for diffuse optical tomography typically relies on numerical approximations (Kim and Hielscher, 2009). In contrast, the Green's function for the diffusion equation is available for homogeneous media in a variety of medium geometries (infinite medium, semi-infinite half-space, infinite slab, sphere, cylinder, etc.) in both the time domain and the frequency domain (Arridge et al., 1992), and for the inhomogeneous case of a spherical inhomogeneity embedded in an infinite scattering medium (Boas et al., 1994).

14.5.1.2 *Finite-difference method (FDM)*

Finite-difference methods (FDMs) refer to a class of numerical techniques to solve partial differential equations by approximating derivatives with differences after having discretized the sample volume into a regular three-dimensional lattice. (In Section 7.3, we introduced the time-domain variant of the FDM, or FDTD, as applied to simulating the scattering properties of a single particle.) In the context of diffuse optical tomography, the spatial variability of the medium optical properties is represented by the individual values of μ_a and μ_s' in each cell of the lattice or, more accurately, at the grid points. After approximating derivatives with differences, FDMs realize a Taylor expansion, typically to the second order in the grid step, of the function of interest (the radiance or the fluence rate in our case) about each grid point. The partial differential equation is thus translated into a system of algebraic equations for the values of the unknown function at the grid points, which can be solved by matrix algebra.

The time-independent diffusion equations for the FD and CW cases (Eqs. (14.16) and (14.17)) are classified as elliptic equations, for which the *multigrid* (MG) method is one choice of solver. The MG method, based on a discrete lattice decomposition with step sizes ΔL_x, ΔL_y, and ΔL_z along the x, y, and z axes, respectively, has been applied to frequency-domain diffuse optical tomography (Pogue et al., 1995). The time-dependent diffusion equation for the TD case (Eq. (14.15)) is classified as a parabolic equation, for which a grid with equal step sizes in the three dimensions and the *alternating direction implicit* (ADI) method is another efficient solver of the resulting set of algebraic equations.

14.5.1.3 *Finite-element method (FEM)*

The *finite-element method* (FEM) is another numerical method to approximate solutions to partial differential equations. The FEM consists of discretizing the volume of interest into small regions (called *elements*) that are usually triangular or tetrahedral, and introducing a set of *nodes* that are typically at the vertices of the elements. The finite elements and the nodes form the *mesh*, which is the basic structure on which the FEM analysis is based. In contrast to the FDM, which requires a regular grid, the power of the FEM rests on its inherent ability to handle irregular grids; in fact, FEMs treat equally well both regular and irregular grids. The idea is to approximately solve the so-called "weak form" of the partial differential equation, which is obtained by multiplying it by an arbitrary weighting function and by integrating it over the volume of interest. (For a description of the general principles of FEM, the reader is referred to the publication by Zienkiewicz and Taylor listed at the end of the chapter under "Further reading" on the finite-element method.) The arbitrary weighting function can be written as a linear combination of so-called *test functions*. One can interpret this weak statement of the problem as a method to seek a solution that, rather than being an exact solution everywhere, satisfies the partial differential equation "on average" (as expressed by the integral in the weak form) over the support of each test function.

The desired approximate solution to the partial differential equation is assumed to be a linear combination of a set of properly selected *basis functions*. The test and basis functions are typically defined in each finite element by linear interpolation of their values at the nodes of the element. (Considering the same set of test and basis functions results in the Galerkin method, named after the Russian mathematician Boris Galerkin [1871–1945].) To solve the FEM representation is to recover the values of the expansion coefficients in the set of basis functions. This is done by applying matrix algebra techniques to the set of algebraic equations in which the FEM converts the weak form of the partial differential equation. Basis functions with compact support produce a sparse coefficient matrix (i.e., a matrix consisting primarily of zero elements). The solution of the diffusion equation with the FEM for diffuse optical measurements has been discussed in Arridge et al. (1993).

14.5.2 Stochastic models and methods: Monte Carlo simulations and random walk theory

The physical processes that form the basis for the deterministic description of the Boltzmann transport equation can also be modeled with stochastic approaches. The most common stochastic approach in biomedical optics is the Monte Carlo method (introduced in Section 8.1.1), which treats the number of photons that make up the optical signal as a stochastic variable whose expectation value represents the physical, measurable optical signal. Individual photons are launched at the source location and their paths in the scattering medium consist of straight line steps whose lengths (l) and directions ($\hat{\Omega}$) are determined statistically by a set of random numbers that describe the probability distributions for possible physical processes experienced by the photon in the medium (absorption, scattering, reflection at an interface, etc.). For example, on the basis of the definition of the absorption (μ_a) and scattering (μ_s) coefficients (see Section 1.3), the cumulative probability of a photon interaction, either absorption or scattering, far from a boundary over a distance l is given by:

$$P_{as}(l) = 1 - e^{-(\mu_a+\mu_s)l}. \tag{14.18}$$

Therefore, $e^{-(\mu_a+\mu_s)l}$ is the probability that a photon is neither absorbed nor scattered over a distance l. By associating this free-path *probability* with a random number ξ_l ($0 \leq \xi_l \leq 1$) one can write:

$$\xi_l = e^{-(\mu_a+\mu_s)l}, \tag{14.19}$$

which can be inverted to yield the following step length associated with a random number ξ (Wang et al., 1995):

$$l = -\frac{\ln(\xi_l)}{\mu_a + \mu_s}. \tag{14.20}$$

In this specific implementation of Monte Carlo methods, absorption is taken into account by introducing, at each step, a weight factor given by the single-scattering albedo, a (given by $\mu_s/(\mu_a + \mu_s)$). (In this representation, photons are not quantized, but are represented by "photon packets" that are allowed to take on fractional values following absorption.) Thus, the weight of each photon packet at its i-th step (W_i) is given by $W_i = W_{i-1}(1 - a)$, where W_{i-1} is the weight at the prior step. It should be noted that in this approach the weight reduction at each step is the same, regardless of the lengths of the step (l) determined randomly by Eq. (14.20).

The direction of the photon path at the i-th step ($\hat{\Omega}_i$) is determined from the photon path direction at the previous step ($\hat{\Omega}_{i-1}$) by translating two additional random numbers (ξ_θ, ξ_φ) into the polar (θ) and azimuthal (φ) scattering angles using the probability distribution of a scattering phase function $p(\hat{\Omega}_{i-1}, \hat{\Omega}_i)$. The photon path is built until the photon is terminated, has escaped the scattering medium through a boundary (a fourth random number ξ_r may determine whether reflection or transmission occurs at a boundary), or has reached the optical detector. The process is repeated for a large number of launched photons, of the order of 10^6 or much more, depending on the source-detector distance, to achieve a relatively small variance for the expectation value of the collected optical signal at the detector. For example, if one wishes the variance in the number of collected photons for a simulation to be $\sim$1%, then 10,000 photons must be collected ($\sqrt{N}/N = 1\%$). Thus, for a given optical geometry and medium properties, if only one out of 10^3 launched photons reaches the detector, then the simulation must be run for 10^7 launched photons.

The high versatility of the Monte Carlo method in treating complex geometries and inhomogeneous media is balanced by the long computation times required to generate the large number of photon trajectories to obtain statistically significant results. With the photon-packet representation, even very small fractional photons can contribute to the collected signal, such that the computational efficiency is modestly improved. More importantly, however, Monte Carlo methods have recently been adapted for massively parallel computation on graphics processing units (GPUs) (Alerstam et al., 2008), achieving computational speed improvement by a factor $\sim$1000.

Random walk theory describes the photon path in terms of a sequence of steps on a regular cubic lattice. This apparently oversimplified approximation of reality is indeed a reasonable approach when the lattice spacing (L) is small enough (of the order of the scattering mean free path) to guarantee that photon paths comprise a large number of lattice steps. The absorption probability per lattice step can be represented in an expression similar to Eq. (14.18), by only considering the absorption coefficient:

$$P_a(L) = 1 - e^{-\mu_a L}. \tag{14.21}$$

Scattering may be considered to be isotropic, an acceptable assumption in the diffusion approximation where large numbers of individual, in general anisotropic, scattering events described by μ_s are collectively characterized in terms of effectively isotropic scattering events described by μ_s' (Bonner et al., 1987). However, anisotropic random walks have also been considered, where the length and direction of each step in the random walk reflect the scattering properties of the modeled medium (Gandjbakhche et al., 1992).

14.6 Diffuse optical tomography: the inverse problem

As illustrated in Figure 14.3, the inverse problem consists of determining the spatial distribution of the optical properties of a scattering medium, given the measured optical signals for a specified arrangement of light sources and optical detectors. The backprojection method described in Section 14.2 is a straightforward imaging approach, even though the quantitative accuracy of diffuse optical images obtained by backprojection is limited. The optical signals resulting from the solution of the Boltzmann transport equation (Eqs. (14.12)–(14.14)) and the diffusion equation (Eqs. (14.15)–(14.17)) feature a nonlinear dependence on the optical properties of the scattering medium. Under some conditions, for example the case of relatively slow temporal changes in the optical properties, it is reasonable to apply perturbation methods to linearize such dependence, but nonlinear imaging methods are in general required for quantitative diffuse optical tomography. Linear and nonlinear image reconstruction methods are described in the next two sections.

14.6.1 Linear methods based on perturbation theory

The basic idea of linear imaging methods is to linearize the forward problem to translate it into a linear system of equations. The inverse problem then consists of solving this linear system of equations, which can be done with established numerical techniques.

14.6.1.1 *Linearization for absorption and diffusion perturbations*

A method to linearize the forward problem is to use a perturbation approach based on expressing the absorption coefficient ($\mu_a(\mathbf{r})$) and the optical diffusion coefficient ($D(\mathbf{r})$) as the sum of spatially uniform terms (μ_{a0} and D_0) and small spatially dependent perturbations ($\delta\mu_a(\mathbf{r}) \ll \mu_{a0}$ and $\delta D(\mathbf{r}) \ll D_0$) ($\delta$ as used here should not be confused with the Dirac delta):

$$\mu_a(\mathbf{r}) = \mu_{a0} + \delta\mu_a(\mathbf{r}), \tag{14.22}$$

$$D(\mathbf{r}) = D_0 + \delta D(\mathbf{r}). \tag{14.23}$$

Invoking these perturbation expressions, the fluence rate ($\phi(\mathbf{r})$) can be written in terms of a Taylor series that, in the perturbation approach, is truncated after the first, linear term:

$$\phi(\mathbf{r}) \cong \phi_0(\mathbf{r}) + \delta\phi(\mathbf{r}) = \phi_0(\mathbf{r}) + \left(\frac{\partial\phi_0(\mathbf{r})}{\partial\mu_a}\right)_{\mu_{a0},D_0} \delta\mu_a(\mathbf{r}) + \left(\frac{\partial\phi_0(\mathbf{r})}{\partial D}\right)_{\mu_{a0},D_0} \delta D(\mathbf{r}), \tag{14.24}$$

where $\phi_0(\mathbf{r})$ represents the fluence rate for the unperturbed, homogeneous optical properties μ_{a0} and D_0, and $\delta\phi(\mathbf{r})$ is the fluence rate perturbation due to the optical perturbations $\delta\mu_a(\mathbf{r})$ and $\delta D(\mathbf{r})$. The approximation of Eq. (14.24) is referred to as the *Born approximation* (named after the German physicist Max Born [1882–1970], who received the Nobel Prize for physics in 1954).

Let's consider the frequency-domain diffusion equation (Eq. (14.16)), which reduces to the CW diffusion equation for $\omega = 0$. We substitute the above expressions for $\mu_a(\mathbf{r})$, $D(\mathbf{r})$, and $\phi(\mathbf{r})$ (which is written $\phi_{\mathrm{FD}}(\mathbf{r}, \omega)$ in the frequency domain) into the frequency-domain diffusion equation (Eq. (14.16)), and take into account that $\phi_{\mathrm{FD0}}(\mathbf{r}, \omega)$ is the solution to the equation for the source term $S_0(\mathbf{r}, \omega)$. Further, if we assume uniform optical properties μ_{a0} and D_0, then ignoring second-order terms leads to the following equation for $\delta\phi_{\mathrm{FD}}(\mathbf{r}, \omega)$:

$$\begin{aligned} &D_0\nabla^2(\delta\phi_{\mathrm{FD}})(\mathbf{r}, \omega) - [c_n\mu_{a0} - i\omega](\delta\phi_{\mathrm{FD}})(\mathbf{r}, \omega) \\ &= -c_n\left[-\delta\mu_a(\mathbf{r})\phi_{\mathrm{FD0}}(\mathbf{r}, \omega) \right. \\ &\quad \left. + \frac{\delta D(\mathbf{r})\nabla^2\phi_{\mathrm{FD0}}(\mathbf{r}, \omega) + \nabla(\delta D)(\mathbf{r}) \cdot \nabla\phi_{\mathrm{FD0}}(\mathbf{r}, \omega)}{c_n}\right], \end{aligned} \tag{14.25}$$

where the subscript "FD" and the argument ω specify that this is the frequency-domain case. Equation (14.25) is formally identical to the diffusion equation for a homogeneous medium with optical properties μ_{a0} and D_0 (indicated in the left-hand side of the equation), with an effective source term ($S_{\mathrm{eff}}(\mathbf{r}, \omega)$) (the term in square brackets that multiplies $-c_n$ on the right-hand side of the equation). This effective source term contains the fluence rate solution for the homogeneous medium ($\phi_{FD0}(\mathbf{r}, \omega)$) and the perturbations in the optical properties ($\delta\mu_a(\mathbf{r})$ and $\delta D(\mathbf{r})$). Equation (14.25) can be solved using Green's functions, i.e., by integrating the product of the Green's function of the frequency-domain (FD) diffusion equation for a homogeneous medium ($G_{\phi_{FD}}(\mathbf{r}', \mathbf{r}, \omega)$ for a point source at $\mathbf{r}'$) times the effective source term $S_{eff}(\mathbf{r}', \omega)$ (again, as given by the expression that multiplies $-c_n$ on the right-hand side of Eq. (14.25)):

$$
\begin{aligned}
\delta\phi_{FD}(\mathbf{r}, \omega) &= \int_V S_{eff}(\mathbf{r}', \omega) G_{\phi_{FD}}(\mathbf{r}', \mathbf{r}, \omega) d\mathbf{r}' \\
&= \int_V \left\{ -\delta\mu_a(\mathbf{r}')\phi_{FD0}(\mathbf{r}', \omega) + \frac{1}{c_n}[\delta D(\mathbf{r}')\nabla^2\phi_{FD0}(\mathbf{r}', \omega) \right. \\
&\qquad \left. + \nabla(\delta D)(\mathbf{r}') \cdot \nabla\phi_{FD0}(\mathbf{r}', \omega)] \right\} G_{\phi_{FD}}(\mathbf{r}', \mathbf{r}, \omega) d\mathbf{r}' \\
&= -\int_V \delta\mu_a(\mathbf{r}')\phi_{FD0}(\mathbf{r}', \omega) G_{\phi_{FD}}(\mathbf{r}', \mathbf{r}, \omega) d\mathbf{r}' \\
&\quad + \frac{1}{c_n} \int_V \nabla \cdot [\delta D(\mathbf{r}')\nabla\phi_{FD0}(\mathbf{r}', \omega)] G_{\phi_{FD}}(\mathbf{r}', \mathbf{r}, \omega) d\mathbf{r}'
\end{aligned}
\tag{14.26}
$$

The second integral can be put in a form that is linear in δD by using the following identity for three-dimensional integration by parts (where u and $\mathbf{v}$ are arbitrary scalar and vector functions, respectively):

$$
\int_V (\nabla \cdot \mathbf{v}) u \, dV = \oint_S u\mathbf{v} \cdot d\mathbf{S} - \int_V \mathbf{v} \cdot \nabla u \, dV. \tag{14.27}
$$

Because the integral on the left-hand side of Eq. (14.27) is carried out over the entire volume extending to infinity, and both $G_{\phi_{FD}}(\mathbf{r}', \mathbf{r}, \omega)$ and $|\nabla\phi_{FD0}(\mathbf{r}', \omega)|$ vanish as $|\mathbf{r}'| \to \infty$, the surface integral is zero. Therefore, Eq. (14.26) becomes:

$$
\begin{aligned}
&\delta\phi_{FD}(\mathbf{r}, \omega) \\
&\quad = -\int_V \delta\mu_a(\mathbf{r}')\phi_{FD0}(\mathbf{r}', \omega) G_{\phi_{FD}}(\mathbf{r}', \mathbf{r}, \omega) d\mathbf{r}' \\
&\qquad - \frac{1}{c_n} \int_V \delta D(\mathbf{r}')\nabla\phi_{FD0}(\mathbf{r}', \omega) \cdot \nabla G_{\phi_{FD}}(\mathbf{r}', \mathbf{r}, \omega) d\mathbf{r}'
\end{aligned}
\tag{14.28}
$$

In practice, the volume integrals in Eq. (14.28) are performed over the volume of the scattering medium, since $\delta\mu_a(\mathbf{r})$ and $\delta D(\mathbf{r})$ can be non-zero only inside the medium.

The solution of the perturbation problem in the time domain can be readily found by recognizing that the product of ϕ_{FD0} and $G_{\phi_{FD}}$ and the dot product of $\nabla\phi_{FD0}$ and $\nabla G_{\phi_{FD}}$ in the frequency-domain expression of Eq. (14.28) become convolution products in the time domain:

$$\delta\phi_{TD}(\mathbf{r},t) = -\int_V \delta\mu_a(\mathbf{r}')d\mathbf{r}' \int_{-\infty}^{+\infty} \phi_{TD0}(\mathbf{r}',t')G_{\phi_{TD}}(\mathbf{r}',\mathbf{r},t-t')dt' +$$

$$-\frac{1}{c_n}\int_V \delta D(\mathbf{r}')d\mathbf{r}' \int_{-\infty}^{+\infty} \nabla\phi_{TD0}(\mathbf{r}',t')\cdot\nabla G_{\phi_{TD}}(\mathbf{r}',\mathbf{r},t-t')dt'. \quad (14.29)$$

We recall that the units of the fluence rate Green's function are those of a fluence rate per unit power (i.e., m^{-2}) in the frequency domain, and fluence rate per unit energy (i.e., $s^{-1}m^{-2}$) in the time domain. Consequently, the left-hand sides of Eqs. (14.28) and (14.29) indeed have units of fluence rate (W/m^2). The integral Eqs. (14.28) and (14.29) provide analytical solutions to the first-order perturbation of the diffusion equation and show a linear dependence on $\delta\mu_a(\mathbf{r})$ and $\delta D(\mathbf{r})$. A comparison with Eq. (14.24) shows that Eqs. (14.28) and (14.29) provide analytical expressions for the Jacobian matrix (i.e., the set of first derivatives) of the fluence rate perturbation with respect to the optical coefficients.

14.6.1.2 *The CW sensitivity function for absorption perturbations in infinite and semi-infinite media*

The absorption perturbation term in Eq. (14.28), in the limiting case of $\omega = 0$, specifies how the product of the two Green's functions (of the CW diffusion equation for the fluence rate) that we have considered in Section 14.1 is related to the sensitivity function of ϕ_{CW} for absorption perturbations. In fact, by considering a point source of power P_{CW} at position $\mathbf{r}_s$ (so that $\phi_{CW0}(\mathbf{r}') = P_{CW}G_{\phi_{CW}}(\mathbf{r}_s,\mathbf{r}')$) and an absorption perturbation confined within a small volume ΔV about $\mathbf{r}'$ (so that $\delta D = 0$, and the volume integral becomes a product by ΔV), Eq. (14.28) provides the following expression for the change in the CW fluence rate measured at detector position $\mathbf{r}_d$ due to an absorption perturbation:

$$\delta\phi_{CW}(\mathbf{r}_s,\mathbf{r}_d) = -P_{CW}G_{\phi_{CW}}(\mathbf{r}_s,\mathbf{r}')G_{\phi_{CW}}(\mathbf{r}',\mathbf{r}_d)\delta\mu_a(\mathbf{r}')\Delta V. \quad (14.30)$$

The relative change in the CW intensity measured in an infinite medium ($I_{CW}^{(inf)}$) at the detector position $\mathbf{r}_d$ can be directly obtained from Eq. (14.30) because $I_{CW}^{(inf)}$ is proportional to the fluence rate:

$$I_{CW}^{(inf)}(\mathbf{r}_s,\mathbf{r}_d) \propto P_{CW}G_{\phi_{CW}}^{(inf)}(\mathbf{r}_s,\mathbf{r}_d). \quad (14.31)$$

Therefore, the relative change in $I^{(\text{inf})}_{\text{CW}}(\mathbf{r}_s, \mathbf{r}_d)$ as a result of a localized absorption perturbation $\delta\mu_a$ over volume ΔV about $\mathbf{r}'$ is given by:

$$\frac{\delta I^{(\text{inf})}_{\text{CW}}(\mathbf{r}_s, \mathbf{r}_d)}{I^{(\text{inf})}_{\text{CW}0}(\mathbf{r}_s, \mathbf{r}_d)} = -\frac{P_{\text{CW}} G^{(\text{inf})}_{\phi\text{CW}}(\mathbf{r}_s, \mathbf{r}') G^{(\text{inf})}_{\phi\text{CW}}(\mathbf{r}', \mathbf{r}_d) \delta\mu_a(\mathbf{r}') \Delta V}{P_{\text{CW}} G^{(\text{inf})}_{\phi\text{CW}}(\mathbf{r}_s, \mathbf{r}_d)}$$

$$= -\frac{\frac{3(\mu_s'+\mu_a)}{4\pi} \frac{e^{-\mu_{\text{eff}}|\mathbf{r}'-\mathbf{r}_s|}}{|\mathbf{r}'-\mathbf{r}_s|} \frac{e^{-\mu_{\text{eff}}|\mathbf{r}_d-\mathbf{r}'|}}{|\mathbf{r}_d-\mathbf{r}'|}}{\frac{e^{-\mu_{\text{eff}}|\mathbf{r}_d-\mathbf{r}_s|}}{|\mathbf{r}_d-\mathbf{r}_s|}} \delta\mu_a(\mathbf{r}') \Delta V, \qquad (14.32)$$

A striking implication of Eq. (14.32) is that the relative change in the optical intensity for absorbing perturbations on the source-detector line is essentially independent of the absorption coefficient of the medium. In fact, for a perturbation along the source-detector line, $|\mathbf{r}_d - \mathbf{r}_s| = |\mathbf{r}_d - \mathbf{r}'| + |\mathbf{r}' - \mathbf{r}_s|$, and the dependence on μ_{eff} disappears in Eq. (14.32); the remaining factor $(\mu_s' + \mu_a)$ is only weakly dependent on μ_a since $\mu_a \ll \mu_s'$ in the diffusion regime.

In a semi-infinite medium, the optical intensity measured at the surface is represented by the component of the gradient of the fluence rate Green's function along the outer normal to the surface, which we take to be $\hat{\mathbf{z}}$. In the case of the $(\mathbf{r}_s, \mathbf{r}')$ term, this results from the difference of two infinite-medium Green's function for sources separated by $2z_0 = 2/\mu_{s0}'$ (under zero boundary conditions), which yields a term $2z_0 \frac{\partial G^{(\text{inf})}_{\phi\text{CW}}(\mathbf{r}_s, \mathbf{r}')}{\partial z}$ (see Eq. (14.A.5) in Appendix 14.A). In the case of the $(\mathbf{r}', \mathbf{r}_d)$ term, the measured CW intensity on a semi-infinite medium boundary, at the plane $z = 0$ ($I^{(\text{seminf})}_{\text{CW}}$), is given by the component of the optical flux along $\hat{\mathbf{z}}$, which in the diffusion approximation is given by:

$$I^{(\text{seminf})}_{\text{CW}}(\mathbf{r}', \mathbf{r}_d) \propto -\frac{P_{\text{CW}}}{3(\mu_s' + \mu_a)} \left(\frac{\partial G^{(\text{seminf})}_{\phi\text{CW}}(\mathbf{r}', \mathbf{r}_d)}{\partial z_d} \right)_{z_d=0}. \qquad (14.33)$$

Consequently, the relative change in the CW intensity measured at the surface of a semi-infinite medium for an absorption perturbation is:

$$\frac{\delta I^{(\text{seminf})}_{\text{CW}}(\mathbf{r}_s, \mathbf{r}_d)}{I^{(\text{seminf})}_{\text{CW}0}(\mathbf{r}_s, \mathbf{r}_d)} = \frac{2z_0 P_{\text{CW}} \frac{\partial G^{(\text{inf})}_{\phi\text{CW}}(\mathbf{r}_s, \mathbf{r}')}{\partial z} \frac{1}{3(\mu_s'+\mu_a)} \left(\frac{\partial G^{(\text{seminf})}_{\phi\text{CW}}(\mathbf{r}', \mathbf{r}_d)}{\partial z_d} \right)_{z_d=0} \delta\mu_a(\mathbf{r}') \Delta V}{\frac{P_{\text{CW}}}{3(\mu_s'+\mu_a)} \left(\frac{\partial G^{(\text{seminf})}_{\phi\text{CW}}(\mathbf{r}_s, \mathbf{r}_d)}{\partial z_d} \right)_{z_d=0}}$$

$$= -\frac{2z_0 \frac{3(\mu_s'+\mu_a)}{4\pi} z' \left(\mu_{\text{eff}} + \frac{1}{|\mathbf{r}'-\mathbf{r}_s|} \right) \frac{e^{-\mu_{\text{eff}}|\mathbf{r}'-\mathbf{r}_s|}}{|\mathbf{r}'-\mathbf{r}_s|^2} \frac{2z'}{4\pi} \left(\mu_{\text{eff}} + \frac{1}{|\mathbf{r}_d-\mathbf{r}'|} \right) \frac{e^{-\mu_{\text{eff}}|\mathbf{r}_d-\mathbf{r}'|}}{|\mathbf{r}_d-\mathbf{r}'|^2}}{\frac{2z_0}{4\pi} \left(\mu_{\text{eff}} + \frac{1}{|\mathbf{r}_d-\mathbf{r}_s|} \right) \frac{e^{-\mu_{\text{eff}}|\mathbf{r}_d-\mathbf{r}_s|}}{|\mathbf{r}_d-\mathbf{r}_s|^2}}$$

$$\times \delta\mu_a(\mathbf{r}') \Delta V$$

$$= -\frac{\frac{3(\mu_s'+\mu_a)}{2\pi}(z')^2\left(\mu_{\text{eff}}+\frac{1}{|\mathbf{r}'-\mathbf{r}_s|}\right)\frac{e^{-\mu_{\text{eff}}|\mathbf{r}'-\mathbf{r}_s|}}{|\mathbf{r}'-\mathbf{r}_s|^2}\left(\mu_{\text{eff}}+\frac{1}{|\mathbf{r}_d-\mathbf{r}'|}\right)\frac{e^{-\mu_{\text{eff}}|\mathbf{r}_d-\mathbf{r}'|}}{|\mathbf{r}_d-\mathbf{r}'|^2}}{\left(\mu_{\text{eff}}+\frac{1}{|\mathbf{r}_d-\mathbf{r}_s|}\right)\frac{e^{-\mu_{\text{eff}}|\mathbf{r}_d-\mathbf{r}_s|}}{|\mathbf{r}_d-\mathbf{r}_s|^2}}$$

$$\times\, \delta\mu_a(\mathbf{r}')\Delta V, \qquad (14.34)$$

where $\rho_{sd} = |\mathbf{r}_d - \mathbf{r}_s|$ is the source-detector distance measured on the surface of the semi-infinite medium, and z' is the depth coordinate of the volume element (ΔV) where the absorption perturbation ($\delta\mu_a$) occurs. It is very important to observe that while μ_{eff} completely determines the spatial dependence of the region of sensitivity (as shown by Eqs. (14.5) and (14.8)), the factor $\mu_s' + \mu_a \approx \mu_s'$ in Eqs. (14.32) and 14.34) introduces a linear enhancement of the sensitivity function for higher values of the reduced scattering coefficient. This may be understood from the fact that photons will spend more time in an absorbing perturbation if they scatter more frequently.

Equations (14.32) and (14.34) yield the percent changes of the measured CW intensity in an infinite and semi-infinite medium, respectively, resulting from an absorption perturbation of a certain strength ($\delta\mu_a$), volume (ΔV), and location ($\mathbf{r}'$). These equations can therefore be used to quantify the effect on the measured CW intensity of a localized absorption perturbation. In Figures 14.4–14.7, we report such effects in an infinite medium (Figures 14.4 and 14.5) and in a semi-infinite medium (Figures 14.6 and 14.7), resulting from an absorption perturbation $\delta\mu_a = 0.005\ \text{cm}^{-1}$ in a volume element $\Delta V = 1\ \text{mm}^3$. As shown in Eqs. (14.32) and (14.34), percent changes in the optical signals scale linearly with $\delta\mu_a$ and ΔV, provided that these quantities are small enough to be treated with perturbation theory.

Figure 14.4 shows the percent change in the CW intensity measured in an infinite medium for a perturbation that is located at different positions along the source-detector line (Figure 14.4(a)) and along the transverse coordinate at the mid-plane between source and detector (Figure 14.4(b)). Longer source-detector distances result in broader regions of sensitivity at reduced sensitivity levels. Figure 14.5 also refers to the infinite medium, but it shows the effects of different optical properties of the medium on the sensitivity levels to an absorbing perturbation along the source-detector line (Figure 14.5(a)) or along the transverse coordinate at mid-plane (Figure 14.5(b)). As observed above, the level of sensitivity to an absorption perturbation along the source-detector line is independent of the absorption coefficient on the medium, and Figure 14.5(b) shows the effect of the absorption properties when the perturbation is away from the source-detector line. Figure 14.5 shows the important role played by the reduced scattering coefficient of the medium in determining the sensitivity of CW optical signals to localized absorption perturbations.

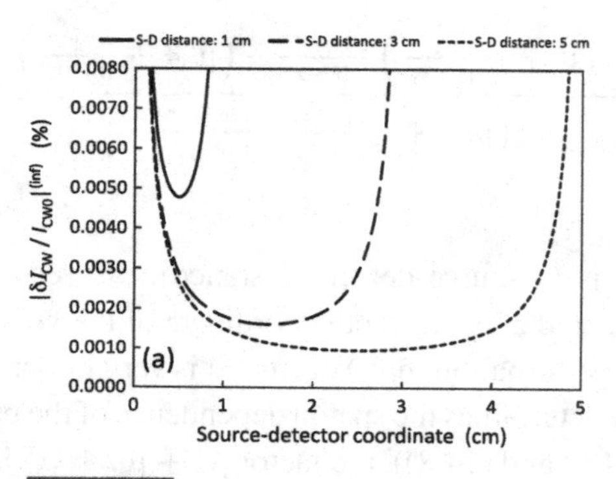

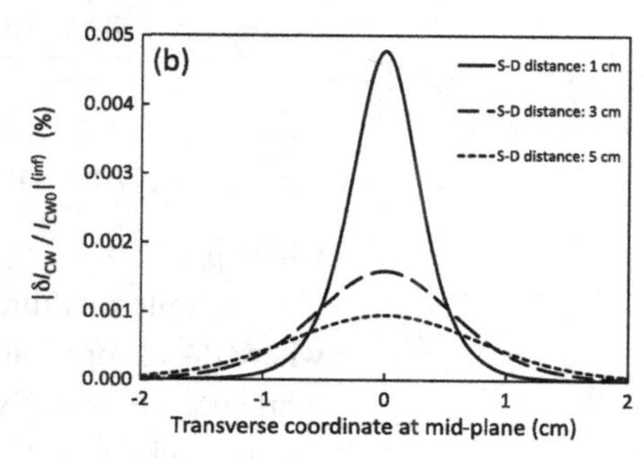

Figure 14.4

Percent changes in the CW intensity measured in an infinite medium for a localized optical perturbation of magnitude $\delta\mu_a = 0.005$ cm^{-1}, within a volume $\Delta V = 1$ mm^3, as a function of the perturbation position. Three source-detector distances of 1, 3, and 5 cm are shown. (a) The perturbation is located along the source-detector line. (b) The perturbation is located along the transverse coordinate, perpendicular to the source-detector line at the mid-plane between source and detector.

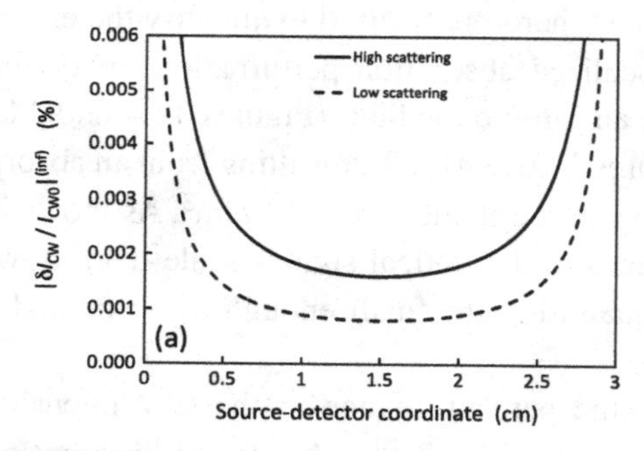

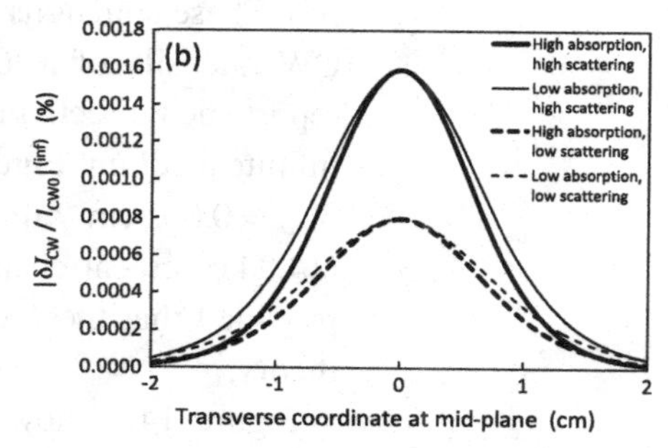

Figure 14.5

Percent changes in the CW intensity measured in an infinite medium for a localized optical perturbation of magnitude $\delta\mu_a = 0.005$ cm^{-1}, within a volume $\Delta V = 1$ mm^3, as a function of the perturbation position. Four combinations of low absorption ($\mu_a = 0.05$ cm^{-1}), high absorption ($\mu_a = 0.10$ cm^{-1}), low scattering ($\mu_s' = 5$ cm^{-1}), and high scattering ($\mu_s' = 10$ cm^{-1}) of the scattering medium are shown. (a) The perturbation is located along the source-detector line (where the percent signal change is independent of the medium absorption coefficient). (b) The perturbation is located along the transverse coordinate, perpendicular to the source-detector line at the mid-plane between source and detector.

Figures 14.6 and 14.7 are the equivalent of Figures 14.4 and 14.5 for a semi-infinite medium. In the case of Figures 14.6(a) and 14.7(a), the horizontal axis reports the projection onto the source-detector coordinate of an absorbing perturbation located at the point of maximal sensitivity $z_{\max}(x)$. Figure 14.6(b) shows that increased source-detector distances result in a deeper region of sensitivity at the

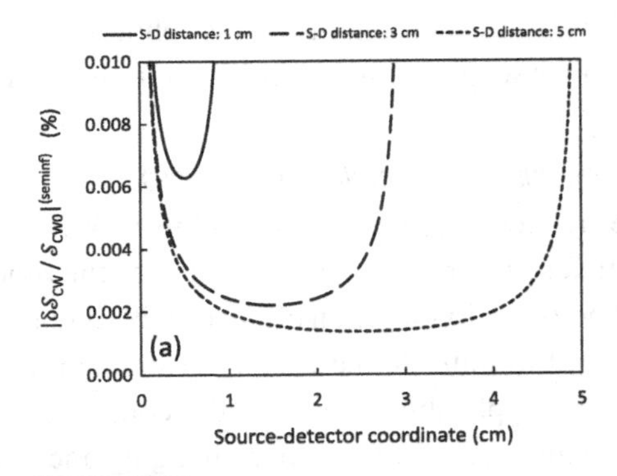

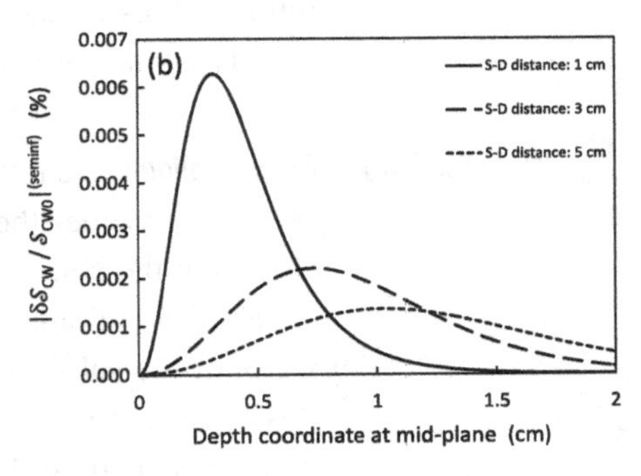

Figure 14.6

Percent changes in the CW intensity measured at the surface of a semi-infinite medium for a localized optical perturbation of magnitude $\delta\mu_a = 0.005$ cm^{-1}, within a volume $\Delta V = 1$ mm^3, as a function of the perturbation position. Three source-detector distances of 1, 3, and 5 cm are shown. (a) The perturbation is located along the line of maximal sensitivity ($z = z_{max}(x)$) and the horizontal axis reports the coordinate of its projection along the source-detector line. (b) The perturbation is located along the depth coordinate, perpendicular to the source-detector line at the mid-plane between source and detector.

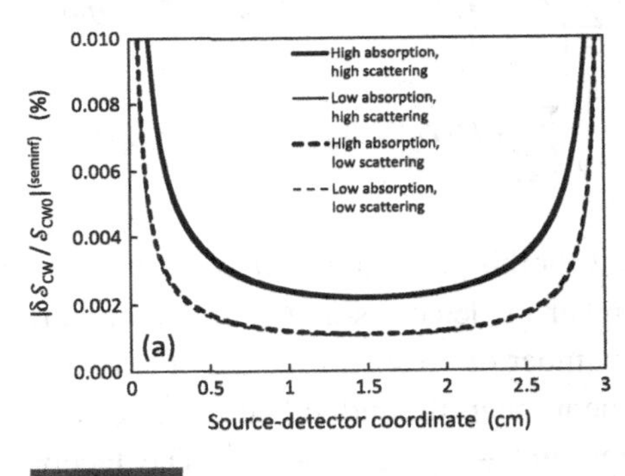

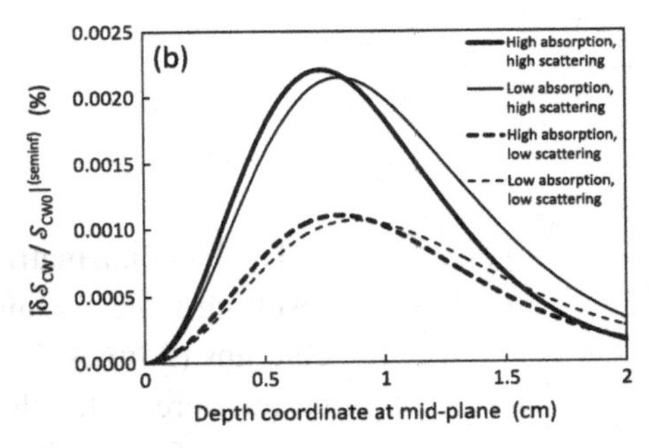

Figure 14.7

Percent changes in the CW intensity measured at the surface of a semi-infinite medium for a localized optical perturbation of magnitude $\delta\mu_a = 0.005$ cm^{-1}, within a volume $\Delta V = 1$ mm^3, as a function of the perturbation position. Four combinations of low absorption ($\mu_a = 0.05$ cm^{-1}), high absorption ($\mu_a = 0.10$ cm^{-1}), low scattering ($\mu_s' = 5$ cm^{-1}), and high scattering ($\mu_s' = 10$ cm^{-1}) of the scattering medium are shown. (a) The perturbation is located along the line of maximal sensitivity ($z = z_{max}(x)$) and the horizontal axis reports the coordinate of its projection along the source-detector line. (b) The perturbation is located along the depth coordinate, perpendicular to the source-detector line at the mid-plane between source and detector.

expense of reduced sensitivity levels. Figure 14.7 confirms, also in the semi-infinite case, the strong influence of the reduced scattering coefficient on the sensitivity of the CW intensity to an absorbing perturbation. The fact that stronger scattering translates into enhanced sensitivity to the perturbation volume ΔV can be intuitively associated with the additional time that photons spend within volume ΔV as a result of a shorter scattering mean free path in the medium.

14.6.1.3 *Discretization of the problem into a linear system of equations*

We now consider the frequency-domain case (and its CW limiting case, $\omega = 0$). However, the conceptual steps described below apply to the time domain as well, once the time-domain perturbation solution of Eq. (14.29) is considered. Therefore, we omit the subscript FD to indicate the frequency-domain fluence rate because this treatment is formally applicable equally to the continuous-wave, frequency-domain, and time-domain cases. In an optical imaging measurement, $\delta\phi(\mathbf{r}, \omega)$ denotes the fluence rate perturbation for a number of source configurations S_i (index i) and a number of detector locations $\mathbf{r}_j$ (index j), while $\delta\mu_a(\mathbf{r}')$, and $\delta D(\mathbf{r}')$ denote the absorption and diffusion coefficient perturbations at all image voxels $\mathbf{r}_k$ (index k). This may be expressed explicitly by discretizing the integral in Eq. (14.28) over the total number of voxels N, each one assumed to have the same volume ΔV:

$$\begin{aligned}\delta\phi_i(\mathbf{r}_j, \omega) = &-\sum_{k=1}^{N} \delta\mu_a(\mathbf{r}_k)\phi_{0i}(\mathbf{r}_k, \omega)G_\phi(\mathbf{r}_k, \mathbf{r}_j, \omega)\Delta V \\ &- \frac{1}{c_n}\sum_{k=1}^{N} \delta D(\mathbf{r}_k)\nabla\phi_{0i}(\mathbf{r}_k, \omega)\cdot\nabla G_\phi(\mathbf{r}_k, \mathbf{r}_j, \omega)\Delta V \end{aligned} \qquad (14.35)$$

The Jacobian matrix $\mathbf{J}$ for this problem has m rows (the number of optical signals resulting from the number of detectors and source configurations used) and $2N$ columns (2 times the number of voxels because of the two optical coefficients to be measured). Its elements map the individual terms of the two summations in Eq. (14.35), which represent $\frac{\partial\phi_{0i}(\mathbf{r}_j)}{\partial\mu_a(\mathbf{r}_k)}$ and $\frac{\partial\phi_{0i}(\mathbf{r}_j)}{\partial D(\mathbf{r}_k)}$. The Jacobian matrix is named after the German mathematician Carl Jacobi [1804–1851]. Equation (14.35) can be written in matrix notation as follows:

$$\boldsymbol{\delta\phi} = \mathbf{J}\,\boldsymbol{\delta\mu}, \qquad (14.36)$$

where $\boldsymbol{\delta\phi}$ is the $(m \times 1)$ column vector of the fluence rate perturbation, $\mathbf{J}$ is the $(m \times 2N)$ Jacobian matrix, and $\boldsymbol{\delta\mu}$ is the $(2N \times 1)$ column vector of the optical coefficient perturbations (δD and $\delta\mu_a$) at the N voxels. Since the elements of the Jacobian matrix are given analytically by Eq. (14.35), the imaging problem is translated into the algebraic problem of computing an approximate solution to Eq. (14.36), which can be done using methods such as singular value decomposition (SVD), algebraic reconstruction technique (ART), and simultaneous iterative reconstruction technique (SIRT). In practice, since the system of equations in

Eq. (14.36) is under-determined, some form of regularization (see Section 14.7 below) is needed to find a plausible physical solution.

The essence of perturbation methods is described by Eq. (14.36):

(a) the optical measurements are difference measurements ($\boldsymbol{\delta\phi}$) between a reference condition of homogeneous optical properties and an experimental condition of perturbed optical properties;
(b) the reconstructed optical coefficients represent the differences ($\boldsymbol{\delta\mu}$) between the optical properties of the experimental and reference conditions.

This difference approach is quite limiting, especially considering the requirement of small optical perturbations, but there are practical cases in biomedical imaging that are within its range of applicability, most notably in the case of functional studies of brain activation.

14.6.2 Nonlinear, iterative methods based on calculation of the Jacobian

The linear relationship between perturbations in the optical properties and changes in the fluence rate (Eq. (14.36)) has a Jacobian matrix that is independent of the reconstructed optical images. In fact, the Jacobian $\mathbf{J}$ in Eq. (14.36) does not depend on $\boldsymbol{\delta\mu}$, but only on the background homogeneous optical properties μ_{a0} and D_0 (and, of course, on the locations of light sources and optical detectors). The general nonlinear relationship between an optical signal ($\mathcal{S}$) and the set of optical coefficients (indicated here as $\boldsymbol{\mu}$, a vector of parameters that describe both absorption and scattering properties) can be written as:

$$\mathcal{S} = F[\boldsymbol{\mu}(\mathbf{r})], \tag{14.37}$$

where F is the nonlinear function that solves the forward problem (say, the solution to the Boltzmann transport equation or the diffusion equation). The elements of the Jacobian matrix for this nonlinear problem are now dependent on $\boldsymbol{\mu}(\mathbf{r})$, and they represent the sensitivity of the measured optical signals to the optical properties at voxel $\mathbf{r}$. Therefore, they coincide with the sensitivity function defined by Eq. (14.1). It is important to observe that the Jacobian as well as the sensitivity function of Eq. (14.1) depend on the measured optical signal ($\mathcal{S}$) and on the nature (say, absorption or scattering) and location (identified here by $\mathbf{r}'$) of the changes in the optical properties. This is explicitly specified by the following expression of the Jacobian (Arridge, 1995), written here for the Jacobian element corresponding to a specific optical signal $\mathcal{S}$ (which is collected by one of the source-detector pairs used for imaging) and optical property μ_i (the i-th component of $\boldsymbol{\mu}$ representing, for example, the absorption or the diffusion coefficient):

$$J_{\mathcal{S},\mu_i}(\mathbf{r}') = \lim_{\delta\mu_i \to 0} \frac{F[\boldsymbol{\mu}(\mathbf{r}) + \mathbf{e}_i \delta\mu_i(\mathbf{r}')] - F[\boldsymbol{\mu}(\mathbf{r})]}{\delta\mu_i(\mathbf{r}')}, \tag{14.38}$$

where, as in Eq. (14.1), $\mathbf{e}_i$ is the unit vector whose components are all zero except for the i-th component, which is 1. Equation (14.38) shows how the Jacobian may be determined, for a given distribution of optical properties $\boldsymbol{\mu}(\mathbf{r})$, by solving the perturbation problem of the effect of a small optical perturbation $\boldsymbol{\delta\mu}(\mathbf{r}')$ around $\mathbf{r}'$ on the optical signal $\mathcal{S} = F[\boldsymbol{\mu}(\mathbf{r})]$.

The method of identifying the Jacobian to solve a linear perturbation problem can be used in each step of nonlinear, iterative reconstruction methods. This is the essence of methods to solve the inverse problem and yield a (previously unknown) spatial distribution of medium properties, given the location and properties of illumination sources and the measured signals at the detectors. Conceptually, *iterative methods involve the determination of the distribution of optical properties that minimizes the difference between the measured optical signals and the solution of the forward model.* An obvious requirement is that the forward model provide an accurate description of light propagation in the scattering medium. The quantity to be minimized, i.e., the parameter that quantifies the difference between the set of optical measurements and the forward model solution, is variously called the *objective function*, *cost function*, *error function*, or *error norm*.

One way to define the objective function (Ψ) is to take the norm of the difference between the vectors of the measured ($\boldsymbol{\mathcal{S}}_{\text{meas}}$) and modeled ($\boldsymbol{\mathcal{S}}_{\text{model}}$) optical signals:

$$\Psi = \|\boldsymbol{\mathcal{S}}_{\text{meas}} - \boldsymbol{\mathcal{S}}_{\text{model}}\|^2 = \sum_{i,j} (\mathcal{S}_{\text{meas}}|_{i,j} - \mathcal{S}_{\text{model}}|_{i,j})^2. \tag{14.39}$$

where the indices i and j enumerate the source configurations and the optical detectors, respectively. The minimization of the objective function involves setting to zero its first derivatives with respect to the optical properties:

$$\frac{\partial \Psi}{\partial \boldsymbol{\mu}} = 2\mathbf{J}^{\mathrm{T}}(\boldsymbol{\mathcal{S}}_{\text{meas}} - \boldsymbol{\mathcal{S}}_{\text{model}}) = 0, \tag{14.40}$$

where the derivative of Ψ with respect to the optical-properties vector field $\boldsymbol{\mu}$ is the column vector (with $2N$ rows) whose components are the derivatives of Ψ with respect to each component of $\boldsymbol{\mu}$. ($\mathbf{J}^{\mathrm{T}}$ is the transpose of the Jacobian matrix $\mathbf{J}$.) At each step of the iterative procedure, one can update the model prediction at the previous step ($\boldsymbol{\mathcal{S}}^{(-1)}_{\text{model}}$) by using the sensitivity to the optical properties given by the Jacobian, thereby considering a linear approximation:

$$\boldsymbol{\mathcal{S}}_{\text{model}} \approx \boldsymbol{\mathcal{S}}^{(-1)}_{\text{model}} + \mathbf{J}\,\boldsymbol{\delta\mu}. \tag{14.41}$$

It is important to observe that the Jacobian in Eq. (14.41) is computed at each step on the basis of the current value of the vector field of optical properties $\boldsymbol{\mu}$.

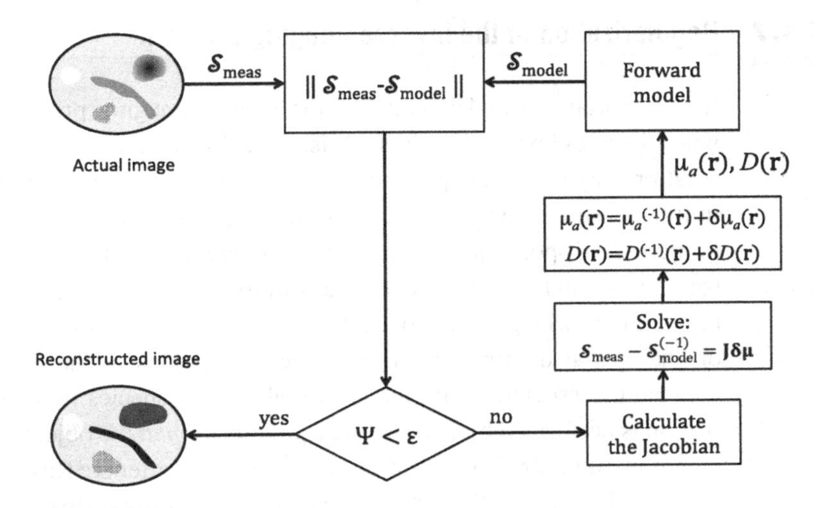

Figure 14.8

General approach of iterative methods for image reconstruction based on repeated applications of a forward model. The set of measured optical signals ($\mathcal{S}_{meas}$) are compared with the set of optical signals predicted by the forward model ($\mathcal{S}_{model}$) for a given spatial distribution of the absorption and diffuse optical coefficients ($\mu_a(\mathbf{r})$, $D(\mathbf{r})$). If the difference between the measured and modeled signals, as measured by an objective function Ψ, is greater than a set tolerance level ε, the optical coefficients are updated, otherwise the iterative procedure stops. The updated optical properties (the previous step is indicated by a (-1) superscript) are based on solving a perturbation problem that, through a calculated Jacobian matrix ($\mathbf{J}$), yields the changes to the optical properties $\delta\mu_a(\mathbf{r})$, $\delta D(\mathbf{r})$ for the next iterative step.

This requirement accounts for the fact that nonlinear iterative methods are computationally expensive. By substituting Eq. (14.41) in the minimization condition of Eq. (14.40) and indicating with ($\boldsymbol{\delta\phi}$) the difference between the measured signal and the modeled signal at the previous step ($\boldsymbol{\delta\phi} = \mathcal{S}_{meas} - \mathcal{S}^{(-1)}_{model}$), one obtains:

$$\mathbf{J}^{\mathrm{T}}\boldsymbol{\delta\phi} = \mathbf{J}^{\mathrm{T}}\mathbf{J}\,\boldsymbol{\delta\mu}. \tag{14.42}$$

Equation (14.42) represents a linear perturbation problem that makes up each step of an iterative procedure to update the optical properties as relates to the step change $\boldsymbol{\delta\mu}$. The required Jacobian to solve this perturbation problem can be obtained from Eq. (14.38) using a proper forward model. For example, Eq. (14.35) gives analytical expressions of the Jacobian elements for the perturbative, frequency-domain fluence rate in the diffusion approximation. In addition to updating the optical properties at each step, the result for $\boldsymbol{\delta\mu}$ is also used to update the Jacobian for a new application of Eq. (14.42) in the next iterative step, and this procedure is repeated until the forward model solution reproduces the measured optical signals to within a given tolerance. This iterative approach is illustrated in Figure 14.8.

14.7 Regularization of the inverse imaging problem

It is important to understand that the inverse imaging problem (for a problem with the complexity of biological tissue) is inherently *ill posed*, meaning that its solution may not be unique and that it may be highly sensitive to small variations in the measured signals or to the starting "guesstimate" of spatial distribution of properties. Ill-posedness is tackled by *regularization* techniques, which impose some constraints on the acceptable solutions. Regularization conditions may be based on knowledge of the noise features of the optical signals, prior information on the spatial distribution of the optical properties in the medium, or spectral constraints associated with the known absorbing species in the sample. Regularization is performed by introducing a *penalty term* in the objective function. In the common Tikhonov regularization scheme (named after the Russian mathematician Andrei Tikhonov [1906–1993]), the contribution of the penalty term is controlled by a regularization parameter $\lambda > 0$, so that the regularized objective function is written:

$$\Psi_{\text{Reg}} = \|\mathcal{S}_{\text{meas}} - \mathcal{S}_{\text{model}}\|^2 + \lambda\|\boldsymbol{\delta\mu}\|^2. \tag{14.43}$$

The role of the penalty term $\lambda\|\boldsymbol{\delta\mu}\|^2$ is to penalize overblown solutions by increasing their objective function. The regularization parameter λ is usually not known a priori and must be determined as part of the solution process. Methods for estimating the regularization parameter include the *discrepancy principle*, the *L-curve method*, and *generalized cross validation*, among others (Vogel, 2002). Minimization of the regularized objective function of Eq. (14.43) is equivalent to solving a modified linear problem with respect to that given by Eq. (14.42):

$$\mathbf{J}^{\mathbf{T}}\boldsymbol{\delta\phi} = (\mathbf{J}^{\mathbf{T}}\mathbf{J} + \lambda\mathbf{I})\boldsymbol{\delta\mu}, \tag{14.44}$$

where $\mathbf{I}$ is the identity matrix and the diagonal terms associated with it regularize the problem.

To take into account a priori spatial information provided, for example, by complementary imaging techniques (e.g., X-ray computed tomography, magnetic resonance imaging, etc.), known spatial constraints, or spatially variant regularization (Pogue et al., 1999), the penalty term may be written in terms of a filter matrix $\mathbf{L}$ that incorporates such a priori information (Brooksby et al., 2005):

$$\Psi_{\text{Reg}} = \|\mathcal{S}_{\text{meas}} - \mathcal{S}_{\text{model}}\|^2 + \lambda\|\mathbf{L}\boldsymbol{\delta\mu}\|^2. \tag{14.45}$$

In this case, the linear problem is expressed as follows:

$$\mathbf{J}^{\mathbf{T}}\boldsymbol{\delta\phi} = (\mathbf{J}^{\mathbf{T}}\mathbf{J} + \lambda\mathbf{L}^{\mathbf{T}}\mathbf{L})\boldsymbol{\delta\mu}, \tag{14.46}$$

and the spatial priors included in **L** can significantly enhance the quality and accuracy of the reconstructed optical images. The approach of Eqs. (14.45) and (14.46) to take into account a priori spatial information is referred to as a *soft-prior* approach (Yalavarthy et al., 2007), as opposed to a *hard-prior* approach that segments the reconstructed region into a set of sub-regions, each with uniform optical properties. The hard-prior approach dramatically reduces the number of unknowns of the inverse problem, but the soft-prior approach is less sensitive to uncertainties or inaccuracies in the a priori information.

Another powerful regularization approach in diffuse optical tomography is one that takes into account spectral constraints. In fact, the optical properties of biological tissues that determine light propagation in the diffusion regime are fully determined by a few known absorbing species and by the reduced scattering coefficient whose wavelength dependence can be estimated. The absorption coefficient results from the superposition of the contributions from each of the absorbing species, which, as shown in Eqs. (2.2) and (10.2), are given by the products of their molar extinction coefficients times their concentrations. Consequently, the absorption spectrum can be fully described by the concentrations of the few absorbing species. The reduced scattering coefficient depends on the sizes and densities of the scattering particles, as described in Chapters 7 and 8, and its wavelength dependence can be adequately described by two parameters, as shown by the power law of Eq. (8.2). Therefore, the spectral imaging problem can be solved by finding the chromophore concentrations and the tissue scattering parameters, rather than solving independently for the absorption and reduced scattering coefficients at all wavelengths. In this approach, the unknowns are no longer the optical properties ($\boldsymbol{\delta\mu}$), but rather the set of chromophore concentrations and scattering parameters (which lead to the optical properties), and we collectively indicate these with the vector $\boldsymbol{\delta c}$. Equation (14.44) is then replaced by a formally identical equation:

$$\mathbf{J}^{\mathrm{T}}\boldsymbol{\delta\phi} = (\mathbf{J}^{\mathrm{T}}\mathbf{J} + \lambda\mathbf{I})\boldsymbol{\delta c}, \tag{14.47}$$

where, however, the Jacobian matrix in Eq. (14.47) contains the derivatives of the fluence rate perturbation (at all measured wavelengths) with respect to the chromophore concentrations and scattering parameters.

14.8 Contrast and resolution in diffuse optical imaging

14.8.1 Optical properties contrast

In diffuse optical imaging, the relevant optical contrast associated with localized tissue inhomogeneities (e.g., tumors, blood vessels, fluid-filled cysts, biochemical imbalances, morphological changes, etc.) is related to their absorption and scattering properties relative to those of the background tissue. This localized optical contrast for an optical inhomogeneity about position $\mathbf{r}'$ is specified by $\delta\mu_a(\mathbf{r}')$ (for absorption) and $\delta\mu_s'(\mathbf{r}')$ (for reduced scattering), which are particular cases of the general optical contrast expressed in Eq. (14.1) by $\delta\mu_i(\mathbf{r}')$, the local change in the i-th component of the optical-property vector $\boldsymbol{\mu}$. The absolute optical contrast represented by $\delta\mu_i(\mathbf{r}')$ may be expressed as a relative *optical property contrast* (dimensionless or percent), C_{μ_i}, by dividing it by the *background value* of the optical property, μ_{i0}, which can be thought of as an average value over a larger surrounding volume:

$$C_{\mu_i}(\mathbf{r}') = \frac{\delta\mu_i(\mathbf{r}')}{\mu_{i0}}. \tag{14.48}$$

The value of diffuse optical imaging in biomedical applications generally rests on this contrast in the optical properties and their associated biological, diagnostic, functional, and structural information. For example, tumor angiogenesis (the formation of new blood vessels), a local hemorrhage, the focal increase in cerebral blood flow as a result of brain activation, or a localized hypoxic tissue region, all result in a spatially localized change in the absorption coefficient due to changes in hemoglobin concentration or oxygen saturation.

Similarly, cancer-induced modifications to the microscopic structure or morphology of tissue, including the sizes and shapes of cell nuclei and other organelles, result in a change in the reduced scattering coefficient. Of course, the specific extent of absorption or scattering contrast depends on the particular case, but it is important to consider that it can be large. One should keep in mind that the near-infrared absorption coefficient (at, say, 800 nm) of normally perfused tissue is of the order of 0.1 cm^{-1}, whereas the absorption coefficient of whole blood at 800 nm is of the order of 5 cm^{-1} (greater by a factor of 50!), which accounts for the large absorption contrast of blood vessels. In the case of scattering, even though the perturbations determined by biological effects on individual scattering events may be small, one must consider that in diffuse imaging every collected photon undergoes thousands of scattering events before being detected. (On average, photons travel a distance equal to the source-detector separation times the differential pathlength factor (DPF; see Section 10.3), i.e., ~20 cm for a source-detector separation of ~3 cm, and suffer a scattering event, on average, every $1/\mu_s \sim 100\ \mu\text{m}$.) As a point of reference, breast cancer can feature an absorption coefficient that is 2–3 times the background absorption, resulting in an absorption contrast $C_{\mu_a} \sim$ 200–300%! (Grosenick et al., 2005). Similarly, the reduced scattering contrast ($C_{\mu_s'}$) of breast tumors can be as high as 30–50% (Grosenick et al., 2005; Cerussi et al., 2006; Choe et al., 2009).

14.8.2 Image contrast

The *image contrast* associated with a given optical inhomogeneity can be defined as the relative difference between the imaged property value X (which may be a derived optical property or a directly measured signal) at the location of the inhomogeneity (X_I) and its background value (X_0):

$$C_X = \frac{X_I - X_0}{X_0}. \tag{14.49}$$

Ideally, if the imaged property is an optical property, the associated image contrast should be a faithful representation of the corresponding optical property contrast considered in the previous section and expressed by Eq. (14.48). This is indeed the case for imaging and tomographic reconstruction methods that quantitatively take into account the different effects on the measured signal of a given optical inhomogeneity, depending on its location relative to the illumination and optical detection points. In some straightforward backprojection methods, however, an optical image may be obtained by spatially combining optical properties that are derived from the measured optical signals, say the CW intensities, for each source-detector pair. When the measured optical signal for each source-detector pair is translated into an optical property, say the absorption coefficient, by assuming optical homogeneity of the medium, the resulting image contrast may be degraded because the optical inhomogeneity occupies only a portion of the optically interrogated volume. This reduction in image contrast is referred to as a *partial volume effect*.

The definition of image contrast of Eq. (14.49) does not take into account the noise associated with the imaged quantity X, which may have both instrumental and biological origins. The noise in the imaged quantity X, which we indicate with σ_X, reduces the visibility of optical inhomogeneities in the image, a fact that is not reflected in the contrast definition of Eq. (14.49). Therefore, a more meaningful quantity to describe the ability to visualize an optical inhomogeneity is the *contrast-to-noise ratio* (CNR), defined as:

$$\text{CNR} = \frac{X_I - X_0}{\sigma_X}. \tag{14.50}$$

The CNR specifies the effect of an optical inhomogeneity on the imaged quantity ($X_I - X_0$) in relation to the noise level (σ_X): it specifies the detectability level of optical inhomogeneities, in that a CNR $\lesssim 1$ is inadequate for the detection of an object in the optical image.

Small optical perturbations ($\delta\mu_a \ll \mu_{a0}$, $\delta\mu_s' \ll \mu_{s0}'$) can be treated with the linear Born approximation of Section 14.6.1.1, which leads to the analytical solutions of Eqs. (14.28) and (14.29) for the associated fluence rate perturbations in the frequency domain and time domain, respectively. Inspection of these equations shows that a localized optical perturbation within a small volume ΔV – here, "small" means that the volume integrals reduce to the multiplication of the integrand by ΔV – results in an absorption term containing $\delta\mu_a \Delta V$ and a scattering term containing $\delta\mu_s' \Delta V$. (Consider that $\delta D = -\frac{c_n}{3(\mu_s'+\mu_a)^2}\delta\mu_s'$.) Therefore, within the Born approximation, it is not possible to discriminate the size and optical contrast of an embedded object, since a smaller object with a stronger optical contrast would cause the same fluence rate change as a larger object with a weaker optical contrast.

Larger perturbations, requiring higher-order terms than those in the Born approximation, are needed for a full characterization of the size and optical properties of optical inhomogeneities. The specific object size and optical property contrast needed for detectability (which is determined by the products $\delta\mu_a \Delta V$ and $\delta\mu_s' \Delta V$) and optical characterization (which is determined by non-negligible higher-order terms than in the Born approximation) depend on the specific conditions of source-detector separation, measured optical signal and associated noise, background optical properties, etc. By considering conditions that pertain to clinical applications of frequency-domain diffuse optical imaging, it was found that the smallest detectable absorbing object with $C_{\mu_a} = 200\%$ is $\sim$3 mm in diameter, whereas the smallest detectable scattering object with $C_{\mu_s'} = 50\%$ is $\sim$4 mm in diameter; a larger diameter of at least $\sim$1 cm is needed for a complete optical characterization and size assessment of the object (Boas et al., 1997).

14.8.3 Spatial resolution

The concept of *spatial resolution* refers to the ability of an optical imaging method to discriminate two small features that are close to each other. It can be defined it terms of the minimum distance between two small objects that allows for their discrimination. This translates to the width of the blurred image of either a point object (*point spread function*, which will be treated in detail in Section 17.1.2) or a linear object (*line spread function*), or the edge steepness of the image of an opaque half-plane bounded by a straight edge (*edge spread function*). (The line spread function is equal to the spatial derivative of the edge spread function along the direction perpendicular to the line and the edge.) The higher the spatial resolution, the greater is the ability of an imaging system to discern finer structures in the investigated sample, provided that the structures offer a suitable optical property contrast.

As is intuitive, the diffusive nature of light propagation in tissues poses intrinsic limitations to the spatial resolution of diffuse optical imaging. The width of the CW regions of sensitivity illustrated in Figures 14.1 and 14.2, which is directly related to the width of the CW point spread function, allows for a qualitative description of the resolving power for optical inhomogeneities embedded in tissues. The resolution is higher for objects that are closer to either the source or the detector, than for objects near the mid-plane, and the resolution improves (perhaps less intuitively but evident from Figures 14.1(b) and 14.2(b)) at higher background absorption and scattering coefficients.

In practice, in diffuse optical imaging the spatial resolution is generally limited to ~1 cm at depths of 3–5 cm (and scaling with the source-detector separation), although a number of techniques (such as time gating in the time domain, high modulation frequencies in the frequency domain, high-density grids of sources and detectors, tomographic reconstruction schemes) can somewhat improve the spatial resolution. Consequently (when compared to, say, MRI or X-ray imaging), spatial resolution is not a strength of diffuse optical imaging, whose major promise for biomedical applications rests instead on the pathological and physiological processes that can be elucidated by quantitative imaging of intrinsic optical contrast associated with biochemical, structural, and functional changes. Because of the complementary contrast and resolution performance of diffuse optical imaging and other medical imaging techniques such as MRI, X-ray, or ultrasound, there is sometimes value in co-registering optical and other imaging modalities to enhance the information content provided by stand-alone diffuse optical imaging.

Appendix 14.A: Fluence rate Green's functions for a semi-infinite medium

The fluence rate Green's function for a semi-infinite medium with CW illumination at the boundary is given by Eq. (10.22), but with $a'P_{CW}$ (the source power) replaced by 1, because the Green's function refers to a point source of unitary power. By considering zero boundary conditions ($z_b = 0$), and by introducing a Cartesian coordinate system with the x and y axes on the medium surface (illumination point at $x = 0$, $y = 0$), and the z axis pointing outward from the medium (with the medium surface at $z = 0$), Eq. (10.22) can be written as follows to yield the fluence rate semi-infinite Green's function:

$$G_{\phi_{CW}}^{(\mathrm{seminf})}(\mathbf{r}_s, \mathbf{r}) = \frac{3(\mu_s' + \mu_a)}{4\pi} \left(\frac{e^{-\mu_{\mathrm{eff}}\sqrt{x^2+y^2+(z+z_0)^2}}}{\sqrt{x^2+y^2+(z+z_0)^2}} - \frac{e^{-\mu_{\mathrm{eff}}\sqrt{x^2+y^2+(z-z_0)^2}}}{\sqrt{x^2+y^2+(z-z_0)^2}} \right), \quad (14.A.1)$$

where μ_a and μ_s' are the absorption and reduced scattering coefficient, respectively, $\mu_{\text{eff}} = \sqrt{3\mu_a(\mu_s' + \mu_a)}$, and $-z_0 = -1/\mu_s'$ is the z coordinate of the effective light source inside the medium. Because diffusion theory holds at distances from the light source that are greater than several times the inverse of μ_s', Eq. (14.A.1) can be further simplified under the diffusion condition $z_0 \ll \sqrt{x^2 + y^2 + z^2}$. Under this condition, the distances from the real and imaginary sources can be approximated as follows:

$$\sqrt{x^2 + y^2 + (z \pm z_0)^2} = \sqrt{x^2 + y^2 + z^2}\sqrt{1 + \frac{z_0^2 \pm 2z_0 z}{x^2 + y^2 + z^2}}, \tag{14.A.2}$$

where:

$$\left|\frac{z_0^2 \pm 2z_0 z}{x^2 + y^2 + z^2}\right| \ll 1. \tag{14.A.3}$$

With the condition of Eq. (14.A.3), Eq. (14.A.1) becomes:

$$G_{\phi\text{CW}}^{(\text{seminf})}(\mathbf{r}_s, \mathbf{r}) \cong -\frac{3(\mu_s' + \mu_a)}{4\pi} 2z_0 z \left(\mu_{\text{eff}} + \frac{1}{\sqrt{x^2 + y^2 + z^2}}\right) \frac{e^{-\mu_{\text{eff}}\sqrt{x^2+y^2+z^2}}}{(x^2 + y^2 + z^2)}, \tag{14.A.4}$$

Another way to find the same result is to consider that, in the limit $z_0 \ll \sqrt{x^2 + y^2 + z^2}$, the semi-infinite medium fluence rate Green's function given in Eq. (14.A.1) becomes proportional to the partial derivative of the infinite medium fluence rate Green's function with respect to z:

$$G_{\phi\text{CW}}^{(\text{seminf})}(\mathbf{r}_s, \mathbf{r}) = 2z_0 \frac{\partial G_{\phi}^{(\text{inf})}(\mathbf{r}_s, \mathbf{r})}{\partial z}, \tag{14.A.5}$$

which yields the same expression of Eq. (14.A.4).

Problems – answers to problems with * are on p. 647

14.1* Consider the sensitivity function of the CW fluence rate to an absorption perturbation in an infinite medium with a source-detector separation $r_{sd} = 4$ cm, and background absorption and reduced scattering coefficients $\mu_a = 0.1\ \text{cm}^{-1}$, $\mu_s' = 10\ \text{cm}^{-1}$. What is the ratio of this sensitivity function for a perturbation at each of the points listed below and for the same perturbation at the mid-point along the source-detector line (i.e., $\rho = 0$, $r_{sd} = 2$ cm using the coordinate system defined in the derivation of Eq. (14.4))? In other words, what is the ratio $s_{\phi\text{CW},\mu_a}^{(\text{inf})}(\rho, z)/s_{\phi\text{CW},\mu_a}^{(\text{inf})}(\rho = 0,\ z = 2\ \text{cm})$, for the following points (ρ, z)?

(a) $(\rho, z) = (0, 1 \text{ cm})$; (b) $(\rho, z) = (0, 3 \text{ cm})$; (c) $(\rho, z) = (0, 0.1 \text{ cm})$; (d) $(\rho, z) = (0.1 \text{ cm}, 0)$; (e) $(\rho, z) = (1 \text{ cm}, 0)$; (f) $(\rho, z) = (1 \text{ cm}, 2 \text{ cm})$; (g) $(\rho, z) = (2 \text{ cm}, 2 \text{ cm})$.

14.2* For a semi-infinite medium of $\mu_a = 0.1 \text{ cm}^{-1}$, $\mu_s' = 10 \text{ cm}^{-1}$ with surface illumination and collection at a source-detector distance $\rho_{sd} = 4$ cm:

(a) find the depth of maximal CW sensitivity to a localized absorption perturbation at the mid-plane between source and detector;

(b) find the difference between the depth of maximal sensitivity at the mid-plane (found in part (a)) and the average depth of sensitivity approximated by Eq. (14.10).

14.3 According to the expression of Eq. (14.10) is the average depth of sensitivity more strongly sensitive to a given percent change (say 5%) in the source-detector separation, absorption coefficient, or reduced scattering coefficient?

14.4* Consider a monochromatic ($\lambda = 720$ nm), directional light source that is pointed toward an optical detector that is at a distance of 0.9 cm from the source. Both the source and the detector are deeply embedded within a scattering medium with $\mu_a = 0.13 \text{ cm}^{-1}$, $\mu_s' = 11.5 \text{ cm}^{-1}$, and anisotropy factor $g = 0.875$. If the light source emits a power of 135 mW, for how long do you need to wait, on average, to detect one ballistic photon? Compare your answer with the age of the universe of $\sim 13.8 \times 10^9$ years.

14.5* Work out Problem 14.4 for a shorter source-detector distance of 400 μm and find the detected power associated exclusively with ballistic photons.

14.6 The perturbation approach expressed in Eq. (14.24) as the Born approximation (namely $\phi(\mathbf{r}) \cong \phi_0(\mathbf{r}) + \delta\phi(\mathbf{r})$ with $\delta\phi \ll \phi_0$) can also be expressed in terms of the *Rytov approximation*, named after the Russian physicist Sergei Mikhailovich Rytov [1908–1996], as follows:

$$\phi(\mathbf{r}) \cong \phi_0(\mathbf{r})e^{\psi(\mathbf{r})},$$

where $\psi(\mathbf{r})$ is a slowly varying (i.e., with limited $|\nabla\psi(\mathbf{r})|^2$) surrogate function to be determined. By considering that, for small perturbations, $\phi(\mathbf{r}) \cong \phi_0(\mathbf{r})$ and $|\psi(\mathbf{r})| \ll 1$, relate the Rytov approximation function $\psi(\mathbf{r})$ to the Born approximation fluence rate perturbation $\delta\phi(\mathbf{r})$ to show that the surrogate function $\psi(\mathbf{r})$ is given by:

$$\psi(\mathbf{r}) = -\frac{1}{\phi_0(\mathbf{r})}\int_V \delta\mu_a(\mathbf{r}')\phi_0(\mathbf{r}')G_\phi(\mathbf{r}', \mathbf{r})d\mathbf{r}'$$

$$-\frac{1}{c_n\phi_0(\mathbf{r})}\int_V \delta D(\mathbf{r}')\nabla\phi_0(\mathbf{r}', \omega)\cdot\nabla G_\phi(\mathbf{r}', \mathbf{r}, \omega)d\mathbf{r}'.$$

14.7 Consider a localized optical perturbation inside an otherwise homogeneous scattering medium at $\mathbf{r} = \mathbf{r}_p$ (say, $\delta\mu_a(\mathbf{r}) = k_a\delta(\mathbf{r} - \mathbf{r}_p)$ for an absorption perturbation, or $\delta\mu_s'(\mathbf{r}) = k_s\delta(\mathbf{r} - \mathbf{r}_p)$ for a scattering perturbation, where $\delta(\mathbf{r} - \mathbf{r}_p)$ is the Dirac delta, and k_a, k_s are constants).

(a) What are the units of k_a and k_s?

(b) Does the introduction of a localized absorption perturbation ($k_a > 0$, $k_s = 0$) result in an increase, decrease, or no change in the CW fluence rate measured at a given detector position $\mathbf{r}_d$ ($\phi_{CW}(\mathbf{r}_d)$) according to the Born approximation? Does the answer depend on the spatial distribution of the light sources, detector position ($\mathbf{r}_d$), perturbation location ($\mathbf{r}_p$), and/or boundary conditions?

(c) Answer part (b) for the case of a scattering perturbation ($k_a = 0$, $k_s > 0$).

(d) If you think that the answers to (b) and (c) depend on the source locations, detector position ($\mathbf{r}_d$), perturbation location ($\mathbf{r}_p$), or boundary conditions find the answers for the specific case of an infinite medium and a point source at position $\mathbf{r}_s$.

14.8 How many rows and columns does the matrix $\mathbf{M} = \mathbf{J}^T\mathbf{J}$ (which appears in Eq. (14.42)) have? (Here, $\mathbf{J}$ is the Jacobian matrix introduced in Eq. (14.36).)

14.9* According to the Born approximation, find the CW intensity contrast (i.e., the relative change in the CW intensity) associated with a small absorbing inhomogeneity (optical property contrast: $C_{\mu_a} = 90\%$, $C_{\mu_s'} = 0$; volume of inhomogeneity: $\Delta V = 3\ \text{mm}^3$) located half-way between a source and a detector that are separated by 3.8 cm within an infinite medium of $\mu_{a0} = 0.062\ \text{cm}^{-1}$ and (a) $\mu_{s0}' = 7.1\ \text{cm}^{-1}$; (b) $\mu_{s0}' = 14.2\ \text{cm}^{-1}$.

References

Alerstam, E., Svensson, T., and Andersson-Engels, S. (2008). Parallel computing with graphics processing units for high-speed Monte Carlo simulation of photon migration. *Journal of Biomedical Optics*, 13, 060504.

Arridge, S. R. (1995). Photon-measurement density functions. Part I: Analytical forms. *Applied Optics*, 34, 7395–7409.

Arridge, S. R., Cope, M., and Delpy, D. T. (1992). The theoretical basis for the determination of optical pathlengths in tissue: temporal and frequency analysis. *Physics in Medicine and Biology*, 37, 1531–1560.

Arridge, S. R., Schweiger, M., Hiraoka, M., and Delpy, D. T. (1993). A finite element approach to modelling photon transport in tissue. *Medical Physics*, 20, 209–309.

Boas, D. A., O'Leary, M. A., Chance, B., and Yodh, A. G. (1994). Scattering of diffuse photon density waves by spherical inhomogeneities within turbid media: analytic solution and applications. *Proceedings of the National Academy of Sciences USA*, 91, 4887–4891.

Boas, D. A., O'Leary, M. A., Chance, B., and Yodh, A. G. (1997). Detection and characterization of optical inhomogeneities with diffuse photon density waves: a signal-to-noise analysis. *Applied Optics*, 36, 75–92.

Bonner, R. F., Nossal, R., Havlin, S., and Weiss, G. H. (1987). Model for photon migration in turbid biological media. *Journal of the Optical Society of America A*, 4, 423–432.

Brooksby, B., Jiang, S., Dehghani, H., Pogue, B. W., and Paulsen, K. D. (2005). Combining near-infrared tomography and magnetic resonance imaging to study in vivo breast tissue: implementation of a Laplacian-type regularization to incorporate magnetic resonance structure. *Journal of Biomedical Optics*, 10, 051504.

Cai, W., Lax, M., and Alfano, R. R. (2000). Analytical solution of the elastic Boltzmann transport equation in an infinite uniform medium using cumulant expansion. *Journal of Physical Chemistry B*, 104, 3996–4000.

Cerussi, A. E., Shah, N., Hsiang, D., et al. (2006). *In vivo* absorption, scattering, and physiologic properties of 58 malignant breast tumors determined by broadband diffuse optical spectroscopy. *Journal of Biomedical Optics*, 11, 044005.

Choe, R., Konecky, S. D., Corlu, A., et al. (2009) Differentiation of benign and malignant breast tumors by *in-vivo* three-dimensional parallel-plate diffuse optical tomography. *Journal of Biomedical Optics*, 14, 024020.

Cuccia, D. J., Bevilacqua, F., Durkin, A. J., Ayers, F. R., and Tromberg, B. J. (2009). Quantitation and mapping of tissue optical properties using modulated imaging. *Journal of Biomedical Optics*, 14, 024012.

Fantini, S., Franceschini, M. A., Walker, S. A., Maier, J. S., and Gratton, E. (1995). Photon path distributions in turbid media: applications for imaging. *Proceedings of SPIE*, 2389, 340–349.

Feng, S., Zeng, F., and Chance, B. (1995). Photon migration in presence of a single defect: a perturbation analysis. *Applied Optics*, 34, 3826–3837.

Fishkin, J. B., and Gratton, E. (1993). Propagation of photon-density waves in strongly scattering media containing an absorbing semi-infinite plane bounded by a straight edge. *Journal of the Optical Society of America A*, 10, 127–140.

Gandjbakhche, A. H., Bonner, R. F., and Nossal, R. (1992). Scaling relationships for anisotropic random walks. *Journal of Statistical Physics*, 69, 35–53.

Gioux, S., Mazhar, A., Lee, B. T., et al. (2011). First-in-human pilot study of a spatial frequency domain oxygenation imaging system. *Journal of Biomedical Optics*, 16, 086015 (10pp).

Grosenick, D., Weibnitz, H., Moesta, K. T., et al. (2005) Time-domain scanning optical mammography: II. Optical properties and tissue parameters of 87 carcinomas. *Physics in Medicine and Biology*, 50, 2451–2468.

Hall, D. J., Hebden, J. C., and Delpy, D. T. (1997). Imaging very-low-contrast objects in breastlike scattering media with a time-resolved method. *Applied Optics*, 36, 7270–7276.

Kim, H. K., and Hielscher, A. H. (2009). A PDE-constrained SQP algorithm for optical tomography based on the frequency-domain equation of radiative transfer. *Inverse Problems*, 25, 015010 (20pp).

Konecky, S. D., Mazhar, A., Cuccia, D., et al. (2009) Quantitative optical tomography of sub-surface heterogeneities using spatially modulated structured light. *Optics Express*, 17, 14780–14790.

Liemert, A., and Kienle, A. (2013). Exact and efficient solution of the radiative transport equation for the semi-infinite medium. *Scientific Reports*, 3, 2018.

Machida, M., Panasyuk, G. Y., Schotland, J. C., and Markel, V. A. (2010). The Green's function for the radiative transport equation in the slab geometry. *Journal of Physics A: Mathematical and Theoretical*, 43, 065402.

Mitic, G., Kölzer, J., Otto, J., et al. (1994). Time-gated transillumination of biological tissues and tissuelike phantoms. *Applied Optics*, 33, 6699–6710.

Paasschens, J. C. J. (1997). Solution of the time-dependent Boltzmann equation. *Physics Review E*, 56, 1135–1141.

Patterson, M. S., Andersson-Engels, S., Wilson, B. C., and Osei, E. K. (1995). Absorption spectroscopy in tissue-simulating materials: a theoretical and experimental study of photon paths. *Applied Optics*, 34, 22–30.

Perelman, L. T., Winn, J., Wu, J., Dasari, R. R., and Feld, M. S. (1997). Photon migration of near-diffusive photons in turbid media: a Lagrangian-based approach. *Journal of the Optical Society of America A*, 14, 224–229.

Pogue, B. W., Patterson, M. S., and Paulsen, K. D. (1995). Initial assessment of a simple system for frequency domain diffuse optical tomography. *Physics in Medicine and Biology*, 40, 1709–1729.

Pogue, B. W., McBride, T. O., Prewitt, J., Österberg, U. L., and Paulsen, K. D. (1999). Spatially variant regularization improves diffuse optical tomography. *Applied Optics*, 38, 2950–2961.

Polishchuk, A. Ya., and Alfano, R. R. (1995). Fermat photons in turbid media: an exact analytic solution for most favorable paths – a step toward optical tomography. *Optics Letters*, 20, 1937–1939.

Sawosz, P., Kacprzak, M., Weigl, W., et al. (2012). Experimental estimation of the photons visiting probability profiles in time-resolved diffuse reflectance measurement. *Physics in Medicine and Biology*, 57, 7973–7981.

Schotland, J. C., Haselgrove, J. C., and Leigh, J. S. (1993). Photon hitting density. *Applied Optics*, 32, 448–453.

Sevick, E. M., Frisoli, J. K., Burch, C. L., and Lakowicz, J. R. (1994). Localization of absorbers in scattering media by use of frequency-domain measurements of time-dependent photon migration. *Applied Optics*, 33, 3562–3570.

Torricelli, A., Pifferi, A., Spinelli, L., et al. (2005). Time-resolved reflectance at null source-detector separation: Improving contrast and resolution in diffuse optical imaging. *Physical Review Letters*, 95, 078101 (4pp).

Tosi, A., Dalla Mora, A., Zappa, F., et al. (2009). Fast-gated single-photon avalanche diode for extremely wide dynamic-range applications. *Proceedings of SPIE*, 7170, 71700K (11 pp).

Vogel, C. R. (2002). *Computational Methods for Inverse Problems*. Philadelphia, PA: SIAM, Ch. 7.

Wang, L., Jacques, S. L., and Zeng, L. (1995). MCML – Monte Carlo modelling of light transport in multi-layered tissues. *Computer Methods and Progress in Biomedicine*, 47 131–146.

Yalavarthy, P. K., Pogue, B. W., Dehghani, H., et al. (2007). Structural information within regularization matrices improves near infrared diffuse optical tomography. *Optics Express*, 15, 8043–8058.

Further reading

Green's functions in diffusion-wave fields

Mandelis, A. (2001). *Diffusion-Wave Fields: Mathematical Methods and Green Functions*. New York: Springer.

Region of optical sensitivity in scattering media

Arridge, S. R. (1995). Photon-measurement density functions. Part 1: Analytical forms. *Applied Optics*, 34, 7395–7409.

Arridge, S. R., and Schweiger, M. (1995). Photon-measurement density functions. Part 2: Finite-element-method calculations. *Applied Optics*, 34, 8026–8037.

Weiss, G. H., and Kiefer, J. E. (1998). A numerical study of the statistics of penetration depth of photon re-emitted from irradiated media. *Journal of Modern Optics*, 45, 2327–2337.

Weiss, G. H., Nossal, R., and Bonner, R. F. (1989). Statistics of penetration depth of photon re-emitted from irradiated tissue. *Journal of Modern Optics*, 36, 349–359.

Imaging by backprojection

Colak, S. B., Papaioannou, D. G., 't Hooft, G. W., et al. (1997). Tomographic image reconstruction from optical projections in light-diffusing media. *Applied Optics*, 36, 180–213.

Lyubimov, V. V., Kalintsev, A. G., Konovalov, A. B., et al. (2002). Application of the photon average trajectories method to real-time reconstruction of tissue inhomogeneities in diffuse optical tomography of strongly scattering media. *Physics in Medicine and Biology*, 47, 2109–2128.

Walker, S. A., Fantini, S., and Gratton, E. (1997). Image reconstruction by backprojection from frequency-domain optical measurements in highly scattering media. *Applied Optics*, 36, 170–179.

Imaging with time-gated optical detection

Alfano, R. R., Liang, X., Wang, L., and Ho, P. P. (1994). Time-resolved imaging of translucent droplets in highly scattering turbid media. *Science*, 264, 1913–1915.

Mazurenka, M., Di Sieno, L., Boso, G., et al. (2013). Non-contact in vivo diffuse optical imaging using a time-gated scanning system. *Biomedical Optics Express*, 4, 2257–2268.

Patel, N., Lin, Z.-J., Rathore, Y., et al. (2010). Relative capacities of time-gated versus continuous-wave imaging to localize tissue embedded vessels with increasing depth. *Journal of Biomedical Optics*, 15, 016015 (9 pp).

Optical imaging with spatially modulated illumination

O'Sullivan, T. D., Cerussi, A. E., Cuccia, D. J., and Tromberg, B. J. (2012). Diffuse optical imaging using spatially and temporally modeulated light. *Journal of Biomedical Optics*, 17, 071311 (14 pp).

Yafi, A., Vetter, T. S., Scholz, T., et al. (2011). Postoperative quantitative assessment of reconstructive tissue status in a cutaneous flap model using spatial frequency domain imaging. *Plastic and Reconstuctive. Surgery*, 127, 117–130.

Finite-element method

Zienkiewicz, O. C., and Taylor, R. L. (2000). *The Finite Element Method: Volume 1*, 5th edn. Oxford: Butterworth-Heinemann.

Monte Carlo simulations

Fang, Q., and Kaeli, D. R. (2012). Accelerating mesh-based Monte Carlo method on modern CPU architectures. *Biomedical Optics Express*, 3, 3223–3230.

Flock, S. T., Patterson, M. S., Wilson, B. C., and Wyman, D. R. (1989). Monte Carlo modeling of light-propagation in highly scattering tissues: 1. Model predictions and comparison with diffusion theory. *IEEE Transactions on Biomedical Engineering*, 36, 1162–1168.

Sassaroli, A. (2011). Fast perturbation Monte Carlo method for photon migration in heterogeneous turbid media. *Optics Letters*, 36, 2095–2097.

Sassaroli, A., and Martelli, F. (2012). Equivalence of four Monte Carlo methods for photon migration in turbid media. *Journal of the Optical Society of America A*, 29, 2110–2117.

Wang, L., Jacques, S. L., and Zheng, L. (1995). MCML-Monte Carlo modeling of light transport in multi-layered tissues. *Computer Methods and Progress in Biomedicine*, 47, 131–146.

Zaccanti, G. (1991). Monte Carlo study of light propagation in optically thick media: point source case. *Applied Optics*, 30, 2031–2041.

Random walk theory

Gandjbakhche, A. H., Nossal, R., and Bonner, R. F. (1993). Scaling relationships for theories of anisotropic random walks applied to tissue optics. *Applied Optics*, 32, 504–516.

Gandjbakhche, A. H., Weiss, G. H., Bonner, R. F., and Nossal, R. (1993). Photon path-length distributions for transmission through optically turbid slabs. *Physics Review E*, 48, 810–818.

Diffuse optical tomography

Arridge, S. R. (1999). Optical tomography in medical imaging. *Inverse Problems*, 15, R41-R93.

Arridge, S. R., and Hebden, J. C. (1997). Optical imaging in medicine: II. Modeling and reconstruction. *Physics in Medicine and Biology*, 42, 841–853.

Arridge, S. R., and Schotland, J. C. (2009). Optical tomography: forward and inverse problems. *Inverse Problems*, 25, 123010.

Dehghani, H., Srinivasan, S., Pogue, B. W., and Gibson, A. (2009). Numerical modelling and image reconstruction in diffuse optical tomography. *Philosophical Transactions of the Royal Society A*, 367, 3073–3093.

Durduran, T., Choe, R., Baker, W. B., and Yodh, A. G. (2010). Diffuse optics for tissue monitoring and tomography. *Reports on Progress in Physics*, 73, 076701.

Hielscher, A. H., Bluestone, A. Y., Abdoulaev, G. S., et al. (2002). Near-infrared diffuse optical tomography. *Disease Markers*, 18, 313–337.

Jiang, H. (2011). *Diffuse Optical Tomography: Principles and Applications* (CRC Press, Boca Raton, FL).

Srinivasan, S., Davis, S. C., and Carpenter, C. M. (2011). Diffuse optical tomography using CW and frequency domain imaging systems. In *Handbook of Biomedical Optics*, ed. D. A. Boas, C. Pitris, and N. Ramanujam. Boca Raton, FL: CRC Press, Ch. 19, pp. 373–394.

Regularization techniques for optical tomography

Belge, M., Kilmer, M. E., and Miller, E. L. (2002). Efficient determination of multiple regularization parameters in a generalized L-curve framework. *Inverse Problems*, 18, 1161–1183.

Katamreddy, S. H., and Yalavarthy, P. K. (2012). Model-resolution based regularization improves near infrared diffuse optical tomography. *Journal of the Optical Society of America A*, 29, 649–656.

Li, A., Miller, E. L., Kilmer, M. E., et al. (2003). Tomographic optical breast imaging guided by three-dimensional mammography. *Applied Optics*, 42, 5181–5190.

Larusson, F., Fantini, S., and Miller, E. L. (2011). Hyperspectral image reconstruction for diffuse optical tomography. *Biomedical Optics Express*, 2, 946–965.

Yalavarthy, P. K., Pogue, B. W., Dehghani, H., and Paulsen, K. D. (2007). Weight-matrix structured regularization provides optimal generalized least-squares estimate in diffuse optical tomography. *Medical Physics*, 34, 2085–2098.

Ye, J. C., Webb, K. J., Bouman, C. A., and Millane, R. P. (1999). Optical diffusion tomography by iterative-coordinate-descent optimization in a Bayesian framework. *Journal of the Optical Society of America A*, 16, 2400–2412.

15 In vivo applications of diffuse optical spectroscopy and imaging

In a sense, quantitative biomedical optics methods seek to enhance and quantify the visual perception of color and brightness of light transmitted through or reflected by biological tissues. This is especially the case for diffuse optical spectroscopy and imaging, which are based on the spectral dependence, spatial distribution, and temporal characteristics of the detected optical signals.

The "color" of tissue is determined by the spectral features of its absorbing and scattering constituents, with hemoglobin playing an important role, given its abundance and the relationship between its oxygen saturation and its absorption spectrum. For example, human skin color is determined by the amount of melanin in the skin, by the hemoglobin in cutaneous circulation, and by other chromophores, such as β-carotene and bilirubin, that may be indicative of pathological conditions. The dilatation or contraction of skin arterioles (for example as a result of warm or cold environments), and the resulting enhancement or reduction in cutaneous blood flow, affects the color of the skin, which, for fair-skinned people, may tend toward white (pallor: vascular constriction), red (erythema: vascular dilatation), or blue/purple (cyanosis: inadequate blood flow). Of note, the blue appearance of superficial veins is due to a combination of the reduced level of venous blood oxygenation, the scattering properties of the skin and blood constituents, and the visual perception of color contrast (Kienle et al., 1996).

The spatial distribution of light transmission through tissue can be affected by an inhomogeneous distribution of chromophores. For example, tumor angiogenesis or hematomas are associated with a focal increase in the concentration of blood in tissue, which results in a localized increase in light absorption.

Temporal characteristics of the detected optical signal may be directly associated with time-varying and dynamic physiological processes (arterial pulsation, respiration-modulated blood pressure, blood flow, etc.), thus yielding potentially valuable diagnostic information, and can also be used to monitor the evolution of metabolic and physiological conditions or drug delivery to tissues.

This chapter describes a representative sampling of in vivo applications of diffuse tissue spectroscopy and imaging, demonstrating how the quantitative methods described in Chapters 10–14 find practical applications in a number of research and clinical areas.

15.1 Oximetry

15.1.1 The oxygen dissociation curve of hemoglobin

Diffuse optical spectroscopy (DOS) of macroscopic tissue volume must be performed in the red to near-infrared (NIR) spectral region (say, in the wavelength range 600–1000 nm) to achieve a suitable optical penetration into tissue; this is why it is identified with the acronym NIRS (near-infrared spectroscopy). This optical "window" for biological tissue is determined by the combination of a low absorption by hemoglobin (see Figure 2.4) and water (see Figure 2.10), the two dominant tissue chromophores in blood-perfused tissues. The strong absorption of hemoglobin at shorter wavelengths (<600 nm) and water at longer wavelengths (>1000 nm) leads to high optical attenuation, which hinders optical measurements over macroscopic tissue volumes with long photon pathlengths typical of DOS methods. This fact can be visualized by placing a hand over a flashlight (best observed in a dark room); only red light makes it through the hand, since all shorter wavelengths of the white spectrum of the incandescent lamp are absorbed by hemoglobin and other tissue constituents.

In the spectral region 600–850 nm, appropriate for diffuse optical spectroscopy and imaging, the absorption of hemoglobin, while relatively low, is typically still dominant over water absorption, which is easily demonstrated by an approximate calculation. The extinction coefficient of hemoglobin (defined in terms of exponential attenuations to base *e* and for a full functional molecule containing four heme groups) is ~ 2000 cm^{-1}M^{-1} at the isosbestic point at ~ 800 nm, and its concentration in blood-perfused tissue is ~ 100 μM. (The concentration of hemoglobin in whole blood is ~ 2.3 mM, and the v/v concentration of blood in tissue is a few %, as discussed in Section 2.2.) The result is a contribution from hemoglobin to the absorption coefficient in tissue of ~ 0.1 cm^{-1} at that wavelength. This is significantly greater than the absorption due to water, as determined by the absorption coefficient of pure water at 800 nm (~ 0.02 cm^{-1}) multiplied by the v/v water concentration in tissue (which can be as high as $\sim 80\%$ in brain, skeletal muscle, and skin, and as low as 10% in fatty tissue).

The fact that the absorption spectra of oxyhemoglobin and deoxyhemoglobin differ from each other significantly in the NIR spectral region (see Figure 2.4) accounts for the feasibility of NIR blood oximetry. It needs to be clear that, in this context, oximetry does not refer to a direct measure of oxygen concentration, but,

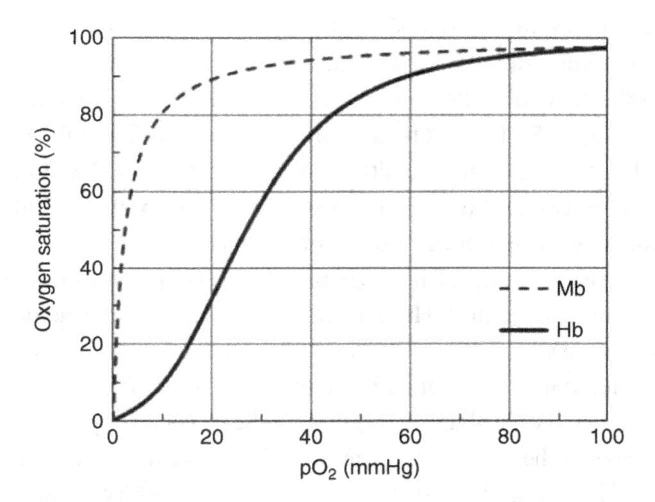

Figure 15.1

Oxygen dissociation curve of hemoglobin (Hb, solid line) and myoglobin (Mb, dashed line). A shift of the dissociation curve to the right indicates a decreased affinity for oxygen, which results in a lower oxygen saturation for a given partial pressure of dissolved oxygen (pO_2). Myoglobin has a higher oxygen affinity than hemoglobin. The dissociation curve for hemoglobin depends on pH (Bohr effect), and both dissociation curves for hemoglobin and myoglobin depend on temperature.

as introduced in Section 2.2, to a measure of the percentage of the heme groups in hemoglobin that bind oxygen, a quantity that is referred to as *oxygen saturation* of hemoglobin, or *hemoglobin saturation.* There is, however, a specific relationship between hemoglobin saturation ($S_{Hb}O_2$) and the partial pressure of oxygen (pO_2) in solution in blood, which is expressed by the *oxygen dissociation curve* of hemoglobin, reported in Figure 15.1 (together with the oxygen dissociation curve for myoglobin, which will be discussed below, in Section 15.2.1). This relationship can be well approximated by the Hill equation (named after the British physiologist Archibald Hill [1886–1977], who received the 1922 Nobel Prize for physiology and medicine):

$$S_{Hb}O_2 = \frac{(pO_2)^n}{(pO_2)^n + (p_{50})^n}, \tag{15.1}$$

where p_{50} is the partial pressure of oxygen for which the oxygen saturation is 50% (which is about 26 mmHg for hemoglobin under normal physiological conditions) and n is the *Hill coefficient*, which specifies the level of cooperative binding of oxygen for the four hemoglobin subunits. The oxygen binding *cooperativity*, which refers to the fact that the oxygen affinity of a subunit depends on whether the other subunits already bind oxygen, is a crucial aspect of the functional role of hemoglobin and accounts for the sigmoidal shape of the oxygen dissociation curve of hemoglobin in Figure 15.1. In the ideal case of complete cooperativity among the four subunits of hemoglobin, the Hill coefficient n would take a value of 4, and Eq. (15.1) would be a rigorous representation of the oxygen dissociation curve. For hemoglobin in healthy red blood cells, $n \sim 2.8$–3.0, which indicates a high level of cooperativity that can be loosely interpreted as indicating a complete cooperativity among three of the four subunits.

The oxygen affinity of hemoglobin, denoted by p_{50} in Eq. (15.1), depends on temperature and acidity. Higher temperature and higher acidity (i.e., lower pH) reduce the oxygen affinity of hemoglobin, increase p_{50}, and shift the oxygen dissociation curve of Figure 15.1 to the right. This effect has physiological implications, since it contributes to regulating the oxygen supply to tissue. For example, an increase in the CO_2 concentration in blood, termed *hypercapnia*, increases the blood acidity. The dependence of the oxygen affinity of hemoglobin on pH is called the *Bohr effect* after the Danish physiologist Christian Bohr [1855–1911], who first described it. (Christian Bohr did not receive a Nobel Prize, but his son Niels Bohr [1885–1962] and grandson Aage Bohr [1922–2009] received Nobel Prizes for physics in 1922 and 1975, respectively.)

The oxygen saturation on the y axis of Figure 15.1 is the percentage of molecules (for myoglobin) or heme complexes (for hemoglobin) that have bound oxygen, and the partial pressure of oxygen on the x axis of Figure 15.1 is directly proportional to the concentration of dissolved oxygen in plasma.

An estimating calculation allows us to realize that the vast majority of oxygen in blood is actually bound to hemoglobin. In fact, in arterial blood at ~100% saturation, the concentration of oxygen bound to hemoglobin is ~9.2 mM (four times the concentration of hemoglobin in blood, 2.3 mM, because of the four heme binding sites for oxygen in the hemoglobin molecule). On the other hand, the corresponding partial pressure of oxygen of ~100 mmHg results in a concentration of dissolved oxygen of ~0.17 mM, as obtained by Henry's law (named after the British chemist William Henry [1775–1836]), with an equilibrium constant for oxygen in water of 584.6 mmHg/mM). Therefore, dissolved oxygen in plasma is of the order of only ~2% of the total oxygen in blood. Nevertheless, it is the dissolved oxygen in plasma that is involved in the oxygen transport from alveoli to pulmonary capillaries and from systemic capillaries to peripheral tissue in the diffusion process, driven by a gradient in the concentration of dissolved oxygen.

The hemoglobin dissociation curve of Figure 15.1 quantifies the relationship between pO_2 in plasma and the saturation of hemoglobin, which is directly measurable with NIRS. The alveolar pO_2 is ~100 mmHg (somewhat lower than the atmospheric air value of ~160 mmHg because of humidification of inspired air and mixing with residual O_2-depleted air in the lungs). Correspondingly, the hemoglobin saturation in the pulmonary capillaries is close to 100% (as evident in Figure 15.1), resulting in the almost complete oxygen saturation of hemoglobin in arterial blood. In fact, if arterial hemoglobin saturation drops even slightly to ≤90% in a clinical setting, this is treated as a warning sign of pulmonary insufficiency.

The pO_2 of interstitial fluid in peripheral tissue may have typical values of 30–40 mmHg, depending on the type of tissue and its metabolic demand, so that, following oxygen transport from systemic capillaries to peripheral tissue, venous blood may have a hemoglobin saturation of about 50–70%. From this discussion, it is apparent that hemoglobin saturation is highly variable spatially in the vasculature, and it is also temporally variable in response to dynamic changes in the blood flow and metabolic demand of individual tissues. Selective measurement of the hemoglobin saturation of arterial blood (SaO_2) is achieved by *pulse oximetry*, whereas global measurement of the hemoglobin saturation in tissue (StO_2) is achieved by *tissue oximetry*, which can be further extended to optical measurement of the redox state of intracellular cytochrome *c* oxidase. Pulse oximetry, tissue oximetry, and the assessment of the redox state of cytochrome *c* oxidase are described below in Sections 15.1.2 and 15.1.3.

15.1.2 Pulse oximetry

Pulse oximetry, among the earliest disseminations of biomedical optics technology, was developed during the 1970s, and by the 1980s it had demonstrated a clinical value that is still recognized today, accounting for the ubiquitous presence of pulse oximeters in clinical and doctors' office settings. The oxygenation measurement of pulse oximetry (SpO_2) refers to the oxygen saturation of hemoglobin in the arterial component of the blood pool (SaO_2), which is typically >95% in healthy individuals. Because arterial saturation reflects the oxygenation of blood upstream of the oxygen transport to tissue, which occurs at the level of small arterioles and systemic capillaries, pulse oximetry does not convey information on the adequacy of oxygen delivery to tissue; rather, it indicates the adequacy of blood oxygenation in the pulmonary capillaries. Because SaO_2 (in individuals with normal red blood cells) is only determined by the arterial partial pressure of oxygen (as indicated by the oxygen dissociation curve of Figure 15.1), it is not dependent on the concentration of hemoglobin in blood, and thus it cannot provide indication of, for example, anemia.

Optical measurements of SpO_2 are based on the fact that arteries and arterioles, as opposed to capillaries and veins, feature a periodic dilatation and contraction (or pulsation) in response to the oscillations in the arterial blood pressure associated with the heartbeat. Such arterial pulsation results in an oscillation of the

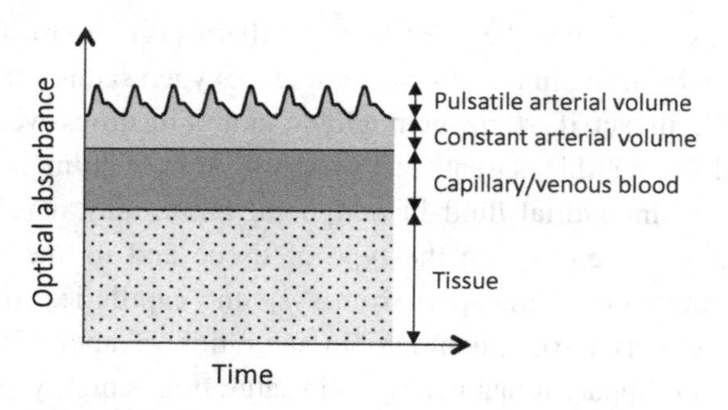

Figure 15.2

Illustration of the constant contributions to the optical absorbance from tissue, capillary/venous blood (non-pulsatile) and a portion of the arterial blood, to which the pulsatile component of arterial blood volume contributes a time-varying oscillation. It is this oscillatory component of the optical absorbance, associated with arterial blood, which is used by pulse oximeters to measure the arterial saturation.

arterial blood volume fraction in the tissue, which leads to a dynamic, pulsatile component of the optical absorbance of tissue. This pulsatile tissue absorbance adds to the constant absorbance from the minimum arterial volume, in addition to non-pulsatile capillary and venous blood, and tissue. This is illustrated in Figure 15.2. Conceptually, pulse oximetry is based on measurements of the pulsatile component of the optical signal at different wavelengths, which can be translated into a measure of the hemoglobin saturation of arterial blood on the basis of the different absorption spectra of oxyhemoglobin and deoxyhemoglobin (see Figure 2.4).

Strictly speaking, an absolute measurement of arterial saturation requires the measurement of the pulsatile component of the absorption coefficient $\Delta\mu_{a|p}(\lambda_i)$ at a minimum of two wavelengths, which can then be translated into measurements of the pulsatile concentrations of HbO and Hb ($\Delta[\mathrm{HbO}]_p$ and $\Delta[\mathrm{Hb}]_p$, respectively, where the square brackets indicate tissue concentrations) according to Eq. (10.2), which can be explicitly written at two wavelengths λ_1 and λ_2 as:

$$\Delta\mu_{a|p}(\lambda_1) = \varepsilon_{\mathrm{HbO}}(\lambda_1)\Delta[\mathrm{HbO}]_p + \varepsilon_{\mathrm{Hb}}(\lambda_1)\Delta[\mathrm{Hb}]_p, \qquad (15.2)$$

$$\Delta\mu_{a|p}(\lambda_2) = \varepsilon_{\mathrm{HbO}}(\lambda_2)\Delta[\mathrm{HbO}]_p + \varepsilon_{\mathrm{Hb}}(\lambda_2)\Delta[\mathrm{Hb}]_p. \qquad (15.3)$$

The extinction spectra of oxyhemoglobin ($\varepsilon_{\mathrm{HbO}}(\lambda)$) and deoxyhemoglobin ($\varepsilon_{\mathrm{Hb}}(\lambda)$) are shown in Figure 2.4 and they are reported (for the wavelength range 650–1042 nm) in tabulated form in Appendix 2.A (Table 2.A.1). Equations (15.2) and (15.3) are a linear system of two equations and two unknowns, which can be solved to yield the pulsatile components of the concentrations of oxy- and deoxyhemoglobin:

$$\Delta[\text{HbO}]_p = \frac{\varepsilon_{\text{Hb}}(\lambda_2)\Delta\mu_{a|p}(\lambda_1) - \varepsilon_{\text{Hb}}(\lambda_1)\Delta\mu_{a|p}(\lambda_2)}{\varepsilon_{\text{HbO}}(\lambda_1)\varepsilon_{\text{Hb}}(\lambda_2) - \varepsilon_{\text{HbO}}(\lambda_2)\varepsilon_{\text{Hb}}(\lambda_1)}, \tag{15.4}$$

$$\Delta[\text{Hb}]_p = \frac{\varepsilon_{\text{HbO}}(\lambda_1)\Delta\mu_{a|p}(\lambda_2) - \varepsilon_{\text{HbO}}(\lambda_2)\Delta\mu_{a|p}(\lambda_1)}{\varepsilon_{\text{HbO}}(\lambda_1)\varepsilon_{\text{Hb}}(\lambda_2) - \varepsilon_{\text{HbO}}(\lambda_2)\varepsilon_{\text{Hb}}(\lambda_1)}. \tag{15.5}$$

The pulsatile saturation (SpO_2) is then obtained as follows:

$$\begin{aligned} \text{SpO}_2 &= \frac{\Delta[\text{HbO}]_p}{\Delta[\text{HbO}]_p + \Delta[\text{Hb}]_p} \\ &= \frac{\varepsilon_{\text{Hb}}(\lambda_2) - \varepsilon_{\text{Hb}}(\lambda_1)\frac{\Delta\mu_{a|p}(\lambda_2)}{\Delta\mu_{a|p}(\lambda_1)}}{\varepsilon_{\text{Hb}}(\lambda_2) - \varepsilon_{\text{HbO}}(\lambda_2) + (\varepsilon_{\text{HbO}}(\lambda_1) - \varepsilon_{\text{Hb}}(\lambda_1))\frac{\Delta\mu_{a|p}(\lambda_2)}{\Delta\mu_{a|p}(\lambda_1)}}. \end{aligned} \tag{15.6}$$

Inasmuch as the pulsatile saturation SpO_2 is associated with in-phase oscillations of [HbO] and [Hb] that are directly associated with a volume oscillation of the arterial vascular compartment, it is indeed a valid estimator of the arterial saturation SaO_2.

The translation of the measured amplitude of the pulsatile optical intensity at wavelength $\lambda_i(\Delta I_p(\lambda_i))$ into an absorption coefficient change $(\Delta\mu_{a|p}(\lambda_i))$ must take tissue scattering into account. This may be done by assuming the value of the differential pathlength factor (DPF) in the modified Beer-Lambert law (Eq. (10.14)) on the basis of reference measurements of absorption and reduced scattering coefficients on similar tissues, or by performing a calibration measurement. Pulse oximetry invokes the latter approach, which initially involves reference measurements of tissue scattering in the absence of blood (by applying a pressure higher than systolic pressure, thus squeezing blood out of the measured tissue) and, currently, relies on a lookup table based on calibration measurements on a healthy population. The idea is to recognize that the wavelength dependence of the pulsatile optical intensity ΔI_p (divided by its average value I to cancel out contributions from the light source emission properties) is essentially only a function of arterial saturation.

A preliminary set of measurements on healthy subjects, where the arterial saturation is varied within the range ~80–100%, provides an empirical calibration lookup table. It specifies the dependence of the ratio (R_p) between the pulsatile optical intensities at a red wavelength (λ_1, at say 660 nm) and a NIR wavelength (λ_2, at say 940 nm) on the arterial saturation for the specific instrument and tissue considered (typically a finger, toe, or ear lobe). The ratio $R_p(\lambda_1, \lambda_2)$ is defined as follows:

$$R_p\,(\lambda_1, \lambda_2) = \frac{\Delta I_p(\lambda_1)/\langle I\rangle(\lambda_1)}{\Delta I_p(\lambda_2)/\langle I\rangle(\lambda_2)}. \tag{15.7}$$

The measurement of $R_p(\lambda_1, \lambda_2)$ is translated into a measurement of SaO_2 by proprietary, empirical lookup tables implemented in the software of commercial pulse oximeters.

15.1.3 Tissue oximetry

15.1.3.1 *Oxygen saturation of hemoglobin in blood-perfused tissue*

The sensitivity to oxygen saturation of the absorption spectrum of hemoglobin, which accounts for the change in color from the bright red of oxyhemoglobin to the dusky burgundy color of deoxyhemoglobin, has been exploited since the 1940s to perform optical measurements of blood oxygenation in tissue (Millikan, 1942; Kay and Coxon, 1957). Relative measurements of blood oxygenation can be based on the optical density of tissue at two wavelengths, as described above for pulse oximetry. A measurement of the absorption coefficient of tissue, in cases where the optical absorption is dominated by HbO and Hb, can be translated into absolute tissue concentrations and oxygen saturation of hemoglobin, by rewriting Eqs. (15.4) and (15.5) in terms of overall tissue, rather than pulsatile, quantities:

$$[\mathrm{HbO}] = \frac{\varepsilon_{\mathrm{Hb}}(\lambda_2)\mu_a(\lambda_1) - \varepsilon_{\mathrm{Hb}}(\lambda_1)\mu_a(\lambda_2)}{\varepsilon_{\mathrm{HbO}}(\lambda_1)\varepsilon_{\mathrm{Hb}}(\lambda_2) - \varepsilon_{\mathrm{HbO}}(\lambda_2)\varepsilon_{\mathrm{Hb}}(\lambda_1)}, \tag{15.8}$$

$$[\mathrm{Hb}] = \frac{\varepsilon_{\mathrm{HbO}}(\lambda_1)\mu_a(\lambda_2) - \varepsilon_{\mathrm{HbO}}(\lambda_2)\mu_a(\lambda_1)}{\varepsilon_{\mathrm{HbO}}(\lambda_1)\varepsilon_{\mathrm{Hb}}(\lambda_2) - \varepsilon_{\mathrm{HbO}}(\lambda_2)\varepsilon_{\mathrm{Hb}}(\lambda_1)}, \tag{15.9}$$

from which one can express the absolute value of tissue hemoglobin saturation:

$$\mathrm{StO_2} = \frac{[\mathrm{HbO}]}{[\mathrm{HbT}]} = \frac{\varepsilon_{\mathrm{Hb}}(\lambda_2) - \varepsilon_{\mathrm{Hb}}(\lambda_1)\frac{\mu_a(\lambda_2)}{\mu_a(\lambda_1)}}{\varepsilon_{\mathrm{Hb}}(\lambda_2) - \varepsilon_{\mathrm{HbO}}(\lambda_2) + (\varepsilon_{\mathrm{HbO}}(\lambda_1) - \varepsilon_{\mathrm{Hb}}(\lambda_1))\frac{\mu_a(\lambda_2)}{\mu_a(\lambda_1)}}, \tag{15.10}$$

where $[\mathrm{HbT}] = [\mathrm{HbO}] + [\mathrm{Hb}]$ is the total hemoglobin concentration in tissue. It is important to note that a measurement of the absolute concentration of total hemoglobin in tissue requires the individual values of the absorption coefficients at two wavelengths (see Eqs. (15.8) and (15.9)), whereas a measurement of the tissue saturation only requires the ratio of the absorption coefficients at two wavelengths (see Eq. (15.10)). This means that a systematic error introduced by an unknown wavelength-independent factor in the measurement of the absorption coefficient would induce errors in the measured total hemoglobin concentration, but it would not affect the accuracy of the measurement of tissue Hb saturation. As a result, absolute measurements of StO_2 are typically more robust than absolute measurements of [HbO] and [Hb].

15.1.3.2 *Redox state of cytochrome c oxidase*

In addition to the oxygen saturation of hemoglobin (located only in red blood cells), which is an intravascular measure of blood oxygenation, diffuse optical spectroscopy can also sense the redox state of the cytochrome *c* oxidase complex (located in the inner mitochondrial membrane of cells), which is a measure of oxygen consumption at the cellular level. Cytochromes are proteins that, similar to hemoglobin, include a covalently bound heme group incorporating an iron atom. The functional role of iron in the heme group of hemoglobin is to carry oxygen, whereas in cytochromes it is to carry electrons. As introduced in Section 2.2, the role of cytochrome *c* oxidase in the transport of electrons is a key element in the energy cycle of cells and in the steps leading to adenosine triphosphate (ATP) synthesis in mitochondria. (ATP hydrolysis is the primary source of energy for cells to do work.) This is why mitochondria are often called the "energy engines" of cells, and the redox state of cytochrome *c* oxidase is a key indicator of that energetic activity. In fact, cythochromes act as electron carriers in the intracellular electron transport chain that transfers electrons from NADH to reduce oxygen (the terminal electron acceptor) and form water with the addition of two hydrogen ions. Therefore, the term *redox state* of cytochromes refers to their stoichiometry resulting from the processes of gaining an electron (reduction) or losing an electron (oxidation) in the electron transport chain.

The NIR optical absorption of cytochrome *c* oxidase is dominated by the copper A complex (Cu_A), a copper-based center that acts as the primary electron acceptor from cytochrome *c* and is in redox equilibrium with cytochrome *a*. Cytochrome *c* is a mobile electron carrier that conveys electrons to the cytochrome oxidase complex consisting of cytochrome *a* and cytochrome a_3, which are the last two steps in the electron transport chain. The NIR absorption spectrum of the Cu_A center features a broad band around ~830 nm, which is assigned to a charge-transfer transition between the two copper atoms in the Cu_A center (Larsson et al., 1995). The absorption-difference spectrum between normoxia and anoxia conditions in the brain of experimental animals after removal of blood (to eliminate the confounding absorption of hemoglobin) (Wray et al., 1988) is representative of the difference spectrum between oxidized and reduced cytochrome oxidase. Such a difference spectrum is illustrated in Figure 15.3 (from Kolyva et al., 2012), which shows the broad differential absorption peak centered at around 830 nm, and is tabulated in Appendix 2.A (Table 2.A.1). The reader should note that the units of the molar extinction coefficient in Figure 15.3 are OD $cm^{-1}M^{-1}$, indicating an attenuation by powers of 10 – the optical density, OD, is defined in terms of Log_{10} – so that (as in the case of Figure 2.4) it is a factor of ln(10) (~2.3) smaller than the corresponding extinction coefficient defined in terms of powers of *e*.

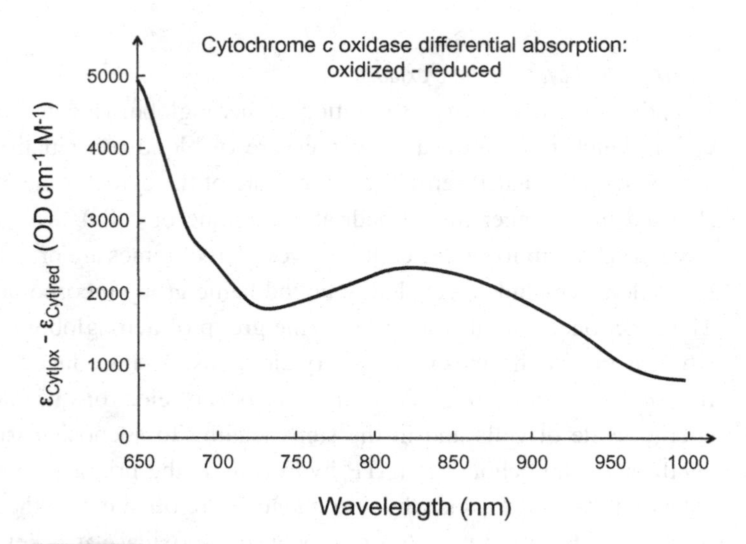

Figure 15.3

Difference extinction spectra of oxidized and reduced cytochrome c oxidase (from tabulated data in Kolyva et al., 2012, also reported in Appendix 2.A, Table 2.A.1). The optical density (OD) is a base-10 logarithmic measure of optical attenuation and it is related to the base-e absorbance (*A*) introduced in Eq. (10.11) by $A = \mathrm{OD} \times \ln(10)$.

By considering a constant total concentration of cytochrome *c* oxidase ([Cyt|tot]), a measurement of the change in concentration of oxidized cytochrome *c* oxidase (Δ[Cyt|ox]) provides a measure of a change in its redox state, which is a marker of mitochondrial oxygen consumption. This measurement can be accomplished by including a cytochrome *c* oxidase contribution in the Beer's law equation for the overall tissue absorption change:

$$\begin{aligned}\mu_a(\lambda) = \varepsilon_{\mathrm{HbO}}(\lambda)[\mathrm{HbO}] + \varepsilon_{\mathrm{Hb}}(\lambda)[\mathrm{Hb}] + \varepsilon_{\mathrm{Cyt|ox}}(\lambda)[\mathrm{Cyt|ox}] \\ + \varepsilon_{\mathrm{Cyt|red}}(\lambda)([\mathrm{Cyt|tot}] - [\mathrm{Cyt|ox}]) \qquad (15.11)\end{aligned}$$

where we have expressed the concentration of cytochrome in the reduced state ([Cyt|red]) as the difference between the constant total cytochrome concentration ([Cyt|tot]) and the concentration of cytochrome in the oxidized state ([Cyt|ox]). Differentiation of Eq. (15.11), assuming [Cyt|tot] to be constant, yields the following expression for tissue absorption changes associated with changes in the concentrations of oxy- and deoxyhemoglobin, as well as the redox state of cytochrome *c* oxidase:

$$\begin{aligned}\Delta\mu_a(\lambda) = \varepsilon_{\mathrm{HbO}}(\lambda)\Delta[\mathrm{HbO}] + \varepsilon_{\mathrm{Hb}}(\lambda)\Delta[\mathrm{Hb}] \\ + [\varepsilon_{\mathrm{Cyt|ox}}(\lambda) - \varepsilon_{\mathrm{Cyt|red}}(\lambda)]\Delta[\mathrm{Cyt|ox}]. \qquad (15.12)\end{aligned}$$

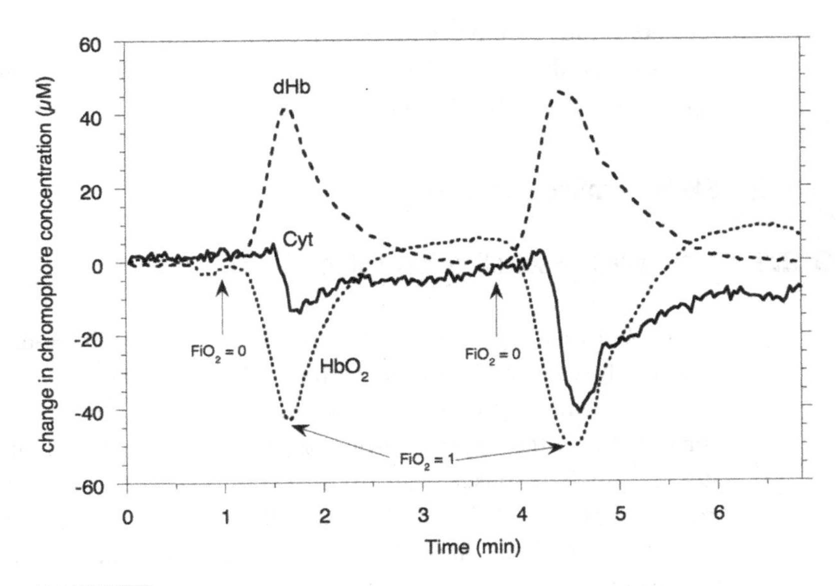

Figure 15.4

Time traces of changes in the concentrations of oxyhemoglobin (Δ[HbO], or HbO_2 in the figure), deoxyhemoglobin (Δ[Hb], or dHb in the figure), and oxidized cytochrome *c* oxidase (Δ[Cyt|ox], or Cyt in the figure) measured in the rat brain during rapid decrease in the fraction of inspired oxygen (FiO_2) at the times indicated by the vertical arrows. (Reprinted from Cooper et al., 1997 (Figure 5); Copyright 1997, Springer and Plenum Press, with permission from Springer Science and Business Media.)

On the basis of Eq. (15.12), spectral measurements at multiple discrete wavelengths or over a continuous band can quantify changes in the concentrations of oxyhemoglobin and deoxyhemoglobin, and in the redox state of cytochrome *c* oxidase. The NIR molar extinction coefficient of cytochrome *c* oxidase is comparable to (and even greater than) that of hemoglobin (for example, the extinction coefficient at 800 nm is ~2000 $M^{-1}cm^{-1}$ for hemoglobin and ~5200 $M^{-1}cm^{-1}$ for cytochrome *c* oxidase). However, the concentration of cytochrome *c* oxidase ([Cyt|tot]) is much smaller, typically 10% or less, than the concentration of hemoglobin ([HbT]) in tissue, resulting in a larger contribution of hemoglobin to the overall NIR absorption of tissue, even in highly cellular tissue with large numbers of mitochondria, as in brain tissue.

As an example of the complementary information provided by hemoglobin saturation (a measure of vascular oxygenation) and the redox state of cytochrome *c* oxidase (a measure of tissue oxygenation at the cellular level), Figure 15.4 shows the time traces of Δ[Hb], Δ[HbO], and Δ[Cyt|ox] measured in the rat brain during modulation of the fraction of inspired oxygen (FiO_2) (Cooper et al., 1997). The delayed decrease of Δ[Cyt|ox] with respect to Δ[HbO] indicates that the cerebral metabolic rate of oxygen does not change until the reduced oxygen delivery associated with reduced FiO_2, as indicated by the desaturation of hemoglobin, reaches a certain critical level.

15.2 Skeletal muscle studies

15.2.1 The oxygen dissociation curve of myoglobin

Myoglobin (already introduced in Section 2.2) is a single-unit protein that is structurally similar to each of the four subunits of the hemoglobin molecule. It contains one heme group, to which molecular oxygen can bind covalently. While hemoglobin is the oxygen-carrier protein in blood, myoglobin is the oxygen-storage protein in muscle cells. Hemoglobin is needed to achieve oxygen concentrations in blood (~9.2 mM) that are well above what may be achieved by oxygen solubility alone (~0.2 mM). Myoglobin is an intracellular protein in muscle tissue that readily binds oxygen and stores it, releasing it only when the oxygen tension in tissue reaches very low values (below ~10 mmHg). The greater oxygen affinity of myoglobin with respect to hemoglobin is illustrated in Figure 15.1. One of the interesting observations that can be made by inspection of Figure 15.1 is that, although myoglobin holds on to its oxygen until local tension is low, it has a much steeper dissociation curve than hemoglobin. That is, when the oxygen tension does drop, myoglobin can quickly release its oxygen. This explains why the locally stored oxygen (in the myoglobin) in skeletal muscle tissue is the source that is used for near-instantaneous and short-term muscle activity, such as lifting a weight or running a short sprint. (Such activity is often referred to as "anaerobic" because it occurs on a time-scale too short for oxygen replenishment by circulation from the lungs.)

The oxygen dissociation curve for Hb is described by the sigmoidal (S-shaped) function of Eq. (15.1), whereas the curve for Mb is described by a rectangular hyperbolic function:

$$S_{Mb}O_2 = \frac{pO_2}{pO_2 + p_{50}}, \tag{15.13}$$

which is formally identical to Eq. (15.1) with a Hill coefficient of 1, indicating the lack of cooperativity effects for the single-unit Mb molecule. The Bohr effect described in Section 15.1.1 for hemoglobin (i.e., the dependence of the oxygen affinity of hemoglobin on pH) is associated with cooperativity effects, and it does not apply to myoglobin. However, similarly to hemoglobin, temperature does affect the oxygen dissociation curve of myoglobin by shifting it to the right at higher temperatures. The value of p_{50} is ~2.4 mmHg for myoglobin, indicating its stronger oxygen affinity compared to hemoglobin (for which p_{50} is ~26 mmHg).

The optical absorption of myoglobin in the red and NIR is associated with charge transfer from the iron to the porphyrin in the heme group, exactly as with hemoglobin. Consequently, the shape of the extinction spectra of oxy-myoglobin and deoxy-myoglobin closely match those of oxyhemoglobin and deoxyhemoglobin. (Of course, as explained in Section 2.2, the molar extinction coefficient of Mb is $\sim 1/4$ that of Hb, given that there is only one heme group per molecule in Mb, versus four in Hb.) The slight deviations between the extinction spectral shapes of hemoglobin and myoglobin (a ~4 nm red shift of Mb vs. Hb spectra) have been exploited to measure myoglobin saturation in vivo on the exposed muscle in animal models using reflectance spectroscopy in the visible (Arakaki et al., 2007) and near-infrared (Schenkman et al., 1999) spectral regions. However, noninvasive optical measurements on muscles are incapable of discriminating Hb and Mb, so that such measurements effectively target the combined contributions (Hb+Mb) rather than either species individually.

Even though the concentration of myoglobin in human muscle may be of the order of 200 μM, twice the concentration of hemoglobin, as mentioned above Mb has a lower extinction coefficient than Hb. Moreover, Mb is localized only in muscle tissue, whereas other neighboring tissues that are also interrogated in noninvasive optical measurements (skin, fat, bone, etc.) do not contain Mb. Consequently, because of the more limited spatial distribution of Mb, and the smaller molar extinction coefficient of Mb vs. Hb, the resulting measurement is likely to result in a dominant contribution of Hb to noninvasive optical measurements of muscle tissue. Even though this is an open research question that still awaits a conclusive answer (a number of studies have reported Mb contributions to NIRS measurements in muscle ranging from $<25\%$ [Chance et al., 1992] to $\sim 80\%$ [Marcinek et al., 2007]), in what follows we do not explicitly consider Mb contributions to optical measurements on muscle tissue. In some cases, such as measurements of changes in hemoglobin concentration resulting from hemodynamic perturbations, this may be well justified. In other cases, such as absolute measurements of hemoglobin concentration and saturation in skeletal muscle, myoglobin contributions may play a role, and the above discussion should be kept in mind. In general, it is appropriate to treat the NIRS measurements of "Hb concentration" and "Hb saturation" in muscle as "(Hb+Mb) concentration" and "(Hb+Mb) saturation" to reflect the combined contributions of Hb and Mb to optical measurements.

15.2.2 Blood volume, blood flow, and oxygen consumption in skeletal muscle

The concentration of hemoglobin in tissue provides an indirect indication of the amount of blood per unit volume of tissue, defined as the *blood volume fraction* (units: ml_{blood}/ml_{tissue} or %), or, often, simply *blood volume*. In fact, the tissue blood volume (BV) is given by the concentration of hemoglobin in tissue ([HbT]) divided by the concentration of hemoglobin in blood (ctHb):

$$\mathrm{BV} = \frac{1}{\mathrm{ctHb}}[\mathrm{HbT}]. \tag{15.14}$$

(The calculation of an estimated concentration of hemoglobin in whole blood was presented in Section 2.2.) Equation (15.14) reflects the definitions of the quantities involved, and therefore provides a general relationship between blood volume and tissue concentration of hemoglobin. In the case of muscle tissue, muscle blood volume is denoted as mBV. The saturation of hemoglobin in tissue, which reflects a combination of the oxygen saturation in the arterial, capillary, and venous compartments, is a measure of the balance of the supply and utilization of oxygen in tissue. The oxygen supply is related to available oxygen in blood (as discussed above) and the *blood flow*, defined as the amount of blood flowing per unit time per unit volume of tissue (units: $\mathrm{ml_{blood}/(ml_{tissue}\text{-}min)}$). The oxygen utilization is quantified by the *oxygen consumption*, defined as the amount of oxygen consumed per unit time per unit volume of tissue (units: $\mathrm{mol_{O2}/(ml_{tissue}\text{-}min)}$). There are cases in which NIR spectroscopy (NIRS) measurements can noninvasively provide a measure of blood flow and oxygen consumption. One such case is the study of skeletal muscle in protocols involving blood circulation perturbations associated with venous or arterial occlusion.

The pulsatile blood pressure in arteries oscillates from a systolic value of ~120 mmHg to a diastolic value of ~80 mmHg, following the heartbeat; and the *mean arterial pressure* (MAP) is a weighted average of systolic and diastolic pressures, with weights of 1/3 and 2/3, respectively. The blood pressure in systemic capillaries and veins is negligibly pulsatile and features typical values of 20–30 mmH in capillaries and ~10 mmHg or less in veins.

The inflation of a pneumatic cuff placed around a subject's arm or leg will affect the circulation in distal tissues. Inflation to a pressure of ~50 mmHg does not affect the arterial circulation but blocks venous return. As a result, distal tissues feature an increase in blood volume at a rate that is initially proportional to the blood flow. This basic principle, used by *venous occlusion plethysmography* to measure forearm blood flow, can also be used for NIRS measurements of muscle blood flow (mBF) on the basis of the initial rate of increase of total hemoglobin concentration during venous occlusion (de Blasi et al., 1994):

$$\mathrm{mBF} = \frac{1}{\mathrm{ctHb}} \left. \frac{d[\mathrm{HbT}]}{dt} \right|_0, \tag{15.15}$$

where the subscript 0 indicates the initial (and highest) value of the time derivative (i.e., the rate of increase) of the total hemoglobin concentration in tissue measured with NIRS immediately following venous occlusion.

Inflation of a cuff to a pressure of 200 mmHg or higher ensures the occlusion of arteries as well (in a healthy subject), and all blood flow to and from the limb is completely blocked. Under such occlusion, the total blood volume in the limb cannot change (although a limited local change may occur as a result of blood redistribution within the limb), and the rate of hemoglobin desaturation provides a direct measure of the muscle oxygen consumption, $\mathrm{mV}_{\mathrm{O}_2}$. In fact, the rate of conversion of oxyhemoglobin to deoxyhemoglobin serves as a measure of the rate of oxygen diffusion from plasma to tissue. Under conditions of constant blood volume with no flow, as is the case for arterial occlusion, the rate of decrease of [HbO] coincides with the rate of increase of [Hb], and they can be combined in the expression for muscle oxygen consumption:

$$\mathrm{mV}_{\mathrm{O}_2} = 4\frac{d}{dt}\left(\frac{[\mathrm{HbO}] - [\mathrm{Hb}]}{2}\right), \tag{15.16}$$

where the factor 4 accounts for the four binding sites for oxygen on the hemoglobin molecule.

Frequency-domain NIRS measurements of absolute [HbO], [Hb], [HbT], and StO_2 in the human forearm (brachioradialis muscle) during venous occlusion at the upper arm are shown in Figure 15.5. In addition to the measurements of muscle blood flow and oxygen consumption associated with venous and arterial occlusions, respectively, Figure 15.5 shows some relevant quantitative results that are worth a closer look. The baseline value of concentration of hemoglobin in muscle tissue ([HbT]) is typically 70–100 μM, indicating a blood volume fraction of ~3–4% (by considering a typical hemoglobin concentration in blood of ~2.3 mM). The baseline value of muscle tissue saturation (StO_2) is typically 70–80%, because it is indicative of a weighted average of the highly oxygenated arterial blood, the capillary blood featuring longitudinal blood deoxygenation, and the deoxygenated venous blood.

During venous occlusion, StO_2 shows a small decrease associated with the accumulation of venous blood, whereas during arterial occlusion StO_2 decreases significantly as a result of the imbalance between oxygen supply (which is interrupted) and oxygen utilization (which continues). StO_2 reaches a plateau, during arterial occlusion, at a value of ~40 mmHg, which is indicative of a tissue oxygen tension of ~25 mmHg, according to the oxygen dissociation curve for hemoglobin of Figure 15.1. (This is an estimate, because StO_2 is determined not only by the blood in the capillary bed from which oxygen diffuses to tissue, but also by arterial and venous blood. Consequently, care should be taken in translating StO_2 into tissue pO_2 by using the oxygen dissociation curve.) Muscle exercise will further reduce the value of StO_2 from the value at plateau by further decreasing the tissue pO_2 (in Mb as well as Hb). The overshoot in StO_2 after release of arterial occlusion

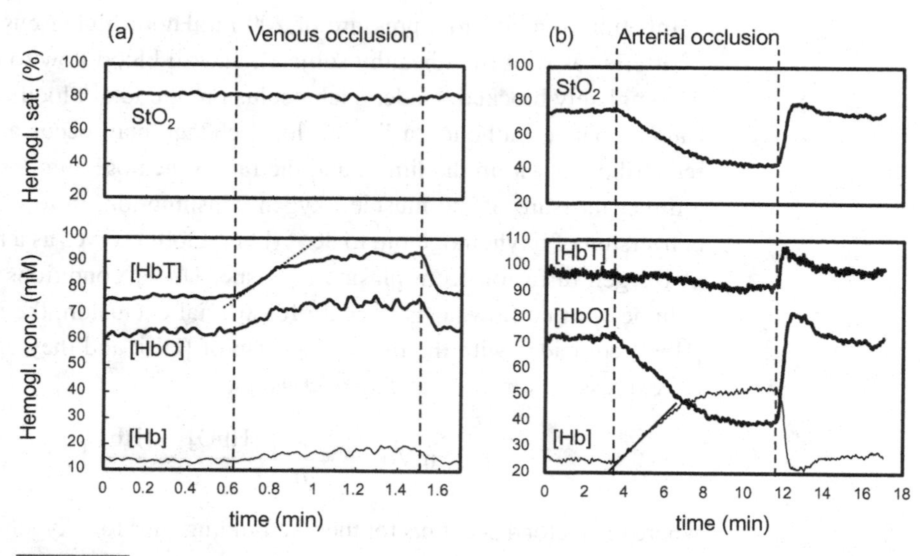

Figure 15.5

Frequency-domain NIR spectroscopy (NIRS) measurements on human skeletal muscle (brachioradialis in the forearm) during a protocol involving the inflation of a pneumatic cuff around the upper arm to achieve (a) venous occlusion (cuff pressure: 50 mmHg) and (b) arterial occlusion (cuff pressure: 220 mmHg). The measured quantities are the absolute values of the concentration of oxyhemoglobin ([HbO]), deoxyhemoglobin ([Hb]), total hemoglobin ([HbT]), and hemoglobin saturation (StO_2) in tissue. The initial rate of increase of [HbT] following venous occlusion (shown by the dotted line in panel (a)) results in a muscle blood flow measurement of 2.5 $ml_{blood}/(100\ ml_{tissue}\text{-min})$. The initial rate of increase of [Hb] during arterial occlusion (shown by the solid line segment in panel (b)) results in a muscle oxygen consumption measurement of 3.2 $\mu mol_{O2}/(100\ ml_{tissue}\text{-min})$.

results from a hyperemic response, which is also reflected in the increase of blood volume, or [HbT], following the relatively constant blood volume regime during arterial occlusion (see Figure 15.5(b)).

15.2.3 NIRS measurements of muscle metabolism during exercise

Near-infrared spectroscopy, with its sensitivity to the oxygen saturation of hemoglobin and myoglobin, and the redox state of cytochrome *c* oxidase, is an ideal noninvasive technique to study muscle activity, which is strongly correlated with oxidative metabolism. A baseline assessment of the concentration and saturation of Hb+Mb in muscle tissue requires absolute optical measurements such as those afforded by diffuse optical spectroscopy in the time domain (see Eqs. (11.26) and (11.27)) or in the frequency domain (see Eqs. (12.13) and (12.14)). By contrast, the effects of temporal perturbations (for example, as induced by vascular occlusion or

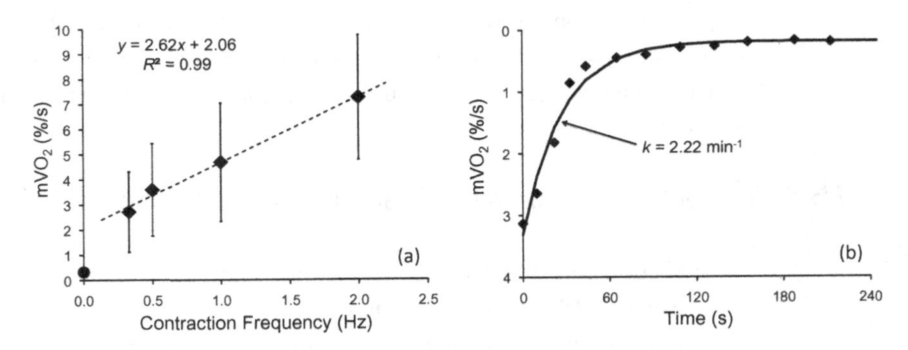

Figure 15.6

NIRS measurements of muscle oxygen consumption (mVo_2) (human medial gastrocnemius) in a plantar-flexion exercise protocol. (a) Increase in mVo_2 as a function of contraction frequency (zero frequency signifies rest conditions); (b) recovery to baseline following exercise. In this case, mVo_2 is obtained from relative measurements of muscle [HbO] (expressed as % of the maximal [HbO] variability recorded in a calibration measurement), so that the expression in Eq. (15.16) results in relative units of %/s for mVo_2. (Reprinted from Ryan et al., 2013, with permission; copyright 2013, The American Physiological Society.)

muscle exercise) may be assessed by measurements of relative changes in the muscle absorption coefficient from a given reference condition. This is a case in which a continuous-wave measurement, analyzed using the modified Beer-Lambert law (Eq. (10.14)), can quantify hemoglobin concentration and saturation changes.

It is important to recall that the application of the modified Beer-Lambert law requires knowledge of the differential pathlength factor (DPF), defined as the ratio between the mean photon pathlength (for the examined tissue at the measured wavelength) and the geometric source-detector separation. In the case of human thigh muscle tissue, the DPF is ~5.2 at 690 nm and ~4.5 at 830 nm, and it decreases by 5–7% during exercise (Ferreira et al., 2007). The measurement of muscle oxygen consumption (mVo_2) with the arterial occlusion protocol and Eq. (15.16) allows for the assessment of mitochondrial capacity by monitoring the recovery of mVo_2 after exercise. This is accomplished using multiple arterial occlusions over time. Such recovery may be described by a single exponential function:

$$mVo_2(t) = mVo_{2|Rest} - \left(mVo_{2|Rest} - mVo_{2|Ex}\right) e^{-kt}, \qquad (15.17)$$

where $mVo_{2|Rest}$ and $mVo_{2|Ex}$ are the muscle oxygen consumption at rest and immediately following exercise (time $t = 0$ defined at end of exercise), respectively, and k is the recovery rate constant, which represents the maximal muscle oxidative capacity, a measure of muscle mitochondrial function.

Figure 15.6(a) shows the linear increase of the oxygen consumption of the medial gastrocnemius (calf) muscle as a function of contraction frequency in a plantar-flexion exercise, whereas Figure 15.6(b) shows the single exponential recovery to baseline following exercise, featuring a rate constant of 2.22 min^{-1}, which can be used as a measure of mitochondrial function (Ryan et al., 2013).

15.2.4 Confounding factors in optical studies of skeletal muscle

In Section 15.2.1, we stated that the quantitative role played by myoglobin in noninvasive NIRS muscle studies is still an open question, which requires an accurate assessment of myoglobin content in muscle. There are at least two additional confounding factors to be considered in noninvasive optical measurements of muscle tissue. The first has to do with possible scattering changes associated with blood volume effects and/or with microscopic morphological tissue changes associated with muscle exercise. This issue can be addressed by quantifying exercise-induced scattering changes in muscle tissue with time-resolved measurements in either the time domain (Torricelli et al., 2004) or the frequency domain (Ferreira et al., 2007).

The second confounding factor is related to the skin and subcutaneous adipose tissue layers, which may provide significant contributions to the NIRS signal. To fully address this problem, the assumption of tissue homogeneity must be abandoned to realize depth-resolved measurements that can specifically target the muscle tissue located underneath the skin and lipid layers. Alternatively, a suitable correction factor, dependent on the thickness of the adipose tissue layer, may be introduced (Niwayama et al., 2000), or optical data collected at multiple source-detector separations may be combined to minimize sensitivity to the most superficial tissue layer (Franceschini et al., 1998).

15.3 Functional brain investigations

15.3.1 Effects of brain activation on hemoglobin and oxidized cytochrome concentrations

The NIR optical signatures associated with the oxygen saturation level of hemoglobin and with the redox state of cytochrome *c* oxidase (discussed above in Section 15.1) were the basis for Frans Jöbsis's pioneering work that showed the feasibility and significance of noninvasive optical spectroscopy of brain tissue in the mid 1970s (Jöbsis, 1977). Near-infrared light can penetrate the intact scalp and skull to probe the human brain cortex just as sunlight can be partially transmitted through the clouds to illuminate the surface of the earth on a cloudy day. Scalp and skull tissues present as strong scatterers to light delivered at the scalp surface, like the cloud layer in the sunlight transmission analogy, but some scattered photons do reach the brain cortex layer and find their diffusive way back (along the "photon banana") to the scalp surface, where they can be detected. These diffusely "reflected" photons carry information about the optical properties of brain tissue. The absorption properties, in turn, depend on hemoglobin concentration (as determined by cerebral blood volume [CBV]), hemoglobin saturation (as determined by arterial saturation [SaO_2], cerebral blood flow [CBF], the cerebral metabolic rate of oxygen [$CMRO_2$]), and the redox state of cytochrome *c* oxidase (as determined by cellular oxygen utilization). As a result, noninvasive NIR spectroscopy (NIRS) can perform regionally sensitive hemodynamic-based functional studies of the brain, with the additional capability of sensing cellular metabolism, something that was initially realized and developed in the early 1990s (Chance et al., 1993; Villringer et al., 1993; Hoshi and Tamura, 1993; Kato et al., 1993). This technique is known as *functional NIRS*, or fNIRS. The qualitative effect of individual physiological parameters on quantities measurable with fNIRS can be intuitively described as follows:

- a decrease in SaO_2 causes a decrease in [HbO], an increase in [Hb], and a decrease in [Cyt|ox] (because of the smaller amount of oxyhemoglobin and larger amount of deoxyhemoglobin in the incoming arterial blood, resulting in a lower tissue oxygen tension);
- an increase in CBV causes an increase in [HbT] (because of the larger amount of blood per unit volume of tissue);
- an increase in CBF causes an increase in [HbO], a decrease in [Hb], and an increase in [Cyt|ox] (because the increased rate of oxyhemoglobin flow results in a wash-out of deoxyhemoglobin and an increase in oxygen delivery and tissue oxygen tension);
- an increase in $CMRO_2$ causes a decrease in [HbO], an increase in [Hb], and a decrease in [Cyt|ox] (because the increased oxygen consumption translates into a larger number of oxyhemoglobin molecules being converted into deoxyhemoglobin, and a decrease in the amount of oxidized cytochrome *c* oxidase);
- an increase in cellular metabolism causes a decrease in [Cyt|ox] (because a lower oxygen tension in tissue moves the redox state of cytochrome *c* oxidase toward its reduced form).

The above observations indicate the opposing effects that various physiological and metabolic changes may have on the quantities measurable with fNIRS. It is therefore often difficult to identify the relative contributions of physiological and metabolic changes to the measured fNIRS signals. Furthermore, it is often not easy to draw indications about the temporal and dynamic relationships between functional processes and fNIRS-measured quantities. The situation is even more

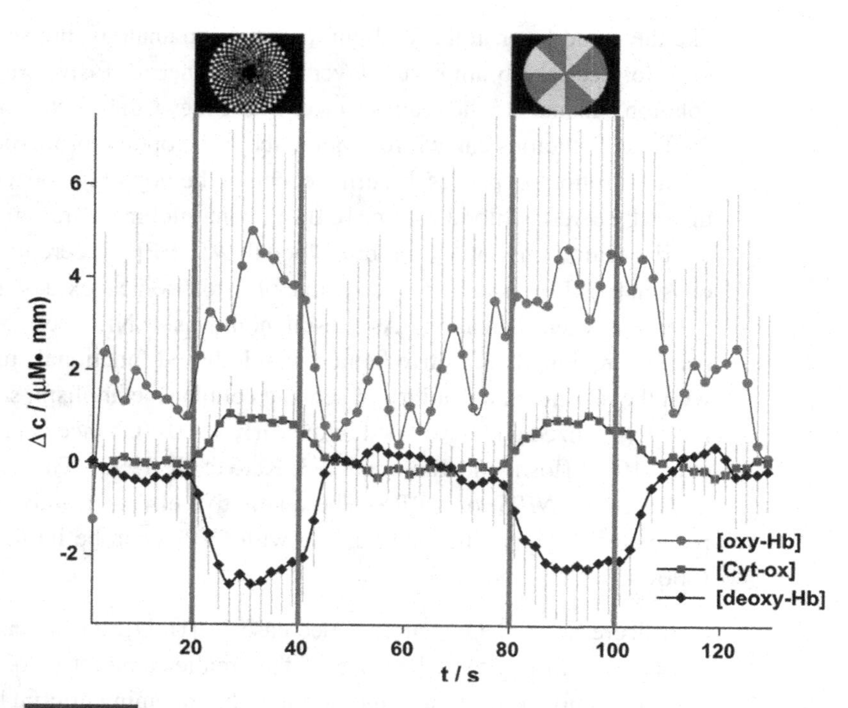

Figure 15.7

Representative time traces of cerebral concentration changes (Δc) of oxyhemoglobin ([oxy-Hb]), deoxyhemoglobin ([deoxy-Hb]) and oxidized cytochrome *c* oxidase ([Cyt-ox]) measured with fNIRS on the human visual cortex (right occipital area) in response to two visual stimuli. The first stimulus, between 20 and 40 s, is a form stimulus achieved by a rotating disk featuring an angular black and white checkerboard. The second stimulus, between 80 and 100 s, is a color stimulus achieved by rotating a disk with alternating red and green sectors. (Reprinted from Uludağ et al., 2004; copyright 2004, with permission from Elsevier.)

complicated for cytochrome *c* oxidase than hemoglobin, since NADH levels can further affect the redox state of cytochrome *c* oxidase. Nevertheless, the qualitative observations above may be used to guide the interpretation of optical responses to brain activation. For example, after data analysis based on Eq. (15.12), the optical signal measured in response to visual stimulation in human subjects results in the hemoglobin and oxidized cytochrome concentration response of Figure 15.7 (Uludağ et al., 2004). The increase in [HbO] and decrease in [Hb] are well-established responses to brain activation, and, for the most part, so is the increase in [Cyt|ox]. However, this latter effect may show some subject- or location-dependent variability (Kolyva et al., 2012). The representative results of Figure 15.7 (increase in [HbO], decrease in [Hb], increase in [Cyt|ox]) indicate that the increase in blood flow elicited by brain activation has a dominant effect over the increase in blood volume and oxygen consumption.

We remind the reader that there is also physiological information available from the measured scattering properties of tissue. The scattering properties depend on factors like cell density, structure of nuclei, condition of the extracellular matrix, cell swelling, and other microscopic morphological features that are often observed in histology. Still, most of the published reports on functional brain imaging with diffuse reflectance measurements have focused on parameters extractable from the absorption properties as described in this section. A notable exception is the reported fast optical signal that has been shown to be directly proportional to the transmembrane potential in cultured neurons (Stepnoski et al., 1991), and that has also been observed noninvasively in human subjects (Gratton and Fabiani, 2003).

15.3.2 Modeling the hemoglobin concentration dependence on CBV, CBF, and $CMRO_2$

As we have seen in the previous section (15.3.1), the basic hemodynamic and metabolic signatures of brain activation, namely cerebral blood volume (CBV), cerebral blood flow (CBF), and cerebral metabolic rate of oxygen ($CMRO_2$), have competing effects on the measurable concentrations of oxyhemoglobin ([HbO]) and deoxyhemoglobin ([Hb]) in cerebral tissue. A quantitative model of the effects of CBV, CBF, and $CMRO_2$ on [HbO] and [Hb] can help identify the physiological and metabolic origins of measured [HbO] and [Hb] changes in response to brain activation. Furthermore, a dynamic model can also give insight into the relative timing of physiological and metabolic processes. One such model is based on conceptually breaking up the cerebral vasculature into three compartments: an arterial compartment (a) and a venous compartment (v) from which no oxygen diffuses out to tissue, and a capillary compartment (c) (including small arterioles), from which oxygen diffuses out to tissue and wherein hemoglobin becomes desaturated. Because CBF and $CMRO_2$ have no effect on the total concentration of hemoglobin (given that any changes in the rate of hemoglobin desaturation would only affect the stoichiometry of Hb and HbO, not their total concentration), [HbT] is a direct measure of CBV:

$$[\mathrm{HbT}](t) = \mathrm{ctHb}\,\{\mathrm{CBV}_0 + \Delta\mathrm{CBV}(t)\}, \qquad (15.18)$$

where ctHb is the concentration of hemoglobin in blood, CBV_0 is the baseline cerebral blood volume, and $\Delta CBV(t)$ is its time-dependent perturbation. Considering that $CBV(t) = CBV_0 + \Delta CBV(t)$, Eq. (15.18) is consistent with the general relationship of Eq. (15.14). Similarly, we indicate with CBF_0 and $CMRO_2|_0$ the baseline cerebral blood flow and metabolic rate of oxygen, and with $\Delta CBF(t)$ and $\Delta CMRO_2(t)$ their time-varying perturbations. After introducing notation for the arterial saturation (SaO_2), the average capillary saturation (ScO_2), and the venous saturation (SvO_2), the following dynamic relationships have been derived (Fantini, 2014a; Fantini, 2014b):

$$\begin{aligned}[\mathrm{HbO}](t) = \mathrm{ctHb}\{&\mathrm{SaO_2}[\mathrm{CBV}_0^{(a)} + \Delta\mathrm{CBV}^{(a)}(t)] + \mathrm{ScO_2}\mathcal{F}^{(c)}\mathrm{CBV}_0^{(c)} \\ &+ \mathrm{SvO_2}[\mathrm{CBV}_0^{(v)} + \Delta\mathrm{CBV}^{(v)}(t)]\} \\ &+ \mathrm{ctHb}\left[\frac{\mathrm{ScO_2}}{\mathrm{SvO_2}}(\mathrm{ScO_2} - \mathrm{SvO_2})\mathcal{F}^{(c)}\mathrm{CBV}_0^{(c)} h_{RC-LP}^{(c)}(t)\right. \\ &\left. + (\mathrm{SaO_2} - \mathrm{SvO_2})\mathrm{CBV}_0^{(v)} h_{G-LP}^{(v)}(t)\right] * \left(\frac{\Delta\mathrm{CBF}(t)}{\mathrm{CBF}_0} - \frac{\Delta\mathrm{CMRO_2}(t)}{\mathrm{CMRO_2}|_0}\right)\end{aligned} \quad (15.19)$$

$$\begin{aligned}[\mathrm{Hb}](t) = \mathrm{ctHb}\{&(1 - \mathrm{SaO_2})[\mathrm{CBV}_0^{(a)} + \Delta\mathrm{CBV}^{(a)}(t)] + (1 - \mathrm{ScO_2})\mathcal{F}^{(c)}\mathrm{CBV}_0^{(c)} \\ &+ (1 - \mathrm{SvO_2})[\mathrm{CBV}_0^{(v)} + \Delta\mathrm{CBV}^{(v)}(t)]\} \\ &- \mathrm{ctHb}\left[\frac{\mathrm{ScO_2}}{\mathrm{SvO_2}}(\mathrm{ScO_2} - \mathrm{SvO_2})\mathcal{F}^{(c)}\mathrm{CBV}_0^{(c)} h_{RC-LP}^{(c)}(t)\right. \\ &\left. + (\mathrm{SaO_2} - \mathrm{SvO_2})\mathrm{CBV}_0^{(v)} h_{G-LP}^{(v)}(t)\right] * \left(\frac{\Delta\mathrm{CBF}(t)}{\mathrm{CBF}_0} - \frac{\Delta\mathrm{CMRO_2}(t)}{\mathrm{CMRO_2}|_0}\right)\end{aligned} \quad (15.20)$$

where $\mathcal{F}^{(c)}$ is the Fåhraeus factor (named after Swedish pathologist Robin Sanno Fåhræus [1888–1968]), which is the small-to-large vessel hematocrit ratio that takes into account the reduced hemoglobin concentration in capillaries with respect to large blood vessels ($\mathcal{F}^{(c)} \sim 0.8$), and $*$ indicates a convolution product (i.e., $f(t) * g(t) = \int_{-\infty}^{\infty} f(\tau)g(t-\tau)d\tau$). The time-dependent functions $h_{RC-LP}^{(c)}(t)$ and $h_{G-LP}^{(v)}(t)$ in Eqs. (15.19) and (15.20) are impulse response functions that describe the low-pass (LP) nature of the capillary and venous contributions, respectively, to the tissue hemoglobin concentration changes in response to blood flow and oxygen consumption perturbations. The capillary response is approximated with a first-order low-pass filter that is used in circuit theory to describe the voltage across the capacitor in an RC (resistor-capacitor) series circuit. The time constant of $h_{RC-LP}^{(c)}(t)$ is $\sim t^{(c)}/e$, where $t^{(c)}$ is the blood transit time in capillaries. The venous response is approximated with a first-order Gaussian (G) low-pass filter with a characteristic time of $\sim(t^{(c)} + t^{(v)})/2$, where $t^{(v)}$ is the venous blood transit time. Because $t^{(c)}$ is ~0.5–1 s and $t^{(v)}$ may extend up to 3 s or more, the time constant associated with the microvascular response function is of the order of ~1 s. This is comparable to the time scale of the hemodynamic response to brain activation (a few seconds) seen in studies, as plotted in Figure 15.7. Equations (15.19) and (15.20) show that it is the difference of the perturbations in CBF and

$CMRO_2$ that determines the concentrations of HbO and Hb measured with fNIRS, so that fNIRS, by itself, cannot discriminate the individual perturbations of blood flow and oxygen consumption associated with brain activation.

15.3.3 Diffuse correlation spectroscopy (DCS) to measure cerebral blood flow

The inability of fNIRS to discriminate the effects of CBF and $CMRO_2$ on the measured concentrations of Hb and HbO accounts for the significance of alternative methods to measure cerebral blood flow and thus complement fNIRS. One such method, described in Sections 8.9.1 and 9.9, is diffuse correlation spectroscopy (DCS), a diffuse optical method that provides noninvasive, continuous measurements of relative cerebral blood flow (rCBF) on the basis of an effective diffusion coefficient (typically based on a Brownian motion model) for the moving scatterers in blood. As described in Section 9.9, the basis for DCS is the correlation diffusion equation for the electric field autocorrelation function, G_1, which, in its normalized form g_1, is linked to the measured normalized intensity autocorrelation function, g_2, by the Siegert relation (Eq. (7.50)). The cerebral blood flow sensed by DCS is mainly associated with the cerebral microvasculature, and DCS is therefore especially suited to complement fNIRS, in which measurements are also mostly affected by microvascular blood volume and blood flow. Figure 15.8(a) illustrates a typical optical probe arrangement for concurrent fNIRS and DCS measurements on the human brain, whereas Figure 15.8(b) shows the cerebral blood flow increase in the motor cortex (measured with DCS) corresponding to the increase in [HbO] and decrease in [Hb] (measured with fNIRS) during a finger-tapping protocol (Durduran et al., 2004).

15.3.4 Optical imaging of intrinsic signals (OIS): high-resolution brain mapping

When high-resolution optical imaging of brain activity is desired, measurements cannot be made through the intact scalp and skull. Higher spatial resolution of neuronal activity, correlated with more finely localized perturbations of tissue optical properties (including hemodynamics and scattering properties), requires exposure of and closer access to the cortex itself.

Optical imaging of intrinsic signals (OIS) is a powerful, albeit invasive (hence, limited to animal models or surgical scenarios), technique for imaging of brain function on the basis of intrinsic optical contrast, as opposed to methods based on extrinsic optical contrast provided, for example, by calcium indicators or voltage-sensitive dyes. OIS, a wide-field imaging version of diffuse reflectance

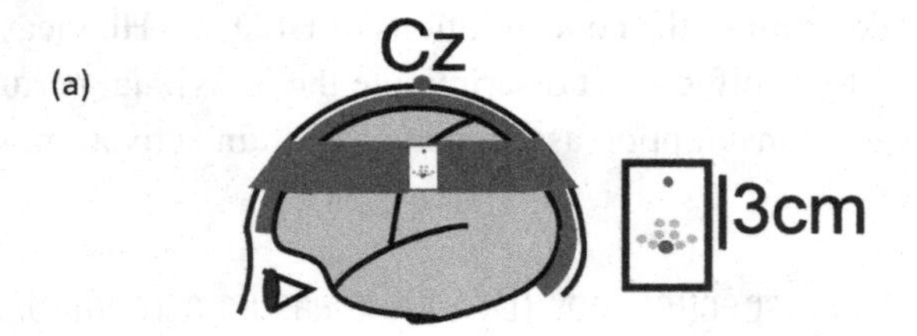

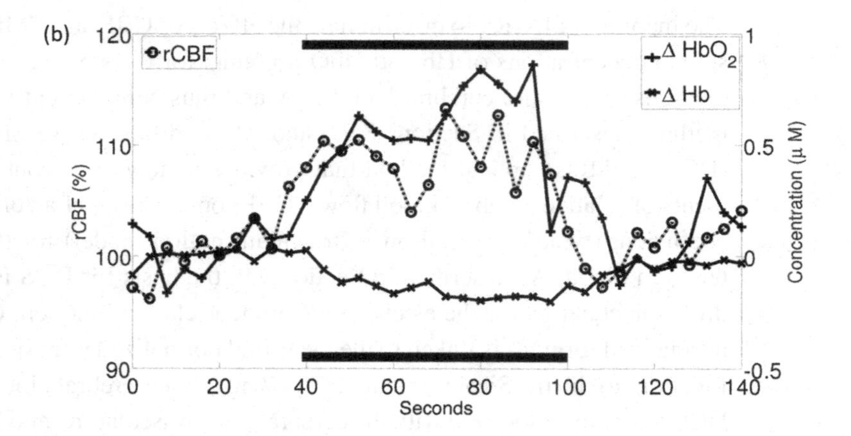

Figure 15.8

(a) Cross section of a human head showing the C_z reference point according to the International 10–20 system ("C" indicates the central location between nasion and inion, and the subscript "z" indicates the midline); the inset shows an enlargement of the optical probe for concurrent diffuse correlation spectroscopy (DCS) and functional NIR spectroscopy (fNIRS) on the motor cortex of a human subject. (b) Results of relative cerebral blood flow (rCBF) (from DCS) and hemoglobin concentration changes (oxyhemoglobin: ΔHbO_2; deoxyhemoglobin: ΔHb) (from fNIRS) during a 60 s finger-tapping period indicated by the thick horizontal lines. (Reprinted from Durduran et al., 2004. with permission; copyright 2004, The Optical Society of America.)

spectroscopy at small source-detector separation (see Chapter 8), was developed in the mid 1980s (Grinvald et al., 1986; Ts'o et al., 1990; Frostig et al., 1990). The method provides a high spatial resolution (10–100 μm) over a large field of view. It is a reflectance imaging technique consisting of broad-beam illumination of the brain cortex after skull thinning or removal (craniotomy).

According to the notation introduced in Section 10.4.1, under plane-wave illumination conditions, methods based on OIS measure the diffuse reflectance $\bar{R}_d$, which is given by the three-flux theory expression of Eq. (10.17) (in terms of the single-scattering albedo $a = \mu_s/(\mu_s + \mu_a)$), and by the diffusion theory expression of Eq. (10.26) (in terms of the reduced single-scattering albedo $a' = \mu_s'/(\mu_s' + \mu_a)$). The dependence of $\bar{R}_d$ on the single-scattering albedo for isotropic scattering (a particular case in which $a = a'$) was illustrated in Figure 10.6 for transport theory, three-flux theory, and diffusion theory, and it is evident that $\bar{R}_d$ decreases with absorption and increases with scattering. (See also extensive analysis of diffuse reflectance at small source-detector separations in Chapter 8.) It is this dependence on the intrinsic optical properties of cortical tissue that is the basis for OIS.

Because the geometry of OIS is not commensurate with analytical methods of the diffusion regime and only the superficial cortical layer is interrogated, OIS can operate at shorter wavelengths in the visible, where optical absorption by hemoglobin is significantly greater than in the NIR. For example, measurements at the isosbestic points of hemoglobin absorption at 545 and 570 nm (where the extinction coefficients of Hb and HbO are equal; see Figure 2.4) are used to maximize sensitivity to blood volume, whereas measurements in the range 600–650 nm, where Hb and HbO absorption differ significantly, maximize sensitivity to blood oxygenation. Wavelengths longer than 800 nm are less commonly used in OIS because, in the NIR, light scattering contributions become more significant relative to the lower absorption by hemoglobin. Therefore, the *intrinsic signals* to which OIS is sensitive are the hemodynamic and metabolic changes associated with neuronal activity, of which Figures 15.4 and 15.7 show some examples, as well as optical scattering changes. While OIS does not directly measure neuronal activity, it is capable of generating perfusion-based functional representations of the brain cortex with high spatial resolution. (We note that invasive optical measurements of neuronal activity, based on perturbations of local scattering properties, with high temporal resolution have been reported, although these are essentially point measurements rather than wide-field images [e.g., Rector et al., 1997].)

The instrumentation for OIS based on absorption changes (illustrated in Figure 15.9(a)) consists of:

- a stable light source, typically a tungsten-halogen lamp with a regulated power supply, which is band-pass filtered to select the wavelength band of interest, and coupled to flexible fiberoptic light guides for cortical illumination; discrete wavelength light sources, such as laser diodes or LEDs, can also be used for the illumination;
- an imaging lens that is focused on the cortical surface for visualization of blood vessels or several hundred microns below the cortical surface for functional mapping;
- a CCD camera detector, with large well capacity ($>2\times10^5$ electrons per pixel) to accommodate the large dynamic range of signals detected in OIS and maximize the signal-to-noise ratio.

One of the major sources of image degradation, cortical movement, is typically minimized by a glass plate placed on top of the cortex and by synchronizing

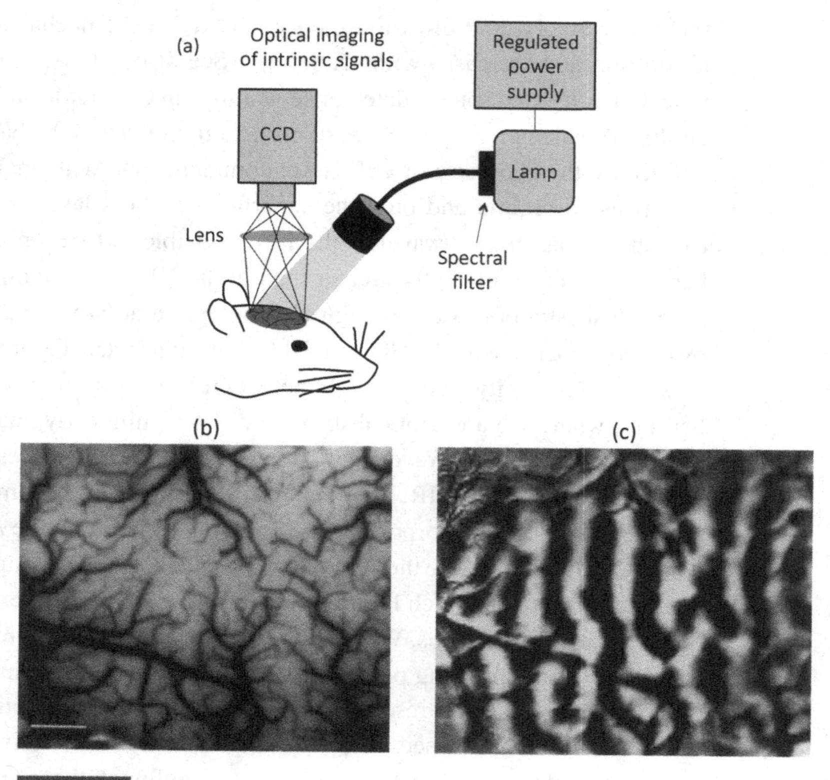

Figure 15.9

(a) Experimental setup for optical imaging of intrinsic signals (OIS) of the exposed brain cortex. (b) Baseline cortical surface imaged in a monkey at 570 nm (an isosbestic point for hemoglobin absorption) showing superficial vasculature (bar = 1 mm). (c) Ocular dominance columns of the monkey cortex obtained by dividing the responses to left eye and right eye stimulations (stimulus was a drifting bar). (Panels (b) and (c) are reprinted from Frostig et al., 1990, with permission.)

the image acquisition with the heartbeat and respiratory cycle (Frostig et al., 1990). Figures 15.9(b) and (c) report a baseline cortical image showing superficial vasculature and ocular dominance columns, respectively, measured on the exposed cortex of a monkey (Frostig et al., 1990). The high spatial resolution of OIS allows for the visualization of functional units having a size of the order of 100 μm on the brain cortex.

15.3.5 Functional near-infrared imaging (fNIRI): noninvasive brain mapping

While optical imaging of intrinsic signals (OIS), described in the previous section, is a powerful method for high-resolution functional mapping of the brain cortex, its obvious drawback for human applications is its invasiveness, since removal of the skull (craniotomy) is required. Analogous to spectroscopic measurements on muscle discussed above (Sections 15.1 and 15.2), functional NIR imaging (fNIRI) is a noninvasive diffuse optical technique that, as with OIS, relies on intrinsic hemodynamic, metabolic, and scattering signals, but operates through the intact scalp and skull. fNIRI is the imaging extension of fNIRS; i.e., it realizes spatially resolved fNIRS measurements. The noninvasiveness of fNIRI, however, comes at the expense of a much reduced spatial resolution (when compared with OIS), typically on a scale of several mm to ~1 cm. Nonetheless, the compact and portable optical instrumentation for fNIRI, together with its noninvasiveness, accounts for the variety of fNIRI applications, including the diagnostics of vulnerable populations, studies of brain development in infants, and functional/cognitive tests in everyday activities. This latter capability renders fNIRI an *ecologically valid* functional imaging technique, where ecological validity is a concept and term developed in the field of psychology to indicate studies that closely simulate real-world conditions.

In the most straightforward approach, fNIRI images may be obtained by a two-step procedure: (1) fNIRS measurements for each source-detector pair are back-projected (see Section 14.2) onto the measured tissue with a spatially dependent weight function given by the relevant sensitivity function (defined by Eq. (14.1)); (2) the spatially weighted backprojections for all source-detector pairs are linearly combined and projected onto a 2D surface to generate the final 2D image. For the case in which the detected signal is the optical intensity (I) and the quantity of interest is the absorption perturbation ($\Delta\mu_a$), the relevant sensitivity function is $s_{I,\mu_a}(x, y)$, as defined in Eq. (14.1), where (x, y) indicates here the generic 2D image pixel (representing the projection of 3D structures at depths z that are optically sensed). By considering a set of N source-detector pairs, labeled by the index i, the absorption perturbation image resulting from this backprojection and 2D rendition approach is given by:

$$\Delta\mu_a(x, y) = \frac{\sum_{i=1}^{N} s_{I,\mu_a}^{(i)}(x, y)\, \Delta\mu_a^{(i)}}{\sum_{i=1}^{N} s_{I,\mu_a}^{(i)}(x, y)}. \tag{15.21}$$

In Eq. (15.21) $\Delta\mu_a^{(i)}$ is the absorption perturbation measured by the i-th source-detector pair, and it can be obtained from the associated intensity perturbation ($\Delta I^{(i)}$) by using the differential form of the modified Beer-Lambert law (derived in Chapter 10 as Eq. (10.14)):

$$\Delta\mu_a^{(i)} = -\frac{1}{r^{(i)}\,\mathrm{DPF}} \frac{\Delta I^{(i)}}{I_0^{(i)}}, \tag{15.22}$$

where $r^{(i)}$ is the i-th source-detector distance, DPF is the differential pathlength factor (which is wavelength dependent) introduced in Section 10.3, and $I_0^{(i)}$ is the

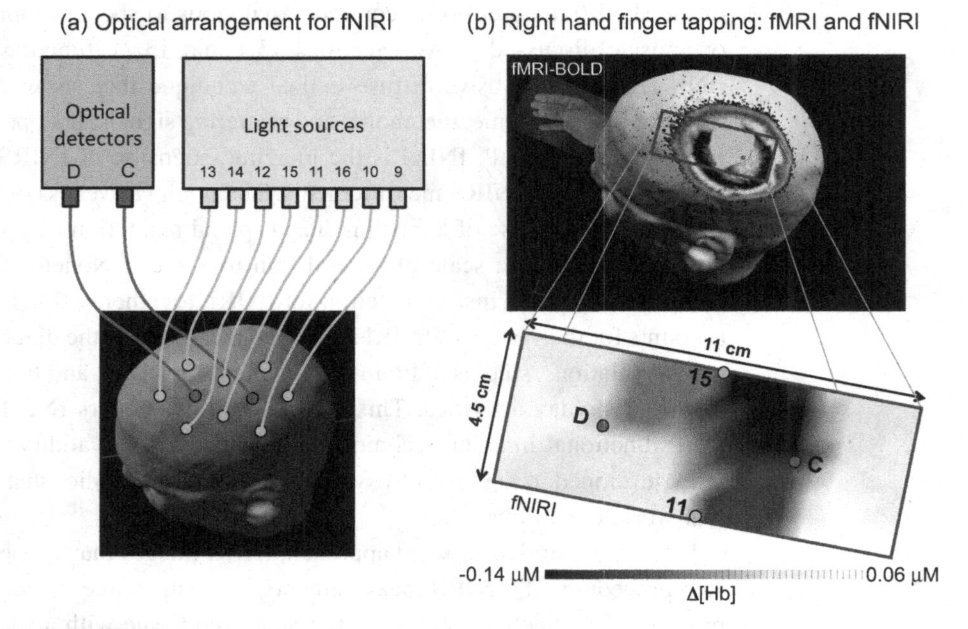

Figure 15.10

(a) Arrangement of a set of illumination optical fibers (labeled 9–16) and collection optical fibers (labeled C and D) on the scalp of a human subject for functional NIR imaging (fNIRI), also referred to as optical topography. (b) Functional brain mapping during a right-hand finger-tapping protocol. The top panel shows the activated cortical area as measured with functional magnetic resonance imaging (fMRI) based on blood oxygenation level dependent (BOLD) contrast. The bottom panel shows the map of deoxyhemoglobin concentration change (Δ[Hb]) measured with fNIRI. (Panel (b) is reprinted from Sassaroli et al., 2006; copyright (2006), with permission from Elsevier.)

reference intensity value for the i-th source-detector pair (typically a baseline or average value). In the NIR, the DPF of brain tissue typically ranges between 4 and 6, with a general decreasing trend for increasing wavelength.

An example of the optical arrangement of sources and detectors for fNIRI is shown in Figure 15.10(a). In this case, eight illumination optical fibers (labeled 9–16) and two detector optical fibers (labeled C and D) are placed on the scalp of a human subject covering an area that includes the primary motor cortex in the rear portion of the frontal lobe. Figure 15.10(b) shows the cortical area activated by right finger tapping as measured with fMRI (functional magnetic resonance imaging: top panel) and with fNIRI (bottom panel) according to the backprojection method described above (Sassaroli et al., 2006). Cortical activation is characterized by a reduced cerebral concentration of deoxyhemoglobin ([Hb]), as shown in Figures 15.7 and 15.8, which is primarily caused by a focal increase in cerebral blood flow, as shown in Figure 15.8. This localized decrease in [Hb] is also responsible for the increase in the *blood oxygenation level dependent* (BOLD) signal that is the basis for the functional MRI (fMRI) image in Figure 15.10(b).

It is worth pointing out that the quantitative results of this backprojection approach for fNIRI are affected by a so-called *partial volume effect* (already introduced in Section 14.8.2), which results from using Eq. (15.22) (based on a homogeneous distribution of absorption changes) to describe a localized absorption change. The fact that the absorption perturbation occurs on a portion of the optically probed volume (a partial volume) leads to an underestimate of the absorption perturbation by Eq. (15.22). In light of this observation, the maximum value of $|\Delta[\text{Hb}]|$ measured by fNIRI in Figure 15.10(b) (namely, 0.14 μM) sets a lower limit for the actual focal cortical change in [Hb]. We also observe the excellent agreement between the spatial location of the cortical activated areas identified by fMRI (top panel) and fNIRI (bottom panel), a finding that is matched by the congruent temporal shapes of the fMRI and fNIRI brain activation signals (Sassaroli et al., 2006).

The approach of fNIRI described above aims to obtain two-dimensional images of the brain cortex surface, and for this reason it is sometimes referred to as *optical topography*. The term "topography" comes from the Greek *topos* (place) and *graphia* (writing), and has developed into meaning the study of the geometrical shape and features of a surface. In the case of brain mapping, the surface of interest is the cerebral cortex, and relevant features include its blood content, blood flow, and metabolic rate of oxygen. In this sense, a two-dimensional brain image obtained with fNIRI realizes a topographical representation of hemodynamic and metabolic features on the cortical surface.

Functional NIRI can also be performed using the more sophisticated approach of diffuse optical tomography (DOT), improving lateral spatial resolution and achieving cross-sectional capabilities (tomography) in addition to two-dimensional mapping (topography). Figure 15.11 shows the application of DOT to functional optical imaging of the human visual cortex in a visual stimulation paradigm (Zeff et al., 2007). In this case, the forward model is based on a finite-element method solution (see Section 14.5.1.3) of the diffusion equation for a two-layered, hemispherical geometry, which simulates the human head. The inverse problem is solved with linear methods (as discussed in Section 14.6.1) and spatially variant regularization (see Section 14.7). The implementation of effective DOT methods for fNIRI, albeit involving necessarily more complex instrumentation and computationally demanding image-reconstruction schemes, can substantially improve upon straightforward backprojection methods in terms of optical contrast (since DOT is not affected by contrast-reducing partial volume effects) and depth discrimination (which is particularly relevant in brain imaging to discriminate signal contributions originating from scalp, skull, and brain tissue).

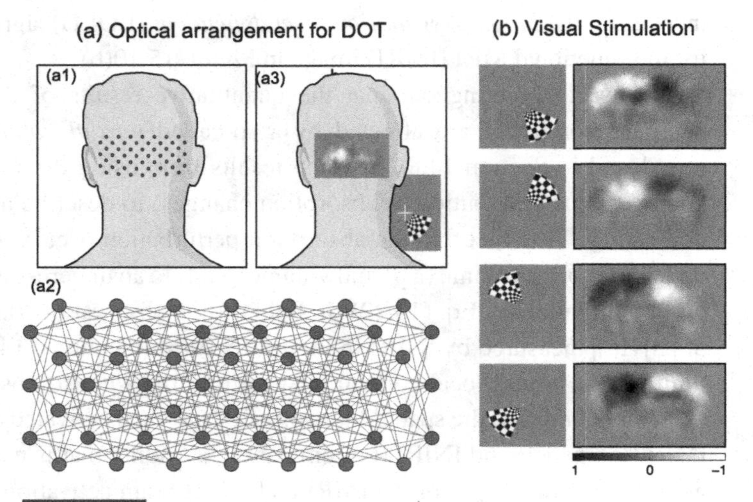

Figure 15.11

(a) Optical arrangement of source and detector locations for functional NIR imaging with high-density diffuse optical tomography (DOT). (a1) 24 sources and 28 detectors are arranged over the occipital lobe for measurements on the visual brain cortex. (a2) The dense grid of source-detector pairs is represented by the set of lines joining each detector with the associated sources, with source-detector separations of 1.3, 3.0, 4.0, and 4.8 cm. (a3) The rectangle in the occipital region shows the optically imaged area (having a size of ~6 cm × 11 cm), and presenting the subject with a white and black checkerboard pattern provides visual stimulation. (b) DOT cortical projection images of normalized oxyhemoglobin concentration changes (Δ[HbO]) in response to angularly swept (10°/s) radial reversing (10 Hz) checkerboard grids for the four stimuli (grid locations) shown. (Reprinted from Zeff et al., 2007, with permission; copyright 2007, National Academy of Sciences, USA.)

15.4 Optical mammography

One of the "holy grails" of diffuse optical tomography has been the prospect of developing imaging methods for detecting breast tumors in the radiographically dense breasts of pre-menopausal women, for whom conventional X-ray mammography is not viable. Like optical oximetry, optical mammography also dates back to the early decades of the 1900s. In the case of oximetry, the early development was, perhaps, instigated by the visible change in blood color as a function of oxygenation. In the case of optical mammography, the relatively low optical absorption by tissue at NIR wavelengths and the optical accessibility of breast tissue account for early attempts of detecting breast tumors with a visual inspection of the breast when transilluminated with red light (Cutler, 1931). Such early studies were based simply on transillumination of the breast with an incandescent light source, in a dark room, coupled with visual observation of darker areas in the breast, presumably attributable to tumors. These early attempts exhibited modest sensitivity, but with very poor specificity.

In more current work, the detection of breast tumors with optical methods is based on imaging a combination of the cancer-induced perturbations to the composition (lipids and water content, scattering properties), hemodynamics (blood content and blood flow), and metabolic features (oxygen consumption) of breast tissue. In particular, it has been established that cancerous breast tissue features a greater blood volume (or total hemoglobin concentration) resulting in stronger optical absorption, greater water content, and reduced lipid content with respect to healthy breast tissue. (A review of results in the field is provided in Fantini and Sassaroli, 2012.)

A simplified *tissue optical index* (TOI) has been introduced to combine these intrinsic sources of tumor contrast into a single indicator of tissue malignancy (Cerussi et al., 2006):

$$\mathrm{TOI} = \frac{[\mathrm{H_2O}]\,[\mathrm{Hb}]}{[\mathrm{Lipids}]}, \tag{15.23}$$

where $[\mathrm{H_2O}]$, [Hb], and [Lipids] are the tissue concentrations of water, deoxyhemoglobin, and lipids, respectively. On the basis of the above observations, cancerous tissue features a stronger attenuation of light (i.e., a greater optical absorbance, as defined in Eq. (10.11)), based on the changes in the concentrations in Eq. (15.23), leading to a greater value of the tissue optical index (TOI) when compared with healthy tissue.

Examples of the appearance of breast cancer in optical images that take advantage of these sources of intrinsic contrast are illustrated in Figure 15.12 for four different approaches to optical mammography. Figure 15.12(a) represents the original method of breast transillumination for a qualitative visual examination of the breast translucence, proposed as a complementary, rather than stand-alone, diagnostic method (Cutler, 1931). In this approach, the examination is performed in a dark room. The examiner places a single red light source of variable optical power beneath the breast lesion of interest, and a solid tumor appears as a localized reduction in optical transmission. This is the basic principle of optical mammography, which has progressed to quantifying the amount and spatial distribution of the cancer-related reduction in optical transmission, but further informed by characterization of its spectral features and physiological origins.

While Figure 15.12(a) may appear to be a remarkably good visualization of a tumor, in practice simple transillumination results in poor specificity, unable to distinguish cancer from a variety of benign morphologies of the breast. Three different approaches for more quantitative and more systematic optical examination of the breast are illustrated in Figure 15.12(b)–(d). Figure 15.12(b) shows a planar

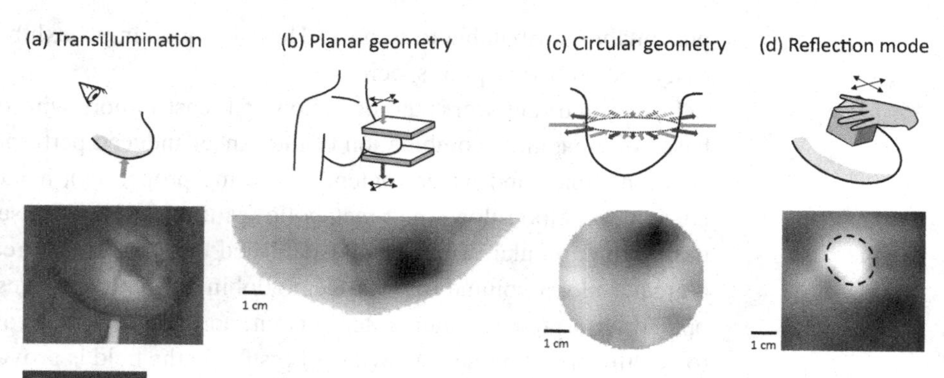

Figure 15.12

Various configurations for optical mammography (top panels) and representative corresponding breast images (bottom panels). (a) The original transillumination approach, in which the examiner visually inspects the spatial pattern of light transmission through the breast using a hand-held light source. The image shows the circular area of reduced light transmission due to a solid tumor (reprinted from Cutler, 1931, with permission; copyright 1931, Lippincott-Raven Publishers). (b) Planar geometry, in which the breast is placed between two parallel glass plates (with minimal compression) and a set of light sources and optical detectors are scanned along the plates on the opposite sides of the breast. The image is an optical absorbance image at 810 nm obtained from frequency-domain data and showing a darker area of increased optical density corresponding to a 3 cm invasive ductal carcinoma (reprinted from Franceschini et al., 1997, with permission; copyright 1997, National Academy of Sciences, USA). (c) Circular arrangement of illumination (gray arrows) and collection (black arrows) optical fibers around the pendulous breast. The image is a map of total hemoglobin concentration, showing a darker area of greater hemoglobin concentration corresponding to a 1 cm invasive ductal carcinoma (reprinted from Pogue et al., 2001, with permission; copyright 2001, Radiological Society of North America). (d) Diffuse reflection configuration realized by scanning a hand-held optical probe containing light sources and optical detectors. The image is a map of tissue optical index (TOI) (which is defined in terms of the concentrations of deoxyhemoglobin, water, and lipids) showing a localized area of greater TOI that corresponds to a 2.1 cm invasive ductal carcinoma (reprinted from Leproux et al., 2013, with permission; copyright 2013, BioMed Central).

scan of a light source and an optical detector that are placed on the opposite sides of a breast, which is slightly compressed between two parallel transparent (glass) plates. In the implementation depicted in Figure 15.12(b), a frequency-domain method yields quantitative optical absorbance measurements that are corrected for the variability of breast tissue thickness, resulting in a reliable, reproducible, and quantitative image that highlights the increased optical absorbance of breast cancer (Franceschini et al., 1997).

Figure 15.12(c) shows a circular arrangement of a set of illumination and collection optical fibers placed around the pendulous breast. This arrangement lends itself to a tomographic reconstruction of the optical properties of breast tissue, based on frequency-domain data, wherein a quantitative cross-sectional image identifies cancer as an area of greater hemoglobin concentration (Pogue et al., 2001).

Figure 15.12(d) shows yet another approach of *diffuse optical spectroscopy and imaging* (DOSI) that is based on successive diffuse reflectance measurements of a hand-held optical probe that is scanned over the breast surface, with source and detector fibers on the same side of the breast. This reflectance mode method offers the potential of accessing tumors that are closer to the chest wall and that may be out of the field of view of the planar (slightly compressed breast) and circular (pendulous breast) geometry approaches. The image shown in Figure 15.12(d) is a tissue optical index (TOI) image, which identifies breast cancer as the region where the TOI shows a local maximum (Leproux et al., 2013).

There are a number of practical considerations in the design of optical mammography instrumentation:

- First, there is the subject position and breast arrangement. In Figure 15.12(a) the subject is sitting and the breast is unconstrained; in Figure 15.12(b) the subject is either standing or sitting and the breast is slightly compressed between glass plates; in Figure 15.12(c) the subject lies on a bed in a prone position, and the breast is pendulous through an opening in the bed; in Figure 15.12(d) the subject is lying on a bed in a supine position and the breast is unconstrained. These different conditions result in different levels of subject comfort, breast stabilization, and range of accessible breast tissue.
- Second, there is the geometrical arrangement of the light sources and optical detectors. There may be as few as one source and one detector in the cases of a tandem scan across the slightly compressed breast in a planar geometry (Figure 15.12b), or with a hand-held optical probe for diffuse reflectance (Figure 15.12d). On the opposite end, there may be as many as tens of sources and tens of detectors in a complete coverage of the pendulous breast along the angular and radial directions. The number of sources and detectors and the acquisition method for spatially resolved measurements affect the spatial sampling rate and the time necessary to collect a complete breast image, thereby resulting in optical mammograms with a range of spatial and temporal resolutions. Furthermore, the mathematical effectiveness of optical tomographic reconstruction may be strongly dependent on the number and geometrical arrangement of the sources and detectors.
- Third, the optical data collection scheme may be based on continuous-wave (CW) methods or time-resolved methods, either in the time domain (TD) or frequency domain (FD). As discussed in Chapters 10–13, time-resolved methods can generate data with a richer information content than CW data, albeit at a limited number of discrete wavelengths, thus enhancing the optical mammography

images in terms of their spatial accuracy and/or optical properties quantification. Of course, the increased information of time-resolved data comes at the price of a more complex instrumentation and data-processing schemes. Furthermore, one should consider that the lack of time-resolved information of CW data may be compensated by the fact that it is usually easier to collect spectrally rich data with CW rather than with time-resolved systems. The relevance of spectral information leads us to the next point.

- Fourth, there is the spectral information content of the optical data. As discussed in Section 10.1, the determination of the concentration of n chromophores requires the measurement of the absorption coefficient at a minimum of n wavelengths. In the case of strongly scattering tissue, the measurement of the absorption coefficient requires discrimination of the contributions to the optical attenuation coming from scattering, which in itself may carry information of diagnostic value. Therefore, targeting the four chromophores of relevance in breast tissue, namely HbO, Hb, water, and lipids, requires measurements at a minimum of four wavelengths that feature selective sensitivities to the four chromophores of interest. However, there may be value in maximizing the spectral information of the optical data, even to the point of measuring a continuous spectrum, to improve the signal-to-noise ratio of concentration measurements or to allow for *derivative spectroscopy*: calculating the first and second derivatives, with respect to wavelength, of the optical absorbance, which may enhance the detection of individual chromophores and minimize scattering contributions.

The above considerations play important roles in the optimization of instrumentation, methods, and measurement protocols for optical mammography. It is worth noting that optical mammography is also being explored as a safe and noninvasive technique to monitor individual responses to neoadjuvant therapy, which is chemotherapy or hormone therapy administered to reduce the size of breast cancer in advance of breast surgery (Cerussi et al., 2007; Enfield et al., 2009; Soliman et al., 2010).

While the intrinsic contrast provided by the tissue content of blood, water, and lipids enables optical mammography to be implemented noninvasively, there is the possibility of enhancing the optical contrast of tumors by administration of exogenous contrast agents. For example, indocyanine green (ICG) (Corlu et al., 2007) and omocyanine (van de Ven et al., 2010), blood-pool agents that have strong absorption/fluorescence bands in the NIR spectral region, have been used for enhanced diffuse optical tomography of the human breast. In this case, fluorescence diffusion theory (see Section 9.8) provides the forward model solver for image reconstruction. The loss of noninvasiveness associated with the introduction of exogenous contrast agents, with the ever-present possibility of side effects, is

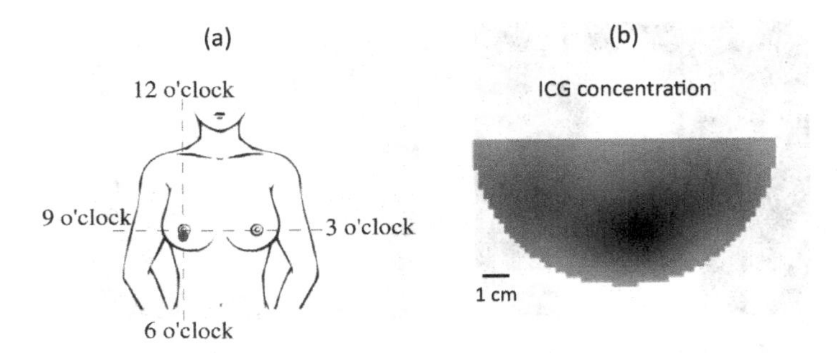

Figure 15.13

Optical mammography following the injection of a fluorescent extrinsic contrast agent (ICG: indocyanine green). (a) Location of an invasive ductal carcinoma in the right breast. (b) Fluorescence image of the indocyanince green (ICG) concentration obtained with diffuse optical tomography (DOT), showing the increased ICG concentration at the cancer location. (Reprinted from Corlu et al., 2007, with permission; copyright 2007, The Optical Society of America.)

compensated by the higher optical contrast and the potential for cancer selectivity of the optical agent, which can translate into higher sensitivity and specificity for breast cancer detection by optical methods. Figure 15.13 shows an example of optical mammography based on extrinsic contrast where, following the intravenous injection of ICG, the ICG is deposited in greater concentrations in cancerous tissue than in healthy breast tissue due to enhanced extravasation through the permeable neo-vasculature of tumors (Corlu et al., 2007).

15.5 Small-animal imaging

Diffuse optical spectroscopy and imaging facilitates a broad range of applications for small-animal studies. We have already shown the applicability of diffuse reflectance measurements to the study of exposed organs, such as the brain cortex in Figure 15.9. That example refers to the monkey brain, but the same technique can be applied to the brains of mice and rats. Exposed organs can be investigated by other optical techniques as well: for example, photoacoustic imaging, which will be presented in Chapter 16, and various forms of optical microscopy and optical coherence tomography, which will be presented in Chapters 17 and 18.

Optical imaging of intrinsic signals (OIS) and other NIR spectroscopy and imaging methods often rely on intrinsic sources of optical contrast (hemoglobin, cytochrome *c* oxidase, scattering properties, etc.). However, the introduction of exogenous sources of contrast opens new opportunities in small-animal imaging, to specifically target proteins or genes, a technique referred to as *molecular imaging*. Optical contrast agents for molecular imaging are based on compounds that provide

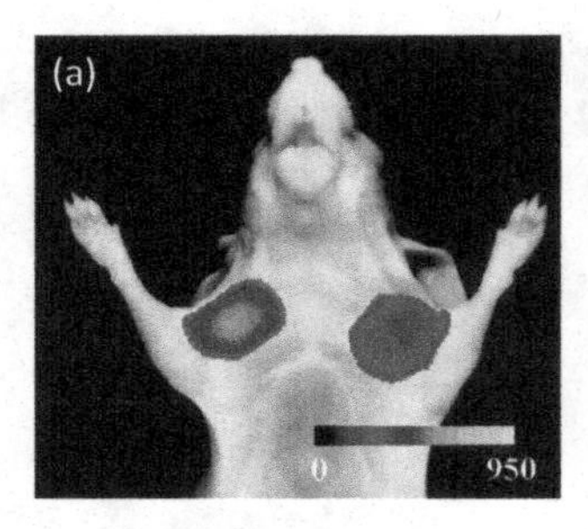

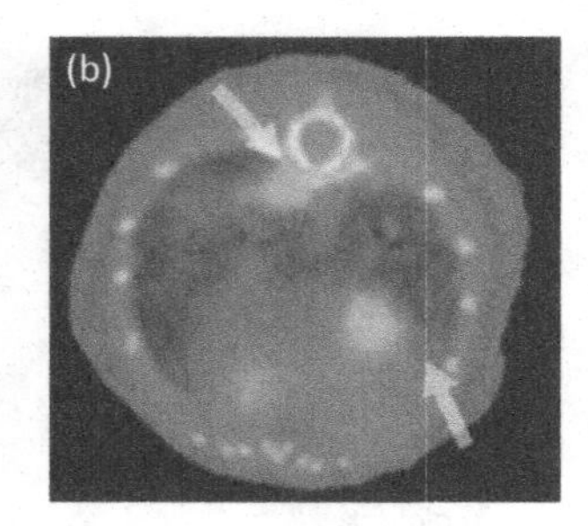

Figure 15.14

(a) Thresholded fluorescence image, superimposed on a light image, of a mouse injected with a fluorescent probe activated by cathepsin-B. The two areas of high fluorescence signal correspond to two tumors. The stronger fluorescent signal from the left tumor (a highly invasive breast carcinoma) reflects a higher cathepsin-B content. Illumination wavelengths: 610–650 nm; detection wavelengths: 680–720 nm; CCD exposure time: 30 s. (Reprinted from Bremer et al., 2002, with permission; copyright 2002, Radiological Society of North America.) (b) Transverse slice of fluorescence molecular tomography of a lung cancer mouse model showing enhanced fluorescence (arrows) from a probe targeting $\alpha_V\beta_3$ integrin, which is expressed in cancer cells. (Reprinted by permission from Macmillan Publishers Ltd: *Nature Methods* [Ale et al., 2012], copyright 2012.)

targeted absorption, fluorescence or bioluminescence (which requires an enzyme to catalyze a chemical reaction that results in light emission). The limited tissue thickness of small animals enables the imaging of highly specific absorption or visualization of the fluorescence or bioluminescence light originating from target cells over volumes as large as complete organs and even from the entire animal.

One straightforward approach consists of placing the animal in a light-tight chamber, providing broad-beam illumination at the fluorescence excitation wavelength (for example, by LEDs or using a halogen lamp with an excitation band-pass filter), and imaging the fluorescence emission with a CCD camera equipped with an emission band-pass filter (as illustrated in Figure 5.10). This method has been used to generate Figure 15.14(a), which shows a fluorescence image, superimposed on a white-light image, of a mouse injected with a fluorescent probe that targets cathepsin-B, a protein that is overexpressed in a number of human cancers (Bremer et al., 2002). The different fluorescent signals from two size- and depth-matched tumors, a highly invasive breast adenocarcinoma and a well-differentiated (hence, less malignant) adenocarcinoma, reflect a significantly higher concentration of cathepsin-B in the former (the one on the left in Figure 15.14(a)) (Bremer et al., 2002).

A tomographic approach to fluorescence diffuse imaging of small animals, called *fluorescence molecular tomography (FMT)*, is based on the general scheme of diffuse optical imaging illustrated in Figure 14.3. This tomographic approach can generate quantitative optical images, and can significantly benefit from regularization techniques based on spatial priors, which were presented in Section 14.7. In this spirit, the combination of FMT with X-ray computed tomography (XCT) has led to the image in Figure 15.14(b), which is a transverse slice of a lung cancer mouse model, showing enhanced fluorescence (arrows) from a probe targeting $\alpha_v\beta_3$ integrin, which is expressed in cancer cells (Ale et al., 2012). It should be noted, nonetheless, that while it is possible to image absolute values for scattering and absorption properties of tissue, it is highly challenging to generate quantitative images of fluorophore concentrations. Hence, such images are generally assumed to provide indications of relative concentrations of targeted vs. background fluorescence.

15.6 Prospects of diffuse optics for biomedical applications

Applications of diffuse optics for spectroscopy and imaging of biological tissue have been explored for almost a century. The relatively low absorption of light in the red to NIR spectral region allows for optical sensing of thick tissues, albeit at a limited spatial resolution. Early investigations aimed at breast cancer detection and oximetry demonstrated some of the most appealing features of NIR spectroscopy and imaging: the high contrast associated with blood content and oxygenation, its noninvasive nature, compact instrumentation, and safety. Pulse oximetry, the first clinical application of such techniques, remains in widespread use in clinics and doctor's offices today. However, diffuse optical spectroscopy and imaging are still mostly research tools aimed at physiological, functional, diagnostic, or prognostic assessment, as well as for monitoring the efficacy of therapeutic treatments. The significant promise of noninvasive diffuse optics in the biomedical field is reflected by a number of companies that have introduced commercial systems that are suitable for laboratory as well as clinical applications involving a broad range of diffuse optical spectroscopy and imaging methods. Nonetheless, broad clinical adoption of diffuse optical methods remains an opportunity for future commercial development and dissemination.

Problems – answers to problems with * are on p. 648

15.1* A measurement of the oxygen saturation of hemoglobin at a partial pressure of oxygen of 45 mmHg gives a reading of 80%. According to the Hill equation, what are the values of $p50$ in the cases of (a) no cooperativity ($n = 1$), (b) partial cooperativity of $n = 2.5$, and (c) complete cooperativity ($n = 4$)?

15.2 Consider the equilibrium reaction $Mb + O_2 \rightleftarrows MbO_2$ that describes the binding of oxygen to myoglobin. The equilibrium constant of this reaction is $k_M = [MbO_2]/\{[Mb][O_2]\}$. Find the expression for p_{50} (the partial pressure of dissolved oxygen at 50% myoglobin saturation) in terms of k_M and of the Henry's law constant H.

15.3* (a) Convert a tissue oxygen consumption of 3.2 $\mu mol_{O2}/(100\ ml_{tissue}$-min), representing a typical resting value for muscle, into units of $ml_{O2}/(kg_{tissue}$-min) by assuming standard temperature and pressure (STP) conditions, and a mass density of muscle tissue of 1.04 kg/l. (b) Compare your result in part (a) with a typical maximal muscle oxygen consumption, a measure of the ability for work output, of ~40 $ml_{O2}/(l_{tissue}$-min) in untrained individuals and ~80 $ml_{O2}/(l_{tissue}$-min) in elite endurance athletes. How much larger is the maximal muscle oxygen consumption in these two cases with respect to the oxygen consumption at rest?

15.4 A calf skeletal muscle has a baseline hemoglobin concentration of 83 μM. A measurement performed 5 s after a sudden venous occlusion of the thigh records a hemoglobin concentration of 89 μM. Assuming a hemoglobin concentration in blood of 2.3 mM, find: (a) the baseline muscle blood volume in %, and (b) the muscle blood flow in units of $ml_{blood}/(100\ ml_{tissue}$-min).

15.5* A forearm skeletal muscle has a hemoglobin concentration of 92 μM. An arterial occlusion of the upper arm determines a decrease in saturation at a rate of 15% per minute, and no change in blood volume. Find the muscle oxygen consumption in units of $\mu mol_{O2}/(100\ ml_{tissue}$-min).

15.6 The oxygen consumption of a given muscle is 4.68 $\mu mol_{O2}/(100\ ml_{tissue}$-min) and its blood flow is 2.6 $ml_{blood}/(100\ ml_{tissue}$-min). If the muscle mass is 2.8 kg (muscle density: 1.04 kg/l) and the oxygen saturation of incoming arterial blood is 98%, what is the oxygen saturation of the venous blood leaving the muscle? (Assume a hemoglobin concentration in blood of 2.3 mM, and ignore the amount of dissolved oxygen in plasma.)

15.7* After intravenous injection of indocyanine green (ICG), an FDA-approved optical contrast agent, its concentration in a tissue of interest is 0.15 mg/dl. The molar extinction coefficient of ICG in plasma at 780 nm (defined in terms of powers of 10) is $10^5\ cm^{-1}M^{-1}$, and its molecular weight is 775 g/mol. What is the ICG absorption coefficient (defined in terms of powers of e) in the examined tissue at 780 nm?

15.8 An optical instrument performs measurements on a living tissue at a source-detector separation of 2 cm and at a wavelength of 800 nm, the isosbestic point of hemoglobin where the molar extinction coefficients of Hb and HbO are both 2 $cm^{-1}mM^{-1}$. This instrument records intensity oscillations having a relative amplitude of 0.9% as a result of arterial pulsation in the examined tissue. Find the pulsatile arterial blood volume, expressed as a percent volume fraction in tissue, using the modified Beer-Lambert law, a differential pathlength factor (DPF) of 4.9, and a hemoglobin concentration in blood of 2.3 mM.

15.9* What is the image contrast in an optical mammogram displaying the TOI defined in Eq. (15.23) if a breast cancer features percent changes, with respect to background tissue, of +24%, +36%, and −15% for [H_2O], [Hb], and [Lipids], respectively?

References

Ale, A., Ermolayev, V., Herzog, E., et al. (2012). FMT-XCT in vivo animal studies with hybrid fluorescence molecular tomography-X-ray computed tomography. *Nature Methods*, 9, 615–620.

Arakaki, L. S. L., Burns, D. H., and Kushmerick, M. J. (2007). Accurate myoglobin oxygen saturation by optical spectroscopy measured in blood-perfused rat muscle. *Applied Spectroscopy*, 61, 978–985.

Bremer, C., Tung, C.-H., Bogdanov, A., and Weissleder, R. (2002). Imaging of differential protease expression in breast cancers for detection of aggressive tumor phemotypes. *Radiology*, 222, 814–818.

Cerussi, A. E., Shah, N., Hsiang, D., et al. (2006). In vivo absorption, scattering, and physiologic properties of 58 malignant breast tumors determined by broadband diffuse optical spectroscopy. *Journal of Biomedical Optics*, 11, 044005 (16pp).

Cerussi, A. E., Hsiang, D., Shah, N., et al. (2007). Predicting response to breast cancer neoadjuvant chemotherapy using diffuse optical spectroscopy. *Proceedings of the National Academy of Sciences USA*, 104, 4014–4019.

Chance, B., Dait, M. T., Zhang, C., Hamaoka, T., and Hagerman, F. (1992). Recovery from exercise-induced desaturation in the quadriceps muscles of elite competitive rowers. *American Journal of Physiology*, 262, C766–C775.

Chance, B., Zhuang, Z., UnAh, C., Alter, C., and Lipton, L. (1993). Cognition-activated low-frequency modulation of light absorption in human brain. *Proceedings of the National Academy of Sciences USA*, 90, 3770–3774.

Cooper, C. E., Cope, M., Quaresima, V., et al. (1997). Measurement of cytochrome oxidase redox state by near infrared spectroscopy. In *Optical Imaging of Brain Function and Metabolism II*, ed. Villringer, A. and Dirnagl, U. New York: Plenum Press.

Corlu, A., Choe, R., Durduran, T., et al. (2007). Three-dimensional *in vivo* fluorescence diffuse optical tomography of breast cancer in humans. *Optics Express*, 15, 6696–6716.

Cutler, M. (1931). Transillumination of the breast. *Annals of Surgery*, 93, 223–234.

de Blasi, R. A., Ferrari, M., Natali, A., et al. (1994). Noninvasive measurement of forearm blood flow and oxygen consumption by near-infrared spectroscopy. *Journal of Applied Physiology*, 76, 1388–1393.

Durduran, T., Yu, G., Burnett, M. G., et al. (2004). Diffuse optical measurement of blood flow, blood oxygenation, and metabolism in a human brain during sensorimotor cortex activation. *Optics Letters*, 29, 1766–1768.

Enfield, L. C., Gibson, A. P., Hebden, J. C., and Douek, M. (2009). Optical tomography of breast cancer-monitoring response to primary medical therapy. *Targeted Oncology*, 4, 219–233.

Fantini, S. (2014a). Dynamic model for the tissue concentration and oxygen saturation of hemoglobin in relation to blood volume, flow velocity, and oxygen consumption: Implications for functional neuroimaging and coherent hemodynamics spectroscopy (CHS). *NeuroImage*, 85, 202–221.

Fantini, S. (2014b). A new hemodynamic model shows that temporal perturbations of cerebral blood flow and metabolic rate of oxygen cannot be measured individually using functional near-infrared spectroscopy. *Physiological Measurement,* 35, N1–N9.

Fantini, S., and Sassaroli, A. (2012). "Near-infrared optical mammography for breast cancer detection with intrinsic contrast,". *Ann. Biomed. Eng. Annals of Biomedical Engineering*, 40, 398–407.

Ferreira, L. F., Hueber, D. M., and Barstow, T. J. (2007). Effects of assuming constant optical scattering on measurements of muscle oxygenation by near-infrared spectroscopy during exercise. *Journal of Applied Physiology*, 102, 358–367.

Franceschini, M. A., Moesta, K. T., Fantini, S., et al. (1997). Frequency-domain techniques enhance optical mammography: initial clinical results. *Proceedings of the National Academy of Sciences USA*, 94, 6468–6473.

Franceschini, M. A., Fantini, S., Paunescu, L. A., Maier, J. S., and Gratton, E. (1998). Influence of a superficial layer in the quantitative spectroscopic study of strongly scattering media. *Applied Optics*, 37, 7447–7458.

Frostig, R. D., Lieke, E. E., Ts'o, D. Y., and Grinvald, A. (1990). Cortical functional architechture and local coupling between neuronal activity and the microcirculation revealed by in vivo high-resolution optical imaging of intrinsic signals. *Proceedings of the National Academy of Sciences USA*, 87, 6082–6086.

Gratton, G., and Fabiani, M. (2003). The event-related optical signal (EROS) in visual cortex: replicability, consistency, localization, and resolution. *Psychophsyiology*, 40, 561–571.

Grinvald, A., Lieke, E., Frostig, R. D., Gilbert, C. D., and Wiesel, T. N. (1986). Functional architecture of cortex revealed by optical imaging of intrinsic signals. *Nature*, 324, 361–364.

Hoshi, Y., and Tamura, M. (1993). Dynamic multichannel near-infrared optical imaging of human brain activity. *Journal of Applied Physiology*, 75, 1842–1846.

Jöbsis, F. F. (1977). Noninvasive, infrared monitoring of cerebral and myocardial oxygen sufficiency and circulatory parameters. *Science*, 198, 1264–1267.

Kato, T., Kamei, A., Takashima, S., and Ozaki, T. (1993). Human visual cortical function during photic stimulation monitoring by means of near-infrared spectroscopy. *Journal of Cerebral Blood Flow & Metabolism*, 13, 516–520.

Kay, R. H., and Coxon, R. V. (1957). Optical and instrumental limitations to the accuracy of oximetry. *Journal of Scientific Instruments*, 34, 233–236.

Kienle, A., Lilge, L., Alex Vitkin, I., et al. (1996). Why do veins appear blue? A new look at an old question. *Applied Optics*, 35, 1151–1160.

Kolyva, C., Tachtsidis, I., Ghosh, A., et al. (2012). Systematic investigation of changes in oxidized cerebral cytochrome c oxidase concentration during frontal lobe activation in healthy adults. *Biomedical Optics Express*, 3, 2550–2566.

Larsson, S., Källebring, B., Wittung, P., and Malmström, B. G. (1995). The CuA center of cytochrome-c oxidase: Electronic structure and spectra of models compared to the properties of CuA domains. *Proceedings of the National Academy of Sciences USA*, 92, 7167–7171.

Leproux, A., Durkin, A., Compton, M., et al. (2013). Assessing tumor contrast in radiographically dense breast tissue using Diffuse Optical Spectroscopic Imaging (DOSI). *Breast Cancer Research*, 15, R89.

Marcinek, D. J., Amara, C. E., Matz, K., Conley, K. E., and Schenkman, K. A. (2007). Wavelength shift analysis: a simple method to determine the contribution of hemoglobin and myoglobin to in vivo optical spectra. *Applied Spectroscopy*, 61, 665–669.

Millikan, G. A. (1942). The oximeter, an instrument for measuring continuously the oxygen saturation of arterial blood in man. *Review of Scientific Instruments*, 13, 434–444.

Niwayama, M., Lin, L., Shao, J., Kudo, N., and Yamamoto, K. (2000). Quantitative measurement of muscle hemoglobin oxygenation using near-infrared spectroscopy with correction for the influence of a subcutaneous fat layer. *Review of Scientific Instruments*, 71, 4571–4575.

Pogue, B. W., Poplack, S. P., McBride, T. O., et al. (2001) Quantitative hemoglobin tomography with diffuse near-infrared spectroscopy: pilot results in the breast. *Radiology*, 218, 261–266.

Rector, D. M., Poe, G. R., Kristensen, M. P., and Harper, R. M. (1997). Light scattering changes follow evoked potentials from hippocampal Schaeffer collateral stimulation. *Journal of Neurophysiology*, 78, 1707–1713.

Ryan, T. E., Brizendine, J. T., and McCully, K. K. (2013). A comparison of exercise type and intensity on the noninvasive assessment of skeletal muscle mitochondrial function using near-infrared spectroscopy. *Journal of Applied Physiology*, 114, 230–237.

Sassaroli, A., Frederick, B. deB., Tong, Y., Renshaw, P. F., and Fantini, S. (2006). Spatially weighted BOLD signal for comparison of functional magnetic resonance imaging and near-infrared imaging of the brain. *NeuroImage*, 33, 505–514.

Schenkman, K. A., Marble, D. R., Burns, D. H., and Feigl, E. O. (1999). Optical spectroscopic method for in vivo measurement of cardiac myoglobin oxygen saturation. *Applied Spectroscopy*, 53, 332–338.

Soliman, H., Gunasekara, A., Rycroft, M., et al. (2010) Functional imaging using diffuse optical spectroscopy of neoadjuvant chemotherapy response in women with locally advanced breast cancer. *Clinical Cancer Research*, 16, 2605–2614.

Stepnoski, R. A., LaPorta, A., Raccuia-Behling, R., et al. (1991). Noninvasive detection of changes in membrane potential in cultured neurons by light scattering. *Proceedings of the National Academy of Sciences USA*, 88, 9382–9386.

Torricelli, A., Quaresima, V., Pifferi, A., et al. (2004). Mapping of calf muscle oxygenation and haemoglobin content during dynamic plantar flexion exercise by multi-channel time-resolved near-infrared spectroscopy. *Physics in Medicine and Biology*, 49, 685–699.

Ts'o, D. Y., Frostig, R. D., Lieke, E. E., and Grinvald, A. (1990). Functional organization of primate visual cortex revealed by high resolution optical imaging. *Science*, 249, 417–420.

Uludağ, K., Steinbrink, J., Kohl-Bareis, M., et al. (2004). Cytochrome-c-oxidase redox changes during visual stimulation measured by near-infrared spectroscopy cannot be explained by a mere cross talk artefact. *NeuroImage*, 22, 109–119.

van de Ven, S., Elias, S., Wiethoff, A., et al. (2010). A novel fluorescent imaging agent for diffuse optical tomography of the breast: First clinical experience in patients. *Molecular Imaging and Biology*, 12, 343–348.

Villringer, A., Planck, J., Hock, C., Schleinkofer, L., and Dirnagl, U. (1993). Near infrared spectroscopy (NIRS): a new tool to study hemodynamic changes during activation of brain function in human adults. *Neuroscience Letters*, 154, 101–104.

Wray, S., Cope, M., Delpy, D. T., Wyatt, J. S., and Reynolds, E. O. R. (1988). Characterization of the near infrared absorption cpectra of cytochrome aa3 and haemoglobin for the non-invasive monitoring of cerebral oxygenation. *Biochimica et Biophysica Acta*, 933, 184–192.

Zeff, B. W., White, B. R., Dehghani, H., Schlaggar, B. L., and Culver, J. P. (2007). Retinotopic mapping of adult human visual cortex with high-density diffuse optical tomography. *Proceedings of the National Academy of Sciences USA*, 104, 12169–12174.

Further reading

Optics of skin color

Kollias, N. (1995). The physical basis of skin color and its evaluation. *Clinics in Dermatology*, 13, 361–367.

Resifeld, P. L. (2000). Blue in the skin. *Journal of the American Academy of Dermatology*, 42, 597–605.

Oxygen dissociation curve of hemoglobin

Gomez-Cambronero, J. (2001). The oxygen dissociation curve of hemoglobin: Bridging the gap between biochemistry and physiology. *Journal of Chemical Education*, 78, 757–759.

Weissbluth, M. (1974). *Hemoglobin: Cooperativity and Electronic Properties*. New York: Springer.

Pulse oximetry

Franceschini, M. A., Gratton, E., and Fantini, S. (1999). Non-invasive optical method to measure tissue and arterial saturation: An application to absolute pulse oximetry of the brain. *Optics Letters*, 24, 829–831.

Mendelson, Y. (1992). Pulse oximetry: theory and applications for noninvasive monitoring. *Clinical Chemistry*, 38, 1601–1607.

Wukitsch, M. W., Petterson, M. T., Tobler, D. R., and Pologe, J. A. (1988). Pulse oximetry: analysis of theory, technology, and practice. *Journal of Clinical Monitoring*, 4, 290–301.

Cytochrome *c* oxidase

Cooper, C. E., Matcher, S. J., Wyatt, J. S., et al. (1994). Near-infrared spectroscopy of the brain: relevance to cytochrome oxidase bioenergetics. *Biochemical Society Transactions*, 22, 974–980.

Kolyva, C., Ghosh, A., Tachtsidis, I., et al. (2014). Cytochrome c oxidase response to changes in cerebral oxygen delivery in the adult brain shows higher brain-specificity than hemoglobin. *NeuroImage*, 85, 234–244.

Matcher, S. J., Elwell, C. E., Cooper, C. E., Cope, M., and Delpy, D. T. (1995). Performance comparison of several published tissue near-infrared spectroscopy algorithms. *Analytical Biochemistry*, 227, 54–68.

Michel, H., Behr, J., Harrenga, A., and Kannt, A. (1998). Cytochrome c oxidase: Structure and spectroscopy. *Annual Review of Biophysics and Biomolecular Structure*, 27, 329–356.

Muscle studies with NIRS and myoglobin contributions

Bhambhani, Y. N. (2004). Muscle oxygenation trends during dynamic exercise measured by near infrared spectroscopy. *Canadian Journal of Applied Physiology*, 29, 504–523.

Ferrari, M., Muthalib, M., and Quaresima, V. (2011). The use of near-infrared spectroscopy in understanding skeletal muscle physiology: recent developments. *Philosophical Transactions of the Royal Society A*, 369, 4577–4590.

Hamaoka, T., McCully, K. K., Quaresima, V., Yamamoto, K., and Chance, B. (2007). Near-infrared spectroscopy/imaging for monitoring muscle oxygenation and oxidative metabolism in healthy and diseased humans. *Journal of Biomedical Optics* 12, 062105 (16pp).

Mancini, D. M., Bolinger, L., Li, H., et al. (1994). Validation of near-infrared spectroscopy in humans. *Journal of Applied Physiology*, 77, 2740–2747.

Modeling optical signals in relation to cerebral hemodynamics and metabolism

Banaji, M., Mallet, A., Elwell, C. E., Nicholls, P., and Cooper, C. E. (2008). A model of brain circulation and metabolism: NIRS signal changes during physiological challenges. *PLoS Computational Biology*, 4, e1000212.

Diamond, S. G., Perdue, K. L., and Boas, D. A. (2009). A cerebrovascular response model for functional neuroimaging including dynamic cerebral autoregulation. *Mathematical Biosciences*, 220, 102–117.

Huppert, T. J., Allen, M. S., Benav, H., Jones, P. B., and Boas, D. A. (2007). A multicompartment vascular model for inferring baseline and functional changes in cerebral oxygen metabolism and arterial dilation. *Journal of Cerebral Blood Flow & Metabolism*, 27, 1262–1279.

Kainerstorfer, J. M., Sassaroli, A., Hallacoglu, B., Pierro, M. L., and Fantini, S. (2014). Practical steps for applying a new dynamic model to near-infrared spectroscopy measurements of hemodynamic oscillations and transient changes: implications for cerebrovascular and functional brain studies. *Academic Radiology*, 21, 185–196.

Optical imaging of intrinsic signals (OIS)

Pouratian, N., Sheth, S. A., Martin, N. A., and Toga, A. W. (2003). Shedding light on brain mapping: advances in human optical imaging. *Trends in Neuroscience*, 26, 277–282.

Yin, C., Zhou, F., Wang, Y., et al. (2013). Simultaneous detection of hemodynamics, mitochondrial metabolism and light scattering changes during cortical spreading depression in rats based on multi-spectral optical imaging. *NeuroImage*, 76, 70–80.

Zepeda, A., Arias, C., and Sendpiel, F. (2004). Optical imaging of intrinsic signals recent developments in the methodology and its applications. *Journal of Neuroscience Methods*, 136, 1–21.

Near-infrared functional brain studies (fNIRS and fNIRI)

Ferrari, M., and Quaresima, V. (2012). A brief review on the history of human functional near-infrared spectroscopy (fNIRS) development and fields of application. *NeuroImage*, 63, 921–935.

Lloyd-Fox, S., Blasi, A., and Elwell, C. E. (2010). Illuminating the developing brain: The past, present and future of functional near infrared spectroscopy. *Neuroscience & Biobehavioral Reviews*, 34, 2690284.

Special Issue. (15 January 2014). Celebrating 20 years of functional near infrared spectroscopy (fNIRS). *NeuroImage*, 85 (Part 1), 1–636.

Wolf M., Morren, G., Haensse, D., et al. (2008). Near infrared spectroscopy to study the brain: an overview. *Opto-Electronics Review*, 16, 413–419.

Optical mammography

Brooksby, B., Pogue, B. W., Jiang, S., et al. (2006). Imaging breast adipose and fibroglandular tissue molecular signatures by using hybrid MRI-guided near-infrared spectral tomography. *Proceedings of the National Academy of Sciences USA*, 103, 8828–8833.

Choe, R., Konecky, S. D., Corlu, A., et al. (2009). Differentiation of benign and malignant breast tumors by in-vivo three-dimensional parallel-plate diffuse optical tomography. *Journal of Biomedical Optics*, 14, 024020.

Demos, S. G., Vogel, A. J., and Gandjbakhche, A. H. (2006). Advances in optical spectroscopy and imaging of breast lesions. *Journal of Mammary Gland Biology and Neoplasia*, 11, 165–181.

Grosenick, D., Wabnitz, H., Moesta, K. T., et al. (2005). Time-domain scanning optical mammography: II. Optical properties and tissue parameters of 87 carcinomas. *Physics in Medicine and Biology*, 50, 2451–2468.

Kukreti, S., Cerussi, A. E., Tanamai, W., et al. (2010) Characterization of metabolic differences between benign and malignant tumors: High-spectral-resolution diffuse optical spectroscopy. *Radiology*, 254, 277–284.

Leff, D. R., Warren, O. J., Enfield, L. C., et al. (2008) Diffuse optical imaging of the healthy and diseased breast: A systematic review. *Breast Cancer Research and Treatment*, 108, 9–22.

Schmitz, C. H., Klemer, D. P., Hardin, R., et al. (2005). Design and implementation of dynamic near-infrared optical tomographic imaging instrumentation for simultaneous dual-breast measurements. *Applied Optics*, 44, 2140–2153.

Taroni, P., Torricelli, A., Spinelli, L., et al. (2005). Time-resolved optical mammography between 637 and 985 nm: Clinical study on the detection and identification of breast lesions. *Physics in Medicine and Biology*, 50, 2469–2488.

Wang, J., Pogue, B. W., Jiang, S., and Paulsen, K. D. (2010). Near-infrared tomography of breast cancer hemoglobin, water, lipid, and scattering using combined frequency domain and cw measurements. *Optics Letters*, 35, 82–84.

Small-animal imaging

Contag, C. H., Jenkins, D., Contag, P. R., and Negrin, R. S. (2000). Use of reporter genes for optical measurements of neoplastic disease in vivo. *Neoplasia*, 2, 41–52.

Weissleder, R., and Ntziachristos, V. (2003). Shedding light onto live molecular targets. *Nature Medicine*, 9, 123–128.

16 Combining light and ultrasound: acousto-optics and opto-acoustics

Measurement techniques based on different physical properties can be combined in a variety of ways to result in multi-modal spectroscopy or imaging methods. One may simply apply them independently and then combine the complementary information content of their results. For example, physicians often combine the information content of X-ray tomography and magnetic resonance imaging (MRI) to provide information about both hard and soft tissues. Alternatively, one may use the results of one technique as prior information to enhance the performance of another technique. Finally, two techniques may be combined at a more fundamental level, yielding a hybrid technology in which both physical mechanisms contribute to the generation of the measured signals. In this chapter, we describe two intimately integrated hybrid imaging methods that combine optical and ultrasound techniques.

In one hybrid implementation, called *acousto-optics*, light and ultrasound are delivered to the tissue and their interaction results in a modulated optical signal that originates from the ultrasound focal region (which can be scanned). Such a modulated optical signal is detected at the tissue surface. This approach is also called *ultrasonic tagging of light* (UTL), *acousto-optic tomography* (AOT), or *ultrasound-modulated optical tomography* (UOT).

In a second hybrid implementation, short laser pulses are delivered to the tissue to generate *opto-acoustic* pressure perturbations as a result of the local thermal expansion/relaxation of optical absorbers within the tissue. These pressure perturbations have a frequency content in the ultrasound range (MHz to tens of MHz), which propagate to the tissue surface where they can be detected with ultrasound imaging methods. This approach is also called *photoacoustic* (PA) *imaging* (with the variants of *PA tomography* [PAT] and *PA microscopy* [PAM]) or *laser optoacoustic imaging* (implemented by *laser optoacoustic imaging systems* [LOIS]).

The two approaches, based on acousto-optics and opto-acoustics, are conceptually illustrated in Figure 16.1. This chapter first introduces general concepts of ultrasound imaging, and then describes the theory, methods, and applications of both acousto-optic and opto-acoustic approaches to the combination of light

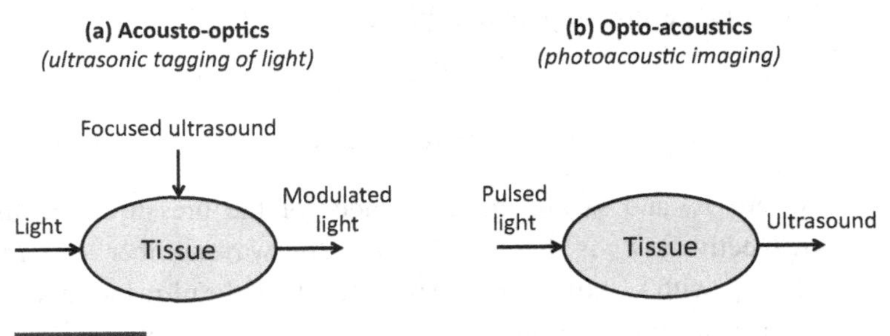

Figure 16.1

Conceptual representation of the implementations of (a) acousto-optics, and (b) opto-acoustics for the study of tissues by combination of light and ultrasound. In acousto-optic methods, light that travels through the ultrasound focal volume is modulated at the ultrasound frequency. In opto-acoustic (or photoacoustic) methods, the absorption of a short pulse of light within the tissue generates an ultrasound wave as a result of a fast thermal expansion.

and ultrasound. These two different techniques (one based on light and one on ultrasound) are joined into integrated systems that exploit their synergistic combination at a fundamental level.

16.1 Basic concepts of ultrasound imaging

16.1.1 The nature of ultrasound

Sound and *ultrasound* are pressure waves that propagate through compressible media such as air, water, or biological tissue. Audible sound refers to the frequency range 20 Hz – 20 kHz, whereas ultrasound refers to frequencies greater than 20 kHz. Conventional diagnostic ultrasound imaging systems operate within the frequency range 1–15 MHz. Higher frequencies achieve better spatial resolution at the expense of a reduced penetration depth in tissue. (Even higher frequencies [up to ~60 MHz] are employed in specialized high-resolution or intravascular applications for imaging at short distances.)

Ultrasound is described in terms of the spatial and temporal dependence of the local pressure perturbation (p) and the displacement (s) of the microscopic constituents of the medium (molecules, cells, etc.). It is important to observe that the pressure perturbation p is the change in pressure associated with ultrasound, which is superimposed on the baseline pressure, so that it is sometimes referred to as *overpressure*. A plane ultrasound wave at angular frequency ω_a (the subscript "a" specifies that sound and ultrasound are acoustic waves) that propagates along the z direction is characterized by the following harmonic expressions for local pressure perturbation and displacement:

$$p(z,t) = p_0 e^{i(k_a z - \omega_a t + \phi_p)}, \tag{16.1}$$

$$s(z,t) = s_0 e^{i(k_a z - \omega_a t + \phi_s)}, \tag{16.2}$$

where p_0 and s_0 are the amplitudes of the pressure and displacement waves, respectively, k_a is the complex acoustic wavenumber – its real part is given by $2\pi/\lambda_a$, with λ_a expressing the wavelength of the ultrasound wave, and its imaginary part determines the attenuation of the ultrasound wave – and ϕ_p and ϕ_s are the phases of the pressure and displacement waves, respectively. The phase velocity of the acoustic wave is $v_a = \omega_a / k_a$. Local maxima and minima (i.e., stationary points) in displacement ($\frac{ds}{dz} = 0$) correspond to no pressure changes ($p = 0$) because the local density of the microscopic medium constituents is unperturbed. As a result, the pressure and displacement waves have a relative phase difference of $\pi/2$: i.e., $\phi_s - \phi_p = \pi/2$. Because the direction of the displacement s is typically along the direction of propagation of the ultrasound wave (the z direction in the case considered here), ultrasound is a longitudinal wave. Therefore, it should be noted that ultrasound differs from light in at least two fundamental ways:

- it is a mechanical wave, which can only propagate in a compressible medium, as opposed to electromagnetic radiation, which can propagate in vacuum;
- it is a longitudinal wave (at least in gases and liquids), in which the particle displacement is along the direction of propagation of the waves, as opposed to optical waves whose electric and magnetic field are orthogonal to the direction of propagation of light.

The amplitudes of the pressure and displacement waves (p_0 and s_0) are not independent, as they are directly proportional to each other through the angular frequency (ω_a) and the characteristic impedance (Z), which is related to the mechanical and elastic properties of the medium:

$$p_0 = Z\omega_a s_0. \tag{16.3}$$

Consequently, ultrasound can be described equivalently in terms of either a pressure or a displacement wave. The characteristic impedance, Z, is defined as the square root of the product of the mass density of the medium (ρ) times the bulk modulus (B), which is the negative ratio of the pressure change to the relative volume change ($-p/(\frac{\Delta V}{V})$) or the inverse of the compressibility:

$$Z = \sqrt{\rho B}. \tag{16.4}$$

Because the physical dimensions of B are those of pressure, the SI units of Z are kg/(m^2-s), which are given the name Rayl after Lord Rayleigh, whose name has also been adopted to refer to scattering from small particles (Rayleigh scattering, derived in Chapters 6 and 7), and to the Rayleigh range of a focused beam (covered in Chapter 17). We also point out that the ultrasound speed is related to the medium density and bulk modulus as follows:

$$v_a = \sqrt{\frac{B}{\rho}}. \tag{16.5}$$

The time-averaged acoustic intensity (I_a), with units of W/m^2, is proportional to the square of the amplitude of the pressure wave with the factor $1/(2Z)$:

$$I_a = \frac{p_0^2}{2Z}. \tag{16.6}$$

The acoustic intensity (I_a) can be specified relative to an arbitrary reference level (I_0), resulting in the dimensionless units of *decibels* (dB). The decibel is one-tenth of a bel (B), a logarithmic unit named after the Scottish-American engineer Alexander Graham Bell [1847–1922], who is aptly mentioned in this chapter, since he is also the father of photoacoustics, having reported the first experiments on the generation of sound by light in 1880. The relative intensity level in decibels is defined as $10\mathrm{Log}_{10}(I_a/I_0)$, so that -3 dB means that $I_a = 0.5I_0$, 0 dB means that $I_a = I_0$, 10 dB means that $I_a = 10I_0$, 20 dB means that $I_a = 100I_0$, etc.

The attenuation of ultrasound intensity in tissue is conveniently expressed in decibels by considering the ratio of the ultrasound intensities at two different tissue locations, $\mathbf{r}_1$ and $\mathbf{r}_2$. Simply, the propagation of ultrasound from $\mathbf{r}_1$ to $\mathbf{r}_2$ results in an intensity attenuation in decibels that is given by $10\mathrm{Log}_{10}[I_a(\mathbf{r}_1)/I_a(\mathbf{r}_2)]$. A 20 dB attenuation means that the ultrasound intensity is decreased to 1% of its original value. It is important to observe that the quadratic relationship between intensity and pressure (Eq. (16.6)) results in the fact that $\mathrm{Log}_{10}(I_a/I_0) = 2\mathrm{Log}_{10}(p_a/p_0)$, where p_a and p_0 are the pressure levels corresponding to I_a and I_0, respectively. To take this into account, the pressure attenuation in decibels is defined as $20\mathrm{Log}_{10}(p_a/p_0)$, so that the two decibel definitions coincide:

$$10\mathrm{Log}_{10}(I_a/I_0) = 20\mathrm{Log}_{10}(p_a/p_0). \tag{16.7}$$

The ultrasound attenuation in soft tissue is of the order of 1 dB/cm at a frequency of 1 MHz, and it increases approximately linearly with frequency, so that it is usually indicated as $\sim$1 dB/(cm-MHz).

16.1.2 The source of contrast in ultrasound imaging

The characteristic impedance Z plays a key role in ultrasound imaging. In fact, a characteristic impedance mismatch at a tissue boundary determines the reflection and refraction of ultrasound waves in a manner similar to the way a refractive index mismatch determines the reflection and refraction of light. It is the reflection and scattering of ultrasound waves within tissue that forms the basis for ultrasound imaging, such that gradients and discontinuities in the characteristic impedance are the features that are displayed in ultrasound images.

In Chapter 7, we discussed the important role played by the size of scattering particles relative to the wavelength of light in determining scattering probabilities. Similarly, the wavelength of ultrasound is the basic reference length to measure the scale of spatial variations in the characteristic impedance. The wavelength of ultrasound is given by the ratio of the ultrasound speed (~1540 m/s in tissue) to the ultrasound frequency (~1–15 MHz in medical imaging applications), and is therefore of the order of 0.1–1.5 mm. Acoustic scattering is most effectively generated by characteristic impedance inhomogeneities or gradients on a length scale that is comparable to the wavelength of the ultrasound. The interfaces of macroscopic tissue structures (having length scales of several millimeters to centimeters) act as specular reflectors for ultrasound, whereas interface roughness and individual acoustic inhomogeneities of sizes comparable to or smaller than the wavelength (submillimeter) determine multidirectional ultrasound scattering.

In *pulse-echo ultrasound* imaging, the time delay between the interrogating ultrasound pulse and the back-reflected echo pulse is translated into the distance of the reflecting structure within the tissue by considering a constant speed of ultrasound (~1540 m/s in soft tissue). The amplitude of the detected echo pulses depends on the gradient of characteristic impedance mismatch at a reflecting tissue boundary, as well as the acoustic intensity that reaches the boundary. The combination of multiple lines of sight (called A-lines) results in the generation of ultrasound images.

16.1.3 Ultrasound transducers

In ultrasound imaging, the ultrasound transducer is the device that generates the ultrasound waves that are sent to the tissue. It also detects the ultrasound waves reflected back from the tissue. The heart of an ultrasound transducer is a *piezoelectric crystal* that converts an electrical voltage into a mechanical stress, and vice versa, by deformation of its polar unit cells or re-orientation of its molecular components, which feature an electric dipole moment. The word piezoelectricity, which indicates the generation of an electrical signal in response to an exerted mechanical pressure, derives from the Greek word *piezein*, which means "to press" or "to squeeze." With the exception of specialized applications (for example, Doppler ultrasound), the same transducer is used for the generation of ultrasound pulses (typically a few microseconds long) and for the detection of the ultrasound echoes (over a time, of the order of 0.5 ms, which is long enough for all the echoes to return to the tissue surface). Therefore, an ultrasound transducer acts as a source for a short time and as a detector for the majority of the time, say ~99.5% of the time, corresponding to a duty cycle of 0.5%.

Ultrasound transducers may consist of single elements or, more commonly, of arrays of individual piezoelectric elements, which are driven individually to control the size, location, direction, and/or focusing of the emitted ultrasound beam. In resonance transducers, the center frequency of the emitted ultrasound is determined by the crystal thickness (the resonance frequency corresponds to a constructive interference of standing ultrasound waves having a wavelength equal to twice the thickness, i.e., the round-trip travel distance within the crystal), whereas the bandwidth is determined by a backing material that dampens the ultrasound wave in the crystal, thereby reducing the quality factor (Q) of the mechanical oscillation. (Effectively, the bandwidth is inversely related, by Fourier transform, to the ultrasound pulse duration.) Ultrasound imaging requires the emission of short pulses (typically a few μs in duration), corresponding to a bandwidth of the order of hundreds of kHz. By contrast, photoacoustic ultrasound methods require ultra-wide bandwidths of tens of MHz.

16.1.4 Spatial resolution in ultrasound imaging

In ultrasound imaging, the spatial resolution along the direction of propagation of ultrasound, the *axial resolution*, is determined by the pulse duration or, equivalently, by its bandwidth. The two closest structures that can be discriminated (along the direction of propagation) by an ultrasound pulse of duration t_p are separated by an axial distance of $\frac{1}{2}v_a t_p$. In fact, the echoes generated by two such structures are separated in time by t_p (taking into account the factor of 2 for the round trip travel of ultrasound in pulse-echo mode), which is the shortest time separation of the echoes from the two structures that allows for them to be resolved. The axial resolution is therefore $\frac{1}{2}v_a t_p$, and, recalling the inverse relationship between the ultrasound pulse duration (t_p) and the acoustic frequency bandwidth (BW_a), it can also be expressed as $\sim\frac{v_a}{2BW_a}$. Therefore, the axial resolution in ultrasound imaging is inversely proportional to the frequency bandwidth. To achieve a high axial resolution, one should use broadband ultrasound, which necessitates high center frequencies of the order of tens of MHz. The axial resolution of conventional clinical diagnostic ultrasound is of the order of a few hundred microns to ~1 mm and it can be as good as ~100 μm for intravascular ultrasound (IVUS) imaging, which operates at center frequencies of 20–40 MHz. The axial resolution is approximately constant as a function of depth in the tissue.

The *lateral resolution* in ultrasound imaging is determined by the shape, in particular the cross-sectional area, of the ultrasound beam. Curved transducers, the introduction of acoustic lenses, or phased arrays of ultrasound transducers achieve a minimal cross-sectional area of the beam at a certain distance, which is identified as the focal distance of the curved transducer, acoustic lens, or transducer phased array. In conventional diagnostic ultrasound, the lateral resolution is typically a few millimeters (down to ~250 μm in intravascular ultrasound), and it degrades outside the focal region of the ultrasound beam. Because higher ultrasound frequencies tend to narrow the ultrasound beam, the lateral resolution improves at higher ultrasound frequencies.

16.2 Acousto-optic spectroscopy and imaging by ultrasonic tagging of light

16.2.1 Mechanisms of ultrasonic modulation of light intensity

The modulated pressure (given in Eq. (16.1)) associated with an ultrasound wave generates a local modulation of the tissue optical properties, namely the absorption coefficient (μ_a), reduced scattering coefficient (μ_s'), and refractive index (n). One may expect that such a modulation of the optical properties would result in a measurable modulation of the optical intensity, if the ultrasound beam intersects the optical region of sensitivity. It turns out that this is not the case, as can be appreciated on the basis of the following two observations. The first observation is that an ultrasound wave consists of alternating compressions and rarefactions, with a spatial period given by the ultrasound wavelength, or ~1 mm. In this pattern, the alternating regions with higher absorption and scattering (compressions) and lower absorption and scattering (rarefactions) tend to compensate each other in their contributions to optical signals that probe tissue volumes that encompass multiple ultrasound wavelengths. The second observation is that even the single compression (or rarefaction) zone that occurs over half a wavelength of ultrasound corresponds to a small change in the optical properties.

How small this change is can be estimated as follows. The tissue absorption and scattering coefficients are proportional to the density of absorbers and scatterers through the molar extinction coefficient (see Eq. (2.1)) and the scattering cross section (see Eq. (7.35)), respectively. Consequently, the absorption and scattering coefficients can be expressed as $\mu \propto \rho$ (where μ indicates either the absorption or scattering coefficient, and ρ indicates the tissue density, which is itself proportional to the density of absorbing or scattering particles), so that relative changes in the optical coefficients are equal to relative changes in tissue density ($\frac{\Delta\mu}{\mu} = \frac{\Delta\rho}{\rho}$). Because the tissue density associated with a given mass element is inversely related to its volume, its relative change is $\frac{\Delta\rho}{\rho} = -\frac{\Delta V}{V}$, which takes a maximum value for the ultrasound wave that can be written in terms of the amplitude of the pressure change (p_0), recalling the definition of the bulk modulus ($B = -p/(\frac{\Delta V}{V})$):

$$\left.\frac{\Delta\mu}{\mu}\right|_{max} = \left.\frac{\Delta\rho}{\rho}\right|_{max} = \frac{p_0}{B}. \tag{16.8}$$

By considering an ultrasound peak pressure of 1 MPa and a typical bulk modulus of soft tissue of $B = 2.9$ GPa, the maximum relative change in the optical coefficients is ~0.034%, too small to be detectable in typical optical measurements.

So, if it is not the local perturbation in the tissue optical properties, what is the mechanism responsible for ultrasonic tagging of light? It is a coherent mechanism associated with perturbations to the optical phase, which result from the ultrasound-induced modulation of the refractive index and displacement of scattering centers. The interaction of an optical wave with the spatially and temporally modulated refractive index, and with collective structural oscillations associated with an acoustic wave, is described in terms of *Brillouin scattering* (which we mentioned in the introduction of Chapter 7). In tissue optics, the general treatment of Brillouin scattering is complicated by the diffusive nature of light propagation over distances of several wavelengths of the acoustic wave, given the submillimeter photon mean free path in tissue.

We have previously described how the motion of red blood cells in the blood stream results in spectral shifts and intensity fluctuations that are used by Doppler flowmetry (Section 7.5.1), dynamic light scattering (Section 7.5.2), and diffuse correlation spectroscopy (Sections 8.9.1 and 9.9) to measure blood flow. In the case of ultrasound, the tissue scatterers undergo a collective, periodic motion at the ultrasound frequency, which is superimposed on any underlying Brownian or flow-related motion. Such collective, periodic motion results in a modulation of the distribution of the optical pathlengths, which in turn modulates the speckle pattern generated by a coherent light source. This process results in the following electric field autocorrelation function $G_1(\tau)$ (Leutz and Maret, 1995):

$$G_1(\tau) = \int_l^\infty p(s)\langle E_s(t)E_s^*(t+\tau)\rangle ds, \tag{16.9}$$

where l is the scattering mean free path (assumed to be much longer than the optical wavelength); s is the photon pathlength in the medium; $p(s)$ is its probability density function; E_s is the electric field contributed by photons that traveled over pathlength s (which is supposed to be uncorrelated with the electric field from different photon pathlengths according to the weak scattering approximation); * indicates the complex conjugate, and the brackets indicate a time average. According to the Wiener-Khinchin theorem (named after the American mathematician

Norbert Wiener [1894–1964] and the Russian mathematician Aleksandr Khinchin [1894–1959]), the intensity power spectrum is the Fourier transform of the field autocorrelation function, such that the spectral intensity (I_n) at a frequency that is a multiple of the acoustic frequency ($n\omega_a$) can be written as follows:

$$I_n \propto \frac{1}{T_a} \int_0^{T_a} \cos(n\omega_a \tau) G_1(\tau) d\tau, \tag{16.10}$$

where T_a is the period of the acoustic wave. The units of $G_1(\tau)$ are those of an electric field squared, which represents an energy density in the Gaussian system of units, whereas in SI units the energy density is the product of the permittivity times the electric field squared. The product of the optical energy density times the speed of light results in an optical intensity. The proportionality sign in Eq. (16.10) is there to avoid the need to consider constant dimensional factors, which do not play a role in the measurements of ultrasonically tagged light. The modulation depth (M) is defined as the ratio of the spectral intensity at the ultrasound frequency (I_1) to the unmodulated intensity (I_0), and is a relevant quantity in acousto-optic measurements.

By considering ultrasound-induced effects related both to the modulation of the refractive index and to the collective motion of scattering particles, the following expression has been derived for the *normalized* field autocorrelation function ($g_1(\tau)$) in the case of optical transmission through a transversely infinite slab of thickness L (the slab is illuminated on one side by a coherent optical plane wave, and a plane ultrasound wave propagates through the slab) (Wang, 2001):

$$g_1^{(\text{slab})}(\tau) = \frac{\frac{L}{l} \sinh\left\{kA\sqrt{\alpha_{l,k_a}\left[1-\cos(\omega_a\tau)\right]}\right\}}{\sinh\left\{kA\frac{L}{l}\sqrt{\alpha_{l,k_a}\left[1-\cos(\omega_a\tau)\right]}\right\}}, \tag{16.11}$$

where sinh is the hyperbolic sine function ($\sinh(x) = \frac{e^x - e^{-x}}{2}$), l is the scattering mean free path, k is the optical wavenumber, A is a measure of the amplitude of the ultrasound wave that is proportional to the acoustic pressure p_0, and α_{l,k_a} is a function of l and of the ultrasound wavenumber (k_a). We recall that the normalized field autocorrelation function is defined as $g_1(\tau) = G_1(\tau)/G_1(0)$, so that the proportionality relation in Eq. (16.10) holds also when $G_1(\tau)$ is replaced with $g_1(\tau)$. In the weak modulation limit expressed by $kA\frac{L}{l}\sqrt{\alpha_{l,k_a}} \ll 1$, and by considering that $L \gg l$ in the strong scattering case considered here, the normalized field autocorrelation function of Eq. (16.11) reduces to (Wang, 2004):

$$g_1^{(\text{slab})}(\tau) \approx 1 - \frac{1}{6}\left(\frac{L}{l}\right)^2 (kA)^2 \alpha_{l,k_a}\left[1-\cos(\omega_a\tau)\right], \tag{16.12}$$

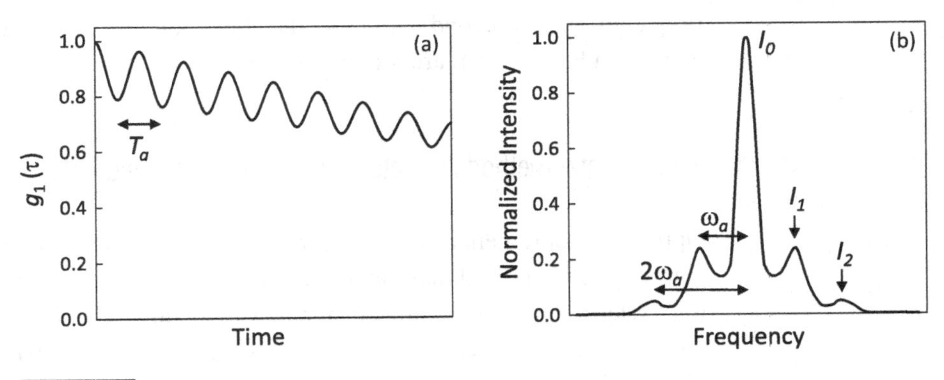

Figure 16.2

(a) Normalized field autocorrelation function ($g_1(\tau)$) as a function of the correlation time; the oscillation has the period of the acoustic wave (T_a) and its decay and modulation dampening are due to particle movement (Brownian motion, blood flow, etc.), which is not considered in Eq. (16.10). (b) Intensity spectrum featuring the Rayleigh peak (I_0) and the sidebands shifted by the acoustic angular frequency $\omega_a(I_1)$ and by twice $\omega_a(I_2)$.

so that, from Eq. (16.10), the modulation depth becomes:

$$M \equiv \frac{I_1}{I_0} \approx \frac{1}{12}\left(\frac{L}{l}\right)^2 (kA)^2 \alpha_{l,k_a} \propto p_0^2 \propto I_a. \tag{16.13}$$

The acoustic modulation, therefore, depends on the square of the acoustic pressure, which in turn is proportional to the acoustic intensity. For this reason, the spatial origin of ultrasonically tagged light is the focal volume of the ultrasound beam, where the acoustic intensity is maximal. Such a localized region where the acoustic waves introduce a modulation in the optical signal acts as a *virtual source of modulated light* that probes the tissue at that particular location, and which can be scanned within the tissue for imaging.

The normalized field autocorrelation function ($g_1(\tau)$) oscillates at the acoustic frequency ω_a, as can be immediately seen in the approximated expression for a slab represented by Eq. (16.12). The treatment above, however, neglects Brownian motion and, most importantly for biological tissues, the hemodynamics associated with blood flow. These dynamic effects tend to decorrelate the speckles, therefore inducing a decay of the autocorrelation function and damping the modulation depth of the optical signal. The effect of speckle decorrelation with time is qualitatively illustrated in Figure 16.2(a), and it can be exploited for blood flow measurements based on ultrasonically modulated light. The intensity spectrum corresponding to the field autocorrelation function of Figure 16.2(a) is shown in Figure 16.2(b) as a function of frequency. The intensity spectrum shows the Rayleigh peak (unmodulated light) and the sidebands associated with amplitude modulation at the acoustic frequency and its second harmonic; the corresponding intensities I_0, I_1, and I_2 (as defined in Eq. (16.10)) are identified in Figure 16.2(b).

16.2.2 Experimental methods to detect ultrasonically tagged light

Optical measurements of ultrasonically tagged light need to detect the signal component that is modulated at the ultrasound frequency, which is superimposed on the background of unmodulated signal. The modulated optical signal results from the ultrasound-induced temporal oscillations in the speckle pattern at the detector plane, which translates into spectral components at the ultrasound frequency. Experimental methods are implemented either in the time domain (to measure the temporal speckle dynamics depicted in Figure 16.2(a)) or in the frequency domain (to measure the spectral intensity shown in Figure 16.2(b)), the two methods constituting a Fourier transform pair.

Time-domain detection consists of direct measurements of the speckle pattern oscillations. While measurements can be done on a single speckle (Leutz and Maret, 1995) or using a single photodetector that collects multiple speckles (Wang and Zhao, 1997), the signal-to-noise ratio is significantly enhanced by parallel detection of a large number of speckles. (The signal-to-noise ratio scales with the square root of the number of speckles measured in parallel, assuming that they are uncorrelated.) Parallel multi-speckle detection can be realized by a CCD camera placed at a proper distance from the sample to match the mean speckle size to the CCD pixel size, so as to take full advantage of the individual speckle modulation (Lévêque et al., 1999).

The experimental setup for such parallel acquisition is shown in Figure 16.3. The fact that the CCD camera's frame rate ($\sim$100 Hz) is not high enough to follow the $\sim$1–15 MHz modulation due to ultrasound is addressed by using a light source that is modulated in synch with the ultrasound. Four illumination sequences are employed to provide illumination during each quarter of the ultrasound period. The resulting four CCD data sets, each corresponding to a specific phase shift between the periodic pulsed illumination and the ultrasound wave, are combined to yield the desired modulation depth at the ultrasound frequency. This approach is based on synchronous illumination rather than synchronous lock-in detection.

Frequency-domain methods employ interferometric techniques to suppress the unmodulated optical signal (I_0 in Figure 16.2(b)) and selectively detect the ultrasound-modulated light. A confocal Fabry-Pérot interferometer (CFPI) has been used because of its ability to provide parallel speckle processing and for its high *étendue*, the product of the acceptance solid angle times the detection area,

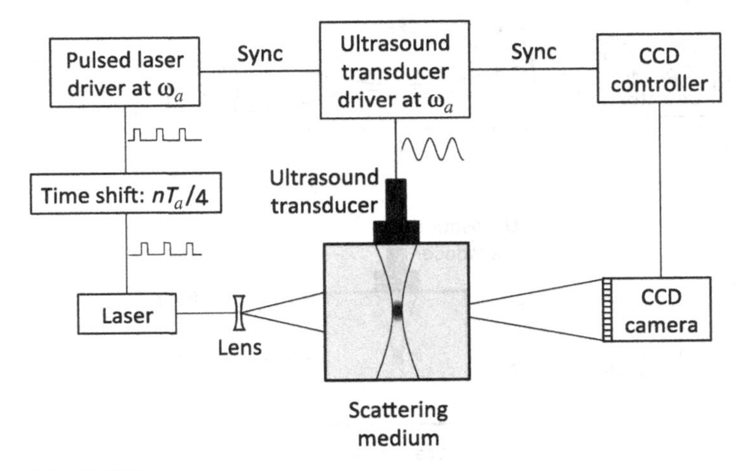

Figure 16.3

Experimental setup for measurements of the dynamic speckle pattern resulting from light modulation by ultrasound. The laser beam is expanded by a diverging lens to illuminate a broad sample area (a few centimeters in diameter), and optical detection is performed with a CCD camera. The ultrasound transducer is driven at the ultrasound frequency ω_a, which is also the repetition frequency of pulsed laser illumination, with pulse duration equal to 1/4 the period of the ultrasound wave (T_a). By time shifting the laser pulse illumination by $nT_a/4$ (with $n = 0,1,2,3$), one achieves synchronous illumination over each quarter of the ultrasound period, so that the speckle pattern dynamics can be measured at the ultrasound frequency (~1–15 MHz) even though the CCD camera acquisition rate is much slower (~100 Hz). The laser driver, ultrasound transducer driver, and CCD controller are all synchronized (Sync).

which is highly desirable to increase the efficiency of optical collection and to optimize the signal-to-noise ratio (Sakadžić and Wang, 2004; Rousseau et al., 2009). A Fabry-Pérot interferometer (named after French physicists Charles Fabry [1867–1945] and Alfred Pérot [1863–1925]) consists of two parallel highly reflective mirrors whose separation can be tuned to select the transmission of specific wavelengths (or frequencies), which are those that realize a constructive interference of the multiple reflections between the two mirrors.

A confocal Fabry-Pérot interferometer uses two spherical mirrors having the same curvature. Their separation, defined as the cavity length (L), is equal to their radius of curvature, so that all optical paths between the mirrors go through a diffraction-limited spot at the common focal point of the mirrors. (The focal length of a spherical mirror is half its radius of curvature.) This feature allows for a relatively large acceptance angle, sensitive area, and, therefore, étendue (of the order of 0.1 mm^2 sr). The *free spectral range* (FSR) of a confocal Fabry-Pérot interferometer, defined as the frequency separation of two successive transmission maxima, is given by the speed of light (c) divided by the round trip optical

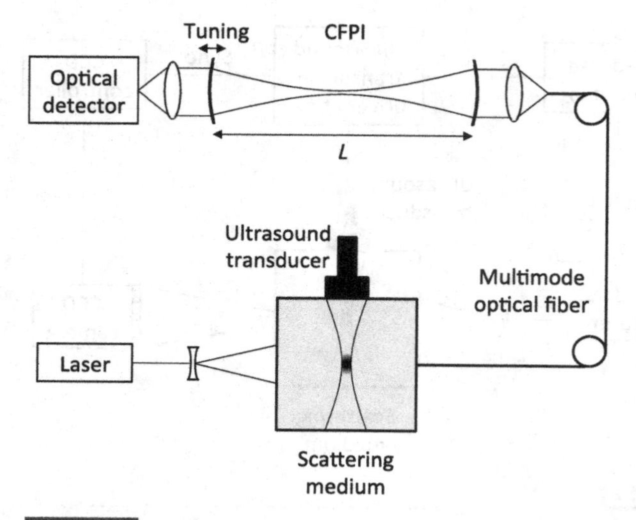

Figure 16.4

Experimental arrangement for interferometric-based measurement of the spectral intensity at a frequency shifted from the laser light frequency by an amount equal to the ultrasound frequency. Light is collected with a multimode optical fiber (~1 mm in core diameter) and coupled to a confocal Fabry-Pérot interferometer (CFPI) with cavity length L (~1 m). The CFPI is tuned to one sideband of the intensity spectrum (I_1 in Figure 16.2(b)) by the fine displacement of one cavity mirror. The optical detector is typically a PIN or avalanche photodiode.

pathlength, which is 4 times the cavity length ($4L$) because of the straight-line, figure-eight shape of the optical paths between the mirrors. For a cavity length $L \sim 1$ m, the FSR is then ~75 MHz. The *finesse* (F) is defined as the ratio of the FSR to the full-width at half-maximum of the transmission frequency peaks, with the latter being a measure of the spectral resolution of the interferometer. A finesse of ~30 yields a frequency resolution of FSR/$F \sim 2.5$ MHz, which is suitable for the detection of the spectral shifts associated with ultrasound modulation of light. This discussion explains the need for long cavity lengths (0.5–1 m) to achieve a suitable spectral resolution. One can view the Fabry-Pérot interferometer as a bandpass frequency filter whose center frequency is tuned away from the laser frequency by an amount equal to the ultrasound frequency, and whose bandwidth is given by its spectral resolution. The experimental setup for the detection of ultrasonically tagged light with a confocal Fabry-Pérot interferometer is shown in Figure 16.4.

A number of variations on the methods described above, as well as other techniques for detection of ultrasonically tagged light, can be found in Li et al. (2011), listed at the end of this chapter under further reading for ultrasonic tagging of light.

16.2.3 Applications of ultrasonically tagged light

Ultrasonically tagged light lends itself to imaging applications, referred to as *ultrasound-modulated optical tomography (UOT)* or *acousto-optic tomography (AOT)*. By modulating the optical signal, the ultrasound focal volume acts as a virtual source of modulated light that can be scanned within the tissue of interest to generate an image. The spatial resolution of this imaging approach is determined by the size of the ultrasound focal volume. The lateral resolution (in the direction orthogonal to the propagation of ultrasound) is determined by the ultrasound focal width, typically ~1 mm, at depths up to a few cm in tissue. The axial resolution (along the direction of propagation of ultrasound) is less precise (~1 cm) as a result of the elongated shape of the ultrasound focal volume, but it may be improved (down to a few mm) by shaping the acoustic wave according to a variety of methods (frequency sweep, phase randomization, pulsed ultrasound, etc.).

The source of contrast of acousto-optic imaging is a combination of the optical and acoustic properties of tissue. This is because the strength of the virtual source of modulated light depends on both the optical fluence rate and the ultrasound intensity at the acoustic focal volume. It is important to observe, however, that it is not just the local optical and acoustic properties at the focal volume that determine the strength of the ultrasonically modulated signal, but there is also a dependence on the overall spatial distribution of the fields in the tissue through which the ultrasound and light propagate. In fact, the ultrasound intensity at the focal volume depends on the attenuation experienced by the acoustic wave in its path from the transducer to the focal volume. Similarly, the optical fluence rate at the focal volume depends on the optical attenuation from the light source to the focal volume. Furthermore, the detected ultrasonically tagged optical signal depends on the optical attenuation between the virtual modulated light source (at the ultrasound focal volume) and the location of the optical detector. For this reason, quantitative acousto-optic imaging requires reconstruction techniques, based on both the modulated and the unmodulated optical signals, similar to those described in Chapter 14 for diffuse optical tomography.

Using multiple laser wavelengths, one can collect spectrally resolved ultrasonically tagged light to perform acousto-optic spectroscopy, which can yield measurements of, for example, hemoglobin oxygen saturation in blood-perfused tissues (Kim and Wang, 2007). It is also possible to take advantage of the speckle decorrelation associated with blood flow to relate the ultrasonically modulated signal to the microvascular blood flow in the ultrasound focal volume (Ron et al., 2012). Ultrasound-modulated fluorescence has also been explored, even though the incoherence of fluorescence emission leads to inefficient modulation, requiring enhancement approaches based on microbubbles (Yuan et al., 2009) or digital time reversing (Wang et al., 2012).

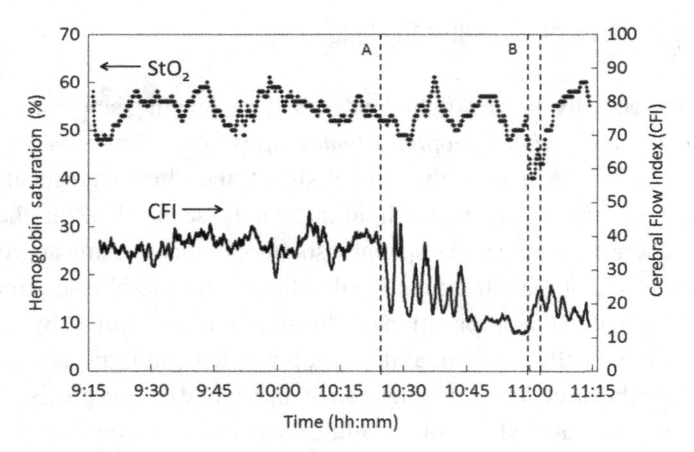

Figure 16.5

Application of ultrasonically tagged light to monitor hemoglobin saturation (StO_2) and cerebral blood flow index (CFI) in the brain of a patient undergoing coronary artery bypass grafting surgery. At time point A, the patient is placed on cardio-pulmonary bypass. Cardiopulmonary bypass parameters are modified at time point B following the measured drops in StO_2 and CFI. (Courtesy of M. Balberg, R. Shechter, N. Racheli, and M. Kamar.)

In vivo applications of ultrasonically modulated light are more challenging than ex vivo measurements because of the strong speckle decorrelation (over a time scale faster than milliseconds) associated with blood flow. However, significant progress is being made, and in vivo applications of this technology have been shown to be feasible in both animal models and human subjects. Pilot clinical studies of regional brain oxygenation with ultrasonically tagged light have been performed on patients with severe traumatic brain injury (Rosenthal et al., 2014). A measurement performed in a clinical setting, reported in Figure 16.5, shows temporal traces of hemoglobin saturation and cerebral blood flow with ultrasonically tagged light collected from the prefrontal cerebral cortex of a patient undergoing coronary artery bypass grafting surgery. In this case, the ultrasound frequency is 1 MHz and the optical wavelengths for spectroscopic determination of hemoglobin saturation are 785, 808, and 830 nm.

16.3 Photoacoustic imaging

16.3.1 Generation of ultrasound by pulsed illumination

The *photoacoustic effect* (also referred to as *optoacoustic effect*) is the generation of a sound-pressure wave as a result of the thermal expansion and relaxation of an object that absorbs light. Such thermal expansion is the result of the conversion of absorbed optical energy into heat by non-radiative relaxation. If the light intensity is periodically modulated, the generated pressure wave is also periodic as a result of cyclic heating and cooling of the absorbing object. Such optical modulation at ultrasound frequencies results in the generation of ultrasound. If the light intensity is a short pulse, the generated pressure perturbation is also pulsed as a result of the fast thermal expansion and relaxation of the absorbing object. The frequency bandwidth of the pressure perturbation is inversely related to its temporal pulse width, so that a ~0.1 μs pulse duration corresponds to a bandwidth of ~10 MHz. Pulsed illumination is most commonly employed in photoacoustic imaging because of the valuable spatial information associated with timing the detected ultrasound pulse. Simply, the distance between an absorbing object and the ultrasound transducer is given by the speed of ultrasound in tissue (~1540 m/s) multiplied by the propagation time of the ultrasound pulse from the object to the transducer.

In describing the photoacoustic effect, there are two relevant characteristic times to consider. The first characteristic time is the acoustic propagation time within the smallest volume element that one wishes to discriminate (the spatial resolution limit). If we denote its volume as V, and its linear dimension as L, ($L \sim V^{1/3}$), then the *acoustic confinement time* is given by:

$$\tau_a = \frac{L}{v_a}, \tag{16.14}$$

with v_a the speed of ultrasound. Since the speed of ultrasound in soft tissue is ~1540 m/s, the acoustic confinement time in a 1 mm^3 volume element is $\tau_a \sim 600$ ns. The second characteristic time is the heat propagation time within the volume element, the so-called *thermal confinement time*, which, in the three-dimensional case considered here, is given by:

$$\tau_{\text{th}} = \frac{L^2}{6\alpha}, \tag{16.15}$$

where α is the *thermal diffusivity* (SI units: m^2/s), not to be confused with α_{l,k_a} in Eqs. (16.11)–(16.13). This relationship between the square of the characteristic distance and the diffusion time coincides with the diffusion expression for Brownian motion reported in Eq. (9.75). (In the literature, one finds the thermal confinement time commonly defined with a factor "4" instead of "6" in the denominator, which would refer to a 2D case, such as the thermal diffusion from axial heating along a collimated laser beam.) The thermal diffusivity is, in turn, defined as:

$$\alpha = \frac{k}{\rho c_p}, \tag{16.16}$$

where k is the *thermal conductivity* (SI units: W/(m-K)), ρ is the mass density (SI units: kg/m^3), and c_p is the *specific heat at constant pressure* (SI units: J/(kg-K)). The thermal diffusivity of biological tissue is similar to that of water, of the order of 1.4×10^{-7} m^2/s, so that the thermal confinement time in a 1 mm^3 tissue volume is $\tau_{th} \sim 1.2$ s. These two characteristic times scale differently with the linear size of the volume element L: τ_a linearly and τ_{th} quadratically. Therefore, a smaller resolution limit of, say, 0.1 mm corresponds to $\tau_a \sim 60$ ns and $\tau_{th} \sim 12$ ms.

In photoacoustic imaging, it is important that the optical irradiation time (τ_{opt}) be much shorter than both τ_a and τ_{th} to maximize the image contrast (as relates to an optimal conversion efficiency of optical into acoustic energy) and resolution (as relates to an optimal spatial confinement for converting thermal into acoustic energy). This condition is easily realized with laser pulse durations of the order of nanoseconds (common for Q-switched solid-state lasers). Even taking into account the photon time-of-flight distribution within the tissue (which may extend for several nanoseconds, as we have seen in Chapter 11), one achieves an optical irradiation time τ_{opt} that is fast compared to τ_a and τ_{th}, thus achieving the desired conditions of *acoustic stress confinement* and *thermal confinement*, respectively. Under these conditions, and under additional conditions typically fulfilled in tissue photoacoustics, the pressure perturbation p obeys the following inhomogeneous wave equation (Sigrist, 1986), known as the *photoacoustic wave equation*:

$$\nabla^2 p(\mathbf{r}, t) - \frac{1}{v_a^2} \frac{\partial}{\partial t^2} p(\mathbf{r}, t) = -\frac{\beta}{c_p} \frac{\partial}{\partial t} [\mu_a(\mathbf{r}) \phi(\mathbf{r}, t)] . \tag{16.17}$$

where β is the *thermal expansion coefficient* (defined as the relative volume change per unit temperature change; SI units: K^{-1}), and the product $\mu_a(\mathbf{r})\phi(\mathbf{r}, t)$ (absorption coefficient times the optical fluence rate) represents the absorbed optical energy density (which is assumed to be completely converted into heat) per unit time.

Over the optical irradiation time (τ_{opt}), the energy absorbed within volume V at position $\mathbf{r}$ is $Q \sim \mu_a(\mathbf{r})\psi(\mathbf{r})V$, where $\psi(\mathbf{r})$ is the optical fluence associated with the illumination pulse. This absorbed optical energy, Q, is converted into thermal energy, which causes a temperature change $\Delta T = Q/(mc_p)$, where m is the volume's mass and c_p is its specific heat. In turn, this temperature change, ΔT, induces a relative volume change $\frac{\Delta V}{V} = \beta \Delta T$, where β is the thermal expansion coefficient. Finally, the relative volume change, $\Delta V/V$, corresponds to a pressure change $p_0 = B\frac{\Delta V}{V}$, where B is the bulk modulus. By combining all of the above conversions, one can write:

$$p_0(\mathbf{r}) = B\frac{\Delta V}{V} = B\beta\Delta T = B\beta\frac{Q}{mc_p} \cong B\beta\frac{\mu_a(\mathbf{r})\psi(\mathbf{r})V}{mc_p} = B\beta\frac{\mu_a(\mathbf{r})\psi(\mathbf{r})}{\rho c_p}, \tag{16.18}$$

where $\rho = m/V$ is the tissue mass density. Given typical values of bulk modulus ($B \sim 2.9$ GPa) and thermal expansion coefficient ($\beta \sim 3 \times 10^{-4}$ K^{-1}) in soft tissues, 1 mK of temperature increase corresponds to a pressure change of almost 1 kPa (~870 Pa). (In photoacoustic imaging, temperature increases are of the order of 10 mK, and the associated pressure waves have amplitudes smaller than 10 kPa, which is much less than peak pressures of ~1 MPa generated by diagnostic ultrasound imaging systems.) Since this temperature increase occurs within volume V, it generates a pressure wave through a surface area $\sim 4\pi r^2$ (where r is the equivalent radius of volume V) that propagates in all directions so that it spreads over an area $4\pi R^2$ at a distance R from volume V. As a result, even ignoring ultrasound attenuation, the acoustic intensity associated with the pressure perturbation reaching the tissue surface at a distance R from volume V will be reduced by a factor $(r/R)^2$. For $r \sim 1$ mm and $R \sim 1$ cm, this results in an intensity reduction by a factor of ~100, which translates into a pressure reduction by a factor of ~10 (recall the quadratic relationship between acoustic intensity and pressure reported in Eq. (16.6)). This geometrical effect accounts for a pressure perturbation of ~100 Pa at the tissue surface as a result of a local temperature increase of ~1 mK in the target volume. Of course, ultrasound attenuation would further decrease this estimate. In any case, a pressure perturbation of 10–100 Pa is within the detection capability of commercial ultrasound transducers, which can be as low as ~1 Pa.

By recalling that the bulk modulus can be written as $B = \rho v_a^2$ (see Eq. (16.5)), the amplitude of the pressure perturbation in Eq. (16.18) becomes:

$$p_0(\mathbf{r}) \cong \frac{\beta v_a^2}{c_p} \mu_a(\mathbf{r}) \psi(\mathbf{r}), \tag{16.19}$$

where the factor $\beta v_a^2/c_p$ is the dimensionless *Grüneisen coefficient* (often denoted by Γ) (named after the German physicist Eduard Grüneisen [1877–1949]). Equation (16.19) shows how the optical fluence associated with the illumination pulse (ψ, which depends on the absorption and scattering properties of tissue) gets converted into a pressure perturbation (p_0) through the optical absorption (μ_a) and some thermodynamic and mechanical properties of tissue (B, c_p, β, ρ). Equation (16.19) may appear to indicate that the pressure perturbation is only determined by the absorption coefficient at the volume element of interest. This is not strictly true, because the scattering coefficient within volume V also plays a role in affecting the local fluence rate, which in turn affects the amount of absorbed energy through scattering-induced absorption (a concept introduced in Section 10.2.1), since a change in the scattering coefficient affects the optical pathlength within the volume V. However, scattering effects are usually much smaller than absorption effects, also because absorption contrast tends to be greater than scattering contrast in biological tissue. Therefore, the source of contrast in photoacoustic imaging of tissue is mainly the absorption coefficient.

It is very important to note, however, that the photoacoustic pressure is not proportional to the absorption coefficient alone, but rather to the absorbed energy

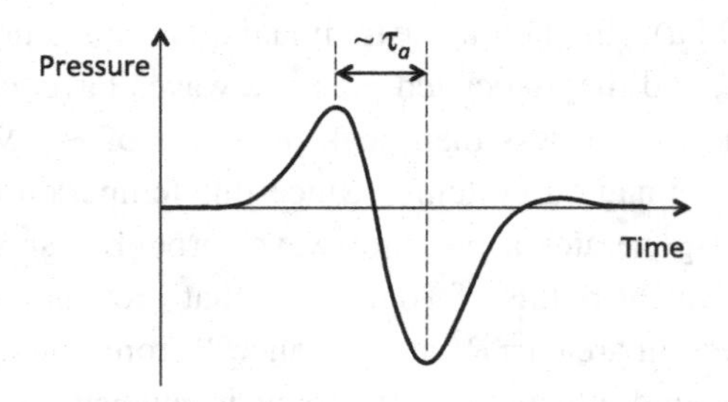

Figure 16.6

Photoacoustic pressure pulse generated by a localized optical absorber in response to pulsed illumination. The time-difference between the peaks of the positive, compression phase and the negative, rarefaction phase is of the order of the acoustic confinement time (τ_a), i.e., the linear size of the absorber divided by the speed of ultrasound.

density, which is given by the product of the absorption coefficient and the optical fluence at position $\mathbf{r}$ ($\mu_a(\mathbf{r})\psi(\mathbf{r})$). Moreover, it is apparent that the photoacoustic pressure at $\mathbf{r}$ depends on both the local absorption coefficient ($\mu_a(\mathbf{r})$) and the global optical properties (absorption, scattering, anisotropy factor, etc.) that affect the optical fluence at position $\mathbf{r}$ ($\psi(\mathbf{r})$). For this reason, performing quantitative measurements and, in particular, quantitative spectroscopy measurements with photoacoustic methods is complicated by the cross-talk between the global optical properties and the local absorption at each voxel.

The time course of the photoacoustic signal depends on a number of parameters, including the spatial distributions of the optical fluence and optical absorption in the investigated tissue. For a localized optical absorber, the photoacoustic pressure pulse is bipolar, with a positive (compression) phase followed by a negative (rarefaction) phase. The temporal separation of the positive and negative pressure peaks is of the order of the acoustic confinement time (τ_a) (Sigrist and Kneubühl, 1978), which represents the propagation time of the acoustic wave within the localized absorber. Such a photoacoustic pressure signal generated by a localized absorber is illustrated in Figure 16.6, for which it is assumed that the pulse duration of the optical fluence is short compared to τ_a.

16.3.2 Instrumentation for photoacoustic imaging

The illumination pulse, nanoseconds in duration, typically has a repetition rate of ~10 Hz, and is commonly provided by a Q-switched laser (Nd:YAG laser: emitting at 1064 nm, frequency-doubled at 532 nm; alexandrite laser: emitting at 757 nm, etc.). A Q-switched laser provides pulsed laser emission by a rapid switch of the optical cavity quality factor (Q). A variable attenuator within the optical cavity allows for a build-up in the population inversion in the laser medium when it acts as an optical absorber, thereby suppressing Q. A rapid increase in the quality factor (a Q-switch), achieved by suddenly removing the optical attenuation within the cavity, allows for a strong laser emission through the rapid depletion of the accumulated population inversion in the gain medium. Spectroscopic applications may use a tunable dye laser, pumped by a frequency-doubled Nd:YAG laser (530 nm) and emitting in the visible-NIR range, to allow for measurements at selected wavelengths within the visible and NIR spectral region of interest in biomedical optics. (Photoacoustic imaging applications typically employ wavelengths in the range 530–900 nm, the shorter wavelengths pertaining to imaging at shallow depths.) Specifically, spectroscopic photoacoustic methods enable measurements of oxygen saturation of hemoglobin according to the principles described in Section 15.1.3.1.

The laser emission may be coupled into a multimode optical fiber and delivered to the tissue surface by a set of lenses at the emitting end of the optical fiber. In photoacoustic imaging of tissues it is often advantageous to distribute the optical fluence over the entire tissue volume of interest, because the image contrast is mainly associated with optical absorption (so that it is desirable to maximize the optical fluence throughout the imaged volume), whereas the spatial resolution is mainly determined by the properties of the weakly scattered thermally generated ultrasound. Thus, spatial localization in photoacoustic imaging is based on the timing of the detected ultrasound and on the focal properties of the ultrasound transducer, rather than on the spatial features of the optical illumination field (with the exception of optical-resolution photoacoustic microscopy described below in Section 16.4.3). The expansion of the illuminated area on the tissue surface also allows for delivery of a larger total optical energy, since safety requirements pose limitations on the radiant exposure, i.e., on the energy per unit area. As described in Section 13.1.4, maximum permissible skin exposure to a nanosecond laser pulse in the red and NIR spectral region is 20 mJ/cm^2.

The ultrasound detectors in photoacoustic imaging must have an ultra-wide frequency response (tens of MHz, or even ~100 MHz in some microscopy applications) to be sensitive to the large frequency content of the photoacoustic pulses, which feature fast pressure transients resulting from sharp or spatially confined tissue structures. Such large bandwidth requirement translates into a high axial resolution in photoacoustic images. Typically, photoacoustic detectors are ultrasound transducers, based on piezoelectric materials, which are also used in ultrasound imaging. Single transducers (mechanically scanned) or transducer arrays can be used. However, an optical method to detect ultrasound, based on a Fabry-Pérot polymer film interferometer (see Section 16.2.2 for the description of a Fabry-Pérot interferometer), has also been described (Zhang et al., 2008).

16.4 Photoacoustic tomography and microscopy

16.4.1 The range of penetration depths afforded by photoacoustic imaging

One of the defining features of photoacoustic imaging is its ability to investigate biological tissues over a wide range of depths, while retaining good spatial resolution. In fact, the combination of excitation by diffuse optical methods (which achieve large penetration depths of several centimeters in the near-infrared optical window) and detection by ultrasound imaging methods (which are intrinsically able to probe tissues at depths of up to ~10 cm using frequencies of a few MHz) affords measurements to the same depths (several centimeters) that are accessible to diffuse optical imaging (as presented in Chapter 14). At the other extreme, at depths of ~1 mm or less, photoacoustic imaging can take advantage of localized excitation by focused illumination and by the high resolution afforded by ultrasound imaging at high frequencies (tens of MHz). Therefore, photoacoustic imaging can operate at depths of several centimeters in the optical diffusion regime, while taking advantage of the better spatial resolution of ultrasound imaging, at depths of hundreds of microns to ~1 mm in a microscopy configuration, as well as at meso-scales that cover the range between microscopic and macroscopic scales. Such a capability for studying tissues over a broad depth range of hundreds of microns to several centimeters is unique to photoacoustic imaging.

As one would expect, the spatial resolution of photoacoustic imaging scales with the penetration depth. Keep in mind that photoacoustic imaging is essentially an ultrasound imaging technique. (It differs from conventional ultrasound imaging in the mechanism for generating the ultrasound waves within the tissue.) Therefore, the bandwidth of the detected ultrasound pulses is directly linked to the achievable image resolution. Even though the bandwidth of the pressure waves generated by the photoacoustic process (induced by nanosecond optical pulses) may have a broad frequency band (as a result of the temporally short photoacoustic pulses), such a broad frequency band is destined to shrink as the pressure waves propagate in the tissue as a consequence of the strong propagation attenuation at high frequencies. Ultimately, this is the limiting factor for spatial resolution, and one can see why the farther an ultrasound pulse has to travel from the imaged tissue volume to the ultrasound transducer, the narrower its bandwidth and the poorer the achievable spatial resolution. To provide some reference values, spatial resolution can be as

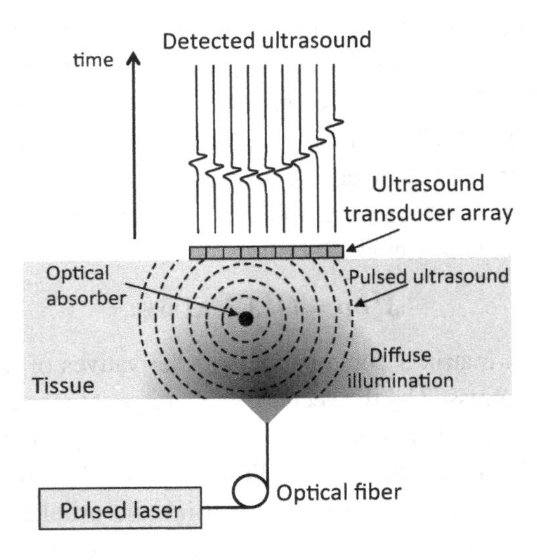

Figure 16.7

Schematic arrangement for photoacoustic tomography (PAT). A nanosecond laser pulse is delivered to the tissue surface by an optical fiber to provide diffuse illumination over an extended tissue volume. A localized optical absorber generates a photoacoustic pulse that is detected at the tissue surface by an array of ultrasound transducers. The time of detection is proportional to the distance between the optical absorber and the transducer element.

good as hundreds of microns at depths of ~1 cm, tens of microns at depths of ~1 mm, and down to microns at depths of ~100 μm.

16.4.2 Photoacoustic tomography (PAT)

In photoacoustic tomography (PAT), the laser pulse is delivered to a broad tissue area, to diffusely distribute optical fluence over an extended tissue volume. The photoacoustically generated ultrasound is then detected at the tissue surface by a mechanically scanned single transducer or by a phase-correlated transducer array. Similarly to ultrasound imaging, the time of arrival of the ultrasound pulses is translated into a distance by assuming an approximately constant speed of ultrasound in tissue. Thus, image sectioning is achieved by choosing different time windows associated with different depths. A schematic diagram of the experimental approach to photoacoustic tomography is shown in Figure 16.7.

The generation of PAT images can also be accomplished using tomographic methods, including backprojection (introduced in Section 14.2 for diffuse optical imaging), which is conceptually based on backprojecting the detected ultrasound signals onto a spherical surface centered at the detection element and having a radius given by the detection time multiplied by the speed of ultrasound (Xu and Wang, 2005). Model-based reconstruction methods use the photoacoustic wave equation (Eq. (16.17)) to represent the optical generation of ultrasound in regularized solutions to the inverse problem such as described in Section 14.7. In the case of photoacoustic tomography, the regularized inverse problem can be expressed as follows (Jiang et al., 2006):

$$\mathbf{J}^{\mathrm{T}}\boldsymbol{\delta}\mathbf{p} = (\mathbf{J}^{\mathrm{T}}\mathbf{J} + \lambda\mathbf{I})\boldsymbol{\delta\chi}, \tag{16.20}$$

where the Jacobian matrix $\mathbf{J}$ contains the first derivatives of the measured ultrasound pressures (expressed by the vector $\mathbf{p}$) with respect to the acoustic and optical properties in each tissue voxel (expressed by the vector $\boldsymbol{\chi}$), λ is the regularization parameter, and $\mathbf{I}$ is the identity matrix. Equation (16.20) is formally identical to Eq. (14.44), which describes the regularized inverse problem in diffuse optical tomography. In nonlinear, iterative reconstruction schemes, $\boldsymbol{\delta}\mathbf{p}$ is the vector of the difference between the measured pressures and the pressure values computed at the previous step, $\boldsymbol{\delta\chi}$ is the vector of the difference between the acoustic and optical properties (at each voxel) associated with the current and previous steps, and the Jacobian $\mathbf{J}$ is updated at each step. The method seeks to minimize the regularized objective function:

$$\psi_{\mathrm{Reg}} = \|\boldsymbol{\delta}\mathbf{p}\|^2 + \lambda\|\boldsymbol{\delta\chi}\|^2. \tag{16.21}$$

16.4.3 Photoacoustic microscopy (PAM)

Rather than relying on tomographic image reconstruction methods (as is the case in PAT), photoacoustic microscopy (PAM) generates images by scanning a focal point within the tissue volume of interest, thus spatially confining the source of the detected ultrasound. The axial resolution is still determined by the bandwidth of the detected ultrasound pulses, whereas the lateral resolution is determined by the spatial extent of the focal region. One may either focus the ultrasound detector (*acoustic-resolution PAM*, or AR-PAM) or the optical illumination (*optical-resolution PAM*, or OR-PAM). The major advantage of focusing ultrasound detection is the larger measurable depth (up to centimeters), which far exceeds the maximum measurable depth (<1 mm) achievable with optical focusing (given the much stronger scattering of light compared to ultrasound). In either case, an image is obtained by combining data from a set of A-lines (where "A" stands for "amplitude"), where each A-line contains information along the depth coordinate. Multiple A-lines are measured by scanning the photoacoustic system along a line

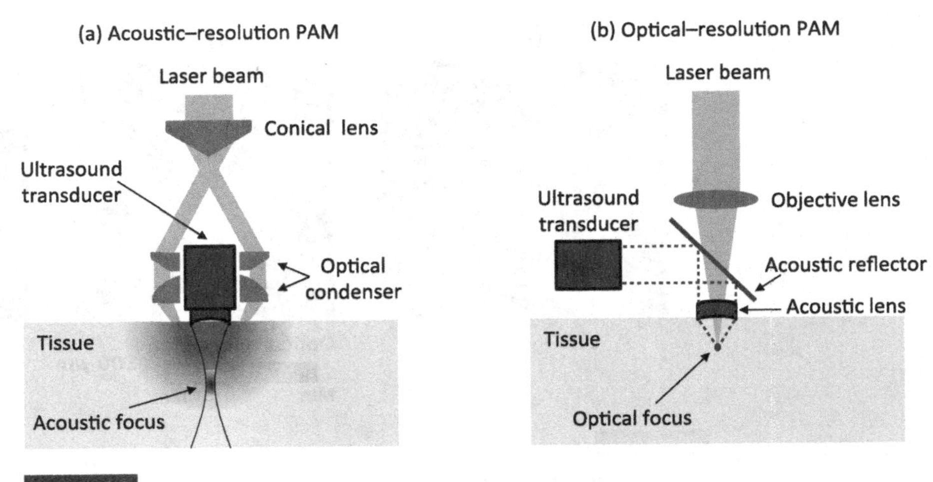

Figure 16.8

Schematic setups of two implementations of (a) acoustic-resolution photoacoustic microscopy (AR-PAM), and (b) optical-resolution photoacoustic microscopy (OR-PAM). While in both methods the axial resolution is determined by the frequency bandwidth of the detected ultrasound, the lateral resolution is determined by the acoustic focus in AR-PAM and by the optical focus in OR-PAM.

(for 2D, or B-mode imaging, where "B" stands for "brightness") or a plane (for 3D imaging) on the tissue surface.

An approach to acoustic-resolution PAM (AR-PAM) that employs dark-field illumination to minimize the strong photoacoustic signal from the tissue surface (Maslov et al., 2005) is illustrated in Figure 16.8(a). Here, dark-field illumination refers to the fact that laser light is delivered to the tissue over a region that does not fall within the field of view of the ultrasound transducer. The lack of light-focusing requirements in acoustic-resolution PAM allows for the delivery of a larger total optical energy, given safety limitations to the delivered energy per unit area, as discussed above. To achieve a high axial spatial resolution, the transducer has a high center frequency (~50 MHz) and a large bandwidth (70–100% of the center frequency), whereas a high lateral resolution is achieved by an acoustic lens with a large numerical aperture (~0.5). Spatial resolution in acoustic-resolution PAM can be of the order of ~15 μm (axial) and ~50 μm (lateral).

Optical-resolution PAM (OR-PAM) is based on focusing the laser beam inside the tissue to depths up to ~1 mm. Since lateral resolution is determined by the size of the focused laser beam, it can be as good as a few microns, as limited by optical diffraction. The axial resolution, however, is still determined by the ultrasound transducer bandwidth, and it is ~10 μm as in acoustic-resolution PAM. While optical-resolution PAM has many points in common with a variety of other optical microscopy techniques (confocal microscopy, nonlinear microscopy,

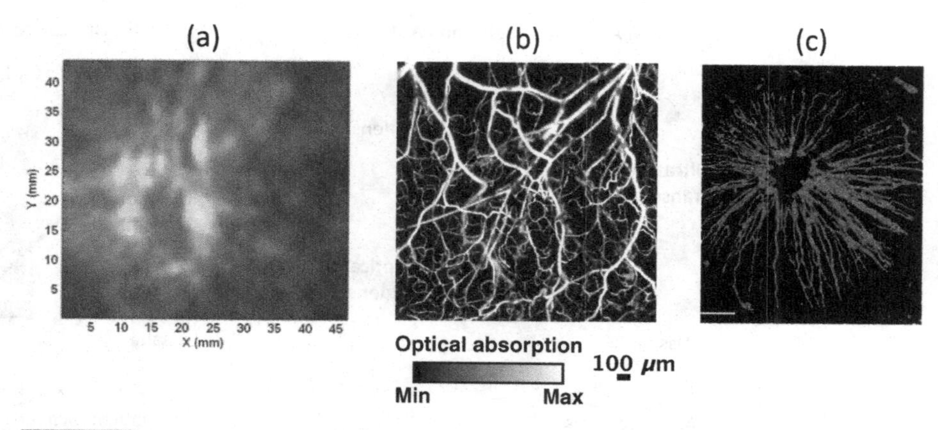

Figure 16.9

(a) Ring pattern of increased vascularization surrounding an invasive ductal carcinoma in a 57-year-old patient measured with photoacoustic tomography (PAT) (this PAT slice is at a depth of 1.2 cm from the breast surface). (Reprinted from Manohar et al., 2007, with permission; copyright 2007, The Optical Society of America.) (b) Microvascular network measured with optical-resolution photoacoustic microscopy (OR-PAM) in the mouse ear. (Reprinted from Hu et al., 2009, with permission; copyright 2009, The Optical Society of America.) (c) Iris vasculature in the mouse eye measured with OR-PAM, where the central void of vasculature corresponds to the location of the pupil (scale bar is 500 μm) (Reprinted from Wu et al., 2014, with permission; copyright 2014, The Optical Society of America.)

optical coherence tomography, etc., which are covered in Chapters 17 and 18) a major difference is the fact that its source of contrast is absorption rather than scattering or fluorescence. The focused laser beam is delivered to the tissue through an optically transparent acoustic reflector, which redirects the photoacoustically generated ultrasound to a broadband, high-frequency (50 MHz or more) transducer. A schematic diagram of optical-resolution PAM is shown in Figure 16.8(b).

16.5 Applications of photoacoustic imaging

16.5.1 Imaging of tissue vascularization and microvasculature

The strong optical absorption of hemoglobin in blood renders the vasculature an ideal target for photoacoustic imaging. At a macroscopic level, the photoacoustic signal relates to the vascular density, or the degree of tissue vascularization, whereas at a microscopic level it can generate high-resolution images of the microvasculature. Figure 16.9(a) shows a photoacoustic tomography (PAT) image of the increased vascularization (due to angiogenesis) around breast cancer, as measured on a patient (Manohar et al., 2007). It is a 2D slice at a depth of 1.2 cm from the breast surface, and it shows the ability of PAT to perform imaging at tissue depths of ≥ 1 cm. Optical-resolution photoacoustic microscopy (OR-PAM) images of the microvasculature in the mouse ear and eye (iris) are shown in Figures 16.9(b) and 16.9(c), respectively. The field of view in these images is several mm^2, and they represent the maximum amplitude projection (MAP) onto the 2D scanning plane. The image in Figure 16.9(b) features an axial resolution of ~ 15 μm and a lateral resolution of ~ 5 μm (Hu et al., 2009). The image in Figure 16.9(c) shows blood vessels of the iris with high contrast, and the central region void of blood vessels corresponds to the location of the pupil (Wu et al., 2014).

16.5.2 Hemoglobin saturation and blood flow

The structural information associated with the visualization of the microvascular network can be complemented by the functional information provided by the oxygen saturation of hemoglobin and the speed of blood flow in each blood vessel. Hemoglobin saturation can be obtained by photoacoustic spectroscopy (i.e., photoacoustic measurements at multiple wavelengths), taking advantage of the dependence of the hemoglobin absorption spectrum on oxygen saturation. Blood flow velocity can be measured by photoacoustic Doppler flowmetry, a technique conceptually similar to Doppler ultrasound, which is based on measuring the bandwidth broadening of photoacoustic ultrasound signals generated by moving red blood cells. The Doppler broadening is linearly dependent on the speed of blood flow (Yao et al., 2010). The combination of structural microvascular information (size, location, and morphology of blood vessels) and functional hemodynamic information (hemoglobin saturation, speed of blood flow) allows for the determination of the rate of oxygen delivery to tissue. This latter measure is directly related to the tissue metabolic rate of oxygen, a critically important physiological parameter.

The capability of metabolic photoacoustic microscopy (or mPAM) to combine structural, oxygenation, and hemodynamic information on the microvasculature of the mouse ear is shown in Figure 16.10 (Yao et al., 2011). Figure 16.10(a) shows a hemoglobin saturation map (based on OR-PAM measurements at 584 and 590 nm) of arterial and venous pairs that feature a typical branching morphology. A common way to classify the microvasculature is to assign a first order to the feeding vessels, and increasing orders (second, third, etc.) to each subsequent branching vessel, so that higher-order vessels are smaller in size. Figure 16.10(b) shows cross-sectional profiles of blood flow speed in artery-vein pairs of first, second, third, and fourth order, revealing the decrease in speed of blood flow from

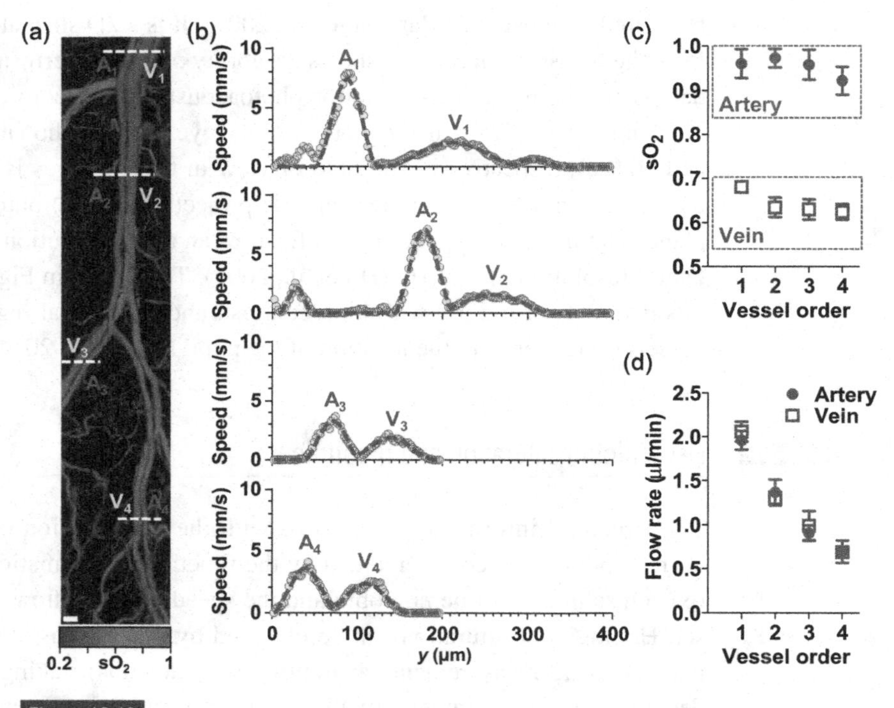

Figure 16.10

Metabolic photoacoustic microscopy (mPAM) of the mouse ear microcirculation. (a) Spatial mapping of hemoglobin saturation (sO_2) (scale bar: 125 μm), (b) speed of blood flow, and (c) hemoglobin saturation at four orders of artery-vein pairs (A_1V_1–A_4V_4). The flow speed and blood vessel diameter are used to compute the flow rate for the four vessel orders (d). (Reprinted from Yao et al., 2011, with permission; copyright 2011, SPIE.)

lower to higher order vessels. Figure 16.10(c) reports the hemoglobin saturation for the four order vessels, showing a range of 90–100% for arterial branches and 60–70% for venous branches. By exploiting knowledge of the vessel size and blood speed, one can calculate the blood flow (in units of microliters of blood per minute) in each vessel order, as reported in Figure 16.10(d).

16.5.3 Other applications

Photoacoustic imaging offers an array of applications that we have briefly summarized here. For more details on applications, the reader is referred to the publications listed under further reading at the end of this chapter. Spectroscopic capabilities, which are exploited in the hemoglobin saturation measurements described in the previous section, can also characterize the biochemical composition of atherosclerotic plaques (by specifying the lipids, elastin, collagen, and water content) in *intravascular photoacoustic imaging* (IVPA). Photoacoustic spectroscopy also finds applications in conjunction with contrast agents such as dyes and nanoparticles. In particular, contrast agents that change their optical properties depending on their biochemical environment or that target specific biomolecular receptors allow for *photoacoustic molecular imaging*. *Photoacoustic thermometry* performs spatially or temporally resolved temperature measurements on the basis of the temperature dependence of the Grüneisen coefficient (defined in Eq. (16.19)).

Clinical translation of photoacoustic imaging is targeting areas such as breast imaging, assessment of skin pathologies, detection of vulnerable atherosclerotic plaques, ophthalmology, functional brain imaging, etc. The photoacoustic imaging field holds great promise, which it is already starting to fulfill, to impact basic research and clinical practice.

Problems – answers to problems with * are on p. 648

16.1* In addition to pressure (p) and particle displacement (s), ultrasound can also be described in terms of the particles' speed (u), which is given by $u = \frac{ds}{dt}$. Similarly to Eqs. (16.1) and (16.2), the particles' speed associated with a plane ultrasound wave is written $u(z,t) = u_0 e^{i(k_a z - \omega_a t + \phi_u)}$. Find the relationships between: (a) u_0 and s_0; (b) ϕ_u and ϕ_s; (c) u_0 and p_0; (d) ϕ_u and ϕ_p.

16.2 Verify that the SI units of Z (Rayl) are kg/(m^2-s).

16.3* Assume that the ultrasound attenuation coefficient is 1 dB/(cm-MHz) in soft tissue. After propagating over what distance in soft tissue does the ultrasound intensity attenuate to 7% of its original value if the ultrasound frequency is (a) 3.5 MHz or (b) 50 MHz?

16.4* By considering that soft tissues have a characteristic impedance (Z) of ~1.6 MRayl, what is the maximum particle displacement associated with a peak overpressure of 1 MPa generated by a 3.3 MHz ultrasound imaging system? (Recall that the angular frequency is 2π times the frequency.)

16.5 Derive Eq. (16.12) from Eq. (16.11) under the approximations $kA\frac{L}{l}\sqrt{\alpha_{l,k_a}} \ll 1$ and $L \gg l$.

16.6 Derive Eq. (16.13) from Eqs. (16.10) and (16.12), recalling the condition $kA\frac{L}{l}\sqrt{\alpha_{l,k_a}} \ll 1$.

16.7 Verify that the photoacoustic wave equation (Eq. (16.17)) is dimensionally consistent by deriving the SI units of each term in the equation and verifying that they are all the same.

16.8* A spherical absorber located at a depth of 1.8 cm from the tissue surface has a radius of 0.6 mm and an absorption coefficient of 4.2 cm^{-1}. Upon receiving a radiant fluence of 1.3 mJ/cm^2 it emits a photoacoustic pressure wave. What is the maximal pressure perturbation that reaches the tissue surface? (Ignore the attenuation coefficient of ultrasound, and assume $B = 2.9$ GPa, $\beta = 3 \times 10^{-4}$ K^{-1}, $\rho = 1.04$ g/cm^3, $c_p = 4.1$ J/(g-K).)

16.9* Assuming that the optical fluence reaching a certain blood vessel is the same at the wavelengths of 650 and 672 nm, find the relative difference of the photoacoustic pressures generated at the two wavelengths ($(p_{672} - p_{650})/p_{650}$) if the hemoglobin saturation in that vessel is: (a) 100%, (b) 75%, (c) 50%. (Use the extinction coefficients of oxyhemoglobin and deoxyhemoglobin tabulated in Appendix 2.A.)

References

Hu, S., Maslov, K., and Wang, L. V. (2009). Noninvasive label-free imaging of microhemodynamics by optical-resolution photoacoustic microscopy. *Optics Express*, 17, 7688–7693.

Jiang, H., Yuan, Z., and Gu, X. (2006). Spatially varying optical and acoustic property reconstruction using finite-element-based photoacoustic tomography. *Journal of the Optical Society of America A*, 23, 878–888.

Kim, C., and Wang, L. V. (2007). Multi-optical-wavelength ultrasound-modulated optical tomography: a phantom study. *Optics Letters*, 32, 2285–2287.

Leutz, W., and Maret, G. (1995). Ultrasonic modulation of multiply scattered light. *Physica B*, 204, 14–19.

Lévêque, S., Boccara, A. C., Lebec, M., and Saint-Jalmes, H. (1999). Ultrasonic tagging of photon paths in scattering media: parallel speckle modulation processing. *Optics Letters*, 24, 181–183.

Manohar, S., Vaartjes, S. E., van Hespen, J. C. G., et al. (2007). Initial results of in vivo non-invasive cancer imaging in the human breast using near-infrared photoacoustics. *Optics Express*, 15, 12277–12285.

Maslov, K., Stoica, G., and Wang, L. V. (2005). In vivo dark-field reflection-mode photoacoustic microscopy. *Optics Letters*, 30, 625–627.

Ron, A., Racheli, N., Breskin, I., et al. (2012). Measuring tissue blood flow using ultrasound modulated diffused light. *Proceedings of SPIE*, 8223, 82232J.

Rosenthal, G., Furmanov, A., Itshayek, E., Shoshan, Y., and Singh, V. (2014). Assessment of a noninvasive cerebral oxygenation monitor in patients with severe traumatic brain injury. *Journal of Neurosurgery*, 120, 901–907.

Rousseau, G., Blouin, A., and Monchalin, J.-P. (2009). Ultrasound-modulated optical imaging using a high-power pulsed laser and a double-pass confocal Fabry-Perot interferometer. *Optics Letters*, 34, 3445–3447.

Sakadžić, S., and Wang, L. V. (2004). High-resolution ultrasound-modulated optical tomography in biological tissues. *Optics Letters*, 29, 2770–2772.

Sigrist, M. W. (1986). Laser generation of acoustic waves in liquid and gases. *Journal of Applied Physics*, 60, R83–R121.

Sigrist, M. W., and Kneubühl, F. K. (1978). Laser-generated stress waves in liquids. *Journal of the Acoustical Society of America*, 64, 1652–1663.

Wang, L. V. (2001). Mechanisms of ultrasonic modulation of multiply scattered coherent light: An analytic model. *Physical Review Letters*, 87, 043903.

Wang, L. V. (2004). Ultrasound-mediated biophotonic imaging: a review of acousto-optical tomography and photo-acoustic tomography. *Disease Markers*, 19, 123–138.

Wang, L., and Zhao, X. (1997). Ultrasound-modulated optical tomography of absorbing objects buried in dense tissue-simulating turbid media. *Applied Optics*, 36, 7277–7282.

Wang, Y. M., Judkewitz, B., DiMarzio, C. A., and Yang, C. (2012). Deep-tissue focal fluorescence imaging with digitally time-reversed ultrasound-encoded light. *Nature Communications*, 3, 928.

Wu, N., Ye, S., Ren, Q., and Li, C. (2014). High-resolution dual-modality photoacoustic ocular imaging. *Optics Letters*, 39, 2451–2454.

Xu, M., and Wang, L. V. (2005). Universal back-projection algorithm for photoacoustic computed tomography. *Physics Review E*, 71, 016706.

Yao, J., Maslov, K. I., Shi, Y., Taber, L. A., and Wang, L. V. (2010). In vivo photoacoustic imaging of transverse blood by using Doppler broadening of bandwidth. *Optics Letters*, 35, 1419–1421.

Yao, J., Maslov, K. I., Zhang, Y., Xia, Y., and Wang, L. V. (2011). Label-free oxygen-metabolic photoacoustic microscopy in vivo. *Journal of Biomedical Optics*, 16, 076003.

Yuan, B., Liu, Y., Mehl, P. M., and Vignola, J. (2009). Microbubble-enhanced ultrasound-modulated fluorescence in a turbid medium. *Applied Physics Letters*, 95, 181113.

Zhang, E., Laufer, J., and Beard, P. (2008). Backward-mode multiwavelength photoacoustic scanner using a planar Fabry-Perot polymer film ultrasound sensor for high-resolution three-dimensional imaging of biological tissue. *Applied Optics*, 47, 561–577.

Further reading

Ultrasound imaging

Aldrich, J. E. (2007). Basic physics of ultrasound imaging. *Critical Care Medicine*, 35, S131–S137.

Lawrence, J. P. (2007). Physics and instrumentation of ultrasound. *Critical Care Medicine*, 35, S314–S322.

Lieu, D. (2010). Ultrasound physics and instrumentation for pathologists. *Archives of Pathology & Laboratory Medicine*, 134, 1541–1556.

Wells, P. N. T. (2006). Ultrasound imaging. *Physics in Medicine and Biology*, 51, R83–R98.

Ultrasonic tagging of light

Li, C., Kim, C., and Wang, L. V. (2011). Photoacoustic tomography and ultrasound-modulated optical tomography. In *Handbook of Biomedical Optics*, ed. D. A. Boas, C. Pitris and N. Ramanujam. Boca Raton, FL: CRC Press.

Marks, F. A., Tomlinson, H. W., and Brooksby, G. W. (1993). A comprehensive approach to breast cancer detection using light: photon localization by ultrasound modulation and tissue characterization by spectral discrimination. *Proceedings of SPIE*, 1888, 500–510.

Selb, J., Lévêque-Fort, S., Dubois, A., et al. (2003). Ultrasonically modulated optical imaging. In *Biomedical Photonics Handbook*, ed. T. Vo-Dinh. Boca Raton, FL: CRC Press, Ch. 35.

Wang, L., Jacques, S. L., and Zhao, X. (1995). Continuous-wave ultrasonic modulation of scattered laser light to image objects in turbid media. *Optics Letters*, 20, 629–631.

Photoacoustic imaging

Bayer, C. L., Luke, G. P., and Emelianov, S. Y. (2012). Photoacoustic imaging for medical diagnostics. *Acoustics Today*, 8, 15–23.

Beard, P. (2011). Biomedical photoacoustic imaging. *Interface Focus*, 1, 602–631.

Ermilov, S. A., Khamapirad, T., Conjusteau, A., et al. (2009). Laser optoacoustic imaging system for detection of breast cancer. *Journal of Biomedical Optics*, 14, 024007.

Kruger, R. A. (1994). Photoacoustic ultrasound. *Medical Physics*, 21, 127–131.

Li, C., and Wang, L. V. (2009). Photoacoustic tomography and sensing in biomedicine. *Physics in Medicine and Biology*, 54, R59-R97.

Zackrisson, S., van de Ven, S. M. W. Y., and Gambhir, S. S. (2014). Light in and sound out: Emerging translational strategies for photoacoustic imaging. *Cancer Research*, 74, 979–1004.

Zhang, H. F., Maslov, K., Stoica, G., and Wang, L. V. (2006). Functional photoacoustic microscopy for high-resolution and noninvasive in vivo imaging. *Nature Biotechnology*, 24, 848–851.

17 Modern optical microscopy for biomedical applications

It can be argued that the earliest example of an optical technology being used for biological applications, in a manner relevant to today's usage, is that of the optical microscope. The word *microscope* comes from the Greek, wherein *micron* means small and *skopein* means to look at or to see. Beyond the simple magnifying glass, compound optical microscopes were first used to observe biological structures invisible to the naked eye in the seventeenth century. The English scientist Robert Hooke [1635–1703] described what he observed with a low-power ($\sim 40\times$) microscope in a popular book called *Micrographia*, which provided the first illustrations of a variety of biological objects, including detail of a fly's eye, and (barely resolved) plant cells. It was the Dutch scientist Anton van Leeuwenhoek [1632–1723], however, who made the first compound lenses that were of high enough quality and short enough focal length to lead to a magnification of $270\times$, powerful enough to be considered a true microscope, rather than a strong magnifying glass. He reported the first observations of bacteria, yeast, blood cells, and various microscopic life forms commonly found in a water droplet. Perhaps surprisingly, it took nearly another 200 years before the French chemist and microbiologist Louis Pasteur [1822–1895] was able to explain the workings of bacteria. It is also interesting to note that the microscopes commonly used today by histopathologists, to examine the cellular structure of tissues for diagnosing disease, do not exhibit dramatically better optical performance than van Leeuwenhoek's early creation.

The general field of optical microscopy is too large to be covered in one chapter, and numerous excellent resources are available to provide ample detail about the optical design and common configurations of "traditional" optical microscopes, including many general textbooks on optics. Some of these are listed under further reading at the end of this chapter. In Section 17.1 we provide an abbreviated review of the basics of traditional microscopy, because these core elements are also found as parts of modern developments in the field, especially those applied to biomedical imaging. Beyond the classical configuration, a variety of more recent inventions in types of microscopy and microscopes offer images with a range of information content and/or resolution beyond the traditional limits. Not all of these are optical microscopes, but many are, and in the subsequent sections of the chapter we introduce some of the important advances in non-traditional optical microscopy.

In some of our presentation of the topics in this chapter, we derive insight from descriptions and formalisms provided in the elegant text by Jerome Mertz, *Introduction to Optical Microscopy* (2010).

17.1 Basic elements and theory of a classical microscope

Whereas a single positive (convex) lens can function as a magnifying glass, when coupled with the lens of your eye in the correct combination of distances, it only takes one additional lens to constitute a microscope. Thus, the two primary elements of an optical microscope are a short focal-length lens that is placed near the object, referred to as the *objective lens* or simply the *objective*, and a longer focal-length lens commonly referred to as the *eyepiece lens* or simply the *eyepiece* (occasionally called the *ocular*). The simple arrangement, as originally assembled by Hooke, van Leeuwenhoek and others of their time, is illustrated in Figure 17.1. (Of course, in this original arrangement, lacking a camera, the lens of the eye plays a role as well, and modern microscopes generally invoke additional optical elements, notably those that accommodate high-resolution digital cameras.)

17.1.1 Magnification

Referring to Figure 17.1, for the objective lens (or for any single positive lens), the *magnification*, M_o, is defined as the ratio of the image size (h_i) to the object size (h_o) in a *real* imaging configuration, and is determined by the ratio of the tube length (L) image distance, d_i, to the objective focal length, f_o:

$$M_o = -\frac{L}{f_o}, \tag{17.1}$$

where the minus sign indicates that the image is inverted with respect to the object. (In fact, any time a *real* image is formed with a single lens, the image is inverted.) Equation (17.1) gives the objective magnification in terms of design standards for microscopes. In common practice, the distances between the components of most microscopes are set, by convention, to standardized lengths. A microscope's *tube length*, L, is the distance from the back focal plane of the objective to the focal plane of the eyepiece, and most manufacturers set this length at 160 mm. This standardized distance enables the manufacturer to engrave the *objective power* or magnification on the lens barrel. Thus, if the objective lens is marked 20×,

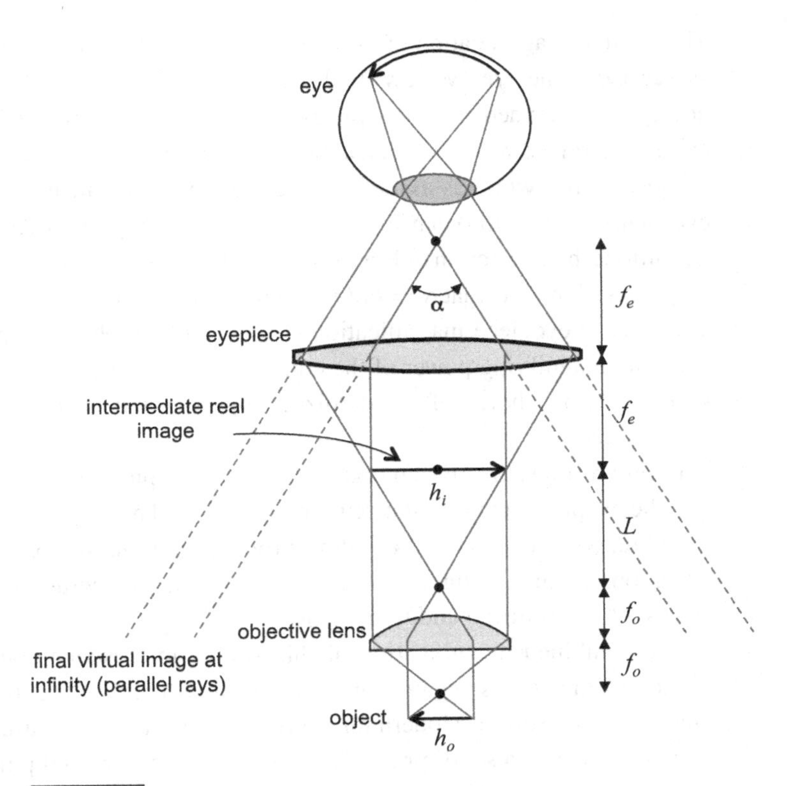

Figure 17.1

Configuration of a simple compound microscope. The virtual image seen by the eye remains inverted with respect to the object. Not shown is a *field stop*, which is effectively a limiting aperture determined by the diameter of the tube holding the assembly together or, more commonly, by the diameter of the eyepiece lens. The field stop affects the field of view, as well as the numerical aperture (NA) of the objective lens. Also indicated are the focal points of the objective lens and of the eyepiece, the focal lengths of both lenses, f_o and f_e, and the tube length, L, which is the distance between the back focal plane of the objective and the front focal plane of the eyepiece.

this indicates that its focal length is 8 mm (160/8 = 20). Since the objective lens generates a real image, the image size-ratio is sometimes referred to as the *transverse magnification*, whereas when an imaging configuration generates a virtual image, the size-ratio is referred to as the *angular magnification*.

In a compound microscope, the second lens, the *eyepiece*, converts the real image at the objective image plane to an enlarged virtual image, and the overall magnification of the combination, called the *magnifying power* of the microscope, is an angular magnification and is the result of the product of the magnifications of the two main elements:

$$\mathrm{MP} = M_o \times M_e. \tag{17.2}$$

The angular magnification of the eyepiece, M_e, is defined as the ratio of the angle subtended at the eye by the virtual image (indicated by α in Figure 17.1) and the angle, α_0, subtended at the eye by an object of the same size (h_i in our case) that is at a standardized reference distance, d_{np}, from the eye. The distance d_{np} is called the *near point*, which is assumed to be the closest distance at which the unaided eye (with good vision) can focus, and it is commonly set at 250 mm. From these definitions, by inspection of Figure 17.1, and in the small-angle approximation (i.e. $\alpha \approx \tan\alpha$), it results that $\alpha = h_i/f_e$ and $\alpha_0 = h_i/d_{\text{np}}$. Consequently, $M_e = d_{\text{np}}/f_e$. Then, if the eyepiece magnification is 10×, and the objective power is 20×, the resulting magnifying power of the combination is 200×, and the microscope user knows this by choice of standardized objective and eyepiece components that he/she selects.

In Figure 17.1, the rays emerging above the eyepiece are collimated, indicating that the eyepiece image distance is set at infinity. Thus, commonly, the observer's eye is relaxed, and focused at infinity for viewing the microscope virtual image. Variations of the configuration can accommodate cameras or bare 2D detector arrays, often requiring additional lenses.

The combined magnification of this simple microscope configuration is still negative (since M_o is negative and M_e is positive), indicating that the final virtual image is still inverted. Modern microscopes, especially those used for histopathology, often invoke a set of prisms in the eyepiece section to flip the image upright, which aids in ease of positioning the sample slide for following moving objects. We also note that, in current technology, the objective lens is composed of several lens elements of various types of glass and surface curvatures, the design goal being to minimize chromatic aberration (variations in focal length for different wavelengths) and spherical aberrations (distortions of the wavefront that blur the focus at the image plane).

17.1.2 Resolving power and the point spread function

In theory, there is no limit to the magnifying power of a microscope, but there *is* a limit to the useful magnification, and that limit is determined by the effective *resolution*, or *resolving power* of the microscope, which is specified as the smallest detail that the microscope can render visible. (In this section we are discussing transverse resolution; axial resolution will be addressed in Section 17.3.2.) A common estimate of the theoretical limit of the transverse resolution, d_{lim}, of a classical microscope is often attributed to Ernst Abbe (German optical physicist [1840–1905]), and is frequently referred to as the "Abbe limit," although a number of optical scientists at the time had reached similar determinations. Abbe's derivation was based on arguments related to light diffraction by an aperture. The value is given as

$$d_{\text{lim}} = \frac{\lambda}{2\text{NA}}, \tag{17.3}$$

where $\text{NA} = n \sin\theta$ is the numerical aperture (first defined by Abbe) of the objective lens, and θ is the half-angle of the convergence from the objective. Here, the refractive index, n, is for the medium between the lens and the object plane, typically air. (This is analogous to the expression for the NA of an optical fiber as discussed in Section 4.8.1.)

A simple way to understand the limiting resolution of a microscope system is to ask the question: what would be the spot size at the focus of the objective lens if a uniform optical field were propagated backwards through the system to the object plane? Assuming aberration-free optical elements, the resulting (resolvable) spot size is commonly defined as the diameter of the first zero of the resulting diffraction-limited *Airy disk* pattern (named after the British astronomer George Biddell Airy [1801–1892]), which is the pattern generated at the focus of a perfect converging spherical wave that is limited in diameter, D, by a finite aperture (e.g., the diameter of the focusing lens). The Airy pattern for the intensity at an angular radius, θ, from the optical axis is given by the expression

$$I(\theta) = I_0\left[\frac{2J_1(x)}{x}\right]^2, \tag{17.4}$$

where I_0 is the intensity at the central peak and J_1 is the Bessel function of the first kind of order 1; $x = ka\sin\theta$, in which $k = 2\pi/\lambda$ is the wavenumber, and $a = D/2$ is the aperture radius. This pattern can be derived simply by taking the 2D Fourier transform of a square "top-hat" function for the intensity distribution at the lens. (At its focal plane, a lens generates a Fourier transform of the spatial distribution of intensity of an object at infinity.) The relative intensity distribution in the focal plane is plotted in Figure 17.2.

Using a series expansion of the Bessel function in Eq. (17.4), the spatial diameter of the first zero at the focal plane, D_0, can be approximated as

$$D_0 \cong \frac{1.22\lambda}{\text{NA}}. \tag{17.5}$$

The reader will note that this is more than double the value of the Abbe limit. That is because the resolvable detail, as discussed further below, is associated with the precision with which one can locate the peak of the Airy disk, which is of the order of the radius rather than the full diameter of the first zero.

In the forward direction, one asks how good a job an optical system (or even a single lens) can do in imaging an infinitesimal object, which can be represented

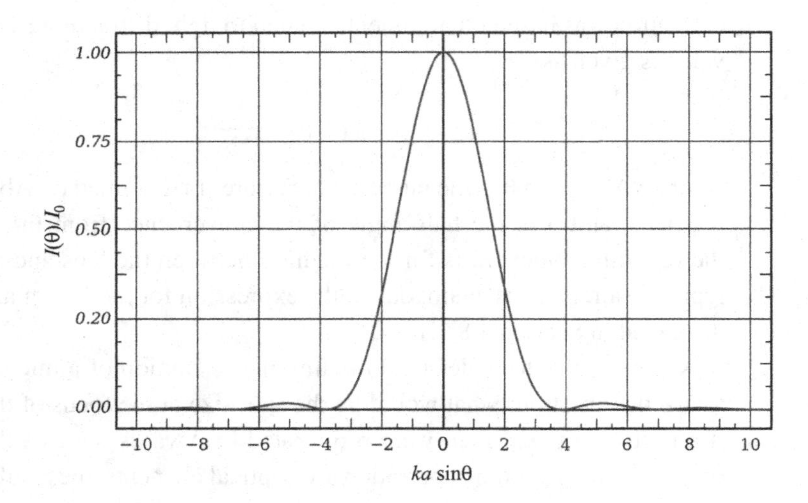

Figure 17.2

Angular radial intensity distribution of an Airy disk pattern (by Inductiveload [public domain]).

as a point source. The *point spread function* (PSF) is a property of the optical system that defines how the system responds to a point source in the object plane, and is related to the system's transverse spatial resolution (determined by the NA of the objective), to the aberrations induced by the optical elements, and to other blurring effects of the system. The PSF is analogous to the temporal *impulse response* of, say, an electrical amplifier to a short electrical pulse. Consequently, the PSF is also sometimes referred to as the *spatial impulse response* of an imaging system.

In a perfect optical system, the coordinates of the image are simply related to the coordinates of the object multiplied by the magnification. Assuming the magnification to be *shift invariant* (i.e., the magnifying power is the same for all locations in the object plane), we can express the coordinates of points in the image plane, $\boldsymbol{\rho}_i = (x_i, y_i)$, as a simple linear transformation ($\boldsymbol{\rho}_i = M\boldsymbol{\rho}_o$) of points in the object plane, $\boldsymbol{\rho}_o = (x_o, y_o)$. The question of image inversion, which depends on the specifics of the optical system, is accommodated by the sign of the magnification, M. Given the limitations of a real optical system, however, a point source in the object plane will appear blurred in the image plane. To express this blurring property of the imaging system more mathematically, if a specific point source in the object plane, located at $\boldsymbol{\rho}'_o$, is represented by a 2D delta function, $\delta(\boldsymbol{\rho}'_o - \boldsymbol{\rho}_o)$, then the point spread function of *any* such point, $\text{PSF}(\boldsymbol{\rho}_i - \boldsymbol{\rho}_o)$, is the blurred image of that point source, after transfer by the optical system to the image plane.

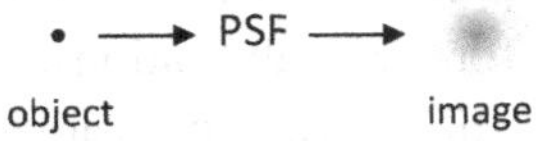

• ⟶ PSF ⟶

object image

Figure 17.3

The point spread function represents the optical system's imperfect response to a spatial "impulse" in the object plane, generating a blurred image in the image plane.

A variety of formats for the argument of the PSF can be found in textbooks and in the literature, seldom with any explanation. We use a common format that incorporates the minus sign between the variables (instead of the sometimes-used semi-colon or comma), because this indicates the assumptions of linearity and shift invariance, and facilitates direct convolution of the PSF with a function that describes the light distribution in the object plane (see below). The PSF can also directly account for the optical magnification, M, if it is expressed as $\text{PSF}(\boldsymbol{\rho}_i - M\boldsymbol{\rho}_o)$, but this is not commonly found in textbooks, and the magnification is generally left to be accounted for separately.

The effect of the PSF in forming the image of an object-plane point source is illustrated in Figure 17.3.

A distribution of light in the object plane (i.e., an object) can be expressed as a weighted distribution of point sources (2D delta functions), which is called the *object function*, $Obj(\boldsymbol{\rho}_o)$. For *incoherent* light in the object plane, the resulting image (the distribution of light intensity in the image plane) can then be computed as the weighted sum over the images (PSFs) of those object point sources, i.e., the convolution of the PSF with the object function:

$$\mathcal{I}(\boldsymbol{\rho}_i) = \iint Obj(\boldsymbol{\rho}_o)\text{PSF}(\boldsymbol{\rho}_i - \boldsymbol{\rho}_o)d^2\boldsymbol{\rho}_o, \tag{17.6}$$

the double integral indicating that the image-plane intensity distribution is a 2D function calculated by integration over $\boldsymbol{\rho}_o = (x_o, y_o)$ in the object plane. The units of $\mathcal{I}(\boldsymbol{\rho}_i)$ and $Obj(\boldsymbol{\rho}_o)$ are those of intensity (W/m^2), and the PSF has units of m^{-2}. The PSF is usually normalized, such that $\iint \text{PSF}(\boldsymbol{\rho}_i - \boldsymbol{\rho}_o)d^2\boldsymbol{\rho}_o = 1$. In essence, the PSF behaves like a "fat" δ-function, or sampling function, in that it describes a narrow distribution whose peak, albeit of finite width, is centered where the argument is zero.

Thus, in this spatial domain representation, the object is represented as a collection of weighted point sources, and the image is a superposition of the images of those point sources. We can understand intuitively that one of the reasons for Eq. (17.6) being expressed as a convolution integral is that, given the blurring effects that lead to the PSF, light originating from any one point in the object plane can contribute to the light detected at many points in the image plane; and, conversely, light detected at one point in the image plane can originate from many points in the object plane. We remember that, even for ideal, aberration-free optics, the PSF is still limited by the finite aperture of the microscope tube and/or the eyepiece, and becomes a function that yields an Airy-disk diffraction image of a point object.

An additional established approach to assess the resolving power of a microscope is to ask what is the smallest separation (generally measured as angular separation) of two point-source-like objects that can be seen as separate in the image plane. This method is often used by astronomers to determine the closest spacing (in angle) of two stars that can still be discriminated by a telescope as being distinct stellar objects. A commonly used measure, called the *Rayleigh criterion* (yes, proposed by the same Lord Rayleigh who was cited so extensively in Chapters 6 and 7, and also in Section 16.1.1), is the point-source object separation that results in the peak of the Airy disk image of one source being located at the first zero of the image pattern of the other source. This criterion is thus related directly to the effective PSF-limited spot diameter of the image of either source. That is, this criterion yields a resolution limit that is half the first-zero diameter, as determined by Eq. (17.5), and close in value to the Abbe limit, Eq. (17.3).

17.1.2.1 *The optical transfer function*

Given that a convolution in the spatial domain is equivalent to a product in the spatial frequency domain, one option to facilitate calculation of the resulting image is to take the Fourier transforms of the elements of Eq. (17.6), and express the result as the product of the transforms:

$$\mathcal{F}\{\mathcal{I}(\boldsymbol{\rho}_i)\} = \mathcal{F}\{Obj(\boldsymbol{\rho}_o)\} \times \mathcal{F}\{\mathrm{PSF}(\boldsymbol{\rho}_i - \boldsymbol{\rho}_o)\}. \tag{17.7}$$

After calculating the product, the inverse Fourier transform of the result, $\mathcal{F}^{-1}\{\mathcal{F}\{\mathcal{I}(\boldsymbol{\rho}_i)\}\}$, gets us back to the image intensity map, $\mathcal{I}(\boldsymbol{\rho}_i)$. This approach leads to another commonly used measure of the optical performance of a microscope (or any optical imaging system), called the *optical transfer function*, $\mathrm{OTF}(\boldsymbol{\kappa}_{\perp i}; \boldsymbol{\kappa}_{\perp o})$, where $\boldsymbol{\kappa}_{\perp}$ represents the transverse wave vector, i.e., the transverse spatial frequency (in the x-y plane) of the image or object, as denoted by the additional subscripts. The OTF is a dimensionless function that defines how an optical system translates spatial frequencies from the object plane to the image plane. The OTF and the PSF are a Fourier transform pair, such that

$$\mathrm{OTF}(\boldsymbol{\kappa}_{\perp i}; \boldsymbol{\kappa}_{\perp o}) = \mathcal{F}\{\mathrm{PSF}(\boldsymbol{\rho}_i - \boldsymbol{\rho}_o)\}. \tag{17.8}$$

Analogous to the Abbe limit in the spatial domain, in the spatial frequency domain it is possible to identify a criterion for the resolution limit as $1/\boldsymbol{\kappa}_{\perp,\mathrm{lim}}$, where $\boldsymbol{\kappa}_{\perp,\mathrm{lim}}$ is the highest spatial frequency for which the OTF is not zero. Rigorous development of these measures and their formulations can be found among the

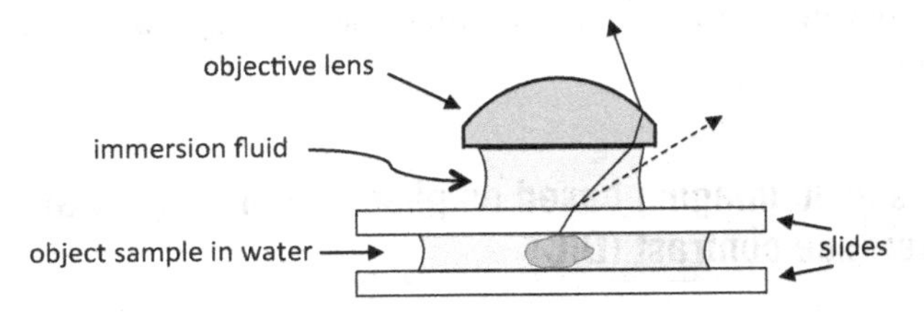

Figure 17.4

Immersion in fluid of higher index allows for collection of a larger range of ray angles, and reduces the effective wavelength in the medium.

resources listed under further reading, with special relevance to the optics of microscopy in Mertz (2010).

We emphasize that the discussion in this section pertains only to *incoherent* imaging, such that each point in the object plane is phase-independent of any other point, and the intensity at a point in the image plane is a weighted superposition of contributions from the points in the object plane. That is, without interference, intensities simply add. For coherent light (or even partially coherent light), however, the complex amplitudes would be combined in a superposition integral, and the PSF would also be a complex function.

17.1.3 Enhancement of resolution with immersion objective lens

One straightforward way to achieve a moderate enhancement of the resolution limit, which has been increasingly employed in recent decades (Seward, 2010), can be deduced by further examination of Eq. (17.3) or Eq. (17.5). For a given wavelength, the only parameter available to play with is the NA of the objective lens. The focal distance of the objective lens and its diameter (hence, its NA) are subject to design constraints that seek to minimize aberrations, resulting, for the current state of the art, in a maximum value for the NA that is ~ 0.95 (in air). If the space between the lens and the object sample, however, is filled with a fluid with higher refractive index than air, then the effective NA is increased by the ratio of the index of that fluid to that of air (~ 1).

An optimal refractive index for the immersion fluid would be approximately equal to that of the glass cover slide and the objective lens. In that case, in addition to increasing the range of angles for rays that are collected from the sample, reflection losses are minimized at the surfaces of the cover slide and the objective lens. An illustration of an immersion objective of this type is shown in Figure 17.4. As depicted here, a ray from the sample (dashed line) that would normally miss the objective lens, due to the refraction at the exit surface of the cover slide, contributes to the collected light for the image when the immersion fluid is present.

17.2 Microscopic imaging based on phase contrast (PC) and differential interference contrast (DIC)

When thin tissue samples are examined with a conventional microscope, there is very little absorption contrast available in the visible wavelength range other than hemoglobin (which is frequently removed in excised tissue), and scattering contrast is also small. Thus, unlabeled cellular structures are difficult to discern, especially with bright-field illumination. This is one of the reasons why histopathologists always use stains that label macromolecules (e.g., proteins and DNA) to facilitate visualization of microscopic structural features. Cellular structures do, however, exhibit higher refractive index than the surrounding cytosol, and this will result in relative phase delays for transmitted light that traverses an organelle, for example, relative to light that only traverses fluid.

Objects that exhibit contrast due to absorption or scattering are called *amplitude objects*, and a transmitted or reflected wave that generates an image of such an object is said to be *amplitude modulated* (spatially). On the other hand, an object that is otherwise transparent, but induces phase shifts in the transmitted light, is called a *phase object*, and the transmitted field is said to be *phase modulated*. Since the eye cannot sense phase differences, it is easy to understand why pathologists use stains to render visible the cellular structures that would be predominantly phase objects. The limitation is the same for standard optical detectors (and cameras), in that they can only sense variations in light intensity. Consequently, the challenge is to convert phase variations into intensity variations over the transverse image plane.

17.2.1 Phase contrast imaging with spatially coherent illumination

If a plane wave (a spatially coherent wave) illuminates a small phase object, a slight phase delay is induced in that part of the wavefront, but not in the majority of the field. This is depicted in Figure 17.5, in which a plane wave illuminates a microscope slide with a small phase object. Most of the illumination wave passes through unimpeded, but a small amount of light interacts (scatters, essentially) with the sample, forming a spreading wave with a slightly distorted and phase-delayed wavefront.

Phase contrast microscopy was first demonstrated in 1934 by the Dutch physicist Fritz Zernike [1888–1966], who received the Nobel Prize for physics in 1953.

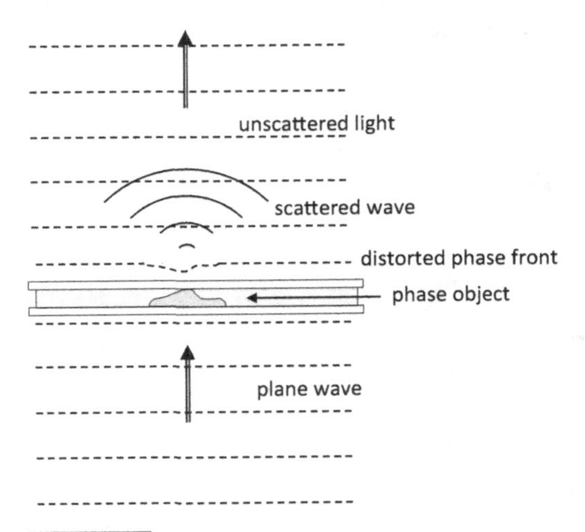

Figure 17.5

A thin phase object creates a phase-lag perturbation for a small portion of an illuminating plane wave transmitted through the sample, while most of the wave propagates through undistorted.

Zernike noted that if a relative phase delay of $\sim\pi$ is induced between the scattered and unscattered waves, then they will mostly interfere destructively, and the slight variations induced by the phase object will show up brightly against the otherwise dark background.

A conceptual diagram of the Zernike microscopy scheme is shown in Figure 17.6. In this incarnation, the unscattered light is focused by the objective through the aperture pupil where a small phase-delaying element (a piece of glass) is placed. The perturbed wave expands from the phase object, is collimated by the objective lens, and mainly misses the phase shifter. The phase shift that is added to the unperturbed wave results in predominantly destructive interference at the image plane (camera sensor) between the unperturbed illuminating beam and the scattered wave with its small phase distortion induced by the phase object. A variable attenuator may also be added at the location of the phase shifter, so that the amplitudes of the perturbed and unperturbed fields are close to equal at the detector plane, yielding optimal destructive interference. Under these conditions, the small phase distortion in the perturbed field is the only thing disrupting the field-cancellation effect of the destructive interference, thus converting small phase differences to intensity variations that appear bright against the mostly dark background.

Zernike phase contrast imaging is sometimes referred to as "phase contrast with coherent illumination," because it requires *spatially* coherent illumination. This also constitutes one of the disadvantages of this method: the requirement of

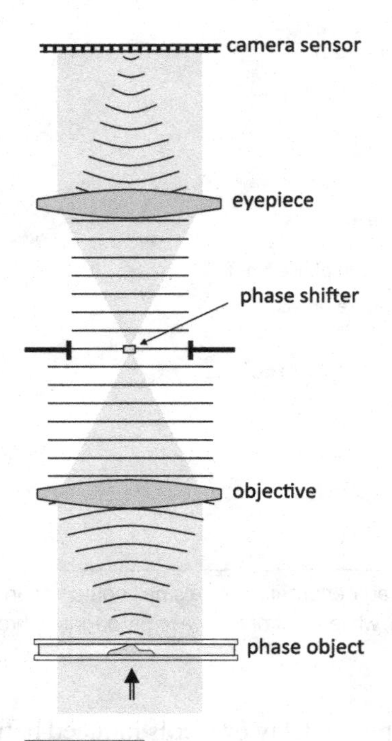

Figure 17.6

Primary components of a Zernike phase contrast microscope. The gray shading represents the propagation of the unperturbed portion of the illuminating plane wave from below. (In this depiction, the lines representing the phase fronts of the perturbed wave do not show the subtle phase distortions introduced by the object.)

spatially coherent illumination, which means that the wavefront must have uniform phase (see Section 4.7.1). If the illumination light source is an incoherent source, such as an incandescent bulb, the spatial coherence (~plane wave) is achieved by focusing the light through a small aperture, which dramatically reduces the power available to illuminate the sample. Of course, a laser source could provide loss-free spatially coherent illumination but, given that this method invokes widefield illumination, the temporal (longitudinal) coherence of laser light would yield the additional image noise of laser speckles.

Additional note: the reader interested in beam optics should note that a plane wave (or a Gaussian beam) propagating through a focal point undergoes a "Gouy phase shift" (named after French physicist Louis Georges Gouy [1854–1926]) of π retardation relative to an unfocused (collimated) beam (or plane wave) that is originally in phase with it. Half of the phase shift, $\pi/2$, occurs for propagation from a large distance up to the focal plane, and the other $\pi/2$ phase shift occurs from the focal plane to the far field, which can be an image plane. In a Zernike phase contrast microscope, this additional phase shift is accounted for when determining the proper amount of phase lag ($\pi/2$ for the conceptual design of Figure 17.6) to be introduced by the phase-shifter element.

17.2.2 Phase contrast imaging with incoherent illumination

Another approach to converting small phase variations in the object plane to intensity variations in the image plane in microscopy is the method called *differential interference contrast* (DIC). DIC microscopy was invented by the Polish-born French physicist Georges Nomarski [1919–1997], and is also sometimes referred to as Nomarski microscopy (Nomarski, 1955). The method is based on the physical separation of orthogonal linear polarization components of the transmitted illumination field, and it works even if the illuminating field is *not* spatially coherent. One of the primary reasons for the popularity of the Nomarski DIC method is that the light source does not need to be focused through a small aperture, because spatial coherence is not required for DIC imaging; thus, much more light is available for generating the image than is the case for Zernike phase contrast microscopy.

If we consider light traveling in the $+z$ direction, the optical electric field in the transverse plane can be represented as comprising two orthogonal polarization components, which (for convenience) we can align with x and y axes. As in phase contrast microscopy, DIC microscopy invokes transmission illumination. For convenience, in Figure 17.7 the illumination and imaging/detection stages of a DIC microscope are separated. Following the light from the light source, the key steps of the method are as follows:

- Unpolarized illumination light passes through a linear polarizer, whose transmission axis is at 45° to the x axis of the coordinate system. With these coordinates, this means that half of the light is x-polarized and half is y-polarized.
- The light next goes through a Nomarski prism (sometimes called a Wollaston prism, after the British scientist William Hyde Wollaston [1766–1828]), which is an optical component incorporating a wedge of birefringent crystal, and its axes are aligned with the x and y coordinates, such that x-polarized and y-polarized light emerge at slightly different angles, separated by $\Delta\theta$.
- The condenser lens then converts the angular separation to a small lateral shift (shear) of $\Delta\rho$ between the two polarization states. The system is typically adjusted so that the two polarizations of a given ray are now two rays separated by approximately the diffraction-limited spot size, ρ_{lim}, at the sample plane. The key point is that, for rays traversing at the edge of a structure in the sample, one polarization (the one going through the denser structure) will be

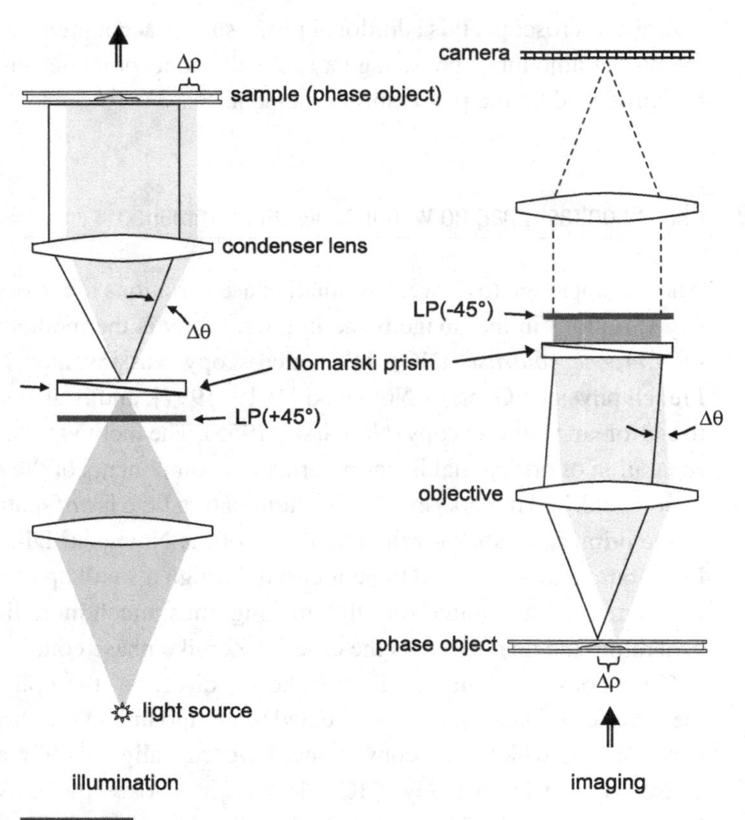

Figure 17.7

Conceptual configuration of a differential interference contrast (DIC) microscope. On the illumination side we illustrate the full field illumination of the object plane, whereas on the imaging/detection side we select the paths of rays from specific points on the phase object towards the detection/image plane. (Images adapted from Mertz, 2010.)

phase-delayed slightly relative to the nearby orthogonally polarized ray that missed the structure. In general, pairs of orthogonally polarized rays traversing any gradient in the optical index of the sample will travel different pathlengths, and hence experience phase shifts with respect to each other.

- The objective lens converts the lateral offset back to a converging angular separation, which is then "undone" by a second Nomarski prism, so that the two polarization states emerge parallel and collinear, and can both be focused by the eyepiece onto the same spot at the detection plane.
- It is important to remember, however, that orthogonal polarizations do not interfere. Consequently, to generate interference between the two polarization components (which were slightly separated at the sample), a second linear

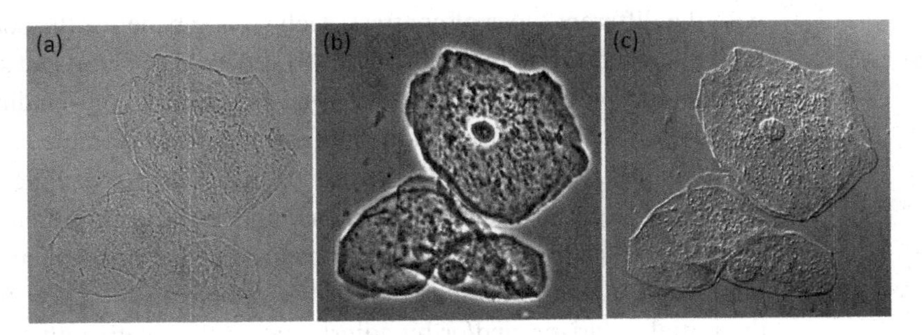

Figure 17.8

Microscope images for hamster cheek cells: (a) standard bright field microscopy, (b) Zernike phase contrast microscopy; and (c) Nomarski differential interference contrast microscopy (courtesy of Mertz, 2010).

polarizer is placed in the field after the second Nomarski prism, with its transmission axis crossed (90°) with respect to the first polarizer, i.e., at −45° relative to the x axis. Half of each polarization component is transmitted through this polarizer, and the end result is that both components are now parallel polarized (at −45°) and do indeed interfere at the image plane, to yield an intensity distribution (against a dark background) that represents the relative phase shifts of the two separated fields at the sample plane.

- The second Nomarski prism can be user-adjusted to introduce an additional relative phase shift between the two polarizations, rendering the total phase shift to be near π, so that the sample-induced phase variations will exhibit greater contrast against a dark background. (The amount of added phase shift will depend on the sample thickness, and can be adjusted to optimize the detail of the specific structures of interest to the user.)

Thus, the result is that DIC microscopy provides an image of the edges of structures and density gradients within the phase object. One consequence of this property is that structural edges that are parallel to the direction of the shear will not yield much contrast, and opposite refractive index gradients will yield either bright or dark contrasts. That is, DIC images can give the false impression that the object is being illuminated obliquely from one side. This effect can be seen in Figure 17.8(c), in which the cell edges facing the upper left appear bright, whereas those facing the lower right appear dark, and edges oriented normal to the strongest contrast edges exhibit lower contrast. (The user can rotate the sample to accentuate different sections of the structures.)

Figure 17.8 compares images of cells for standard bright-field microscopy, phase contrast microscopy, and DIC microscopy. While the phase contrast and DIC images (Figures 17.8(b) and 17.8(c)) provide a great deal of structural detail with the different types of contrast, each having benefits, the conventional bright-field microscope image (Figure 17.8(a)) is devoid of detail and structural contrast.

It should be noted that an additional potentially confounding factor for DIC microscopy, however, is that any birefringence in the sample itself will mix polarization states, and will either distort or blur the image. Since many biological structures (e.g., structural proteins) exhibit significant structural birefringence, this factor must be understood when examining the image, and can sometimes be minimized by careful orientation of the sample with respect to the direction of the initial polarizer, and/or by adjustment of the relative phases with the second Nomarski prism. (See, for example, Preza et al., 1999.)

17.2.3 Applications of phase contrast and DIC microscopy

Phase contrast microscopy and DIC microscopy helped usher in the era of modern cellular biology by enabling, for the first time, high-resolution imaging (with adequate contrast) of living cells that have not been fixed or stained (Pluta, 1989). Consequently, such microscopes are ubiquitous in bioscience research labs and are especially valuable tools for monitoring dynamic processes in living cells and small organisms. Label-free high-resolution microscopy is an essential tool for micromanipulation of cells and other small living biological organisms, and applications abound in areas such as *in vitro fertilization* (IVF, also to be discussed in Chapter 19, Section 19.3.3.1), transgenic techniques, stem cell research, and developmental biology. As one example, it is common to use phase contrast or DIC microscopy in neuroscience to visualize the insertion of micro-electrodes to study nerve cell electrical activations (Carter and Shieh, 2010). Similarly, modern methods of in vitro fertilization would be impossible without high-resolution visualization of the egg cells, sperm cells, and the micromanipulation tools that bring the two together. Figure 17.9 shows a DIC image of IVF in action, as a micropipette is used to inject spermatazoa into the cytoplasm of an ovum (egg cell).

17.3 Optical sectioning: confocal microscopy

One of the primary limitations of applying conventional microscopy to thick biological samples, such as tissue, is that scattered light (or emitted fluorescence, in the case of fluorescence microscopy) from layers of tissue above or below the focal plane adds noise to the images, and conventional microscopes are not effective at rejecting background from outside the intended focal plane. In fact, it is generally

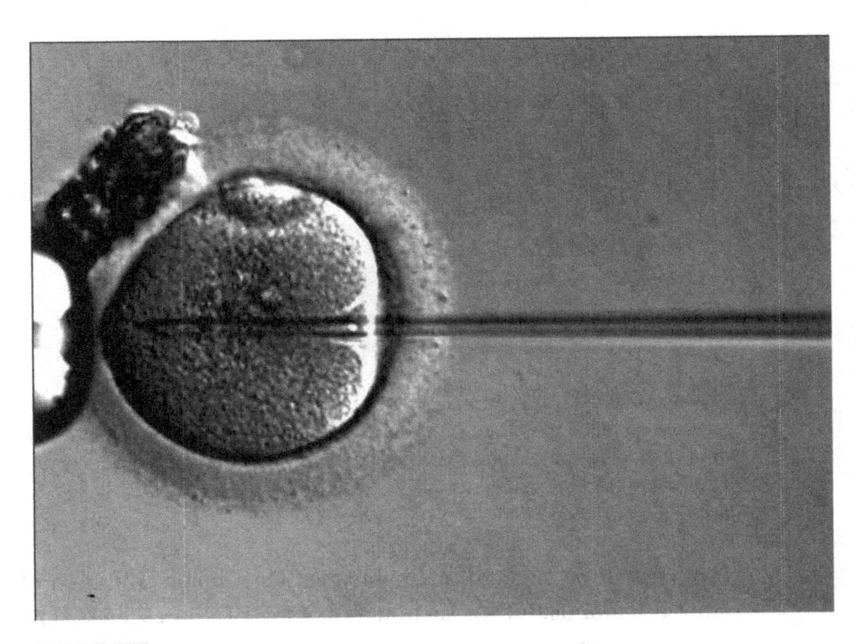

Figure 17.9

Viewed under phase contrast microscopy, an ovum is held by a suction pipette, while spermatozoa are injected into the cytoplasm by an ultrafine micropipette that has pierced the ovum membrane. The protective outer layer of the ovum (called the *zona pelucida*) can be seen as the halo around the dark cell membrane. (Wikimedia Commons [public domain].)

impossible to achieve a sharp image of tissue at a depth of more than one or two cell thicknesses beneath the tissue surface. (Phase contrast and DIC microscopy also suffer from this limitation.) This is one of the reasons why histology slides are typically prepared from very thin, commonly ~5 μm, slices of tissue, typically cut from a paraffin-embedded sample of tissue. As can be seen in Figure 17.10, layered epithelial tissues present different sub-structures at different depths, inviting sub-surface imaging that is not accessible by conventional microscopy.

The ability to reject out-of-focus light in microscopic imaging is referred to as *optical sectioning*, or *axial resolution*, the primary intent of which is to enable imaging of a sub-surface plane in a biological sample, while minimizing the blurring effects of light from the intervening (shallower) or from deeper tissues, but especially from the shallower depths.

There are two primary classes of imaging methods that enable rejection of out-of-plane light: temporal (or phase) gating and spatial gating. With temporal gating, one seeks to selectively detect photons that arrive at precisely the time-of-flight corresponding to the direct propagation from the intended focal plane to the detector. Temporal gating is the basis of optical coherence tomography (OCT),

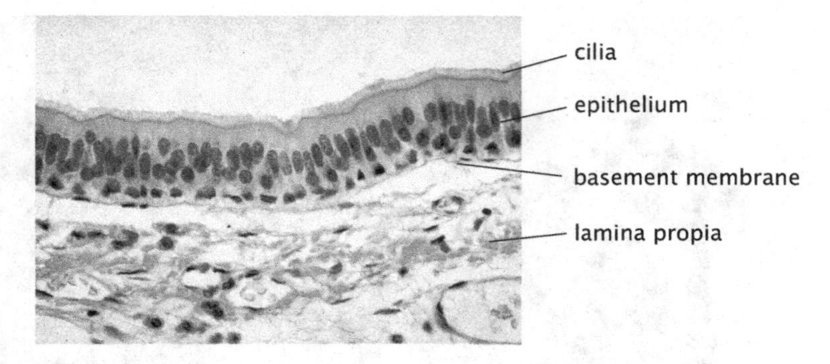

Figure 17.10
Illustration of the cross section of columnar-cell epithelium (found in the entire gastrointestinal tract below the esophagus). It is clearly evident that sectional imaging at different depths would yield images of different structures.

which will be covered in Chapter 18. One major limitation of OCT, however, is that it is not useful for sectional imaging of fluorescence. That is a consequence of the uncertainty of the timing of fluorescence emission. Microscopic resolution (microns) in the axial dimension (depth) would require temporal resolution of the order of femtoseconds, whereas the uncertainty in the timing of fluorescence emission is comparable with the fluorescence lifetime, which is of the order of nanoseconds, or six orders of magnitude too long for precise temporal gating. (See Section 5.6.)

With confocal microscopy, however, axial resolution is achieved by spatial gating, which is not susceptible to the uncertainty of fluorescence emission timing and is equally effective at rejecting out-of-focus scattered illumination light or emitted fluorescence. Imaged fluorescence is most commonly from exogenous fluorophores that can serve as molecular biomarkers. In fact, it is the power to generate axially resolved imaging of fluorophore-tagged cellular and tissue structures that has accounted for the ubiquitous presence of confocal microscopes in bioscience research labs.

17.3.1 Basic design of a confocal microscope

The basic concept of the confocal microscope, as originally conceived by Minsky (1961), is illustrated in Figure 17.11. Instead of broadly illuminating the object (*widefield* illumination) with uniform light, as would be done in conventional microscopy, the sample is illuminated with a tightly focused point of light. The point of illumination at the focal plane is imaged through a conjugate pinhole at

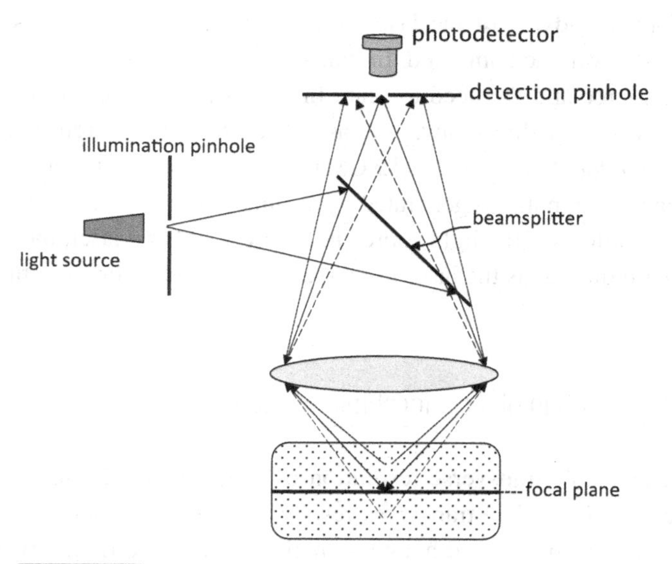

Figure 17.11

Schematic of a basic confocal microscope. Axially resolved (sectional) imaging is achieved by scanning the illumination spot across the focal plane within the thick sample. The illumination pinhole, the focal spot at the sectioned plane, and the detection pinhole are conjugates.

the detector, such that light from axial locations above or below the focal plane is defocused at the detection pinhole, and is mostly blocked. (Similarly, light from other points that are translated laterally from the focal point, but within the focal plane, is also blocked.) We recall here that *conjugate points* in an optical system are such that if one of them is an object, the other is its image generated by the optical system. By transferring the image of the small illumination spot to the detection pinhole, the detection pinhole is a conjugate that is effectively *confocal* with the illumination focal spot, hence the term *confocal microscopy*. Of course, this constitutes measurement of light from only one point in the sample, and, to generate a sectional image, either the sample or the focal point must be scanned in two dimensions laterally. Taking this further, a beautiful aspect of confocal imaging is that by adding axial scanning to the transverse scanning, sectional images can be recorded for a range of depths, and when those multiple sections are combined, a three-dimensional microscopic tomograph can be generated.

To achieve focused illumination, the light source must have good spatial coherence, the same requirement (but for a different reason) as for phase contrast microscopy. Good spatial coherence is inherent in a laser beam, whereas light from an incoherent lamp must be passed through a small pinhole to limit the transverse phase variance, which of course dramatically reduces the amount of light available for the illuminating the sample. The confocal microscope was developed before the advent of the laser, and early incarnations were slow to generate an image, given the combined limitations of low radiance of the incoherent illumination source and the need to scan. In modern confocal microscopes, however, the light source is almost always a laser, and the system is referred to as a *laser scanning confocal microscope* (LSCM). (Laser speckle is not a problem with LSCM, because the image is generated by scanning a small spot, rather than by widefield illumination.) If the light source in Figure 17.11 is a laser, then the source pinhole is not required, as the laser field itself acts as light from a virtual pinhole.

17.3.2 Axial resolution of a confocal microscope

The available transverse resolution of a conventional microscope was treated in Section 17.1.2. The transverse resolution limit of a confocal microscope follows the same principles, although sometimes there is a trade-off between resolution and available light (for scanning speed) that is manipulated by adjusting the size of the detection pinhole (Solberg, 2000). To understand the axial resolution of a confocal microscope, we need to examine the *depth of focus* of a laser beam. We present here, without derivation, the parameters associated with the focal region of a Gaussian laser beam, which is a solution to the homogeneous wave equation for an electromagnetic field in a dielectric medium. (See, for example, the detailed development of Gaussian beams in Saleh and Teich, 2007.) The resulting depth of focus for an aperture-limited plane wave is close in value to that of a Gaussian beam whose $1/e^2$ diameter is equal to the plane wave's aperture diameter, but the mathematics of a Gaussian beam are more convenient.

Suppressing the polarization and propagation phase factors for the moment, the (transverse) electric field amplitude of a Gaussian beam is given as

$$E(\rho, z) = E_0 \frac{w_0}{w(z)} e^{-\rho^2/[w(z)]^2} \tag{17.9}$$

where $w(z)$ is the *beam radius* at axial position z, $\rho = \sqrt{x^2 + y^2}$ is the radial coordinate, w_0 is the beam radius at the focal plane, and E_0 is the electric field amplitude on axis at the focal plane. The z-dependence of the beam radius is expressed as

$$w(z) = w_0 \sqrt{1 + (z^2/z_R^2)}, \tag{17.10}$$

where z_R is the Rayleigh range, to be defined below.

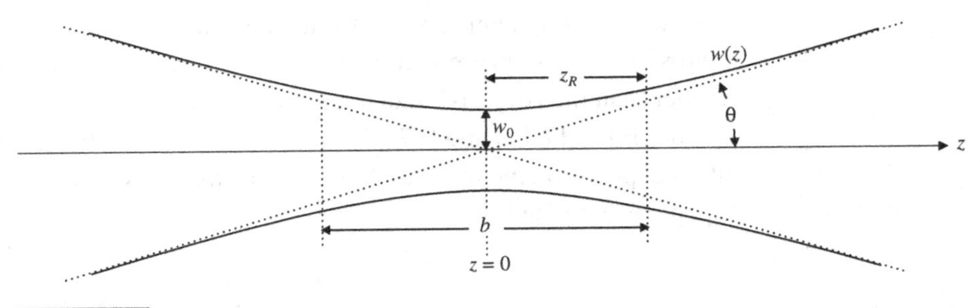

Figure 17.12

The Gaussian beam radius $w(z)$ as a function of the distance along the axis of propagation, z. The parameters indicated around the beamwaist are for a focused Gaussian beam with a far-field convergence angle of θ.

Figure 17.12 is a plot of $w(z)$ for a focused Gaussian beam in the vicinity of the focal plane. Fresnel diffraction (self-diffraction) of the beam limits the beam diameter at the focal plane to a minimum *beamwaist*, w_0, as depicted.

The distance b, often called the *confocal parameter*, is a measure of the depth of focus, and is a good choice as the criterion for the axial resolution for a confocal microscope. It is the distance between points on either side of the beamwaist where $w(z_R) = \sqrt{2}w_0$, in other words, the axial locations where the area of the beam has doubled from its minimum at the beamwaist. The distance b is twice the distance referred to as the *Rayleigh range*, z_R (yes, Lord Rayleigh again!). The relationships among these parameters can be expressed as

$$b = 2z_R = \frac{2\pi w_0^2}{\lambda}, \tag{17.11}$$

where we keep in mind that λ is the optical wavelength in the propagation medium. While the relationship between the confocal parameter and the minimum spot size, defined by the beamwaist radius w_0, is provided by Eq. (17.11), w_0 itself is ultimately determined by two factors only: the wavelength of the light in the medium and the far-field convergence angle, θ.

$$w_0 \cong \frac{\lambda}{\pi n \sin\theta} \tag{17.12}$$

Remembering that an optical detector measures intensity, not field amplitude, if we define the spot diameter of a Gaussian beamwaist by the radius ρ at the $1/e^2$ intensity, then the minimum Gaussian spot diameter is

$$d_{0,G} = 2w_0 \cong \frac{2\lambda}{\pi \mathrm{NA}}, \tag{17.13}$$

which is very close to the values for the transverse resolution limit given in Eqs. (17.3) and (17.5). In practice, manufacturers of confocal microscopes will sometimes claim better (up to 40% smaller) lateral resolution than a conventional microscope, the improvement being achieved by using a detection pinhole that is smaller than the Airy disk size at the image plane.

From Eqs. (17.11) and (17.12) it can be seen that, whereas the transverse resolution depends on the inverse of the NA, the axial resolution has an inverse-square dependence on the NA:

$$b = \frac{2\lambda}{\pi \mathrm{NA}^2} \tag{17.14}$$

Another definition of the axial resolution of a confocal microscope, based on the system response to axial spatial frequencies, yields a similar value (Nakamura and Kawata, 1990; Sheppard, 1986):

$$\Delta z = \frac{n\lambda}{2\mathrm{NA}^2} \tag{17.15}$$

Thus, axial resolution is strongly governed by the NA, and significant effort in optical engineering is expended to maximize the NA of objective lenses used in confocal microscopes.

Both the improved lateral resolution sometimes claimed for confocal microscopes and the axial resolution given by Eq. (17.15) tend to be overly optimistic for a variety of practical reasons, including the effect of the detection aperture size on the optical signal strength at the detector (Mertz, 2010).

We note that in the far field on either side of the focal plane, the wavefront of a Gaussian beam is spherical, whereas close to the beamwaist the wavefront is approximately planar. Additionally, the reader is reminded that the phase of the beam undergoes a delay of π, relative to that of an unfocused plane wave, while propagating from $-\infty$ to $+\infty$, half of which occurs on either side of the beamwaist. A phase shift of $\pi/4$ occurs (on each side of the beamwaist) between $z = 0$ and $z = \pm z_R$.

An example of the advantage that sectioning provides for imaging in a thick sample is provided in Figure 17.13. Rather than comparing the confocal image and its sectioning capability to a conventional image, what is shown is simply the effect of opening the detection aperture to 6× the diameter of the PSF of the microscope. (Of course, the limit of opening the detection aperture yields an image equivalent to that from a conventional microscope, despite the scanned illumination spot.)

17.3.3 Application example of in vivo confocal microscopy

At the time of this writing, intravital applications of fluorescence confocal microscopy are emerging as potentially valuable clinical research tools, enabled

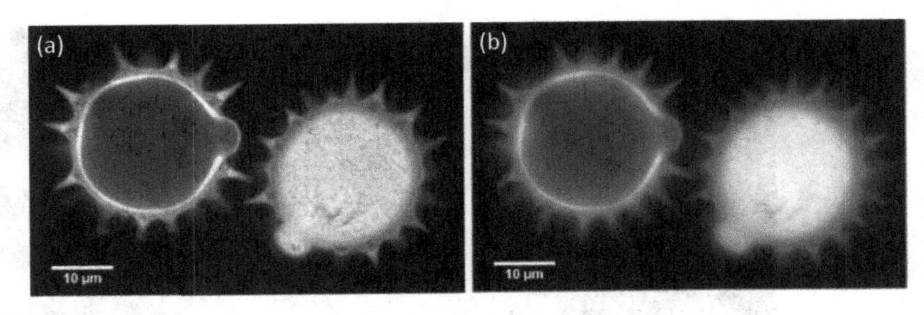

Figure 17.13

Comparison of confocal images of fluorescent pollen grains with the detection aperture set at (a) the diameter of the PSF and (b) 6× the PSF diameter. (Courtesy of Mertz, 2010; permission by Roberts & Co. Publishers.)

by technology advances for size reduction of optical elements, and permitting passage of a confocal microscopy probe through the working channel (typically ~2.5 mm diameter) of a standard endoscope. The instrumentation invokes a "coherent" imaging optical fiber bundle (see Section 4.8.5) to convey light in both directions between the system (with its source and detection components) and the tissue at the end of the endoscope. A focused laser beam at the excitation wavelength is scanned over the proximal ends of the fibers in the instrument, such that one fiber at a time illuminates the tissue, focused to the sectioning depth by a small lens at the distal end of the fiber bundle. Thus, each individual fiber serves as the confocal aperture, both for illumination and for collection of the fluorescence emitted from the tissue. While originally conceived in the 1990s (Gmitro and Aziz, 1993), only recently has microfabrication of optical elements enabled mediation through an endoscope. Transverse resolution is determined by the number (tens of thousands) and individual diameters (typically ~1.5 μm) of fibers in the bundle. A commercial version of this type of implementation has recently become available (CellVizio™, Mauna Kea Technologies, Inc.), opening the door to clinical applications.

One such clinical application was aimed at the micro-physiological origins of optical signatures of pre-malignant polyps in the colon. Figure 17.14 is an endoscopic confocal fluorescence image of the mucosal epithelium of an adenomatous colon polyp, at a sectioning depth of ~65 μm, taken during an otherwise routine colonoscopy screening procedure, following systemic administration of fluorescein. The structures of four mucosal crypts are seen, as are some of the columnar-cell structures within the crypts. Microcirculation vessels are also seen surrounding the crypts. By studying in vivo (with full perfusion) the differences in epithelial architecture and microcirculation between adenomatous (pre-malignant) and hyperplastic (benign) polyps, the researchers are able to identify the potential elements of the underlying basis for real-time "optical biopsy" applications.

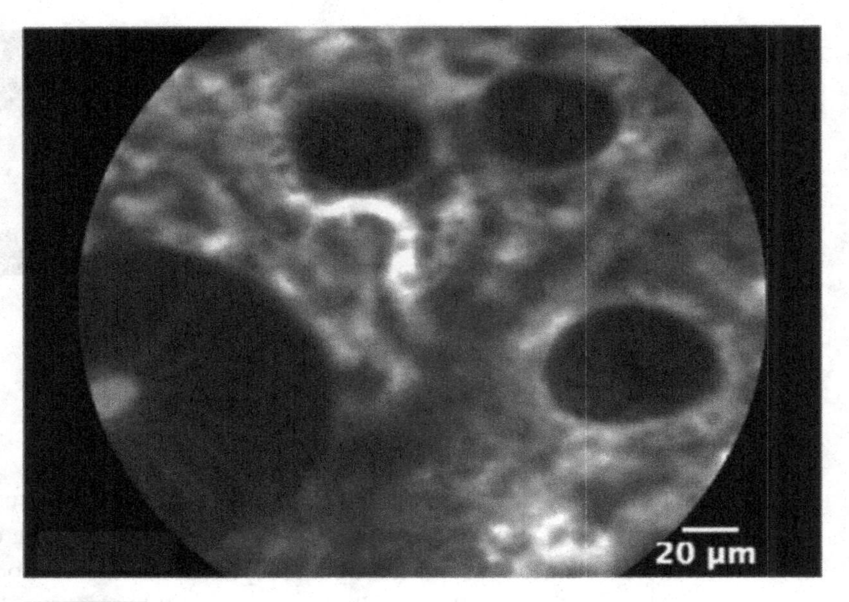

Figure 17.14

A confocal microscopy image, mediated through a standard colonoscope, showing the subsurface epithelial morphology of an adenomatous colonic polyp at a depth of ~65 μm. (Courtesy S.K. Singh and E. Rodriguez-Diaz, Boston University School of Medicine.)

17.4 Nonlinear optical microscopy

More recently developed approaches for providing axial sectioning capability in microscopy include methods that have been enabled by the recent availability of short-pulse laser sources, most commonly *mode-locked* lasers (see, e.g., Siegman, 1986). Mode-locked lasers allow for methods that invoke the nonlinear response of the sample to the illumination light. For a given average power, a mode-locked laser emits much higher peak powers than a continuous-wave laser, because the mode-locked emission is a series of very short pulses (often 1–10 ps duration) separated by longer times (several ns), such that the duty cycle is low, but the peak power of the individual pulses is high. To avoid photodamage to samples, pulse energies may be limited to ~1 μJ, resulting in peak powers of ~100 kW.

Prominent among these new nonlinear microscopy methods are multi-photon excitation of fluorescence, predominantly two-photon excitation fluorescence, and nonlinear optical harmonic generation. For these methods, the axial resolution is achieved as a consequence of the nonlinear process in the illumination stage. Whereas in confocal microscopy the 3D resolution volume (*lateral resolution* ×

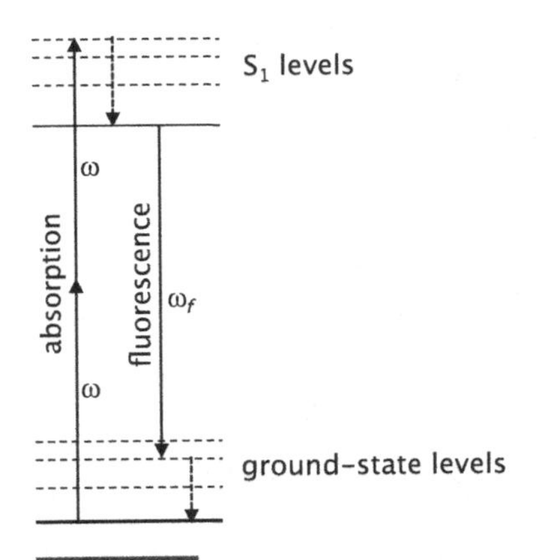

Figure 17.15

Jablonski diagram of the two-photon excitation process for fluorescence. The events happen sequentially, left to right, and the dashed arrows represent internal energy conversion, generally resulting in heat.

axial resolution) is governed by both the illumination optics and the detection optics, in nonlinear optical microscopy the axial resolution (and hence the 3D resolution volume) is determined solely by the illumination optics.

17.4.1 Multi-photon excitation fluorescence microscopy

If the photon energy of an applied optical field is significantly less than the energy level of a molecular excited state, then the absorption probability becomes vanishingly small for typical illumination intensity. On the other hand, if the density of photons is very high, the probability of absorbing two photons simultaneously becomes large enough to yield a useful excited-state population. This effect was theoretically predicted by the German-born American physicist Maria Goeppert-Mayer [1906–1972] in her doctoral dissertation of 1930 (she received the Nobel Prize for physics in 1963), and was experimentally demonstrated soon after the invention of the laser in 1961 (Kaiser and Garrett, 1961). This two-photon effect occurs during the very short pulses of a mode-locked laser, in which the temporal photon density is high. The transition process is represented in the energy-level diagram of Figure 17.15. Since the energy of a single excitation photon is not near to that of any excited state of the molecule, there is no intermediate absorption step, and absorption only occurs by interaction of two photons simultaneously with the molecule.

There are a variety of quantum mechanical constraints, selection rules and transition probability factors that govern the probability of two-photon excitation, and

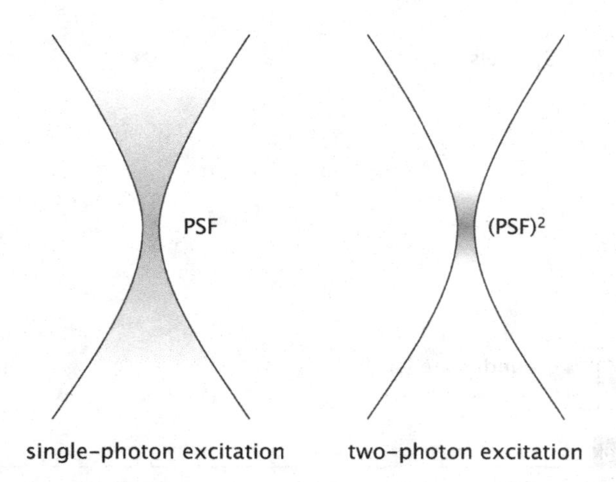

Figure 17.16

Depiction of the relative fluorescence emission volumes for the two cases of single-photon excitation fluorescence (SPEF) and two-photon excitation fluorescence (TPEF). (The axial PSF is defined similar to the transverse PSF.) Due to the quadratic response to excitation light, there is also potential for enhanced resolution in TPEF.

the subsequent emission, but most dipole-allowed electronic absorption transitions in large fluorescent dye molecules have usable cross sections for two-photon excitation. What matters for the discussion here is that the fluorescence emission power, P_f, is proportional to the square of the excitation intensity I_e.

$$P_f \propto \sigma_{2f} I_e^2, \tag{17.16}$$

where σ_{2f} takes the place of a cross section for two-photon excitation. In this expression, the two-photon "cross section" has odd units of m^4/W, rather than units of area as would the cross section for *single-photon excitation fluorescence* (SPEF), or any single-photon interaction probability. (Three-photon excitation fluorescence microscopy has also been demonstrated, and the principles are analogous to those discussed here.) Figure 17.16 illustrates the reduced volume from which fluorescence is emitted in a *two-photon excitation fluorescence* (TPEF) microscope, compared with SPEF microscopy. Given that the PSF (for exciting the fluorescence) is squared for TPEF, two-photon fluorescence microscopy has the potential to yield higher-resolution fluorescence images than conventional fluorescence microscopy, apart from any sectioning benefit (i.e., even when comparing fluorescence imaging of the sample surface).

An additional benefit is that TPEF microscopy images in turbid (scattering) samples exhibit less (darker) background, because scattered excitation light outside the focal volume is not intense enough to generate two-photon absorption. Effectively,

TPEF microscopy images only fluorescence generated by ballistic excitation light, and scattering of the emitted fluorescence is still useful signal (if it can be collected by a large detector) because the image is generated by the scanning of the excitation spot. The first demonstration of practical TPEF microscopy was enabled by the use of a scanning focused laser spot from a sub-picosecond laser (with high peak intensity), with the scanned image generated in a manner similar to confocal microscopy, but with appropriate wavelength-selective filters (Denk et al., 1990).

As with SPEF, TPEF is an incoherent process. Due to the uncertainty of the timing of the fluorescence emission, there is no phase relationship between the excitation fields and the emission field. Additionally, most of the polarization and orientation factors of the fluorescence emission that were discussed in Chapter 5 apply here as well for TPEF. Moreover, as a result of how the selection rules of quantum mechanics apply in multi-photon absorption, the absorption is most probable to a high vibrational level in the excited electronic state, and the energy lost to internal molecular processes (heat) associated with the TPEF interaction, $\Delta E = \hbar(2\omega - \omega_f)$, is typically much larger than that generally associated with the Stokes shift for single-photon excitation fluorescence.

TPEF microscopy has become popular for applications that image fluorescently labeled structures and molecules in thicker samples, especially tissue, for which there might be strong absorption at the shorter wavelengths required for excitation sources of SPEF microscopy (Denk, 1996). For example, if a fluorescent tag emits in the red wavelength range, the excitation source that would be required for single-photon excitation would likely be green light (which is strongly absorbed by hemoglobin), whereas the NIR wavelengths that would work for TPEF would be in a range of minimal absorption in tissue. As seen in Section 7.2.2, longer wavelengths also scatter less, so there is less defocusing of the excitation light on its way to the focus, and more is available for sectional imaging by the small scanned focal volume. The end result is that the typical maximum sectioning depth for TPEF microscopy in thick turbid samples (tissue) is greater for TPEF imaging (~500 μm) than for SPEF confocal microscopy (~100–200 μm) (Helmchen and Denk, 2005). One additional advantage of TPEF imaging is that the very large wavelength separation between the excitation and emission light facilitates spectral separation of the two in the detection stage.

Given the advantages of TPEF microscopy, and commercial availability of TPEF microscopes with combined laser light source, the method has enjoyed broad adoption in various areas of biomedical research. One example of the impressive enabling capability of TPEF microscopy is in the area of neurophysiology, for imaging neuronal activity in vivo. Ohki et al. (2005) labeled neuronal cells in living animals with calcium-indicator dye to enable two-photon microscopy of neural network activity in the cerebral cortex. By sectional imaging with TPEF microscopy, they imaged somatic and dendritic calcium transients, representing spontaneous and evoked activity patterns in the neocortex with cellular resolution. In a similar technique Kerr et al. used the advantages of the pulsed laser to achieve single-spike resolution (also at the cellular spatial resolution).

For a review of a broader array of applications of TPEF microscopy, see Masters and So (2008) in the further reading list.

17.4.2 Harmonic generation microscopy

In contrast to nonlinear fluorescence excitation, microscopy by nonlinear harmonic generation is a coherent process. This means that there is a precise phase relationship between the applied field and the emitted field, through the induced polarization which instantly reradiates without phase lag. The illumination optics are, essentially, the same as for TPEF microscopy, with 3D (sectioning) imaging enabled by the small excitation volume as a consequence on the supralinear response of the medium to the illumination field. In this sense, harmonic-generation microscopy is very similar to TPEF microscopy, in that the image is generated by scanning of the illumination spot relative to the sample, and collecting the harmonic light with a single detector (typically a photomultiplier tube).

In Chapter 6 (Eq. (6.20)) we introduced the concept of the nonlinear components of the polarizability of a medium. Here we make the simplifying assumption that all the applied fields are from the same (laser) source; also, for now, we suppress the subscripts that do the accounting for the tensor relationship, and note that all the applied fields are identical:

$$p = p_0 + \alpha E + \alpha^{(2)} E^2 + \alpha^{(3)} E^3 + \cdots \tag{17.17}$$

For the discussion in this section, we are interested in the nonlinear processes represented by third and fourth terms on the right-hand side of Eq. (17.17), which correspond to *second-harmonic generation* (SHG) and *third-harmonic generation* (THG), respectively. We will mostly address SHG microscopy, for which more applications of nonlinear optical microscopy have been demonstrated. For SHG the relevant term for the induced second-harmonic polarization is given by:

$$p_{2\omega} = \alpha^{(2)} E^2 \tag{17.18}$$

It is easy to see how the second-order hyperpolarizability yields second-harmonic generation. If the fundamental applied field has a time dependence proportional to sin (ωt), then the induced dipole moment, which is proportional to

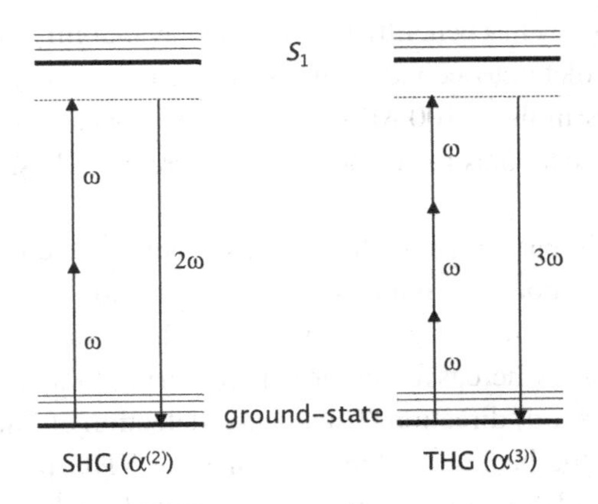

Figure 17.17

Jablonski diagrams of SHG and THG. The dashed line represents a virtual state below the level of the first electronic excited state.

the square of the fundamental field, will have a time dependence provided by the identity $\sin^2(\omega t) = \frac{1}{2}[1 + \cos(2\omega t)]$, and will thus reradiate at frequency 2ω. (The DC component leads to an additional nonlinear effect called *optical rectification*.)

The processes are illustrated by the Jablonski diagrams in Figure 17.17. The induced polarization in the sample is at a frequency twice (SHG) or three times (THG) that of the applied fundamental field, and induced polarization reradiates instantaneously, in phase with the applied field.

In general, the sum of the applied photon energies must be below the lowest excited state of the molecule, to avoid direct absorption, but if the sum is close, then resonance enhancement of the harmonic generation is possible. Unlike fluorescence excitation (linear or nonlinear), harmonic generation invokes no Stokes shift. In other words, the energy of the reradiated photon is exactly equal to the sum of the energies of the applied photons, and there is no internal conversion of any excess energy in the molecule, hence no heat generation. It is generally true that for any nonlinear process that results in emission directly to the state where things started (in Figure 17.17, the ground state) the process is coherent.

Resonance enhancement is not commonly (or often not conveniently) available; thus, the second-order polarizability, $\alpha^{(2)}$, is typically very small, and SHG is generally not measurable with the power available from a continuous-wave laser. The induced polarization, however, is proportional to the square of the applied optical field, which suggests the benefits of ultrashort laser pulses with high peak powers. Just as TPEF benefitted from the development of picosecond mode-locked lasers, SHG microscopy has benefitted from recent availability of lasers with pulse lengths in the tens of femtosecond range. In this case, pulse energies of $\sim 1\ \mu J$ lead to peak powers in the ~ 100 MW range. When focused to a small spot, the optical electric field becomes large enough to generate a robust second-harmonic signal.

Another clear difference between harmonic generation and fluorescence is that fluorescence (single-photon or multi-photon), having no phase relation with the applied field, can radiate in all directions, with directional variations determined by molecular orientations, whereas harmonic generation predominantly radiates in the forward (or near-forward) direction. Thus, while detection of fluorescence images can be achieved in the backward direction, SHG or THG images are generally detected in the forward direction and are thus best suited for thin samples (although aided in thicker samples by the sectioning capability).

17.4.2.1 *Phase shifts and radiation patterns*

In Sections 17.2.1 and 17.3.2 we noted that a focused Gaussian beam undergoes a phase retardation of π (compared with a plane wave) while propagating from $-\infty$ to $+\infty$, half of that retardation occurring while propagating through the focal zone from $z = -z_R$ to $z = +z_R$. At every point along the path, the induced polarization is coherent (phase-locked) with the applied field. The second-harmonic field that is radiated by the nonlinear polarization, however, propagates as a free wave. Thus, second-harmonic light that was generated before the focus undergoes its own phase retardation; but since its frequency is twice that of the fundamental, there will be a growing phase difference between the free propagating SHG wave that was generated before the focus and that which is generated later in the propagation, which is locked to the phase of the fundamental. (Over the focal zone, that difference is $\pi/2$ at the second-harmonic frequency.)

Thus, the "early-generated" 2-ω wave can interfere with the "late-generated" 2-ω wave. Both energy and momentum must be conserved, and the end result is that the second-harmonic emission may propagate in near-forward but off-axis lobes of constructive interference, whose shapes and angles are determined by the conditions of the medium geometry and focus parameters. Fortunately, as for the case of TPEF microscopy, with SHG microscopy the imaging resolution depends only on the optics of the scanned illumination beam, and is independent of the optics or shape of the waves of emitted harmonic light.

Although the harmonic generation is radiated mostly in the near-forward direction, backward detection is possible if the sample plane that is being imaged is in a thick turbid medium, because some of the second-harmonic light will be diffusely backscattered and can be collected, without loss of image resolution (remembering that the resolution is governed only by the illumination optics and the nonlinear process).

17.4.2.2 *Symmetry of the sample and enabled applications*

The symmetry rules of harmonic generation offer the opportunity to selectively image specific biological structures, leading to some of the most interesting applications of SHG microscopy. The components of the polarizability tensor, $\alpha^{(2)}$, must relate to the physical properties of the medium, including its symmetry. If a medium (and hence its polarizability) possesses *inversion symmetry* (a medium property also referred to as *centrosymmetry*), then it cannot generate even-order harmonics without the symmetry being removed. The reason is that if the dipole moment (or local polarization) describes in time a waveform that is symmetrical about the equilibrium point, then the Fourier transform of that waveform will be composed purely of odd harmonics, which appropriately change sign upon inversion. Thus, in order to generate even harmonics a medium must lack inversion symmetry, or be *noncentrosymmetric*.

Many biological structures are noncentrosymmetric, and/or the symmetry breaking can be introduced or accentuated with molecular labels. One of the frequent applications of SHG microscopy has been for imaging the membranes of cells and organelles. Inversion symmetry of the polarizability is intrinsically broken in cell and organelle membranes, because they maintain an ion gradient (and hence a potential difference) across the membrane. Moreover, researchers have designed noncentrosymmetric molecules that are both nonsymmetric and have strong polarizabilities and that intercalate in or bind to the surface of membranes, oriented normal to the membrane plane (Moreau et al., 2000). If the electric field direction of the focused laser beam is oriented parallel to the labeling molecules (normal to the membrane plane), then second-harmonic light will be preferentially generated from those locations.

Figure 17.18 shows an application of SHG microscopy, for imaging of neuronal cells with a membrane biomarker, clearly revealing (otherwise difficult to image) cellular membranes at high resolution in a turbid sample, thus also demonstrating the 3D sectioning capability of SHG microscopy. This type of information would be difficult to visualize by any other imaging method.

We note that, by definition, even an isotropic and centrosymmetric medium with a boundary limit is noncentrosymmetric at its boundary, for any reflection other than across a plane that is precisely normal to the boundary. Thus, SHG microscopy has proven valuable for imaging boundaries of organelles and other biological structures that are denser than their surroundings, although the orientation of the

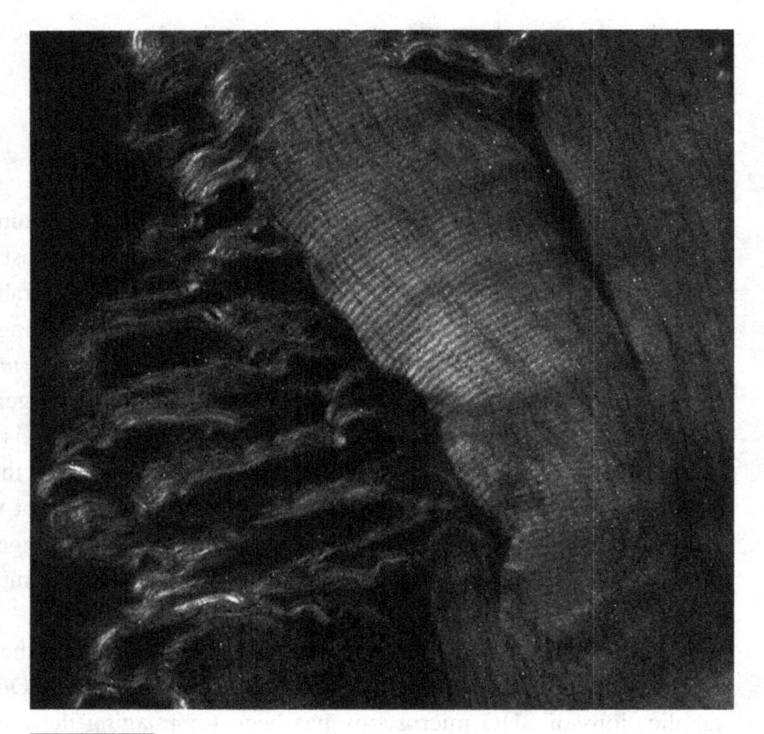

Figure 17.18
Second-harmonic generation (SHG) microscopy image of collagen and myosin on the surface of muscle. Full width of image is 150 μm. (Public domain, courtesy of Steffen Dietzel.)

boundary surface with respect to the optical field polarization affects the strength of the effect.

17.5 Super-resolution microscopy

Finally, we conclude this chapter with a brief introduction to methods of *super-resolution microscopy*, a term that generally refers to optical imaging that achieves significantly finer resolution than the Abbe limit. Full development of the various methods of super-resolution microscopy is beyond the scope of this text, and while this topic is more in the domain of optical physics, with applications in biophysics and biochemistry, the impact on biomedical science is bound to grow. Moreover, three of the key developers of methods for super-resolution microscopy were awarded the 2014 Nobel Prize in chemistry, and all three have addressed biomolecular imaging in their research: Stefan Hell (Germany), Eric Betzig (USA), and William Moerner (USA).

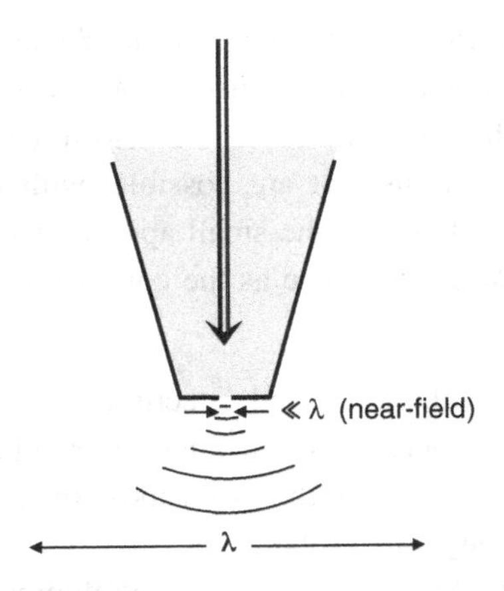

Figure 17.19

The scanning probe tip of an NSOM microscope. The reflective coating (black line) on the small pulled optical fiber tip has a nano-scale hole, permitting leakage of the optical field that, in the near field, is comparable in size to the aperture.

17.5.1 Direct super-resolution: near-field scanning optical microscopy (NSOM)

Prior to the development of indirect methods of achieving super-resolution, such as those related to the 2014 Nobel Prize, optical imaging at a resolution on the scale of tens of nanometers was achieved more directly by avoiding the diffractive limitations of far-field imaging. In the method called *near-field scanning optical microscopy* (NSOM), a small optical probe scans the sample within nanometers of the surface, in a manner similar to the function of an atomic-force microscope (Pohl et al., 1984; Betzig et al., 1986). (Note that one of the 2014 Nobel laureates also played an important role in the earlier development of NSOM.) The most common design for an NSOM probe utilizes an optical fiber that has been drawn down to a small tip, and the outside surface of the probe is coated with a reflective layer, typically vapor-deposited metal. As illustrated in Figure 17.19, there is a very small hole in the reflective coating at the tip of the probe, through which the optical field can leak. Objects close to the tip will interact, in that very small spatial extent, with the electromagnetic field that oscillates at the optical frequency. The imaging capability of NSOM is achieved by raster scanning the small sensitive area over the sample surface.

With NSOM there is a flexible range of the sources of contrast in the sample, which can be based on reflectivity, differential interference, absorption (especially with applied stains) or fluorescence (also with applied biomarkers). Given the exceedingly small interrogation volume, signals for any of the contrast mechanisms tend to be small. As such, NSOM imaging of fluorescence generally yields the best signal-to-noise ratio, because detection of fluorescence is a dark-background type of measurement, and the sensitivity for detection of a fluorescent signal can be at the single photon level. There are also three different types of illumination-detection configurations that are possible, with the choice dependent on sample presentation and size of the small aperture (which relates to the desired spatial resolution). For fluorescence as the contrast, these configurations are:

- illumination of the sample by the probe and detection of fluorescence light that is emitted in any direction (generally in transmission or reflection);
- illumination of the sample externally, either from above or below, and collection of emitted fluorescence through the probe;
- illumination through the probe and collection of the fluorescence through the probe.

An advantage of NSOM over some of the more recent methods for super-resolution is the availability of different sources of contrast, whereas the prominent newer super-resolution methods (see below) are primarily effective for imaging fluorescent objects. On the other hand, the challenge of NSOM is the delicate process of maintaining the tip within a few nanometers of the sample surface while raster-scanning the intended field of view.

17.5.2 Point-source localization techniques for super-resolution

NSOM can be regarded as a direct form of super-resolution, in that a small object is resolved directly through the scanning of the nanometer-scale near-field probe. The methods to be summarized in the next two sections, however, are far-field imaging schemes that indirectly resolve a small object by inferring its location with high precision. That is, methods that work in the far field are subject to the Abbe limit for direct resolution, but the location of a single sub-resolution object can be determined with much greater precision, even if it cannot be resolved directly. This can be understood by examination of Figure 17.2, which represents the shape of the PSF of a distortion-free optical system. As discussed in Section 17.1.2, the closest separation of two point objects that can be distinguished occurs when they are separated by the distance from the axis to the first-zero of the pattern. But the location of the peak of the PSF (for a single point object) can be determined with a higher precision, d_{loc}, that depends, statistically, on the number of photons detected from that point source:

$$d_{\text{loc}} \approx d_{\text{lim}}/\sqrt{N}, \tag{17.19}$$

where N is the number of detected photons from the single point source. The detected photons will gradually fill in the form of the PSF, and its center is determined as the statistical centroid of the distribution of detected photons.

17.5.3 Stimulated emission depletion (STED) microscopy

First proposed by Hell and Wichmann (1994), *stimulated emission depletion microscopy* (STED microscopy) is a method that causes a fluorescent object to appear as a small point source by turning off the fluorescence emission in an annular area surrounding a small central point. The resulting point emission can be located with high precision, as discussed above, and an image can then be generated by scanning the point detection as is done with confocal microscopy. The approach is said to be deterministic, because the laser beams are aimed at the point being detected, and other nearby emitters are suppressed.

The key "trick" of STED microscopy is the "turning off" of the fluorescence from the annular area surrounding the detection point. Following illumination of the area with a beam at the excitation wavelength, the annular turn-off is accomplished by illuminating the donut area with a strong beam at a wavelength in the emission band of the fluorophore. This induces stimulated emission, which depletes the excited molecules much faster than their spontaneous emission rate. Thus, the steps in the process are as follows:

- The spot being scanned at the moment is illuminated with the excitation wavelength for the fluorophores, and emission begins with a decay determined by the excited-state lifetime. This source is generally a picosecond pulse length, much shorter than the fluorescence lifetime.
- Immediately after the excitation pulse, the depletion pulse illuminates a ring surrounding the scanned spot with a strong field at the emission wavelength, inducing depletion by stimulated emission. This pulse is also picoseconds in duration.
- The very small sample spot in the center of the depletion ring is not depleted, and continues to emit fluorescence over a longer time, and is detected, denoting the emission source as the small central spot, which can be located with greater precision than the size of the PSF.

Such donut-shaped beams can be generated by specialized phase gratings (Klar et al., 2001) or by using a laser that emits in the second Laguerre-Gaussian mode (TEM^*_{01}, where TEM stands for transverse electromagnetic, the "donut" shaped transverse mode with zero on-axis intensity [Török and Munro, 2004]). (The modes are named for the French mathematician Edmond Laguerre [1834–1886]

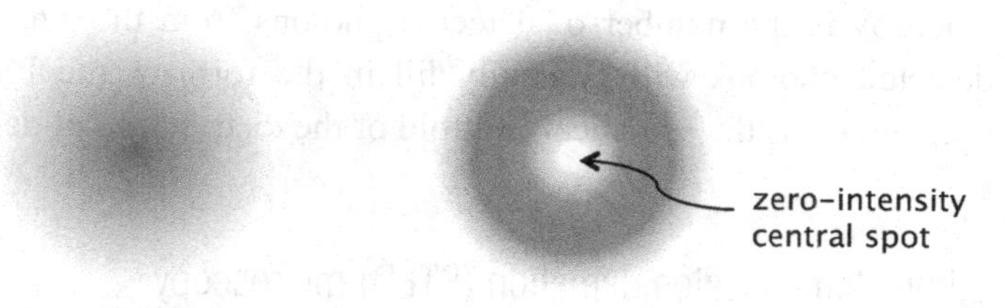

Figure 17.20

Left: a lowest-order Gaussian beam. Right: a second Laguerre-Gaussian mode, used as a depletion beam for STED microscopy, has a small central spot of zero intensity (much smaller than the PSF). The resulting fluorescence emission from the sample is suppressed except in that spot.

and the same Carl Friedrich Gauss of the Gaussian units and distribution.) A qualitative depiction of such a beam is shown in Figure 17.20, alongside a lowest-order Gaussian beam of the same size. To render the central zero spot as small as possible, the depletion beam is strong enough to oversaturate the stimulated emission process of the sample fluorophore.

In practice, the wavelength of the depletion beam is chosen to be in the longer-wavelength range of the emission band of the fluorophore, whereas detection of the fluorescence is performed at a shorter emission wavelength, thus enabling filters to block detection of the strong depletion light. Image generation by scanning of the field of view can be slow, especially if the fluorescence emission rate is low from the non-depleted point center, and care must be taken with biological samples to limit damage by the strong depletion beam.

17.5.4 Stochastic methods of super-resolution

While STED microscopy is a deterministic approach to far-field super-resolution imaging, stochastic methods gradually build an image from randomly located point emitters, whose locations, as discussed above, can be determined with precision greater than they can be resolved. The basic approach was enabled when Moerner and Kador (1989) demonstrated that emission from single molecules could be detected and localized by controlling activation such that emitting molecules at any given time are very sparse, and separated by $\gtrsim 2\times$ the diffraction-limited resolution (or PSF diameter). The sparseness enables localization of each emitting molecule with precision that is a fraction of the PSF, as explained in Section 17.5.2.

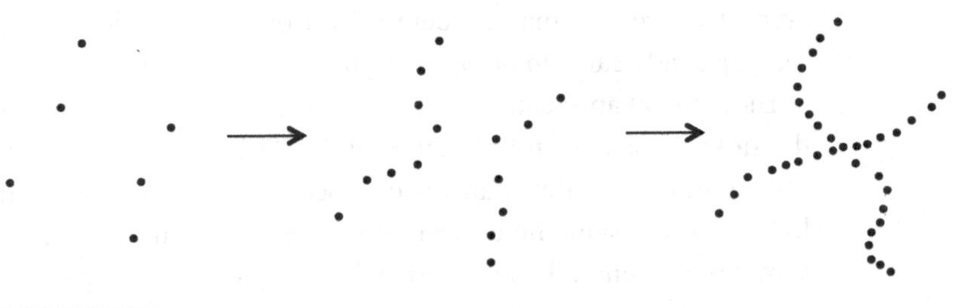

Figure 17.21

Two nano-scale fibrils are imaged by STORM in a succession of sparse single-point recordings that are added. Initially, on the left, the sparse random distribution of points does not yet reveal any structure, but as sets of points are added, the image gradually appears.

A few years later, three groups independently demonstrated super-resolution imaging based on the principles of statistical single-molecule detection (Rust et al., 2006; Betzig et al., 2006; Hess et al., 2006). The term that is commonly used for this approach is *stochastic optical reconstruction microscopy* (STORM). Closely related methods are called *photoactivated localization microscopy* (PALM), and *fluorescence photoactivation localization microscopy* (FPALM), among others.

In its simplest incarnation, the process of STORM proceeds as follows:

- A weak excitation pulse illuminates the entire field of view. A sparse distribution of fluorescent molecules is excited, and all are imaged. Localization of each point source to a fraction of the PSF is calculated and the locations are stored.
- The fluorophores are then all deactivated (e.g., by a broad depletion beam) and the excitation step is repeated, generating a new, and different, sparse distribution of emitters, whose locations are added to those previously recorded.
- The process is repeated as many as thousands of times, eventually building up a full image made of high-precision point locations.

The result is something like a pointillist painting, except that the points are painted in random order on the "canvas." A schematic illustrating the process is shown in Figure 17.21.

In practical terms, the resulting point localization for STORM is limited to a precision of the order of $\sim d_{\text{lim}}/10$, because as the number of collected photons from one point increases, the emitters become less sparse, and there is an increasing probability of more than one emitter being detected within $2 \times$ PSF.

Among far-field super-resolution methods an advantage of the STORM method over STED microscopy is that scanning is not required for STORM, as the entire field of view builds up with the addition of the multiple sparse images. On the other hand, the scanned area for STED microscopy can be limited to a specific area of interest as may be determined by a conventional-resolution image, such that a partial scan can provide all the information that is interesting.

Biomedical applications of super-resolution techniques are generally limited by the slow frame rates for imaging. With STED, for example, given that the effective PSF is much smaller than for confocal microscopy, the scan time can be much longer for the same field of view. Consequently, useful frame rates for biological samples are generally achieved by limiting to a small region of interest in a sample or a small field of view. Moreover, super-resolution techniques generally require use of switchable fluorescent probes, which can be cytotoxic. Nonetheless, in addition to imaging of fixed proteins and other small structures, gradual improvement of the techniques and the increasing availability of improved fluorescent probes are leading to super-resolution imaging of structure and dynamics of live cells (Watanabe et al., 2007; Huang et al., 2008; Shroff et al., 2008). For example, Watanabe et al. (2007) provided the first direct visualization of the transport of nanoscale vesicles in living cells.

Problems – answers to problems with * are on p. 648

17.1 In microscopy, the *Rayleigh criterion* is a measure of the resolving power of the instrument, determined by how close together (as an angular distance) two points in an object can be resolved when imaged by the system. Remembering that intensities simply add, for a given aperture diameter, d, and wavelength, λ, plot a cross-section of the image produced from two point sources in the object plane when the peak of the image of one point source is at the first zero of the Airy disk of the other point source.

17.2* If the PSF of a microscope is an Airy disk, what is the functional shape of the OTF? What does that shape represent in terms of the design of a microscope?

17.3* For a distortion-free microscope whose objective lens has an NA $= 0.95$ in air, what is the resolution limit at the wavelength $\lambda = 500$ nm when an immersion fluid with refractive index $n = 1.5$ is used?

17.4 An isolated nucleus from a cell is on a microscope slide and is surrounded by growth medium. The nucleus has a refractive index of 1.42, while the surrounding fluid has an index of 1.35. For a wavelength $\lambda = 500$ nm, what will be the relative phase shift between light that is transmitted through the center of the nucleus vs. light that goes through the fluid? Can you use this information to explain why the phase of the unretarded light needs to be shifted by $\sim\pi/2$ to accomplish good phase contrast imaging?

17.5* For a Gaussian beam propagating through a focus, the optical electric field amplitude is E_0 at $\boldsymbol{\rho} = 0$, $z = 0$. What is the on-axis optical electric field amplitude at a distance $2z_R$ from the focal plane? What is the intensity, relative to that the focal plane?

17.6* In a confocal microscope, the detection aperture is set to the first zero of the Airy disk of the PSF of that instrument. If the objective lens has an NA = 0.7, what is the axial resolution at $\lambda = 500$ nm?

17.7* In a TPEF microscope, the objective/illumination lens has an NA = 0.7. What is the diameter of the fluorescence source at the focal plane if the excitation laser wavelength is $\lambda = 800$ nm?

17.8 Explain why TPEF microscopy, for imaging of biological samples, is generally carried out with a laser source whose wavelength is in the NIR, rather than in the visible.

17.9 In Figure 17.18, what is the likely polarization direction of the laser field that was used to generate the second-harmonic signal? How do you come to that conclusion?

References

Betzig, E., Lewis, A., Harootunian, A., Isaacson, M., and Kratschmer, E. (1986). Near field scanning optical microscopy (NSOM). *Biophysical Journal*, 49, 269.

Betzig, E., Patterson, G. H., Sougrat, R., et al. (2006). Imaging intracellular fluorescent proteins at nanometer resolution. *Science*, 313, 1642–1645.

Carter, M., and Shieh, J. (2010). *Guide to Research Techniques in Neuroscience*. New York: Academic Press.

Denk, W., Strickler, D., and Webb, W. W. (1990). Two-photon laser scanning fluorescence microscopy. *Science*, 248, 73–76.

Denk, W. (1996). Two-photon excitation in functional biological imaging. *Journal of Biomedical Optics*, 1, 296–304.

Gmitro, A. F., and Aziz, D. (1993). Confocal microscopy through a fiber-optic imaging bundle. *Optics Letters*, 18, 565–567.

Hell, S. W., and Wichmann, J. (1994). Breaking the diffraction resolution limit by stimulated emission: stimulated-emission-depletion fluorescence microscopy. *Optics Letters*, 19, 780–782.

Hell, S. W., Bahlmann, K., Schrader, M., et al. (1996). Three-photon excitation in fluorescence microscopy. *Journal of Biomedical Optics*, 1, 71–74.

Helmchen, F., and Denk, W. (2005). Deep tissue two-photon microscopy. *Nature Methods*, 2, 932–940.

Hess, S. T., Girirajan, T. P. K., and Mason, M. D. (2006). Ultra-high resolution imaging by fluorescence photoactivation localization microscopy. *Biophysical Journal*, 91, 4258–4572.

Huang, B., Jones, S. A., Brandenburg, B., and Zhuang, X. (2008). Whole-cell 3D STORM reveals interactions between cellular structures with nanometer-scale resolution. *Nat. Methods*, 5, 1047–1052.

Kaiser, W., and Garrett, C. G. B. (1961). Two-photon excitation in $CaF2:Eu2+$. *Physical Review Letters*, 7, 229–231.

Kerr, J. N., Greenberg, D., and Helmchen, F. (2005). Imaging input and output of neocortical networks in vivo. *Proceedings of the National Academy of Sciences USA*, 102, 14063–14068.

Klar, T. A., Engel, E., and Hell, S. (2001). Breaking Abbe's diffraction resolution limit in fluorescence microscopy with stimulated emission depletion beams of various shapes. *Physics Review E*, 64, 066613.

Mertz, J. (2010). *Introduction to Optical Microscopy*. St. Catharines, ON: Roberts Press.

Minsky, M. (Dec. 1961). *Microscopy Apparatus*, U.S. Patent 3,013,467.

Moerner, W. E., and Kador, L. (1989). Optical detection and spectroscopy of single molecules in a solid. *Physical Review Letters*, 62, 2535.

Moreau, L., Sandre, O., and Mertz, J. (2000). Membrane imaging by second-harmonic generation microscopy. *Journal of the Optical Society of America B*, 17, 1685–1694.

Nakamura, O., and Kawata, S. (1990). Three dimensional transfer-function analysis of the tomographic capability of a confocal fluorescence microscope. *Journal of the Optical Society of America A*, 7, 522–526.

Nomarski, G. (1955). Microinterféromètre differential à ondes polarisées. *Journal de Physique et le Radium*, 16, S9.

Ohki, K., Chung, S., Ch'ng, Y., Kara, P., and Reid, R. (2005). Functional imaging with cellular resolution reveals precise micro-architecture in visual cortex. *Nature*, 433, 597–603.

Pluta, M. (1989). *Advanced Light Microscopy*, Amsterdam: Elsevier.

Pohl, D. W., Denk, W., and Lanz, M. (1984). Optical stethoscopy: image recording with resolution $\lambda/20$. *Applied Physics Letters*, 44(7), 651.

Preza, C., Snyder, D. L., and Conchello, J. A. (1999). Theoretical development and experimental evaluation of imaging models of differential-interference-contrast microscopy. *Journal of the Optical Society of America A*, 16, 2185–2199.

Rust, M. J., Bates, M., and Zhuang, X. (2006). Sub-diffraction-limit imaging by stochastic optical reconstruction microscopy (STORM). *Nature Methods*, 3, 793–795.

Saleh, B. E. A., and Teich, M. C. (2007). *Fundamentals of Photonics*, 2nd edn. Hoboken, NJ: Wiley.

Seward, G. H. (2010). *Optical Design of Microscopes*, SPIE Tutorial Text Vol. TT88. SPIE Publications.

Sheppard, C. J. R. (1986). The spatial frequency cutoff in three dimensional imaging. *Optik*, 72, 131–133.

Shroff, H., Galbraith, C. G., Galbraith, J. A., and Betzig, E. (2008). Live-cell photoactivated localization microscopy of nanoscale adhesion dynamics. *Nature Methods*, 5, 417–423.

Siegman, A. E. (1986). *Lasers*. Herndon, VA: University Science Books.

Solberg, J. K. (2000). *Light Microscopy*. Trondheim: Tapir Trykk.

Török, P., and Munro, P. R. T. (2004). The use of Gauss-Laguerre vector beams in STED microscopy. *Optics Express*, 12, 3605–3617.

Watanabe, T. M., Sato, T., Gonda, K., and Higuchi, H. (2007). Three-dimensional nanometry of vesicle transport in living cells using dual-focus imaging optics. *Biochemical and Biophysical Research Communications*, 359, 1–7.

Further reading

Basic optics of conventional microscopes

Hecht, E. (2002). *Optics*, 4th edn. Reading, MA: Addison-Wesley.

Seward, G. H. (2010). *Optical Design of Microscopes*, SPIE Tutorial Text Vol. TT88. SPIE Publications.

General biomedical optical microscopy

Cox, G. (2007). *Optical Imaging Techniques in Cell Biology*, Boca Raton, FL: CRC Press.

Fujimoto, J., and Farkas, D., eds. (2009). *Biomedical Optical Imaging*. Oxford: Oxford University Press.

Liang, R., ed. (2013). *Biomedical Optical Imaging Technologies: Design and Applications*. Berlin: Springer.

Mertz, J. (2010). *Introduction to Optical Microscopy*. St. Catharines, ON: Roberts Press.

Phase contrast and differential interference contrast

Pluta, M. (1989). *Advanced Light Microscopy*, Amsterdam: Elsevier.

Preza, C., Snyder, D. L., and Conchello, J. A. (1999). Theoretical development and experimental evaluation of imaging models of differential-interference-contrast microscopy. *Journal of the Optical Society of America A*, 16, 2185–2199.

Shaked, N., Zalevsky, Z., and Satterwhite, L., eds. (2013). *Biomedical Optical Phase Microscopy and Nanoscopy*. Amsterdam: Elsevier.

Confocal combined with DIC microscopy

Cogswell, C. J., and Sheppard, C. J. R. (1992). Confocal differential interference contrast microscopy: including theoretical analysis of conventional and confocal DIC imaging. *Journal of Microscopy*, 165, 81–101.

Gaussian optics (and laser beams)

Hecht, E. (2002). *Optics*, 4th edn. Reading, MA: Addison-Wesley.

Saleh, B. E. A., and Teich, M. C. (2007). *Fundamentals of Photonics*, 2nd edn. Hoboken, NJ: Wiley.

Two-photon fluorescence microscopy and nonlinear optical microscopy

Masters, R., and So, P., eds. (2008). *Handbook of Biomedical Nonlinear Optical Microscopy*. Oxford: Oxford University Press.

Short-pulse laser sources (for two-photon and nonlinear optical microscopy)

Siegman, A. E. (1986). *Lasers*. Herndon, VA: University Science Books.

Super-resolution microscopy

Huang, B., Bates, M., and Zhuang, X. (2009). Super-resolution fluorescence microscopy. *Annual Review of Biochemistry*, 78, 993–1016.

Leung, B. O., and Chou, K. C. (2011). Review of super-resolution fluorescence microscopy for biology. *Applied Spectroscopy*, 65, 967–980.

18 Optical coherence tomography

In Chapter 17 we noted that conventional microscopes are ineffective at rejecting out-of-focus light, either scattered or emitted as fluorescence, and are consequently limited to imaging thin samples. We also reviewed some methods to achieve optical sectioning in microscopic imaging by spatial gating, which requires the illumination source to be spatially coherent, and, for some methods, may also require phasefront-preserving optics for the detection stage. The other general class of methods that can effect sectional imaging is based on temporal gating of back-reflected light. Conceptually, this is simple: if one can generate a very short pulse of light – a 10-fs pulse, for example, would span $\sim$3 μm in the axial direction – and measure precisely the arrival time of a back-reflected pulse, a time gate could (theoretically) reject light that reflects from shorter or longer distances. In principle, this is what pulse-echo ultrasound imaging achieves: an ultrasound imaging system determines the depth of imaged objects by timing the arrival of sound pulses that have reflected from discontinuities in the tissue density (or acoustic characteristic impedance) associated with structures in the body (see Section 16.1.2). Since the speed of sound (in water) is $\sim$1500 m/s, the timing precision needed for, say, 100 μm in axial resolution would be of the order of 100 ns, which is comfortably within the capability of common electronic timing circuitry. In Chapter 16 (Section 16.4.3), we saw that for hybrid photoacoustic imaging, the spatial (lateral) resolution of the ultrasound imaging component can approach microscopic dimensions.

In a purely optical system, however, even though ultrashort laser pulses are available, microscopic axial resolution of, say, 3 μm would require a temporal resolution for detection of the order of 10^{-14} s, which is far shorter than the response time of any electronic sensor. Optical methods, nonetheless, offer us ways to get around this limitation. Early approaches, in a manner conceptually analogous to the sectioning capability of second-harmonic-generation microscopy (as detailed in Section 17.4.2), invoked time gating by correlating the overlap of the sample signal with a reference signal in a nonlinear medium. For example, Fujimoto et al. (1986) first demonstrated the potential for precision "time-of-flight" ranging of distances in biological structures by use of a nonlinear-optical cross-correlation time-gate based on second-harmonic generation, with the harmonic signal separated angularly between the crossed beams from the sample and reference. Nonlinear-optical gates, however, require at least the reference beam to be of high peak power; such methods are not in current use for biological imaging. Nonetheless, one can use a sample of the pulse, delay it in a reference arm, and *interferometrically* compare the *phase* of the light back-reflected from the sample with the phase of the reference light, rather than timing the envelopes of the pulses. This is the basis of *optical coherence tomography* (OCT), a form of *low-coherence interferometry* that is typically carried out in the configuration of a *Michelson interferometer* (named for the American physicist, Albert A. Michelson [1852–1931], who received the Nobel Prize in physics in 1907), adapted to microscopic imaging. We will see in Section 18.1.2 that this method can work even without requiring ultrashort pulses.

18.1 The coherence length of light

To understand the concepts in the sections that follow, we first need to establish a basic formalism for describing the temporal (longitudinal) coherence of a light source. In Section 4.7.1 we briefly introduced the concepts of spatial (lateral) and temporal (axial) coherence of light, but now a more quantitative representation for the latter will be helpful. The reader will also note that an abbreviated development of the formalism presented here for temporal coherence was provided in Section 7.5.2, as it was relevant to the topic of dynamic light scattering. The principles and full derivations of the mathematics of optical coherence can be found in most optics textbooks: good options include Hecht (2002) and Saleh and Teich (2007).

The phase of an ideal monochromatic wave can be known at any point in time. In other words, if one knows the phase at an initial time point, the phase can be calculated for any later (or earlier) time point, which means that the wave would interfere in a predictable way with a later (or earlier) sample of itself. This is referred to as a condition of perfect *autocoherence*. If two monochromatic waves have different frequencies, they have perfect *mutual coherence*, and they will beat against each other in a predictable way, because their relative phases can be known at any point in time or at any location along the axis of propagation.

We are more interested, however, in the coherence (phase) properties of a single and realistic light source. In the non-ideal world, a wave with a center angular frequency ω_0 has some degree of random phase fluctuations, and this will translate to a frequency spread, or bandwidth, about that center frequency; the phase of the wave will not be knowable after a specific length of time, which depends on that bandwidth. The length of time over which a wave oscillates in a predictable manner (i.e., at a predictable phase) is called the *coherence time*, τ_c; analogously,

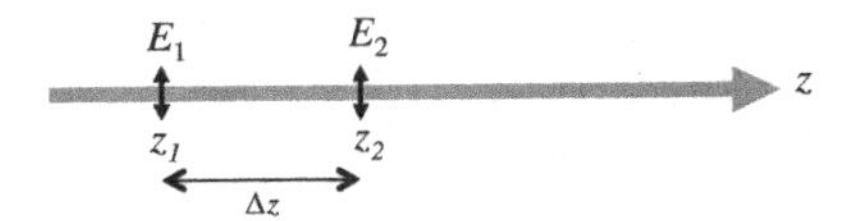

Figure 18.1

The electric field at two locations along the axial (propagation) direction of a wave can be correlated if the coherence length, L_C, is longer than their separation, Δz.

the distance corresponding to that time interval (when multiplied by c) is called the *coherence length*, L_c.

Figure 18.1 symbolizes a beam of light, perhaps a laser beam, propagating in the z direction, and we ask what the degree of correlation (in phase) is between the field at two axial locations.

For a single beam, we can write a *field autocorrelation function* as the average, over z, of the product of the optical field, $E(z)$ and the complex conjugate of the field at a point displaced by Δz:

$$G_1(\Delta z) = \langle E(z)E^*(z + \Delta z)\rangle, \tag{18.1}$$

or, equivalently, if the autocorrelation is examined at a specific z position, but as a function of time, then the *temporal autocorrelation function* is

$$G_1(\Delta t) = \langle E(t)E^*(t + \Delta t)\rangle = \lim_{T\to\infty} \frac{1}{2T} \int_{-T}^{T} E(t)E^*(t + \Delta t)dt \tag{18.2}$$

where $\Delta t = (\Delta z)/c$, and the angular brackets indicate average over time. Here we are using the complex exponential notation for the optical field amplitude, $E(t) = E_0 e^{-i\omega t}$, as a function of the angular frequency, ω, suppressing the spatial dependence, and the asterisk indicates the complex conjugate of the optical electric field term. (If we were asking about the correlation [degree of coherence] between two different beams, we would be interested in the *cross-correlation function*. This will come into play in Section 18.1.2.)

It is useful to write an expression for the autocorrelation function as normalized, represented by $g_1(\Delta t)$, such that it is independent of the intensity:

$$g_1(\Delta t) \equiv \frac{G_1(\Delta t)}{G_1(0)} = \frac{\langle E(t)E^*(t + \Delta t)\rangle}{\langle E(t)E^*(t)\rangle}, \tag{18.3}$$

where the denominator in Eq. (18.3), $G_1(0)$, is effectively the time-averaged intensity. This expression is referred to as the *complex degree of coherence*, and is also called the *temporal coherence function* in some texts. Thus, the value of $|g_1(\Delta t)|$ ranges from 1 (at $\Delta t = 0$) to 0 at any time point (or distance) at which there is no correlation.

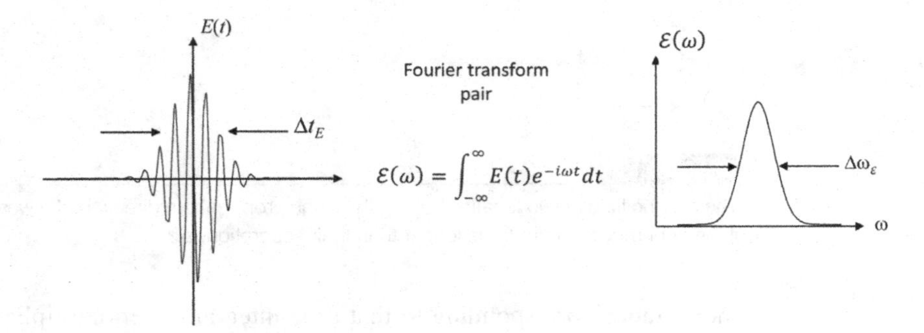

Figure 18.2
The temporal pulse shape of the electric field, $E(t)$, of a short pulse and its amplitude spectrum, $\mathcal{E}(\omega)$, form a Fourier transform pair.

The coherence time, τ_c, is most commonly defined as the separation between time points on either side of $\Delta t = 0$ for which the value of $|g_1(\Delta t)|$ drops to $1/2$ (although some references use the value $1/e$). It is interesting to note that, for a monochromatic wave, the complex degree of coherence becomes $g_1(\Delta t) = e^{i\omega\Delta t}$, whose magnitude, $|g_1(\Delta t)|$, is 1 for all values of Δt.

The coherence length and the coherence time are simply related through the speed of light:

$$L_c = c\tau_c \tag{18.4}$$

18.1.1 Coherence length of a short laser pulse

As discussed above, an ultrashort laser pulse could theoretically offer the timing resolution for optical sectioning in microscopy, but no optical detectors are available to time the arrival of the reflected pulse with adequate precision. One the other hand, the coherence length of an ultrashort pulse is also ultrashort, enabling temporal gating based on phase coherence. The key to this feature is the relationship between the bandwidth of a light beam and its coherence length. Ultrashort pulses are generally referred to as being *transform-limited*, which means that the width of the shortest possible pulse is limited by the frequency bandwidth through the Fourier transform of the power spectrum of the beam.

This relationship between pulse duration and bandwidth is illustrated in Figure 18.2, wherein the temporal history of the field amplitude, $E(t)$, of an ultrashort pulse and its spectrum, $\mathcal{E}(\omega)$, form a Fourier transform pair. Similarly, the temporal history of the field *amplitude* of the pulse (at one location in space), $E(t)$, is the *inverse* Fourier transform of the amplitude spectrum. In this circumstance, the pulse duration is often referred to as being transform-limited. There is an inverse relationship between the width of a pulse in the time domain, Δt_E in Figure 18.2, and the bandwidth of its Fourier transform in the frequency domain, $\Delta\omega_{\mathcal{E}}$ in Figure 18.2:

$$\Delta t_E = \frac{A}{\Delta\omega_{\mathcal{E}}}, \tag{18.5}$$

where A is a constant factor that depends on the pulse shape and on the choice of the measures for the pulse width Δt_E and bandwidth $\Delta\omega_{\mathcal{E}}$. For Gaussian pulse shapes and for pulse widths specified by the standard deviation (σ), one finds $A = 1$. In the more common case where the pulse width is defined by the full-width-half-maximum (FWHM), after considering that, for a Gaussian pulse, $\text{FWHM} = 2\sqrt{2\ln 2}\,\sigma \cong 2.355\,\sigma$, one finds $A = 8\ln 2$.

Now, the optical pulse duration is really defined by the width of the pulse intensity, not the pulse electric field. Because the intensity is proportional to the square of the electric field, in the case of a Gaussian shape the pulse duration and the intensity bandwidth are given by $\Delta t_E/\sqrt{2}$ and $\Delta\omega_{\mathcal{E}}/\sqrt{2}$, respectively. For the case of an ultrashort pulse, the coherence time, τ_c, is of the order of the pulsewidth itself, so that $\tau_c = \Delta t_E/\sqrt{2}$. By indicating with $\Delta\omega_{\text{FWHM}}$ the FWHM intensity bandwidth, Eq. (18.5) yields:

$$\tau_c = \frac{A}{2\Delta\omega_{\text{FWHM}}}, \tag{18.6}$$

and the coherence length is directly obtained from Eq. (18.6) given that $L_c = c\tau_c$.

We note, however, that it is easier to measure wavelength than optical frequency, and the bandwidth of light is more commonly quoted as a range of wavelengths, $\Delta\lambda$. Translation of a frequency range, $\Delta\omega$, centered at angular frequency, ω_0, into a wavelength range, $\Delta\lambda$, depends on the center frequency, ω_0:

$$\Delta\lambda = 2\pi c \frac{\Delta\omega}{\omega_0^2 - \left(\frac{\Delta\omega}{2}\right)^2} \tag{18.7}$$

Moreover, the most commonly used expression for the coherence length (for a Gaussian-shaped bandwidth) is based on the wavelength range associated with the bandwidth. Ignoring the $(\Delta\omega)^2$ term in Eq. (18.7) and considering Eq. (18.6), the coherence length is given by:

$$L_c = \frac{Ac}{2\Delta\omega} = \frac{A2\pi c^2}{2\omega_0^2\Delta\lambda} = \frac{2\ln 2\,\lambda_0^2}{\pi\Delta\lambda} \cong \frac{0.44\lambda_0^2}{\Delta\lambda} \tag{18.8}$$

where we have used $A = 8\ln 2$ for a Gaussian pulse, and $\lambda_0 = 2\pi c/\omega_0$ is the wavelength corresponding to the center frequency, which is *not* the arithmetic center of the wavelength range. We observe that ignoring the $(\Delta\omega)^2$ term in Eq. (18.7) is usually appropriate, because the condition $(\Delta\omega)^2 \ll \omega_0^2$ is typically fulfilled considering that $\omega_0 \approx 2 \times 10^{15}$ rad/s (for $\lambda_0 \approx 900$ nm) and $\Delta\omega \approx 10^{14}$ rad/s (for $\Delta\lambda \approx 50$ nm). However, wide wavelength bandwidths of hundreds of nanometers, achieved for example with halogen lamps or supercontinuum sources, may yield frequency bandwidths $\Delta\omega$ that may approach 10^{15} rad/s, and the approximation would no longer apply.

18.1.2 Coherence length of a long-pulse or continuous light source

While a broad bandwidth (and, hence, a short coherence length) is required for an ultrashort pulse duration, the inverse is not true. A long-duration or continuous light source can have a broad bandwidth, resulting in a short coherence length. For example, an incandescent bulb or a xenon arc lamp can run continuously, and both have very broad spectral bandwidths. Of course, if one of those sources were to be used for illumination in microscopy, and if the type of microscopy required good spatial coherence of the illumination, then the spatial coherence would have to be obtained, at the expense of severe loss of available light, by focusing the light through a small pinhole.

Certain types of lasers, and other light sources with good spatial (transverse) coherence, can also have broad bandwidths. For a beam (or pulse) of arbitrary duration, the coherence time is generally defined in terms of its normalized auto-correlation function:

$$\tau_c = \int_{-\infty}^{\infty} |g_1(\Delta t)|^2 d\Delta t \tag{18.9}$$

and the *power spectral density,* $S(\omega)$ (also called *spectral density* or *power spectrum*), can be calculated from the field autocorrelation function:

$$S(\omega) = \int_{-\infty}^{\infty} G_1(\Delta t) e^{-i\omega\Delta t} d\Delta t \tag{18.10}$$

Analogous to the relation depicted in Figure 18.2, Eq. (18.10) shows that $S(\omega)$ and $G_1(\Delta t)$ form a Fourier transform pair for any duration of light emission. Remembering that $G_1(0)$ represents the time-averaged intensity, $S(\omega)$ can also be called the *intensity spectral density*. As mentioned in Section 16.2.1, Eq. (18.10) is known as the *Wiener-Khinchin theorem*.

The Michelson interferometer depicted in Figure 18.3 offers a simple way to measure the coherence length of any light source, or to measure the mutual

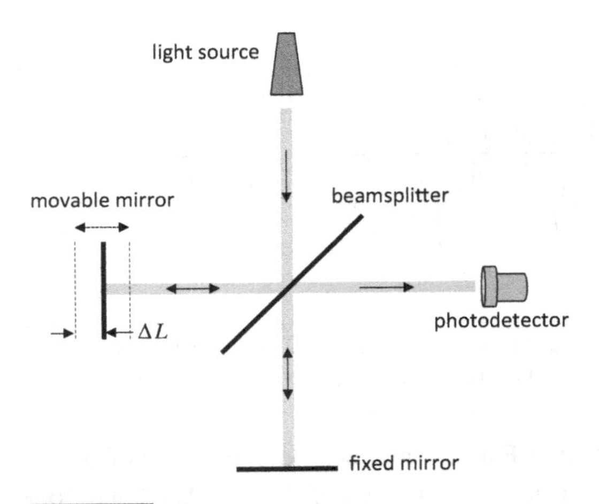

Figure 18.3

Michelson interferometer. By translating the movable mirror, the autocorrelation function of the light source is generated temporally at the detector.

coherence of two or more sources, if those sources are injected collinearly. The difference in lengths of the two arms depends on the position of the movable reference mirror and is denoted as ΔL; thus, the difference in total pathlengths from the source to the detector for the two arms is $2\Delta L$. This pathlength difference translates into a time delay $\Delta t = 2\Delta L/c$, where c is the speed of light.

If two monochromatic waves, E_1 and E_2, of different frequencies, ω_1 and ω_2, are injected collinearly into the interferometer from the source direction, they will simply beat against each other, both in time and as a function of ΔL (or, equivalently, Δt), and the amplitude of that interference can be traced out by varying the position of the movable mirror, hence varying ΔL. Thus, if the two waves are superposed at the detector, the intensity at the detector, I_d, of their time-averaged sum is given by

$$\begin{aligned} I_d(\Delta t) &= \langle |E_1(t) + E_2(t+\Delta t)|^2 \rangle \\ &= \langle |E_1|^2 \rangle + \langle |E_2|^2 \rangle + \langle E_1(t)E_2^*(t+\Delta t) \rangle + \langle E_1^*(t)E_2(t+\Delta t) \rangle \\ &= I_1 + I_2 + 2\mathrm{Re}\{G_{12}(\Delta t)\}, \end{aligned} \tag{18.11}$$

where $\Delta t = 2\Delta L/c$ and $G_{12}(\Delta t)$ is the field cross-correlation function. Equation (18.11) leads to what is sometimes called the *interference equation*:

$$I_d(\Delta t) = I_1 + I_2 + 2\sqrt{I_1 I_2}\,|g_{12}(\Delta t)| \cos \varphi(\Delta t), \tag{18.12}$$

where $\varphi(\Delta t)$ is the phase of $g_{12}(\Delta t)$, the normalized field cross-correlation function, and the third term on the right-hand-side is the interference term. (Expressing the phase term as $\cos \varphi$ was facilitated by taking the real part of the cross-correlation

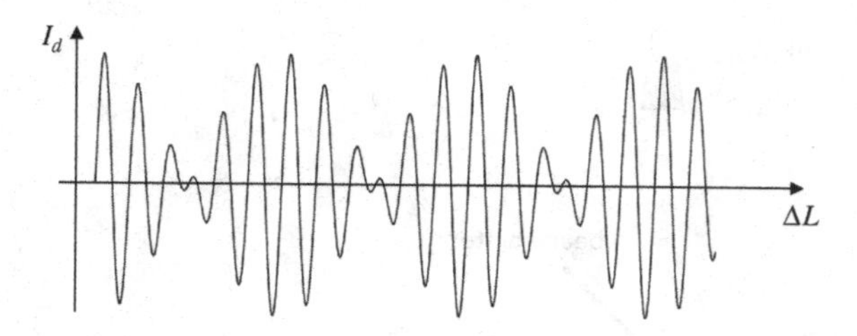

Figure 18.4

Temporal interference of two monochromatic waves is manifest as a beat pattern in the measured intensity as a function of the pathlength difference in a Michelson interferometer.

function, $G_{12}(\Delta t)$.) For two monochromatic waves of different frequencies, the output of the detector, as a function of $\Delta L = c\Delta t/2$, will look like Figure 18.4.

For a single light source, the beam can be split and compared to itself with a variable time delay. We consider the case of Figure 18.3 with a 50/50 beamsplitter, sending half of the source intensity, I, into each arm, denoted as $I_{1/2}$. When the optical pathlengths to the detector for the two arms with the fixed and variable mirrors are precisely equal, then interference will occur at the detector for any coherence length, and the measured strength of the interference will vary with the pathlength difference, which can be described as a special case of Eq. (18.11):

$$I_d(\Delta t) = 2I_{1/2} + 2\text{Re}\{G_1(\Delta t)\} = I\left[1 + |g_1(\Delta t)| \cdot \cos\left[\varphi(\Delta t)\right]\right], \quad (18.13)$$

where the phase part of $g_1(\Delta t)$ is now the phase difference, $\varphi(\Delta t)$, and is a function of the relative time delay (resulting from the pathlength difference, ΔL).

Finally, we consider the example of a light source that has a bandwidth of 53 nm, with a peak wavelength of 600 nm, resulting in a coherence length, L_c, of ~3 μm. Thus, only when the pathlengths of the two arms are equal to within ~3 μm is the interference detected. Fortunately, this level of precision is easily available with precision translation stages. The detector output as a function of the position of the movable mirror will look like Figure 18.5.

18.2 Time-domain optical coherence tomography

18.2.1 Sources of contrast and basics of OCT

The measured signal in OCT is direct backscattering from the sample, and therefore the main sources of *contrast* in OCT imaging are gradients and discontinuities in the refractive index, which are associated with the scattering cross sections of a variety of tissue constituents. Just as in the case of single-scattering events, as

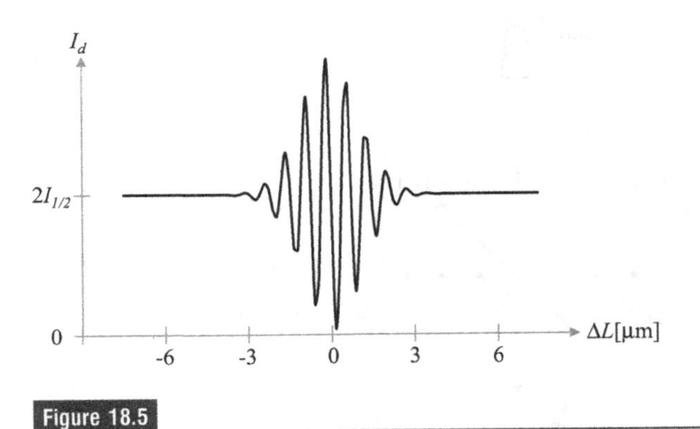

Figure 18.5

Intensity of the light measured at the detector of a Michelson interferometer, as a function of the variable mirror position, for a light source with a short coherence length.

studied in Chapter 7, the relative strength of a backscattered signal from a specific structure in a sample depends on its relative refractive index and the dimensions of the structure. Strongly scattering structures appear bright in an OCT image, but an especially strongly scattering object can "shadow" beyond it because less light reaches deeper objects, and less of the light backscattered from deeper objects is able to reach the detector. Of course, optical absorption also affects the OCT signal since absorbing structures in tissue attenuate the optical signal, thus decreasing the amount of backscattering signal that reaches the detector. Furthermore, localized absorption can result in shadowing light from deeper structures, again reducing the signals therefrom. However, the contrast in OCT is primarily a scattering contrast, so that OCT typically provides structural and morphological information on the investigated tissue. Efforts are ongoing to expand the sensitivity of OCT techniques to optical absorption (differential absorption OCT), mechanical properties (OCT elastography), and blood flow (Doppler OCT) (see Section 18.4).

We now have the basis to understand the scheme of OCT in the time domain. In the Michelson interferometer that was depicted in Figure 18.3, the fixed mirror presents a single reflection plane, which results in a single arrival time for a short pulse (or for the phase of a long pulse) at the detector. If that fixed mirror is replaced by a thick, layered sample (tissue, for example), which exhibits partial reflections from interfaces of layers or structures within the tissue with different scattering strengths, then we get the basic setup of a time-domain OCT (TD-OCT) system, as depicted in the similar Figure 18.6.

A representation of what the detector signal might look like, as a function of ΔL, is shown in Figure 18.7. Here we note that the different reflectivities of layer interfaces or structures within the thick sample are manifest as variations in the

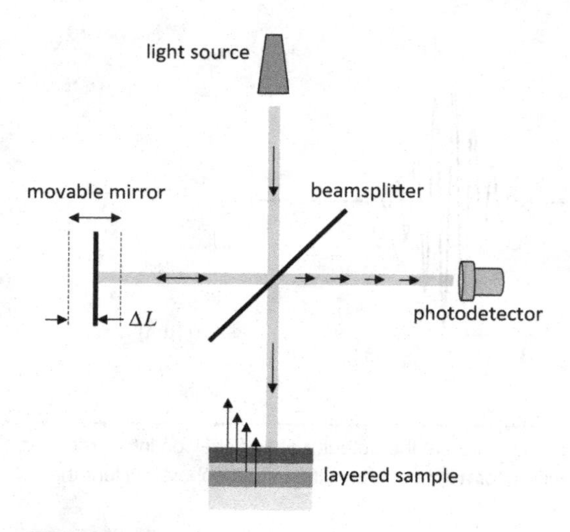

Figure 18.6

Basic configuration of TD-OCT measurement. Reflections in the sample arm can come from structures at different depths in the sample, with various reflectivities.

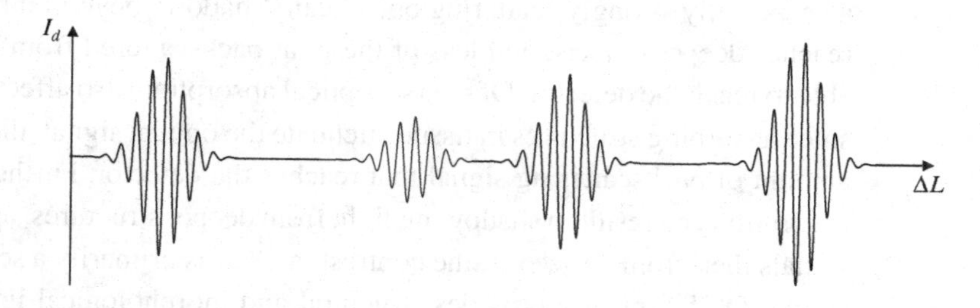

Figure 18.7

Detector output from an OCT interferometer, as a function of ΔL, for a thick sample with structures of various reflectivities at different depths, such as depicted in Figure 18.6.

strength of the interference signal. (We should also keep in mind that the signal generally degenerates with depth, as light is progressively lost to scattering and/or absorption, and less of the reflected light is ballistic.) To provide transverse resolution, a microscope objective is used to focus the beam in the sample arm.

For reflection from a specific surface at a given depth, the signal of interest (one of the "blips" in Figure 18.7) is the interference term in Eq. (18.13), for which the correlation function between the reference and sample beams can be written as:

$$g_1(\Delta t) = e^{-(\Delta\omega\Delta t/4\sqrt{2\ln 2})^2} e^{-i\omega_0\Delta t} \tag{18.14}$$

where $\Delta\omega$ is the FWHM spectral width of the (assumed) Gaussian spectral distribution of the source, whose center frequency is ω_0, and $\Delta t = 2\Delta L/c$. This has a Gaussian shape that is amplitude-modulated by the optical frequency.

Analogous to Eq. (18.13), we can write an expression for the detector signal as:

$$I_d(\Delta t) = I_s + I_r + 2\sqrt{I_s I_r}|g_1(\Delta t)|\cos[\varphi(\Delta t)], \tag{18.15}$$

where the subscripts s and r refer to the light from the sample and reference arms, respectively. The location of each peak of a Gaussian envelope as a function of the scanned distance, ΔL, gives information on the depth of the corresponding reflecting surface.

18.2.2 Detection of the TD-OCT signal

Another important aspect of the moving mirror in the reference arm of TD-OCT is that, in addition to accomplishing the depth scan, the moving mirror also induces a Doppler shift on the reference beam, with the frequency shift determined by the speed of the mirror scan. We recall Eq. (7.46) (here expressed in angular frequency), which leads to an expression for the Doppler frequency shift:

$$\Delta\omega_D \cong 2\omega_0 \frac{v_{\text{ref}}}{c}, \tag{18.16}$$

where ω_0 is the center optical frequency of the light source, and v_{ref} is the scanning velocity of the reference mirror. As in laser Doppler velocimetry (see Section 7.5.1), the factor 2 derives from the round-trip effect of the Doppler shift on the reference beam. When the reference beam and the sample beam are combined in the detection arm, this Doppler shift leads to a modulation frequency at $\Delta\omega_D$, i.e., a beat signal at that frequency (when the pathlengths are approximately equal), enabling heterodyne detection. To facilitate the heterodyne detection, the driver for the reference mirror is typically designed to produce a constant scanning speed, which is accomplished with a triangle-wave drive for the piezoelectric scanning element. (That is, v_{ref} is constant except near the ends of the stroke.) Thus, the depth profile can be generated by demodulation of the heterodyne signal from the detector, by standard heterodyne methods (see below). This depth profile is commonly referred to as an *A-scan*, where "A" stands for "amplitude" as in the amplitude-based scans of A-mode ultrasound.

For thick samples with distributed reflectivity (as compared to a sample with simple flat surfaces) the heterodyne detection affords an improvement in the signal-to-noise ratio of the detected signal. While the reference and sample beams will only interfere when the pathlengths are equal to within L_c, light from nearby

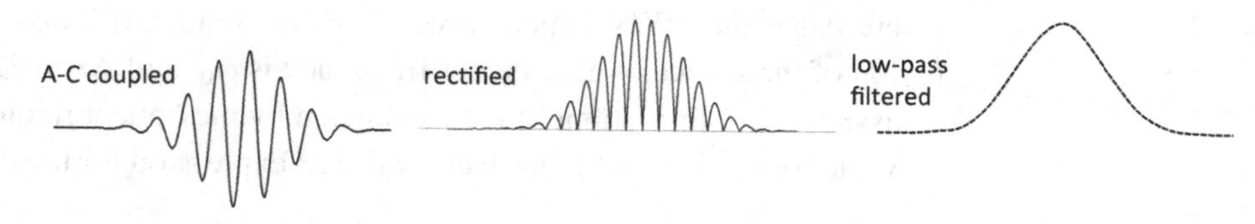

Figure 18.8

Demodulation steps for the interference signal of a TD-OCT system. The initial detector output has a DC offset, due to the non-interfering terms from the sample and reference arms. That DC offset is removed by AC-coupling the detector to the demodulation circuit. The main steps of demodulation include full-wave rectification (i.e., taking the absolute value), followed by a low-pass filter to yield the envelope. This can then be read by an analog-to-digital converter for digital management of the image elements.

sample depths within the sample arm can interfere with each other, effectively generating a homodyne-like signal, which constitutes background "noise" at the detector. That background signal, however, can be blocked by heterodyne detection if the heterodyne (i.e., Doppler) frequency is greater than the self-modulation frequencies from the distributed sample signals. This can be seen mathematically in Eq. (18.15), wherein only the third term contributes to the demodulated heterodyne signal. Demodulation of the rectified detector signal provides the pathlength-dependent amplitude of the heterodyne signal, yielding the A-scan that is the depth profile of the sample reflectivity, without the spurious signal from the sample.

The main steps in demodulating the signal are represented in Figure 18.8, representing the signal from one surface in the sample. There could be a continuous variation of the signal for a range of depth, with multiple structures, as illustrated in Figure 18.7. The peaks and variations of the resulting envelope, as a function of the length of the mirror scan, constitute the reflectivity of the sample vs. depth.

18.2.3 Scanning of TD-OCT to achieve an image

If the axial direction in the sample is designated as the z axis, then the information in Figures 18.7 and 18.8 represents an A-scan at one transverse (x, y) coordinate in the sample, constituting a reflectivity profile of structures as a function of depth for one specific x-y position. A two-dimensional depth cross section can be generated by translating the location of the A-scan laterally, yielding a *B-scan* of the sample, where the signal amplitude is mapped into brightness levels in the image. This terminology is analogous to that of ultrasound imaging, where the "B" in B-mode ultrasound also stands for "brightness." (See an example of an OCT B-scan in Figure 18.11.) Further, if one repeatedly analyzes the detector signal as the movable mirror passes through a specific position, while the focal spot is 2D-scanned across the x-y plane, this generates a depth-sectioned *en face* image. Finally, if multiple *en face* images from a series of different depths are combined, or if the A-scan itself is scanned in both x and y coordinates, a three-dimensional tomographic image of the sample can be generated. While tomographic information can also be achieved with a confocal microscope by axially translating either the sample or the focus of the microscope objective, those approaches are generally much slower than the axial translation of the reference mirror in OCT imaging. The speed advantage of OCT is further enhanced, dramatically, in the frequency domain, as will be described in Section 18.3.

18.2.4 Maximum sectioning depth and transverse resolution

A frequently asked question addresses the maximum sectioning depth at which OCT can generate an acceptable image. There are a number of factors that affect this determination, including the scanning speed, the coherence length (hence the depth resolution that is sought), and, of course, the sample optical properties. For a medium with a low absorption coefficient in the spectral range of the light source, a reasonable "rule of thumb" provides an estimate of the maximum depth of imaging that is in the $\approx 1/\mu_s'$ range. The logic behind this criterion is that beyond that depth, the large majority of the light in the sample beam has diffusely scattered, and since multiply scattered light from shallower depths can have the same total pathlength as ballistic light from a deeper image plane, the diffuse light will increasingly add background noise to the OCT signal, while the proportion of ballistic light becomes small. Thus, for example, for a typical reduced scattering coefficient in tissue of 10 cm^{-1}, a practical limit of section depth, for an image with adequate signal-to-noise ratio, is $\sim$1 mm. We note that this is approximately 10 times deeper than is commonly achieved with confocal microscopy.

One disadvantage of OCT is that the transverse resolution of the resulting image is not, typically, as good as for conventional or confocal microscopy. This is due to the fact that the sample illumination must be less tightly focused, as a compromise, to yield a longer confocal parameter, so that an optimum (as small as possible) focal spot is maintained over the desired axial scan range. From Eq. (17.11) we can see that the beam waist diameter of the focus increases as the square root of the confocal parameter (depth of focus), so the compromise is often acceptable if the desired length of the A-scan is not too large. Nonetheless, the numerical aperture of the focusing element in the sample arm of an OCT system is typically smaller than that of, say, a confocal microscope, and transverse resolution is commonly of the order of 2–3 μm.

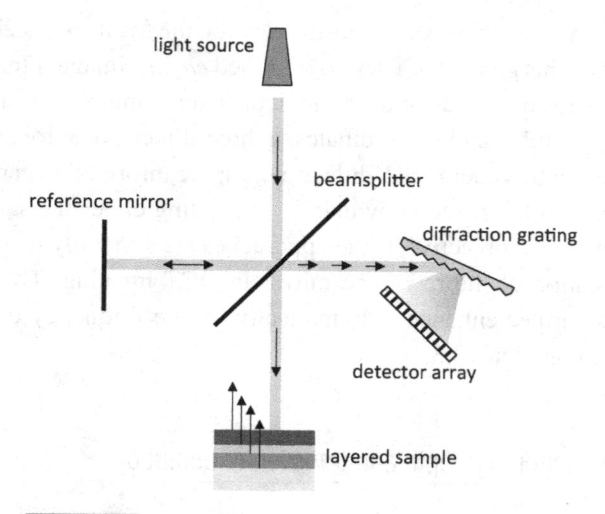

Figure 18.9

Schematic of spectral-domain OCT. The reference mirror is fixed, but the single detector is replaced by a spectrometer with parallel detection.

18.3 Frequency-domain optical coherence tomography

Frequency-domain OCT (FD-OCT) takes a different approach to extracting information about the interference between the sample beam and the reference beam. As the name suggests, FD-OCT works in the Fourier transform "space" of the light reaching the detector. We present the two main classes of FD-OCT.

18.3.1 Spectral-domain OCT

One class of FD-OCT systems is shown in Figure 18.9. Instead of detecting the combined light with a single detector while sweeping the length of the reference arm, with *spectral-domain FD-OCT* the reference mirror is kept fixed, and the different wavelength components of the broadband light are dispersed to separate detectors of a linear array and are measured simultaneously. Essentially, the single detector is replaced by a spectrometer with parallel detection. Given that the temporal autocorrelation of the beam and its spectral power density are related by Fourier transform, as discussed earlier (Eq. (18.10)), the A-scan can be generated numerically by taking the Fourier transform of the spatial distribution of the detected spectrum across the array (without translating the reference mirror). This type of configuration is also called *spatially encoded FD-OCT*.

The axial resolution is still of the order of L_c, which is now an effective coherence length, determined by the total bandwidth (or wavelength range, $\Delta\lambda$) that is dispersed across the detector array. The axial scanning range (length of the A-scan), however, is now determined by the spectral resolution of the array detection, $\delta\lambda$ (as relates to a frequency resolution, $\delta\omega$). Since the A-scan is detected in parallel, this approach can dramatically speed up the rendering of a B-scan in an OCT image. Invoking Eq. (18.8) for a Gaussian spectral distribution, an approximation of the useful length of the A-scan is provided by the expression:

$$L_A \approx L_c \frac{\Delta\lambda}{\delta\lambda} \approx \frac{0.44\lambda_0^2}{\delta\lambda}, \tag{18.17}$$

Thus, at a peak wavelength of, say, 1000 nm, and a spectral resolution of 0.2 nm, the length of the A-scan can be as long as ~2 mm.

18.3.2 Swept-source OCT

An alternative version of FD-OCT can be accomplished by using the configuration of Figure 18.6, but with a fixed mirror in the reference arm and with a light source that has a narrow bandwidth, $\delta\lambda$, whose center frequency is swept across a broad spectral range, $\Delta\lambda$. This approach is often called *swept-source-OCT* (SS-OCT) or *temporally encoded FD-OCT*. It may seem counterintuitive that a narrowband light source can be used for low-coherence interferometry, but in this case the Fourier transform is performed on the temporal output of the single detector, over the entire time that it takes for the frequency sweep over the wavelength range of the light source. As in spectral-domain OCT, the axial resolution is determined by $\Delta\lambda$, with L_c representing the effective coherence length if the dynamic wavelength is swept over the entire gain bandwidth. Here, the depth range of the A-scan is again determined by $\delta\lambda$ (and by limitations on the signal-to-noise ratio at greater depths). While swept-source OCT does invoke scanning of an optical element to achieve the wavelength sweep, as will be discussed in Section 18.5.2, the frequency sweep in SS-OCT can be much faster than the mechanical sweeping of a mirror in TD-OCT.

18.3.3 The SNR advantage of FD-OCT

Interestingly, the major advantage of FD-OCT over TD-OCT is not the faster rendering of a B-scan. An even more important benefit of FD-OCT over TD-OCT relates to the amount of light that contributes to the useful signal, and hence the signal-to-noise ratio (SNR) of the general image-rendering process. In TD-OCT, the light reaching the detector during most of the scan of the reference mirror is wasted, since only the instances of precisely matched pathlengths (to locations of structural surfaces in the sample) yield useful signal. With spatially encoded

FD-OCT, however, the light at all frequencies (i.e., corresponding to all depths of the axial scan) is detected in parallel on the individual pixels of the array detector, and all is useful signal. Similarly, in the case of SS-OCT, all of the instantaneous light at all frequencies during the wavelength scan constitutes useful signal. The overall improvement in SNR of FD-OCT compared to TD-OCT is of the order of the ratio $\Delta\lambda/\delta\lambda$. With modern instrumentation this ratio is of the order of 1000. This dramatically improved SNR is what enables the faster image generation for a given image quality.

18.4 Doppler OCT

The general concepts of Doppler flowmetry were introduced in Chapter 7. In Doppler flowmetry, homodyne detection is used for the interferometry with a single laser beam (i.e., without the use of a reference arm), and consequently a long coherence length is required for the light source. With OCT, the coherence length is intentionally short, but heterodyne detection against a reference arm of equal pathlength enables retention of phase information over a distance much longer than the coherence length. Considering, for the moment, TD-OCT, if, in addition to the mirror motion, objects in the sample are also moving, this can induce an additional Doppler shift, such that the modulation frequency is altered, and the new frequency is

$$\omega_D' = \omega_D \pm 2\omega_0 \frac{n v_s \cos\alpha}{c}, \tag{18.18}$$

where n is the refractive index in the sample, and $v_s \cos\alpha$ is the component of the sample velocity that is parallel to the z axis. (The sign option relates to the relative direction of the mirror motion at the time of measurement.) If the signal is analyzed to extract the information about the sample motion, then the resulting OCT image becomes a spatial map of objects *and* their motion, and is called *Doppler OCT* (DOCT).

Doppler OCT is of special interest for imaging blood flow and, given the strongly scattering nature of red blood cells, the method is well suited for such. It should be noted that this additional frequency shift is generally a small perturbation of the modulation frequency due to the mirror scanning, because the mirror scan speed is generally much faster than blood flow speeds. Consequently, it is difficult to directly measure the slight changes in the modulation frequency for different lateral positions in an image, given that only a few beat cycles are available for a measurement at a given depth. An alternative method, which has been generally adopted, is to measure the phase differences between adjacent A-scans, which are measured at two successive times (Zhao et al., 2002). This approach is especially well suited to FD-OCT methods.

If the time difference between successive scans (i.e., the inverse of the A-scan *rate*) is denoted as t_d, the relative phase, $\Delta\varphi$, between the adjacent A-scans is then calculated as:

$$\Delta\varphi(x, y) = 2\pi n t_d \omega_0 \frac{v_s(x, y)}{c}, \tag{18.19}$$

where we have suppressed the factor $\cos\alpha$, such that $v_s(x, y)$ are axial components of velocities for different lateral (x, y) positions. Thus, a spatial map of relative phase differences yields a velocity map at a range of depths. Various methods for rapid extraction of the phase map have been demonstrated (Zhao et al., 2002; Westphal et al., 2002; Yang et al., 2003).

For the velocities extracted by Doppler OCT to be unambiguous, measurement is limited to determining velocities corresponding to phase shifts that must be within the range $-\pi$ to $+\pi$, for adjacent A-scans. This sets an upper limit to the flow velocities that can be extracted. As such, there is a tremendous advantage to FD-OCT, in that the time between adjacent A-scans can be much shorter, allowing for larger sample velocities for a given phase shift, as can be seen by examination of Eq. (18.19). The range of blood flow velocities encountered in mammalian systems (from $\sim$0.03 cm/s in capillaries to $>$30 cm/s in large arteries) is readily accommodated by the performance parameters of current FD-OCT systems.

Doppler-OCT images are generally displayed similarly to static OCT images, such as that in Figure 18.11, but with an overlay of false color in locations of blood vessels or other locations of scatterer motion.

18.5 Instrumentation for OCT

Given the commercial success of OCT for a number of clinical applications, especially in ophthalmology, a significant amount of engineering development has led to systems that are compact, robust, and user-friendly. Figure 18.10 shows the major elements of a contemporary OCT system.

18.5.1 Optical components

For the interferometer itself, the open-air Michelson configuration (see Figures 18.6 and 18.9) has been replaced by optical fibers, and the beamsplitter has been replaced by a two-way fiber coupler. These are single-mode fibers,

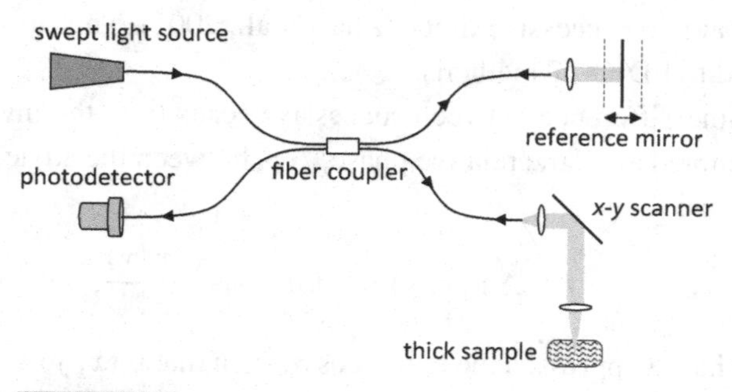

Figure 18.10

Basic elements of a contemporary FD-OCT system. The reference mirror is not scanned, but can be adjusted to facilitate setting the location of the A-scan relative to the sample. The light source is a swept-source fiber laser with semiconductor optical amplifier(s).

as necessitated by the requirement for preservation of precise phase information, which would be lost with multimode fibers. In practice, given that the sample reflectivity is generally less (or much less) than 100%, the fiber coupler (serving the role of beamsplitter) is typically designed to send more light to the sample than to the reference arm, with the intent that the intensities of the reference beam and the sample beam be approximately equal at the detector for optimum visibility of the interference.

Almost all modern OCT systems are of the frequency-domain type, given the clear advantages in speed and signal-to-noise ratio. As such, the mirror in the reference arm is not scanned, as would be the case for TD-OCT, but, nonetheless, is often adjustable to enable setting the pathlength to achieve the desired detection position in the sample, in the middle of the desired axial A-scan range.

18.5.2 Light sources and detectors

In the early implementations of TD-OCT, ultrashort (femtosecond) lasers were used to provide the broad bandwidth and consequent short coherence time. These were often based on mode-locked Ti-sapphire lasers, which have a broad gain bandwidth. Mode-locking, combined with pulse-compression methods (such as chirped pulse amplification), yields pulse durations in the tens of femtoseconds range, with coherence lengths of a few microns. These light sources were large, complex, and expensive laser systems.

Given that what matters for OCT is the coherence time, not the duration time, of the pulse, the development of superluminescent diodes (SLD) enabled the development of lower-cost and less complex OCT systems. The SLD is a continuous quasi-laser source, based on a high-gain semiconductor device, similar to high-gain diode lasers, but faceted so that there is no cavity feedback to achieve full laser action. SLDs are well suited to OCT because they have the high radiance (i.e., good transverse coherence) of a diode laser but the broad bandwidth and low temporal coherence of a light-emitting diode (LED) (Ko et al., 2004).

In recent years, among the approaches to FD-OCT, configurations using a wavelength-swept laser have gained favor (Choma et al., 2003), and swept-source OCT systems now dominate the commercial market. Swept-wavelength tunable lasers typically have spectral gain bands that are centered at 980 nm to 1310 nm, with tuning ranges, $\Delta\lambda$, as wide as >110 nm (e.g., Yun et al., 2004), and instantaneous bandwidths, $\delta\lambda$, of 0.05 nm. In recent developments, these sources are capable of sweep rates as high as 200 kHz (Yamashita and Asano, 2006). These devices are generally compact fiber lasers with a semiconductor optical amplifier (SOA) as the gain medium. Even wider tuning ranges can be achieved using two SOAs in the same cavity. Rapid wavelength tuning was first achieved with a rotating multifaceted mirror for the feedback element (Oh et al., 2005), and, more recently, RF-modulated dispersion (Yamashita and Asano, 2006) and Fourier-domain mode-locking (Huber et al., 2006) have provided the fast sweep rates without any mechanically moving components.

Given the trend towards swept-source OCT, a single photodetector can be used. The desired attributes of good quantum efficiency, high gain and high frequency response have led to the adoption of the *avalanche photodiodes* (APD) as the detector of choice. APDs, presented in Section 13.3, are silicon semiconductor photodetectors that invoke a high (100–1000 V) reverse-bias voltage yielding current gains of 100–1000, with sufficient sensitivity for single-photon detection. Due to the intrinsic small dimensions and low capacitance, frequency response can be as high as 1 GHz, with quantum efficiency of 75% and broad spectral response.

18.5.3 Transverse scanning

As discussed above, the axial scan (A-scan) in FD-OCT is accomplished computationally by the Fourier transform of the detected frequency range. Transverse (x-y) scanning, however, is still accomplished in a manner similar to that of TD-OCT systems, commonly with piezoelectric crystals used to tilt turning mirrors in the sample arm, as represented in Figure 18.10. This can still be fast enough to yield video-rate imaging, but often the mechanism and pattern of the transverse scanning is specific to the imaging application, as will be noted below.

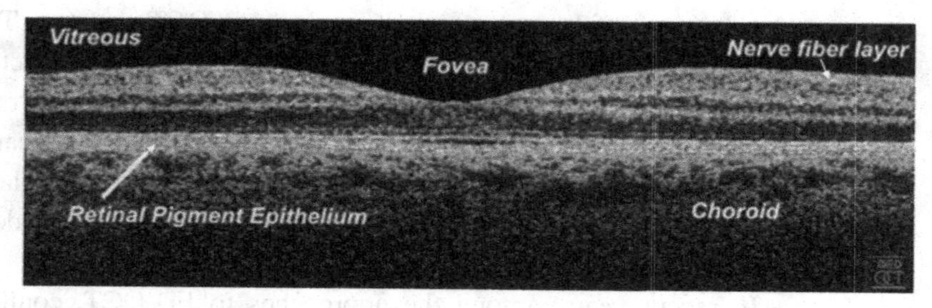

Figure 18.11

An OCT B-scan of the human retina, revealing layers that are not observable by other imaging methods. Image sharpness is degraded by laser speckle. (Public domain.)

18.6 Speckle in OCT images

Figure 18.11 is a B-scan of a human retina, imaged in vivo with a commercial FD-OCT system. One of the features that is immediately obvious is the "graininess" of the image. Any type of coherent imaging suffers from the noise effects of laser speckle, and this is manifest in Figure 18.11 as the random distribution of small dark and light spots throughout the image. Speckle is a locally random interference effect at the detector, arising from phase differences for slightly different pathlengths of two or more backscattered rays reaching a single detection point. When imaging with coherent light from an opaque surface, the speckle is an indication of microscopic surface roughness. When using OCT to image at a depth in a thick turbid medium (tissue), the predominant source of speckle (in, say, an A-scan) is the result of multiple backscatter from scattering points located at nearby depths that are still within the coherence length of the source (Schmitt et al., 1999). Tissue comprises microscopic structures ranging in size from microns to nanometers, and much of the structure is smaller than 1 μm. Thus, within a coherence length of, say, 3 μm, there can be random backscattering from different scattering structures that interfere either constructively or destructively, resulting in bright or dark spots, respectively. The deleterious effects of speckle noise are the reason why laser light sources are not generally used for illumination in conventional microscopy or phase contrast microscopy, despite the potential benefits of good transverse coherence.

Various methods have been investigated to reduce speckle noise in OCT images (Xiang et al., 1997). Most of these attempts to reduce speckle, however, result in unacceptable concomitant degradation of spatial resolution. Among the more promising approaches are *frequency compounding* and *digital signal-processing* methods. Frequency compounding, which has also been used to reduce speckle in ultrasound imaging, is based on adding incoherently the magnitudes of two independent OCT signals that are recorded at two different center wavelengths simultaneously (Pircher et al., 2003). To avoid loss of axial resolution, the wavelength ranges should not overlap significantly. Then each source will generate its own random speckle, resulting in a reduction in the speckle contrast at the detector by a factor $\approx 1/\sqrt{2}$ (for two sources). This modest reduction is, of course, at the expense of increased system complexity, requiring two independent but collinear light sources, at different wavelength ranges.

Digital signal processing, by various numerical methods, can reduce visible speckle by applying various types of spatial-frequency filters (Ozcan et al., 2007). While this reduction in "noise" will not enhance spatial resolution or increase image information content, careful filter design can reduce the appearance of speckle while minimizing loss of spatial resolution. The main advantage of the result is a reduction in the psycho-visual "annoyance" of the grainy appearance.

18.7 A sampling of OCT applications

The first applications of OCT with clinical potential gained acceptance for imaging the layers of the retina of the eye (Fercher et al., 1988; Huang et al., 1991); OCT has since become widely used in ophthalmology and has been expanding to several other areas of medicine. (See, for example, Zysk et al., 2007.)

More than 20 years since its first clinical introduction, by far the broadest dissemination of OCT technology has been in the ophthalmology clinic (Gabriele et al., 2011). This is a natural application for OCT, because the sample arm projects through the non-scattering cornea, aqueous humor, lens, and vitreous humor, to reach the layers of the retina with minimal loss. A number of companies now market FDA-approved systems for clinical use, and most ophthalmology referral centers employ OCT systems in daily practice. (There are also a number of industrial applications of OCT, aimed at detailed imaging of subsurface structures, which are not covered here.)

An example of the diagnostic potential of OCT in ophthalmology is shown in Figure 18.12, which is a B-scan through the fovea, clearly showing retinal detachment on the right side, while the fovea itself is spared. Such timely information enables the physician to offer the most appropriate treatment and management.

Beyond ocular imaging, the most prominent development and clinical dissemination of OCT has been for applications of intravital imaging, i.e., OCT that is conducted through catheters or endoscopes. Intravital imaging, especially intravascular imaging, had previously been dominated by intravascular ultrasound (IVUS). In the small confines of blood vessels, however, OCT offers some distinct advantages over IVUS imaging, chief among them an order-of-magnitude finer spatial

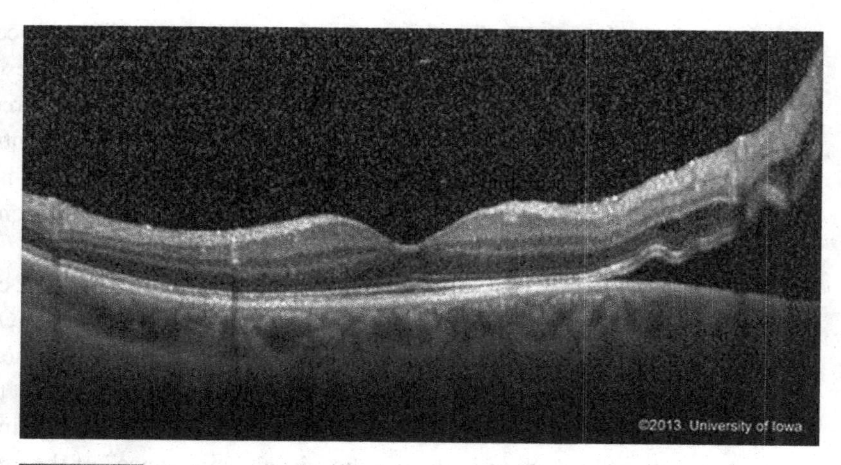

Figure 18.12

An OCT image used in training medical residents in ophthalmology. The detachment of the photoreceptor layers from the retinal pigmented epithelium to the right is readily observed. The OCT image enables determination of microscopic detachments and assessment of whether the fovea is affected, helping to guide appropriate clinical management. (Courtesy, University of Iowa Health Care [2013].)

resolution, and the ability to function without the need for any electrical conductors through the catheter, as would be needed for ultrasound to carry power to, and signal from, the distal end.

For intravital OCT imaging applications, attention has focused on miniaturizing the optomechanical components at the distal end of the sample arm, and significant efforts have addressed methods to accomplish the transverse scanning *in situ*. Of the various approaches to transverse scanning, the systems that have emerged as viable for clinic use invoke the simplest form of scanning, mimicking the methods developed for intravascular ultrasound imaging. In this approach, the tip of the OCT probe incorporates a miniature mirror to direct the sampling beam sideways, and the B-scan is performed azimuthally by simply rotating the OCT probe through 360° about the axis of the catheter. Rotational B-scans can be executed at different locations along the length of the vessel by simply advancing or retracting the probe with the catheter. Recent applications in cardiology have demonstrated the ability of OCT to provide *in situ* images of atherosclerotic plaque (Tearney et al., 2003); Bezerra et al., 2009), an example of which is shown in Figure 18.13.

On a larger intravital scale, commercial application of OCT has begun to impact specific applications in gastroenterology, especially for the diagnosis and management of neoplasia (premalignant growth) in the esophagus, a condition called Barrett's esophagus (after the Australian-born British thoracic surgeon Norman

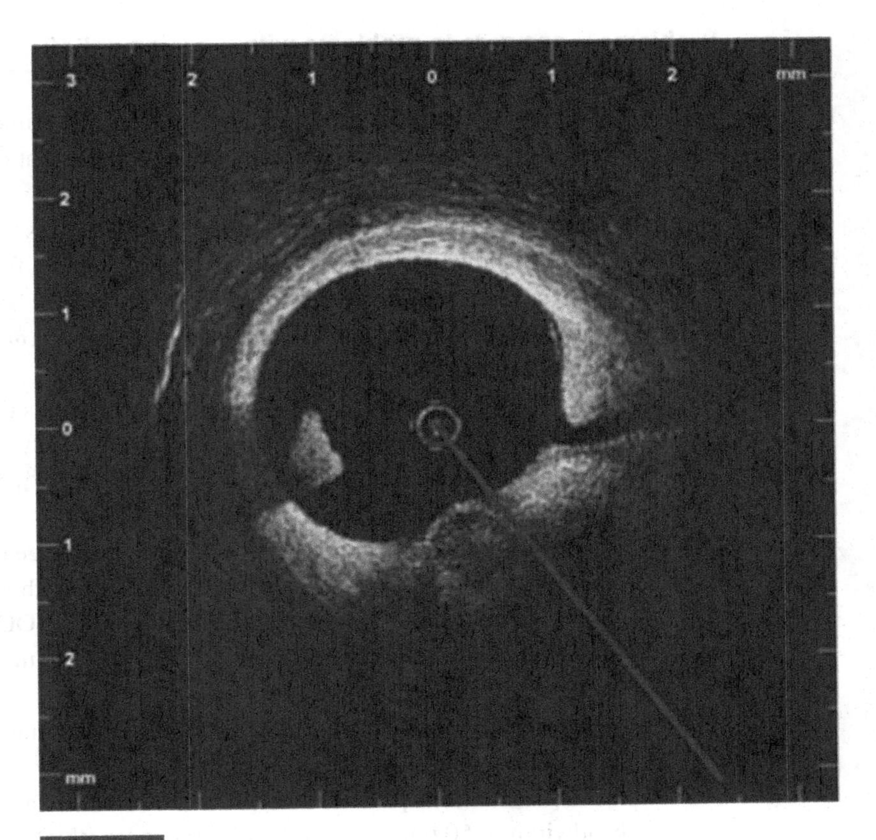

Figure 18.13

Intravascular OCT image of a coronary artery. A heterogeneous plaque is visible at ~6 o'clock, a thrombus (blood clot) at ~9 o'clock and a smaller branch vessel at ~3 o'clock. The diagonal line indicates the direction of an individual A-scan, and such A-scans can be analyzed to provide quantitative diagnostic information. (van Soest et al., 2009, courtesy Springer.)

Barrett [1903–1979]), which can arise as a result of chronic acid reflux, and can lead to esophageal cancer (Suter et al., 2008; Tsai et al., 2012). Here, the power of OCT is, again, its ability to image beneath the surface, in this case to identify cancers that can arise *beneath* the epithelium. In treatment for premalignant Barrett's esophagus, the epithelial layer is ablated (often with radio-frequency heating), and a new epithelium grows on the surface during the healing process. Malignancy can still recur, however, due to any remaining premalignant tissue underneath the new epithelium, which can be seen by scanned OCT imaging.

In the rapidly developing area of clinical applications of OCT, the reader will find new studies reported in publications almost every month.

Problems – answers to problems with * are on p. 648

18.1* Assuming tissue to be mostly water, compare the round-trip travel time for a short acoustic pulse and a short optical pulse that reflect off a tissue structure at a depth of 1 cm.

18.2* Derive Eq. (18.7) for the wavelength range, $\Delta\lambda$, associated with a frequency bandwidth where ω_0 is the center angular frequency and the frequency bandwidth extends from $\omega_0 - \frac{\Delta\omega}{2}$ to $\omega_0 + \frac{\Delta\omega}{2}$.

18.3 Consider a frequency band that extends from ω_{lo} to ω_{hi}, so that its frequency bandwidth is $\Delta\omega = \omega_{hi} - \omega_{lo}$.

(a) Find the wavelength range, $\Delta\lambda$, that corresponds to $\Delta\omega$ by expanding the relationship between λ and ω ($\lambda = 2\pi c/\omega$) in an infinite Taylor series. (Hint: recall that, for $|x| < 1$, $\sum_{n=0}^{\infty} x^n = \frac{1}{1-x}$.)

(b) Use the result in part (a) to derive Eq. (18.7).

18.4 Equation (18.8) shows that there is a distinct advantage to using a shorter center wavelength to achieve a short coherence length with a broadband source. Nonetheless, almost all of the commercial OCT systems have center wavelengths in the NIR, around 1000 or 1300 nm. What are some reasons for this?

18.5* What would be the shape of the spectral density function for a short rectangular pulse of light?

18.6 A laser beam has a peak wavelength of 700 nm, and a Gaussian FWHM bandwidth of 50 nm. What is the coherence length of this beam?

18.7* The reference-arm mirror of a TD-OCT system is translated at a speed of 5 m/s. If the center wavelength is 800 nm, what would be the minimum required frequency response for the photodetector?

18.8 Figure 18.8 depicts the main steps needed to decode the detector signal in a TD-OCT system. What additional steps are needed to decode the signal of a swept-source FD-OCT system?

18.9 The extraction of the A-scan profile in FD-OCT is accomplished by Fourier transform of the detector signal. Explain how the Fourier transform operation for spatially encoded FD-OCT compares with that for temporally encoded (swept-source) FD-OCT.

18.10 A swept-source FD-OCT system operates with a peak wavelength of 1310 nm, and has a useful gain bandwidth of 120 nm. The dispersive feedback mechanism of the laser yields an instantaneous (dynamic) bandwidth of 0.15 nm.

(a) What is the effective coherence length?

(b) What is the maximum range of the A-scan?

18.11* The optics of an OCT imager are set so that the confocal parameter of the focusing lens is equal to the length of the A-scan. If the A-scan range is 300 μm, what is the minimum transverse spot diameter for a peak wavelength of 780 nm?

18.12 In a Doppler OCT system operating with a center wavelength of 1310 nm, the A-scan repetition rate is 100 kHz.

(a) If the phase difference between adjacent A-scans is measured to be $\pi/2$, what is the relative axial velocity of the scattering structures between the two transverse locations?

(b) What is the largest velocity difference between two adjacent A-scans that can be measured unambiguously with this system?

References

Bezerra, H. G., Costa, M. A., Guagliumi, G., Rollins, A. M., and Simon, D. I. (2009). Intracoronary optical coherence tomography: a comprehensive review. *JACC: Cardiovascular Interventions*, 2(11), 1035–1046.

Choma, M. A., Sarunic, M. V., Yang, C., and Izatt, J. A. (2003). Sensitivity advantage of swept source and Fourier domain optical coherence tomography. *Optics Express*, 11, 2183–2189.

Fercher, A. F., Mengedot, K., and Werner, W. (1988). Eye length measurement by interferometry with partially coherent light. *Optics Letters*, 13, 1867–1869.

Fujimoto, J. G., De Silvertri, S., Ippen, E. P., et al. (1986). Femtosecond optical ranging in biological systems. *Optics Letters*, 11, 150–153.

Gabriele, M. L., Wollstein, G., Ishikawa, H., et al. (2011). Optical coherence tomography: history, current status, and laboratory work. *Investigative Ophthalmology & Visual Science.*, 52, 2425–2436.

Hecht, E. (2002). *Optics*, 4th edn. Reading, MA: Addison-Wesley.

Huang, D., Swanson, E. A., Lin, C. P., et al. (1991). Optical coherence tomography. *Science*, 254, 1178–1181.

Huber, R., Wojtkowski, M., and Fujimoto, J. G. (2006). Fourier Domain Mode Locking (FDML): a new laser operating regime and applications for optical coherence tomography. *Optics Express*, 14, 3225–3237.

Ko, T. H., Adler, D. C., and Fujimoto, J. G. (2004). Ultrahigh resolution optical coherence tomography imaging with a broadband superluminescent diode light source. *Optics Express*, 12, 2113–2119.

Oh, W. Y., Yun, S. H., Tearney, G. J., and Bouma, B. E. (2005). 115 kHz tuning repetition rate ultrahigh-speed wavelength-swept semiconductor laser. *Optics Letters*, 30, 3159–3161.

Ozcan, A., Bilenca, A., Desjardins, A. E., Bouma, B. E., and Tearney, G. J. (2007). Speckle reduction in optical coherence tomography images using digital filtering. *Journal of the Optical Society of America* A24, 1901–1910.

Pircher, M., Götzinger, E., Leitgeb, R., and Fercher, A. F. (2003). Speckle reduction in optical coherence tomography by frequency compounding. *Journal of Biomedical Optics*, 8, 565–569.

Saleh, B. E. A., and Teich, M. C. (2007). *Fundamentals of Photonics*, 2nd edn. Hoboken, NJ: John Wiley.

Schmitt, J. M., Xiang, S. H., and Yung, K. M. (1999). Speckle in optical coherence tomography. *Journal of Biomedical Optics*, 4, 95–105.

Suter, M. J., Vakoc, B. J., Yachimski, P. S., et al. (2008). Comprehensive microscopy of the esophagus in human patients with optical frequency domain imaging. *Gastrointestinal Endoscopy*, 68, 745–53.

Tearney, G. J., Yabushita, H., and Houser, S. L. (2003). Quantification of macrophages content in atherosclerotic plaques by optical coherence tomography. *Circulation*, 107, 113–119.

Tsai, T. H., Zhou, C., Tao, Y. K., et al. (2012). Structural markers observed with endoscopic 3-dimensional optical coherence tomography correlating with Barrett's esophagus radiofrequency ablation treatment response. *Gastrointestinal Endoscopy*, 76, 1104–1112.

van Soest, G, Goderie, T. P. M., Gonzalo, N., et al. (2009). Imaging atherosclerotic plaque composition with intracoronary optical coherence tomography. *Netherlands Heart Journal*, 17, 448–450.

Westphal, V., Yazdanfar, S., Rollins, A. M., and Izatt, J. A. (2002). Real-time, high velocity-resolution color Doppler optical coherence tomography. *Optics Letters*, 27, 34–36.

Xiang, S. H., Zhou, L., and Schmitt, J. M. (1997). Speckle noise reduction for optical coherence tomography. *Proceedings of SPIE*, 3196, 79–88.

Yamashita, S., and Asano, M. (2006). Wide and fast wavelength-tunable mode-locked fiber laser based on dispersion tuning. *Optics Express*, 14, 9299–9306.

Yang, V., Gordon, M., Qi, B., et al. (2003). High-speed, wide velocity range Doppler optical coherence tomography: system design, signal processing and performance. *Optics Express*, 11, 794–809.

Yun, S. H., Boudoux, C., Pierce, M. C., et al. (2004). Extended-cavity semiconductor wavelength-swept laser for biomedical imaging. *IEEE Photonics Technology Letters*, 16, 293–295.

Zhao, Y., Chen, Z., Ding, Z., Ren, H., and Nelson, J. S. (2002). Real-time phase-resolved functional optical coherence tomography by use of optical Hilbert transformation. *Optics Letters*, 27, 98–100.

Zysk, A. M., Nguyen, F. T., Oldenburg, A. L., Marks, D. L., and Boppart, S. A. (2007). Optical coherence tomography: a review of clinical development from bench to bedside. *Journal of Biomedical Optics*, 12(5), 051403.

Further reading

Principles of temporal coherence

Hecht, E. (2002). *Optics*, 4th edn. Reading, MA: Addison-Wesley.
Saleh, B. E. A., and Teich, M. C. (2007). *Fundamentals of Photonics*, 2nd edn. Hoboken, NJ: John Wiley.

Ultrashort laser pulses

Weiner, A. M. (2009). *Ultrafast Optics*, Hoboken, NJ: John Wiley.

General topics in OCT

Bouma, B. E., and Tearney, G. J., eds. (2002). *Handbook of Optical Coherence Tomography*, New York: Marcel Dekker.
Fercher, A. F. (1996). Review paper: Optical coherence tomography. *Journal of Biomedical Optics*, 1(2), 157–173.
Brezinski, M. E. (2006). *Optical Coherence Tomography: Principles and Applications*, New York: Academic Press
Fercher, A. F., Drexler, W., Hitzenberger, C. K., and Lasser, T. (2003). Optical coherence tomography – principles and applications. *Reports on Progress in Physics*, 66, 239–303.

Doppler OCT

Leitgeb, R. A., Werkmeister, R. M., Blatter, C., and Schmetterer, L. (2014). Doppler optical coherence tomography. *Progress in Retinal and Eye Research*, 41, 26–43.

19 Optical tweezers and laser-tissue interactions

In the years following the demonstrations of the first working lasers in the early 1960s a flurry of speculation emerged, predicting medical applications of lasers that were both fanciful and unjustifiably optimistic. As the laser industry emerged and grew, a number of the companies opened medical laser divisions, received FDA approval for surgical lasers and began marketing them to hospitals and surgical group practices. These were generally high-power continuous lasers, often Nd:YAG lasers operating in the NIR at 1.06 μm, or their frequency-doubled version at 532 nm. Little in the way of statistically powered research had been carried out to determine whether these "laser scalpels" could actually achieve better patient results than traditional surgical tools. While some laser wavelengths, at appropriate power levels, can indeed achieve improved degrees of hemostasis (stoppage of bleeding), this can also be accomplished with an electrically heated scalpel blade, which cauterizes as it cuts, and at much lower cost. The introduction of a new technology is at its best when it enables a process or procedure that *cannot* be accomplished by other means, or that accomplishes the objective more efficaciously or at lower cost. This was not always the case in those early days of medical lasers, and within a few years many hospitals had large, expensive laser systems gathering dust in closets.

One example of an application that originally inspired optimism but has fallen out of favor is *laser angioplasty*: the use of high-power pulsed lasers to ablate atherosclerotic blockages, thereby restoring blood flow in coronary arteries. Initial studies in the 1980s indicated that acute response was promising, with immediately restored blood flow. Restenosis was found to occur within months, however, in almost all cases, often with blockages worse than the pre-procedure condition (Sanborn, 1996). It was eventually determined that the new blockage was a result of the local infiltration (and proliferation) of smooth muscle cells, released due to the absence of the protective *endothelium* (the layer of cells that line the inner surface of blood vessels), which of course was also ablated while removing the plaque (Köster et al., 2002).

Low-power lasers, however, have facilitated novel methodologies in research for cellular biology, with an explosion of applications enabled by the development of optical tweezers, which offer precision manipulation of particle location and measurement of piconewton forces. This advantage, along with other properties of laser light, has enabled a number of microscopy-based applications that invoke manipulation and alteration of cells and smaller biological objects. Also, more recently, lasers in clinical applications have enjoyed a "reboot," and have achieved broad commercial success in specific clinical settings, including, for example, applications in ophthalmology and for cosmetic or elective procedures. These applications, in both the microscopy and clinical settings, utilize the special capabilities of lasers in more subtle ways than did the early surgical lasers.

In this chapter we introduce the basics of optical tweezers and laser-tissue interactions, and we describe a sampling of the more successful applications of lasers in biomedical research and clinical applications, which are enabled by application-specific performance properties of the lasers.

19.1 Optical tweezers

In the study of cellular dynamics, and a variety of intercellular and sub-cellular interactions, a tool that can mechanically manipulate tiny biological structures (cells or smaller), without destroying them, can be highly enabling in research at those scale lengths. The question is: how can we hold a cell, or part of a cell, without breaking it? Such a capability can be valuable in applications as diverse as genetic engineering, artificial insemination, and cell or genetic repair. A key tool that has become a workhorse for such applications is the *optical tweezers*, also called *laser tweezers*, which uses the momentum of photons to exert small but precise forces on cells and other small dielectric structures.

Photons have no rest mass, but they do carry energy and momentum. The approach is to exert force by exchanging momentum between a light beam and a microscopic object, without exchanging a significant amount of energy. (That is, it is often desirable to avoid depositing energy in the cells so as to avoid heating and damaging them.) We recall from Chapter 7 that light can scatter elastically from cells, a mechanism by which photons will change direction without depositing energy. We also know from classical optics that a transparent dielectric medium (e.g., a lens) can refract light at a boundary with a change in optical refractive index.

19.1.1 Forces resulting from radiation pressure

The mechanical effects of light on small particles, or *radiation pressure*, were first explained by Ashkin (1970), although the fundamental concepts of photon

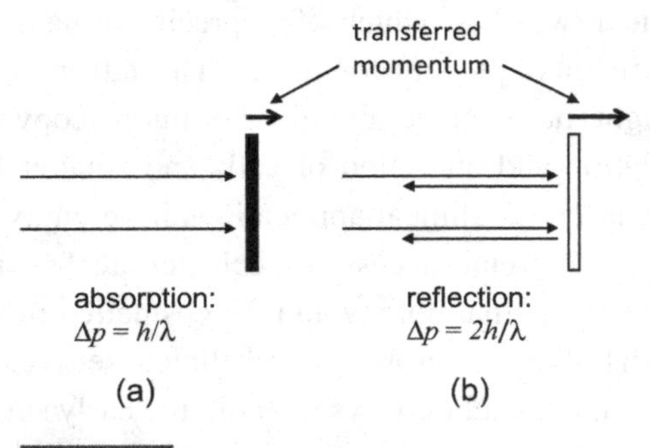

Figure 19.1

Absorption of a photon (a), and reflection of a photon (b), with the lengths of the heavy arrows above representing the relative amounts of momentum imparted to the absorbing or reflecting object.

momentum exchange were known since Einstein. On a scale larger than the photon wavelength, an exchange of photon momentum can be easily represented by classical absorption, reflection or refraction, although we will treat the individual photon momentum from a quantum theory perspective. The scalar momentum, p, of a photon is proportional to its angular frequency, ω (and energy):

$$p = \frac{\hbar\omega}{c_n} = \frac{h}{\lambda}, \tag{19.1}$$

where $\hbar$ is the reduced Planck's constant (h) divided by 2π, and c_n is the speed of light in the medium. The photon vector momentum can also be conveniently expressed, using the wave vector **k** (whose magnitude is $2\pi/\lambda$ and whose direction is along the photon direction of propagation):

$$\mathbf{p} = \hbar\mathbf{k}. \tag{19.2}$$

We can examine the momentum exchange for the photons undergoing absorption and reflection (at normal incidence), as represented in Figure 19.1.

Characterization of the momentum exchange follows the rule of conservation of total momentum of the system. If a photon is simply absorbed (and remembering that we will try to avoid absorption), its momentum (as well as its energy) is transferred to the absorbing object, which experiences a mechanical impulse in the direction of the photon propagation. If the photon is reflected, however, its wave vector reverses direction, and the exchange of momentum with the reflecting object is twice the momentum of the original photon. The tiny "kick" experienced by the object is like an instantaneous or momentary force, which actually induces a mechanical impulse of acceleration.

While the dynamic effect of a single photon on a small object can be calculated quantum mechanically, we can use a semi-classical approach to understand the mechanical effect of a laser beam on a cell, because the effect of a continuous stream of photons interacting with the cell can be averaged, even over a short time span. Under this condition, we can consider the momentum flux of the stream of photons. From classical mechanics we know that an exchange of momentum, $\Delta\mathbf{p}$, over a time span, Δt, translates to a (time-averaged) force as given by:

$$\mathbf{F} = \frac{-\Delta\mathbf{p}}{\Delta t}, \qquad (19.3)$$

where $\mathbf{F}$ is the force on the object and $\Delta\mathbf{p}$ is the change in momentum of the photons during time Δt. The vector notation facilitates keeping track of the direction of the force. In a manner similar to classical mechanics (where *power* = *force* × *velocity*), we can express the energy applied by the laser beam on the object per unit time, i.e., the power P exerted by the radiation pressure of the light on the object, in terms of the momentum change of the photons. In fact, the force exerted by the photons on the object (as given in Eq. (19.3)) is equal and opposite to the force that the object exerts on the photons, which travel at the speed of light in the medium, c_n. Therefore: $P = |\Delta p/\Delta t|\, c_n = F c_n$. We will soon address directional information in some detail. Thus, for a collimated light beam (at normal incidence) that is fully absorbed (i.e., all of the light intensity, I, incident on the particle of cross-section area, A_s) the force exerted by the light is simply:

$$F = \frac{I A_s}{c_n} = \frac{P}{c_n}; \qquad (19.4)$$

here P is the total optical power incident on the particle. (If the particle size is larger than the beam diameter, then this is simply the total beam power.) For reflection, the force is twice that amount. The SI unit of force is the newton, N ($\mathrm{N = kg\text{-}m/s^2}$). In terms of the parameters of Eq. (19.4), N = W-s/m.

But what about light that is neither absorbed, nor simply reflected, but is refracted by a non-absorbing dielectric object? This circumstance is exemplified in Figure 19.2, wherein a ray-trace representation depicts refraction of a collimated beam of light by a dielectric sphere larger than the wavelength, with refractive index greater than that of the surrounding medium. (We temporarily ignore reflections from the surface.) As can be seen from the wave vector diagrams for the two depicted rays, the net force of the radiation pressure on the sphere that results from refraction is in the direction away from the source of the light: that is, a repulsive force.

19.1.2 Trapping force for particles larger than the wavelength

A similar analysis can be performed for a tightly focused laser beam, as would be generated by a microscope objective with a large numerical aperture, NA. Figure 19.3 illustrates the net force direction on a dielectric sphere illuminated by

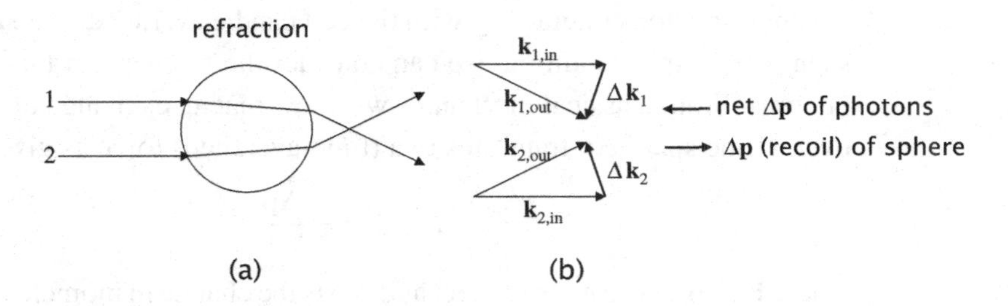

Figure 19.2

(a) A collimated beam of light is refracted by a dielectric, non-absorbing sphere; (b) even if we ignore the effects of back-reflected light, the beam exerts a force on the particle due to the refraction-induced change in momentum, as can be seen from the wave-vector diagrams of the momentum differences for the two depicted rays.

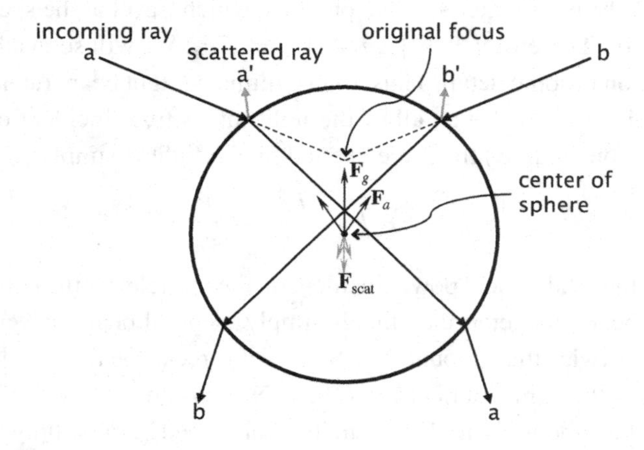

Figure 19.3

A ray-trace illustration of the refraction of a tightly focused laser beam by a spherical object with higher refractive index than the surrounding medium. The force vector arrows indicate the force components for the two rays, exemplified by $\mathbf{F}_a$ and the net upward force due to refraction (in a field gradient), $\mathbf{F}_g$. Gray arrows indicate the backscattered (reflected) rays and the consequent small repulsive force, $\mathbf{F}_{scat}$.

a tightly focused laser beam, whose original focal point is inside and above the center of the sphere. This time, a wave vector analysis similar to that in Figure 19.2 yields an upward, "attractive," force on the sphere (i.e., towards the original focal point of the beam), due to the refraction. We denote this force as $\mathbf{F}_g$, because it is commonly called the *gradient force*, referring to the fact that the force is ultimately due to the gradient of the focused optical field. (In Figure 19.2(a) the gradient results from the refraction by the sphere itself.) Also indicated in

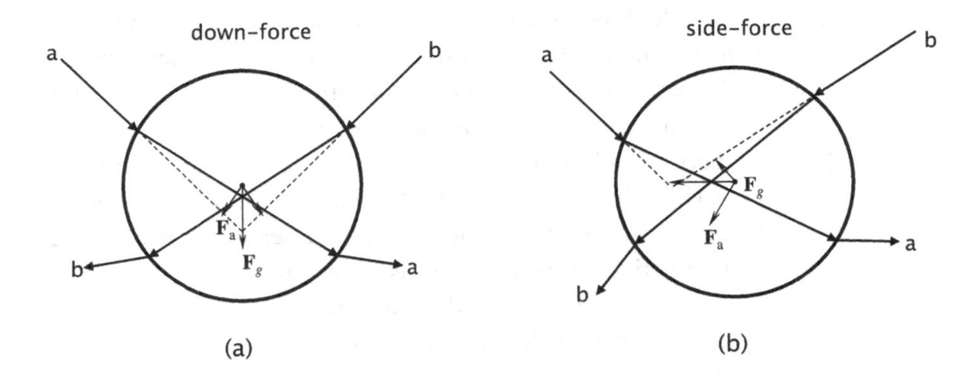

Figure 19.4

Trapping force for a dielectric spherical particle (larger than the wavelength of the light) is directed towards the original focal point of the laser beam: (a) downward force when the sphere center is above the focal point; (b) leftward force when the sphere center is to the right of the focal point.

Figure 19.3 is the fact that there is still a small repulsive force resulting from a small amount of light that is reflected (scattered) back from the surface because of the refractive index mismatch. (The *power reflectivity* or *reflectance*, R, at normal incidence at a dielectric boundary is given by $R = (n_1 - n_O)^2/(n_1 + n_O)^2$.) Thus, for the net force to be in the upward direction, the gradient force (due to refraction) must exceed the force due to scattering (reflection). This condition generally holds for a tightly focused beam, as the reflectivity of biological objects in water is typically less than $\sim 0.06\%$, and even for polystyrene spheres ($n \cong 1.58$) in water it is only $\sim 0.7\%$.

If the focal point of the laser is below the center of the sphere, or to its side, the net force is downward or sideways (towards the focal point), respectively. These circumstances are illustrated in Figure 19.4, where, for clarity, we have suppressed the backscattered rays. Thus, regardless of where the original laser focal point is located, the net force on the particle "attracts" it towards the focal point. The gradient force for a spherical particle is always in the direction of the optical field gradient: i.e., in the direction towards stronger field. In 1986 Ashkin and colleagues reported the basic concepts of optical tweezers and provided the first demonstration of a single-beam optical trap for dielectric particles (Ashkin et al., 1986), and this has since become a widely used tool in cellular biology.

We can make a "ballpark estimate" of the magnitude of the force of a focused laser beam on a cell by including a geometric factor, A, in Eq. (19.4):

$$F_g = AP/c_n, \tag{19.5}$$

where A is dimensionless and accounts for factors that include the convergence cone angle of the beam, the particle size and shape, the relative locations of the particle center and the beam focus, and the refractive indices of the particle and medium. To begin our estimate, we can choose an approximate value for A of 0.1, meaning that 10% of the forward momentum of the laser beam is exchanged with the particle as a consequence of the refraction of the beam. To estimate the speed of light in the medium, we can assume a refractive index of 1.35 (that of water with some dissolved salts and sugars). A reasonable value for the power of the laser beam, representative of the laser power used in many optical tweezers accessories for microscopes, is of the order of 100 mW (Smith et al., 1999). For wavelengths in the red-NIR range, this power results in minimal damage to cells. Inserting these values into Eq. (19.5) yields a trapping force of $\sim 4.5 \times 10^{-11}$ N, or 45 pN (piconewtons).

In this estimate, we have ignored the force due to backscattering. (We leave comparison of the trapping force vs. the backscattering force to a problem at the end of the chapter.) In some circumstances (e.g., with variously shaped particles, or when the laser beam must be less-tightly focused due to concern about cell damage), it may be difficult to generate a trapping force that exceeds the repulsive force due to backscattering. In that circumstance, one option is to split the laser beam into two equal-power, counter-propagating beams that are focused with coincident focal points. In such a geometry the repulsive forces from the two beams would cancel, and the trapping forces would add.

Another way to achieve a better balance between the gradient force and the scattering force can be understood by further examination of Figure 19.4. In this geometry, the depicted pair of rays are large-angle rays, on the outside of the convergent cone from the focusing objective. Of course, there are also paraxial rays (at small angles near the center of the cone) that are not significantly refracted, and therefore do not contribute much to the gradient force, but do contribute maximally to the scattering force. Moreover, commonly used laser beams are lowest-order (Gaussian) mode (TEM_{00}) and have Gaussian transverse intensity profiles (as described in Section 17.3.2), so the most intense light is in the center of the cone. It is possible, however, to utilize laser beams with the second Laguerre-Gaussian mode (TEM^*_{01}), the "donut"-shaped transverse mode, with zero on-axis intensity, as mentioned in Section 17.5.3. (See, for example, Siegman, 1986.) It has been shown, however, that the donut-mode beam only confers an enhancement of axial trapping forces, not of lateral forces, and only for particles significantly larger than the wavelength (O'Neil and Padgett, 2001).

Another insight that can be gleaned from further examination of Figure 19.4 (in the case of spheres larger than the wavelength) is that when the center of the sphere is located at precisely the laser focal point, there is no net gradient force because the phasefront of the beam is spherical and confocal with the sphere center, such that all rays are normal to the surface. For small displacements from center, the restoring force is directly proportional to the magnitude of the displacement, as though the particle were connected to the focal point through a spring (as also described in Section 19.1.6 below). Thus, the equilibrium position is generally on axis, and the balance between the gradient and scattering forces results in the particle center being slightly offset beyond the focal point (away from the light source).

19.1.3 Trapping of particles much smaller than the wavelength

Particles much smaller than the optical wavelength can also be trapped, but we cannot readily explain this with the ray-trace analysis of momentum transfer by refraction as we did for larger particles. Moreover, we cannot assume that all of the light from the laser beam impinges on the particle, since the particle will be smaller than the beamwaist of the focal region. Thus, we will use the incident radiant intensity, I_0, rather than the total power of the optical beam to assess forces. For small particles, we can utilize some of the formalism for Rayleigh scattering developed in Chapter 7 to describe the forces of a focused optical field on small particles.

By using Eq. (7.32) and the first term of Eq. (7.33) (the Rayleigh scattering term), we can rewrite Eq. (19.4) for the radiation pressure force due to scattering by a small particle in terms of the local incident intensity, I_0, and the scattering cross section, σ_s:

$$F_s = \frac{I_0 \sigma_s}{c_n} = \frac{I_0}{c_n} \frac{128\pi^5 a^6}{3\lambda^4} \left(\frac{m^2 - 1}{m^2 + 2} \right)^2 \tag{19.6}$$

where a is the particle radius and m is the relative refractive index (the ratio of the particle index to the medium index). F_s is in the direction of the incident light propagation.

To understand the gradient force on a small dielectric particle, we recall from Chapter 6 that the induced dipole moment in a small particle, or its polarization, $\mathbf{p}$, is equal to *charge* × *separation*, or to the polarizability (α) times the applied electric field: $\mathbf{p} = q d\mathbf{r} = \alpha \mathbf{E}$, where q is the charge, and $d\mathbf{r}$ represents the charge separation vector. If the field is inhomogeneous, i.e., if it has a gradient, then one can think of the Lorentz force on each charge ($\mathbf{F} = q\mathbf{E}$) as being stronger for the positive charge, which is in a location where the field is slightly stronger than where the negative charge is located. (Hendrik Lorentz [1853–1928] was a Dutch physicist who received the Nobel Prize for physics in 1902.) In the small-particle limit, the gradient force (in the direction of the gradient) on a small dielectric sphere (treated like a "point" dipole) is given by (see, for example, Gordon, 1973):

$$\mathbf{F}_g = -\frac{n}{2} \alpha \nabla E^2 \equiv -\frac{2\pi\alpha}{c_n n^2} \nabla I_0 \tag{19.7}$$

where the middle expression (in terms of electric field amplitude) is in Gaussian-cgs units. The equivalent term on the right (in terms of intensity) is in SI units, for which the polarizability, α, of a small dielectric particle is given by:

$$\alpha = n^2 a^3 \left(\frac{m^2 - 1}{m^2 + 2} \right), \tag{19.8}$$

where n is the refractive index of the medium. Both expressions are provided in (19.7) because they appear almost equally in the literature. Since we typically measure intensity, the SI expression is easier to use for calculating forces. The condition for trapping, regardless of the particle size, is that the gradient force must exceed the scattering force:

$$\frac{F_g}{F_s} > 1 \tag{19.9}$$

While this ratio does depend on the intensity gradient, it is dimensionless and does not depend on the intensity itself (i.e., the laser power). This does not mean, of course, that trapping can be achieved with an arbitrarily weak laser. Another condition for trapping is that the net trapping force must exceed forces due to Brownian motion, fluid flow, etc.

To get a feel for the scale of forces relevant to trapping of small biological particles, we can consider a 100 mW Gaussian laser beam, at a wavelength of 500 nm, that is focused to a spot size of 5 μm diameter. (The beam could be focused more tightly, of course, but in practice a larger beamwaist facilitates easier "finding" of the particle with manual manipulation under microscopy.) If a small organelle has a radius of, say, 50 nm, and its relative refractive index, m, is 1.02, then, using Eq (19.6), we can estimate the scattering force in the focal zone to be of the order of 10^{-17} N. When the particle is located, say, in the focal plane, but displaced laterally from the axis by half the beamwaist radius, and using Eqs. (19.7) and (19.8), we can estimate the gradient force on the particle to be $\sim 3 \times 10^{-16}$ N. Thus, the trapping criterion of expression (19.9) is readily satisfied. To accomplish these estimates, we have made the following assumptions: the relative refractive index, m, for this particle is 1.02 and the medium index is 1.35; the average focal plane intensity is simply the laser power divided by the focal spot area; and the magnitude of the intensity gradient across the particle is estimated by $\nabla I \approx \Delta I/(2a)$, where ΔI is difference in intensity across the particle diameter and is taken to be $\sim 1/10$ of the average intensity in the focal plane of a Gaussian beam.

As can be seen from these estimates, the trapping forces on small particles are much weaker than the forces on larger particles. For very small particles, as mentioned above, we want the trapping force to significantly exceed (by, say, a factor 5–10) forces related to Brownian motion, so as to comfortably exceed the maximum stochastic fluctuations of the Brownian forces. This can become important if one wants to trap particles as small as, say, a globular protein molecule. To determine a criterion for stable trapping of very small particles, we note that the scalar potential, U, of a particle in a force field is related to the force through the expression $\mathbf{F} = -\nabla U$, and we see from Eq. (19.7) that the potential of the gradient force is $U_g = n\alpha E^2/2$. Using that expression, we can write the requirement for stable trapping of a very small particle in a fluid at temperature, T, by using the Boltzmann factor in the following inequality (Ashkin, 1978):

$$e^{-\frac{U_g}{k_B T}} = e^{-\frac{n\alpha E^2}{2k_B T}} \ll 1 \tag{19.10}$$

where k_B is Boltzmann's constant, introduced in Chapter 5. (Another way of stating this condition is that the trapping time, i.e., the time it takes to pull the particle close to the focal spot, must be shorter than the diffusion time of the particle in the medium.) At room temperature ($T \cong 295$ K), and with a laser beam focused to 2 μm diameter, this translates to a laser power of the order of >10 mW.

One experimental limitation for trapping of Rayleigh particles is that they are generally too small to observe with standard optical microscopy. So, confirmation of trapping and manipulation of interactions with other biological particles can be challenging. One mechanism that can be used to visualize the particles is fluorescence imaging, if the particle is intrinsically fluorescent, or if it is tagged with a fluorophore. Of course, the fluorescence emission must be separated in wavelength from the trapping light, to enable adequate filtering of the image. Given that the number of photons involved in the absorption and emission process will be much smaller than the number of photons involved in the scattering/trapping, the fluorescence emission will constitute a minimal perturbation of the trapping forces.

19.1.4 Trapping of particles similar in size to the wavelength

But what about the "Goldilocks" particles, which are neither large nor small but are just right (i.e., the size range 0.1–2 μm, which covers many organelles and other biological structures of interest)? For analyzing the forces, we note that neither the ray-trace analysis for large particles nor the point-dipole-in-a-gradient analysis for small particles will provide an unambiguous assessment of trapping when the particle size is of the order of the wavelength; and direct calculation of trapping forces for this size range is complicated and problematic. Fortunately, optical trapping works equally well for this size range, even if calculating the forces is difficult! To get a feel for the complexity, we turn to the Mie theory of scattering, as developed in Chapter 7, and we recall the angle-dependent functions in the scattering matrix elements (Eqs. (7.28) and (7.29)), which lead to the angular dependence of scattering probability, called the scattering phase function, as illustrated in Figures 7.8 and 7.9. By using the phase function, calculation of the angular scattering distribution would indeed enable an estimation of the repulsive scattering force. The formalism of Mie theory, however, assumes a uniform (plane-wave) illumination field, which is gradient-free and cannot generate a trapping force. An analysis of trapping forces on Mie particles for an inhomogeneous (focused) illuminating field is more complex than the already complex Mie theory, and generally invokes various approximations and a limited range of applicable parameter values (Ren et al., 1994; Rohrback and Stelzer, 2001; Rohrbach and Stelzer, 2002; Zemánek and Jonás, 2002).

The complexity of analysis for particle radii in the range approximately 0.2λ–5λ is unfortunate, since this size range covers many of the interesting biological particles and organelles, and is also the size range of dielectric spheres that are often used for studying the biomechanical properties of biological structures. Given the inherent complexity of the calculations, the limited applicability of the approximated conditions, and the fact that many biological structures are not spherical, empirical calibration of trapping forces is most commonly employed. (See Section 19.1.6 below.)

19.1.5 Generic instrumentation for optical tweezers

Although sophisticated, multi-beam, optical tweezers systems can be rather expensive, several companies sell modest-cost optical trapping kits as add-ons for conventional microscopes. The core elements are illustrated in Figure 19.5. In this simple format, the laser trapping beam is introduced into the optical train by a dichroic mirror (which can also serve for translating the beam) that is located in the "barrel" section of the microscope, where the rays are approximately collimated. The microscope objective focuses the laser beam and also images the particle.

The laser should have single-mode (Gaussian) output and stable power in the range 20–200 mW. Semiconductor diode lasers are inexpensive and convenient for this parameter range, and they come in a variety of wavelengths, but the low-cost versions generally do not have single-mode output. Focusing the beam through a pinhole, as depicted, and having it go through the back aperture of the objective lens accomplishes the beam "cleanup" and produces an *apodized* beam, i.e., a beam with no tails (apodized literally means "without feet") with a soft-edged

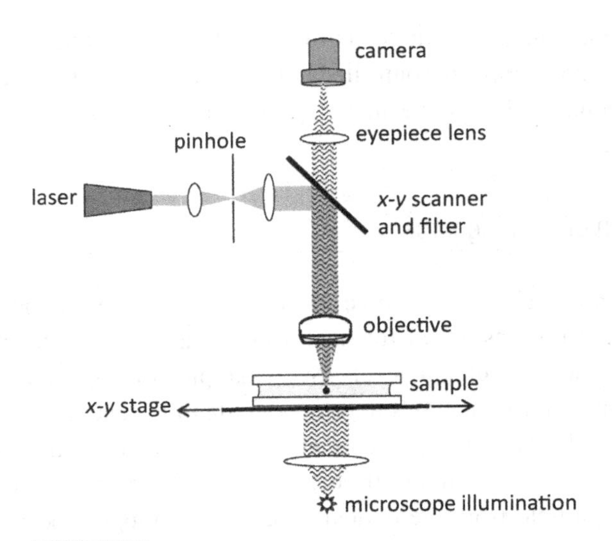

Figure 19.5

Basic elements of a generic optical tweezers setup in a conventional microscope.

top-hat-shaped intensity profile. This apodized beam results in a wider transverse intensity distribution at the conjugate focal plane. For particles larger than the wavelength, such a profile is modestly better for trapping than a standard Gaussian distribution, due to an increased ratio of the gradient force to the scattering force (see, under "Further reading," Neuman and Block, 2004).

A modern CCD camera can serve well as an imaging device, and can also double as position sensor to monitor the particle displacement and transverse motion, by invoking appropriate image analysis software with fast computer response. Once the trapping force is calibrated (see Section 19.1.6), the software can also provide correlation between the particle displacement and forces acting on the particle, although response time is limited by the camera frame rate. Consequently, when faster response times are needed, optical tweezers also often invoke the use of quadrant detectors (photodiodes with four individual sensor areas) to sense particle displacement and motion. The other key element is a mechanism to manipulate and control motion of the trapped particle relative to the rest of the sample. This can be achieved by transverse scanning of either the laser beam or the sample stage, or both. Such motion and positioning control is often accomplished with piezoelectric drivers, which provide greater precision than stepper motors.

We note that optical tweezers systems are not limited to a single laser beam trapping one particle. A multi-beam optical tweezers consists of two or more focused laser beams that are used to trap different objects and manipulate them independently, and large numbers of individually controlled particles have been achieved using computer-controlled holographic gratings to form an array of individually controlled laser "beamlets" (Curtis et al., 2002).

19.1.6 Calibration of trapping force

Calibration of the trapping force of an optical tweezers is necessary when the tweezers are used as a tool for measuring a variety of biomechanical forces at a cellular and sub-cellular level. (See examples of applications below.) Absolute quantitation (by methods developed to date) is generally limited to spherical particles, in the Mie-theory size range or larger. There are three methods that are commonly used to accomplish the calibration, all based on measuring the displacement of a particle from the optical focus of the trapping beam, in the presence of effects from the surrounding fluid. The goal is to calibrate the *trap stiffness constant*, K_{trap}, which, for small displacements, acts like a Hooke's-law spring constant, describing how strongly the trap "pulls" the particle towards the laser focus. This formalism applies because the trapping force is approximately linear with particle displacement, $\mathbf{d} \equiv \mathbf{r}_p - \mathbf{r}_T$, where $\mathbf{r}_p$ is the location of the particle center, and $\mathbf{r}_T$ is the location of the trap focus. Thus, the trapping force acting on the particle, $\mathbf{F}_{\text{trap}}$, can be written as

$$\mathbf{F}_{\text{trap}} = -K_{\text{trap}}\mathbf{d}. \tag{19.11}$$

If the particle is in motion with respect to the fluid, then the external forces acting on the particle include the fluid drag, and the net external force, $\mathbf{F}_r$, can be written

$$\mathbf{F}_r(t) = K_{\text{trap}}\mathbf{d} + b\frac{d\mathbf{r}_p}{dt} \tag{19.12}$$

where b is the damping factor for the viscous drag.

The simplest calibration method, then, is to measure the particle displacement, $\mathbf{r}_T - \mathbf{r}_p$ in a constant flowing fluid of known viscosity with controllable flow rate (Simmons et al., 1996). Once the particle is trapped, the relative flow can be induced by translating the sample stage, or by scanning the trapping beam, at a constant rate. To assess the linear range of the trap stiffness constant, the particle displacement can be measured for a range of flow velocities. The damping factor, b, in Eq. (19.12) can be determined by calculating the drag force using Stokes' law (named after the same Irish-British physicist introduced in Chapter 6 for the named wavelength shift in scattering), which gives the drag force of a slowly flowing fluid on a spherical particle that is not near a surface as:

$$\mathbf{F}_{\text{drag}} = 6\pi\eta a\mathbf{v}, \tag{19.13}$$

where η is the fluid viscosity and a is the particle radius. (Note that $b = 6\pi\eta a\mathbf{v}$.)

Under steady-state conditions (not oscillating) the net external force is zero, and we have $\mathbf{F}_{\text{trap}} = -\mathbf{F}_{\text{drag}}$; thus, we can write (as a scalar relation, for simplicity):

$$K_{\text{trap}} = \frac{6\pi\eta a|\mathbf{v}|}{|\mathbf{d}|} \tag{19.14}$$

A variation of this approach is to oscillate the trap position relative to the fluid. The position of the trap (or sample stage) is typically driven with a triangular wave (of position vs. time), yielding a square wave for its velocity. Under these conditions of a step response, the trapped particle behaves like a highly damped harmonic oscillator (Simmons et al., 1996), which, in one dimension, can be represented by

$$\mathbf{r}_p = \mathbf{r}_T\left(1 - e^{-K_{\text{trap}}t/b}\right). \tag{19.15}$$

Measuring the particle position, $\mathbf{r}_p$, as a function of time yields the rate constant (K_{trap}/b), where b can be determined from Stokes' law, hence determining K_{trap}.

The third approach is based on measuring the thermal Brownian motion by looking at the average particle displacement or the scattering power spectrum (Berg-Sorensen and Flyvbjerg, 2004), in a manner similar to measurement of dynamic light scattering, which was introduced in Chapter 7. From statistical mechanics the equipartition theorem teaches that, in thermal equilibrium, the potential and inertial components of the energy are equal. Thus, for each degree of freedom, say $x \equiv d_x$, the average of each form of the energy of the trapped particle is 1/3 of the thermal energy ($\frac{3}{2}k_BT$), leading to:

$$\frac{1}{2}mv_x^2 = \frac{1}{2}K_{\text{trap}}\langle x^2\rangle = \frac{1}{2}k_BT \tag{19.16}$$

Thus, by measuring the average displacement (or the average particle velocity, which is related through the scattering power spectrum), and given that the temperature is known, the trap stiffness (spring constant) can be calculated.

19.2 A sampling of applications of optical tweezers

As an elegant example of an enabling application of optical tweezers, researchers have measured the stiffness of cochlear hair cells, which are the key transducers of acoustic waves to nerve impulses in the ear (Li et al., 2002; Schumacher et al., 2008). The approach, as depicted schematically in Figure 19.6, is to use a

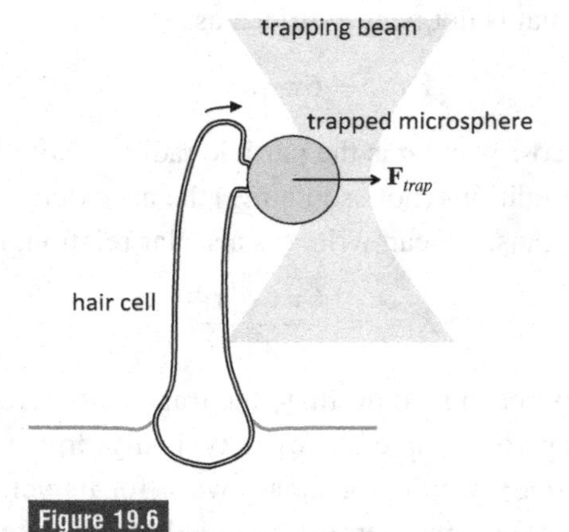

Figure 19.6

Schematic of a trapped sphere tethered to the membrane wall of an auditory hair cell. The sphere is pulled to the side, and the force necessary to pull the membrane tether is measured. The mechanical stiffness of the hair cell is related to the structural integrity of the cytoskeleton bound to the membrane.

spherical bead that is coated with a specific protein, so as to bind strongly to the cell membrane. The bead is held in an optical trap, brought into contact with the membrane on the side of the cell, then pulled sideways to generate a membrane tether between the two, causing the hair cell to bend. Thus, the optical tweezers enable measurement of the forces that are related to the integrity of the cytoskeleton, which forms the internal structural scaffold for the membrane (Ermilov et al., 2006). Studies of this type lead to improved understanding of the mechanisms of hearing loss.

Another broadly used area of application for optical tweezers is the measurement of biophysical properties and forces of molecular motors (e.g., Nishikaza et al., 1995; Arsenault et al., 2009). A large variety of physiological processes related to movement and transport invoke molecular motors, ranging from mitosis to muscle contraction to the swimming of spermatozoa. Optical tweezers have enabled a number of studies that have addressed the mechanical forces and motion of important motor proteins, including kinesin (important to cellular mitosis), myosin (active in muscle contraction), and RNA polymerase (relevant to the mechanics of DNA "unzipping" and transcription). As an example, Lang and coworkers studied the stepping mechanics of kinesin as it walks 100-nm distances along a microtubule, by measuring the step changes in the piconewton-level force on a microsphere that was bound to one end of a kinesin molecule (Lang et al., 2002). A representation of the setup described in their report is shown in Figure 19.7.

On a larger scale, researchers have used optical tweezers to measure the motility (swimming strength) of sperm cells (Nacimento et al., 2006). The swimming

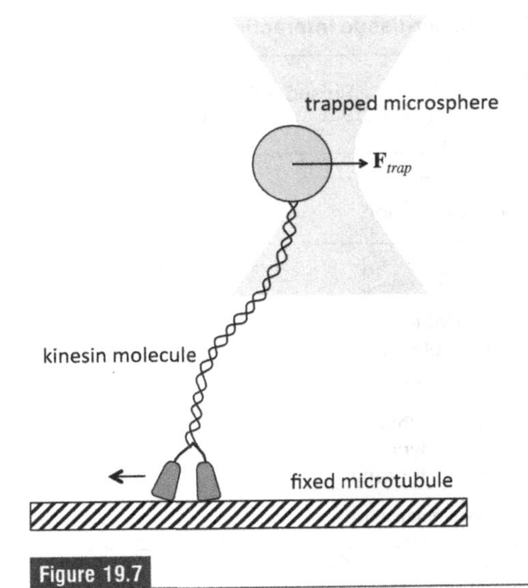

Figure 19.7

A cartoon representation of a kinesin molecule "walking" along a microtubule. An optical trap holds a microsphere that is attached at one end of the molecule, and displacements of the sphere from the trap center enable monitoring of the motion and forces related to the walking.

strength is directly related to the likelihood of a sperm penetrating the outer layer of an oocyte, and achieving insemination. Such studies of sperm motility help to understand loss of fertility in older males.

19.3 Sub-thermal, thermal, and ablative applications of lasers

The other broad area we cover here, for applications of laser beams in biomedicine on a small or microscopic scale, does not address the use of gentle optical forces to hold things, but rather the use of lasers, mainly at higher powers, to heat, melt, burn or ablate (vaporize!) cells or tissue. Nonetheless, and unlike the case of early surgical lasers, the intent is the subtle use of power, when it can be precisely applied, utilizing special properties of laser light, and when the interactions are dominated by absorption, rather than refraction or scattering. The chart shown in Figure 19.8 displays the parameter spaces for different classes of light-tissue interaction, plotted against two parameters: the spatial density of power deposition (Watts per unit volume) and the photon energy (or light frequency).

The zones denoted on the y axis of Figure 19.8 are not marked with hard quantitative boundary points between zones, because the specific demarcation between, say, thermal and ablative processes depends on a number of parameters,

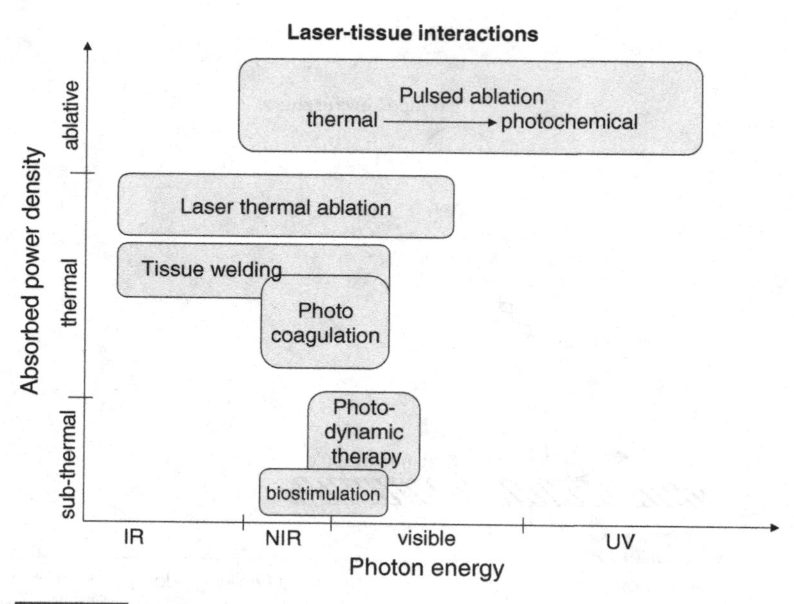

Figure 19.8

Different classes of laser-tissue interactions, mostly dominated by absorption, are plotted against two key parameters that dominate in governing the nature of the interaction: the volumetric density of power deposition, and the energy of the photons (or the optical frequency).

including, importantly, the duration of exposure (hence the total energy deposited). In general, the lower values of power density correspond to continuous-wave lasers with low-to-moderate average powers (milliwatts to watts), whereas the ablative zone generally corresponds to short-pulse (nanoseconds to picoseconds) lasers with high peak powers (≥megawatts). The actual local power density, of course, also depends on the degree of focusing of the beam and the volume over which the light is absorbed.

19.3.1 Sub-thermal irradiation

The term "sub-thermal" refers to the parameter range where power density of deposition is sufficiently low, compared to heat dissipation rates, such that there is only a small or negligible increase of temperature. Thus, if anything significant actually happens, the interaction must be photochemical or photobiological (catalytic) in nature.

The most familiar example of a sub-thermal (or moderately thermal) effect of light on tissue is the common consequence of suntanning, which results from the photochemical effects of UV radiation on cells in the skin (Parrish et al., 1982). The process requires sufficient photon energy (≥2.5 eV), such that no tanning occurs for wavelengths longer than ~420 nm. UVA wavelengths (320–400 nm) oxidize existing melanin, rendering it darker and inducing a short-term tan (and also inducing an inflammatory response). The UVA photon energy is not enough to directly damage DNA, but is sufficient to create reactive oxidative species, which can lead indirectly to DNA damage. On the other hand, UVB wavelengths (280–320 nm) cause direct DNA damage and stimulate production of additional melanin in the melanocytes, by a process called *melanogenesis*, which leads to a delayed but longer-lasting tan and added protection from future exposure (at the expense of the potentially dangerous direct DNA damage). Topical sunscreens are applied in an attempt to block skin exposure, by relying on the absorption of UV wavelengths by the sunscreen on the surface.

A controversial topic in photobiology is a method called *low-level laser therapy* (LLLT), wherein milliwatt levels of red laser irradiation are thought to induce pain reduction, accelerated wound healing, and reduced inflammation, among other health benefits. Unfortunately, there have been a limited number of published controlled studies of LLLT efficacy, and fewer that were double-blinded. One published meta-analysis addressed collected reports of the pain-relieving effect of shining low-power red (~650 nm) laser light onto the skin over locations of neck pain, with apparent statistical significance over placebo (Chow et al., 2009). Determination of pain relief efficacy, however, depends on patient subjective interpretations. More importantly, there is not a well-established explanation for the potential mechanisms of action of LLLT.

In contrast to LLLT, the mechanisms of *photodynamic therapy* (PDT) have been rigorously studied and well explained; and numerous double-blinded studies have been published. (See listed resource under "Further reading.") PDT is a specialized version of chemotherapeutic treatment for cancer, wherein a (hopefully) site-selective drug is photoactivated with an appropriate wavelength of light. That is, absorption of a photon leads to a chemical reaction or transfer of energy to a different species. For many of the PDT agents that have been studied, the induced photochemical reaction results in the production of an excited triplet state of the drug molecule, which then transfers energy to locally available dissolved oxygen, yielding oxygen in its excited singlet state. Singlet oxygen is highly reactive, and causes cytotoxic oxidation, leading to rapid cell death. The promise of PDT is that, in addition to any specific molecular targeting by the drug itself, additional spatial selectivity of the photochemical reaction (treatment) can be achieved by only irradiating the target tissue with the appropriate wavelength of light. Non-activated drug in other parts of the body is eventually metabolized and leaves the body without causing cytotoxicity. Thus, the spatial selectivity of therapy reduces the probability of side effects, which are more common for systemically administered chemotherapy agents.

Most of the research on PDT is about the complex photochemistry and biochemistry of the agents being developed, addressing concerns about biodistribution, targeting, and potential side effects. The optical science aspects of PDT, on the other hand, are relatively straightforward. The activation light is generally in the red region of the optical spectrum, which facilitates diffusion into the tissue, and the light is simply radiated directly onto tissue if the target tissue is near any accessible surface (e.g., directly or through an endoscope). For deeper tissues and solid organs, the light can be delivered through an optical fiber that is passed through a small needle inserted into the tissue. The light source does not need to be coherent, and high-power LEDs can be used for superficial illumination. When optical fibers are used for light delivery, however, lasers (frequently diode lasers) are used due to the relative ease of efficiently coupling laser light into a fiber.

The power level of light required for adequate photoactivation of the PDT agent is generally sub-thermal, and the light distribution follows the laws of diffuse optics, since the tissue volumes irradiated are large enough that the diffusion approximation for transport theory is applicable. Thus, the local light "dosage" (fluence) can be calculated from formalisms of light diffusion that were fully developed in Chapter 10, while accounting for the optical geometry of the specific illumination setup. For example, for the geometry of an optical fiber inserted deep into a tissue volume, Figure 10.1(b) provides a reasonable approximation for the dependence of the fluence rate as a function of distance from the fiber tip, for distances greater than a few times the reduced mean free path for scattering in the tissue. (The directionality of the source is lost at a distance that is consistent with the diffusion regime.) In addition to the optical dosage, the other two factors that govern the treatment efficacy are the tissue concentration of the PDT agent, and the local availability of oxygen to be excited by the activated drug.

19.3.2 Non-ablative thermal effects

When the rate of energy deposition exceeds the local rate of thermal diffusion, the temperature can rise significantly. The rate of heat transfer between two volumes at different temperatures is proportional to their temperature difference. Consequently, in the absence of phase transitions, the temperature of an absorbing volume that is irradiated rises until the rate of heat conduction away (diffusion) equals the energy deposition rate, $\Delta Q/\Delta t$, into that volume. One can write a simplified expression for the steady-state heat flow between two volumes (in this case, an irradiated volume and its surrounding tissue) at different temperatures:

$$P_{abs} = \frac{\Delta Q}{\Delta t} = k\mathcal{A}\frac{\Delta T}{L} \tag{19.17}$$

where P_{abs} is the optical power absorbed in the volume, k is the thermal conductivity, ΔT is the temperature difference between the two volumes, $\mathcal{A}$ is the effective surface area between the two volumes, and L is the characteristic dimension of the irradiated volume. If we approximate the surface area by assuming a spherical volume, we can replace the area with $4\pi L^2$. Then the temperature rise of the irradiated volume can be approximated as:

$$\Delta T = \frac{P_{abs} L}{k\mathcal{A}} \cong \frac{P_{abs}}{4\pi k L}. \tag{19.18}$$

Note that under steady-state conditions, the temperature rise does not depend on the specific heat or density of the medium, but only on its thermal conductivity. It should also be noted that shapes of heated volumes with different surface-to-volume ratios will invoke different multiplicative factors. For example, if a large surface area of, say, skin is irradiated, and if the absorption coefficient at the laser wavelength is large, then the absorption (and heating) occurs in a thin depth ($\sim 1/\mu_a$), and we must use the more-accurate middle term in Eq. (19.18), invoking the actual illuminated area for $\mathcal{A}$, but using the thickness of the absorbing layer for L.

If the temperature of tissue rises above ~60°C for more than a few seconds, *thermal denaturation* begins to occur. Another term for thermal denaturation is *photocoagulation*. Thermal denaturation constitutes a constellation of effects that include cell death (both necrotic and apoptotic) and remodeling or breakdown of the extracellular matrix (ECM), the structural-protein scaffold that gives tissue its structure and mechanical properties, made mostly of collagen and elastin. The degree of denaturation is a nonlinear cumulative effect, in which higher temperatures cause faster denaturation (Pearce and Thomsen, 2011). Thus, significant denaturation can be induced in tens of milliseconds at temperatures greater than ~80°C, whereas it can take several minutes at 55°C.

19.3.2.1 *Laser tissue welding*

One interesting set of applications of thermal denaturation centers on the "melting" of collagen by localized laser-induced heating (Vogel and Venugopalan, 2003). Collagen is the most abundant protein in the body, constituting ~30% of all protein in mammals, and is key to most of the body's mechanical architecture. Collagen is a long fibrillar protein that is found in various tertiary structures, but mostly as a triple helix of three fibrils twisted together. When thermally denatured (melted), the collagen fibers unravel as the crosslink bonds are broken, but the strong molecular covalent bonds maintain the structure of individual strands. Upon cooling, multiple fibrils entangle and reform fibers or globular protein complexes, effectively acting like a hardening glue. This phenomenon has been used to effect *tissue welding*, for sutureless joining or bonding of tissues. One such application was broadly practiced in ophthalmology for some years for treating detached retinas, by "spot-welding" the retina to the vascular layer beneath, using a pulsed laser focused with the lens of the eye (Haller et al., 1997). In recent years, however, this has been replaced by methods that do not leave the consequent small blind spots.

Another application of laser tissue welding is bowel *anastomosis*, which has been the subject of numerous research studies. The promise is that tissue welding can be used to generate a continuous seal of, say, a resected bowel, and the seam will be stronger than for a sutured closure, and with less chance of leakage (e.g., Spector et al., 2009). Despite impressive demonstrations, however, the problem that has limited widespread clinical adoption of this method is that it works well only for a carefully controlled narrow range of procedure parameter values (rate of energy deposition, achieved temperature, and time of elevated temperature), which are difficult to control under standard clinical circumstances.

19.3.2.2 *Interstitial laser thermotherapy*

Beyond tissue welding, the other main area of clinical applications for photocoagulation is simply the killing of a volume of tissue, without physically removing the tissue. The body's immune system does that job, eventually metabolizing the dead cells and debris, and the metabolites are then excreted. This treatment method has been called *interstitial laser thermotherapy* (ILT), and other similar monikers. Commonly, the laser wavelength is chosen to provide a balance between the absorption and scattering coefficients of the tissue, so as to allow some diffusion of the light into a volume of the tissue while being absorbed and heating the tissue. Diode lasers operating at NIR wavelengths generally meet this criterion. With ILT the delivery is often by an optical fiber that is inserted into the bulk of the tissue to be treated, and sufficient laser light power is delivered to cause cell death over a significant volume. The intent is often to heat several cubic centimeters of tissue volume, to temperatures in the range 60–80°C for a few minutes, and this generally translates to laser powers of a few tens of watts.

One example of a clinical application of ILT is the treatment of metastatic cancer in the liver, which is generally not resectable surgically, and is often refractory to chemotherapy. Breast cancer, for example, commonly metastasizes to the liver. For this application ILT has been used clinically and has demonstrated advantages compared to radio-frequency heat treatment (Vogl et al., 2013). Figure 19.9 illustrates how multiple metastatic tumor sites can be treated simultaneously with one

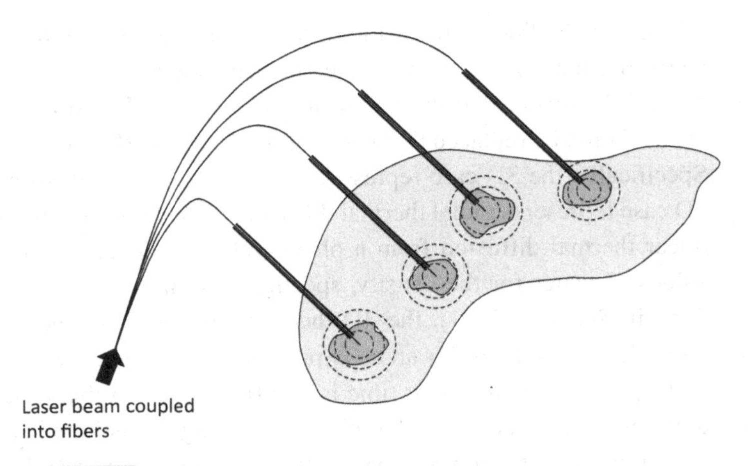

Figure 19.9

Several optical fibers are introduced into cancerous regions of the liver, through small needles that are guided into the lesions under ultrasound imaging. Light diffuses from each fiber tip to heat larger tissue volumes at each site.

laser, by coupling the light to several optical fibers that are introduced into the tumors individually.

19.3.3 Tissue ablation and microsurgery

When the rate of heat deposition in a mass greatly exceeds the rate of heat dissipation into the surrounding media, the temperature can rise dramatically and greatly exceed the boiling temperature, leading to explosive vaporization. Very high optical field densities, especially with shorter wavelength light (UV-visible), can result in instantaneous breaking of chemical bonds or even ionization. Under those circumstances, the optical electric field is converted into electronic energy, more than thermal (vibrational and kinetic) energy.

As we saw in Chapter 16 (Section 16.3.1), the thermal confinement time for a small volume is much longer than the energy deposition time of a short-pulse laser, and steady-state conditions, as discussed in Section 19.3.2, are not applicable. In Section 16.3.1, the *thermal diffusion time* was called the thermal confinement time (which is appropriate when discussing a short duration of irradiation by pulsed laser). Combining Eqs. (16.15) and (16.16), we can use the same term in an expression for the thermal diffusion time, τ_{th}, as relates to the heating dynamics of tissue ablation under much higher peak optical powers:

$$\tau_{\text{th}} = \frac{L^2}{6\alpha} = \frac{L^2 \rho c_p}{6k}, \tag{19.19}$$

where L is the linear dimension of the absorbing volume, α is the thermal diffusivity, ρ is the density, c_p is the specific heat, and k is the thermal conductivity. We remind the reader that the factor of $\frac{1}{6}$ obtains for diffusion in three dimensions, and it should be replaced by factors of $\frac{1}{4}$ or $\frac{1}{2}$ in the 2D and 1D cases, respectively. Specifically, the 3D case represents isotropic thermal diffusion from a point, the 2D case represents radial thermal diffusion from a line, and the 1D case represents linear thermal diffusion from a plane. Taking a linear dimension of 1 mm, and values of water for the density, specific heat, and thermal conductivity (as was done in Section 16.3.1), the 3D thermal confinement time of Eq. (19.19) is of the order of 1 s. Even for an absorption volume with a linear dimension as small as 10 μm, the confinement time is $\sim$100 μs, and, on that time scale, laser pulse durations in the nanosecond to picosecond range constitute instantaneous sources of heating, which simplifies calculation of temperature rise. If we know the laser pulse energy that is deposited in the targeted volume, then we can calculate the temperature rise as

$$\Delta T = \frac{Q_{\text{abs}}}{c_p \rho V}, \tag{19.20}$$

where Q_{abs} is the radiant energy absorbed in the volume, V. To get a feel for the values, if, for example, the energy of a 100-mJ pulse is absorbed in a tissue volume of 1 mm^3, the semi-instantaneous temperature rise will be of the order of 25,000°C. Of course, such a temperature is not reached, and there is nothing gradual about the resulting phase transition; hence, the concept of temperature itself, which implies a form of equilibrium, becomes nebulous. The result is explosive vaporization, with much of the deposited energy being carried away from the site as kinetic energy of ejected tissue.

19.3.3.1 *Laser-assisted in vitro fertilization*

One of the more interesting (and among the most profitable!) applications of laser micro-ablation has been in the area of *in vitro fertilization* (IVF) (e.g., Woods et al., 2014; Woodward and Campbell, 2014). The egg cell, *ovum*, in humans (and in all mammals) is protected by a glycoprotein layer outside the main cell membrane. This additional layer, called the *zona pellucida* (literally, the hairy zone), attracts and binds spermatozoa, but also protects the ovum, resisting penetration, such that the strongest sperm cells are more likely to inseminate. (The term *pellucida* derives from the numerous follicle cells that adhere to the *zona* until induced to disperse by enzymes emitted from sperm cells.) The *zona* is $\sim$15 μm thick and becomes tougher with the age of the female, especially for eggs previously frozen, whereas the swimming strength of spermatozoa becomes weaker with the age of the male. To aid in the process of fertilization, especially for older couples, a

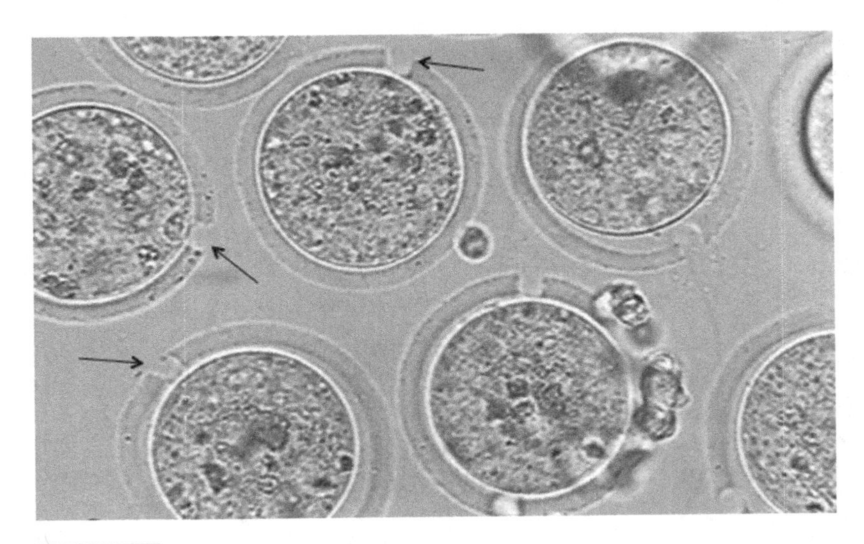

Figure 19.10

Mouse oocytes in which notches (examples denoted by arrows) have been laser drilled through the zona pellucida outer layer, facilitating easier penetration by spermatozoa. (Image adopted from Woods et al., 2014, and converted to grayscale.)

successful method is to drill a tangential slot through the *zona*, without damaging the ovum membrane itself, in a method called *laser-assisted IVF*. Figure 19.10 is a phase-contrast microscope image of ova showing notches drilled by a focused laser beam through the zona pellucida layer.

The laser commonly used in laser-assisted IVF is a 1.48 μm diode laser, chosen because this is a water-absorption peak ($\mu_a = \sim 30 \text{ cm}^{-1}$). There is not excessive loss due to transmission through the aqueous cell-maintenance medium, because the egg cells are drilled while in a monolayer on a microscope slide. Moreover, there is no risk of DNA damage by scattered light (Hammoud et al., 2010), as would be the case with a UV laser. The laser is propagated through the illumination arm of a microscope, in a manner similar to that of some optical tweezers designs, and is focused with a moderate NA lens. Typical optical parameters of the drill laser are: power ~200 mW; focal spot diameter ~10 μm; pulse duration ~100 μs.

19.3.3.2 *Laser refractive surgery*

The application of laser tissue ablation that is best known to the general public, probably because of extensive commercial promotion, is *laser refractive surgery*, also called *photorefractive keratectomy* (PRK), referring to reshaping of the surface of the cornea to correct visual refractive deficiencies, such as myopia and/or astigmatism. The ideal patient is one who will emerge with 20/20 visual acuity,

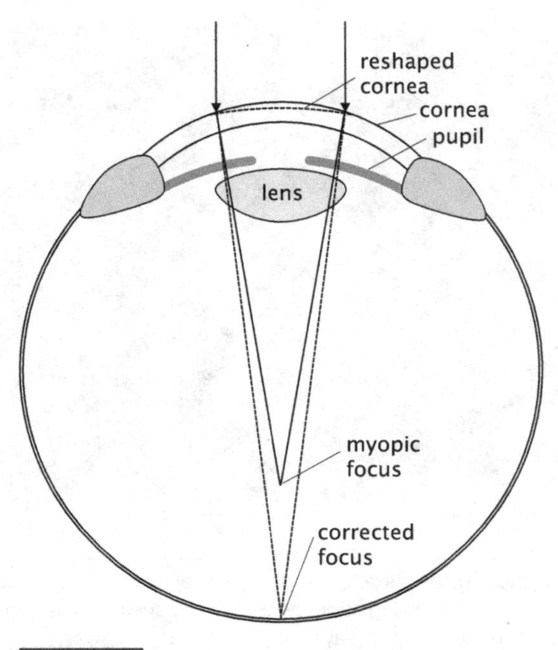

Figure 19.11

Laser ablative reshaping of the cornea to correct a case of myopia. The resulting reduced curvature of the corneal surface refracts less strongly, so that the corrected focal distance is at the corneal surface.

and no longer need to wear corrective eyeglasses. Due to the large mismatch of refractive index between air and the cornea, most of the refractive power of the eye is at the anterior corneal surface. The lens accounts for only a fraction of the total focusing power, and is mainly for adjusting the focal length to accommodate imaging objects at different distances.

The general results of laser refractive surgery are represented in Figure 19.11, which exemplifies correction of myopia, the condition of inability to focus on distant objects. For this condition the combination of cornea and lens results in too much refraction, such that, for objects at infinity, the image plane is short of the retina surface, where a blurred post-focal image appears. This is represented by the shorter (solid-line) focus corresponding to the original corneal shape. After laser ablative reshaping of the surface of the cornea, the central zone of the cornea is flatter, with a larger radius of curvature and, hence, less refractive power. The result, with rays indicated by the dashed lines, is a focal plane at the retina.

Various types of laser procedures for alteration of corneal refractive power have gone through phases of implementation and limited commercial dissemination, but in recent years a version called LASIK has enjoyed widespread adoption. LASIK stands for laser-assisted *in situ* keratomileusis. The LASIK procedure starts with precision cutting and folding back a thin flap of the corneal epithelium (plus a small amount of stroma) over the central portion of the cornea. The exposed stroma is then reshaped by laser ablation to a new curvature, thus changing the refractive power of the eye after the flap is replaced.

The laser used for these procedures is the argon-fluoride (ArF) *excimer laser*, a gas laser that emits ~10-ns pulses at 193 nm, with pulse energies around 1 mJ. Given that the absorption coefficient, at that wavelength, of the combined constituents of the cornea is about 3000 cm^{-1}, the absorption depth is only 3–5 μm. The surface area that is ablated with each pulse is about 0.1–0.5 cm^2, so the absorbing volume for each pulse is of the order of 1×10^{-4} cm^3. Plugging these numbers into Eq. (19.20) yields a temperature rise of only 1–5°C, so the ablative process is not thermal, and there is essentially no thermal damage to stroma beneath the absorbing surface. However, *photochemical ablation*, also called *ablative photodecomposition*, occurs readily, because the energy of a single photon at this wavelength is 6.4 eV, which is enough to break many molecular bonds. The dissociation energies of some of the key bonds in biomolecules are listed here:

C–C bond: 3.6 eV
C=C bond: 6.4 eV
C–H bond: 4.25 eV
C–O bond: 3.6 eV
O–H bond: 4.8 eV

These values are approximate, because they are affected by the molecular environment of the bond, but the 6.4 eV photons from the ArF laser are sufficient to cause dissociation. The result is ejection of low-molecular-weight fragments, with little or no thermal effect on the tissue below. Thus, a very thin layer (≤5 μm) is photochemically ablated with each pulse in a finely controlled manner without damaging the adjacent stroma. Accurate control of the resulting surface curvature is achieved by computer-controlled variation of rings of illumination.

19.3.3.3 *Some cosmetic applications in dermatology*

Finally, we mention briefly some of the laser applications in dermatology, which are predominantly cosmetic. Many of these can be found described in detail in resources listed under "Further reading," especially Vogel and Venugopalan (2003).

One therapeutic application is the laser treatment of *port wine stains*, which can be psychologically burdensome for some people, especially children, since they most frequently occur in facial areas. Port wine stains are a vascular abnormality of excessive and large capillaries in the skin, and are almost always a birthmark. Laser photocoagulation of the blood vessels has been the most successful treatment modality, because selective absorption by the hemoglobin in the blood, for the correct wavelength laser, can destroy the blood vessels with minimal damage to the surrounding dermis. Dye lasers, which can be tuned to the absorption peak of oxyhemoglobin at 580 nm, are often coupled with surface cooling of the skin by various means to minimize thermal effects on the more superficial layers. Laser energies are typically hundreds of millijoules, with pulse durations of ~10–20 ns, and a skin area of ~1 cm^2 irradiated with each pulse. The parameters are controlled so as to heat the capillaries to 70–90°C, and to avoid boiling.

A similar application is the use of lasers to remove tattoos. As with port wine stain treatment, laser tattoo removal has been more successful than other methods because of the selective nature of the interaction of the laser light with the pigment particles in the skin. Tattoo pigments are particles that have been injected beneath the skin surface and are too large for the body's normal methods for eliminating foreign material. Laser treatment constitutes pulsed heating of the pigment particles, such that they break up into smaller particles, which can then be removed by the body's scavenging mechanisms. This process is sometimes called *selective photothermolysis*. Optimal laser treatment happens when the laser wavelength is chosen such that it is selectively absorbed by the specific pigment color. Thus, several different lasers, or widely tunable lasers, are used, so that a different wavelength can be used for each pigment color. Generally, dark or black pigments are most successfully removed. The laser parameters are similar to those for treatment of port wine stains, with the addition of the various wavelengths.

Other cosmetic applications of lasers in dermatology include the reduction of scars, stretch marks, and wrinkles, hair removal, and the removal/reduction of visible veins in the legs. These can also be found under "Further reading."

Problems – answers to problems with * are on p. 648

19.1* Consider an application of laser tweezers.

(a) Assume that the power of the tightly focused laser beam is 100 mW, and that a particle (in water) to be trapped is larger than the focal spot diameter. Also assume that 1% of the laser light from the focused beam is scattered directly backward from the surface of the cell. What is a minimum value for the geometric factor, A, to achieve a net trapping force greater than zero? While the focused laser beam is obviously not collimated (parallel), for the purposes of this problem the force due to backscattering of the focused beam can be approximated to be 1/2 of what it would be if the incoming beam were collimated.

(b) Aside from scattering and absorption, what other property of the cell (or particle) itself can affect the trapping force?

19.2 For a spherical particle larger than the wavelength, must the focal point of the laser beam be located inside the sphere for an attractive (trapping) force to be exerted on the particle? Consider a tightly focused beam incident from above, and with its focal point above the surface of a sphere.

(a) Draw a ray-trace diagram to analyze the net forces on the particle and determine whether they are attractive or repulsive.

(b) How does the repulsive force generated by reflection of a portion of light off the sphere surface compare to that force generated when the focal point is inside the sphere?

19.3 (a) Explain why micro-bubbles of air in water are repelled from the focal volume of a laser beam. Draw a ray-trace diagram to explain your answer.

(b) Calculate the net force on a 5-μm-diameter air bubble in water, for a 50-mW laser beam at a wavelength of 650 nm. Assume a geometric factor, A, of 0.1.

19.4* A lipid sphere, with an effective radius of 20 nm, is suspended in saline solution, with a refractive index of 1.35. The refractive index of the lipid is 1.45. A focused laser beam with a focal diameter of 2 μm and an optical power of 100 mW at a wavelength of 600 nm is used to trap the lipid sphere.

(a) Estimate the scattering force on the sphere when it is near the focal point.

(b) Estimate the gradient force, assuming that the intensity gradient across the diameter of the particle is 0.1 times the average intensity at the focal plane.

(c) Determine whether the inequality of Eq. (19.10) is satisfied for the particle in water at 37°C.

19.5* A quadrant detector with high-frequency response is used (in the microscope image plane) to monitor the position of a cell in an optical trap. The cell is in water at a temperature of 37°C, and the quadrant detector output voltages indicate that the time-averaged random displacement of the particle from trap center, due to Brownian forces, is 200 nm. What is the spring constant of the trap?

19.6 **Photodynamic therapy:** a patient is to be treated for cancer with PDT. The malignant tissue is located within the superficial first centimeter of the organ, and that is the volume that needs to be treated. A dosage of a PDT drug has been administered to the patient at 10 mg per kg of body weight.

The drug has a molecular weight of 100, and its optical absorption peak is at 660 nm. In order to achieve therapeutic treatment, 50% of the PDT molecules must be photoactivated in the illuminated zone. To minimize heating of the tissue, the incident illumination intensity is limited to 100 mW/cm^2 at the surface. Assume that the tissue scattering and absorption coefficients of drug and tissue are such that 20% of the excitation light is diffusely scattered back out of the tissue, and 80% is absorbed within a depth of 1 cm from the tissue surface. Further, assume that 10% of the absorbed excitation light is actually absorbed by the PDT molecules.

For the sake of these estimates, the following additional assumptions can be made:

- the drug is uniformly distributed in the body;
- as drug molecules get photoactivated, the absorption coefficient of the tissue does not change;
- the wavelength of the light source is matched to the absorption peak of the drug, and one absorbed photon is sufficient to excite the drug molecule;
- the mass density of tissue is the same as water (1 g/cm^3).

(a) Estimate the length of time that the light source must be turned on to achieve the desired treatment.

(b) If one oxygen molecule is required for each activated drug molecule in order to produce the singlet oxygen that does the cytotoxic job, and if 10% of the oxygen in the tissue is available to react with the drug molecules, what is the necessary molar concentration of oxygen?

19.7* **Interstitial laser thermotherapy:** when interstitial laser thermotherapy is used for treatment of a tumor, we want to heat a volume of tissue to about 60–80°C for several minutes (steady state). An optical fiber is inserted into the tumor (with the fiber tip placed at the center of the tumor) and laser light from a continuous-wave laser is used to heat the tissue. Assume that the tumor is a spherical volume of tissue 1 cm in radius, and that we want it to be heated so that the temperature at its edge reaches 60°C. (It will be higher in the center.) Also assume that the tissue at a distance 1 cm further out from the edge of the tumor is kept at a constant body temperature of 37°C by the blood flow. A steady-state equilibrium condition is reached within a few tens of seconds after turning on the laser, and the edge of the tumor is at 60°C. What is the laser power (in watts) required to maintain this temperature after reaching equilibrium? (Assume that all light is absorbed within the tumor and that the thermal conductivity of tissue is similar to that of water.)

19.8 For the conditions described in Section 19.3.3.1 relating laser drilling of the zona pellucida of an egg cell:
   (a) Estimate the light intensity at the focal plane.
   (b) If 10% of the laser pulse is absorbed in the zona, calculate the short-term temperature rise in the target mass.
   (c) How does the irradiation time, in this application, compare with the thermal confinement time?
   (d) Explain why higher laser powers or larger deposited energies are not used for this application.

References

Arsenault, M. E., Sun, Y., Baua, H. H., and Goldman, Y. E. (2009). Using electrical and optical tweezers to facilitate studies of molecular motors. *Physical Chemistry Chemical Physics*, 11, 4834–4839.

Ashkin, A. (1970). Acceleration and trapping of particles by radiation pressure. *Physical Review Letters*, 24(4), 156–159.

Ashkin, A. (1978). Trapping of atoms by resonance radiation pressure. *Physical Review Letters*, 40, 729.

Ashkin, A., Dziedzic, J. M., Bjorkholm, J. E., and Chu, S. (1986). Observation of a single-beam gradient force optical trap for dielectric particles. *Optics Letters*, 11(5), 288–290.

Berg-Sorensen, K., and Flyvbjerg, H. (2004). Power spectrum analysis for optical tweezers. *Review of Scientific Instruments*, 75, 594–612.

Chow, R., Johnson, M., Lopes-Martins, R., and Bjordal, J. (2009). Efficacy of low-level laser therapy in the management of neck pain: a systematic review and meta-analysis of randomized placebo or active-treatment controlled trials. *Lancet*, 374, 1897–1908.

Curtis, J. E., Koss, B. A., and Grier, D. G. (2002). Dynamic holographic optical tweezers. *Optics Communications*, 207, 169–175.

Ermilov, S., Murdock, D. R., Qian, F., Brownell, W. E., and Anvari, B. (2006). Studies of plasma membrane mechanics and plasma membrane-cytoskeleton interactions using optical tweezers and fluorescence imaging. *Journal of Biomechanics*, 40, 476–480.

Gordon, J. P. (1973). Radiation forces and momenta in dielectric media. *Physical Review A*, 8, 14.

Haller, J. A., Blair, N., de Juan, E., et al. (1997). Multicenter trial of transscleral diode laser retinopexy in retinal detachment surgery. *Transactions of the American Ophthalmological Society*, 95, 221–230.

Hammoud, I., Molina-Gomes, D., Albert, M., et al. (2010). Are zona pellucida laser drilling and polar body biopsy safe for in vitro matured oocytes? *Journal of Assisted Reproduction and Genetics*, 27(7), 423–427.

Köster, R., Kähler, J., Brockhoff, C., Münzel, T., and Meinertz, T. (2002). Laser coronary angioplasty: history, perspective and future. *American Journal of Cardiovascular Drugs*, 2, 197–207.

Lang, M. J., Asbury, C. L., Shaevitz, J. W., and Block, S. M. (2002). An automated two-dimensional optical force clamp for single molecule studies. *Biophysical Journal*, 83, 491–501.

Li, Z., Anvari, B., Takashima, M., et al. (2002). Membrane tether formation from outer hair cells with optical tweezers. *Biophysical Journal*, 82, 1386–1395.

Nacimento, J. L., Botvinick, E. L., Shi, L. Z., Durrant, B., and Berns, M. W. (2006). Analysis of sperm motility using optical tweezers. *Journal of Biomedical Optics*, 11, 044001.

Nishizaka, T., Miyata, H., Yoshikawa, H., Ishiwata, S., and Kinosita, K. (1995). Unbinding force of a single motor molecule of muscle measured using optical tweezers. *Nature*, 377, 251–254.

O'Neil, A. T., and Padgett, M. J. (2001). Axial and lateral trapping efficiency of Laguerre–Gaussian modes in inverted optical tweezers. *Optics Communications*, 192, 45–50.

Parrish, J. A., Jaenicke, K. F., and Rox Anderson, R. (1982). Erythema and melanogenesis action spectra of normal human skin. *Photochemistry and Photobiology*, 36, 187–191.

Pearce, J., and Thomsen, S. (2011). In *Optical-Thermal Response of Laser-Irradiated Tissue*, 2nd edn., ed. A. J. Welch and M. J. C. van Gemert. Berlin: Springer.

Ren, K. F., Gréha, G., and Gouesbet, G. (1994). Radiation pressure forces exerted on a particle arbitrarily located in a Gaussian beam by using the generalized Lorenz-Mie theory, and associated resonance effects. *Optics Communications*, 108, 343–354.

Rohrbach, A., and Stelzer, E. H. K. (2001). Optical trapping of dielectric particles in arbitrary fields. *Journal of the Optical Society of America A*, 18, 839–853.

Rohrbach, A., and Stelzer, E. H. K. (2002). Trapping forces, force constants, and potential depths for dielectric spheres in the presence of spherical aberrations. *Applied Optics*, 41, 2494–2507.

Sanborn, T. A. (1996). Laser angioplasty: historical perspective. *Seminars in Interventional Cardiology*, 1, 117–119.

Schumacher, K. R., Popel, A. S., Anvari, B., Brownell, W. E., and Spector, A. A. (2008). Modeling the mechanics of tethers pulled from the cochlear outer hair cell membrane. *Journal of Biomechanical Engineering*, 130(3), 031007.

Siegman, A. E. (1986). *Lasers*, new edn. Herndon, VA: University Science Books.

Simmons, R. M., Finer, J. T., Chu, S., and Spudich, J. A. (1996). Quantitative measurements of force and displacement using an optical trap. *Biophysical Journal*, 70, 1813–1822.

Smith, S. P., Bhalotra, S. R., Brody, A. L., et al. (1999). Inexpensive optical tweezers for undergraduate laboratories. *American Journal of Physics*, 67, 27–35.

Spector, D., Rabi, Y., Vasserman, I., et al. (2009). In vitro large diameter bowel anastomosis using a temperature controlled laser tissue soldering system and albumin stent. *Lasers in Surgery and Medicine*, 41, 504–508.

Vogel, A., and Venugopalan, V. (2003). Mechanisms of pulsed laser ablation of biological tissues. *Chemical Reviews*, 103, 577–644.

Vogl, T. J., Farshid, P., Naguib, N. N., and Zangos, S. (2013). Thermal ablation therapies in patients with breast cancer liver metastases: a review. *European Radiology*, 23, 797–804.

Woods, S. E., Qi, P., Rosalia, E., et al. (2014). Laser-assisted *in vitro* fertilization facilitates fertilization of vitrified-warmed C57BL/6 mouse oocytes with fresh and frozen-thawed spermatozoa, producing live pups. *PLoS ONE*, 9(3), e91892: http://www.plosone.org/article/info:doi/10.1371/journal.pone.0091892.

Woodward, B., and Campbell, K. (2014). Laser assisted IVF: a simplified technique to improve zygote production rates without the need for ICSI. *IVF Lite*, 1, 12–66.

Zemánek, P., and Jonás, A. (2002). Simplified description of optical forces acting on a nanoparticle in the Gaussian standing wave. *Journal of the Optical Society of America*, 19, 1025–1034.

Further reading

Optical tweezers, general

Neuman, K. C., and Block, S. M. (2004). Optical trapping. *Review of Scientific Instruments*, 75, 2787–2809.

Bioscience applications of optical tweezers

Grier, D. G. (2003). Review: a revolution in optical manipulation. *Nature*, 424, 810–816.

Moffitt, J. R., Chemla, Y. R., Smith, S. B., and Bustamante, C. (2008). Recent advances in optical tweezers. *Annual Review of Biochemistry*, 77, 205–228.

New capabilities in optical tweezers

Dholakia, K., Spalding, G., and MacDonald, M. (Oct. 2002). Optical tweezers: the next generation. *Physics World*, 31–35.

Photodynamic therapy

Gomer, C., ed. (2010). *Photodynamic Therapy: Methods and Protocols*. New York: Humana Press.

General laser-tissue interactions

Welch, A. J., and van Gemert, M. J. C., eds. (2011). *Optical-Thermal Response of Laser-Irradiated Tissue*, 2nd edn. New York: Springer.

Niemz, M. H. (2007). *Laser-Tissue Interactions – Fundamentals and Applications*, 3rd ed. (Springer, Berlin).

Vogel, A., and Venugopalan, V. (2003). Mechanisms of pulsed laser ablation of biological tissues. *Chemical Reviews*, 103, 577–644.

Applications in ophthalmology

Fankhauser, F., and Kwasniewska, S., eds. (2003). *Lasers in Ophthalmology: Basic, Diagnostic and Surgical Aspects*. Amsterdam: Kugler Publications.

Answers to selected problems identified by *

Chapter 1

1.1 1.026×10^{16} photons/s.
1.3 (a) 60°; (b) 2.356 m^2; (c) 46.81 W; (d) 3.76 s.
1.5 (a) 2.5 W; 2.165 W.
1.8 26.2 mW/cm^2.
1.10 (a) A, $A\hat{\mathbf{x}}$; (b) A, 0; (c) A, 0; (d) A, $-0.25A\hat{\mathbf{z}}$.
1.11 4.5%.
1.13 0.795.
1.15 $\mu_s' = 3.79\mu_a$.

Chapter 2

2.1 43 μM.
2.3 $\mu_a \approx 92\ cm^{-1}$; $mfp_a \approx 108\ \mu m$.
2.6 (a) $5.7 \times 10^9\ cm^{-3}$; (b) 0.8.

Chapter 3

3.1 (a) TP = 93, FN = 16, TN = 106, FP = 5; (b) $Se = 0.853$; $Sp = 0.955$.
3.3 (a) 0.950, 0.867; (b) 0.678, 0.983; (c) 0.161, 0.998; (d) PPV increases, NPV decreases.
3.5 (a) ~92%; (b) ~62%.
3.7 ~2770.

Chapter 4

4.2 ~11.54°.
4.3 25.
4.5 air: 0.24; water: 0.18.

Chapter 5

5.1 $\sim$0.25%.
5.2 6 ns.
5.5 $\sim$25%.
5.7 $\sim$1.81 nm.

Chapter 6

6.2 1.1×10^3 N/m.
6.5 1610 $\text{cm}^{-1} \Rightarrow 6.21$ μm.
6.6 Deuteration, because of a larger change in the reduced mass.

Chapter 7

7.2 Normal.
7.3 $I_{s,1} = I_s - 2I_{s,m}^{\perp}$.
7.4 $\sim$34 cm/s.

Chapter 8

8.1 cm^{-1}.
8.6 (a) $C_{\text{corr}} = 0.16$; (b) $C_{\text{corr}} = 0.78$; (c) reduced difference in the absorption peaks.
8.7 Take the ratio of Eq. (8.35) written for t_1 and t_0, respectively.
8.8 Generalization of the formalism, regardless of scale length.
8.9 Silica spheres and water with sugar ($m = 1.043$), closest to $m = 1.044$ for organelles.

Chapter 9

9.2 (a) $1/(4\pi)$, 0; (b) $1/(4\pi)$, g; (c) $1/(4\pi)$, g; (d) $3/(16\pi)$, 0.
9.4 $u_{00} = 3A\sqrt{4\pi}$; $u_{1,-1} = A\sqrt{8\pi/3}$; $u_{1,1} = -A\sqrt{8\pi/3}$.
9.8 $\sim$1–2 GHz.
9.9 2.1%.
9.10 (a) 0.89; (b) 1.74; (c) 0.297; (d) 0.223.
9.13 $c_n \mu_{ax}^{(f)} U_x(\mathbf{r}, t - \tau)$.
9.15 $D = 6.234 \times 10^8\ \text{cm}^2/\text{s}$; $D_0 = 6.266 \times 10^8\ \text{cm}^2/\text{s}$; $D_\alpha = 6.247 \times 10^8\ \text{cm}^2/\text{s}$.

Chapter 10

10.1 (a) 35.3 μM, 1.46 μM; (b) 2.72 cm^{-1}, 3.30 cm^{-1}.
10.3 10.2 cm^{-1}.
10.5 0.10%.
10.7 22.9 cm, 33.6 times longer.
10.9 0; 0.139; 1.
10.11 0.090.

Chapter 11

11.1 0.627 nJ.
11.4 29.4 pJ.
11.6 Q_{TD}.

Chapter 12

12.3 (a) 0.247, 0.138; (b) −33.2°.
12.5 0.142 cm^{-1}; 13.7 cm^{-1}.
12.6 0.003 cm^{-1}; 0.3 cm^{-1}.
12.7 18.5 cm, 2.22×10^9 cm/s, 0.586 cm.
12.9 0.26 mW/cm^2.

Chapter 13

13.1 0.4377 W/cm^2.
13.3 (a) 4 mW; (b) 0.36 W/cm^2, above the MPE of skin.
13.5 23 μA.
13.7 1.15 pW.

Chapter 14

14.1 (a) 1.33; (b) 1.33; (c) 10.3; (d) 8.38; (e) 0.137; (f) 0.352; (g) 0.0279.
14.2 (a) 0.892 cm; (b) 0.134 cm.
14.4 63.7×10^9 y (~4.6 times the age of the universe).
14.5 3.375 mW.
14.9 (a) −0.03%; (b) −0.06%.

Chapter 15

15.1 (a) 11.25 mmHg; (b) 25.85 mmHg; (c) 31.82 mmHg.
15.3 (a) 0.689 $ml_{O2}/(kg_{tissue}$-min); (b) 56, 112 times.
15.5 5.52 $\mu mol_{O2}/(100ml_{tissue}$-min).
15.7 0.446 cm^{-1}.
15.9 98.4%.

Chapter 16

16.1 (a) $u_0 = \omega_a s_0$; (b) $\phi_u = \phi_s - \frac{\pi}{2}$; (c) $u_0 = p_0/Z$; (d) $\phi_u = \phi_p$.
16.3 (a) 3.3 cm; (b) 0.23 cm.
16.4 0.0301 μm.
16.8 37.13 Pa.
16.9 (a) 1.830; (b) 1.800; (c) 1.794.

Chapter 17

17.2 A "top-hat" (or rect) function; it represents the aperture limit of a microscope.
17.3 ~175 nm.
17.5 $E_0/\sqrt{5}$; $I_0/5$.
17.6 0.65 μm.
17.7 0.40 μm.

Chapter 18

18.1 Sound: ~13 μs; light: ~89 ps.
18.2 Given that $\lambda = 2\pi c/\omega$, write $\Delta\lambda$ as the difference between the wavelengths corresponding to $\omega_0 - \Delta\omega/2$ and $\omega_0 + \Delta\omega/2$.
18.5 A sinc function.
18.7 2× the beat frequency = 25 MHz.
18.11 12.2 μm.

Chapter 19

19.1 (a) 0.01; (b) shape, refractive index.
19.4 (a) $\sim 2.2 \times 10^{-18}$ N; (b) $\sim 8.8 \times 10^{-16}$ N; (c) no.
19.5 $\sim 1.1 \times 10^{-7}$ N/m.
19.7 ~1.7 W.

Table of symbols

Symbol	Name	SI Units
a	Single-scattering albedo	–
a'	Reduced single-scattering albedo	–
B	Bulk modulus	Pa
$\mathbf{B}$	Magnetic induction or magnetic flux density	T
BF	Blood flow	$m^3_{blood}/(m^3_{tissue}s)$
BV	Blood volume	m^3_{blood}/m^3_{tissue}
BW	Frequency bandwidth	Hz
c	Speed of light in vacuum	m/s
c_n	Speed of light in a medium with refractive index n	m/s
c_p	Specific heat at constant pressure	J/(kg-K)
C	Molar concentration	mol/m^3
C	Optical or image contrast	–
d_{np}	Near-point distance	m
D	Optical diffusion coefficient	m^2/s
D_B	Brownian diffusion coefficient	m^2/s
$\mathcal{D}_{TOF}$	Photon time-of-flight distribution	s^{-1}
E	Energy of electronic, vibrational, rotational states	J
$\mathbf{E}$	Electric field	V/m
f	Frequency	Hz
f	Focal length	m
F	Finesse	–
$\mathbf{F}$	Net flux	W/m^2
$\mathbf{F}$	Vector force	N
g	Average cosine of the scattering angle	–
G	Green's function	Variable

(*cont.*)

Symbol	Name	SI Units
g_1	First Legendre moment of the phase function (same as g)	–
$g_1(\tau)$	Normalized electric field autocorrelation function	–
g_2	Second Legendre moment of the phase function	–
$g_2(\tau)$	Normalized intensity autocorrelation function	–
$G_1(\tau)$	Electric field autocorrelation function	V^2/m^2
$G_2(\tau)$	Intensity autocorrelation function	W^2/m^4
h	Planck's constant	J-s
$\hbar$	Reduced Planck's constant ($h/2\pi$)	J-s/rad
H	Radiant exposure	J/m^2
$\mathbf{H}$	Magnetic field	A/m
I	Intensity	W/m^2
$\mathbf{I}$	Identity matrix	–
$\mathcal{J}$	Radiant angular intensity	W/sr
$\mathbf{J}$	Jacobian matrix	Variable
k	Wavenumber	m^{-1}
k	Thermal conductivity	W/(m-K)
k_B	Boltzmann's constant	J/K
k_{nr}	Rate constant for non-radiative decay	s^{-1}
k_r	Rate constant for radiative decay	s^{-1}
K	Kubelka-Munk absorption coefficient	m^{-1}
L	Radiance	$W/(m^2\text{-sr})$
$\langle L \rangle$	Mean photon pathlength	m
L_c	Coherence length	m
L_D	Diffusion length	m
L_{PDW}	Attenuation length of photon-density waves	m
$\mathbf{L}$	Elemental flux vector	$W/(m^2\text{-sr})$
m	Modulation depth	–
m	Relative refractive index	–
M	Optical magnification	–
M	Number of modes in an optical fiber	–
mfp	Optical mean free path	m
mfp_a	Absorption mean free path	m
mfp_s	Scattering mean free path	m
n	Index of refraction	–
NEP	Noise equivalent power	$W/\sqrt{Hz}$
NPV	Negative predictive value	–

Symbol	Name	SI Units
p	Scattering phase function	sr^{-1}
p	Overpressure due to ultrasound	Pa
p_θ	Scattering probability per unit angle	rad^{-1}
$\mathbf{p}$	Dipole moment	C-m
$\mathbf{p}$	Momentum	kg-m/s
$\hat{\mathbf{p}}$	Polarization unit vector	–
P	Radiant power	W
PPV	Positive predictive value	–
pO_2	Partial pressure of oxygen	Pa
Q	Radiant energy	J
Q	Electric charge	C
Q	Quality factor	–
Q_s	Scattering efficiency	–
r	Source-detector distance in infinite media	m
R	Responsivity	A/W
R_F	Fresnel reflection coefficient	–
$\dot{R}_d$	Total diffuse reflectance for point illumination	–
$\bar{R}_d$	Total diffuse reflectance for plane-wave illumination	–
$\ddot{R}_d$	Total diffuse reflectance for diffuse illumination	–
$\tilde{R}_d$	Total diffuse reflectance for spatially modulated illumination	–
$R_{CW}(\rho)$	Single-distance diffuse reflectance in continuous wave	m^{-2}
$R_{FD}(\rho)$	Single-distance diffuse reflectance in the frequency domain	m^{-2}
$R_{TD}(\rho)$	Single-distance diffuse reflectance in the time domain	$m^{-2}s^{-1}$
s	Particle displacement due to ultrasound	m
$S_{\mathcal{S},\mu}$	Sensitivity function of optical signal $\mathcal{S}$ to optical property μ	Variable
S	Kubelka-Munk scattering coefficient	m^{-1}
S_i	Elements of the scattering matrix	–
Se	Sensitivity	–
Sp	Specificity	–
$S_{Hb}O_2$	Oxygen saturation of hemoglobin	–
SaO_2	Arterial saturation	–

(cont.)

Symbol	Name	SI Units
SpO_2	Pulsatile saturation	
StO_2	Tissue saturation	–
SvO_2	Venous saturation	–
$\mathcal{S}$	General optical signal	Variable
T	Period of a wave	s
T	Temperature	K
u	Angular energy density	$J/(m^3\text{-sr})$
u	Particle speed due to ultrasound	m/s
U	Energy density	J/m^3
v_a	Speed of ultrasound	m/s
v_{PDW}	Phase velocity of photon-density waves	m/s
V	"V number" for an optical fiber	–
Vo_2	Oxygen consumption	$mol/(m^3\text{-s})$
w	Laser beam radius	m
Z	Acoustic characteristic impedance	$kg/(m^2\text{-s})$
α	Polarizability	$C\text{-}m^2/V$
α	Thermal diffusivity	m^2/s
β	Thermal expansion coefficient	K^{-1}
γ	Similarity parameter for the scattering phase function	–
γ	Skewness	–
Γ	Grüneisen coefficient	–
ε	Molar extinction coefficient	m^2/mol
η	Viscosity	Pa-s
θ	Polar angle	rad
θ	Phase of frequency-domain fluence rate	rad
θ, Θ	Scattering angle	rad
κ	Excess kurtosis	–
λ	Regularization parameter	Variable
λ	Wavelength	m
λ_{PDW}	Wavelength of photon-density waves	m
μ_a	Absorption coefficient	m^{-1}
μ_{eff}	Effective attenuation coefficient	m^{-1}
μ_s	Scattering coefficient	m^{-1}
μ_s'	Reduced scattering coefficient	m^{-1}
μ_t	Total attenuation coefficient	m^{-1}
μ_t'	Total reduced attenuation coefficient	m^{-1}
$\bar{\nu}$	Wavenumbers	m^{-1}

Symbol	Name	SI Units
$\Delta\bar{\nu}$	Raman shift	m^{-1}
ρ	Mass density	kg/m^3
ρ	Source-detector distance on boundary of semi-infinite media	m
σ_s	Scattering cross section	m^2
τ_a	Acoustic characteristic (confinement) time	s
τ_c	Coherence time	s
τ_F	Fluorescence lifetime	s
τ_{opt}	Optical irradiation time	s
τ_{th}	Thermal diffusion (confinement) time	s
φ	Azimuthal angle	rad
ϕ	Fluence rate	W/m^2
Φ_F	Fluorescence quantum yield	–
ψ	Fluence	J/m^2
Ψ	Objective function	Variable
ω	Angular frequency	rad/s
$\hat{\Omega}$	Unit directional vector	–
$d\Omega$	Infinitesimal solid angle	sr

Table of acronyms

Acronym	Spelled-out form
A/D	Analog-to-Digital
AC	Alternating Current
ALA	AminoLevulinic Acid
AM	Amplitude Modulation
ANSI	American National Standards Institute
AOM	Acousto-Optic Modulator
AOT	Acousto-Optic Tomography
APD	Avalanche PhotoDiode
AR-PAM	Acoustic-Resolution PhotoAcoustic Microscopy
ART	Algebraic Reconstruction Technique
ATP	Adenosine TriPhosphate
AUC	Area Under the Curve
BLL	Beer-Lambert Law
BOLD	Blood Oxygenation Level Dependent
BTE	Boltzmann Transport Equation
BV	Blood Volume
CARS	Coherent Anti-Stokes Raman Spectroscopy
CBF	Cerebral Blood Flow
CBV	Cerebral Blood Volume
CCD	Charge-Coupled Device
CFD	Constant Fraction Discriminator
CGS	Centimeter-Gram-Second
CI	Confidence Interval
$CMRO_2$	Cerebral Metabolic Rate of Oxygen
CNR	Contrast-to-Noise Ratio
CRC	ColoRectal Cancer
CRT	Cathode Ray Tube

Acronym	Spelled-out form
CW	Continuous Wave
DC	Direct Current
DCS	Diffuse Correlation Spectroscopy
DIC	Differential Interference Contrast
DLS	Dynamic Light Scattering
DNA	DeoxyriboNucleic Acid
DOI	Diffuse Optical Imaging
DOS	Diffuse Optical Spectroscopy
DOSI	Diffuse Optical Spectroscopy and Imaging
DOT	Diffuse Optical Tomography
DPF	Differential Pathlength Factor
DR	Dynamic Range
DRS	Diffuse Reflectance Spectroscopy
DWS	Diffusing Wave Spectroscopy
ECM	ExtraCellular Matrix
EEM	Excitation-Emission Matrix
EOM	Electro-Optic Modulator
ESS	Elastic-Scattering Spectroscopy
FAD	Flavin Adenine Dinucleotide
FADH	reduced Flavin Adenine Dinucleotide
FD	Frequency Domain
FDA	Food and Drug Administration
FDM	Finite-Difference Method
FD-OCT	Frequency-Domain Optical Coherence Tomography
FDTD	Finite-Difference Time Domain
FEM	Finite-Element Method
FMN	Flavin MonoNucleotide
fMRI	functional Magnetic Resonance Imaging
FMT	Fluorescence Molecular Tomography
FN	False Negative
fNIRI	functional Near-InfraRed Imaging
fNIRS	functional Near-InfraRed Spectroscopy
FP	False Positive
FPALM	Fluorescence PhotoActivation Localization Microscopy
FRET	Förster Resonance Energy Transfer
FSR	Free Spectral Range

(*cont.*)

Acronym	Spelled-out form
FWHM	Full Width at Half Maximum
GPU	Graphics Processing Unit
HG	Henyey-Greenstein
I/Q	In-phase/Quadrature
ICCD	Intensified Charge-Coupled Device
ICG	IndoCyanine Green
IEC	International Electrotechnical Commission
IF	Intermediate Frequency
ILT	Interstitial Laser Thermotherapy
IR	InfraRed
IVF	In Vitro Fertilization
IVPA	IntraVascular PhotoAcoustic
IVUS	IntraVascular UltraSound
LASIK	Laser-Assisted *in SItu* Keratomileusis
LCI	Low-Coherence Interferometry
LD	Laser Diode
LDF	Laser Doppler Flowmetry
LDV	Laser Doppler Velocimetry
LEBS	Low-coherence Enhanced BackScattering
LIBS	Laser-Induced Breakdown Spectroscopy
LLLT	Low-Level Laser Therapy
LO	Local Oscillator
LOIS	Laser Optoacoustic Imaging System
LSCM	Laser Scanning Confocal Microscopy
MAP	Mean Arterial Pressure
mBF	muscle Blood Flow
mBLL	modified Beer-Lambert Law
mBV	muscle Blood Volume
MC	Monte Carlo
MCP	MicroChannel Plate
MCP-PMT	MicroChannel Plate PhotoMultiplier Tube
MG	MultiGrid
MHG	Modified Henyey-Greenstein
MKS	Meter-Kilogram-Second
MOS	Metal Oxide Semiconductor
MP	Magnifying Power
mPAM	metabolic PhotoAcoustic Microscopy
MPE	Maximum Permissible Exposure

Acronym	Spelled-out form
MRI	Magnetic Resonance Imaging
mTHPC	m-TetraHydroxyPhenylChlorin
NA	Numerical Aperture
NAD	Nicoatinamide Adenine Dinucleotide
NADH	reduced Nicoatinamide Adenine Dinucleotide
NEP	Noise Equivalent Power
NIR	Near InfraRed
NIRS	Near-InfraRed Spectroscopy
NPV	Negative Predictive Value
NSOM	Near-field Scanning Optical Microscopy
OCT	Optical Coherence Tomography
OIS	Optical imaging of Intrinsic Signals
OR-PAM	Optical-Resolution PhotoAcoustic Microscopy
OTF	Optical Transfer Function
PA	PhotoAcoustic
PALM	PhotoActivated Localization Microscopy
PAM	PhotoAcoustic Microscopy
PAT	PhotoAcoustic Tomography
PC	Phase Contrast
PCF	Photonic Crystal Fiber
PCS	Photon Correlation Spectroscopy
PD	PhotoDiode
PDT	PhotoDynamic Therapy
PDW	Photon-Density Wave
PIN PD	p-i-n PhotoDiode
PMT	PhotoMultiplier Tube
PPV	Positive Predictive Value
PRK	PhotoRefractive Keratectomy
PSF	Point Spread Function
QD	Quantum Dot
QELS	Quasi-Elastic Light Scattering
QERS	Quasi-Elastic Rayleigh Scattering
RBC	Red Blood Cell
RC	Resistor-Capacitor
rCBF	relative Cerebral Blood Flow
RF	Radio Frequency
RMS	Root Mean Square

(cont.)

Acronym	Spelled-out form
RNA	RiboNucleic Acid
ROC	Receiver Operating Characteristic
RRS	Resonance Raman Scattering
RTE	Radiative Transfer Equation
SE	Standard Error
SERS	Surface-Enhanced Raman Scattering
SFDI	Spatial Frequency-Domain Imaging
SHG	Second-Harmonic Generation
SI	International System
SiPMT	Silicon PhotoMultiplier Tube
SIRT	Simultaneous Iterative Reconstruction Technique
SLD	SuperLuminescent Diode
SNR	Signal-to-Noise Ratio
SOA	Semiconductor Optical Amplifier
SPAD	Single-Photon Avalanche Diode
SPEF	Single-Photon Excitation Fluorescence
SS-OCT	Swept-Source Optical Coherence Tomography
STED	Stimulated Emission Depletion microscopy
STORM	Stochastic Optical Reconstruction Microscopy
TCSPC	Time-Correlated Single-Photon Counting
TD	Time Domain
TD-OCT	Time-Domain Optical Coherence Tomography
TEM	Transverse Electro-Magnetic
THG	Third-Harmonic Generation
TN	True Negative
TOI	Tissue Optical Index
TP	True Positive
TPEF	Two-Photon Excitation Fluorescence
UOT	Ultrasound-modulated Optical Tomography
UTL	Ultrasonic Tagging of Light
UV	UltraViolet
UVA	UltraViolet A
UVB	UltraViolet B
XCT	X-ray Computed Tomography

Index

Printed in the United States
by Baker & Taylor Publisher Services